Periodic Table of the Elements

Main groups — Transition metals — Main groups

1 1A	2 2A	3 3B	4 4B	5 5B	6 6B	7 7B	8 8B	9 8B	10	11 1B	12 2B	13 3A	14 4A	15 5A	16 6A	17 7A	18 8A
1 H 1.00794																	2 He 4.00260
3 Li 6.941	4 Be 9.01218											5 B 10.81	6 C 12.011	7 N 14.0067	8 O 15.9994	9 F 18.998403	10 Ne 20.1797
11 Na 22.98977	12 Mg 24.305											13 Al 26.98154	14 Si 28.0855	15 P 30.97376	16 S 32.066	17 Cl 35.453	18 Ar 39.948
19 K 39.0983	20 Ca 40.078	21 Sc 44.9559	22 Ti 47.88	23 V 50.9415	24 Cr 51.996	25 Mn 54.9380	26 Fe 55.847	27 Co 58.9332	28 Ni 58.69	29 Cu 63.546	30 Zn 65.39	31 Ga 69.72	32 Ge 72.61	33 As 74.9216	34 Se 78.96	35 Br 79.904	36 Kr 83.80
37 Rb 85.4678	38 Sr 87.62	39 Y 88.9059	40 Zr 91.224	41 Nb 92.9064	42 Mo 95.94	43 Tc (98)	44 Ru 101.07	45 Rh 102.9055	46 Pd 106.42	47 Ag 107.8682	48 Cd 112.41	49 In 114.82	50 Sn 118.710	51 Sb 121.757	52 Te 127.60	53 I 126.9045	54 Xe 131.29
55 Cs 132.9054	56 Ba 137.33	57 *La 138.9055	72 Hf 178.49	73 Ta 180.9479	74 W 183.85	75 Re 186.207	76 Os 190.2	77 Ir 192.22	78 Pt 195.08	79 Au 196.9665	80 Hg 200.59	81 Tl 204.383	82 Pb 207.2	83 Bi 208.9804	84 Po (209)	85 At (210)	86 Rn (222)
87 Fr (223)	88 Ra 226.0254	89 †Ac 227.0278	104 Rf (261)	105 Db (262)	106 Sg (263)	107 Bh (262)	108 Hs (265)	109 Mt (266)	110 (269)	111 (272)	112 (277)						

*Lanthanide series

58 Ce 140.12	59 Pr 140.9077	60 Nd 144.24	61 Pm (145)	62 Sm 150.36	63 Eu 151.96	64 Gd 157.25	65 Tb 158.9254	66 Dy 162.50	67 Ho 164.9304	68 Er 167.26	69 Tm 168.9342	70 Yb 173.04	71 Lu 174.967

†Actinide series

90 Th 232.0381	91 Pa 231.0359	92 U 238.0289	93 Np 237.048	94 Pu (244)	95 Am (243)	96 Cm (247)	97 Bk (247)	98 Cf (251)	99 Es (252)	100 Fm (257)	101 Md (258)	102 No (259)	103 Lr (260)

Operational Organic Chemistry

A Problem-Solving Approach to the Laboratory Course

THIRD EDITION

John W. Lehman

Lake Superior State University

Prentice Hall, Upper Saddle River, New Jersey 07458

Library of Congress Cataloging-in-Publication Data

Lehman, John W.
  Operational organic chemistry : a problem-solving approach to the
laboratory course / John W. Lehman.—3rd ed.
    p. cm.
  Includes bibliographical references and index.
  ISBN 0–13–841917–5
  1. Chemistry, Organic—Experiments. I. Title.
QD261.L39. 1999
547′.0078—DC21   98–7010
                                                        CIP

> *To my wife, Maureen,*
> *our two abys, and the abbess,*
> *Max, Deli, and Hildegard of Bingen*

Acquisitions Editor: *Matthew Hart*
Executive Managing Editor: *Kathleen Schiaparelli*
Assistant Managing Editor: *Lisa Kinne*
Art Director: *Jayne Conte*
Cover Designer: *Bruce Kenselaar*
Art Editor: *Karen Branson*
Manufacturing Manager: *Trudy Pisciotti*
Assistant Editor: *Mary Hornby*
Production Supervision/Composition: *WestWords Inc.*

Spectra © Sigma-Aldrich Co.

Printed in the United States of America

10 9 8 7 6 5 4 3 2

ISBN 0-13-841917-5

Prentice-Hall International (UK) Limited, *London*
Prentice-Hall of Australia Pty. Limited, *Sydney*
Prentice-Hall Canada Inc., *Toronto*
Prentice-Hall Hispanoamericana, S.A., *Mexico*
Prentice-Hall of India Private Limited, *New Delhi*
Prentice-Hall of Japan, Inc., *Tokyo*
Pearson Education Asia Pte. Ltd., *Singapore*
Editora Prentice-Hall do Brasil, Ltda., *Rio de Janeiro*

Contents

Part I Mastering The Operations   19

Part II Correlated Laboratory Experiments   101

Part III Minilabs 473

Part IV Qualitative Organic Analysis   529

Part V The Operations   573

Appendixes and Bibliography   762

Preface

This book is based on the first and second editions of *Operational Organic Chemistry: A Laboratory Course*, but every page of the second edition has been revised or rewritten and the focus of the book had been altered, so it is being issued with a new subtitle. Part I contains experiments designed to teach basic laboratory techniques, here called "operations" and identified by numbers preceded by "OP." Part II contains experiments that are correlated to lecture topics covered in most sophomore-level organic chemistry courses. Four experiments were deleted and 12 new ones were added, including a new techniques experiment (12); two spectral analysis experiments (13, 33); analysis of an analgesic drug by TLC (15); a nucleophilic substitution reaction of an ambident nucleophile (20); the synthesis of ethanol by fermentation (27); a multistep synthesis involving a ring contraction (30); a Wittig synthesis (40); a haloform oxidation (42); and two qualitative analysis experiments (39, 50). In addition, the Advanced Projects section (previously Part IV) has been extensively revised and converted to an independent research experiment (58).

The Minilabs have been removed from the Experiments sections (Parts I/II) and put in a section of their own (Part III). Eleven Minilabs were deleted or converted to Experiments, 18 new Minilabs were added, and the rest were extensively revised.

The Qualitative Organic Analysis section (Part IV) was completely reorganized to consolidate the former "Methodology" and "Procedures" sections, and the material relating to IR and NMR spectral analysis was moved to the corresponding Operations (OP-34, OP-35). The Operations (Part V) have been updated to incorporate advances in instrumentation and techniques. A separate HPLC operation (OP-33) has been added, and new or expanded sections on flash chromatography, reversed-phase column chromatography, capillary gas chromatography, Fourier-transform IR and NMR spectrometry, ^{13}C NMR spectrometry, GC-MS, and modern mass spectrometers are included.

Appendix V, Planning an Experiment, now includes directions for writing flow diagrams, and Appendix VII, The Chemical Literature, has been revised and updated to include a more detailed description of literature sources and information about searching the printed and online versions of *Beilstein* and *Chemical Abstracts*. The Bibliography has been updated to include new works and revisions since 1988, most of the pre-1980 material has been deleted, and a new section on software has been added.

Many people were involved in the preparation of this edition and have earned my sincere gratitude for their contributions. I am grateful to David Todd of Pomona College, whose correspondence about the "Evelyn effect" inspired my revision of Experiment 21; William Haag of Lake Superior State University, who reviewed the new HPLC operation; Mike D'Agostino, who tested some of the the new procedures; and my wife Maureen, for her exemplary patience and moral support. I am indebted to Matthew Hart, John Challice, and Lisa Kinne of Prentice Hall, for their assistance and encouragement throughout the project; and Pat McCutcheon, Julie Hollist, and the staff of WestWords, for their skillful production of the book. Most of the spectra in this book and the *Instructor's Manual* are reproduced from the spectral libraries of the Aldrich Chemical Company, whose generosity is gratefully acknowledged. I would also like to thank the following reviewers, whose suggestions and comments were invaluable: Philip J. Chenier (University of Wisconsin, Eau Claire), Gary W. Earl (Augustana College), Barbara L. Gaffney (Rutgers University), John C. Gilbert (University of Texas at Austin), Philip D. Hampton (University of New Mexico), Ulrich Hollstein (University of New Mexico), Floyd Kelly (Casper College), Jhong K. Kim (University of California, Irvine), Claire Olander (Appalachian State University), John P. Richard (State University of New York at Buffalo), Kerri Scott, (University of Mississippi), Jason Stenzel (University of Idaho), James M. Takacs (University of Nebraska at Lincoln), and Darrell J. Woodman (University of Washington).

To the Instructor

My thirty-some years of teaching organic chemistry have convinced me that many science students don't really know how science *works*. They attend class, absorb some knowledge, and learn some problem-solving techniques, but all too often they see little relationship between what they have learned in class and what they do in the laboratory. They approach the organic chemistry lab as if its sole purpose were to *make* something and not to *learn* something. That is not necessarily their fault. If we, as lab instructors and lecture professors, fail to convey to students that the essence of science is careful, dispassionate observation coupled with the critical thinking skills needed to discover the meaning of what is observed, it is not surprising that they fail to gain much of value from their laboratory experiences.

I agree with those chemists who have championed the idea that organic lab experiments should be designed not merely to synthesize a product but to solve a scientific problem (see, for example *J. Chem. Educ.* **1991**, *68*, 232). This book uses problem-based experiments to help students develop the observational and critical thinking skills that are essential prerequisites for a successful career in science—or in virtually any professional field. The expressed purpose of each major experiment in this book is the solution of a problem through the application of scientific methodology, with the student as the problem solver. Before each experiment the student must first define the problem based on information provided in a hypothetical scenario. After a preliminary reading of the experiment, the student is expected to develop a working hypothesis regarding its outcome. During the experiment, the student gathers and evaluates evidence bearing on the working hypothesis and, as necessary, re-evaluates and revises the hypothesis based on experimental observations and data. Finally, the student tests the hypothesis by obtaining a melting point, a spectrum, a gas chromatogram, or by some other means, and arrives at a conclusion. In reporting the experiment, the student is expected to describe how he or she applied scientific methodology in solving the problem, as illustrated by the sample report in Appendix III.

To stimulate interest and provide a frame of reference for the problems, students are asked to view themselves as "consulting chemists" working for an institute operated by their college or university. Various individuals and organizations come to the institute with their scientific problems, and the problems are relayed to the "project group" comprising each lab section, to be solved individually or by collaboration. Although some of the scenarios may appear a bit contrived, most are designed to suggest the kinds of tasks a practicing chemist might be called upon to perform.

To implement this problem-solving approach in the organic chemistry lab, I have rewritten and reorganized every experiment in Parts I and II. Each experiment now includes "Applying Scientific Methodology," a new section intended to help the student understand the problem, formulate a meaningful hypothesis, and solve the problem. Experiment 1 describes in some detail how the student can apply scientific methodology to the solution of a problem, so students should be asked to read the relevant parts of that experiment even if you choose not to assign it. The section "Understanding the Experiment" replaces the old "Methodology" section of each experiment, but has been revised to clarify the purpose of the experiment. Students now follow "Directions" rather than carry out a "Procedure"; while this is a minor change, it is intended to help students see the experiment as describing a course of action rather than providing a recipe to be followed mechanically. The Directions also include hazard warnings, directions for disposal of chemicals, questions to test the student's understanding of the experiment, and occasional reminders to observe and describe a phenomenon. Many of the old collateral projects and library topics have been combined in a new section, "Other Things You Can Do," which may also refer to one or more Minilabs that are related to the experiment.

Because each experiment in this book is formulated as a problem for the student to solve, the outcome of an experiment is not explicitly stated in the experiment. In some experiments such as Experiments 32 and 38, the identity of an organic reactant is not specified. Therefore it is imperative that the instructor or laboratory coordinator obtain a copy of the *Instructor's Manual* (ISBN: 0-13-919267-0), which is provided free of charge by Prentice Hall to adopters of this book.

John W. Lehman

Introduction

Problem Solving in the Organic Chemistry Laboratory

Organic chemistry is not most people's idea of a "fun" course, but that is no reason for failing to enjoy your organic chemistry lab experience. Many kinds of work can be enjoyable if you have the opportunity to use your imagination and to test your mental and manual skills. During this lab course you will play the role of a consultant in a Consulting Chemists Institute operated by your college or university. When D. K. Little wants to know what happens to his company's food preservative in stomach acid, when Rusty Tappet accidentally pours diesel fuel into a barrel of racing fuel, when Gilda Lilly wants to know the color of a triphenylmethane dye, or when the Olfactory Factory needs a way to convert an oversupply of anisole to a perfume ingredient, you and the other members of your project group will be called upon to solve their problems.

To solve such a problem you must *think* before you *act*. In other words, you will need to read the experiment, understand the problem, and try to predict a likely outcome of the experiment before you actually carry out the experiment in the laboratory. Your prediction, stated clearly in writing, will become your *working hypothesis*. In many cases the most likely outcome will become apparent after you read the experiment, especially if you apply the concepts you have learned in the organic chemistry lecture. In other cases there may be several reasonable outcomes and you will have to make an "educated guess" as to the most likely outcome. During the experiment you will need to gather evidence that may support your hypothesis— or prove it wrong. That means making careful observations and gathering data that relate to the problem. As you evaluate the evidence, you may decide that your original hypothesis was wrong—or at least incomplete— and needs to be revised. By the time you finish the experiment you will have the opportunity to test your original or revised hypothesis and arrive at a conclusion. In this way each experiment will help you apply your observational and critical thinking skills, as well as your lab skills, to the solution of each problem. The next section will tell you, in more detail, how to approach and solve a scientific problem.

Scientific Methodology

Anyone expecting to start a career in some field of science or technology, or a field that is based on scientific knowledge and principles (such as medicine), must learn and apply sound scientific methodology. Although there is no universal "scientific method" that all scientists adhere to rigorously, most of them follow at least some of the following steps when dealing with a scientific problem:

1 Define the problem.
2 Plan a course of action.
3 Gather evidence.
4 Evaluate the evidence.
5 Develop a hypothesis.
6 Test the hypothesis.
7 Reach a conclusion.
8 Report the results.

Defining the Problem. To a scientist, a "problem" is not so much a perceived difficulty as an *opportunity* for exploring and learning more about some aspect of the physical world. A problem may be inherent in an assigned task, or it may arise from anything that the scientist is curious about, such as an unexplained phenomenon or an unexpected observation. A problem is often defined in the form of a question: What is the identity of the liquid my instructor gave me? What is the mercury concentration in a Lake Superior salmon? How do fireflies generate light? In this laboratory course the problem associated with each experiment will be described in the *Scenario* that leads off the experiment. While performing an experiment, you may be able to formulate additional problems worth investigating.

Planning a Course of Action. A scientist must plan his or her own course of action for solving a scientific problem. This often requires that the scientist carry out a literature search to glean information and data relating to the problem, decide which experimental methods and instruments to use, and develop a detailed procedure to be followed. In most lab courses the procedure is "in the book" and the student simply follows the procedure as if it were a recipe for baking a cake. That is true of a few basic experiments in this textbook, but for most of them you will have to perform some calculations and develop your own experimental plan based on the information and directions given in the experiment. In a few cases you will be required to develop a procedure of your own.

Gathering Evidence. Scientists will gather as much evidence as they feel is needed to solve a problem and convice other scientists that their solution is correct. Evidence is gathered by making careful *observations* and *measurements*. Making accurate measurements, such as measuring the mass or melting point of the product of a chemical synthesis, requires a certain amount of skill and know-how. You can obtain such skills by, for example, watching your instructor demonstrate the operation of an instrument, then practicing on the instrument until you obtain consistent and accurate results *before* you use it to make a measurement you intend to report. If you are not sure how to use an instrument properly, ask the instructor to show you.

To make valid observations you must be *objective*, reporting only what you actually saw and not what you expected to see. A wildlife biologist who expects wild chimpanzees to behave just like chimpanzees at the zoo is not likely to make any important discoveries about chimpanzees! Keep the following in mind when you make observations.

1 Don't confuse an *observation* with an *inference*. An observation is whatever you perceive with your senses (sight, smell, touch, taste, or hearing) during an event. An inference is a guess about the *cause* of the event. Writing "The solution turned brown upon addition of 0.1 M $KMnO_4$" records an observation. Writing "The solution must have contained an alkene because it turned brown upon addition of 0.1 M $KMnO_4$" is an inference.

2 Be prepared to be surprised. If you observe something you didn't expect, don't simply disregard the observation or report what you thought you should have seen. Regard an unexpected observation as an opportunity to learn something you didn't know before—something that might lead to a new discovery.

3 Write down your observations as you make them or shortly afterward. If you wait too long you are likely to leave out important details.

4 Record your observations clearly, completely, and systematically. For example, if you are carrying out the same test on a series of samples, you should record your observations in a table.

Evaluating the Evidence. Evaluating the evidence involves assessing the reliability of your experimental results and looking for clues among your results that may point to a solution for the problem. For example, searching the infrared spectrum of an unknown liquid for evidence of a specific functional group and comparing its boiling point with the boiling points of known compounds may help you solve the problem "What is the identity of the liquid my instructor gave me?" Solving such problems often requires clear and logical thinking; in other cases a more creative, intuitive approach can be valuable. In either case, all of your reasoning and intuition may be fruitless if your experimental results are unreliable. The validation of experimental results requires first asking yourself whether the results make sense physically; in some kinds of experiments (but none in this book) it may also require a statistical analysis of experimental data. If you obtain a melting point that is much lower than the literature value, a product mass that is higher than the theoretical yield for a synthesis, or any other result that seems suspect, then you need to find out whether the result is, in fact, erroneous. You should review everything you did that led to that result, using notes from your lab notebook to jog your memory as necessary. Perhaps you only need to repeat a melting point or dry a product longer, but in any case you should find out what you did wrong and correct it.

Developing and Testing Hypotheses. A hypothesis can be regarded as an "educated guess" made to explain the results of one or more experiments. Hypotheses help us see the significance of an object or event that would otherwise mean little. For example, the movements of the planets seemed erratic and mysterious before Copernicus developed his hypothesis that the earth revolves around the sun. A hypothesis must be *testable* to have validity; that is, it must be formulated in such a way that experiments can be devised whose outcome might prove the hypothesis *wrong*. John Dalton's hypothesis that atoms are indivisible was proved wrong after it was shown that bombarding uranium atoms with neutrons caused them to split into smaller atoms. But Dalton's more fundamental hypothesis that all matter is made of atoms has been tested repeatedly over the years and never proved wrong, so it is generally accepted as true.

Most of the experiments in this book require you to formulate and test a hypothesis based on a problem outlined in the Scenario. A hypothesis can be proposed and revised at any time during the course of an experiment. Many scientists develop a working hypothesis as soon as a problem has been defined, since they often have some idea how the experiment may turn out—or how they hope it will turn out. An appropriate working hypothesis and the method of testing it may become apparent upon reading the experiment. For example, after reading Experiment 9 you should be able to develop a working hypothesis such as "The red pigment in Brand X tomato paste was (or was not) isomerized during processing" and then test your hypothesis by recording the ultraviolet-visible spectrum of the pigment in solution, as directed. A working hypothesis often turns out to be wrong, in which case the scientist must be willing to discard it and formulate a new hypothesis. One drawback of a working hypothesis is that the scientist may become so attached to it that he or she will overlook or ignore evidence that contradicts it. For this reason some scientists prefer to explore a problem without any preconcieved ideas about the outcome—which is not always easy to do. In most of the experiments in this book you can, after reading the experiment, formulate a working hypothesis, but you should be ready to revise or abandon it without regret if the evidence does not support it.

Reaching Conclusions. If a hypothesis passes all the tests you carry out, then you are ready to state a conclusion, such as "The liquid my instructor gave me is benzaldehyde." This does not necessarily mean that your conclusion is correct. You may have not carried out enough tests or carried them out accurately enough to justify your conclusion. But if you have performed an experiment carefully and reasoned logically, you will probably arrive at a valid conclusion.

Reporting Results. You should report all results of an experiment as clearly, completely, and unambiguously as possible. Write up your results in correct English using complete sentences and correct spelling. Be as specific as you can, avoiding such generalities as "My yield was lower than expected due to human error." Label any tables clearly, giving names or standard abbreviations for all physical properties and the units in which they are measured. Construct graphs using accurately ruled graph paper (never notebook paper or paper you have ruled yourself), with the dependent variable plotted on the y axis and the independent variable on the x axis. Label both axes with the physical quantity being graphed and its units, if any. Select appropriate uniform scale intervals so that your data points extend most of the way up and across the graph paper. If the relationship you are graphing is linear, draw the straight line that best fits the data points.

Appendixes II and III suggest ways of recording and reporting experimental results. Your instructor will let you know what kind of report he or she prefers.

Organization of this Book

Operational Organic Chemistry is divided into five parts. Part I contains 12 experiments whose purpose is to help you learn basic laboratory techniques (here called "operations") by applying them to the solution of a problem. For example, you will measure a melting point not just to learn how to measure melting points, but also to establish the identity of a substance you have isolated or synthesized. Once you have mastered the basic operations, you will apply them in the experiments of Part II, which are correlated with topics from your lecture course. These experiments will help you apply what you have learned in the lecture course and increase your proficiency in the laboratory.

Each experiment in Parts I and II contains a list of lecture-course topics and information under most or all of the following headings.

Operations. This heading is followed by a list of the operations to be used in the experiment, each preceded by an operation number such as OP-5. The description for each operation can be located quickly by using the large operation numbers in the right-hand corners of the odd-numbered pages in

Part V. Operations being used for the first time are in boldface type; you should read their descriptions thoroughly before you come to the laboratory. Once you have used an operation, you should not have to reread the entire description the next time you use it, but you should at least read the Summary and review the General Directions (if provided) to refresh your memory. Eventually you should have to refer to the operation descriptions only if you encounter experimental difficulties or are applying an operation in a new situation.

Before You Begin. Under this heading you will find a list of things to do before you come to the laboratory, which always includes reading the experiment and reading or reviewing the operations. Starting with Experiment 5 you will be expected to write an experimental plan for each experiment, as described in Appendix V, and you will usually need to carry out some calculations. Your instructor may require that you have your experimental plan and calculations approved before you begin an experiment.

Scenario. The Scenario presents a hypothetical situation involving the scientific problem you are to solve. A typical Scenario will describe a chemistry-related problem being experienced by a company or individual, and tell what role you will play in its solution.

Applying Scientific Methodology. This section is intended to help you formulate and solve the problem posed in the Scenario by using the methods described under *Scientific Methodology* on page 2.

Background Essay. Each background essay appears under a different descriptive heading. The essay will often show the relation between the lab work and related concepts from the lecture course. It may also relate historical sidelights or interesting facts that show the "real-world" relevance of the experiment.

Understanding the Experiment. This section describes the main purpose of the experiment, explains the theoretical basis of the experiment as necessary, and helps you understand the experimental methodology. It may also provide information that will help you interpret your results or cope with unexpected complications as they arise.

Reactions and Properties. For most experiments, this section gives balanced equations for synthetic reactions and tabulates the relevant physical properties of reactants, products, and other chemicals. The Properties Table contains the data needed for most of the pre-lab calculations.

Directions. This section describes the course of action you will follow to carry out the experiment. You should not follow it mechanically, as you would a recipe, but try to understand the purpose of each operation you are performing. The "Understanding the Experiment" section will help you do that. The directions for most Part I experiments are more detailed than those for Part II. By the time you get to Part II, you will be expected to know how to perform most laboratory operations proficiently without the aid of frequent reminders.

Safety Notes. Characteristics of some hazardous chemicals and precautions for their use are described in the Directions under this heading. See *Laboratory Safety* (p. 9) for general information about laboratory hazards.

Report. This section describes specific calculations, analytical data, or other information that should be included in your laboratory report. Your instructor will describe the format you should follow and indicate any additional material that should be included.

Exercises. Your instructor will assign exercises to be completed and turned in with your laboratory report.

Other Things You Can Do. The "other things" may include additional experiments or Minilabs you can perform, or library research projects you can do. You must have your instructor's permission to start any project marked by an asterisk. You should read the relevant sections of Appendix VII, *The Chemical Literature*, and scan the Bibliography for possible sources before you begin a library research project.

Part III of the text contains 46 short experiments called Minilabs. Minilabs take only part of a lab period and do not require extensive write-ups. They are designed to add flexibility to the lab course and "fill in the gaps" when, for example, a two-week experiment can be finished in one-and-a-half lab periods.

Part IV is a comprehensive, self-contained introduction to qualitative organic analysis that will help you learn how to identify organic compounds by using chemical and spectral methods.

Part V contains descriptions of all the operations, which are referred to in the Directions by number, in the form [OP-5]. Appendixes I through V contain illustrations of laboratory equipment and information about laboratory notebooks, reports, experimental plans, and calculations. Appendix VI contains tables of properties for qualitative analysis. Appendix VII is a guide to the chemical literature that describes some of the most important works in chemistry and, in some cases, tells you how to use them. The Bibliography lists a large number of useful works in chemistry, from enormous multi-volume sets such as *Chemical Abstracts* to short papers from the *Journal of Chemical Education.* References to entries listed in the Bibliography are made throughout the text in the form [Bibliography, F20], where the letter refers to a category and the number to a location within that category; for example, F20 is the 20th book listed under category F, Spectrometry.

Getting Along in the Organic Chemistry Laboratory

Because of wide variations in individual working rates, it is usually not possible to schedule experiments so that everyone can be finished in the time allowed. If all labs were geared to the slowest student, the objectives of the course could not be accomplished in the limited time available. If you get behind in the lab, you will probably be required to put in extra hours outside your scheduled laboratory period in order to complete the course. The following suggestions should help you work more efficiently and finish each experiment on time.

1 *Be prepared to start the experiment the moment you reach your work area.* Don't waste the precious minutes at the start of a laboratory period doing calculations, reading the experiment, washing glassware, or carrying out other activities that should have been performed at the end of the previous period or during the intervening time. The first half hour of any lab period is the most important—if you can use it to collect your reagents, set up the apparatus, and get the initial operation (reflux, distillation, etc.) under way, you should have no trouble completing the experiment on time.

2 *Organize your time efficiently.* Set up a regular schedule that requires you to read the experiment and operation descriptions and complete the prelab assignments a day or two before the laboratory period—an hour before the lab is too late! Plan ahead so that you know approximately what you will be doing at each stage of the experiment. A written experimental plan, prepared as described in Appendix V, is invaluable for this purpose.

3 *Organize your work area.* Before performing any operation, arrange all of the equipment and supplies you will need during the operation neatly on your bench top, in approximately the order in which they will be used. Place small objects such as spatulas and items that might be contaminated by contact with the bench top on a paper towel, laboratory tissue, or mat. After you use each item, move it to an out-of-the-way location (for example, dirty glassware to a washing trough in the sink) where it can be cleaned and then returned to its proper location when time permits. Keep your locker well organized, placing each item in the same location after use so that you can immediately find the equipment you need. This will also help you notice missing items, so they can be found or reported to the instructor without delay.

You will get along much better in the laboratory if you can maintain peace and harmony with your coworkers—or at least keep from aggravating them—and stay on good terms with your instructor. Following these common-sense rules will help you do that.

1 *Leave all chemicals where you found them.* You will understand the reason for this rule once you experience the frustration of hunting high and low for a reagent, only to find it at another student's station in a far corner of the lab. Containers should be taken to the reagent bottles to be filled; reagent bottles should never be taken to your lab station.

2 *Take only what you need.* Unless they are obtained from calibrated dispensers, liquids and solutions should ordinarily be measured into graduated containers so that you will take no more than you expect to use for a given operation. Solids can usually be weighed directly from their containers or measured from a special solids dispenser.

3 *Prevent contamination of reagents.* Don't use pipets or droppers to remove liquids from reagent bottles, or return unused reagent to a stock bottle. Be sure to close all bottles tightly after use—particularly those containing anhydrous chemicals and drying agents.

4 *If you must use a burner, inform your neighbors*—unless they are already using burners. This will allow them to cover any containers of flammable solvents and take other necessary precautions. In some circumstances you may have to use a different heat source, move your operation to a safe location (for instance, under a fume hood), or find something else to do while flammable solvents are in use.

5 *Return all community equipment to the designated locations.* This may include ring stands, steam baths, lab kits, clamps, condenser tubing, and other items that are not in your locker. Since such items will be needed by students in other lab sections, they should always be returned to the proper storage space at the end of the period.

6 *Clean up for the next person.* There are few experiences more annoying than finding that the lab kit you just checked out is full of dirty glassware, or that your lab station is cluttered with paper towels, broken glass, and spilled chemicals. The last 15 minutes or so of every laboratory period should be set aside for cleaning up your lab station and the glassware used during the experiment. Put things away so that your work station is uncluttered. Clean off the bench top with a towel or wet sponge, remove debris (including condenser tubing and other community supplies) from the sink, and thoroughly wash any dirty glassware that is to be returned to the stockroom, as well as that from your locker. Clean up any spills and broken glassware immediately. If you spill a corrosive or toxic chemical such as sulfuric acid or aniline, inform the instructor before you attempt to clean it up.

7 *Heed Gumperson's Second Law,* which advises you to maximize the labor and minimize the oratory in the **labor**atory. This does not mean that all conversation must come to a halt. Quiet conversation during a lull in the experimental activity is okay, but a constant stream of chatter directed at a student performing a delicate operation is distracting and can lead to an accident. For the same reason, radios and tape or CD players should not be used in the laboratory.

Laboratory Safety

Preventing Laboratory Accidents

Some of the experiments in this book contain many hazard warnings and safety precautions, which may give you the impression that an organic chemistry laboratory is a dangerous place to work. Actually, most organic lab courses are completed without incident, aside from minor cuts or burns, and serious accidents are rare. Nevertheless, the potential for a serious accident is always there. To reduce the likelihood of an accident, *you must learn the following safety rules and observe them at all times.* Anyone failing to do so may be expelled from the laboratory.

1 *Wear approved safety goggles or safety glasses in the laboratory at all times.* Even when you are not working with hazardous materials, another student's actions could endanger your eyes, so never remove your safety goggles or glasses until you leave the lab. Don't wear contact lenses in the laboratory because chemicals splashed into an eye may get underneath a contact lens and cause damage before it can be removed. Learn the location of the eyewash fountain nearest to you at the first laboratory session, and learn how to use it.

2 *Never smoke in the laboratory or use open flames in operations involving low-boiling flammable solvents.* Anyone found smoking in an organic chemistry laboratory is subject to immediate expulsion. Before you light a burner or even strike a match, inform your neighbors of your intention to use a flame. If anyone nearby is using flammable solvents, either wait until they are finished or move to a safer location, such as a fume hood. Ethyl ether and petroleum ether are extremely flammable, but other common solvents such as acetone and ethanol can be dangerous as well. When ventilation is inadequate, the vapors of ethyl ether and other highly volatile liquids can travel a long distance across a benchtop, so igniting a flame at one end of a lab bench may start an ether fire at its opposite end! Learn the location and operation of the fire extinguishers, fire blankets, and safety showers at the first laboratory session.

3 *Consider all chemicals to be hazardous and minimize your exposure to them.* Never taste chemicals, do not inhale the vapors of volatile chemicals or the dust of finely divided solids, and prevent contact between chemicals and your skin, eyes, and clothing. Many chemicals can cause poisoning by ingestion, inhalation, or absorption through the skin. Strong acids and bases, bromine, thionyl chloride, and other corrosive materials can produce severe burns and require special precautions such as wearing gloves and lab aprons. Some chemicals cause severe allergic reactions, and others may be carcinogenic (tending to cause cancer) or teratogenic (tending to cause birth defects) by inhalation, ingestion, or skin absorption. To prevent accidental ingestion of toxic chemicals, do not bring food or drink into the laboratory or use mouth suction for pipetting, and wash your hands thoroughly after handling any chemical. Clean up chemical spills immediately, using a neutralizing agent and plenty of water for acids and bases, and an absorbent for solvents. In case of a major spill, or if the chemical spilled is very corrosive or toxic, notify your instructor before you try to clean it up. Read the *Safety Notes* section for each experiment, and use protective gloves or a fume hood when directed.

4 *Exercise great care when working with glass and inserting or removing thermometers and glass tubing.* Among the most common accidents in a chemistry lab are cuts from broken glass and burns from touching hot glass. Protect your hands with gloves or a towel when removing or inserting glass tubes and thermometers; grasp the glass close to the stopper or thermometer adapter and gently twist it out, or in. Remember that hot glass remains hot for some time, so after you have fire polished a glass rod or done another glassworking operation, give the glass plenty of time to cool before you touch it. Refer to OP-3 for more information about safe glassworking procedures.

5 *Wear appropriate clothing in the laboratory.* Wear clothing that is substantial enough to offer some protection against accidental chemical spills; long-sleeved shirts or blouses and long pants or dresses are preferable. Wear shoes that provide adequate protection against spilled chemicals and broken glass, not open sandals or cloth-topped athletic shoes. Human hair is very flammable, so wear a hair net while using a burner if you have long hair. To protect your clothing (and yourself) from chemical spills, it is a good idea to wear a lab jacket or apron in the lab.

6 *Dispose of chemicals properly.* For reasons of safety and environmental protection, most organic chemicals should not be washed down the drain. Except when an experiment's directions or your instructor indicate otherwise, place used organic chemicals and solutions in designated waste containers. Some aqueous solutions can be safely poured down the drain, but consult your instructor if there is any question as to the best method for disposing of a particular chemical or solution. See "Disposal of Hazardous Wastes" (p. 17) for additional information.

7 *Never work alone in the laboratory or perform unauthorized experiments.* If you wish to work in the laboratory when no formal lab period is scheduled, you must obtain written permission from the instructor and be certain that others will be present while you are working.

Reacting to Accidents: First Aid

If you have or witness a serious accident involving poisoning or injury, report it to the instructor as soon as possible. Serious accidents should be treated by a competent physician, but applying some basic first aid procedures before a physician arrives can help minimize the damage.

If you have an accident that requires quick action to prevent permanent injury, take the appropriate action as described here, if you can, and see that the instructor is informed of the accident. If you *witness* an accident, call the instructor immediately and leave the first aid to him or her unless: (1) no instructor or assistant is in the laboratory area; (2) the victim requires immediate attention because of stopped breathing, heavy bleeding, etc.; (3) you have had formal emergency response training.

If an accident victim stops breathing or goes into shock as a result of any kind of accident, standard procedures for artificial respiration and treating shock should be applied. Descriptions of these procedures can be found in Chapter 2 of *The CRC Handbook of Laboratory Safety*, 4th ed. [Bibliography, C1].

Eye Injuries

If any chemical enters your eyes, flush them *immediately* with water from an eyewash fountain while holding your eyelids open. If you are wearing contact lenses, remove them first. Continue irrigation for at least 15 minutes (or until a nurse or physician arrives), then have your eyes examined by a physician. If foreign bodies such as glass particles are propelled into an eye, get immediate medical attention. Removal of such particles is a job for a specialist.

Chemical Burns

Remove any contaminated clothing and flush the affected area *immediately* with a large amount of water until the chemical is completely removed, using a safety shower if the area of injury is extensive or if it is inaccessible to washing from a tap. Speed and thoroughness in washing are the most important factors in reducing the extent of injury. Dry the area gently with a clean, soft towel. If the victim experiences pain or if the skin is red or swollen, immerse the burned area in cold water or apply cold, wet dressings. Do not use neutralizing solutions, ointments, or greases on chemical burns unless they are specifically called for in a first aid procedure. Unless the skin is only reddened over a small area, a chemical burn should be examined by a nurse or physician. First aid procedures for burns caused by specific chemicals, such as bromine, are given in the *Sigma-Aldrich Library of Chemical Safety Data* [Bibliography, C4] and in the *First Aid Manual for Chemical Accidents* [Bibliography, C3].

If a chemical burn is very extensive or severe, the victim should lie down with the head and chest a little lower than the rest of the body. If the injured person is conscious and able to swallow, he or she should be provided with plenty of nonalcoholic liquid to drink (water, tea, coffee, etc.) until a physician or ambulance arrives.

Thermal Burns

For a *first-degree burn*, when the skin is reddened but there are no blisters or broken skin, immerse the affected area in clean, cold water or apply ice to reduce the pain and facilitate healing. For a *second-degree burn,* when blisters are raised, immerse the burned area in clean, cold water or apply ice, then cover the area with sterile gauze or another clean dressing. If legs or arms are burned, keep them elevated above the trunk of the body. Never puncture blisters raised by a second-degree burn. For a *third-degree burn,* when the skin is broken and underlying tissues are damaged, place a thick sterile dressing or clean cloths over the affected area and have the burn examined by a nurse or physician as soon as possible. Do not remove burned clothing, immerse the burned area in cold water, or cover the burn with greasy ointment. If legs or arms are burned, keep them elevated above the trunk of the body.

In case of an extensive thermal burn, the burned area should be covered with the cleanest available cloth material and the victim should lie down, with the head and chest lower than the rest of the body, until a physician or ambulance arrives. If the injured person is conscious and able to swallow, he or she should be provided with plenty of nonalcoholic liquid to drink (water, tea, coffee, etc.).

Bleeding, Cuts, and Abrasions

In case of a cut or abrasion that does not involve heavy bleeding, cleanse the wound and surrounding skin with soap and lukewarm water, applying it by wiping away from the wound. Try to remove imbedded glass shards, if there are any, by using tweezers if necessary. Hold a sterile gauze pad over the wound until bleeding stops. If the damage is confined to the skin, apply an antibacterial cream such as Neosporin to prevent infection. Place a fresh gauze pad over the wound and secure it loosely with a triangular or rolled bandage. Replace the pad and bandage as necessary with clean, dry ones. Avoid contact between the wound and the mouth, fingers, handkerchiefs, or other unsterile objects. If the wound is deep or extensive, it should be treated further by a nurse or physician.

In case of a major wound that involves heavy bleeding, *immediately* apply pressure directly over the wound with a cloth pad (such as a clean handkerchief or other clean cloth) or sterile dressing, pressing firmly with one or both hands to reduce the bleeding as much as possible. Then call for assistance. A compression bandage should be applied *on top of the original dressing* if profuse bleeding continues. Removing the original dressing may delay clotting. If no compression bandage is available you can roll a clean cork in sterile gauze and bind it to apply pressure directly over the wound. The victim should lie down with the bleeding part higher than the heart, and the dressing should be held in place with heavy gauze or other cloth strips. A physician or ambulance should be called as soon as possible, and the victim should be kept warm with a blanket or coat in the meantime. If the injured person is conscious and able to swallow, he or she should be provided with plenty of nonalcoholic liquid to drink (water, tea, coffee, etc.) until a physician arrives.

Poisoning

If, after contact with, inhalation of, or accidental ingestion of a chemical, you experience a burning sensation in the throat, discoloration of lips or mouth, stomach cramps, nausea and vomiting, or confusion, you should seek treatment for chemical poisoning.

If a poison is swallowed, an ambulance and a poison control center should be called immediately. If the victim is conscious, loosen tight clothing around the neck and waist, have the victim rinse his or her mouth several times with cold water and spit it out, then give the victim 1–2 cups of water or milk to drink. Unless otherwise advised by the poison control center, induce vomiting by tickling the back of the throat, or by giving 2 tablespoons (~30 mL) of ipecac syrup followed by a cup of water. Vomiting should not be induced if the patient is unconscious, in convulsions, or has severe pain and burning sensations in the mouth or throat, or if the poison is a petroleum product or a strong acid or alkali. When vomiting begins, lower the victim's head to prevent vomit from reentering the mouth, and collect a sample of the vomit for possible analysis. If convulsions occur, remove any objects that might cause injury or move the victim away from such objects. Watch out for and remove any obstructions in the victim's mouth (including dentures). If necessary, insert a soft pad between the victim's teeth to protect the tongue from being bitten. If the victim has stopped breathing, clear the airway and administer artificial respiration. When there is time, the poison should be identified (if possible) and an appropriate antidote given (see the *Merck Index*, 9th ed., pages MISC-22ff., or call a poison control center). If the poison cannot be identified and a medical professional is not present, a heaping tablespoon (15 g) of a universal antidote (consisting of 2 parts activated charcoal, 1 part magnesium oxide, and 1 part tannic acid) can be given in a half glass of warm water. A sample of the poison should be saved for the physician, if it is known.

If a poison has been inhaled, the victim must be taken to fresh air and a physician called immediately. Loosen any tight clothing around the neck and waist, and if necessary, use a tongue depressor or other device to keep the victim's airway open. Administer manual (not mouth-to-mouth) artificial respiration if the victim has stopped breathing. Keep the victim warm and as quiet as possible until the physician arrives. If the poison is a highly toxic gas such as hydrogen cyanide, hydrogen sulfide, or phosgene, the persons attempting to rescue the victim should wear self-contained respirators while they are in contact with the vapors.

In case of skin contamination by a toxic substance, the same general procedure as for chemical burns should be followed. The *Sigma-Aldrich Library of Chemical Safety Data* [Bibliography, C4] or another appropriate source should be consulted for specific procedures to be used for certain substances.

Reacting to Accidents: Fire

In case of fire, your first response should be to *get away* as quickly as possible. If a fire is small and confined to a container such as a flask or beaker, you may be able to extinguish it by placing a watch glass over the mouth of the container. Otherwise, let the instructor deal with the fire. If no instructor is in the laboratory, obtain a fire extinguisher of the appropriate type and attempt to put out the fire by aiming the extinguisher at the base of the fire, while maintaining a safe distance. Most labs should be equipped with class BC or ABC dry chemical extinguishers, which are effective against solvent and electrical fires. A class D fire, which involves burning metals or metal hydrides such as sodium metal and lithium aluminum hydride, can be extinguished with an appropriate class D fire extinguisher or by smothering the fire in dry sand, sodium chloride, or sodium carbonate. If a fire is too large to be extinguished by a fire extinguisher, sound the nearest fire alarm and evacuate the area.

If your hair or clothing catches on fire, *don't panic. Walk* (don't run) directly to the nearest fire blanket or safety shower and attempt to extinguish the fire. Don't wrap yourself in a fire blanket while you are standing, as that may direct the flames around your face; instead *drop* to the floor and *roll* as you wrap the blanket around your body. To use a safety shower, get directly under the shower head and pull the chain. If another person's hair or clothing has caught on fire, try to prevent panic as you lead him or her to a safety shower or fire blanket.

Chemical Hazards

For your own health and safety, it is essential that you *exercise caution while handling chemicals* and *minimize your exposure to them*. Most academic chemistry departments have developed their own policies and procedures for dealing with hazardous chemicals. Your instructor will inform you of any departmental or institutional rules relating to the safe handling and disposal of such chemicals. The experiments in this book describe specific chemical hazards and handling precautions under the heading *Safety Notes*. The hazard descriptions are meant to inform you of the potential danger posed by certain chemicals when they are not handled properly. If you take reasonable precautions and follow the directions in the Safety Notes, there is little likelihood that you will suffer ill effects from working with any chemical.

In many experiments, specific hazard information on some chemicals is provided in the form of octagonal hazard signs containing index numbers and/or letters. Each hazard index is a number from 0 to 4, where 0 indicates no known hazard and 4 indicates a very great health or safety hazard. The letters are used to warn of other hazards, such as reactivity with water and carcinogenic potential. The significance of the numbers and letters is explained in Table 1.

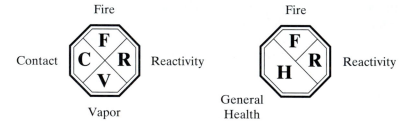

You should learn the location of each category on the hazard signs: Going clockwise from the top, the order is Fire → Reactivity → Vapor → Contact. The *fire* index number in the upper quadrant rates the fire danger posed by the chemical. The *reactivity* index number in the right quadrant assesses the danger of a violent reaction or explosion. The *vapor* index number in the lower quadrant rates the potential health effect of inhaling the vapors of the chemical. The *contact* index number in the left quadrant rates the adverse health effects that may result from skin contact (the eye contact hazard, which is not rated, may be even greater in some cases). For some chemicals the vapor and contact index numbers are replaced by a single

Table 1 Meanings of numbers and letters on hazard symbols

Symbol	Health (Contact or Vapor)	Fire	Reactivity
0	No known hazard	Will not burn	Stable
1	May cause irritation if not treated	Ignites after strong preheating	Unstable only at high temperature and pressure
2	May cause injury; requires treatment	Ignites after moderate heating	Unstable, but won't detonate
3	May cause serious injury despite treatment	Ignites at normal temperatures	Detonates or explodes with difficulty
4	May cause death or major injury despite treatment	Very flammable	Readily detonates or explodes
CA	Carcinogen (cancer-causing agent)		
OX			Strong oxidant; may react violently with combustible material
P			Polymerizes readily
W			Reacts violently with water

health index number that represents the overall health hazard of the chemical. If any quadrant is left blank, it means that no hazard index number was reported in the sources consulted; it does not mean that no hazard exists.

The meaning of the hazard signs is illustrated by these examples.

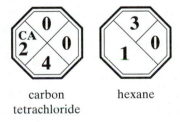

carbon
tetrachloride   hexane

Carbon tetrachloride is nonflammable and thermally stable, so its Fire and Reactivity hazard index numbers are both zero. Inhaling its vapors may be extremely harmful, and there is a moderate risk of injury from skin contact. Carbon tetrachloride has also been identified as a carcinogen in tests with laboratory animals and is suspected of causing cancer in humans. For these reasons, carbon tetrachloride is not used in the experiments in this book. The second hazard sign shows that hexane is highly flammable but very stable and its health risk is comparatively low.

Chemicals with a high fire hazard number should obviously be kept away from ignition sources including flames, sparks, and hot surfaces. For example, ethyl ether will ignite if it is spilled on a hot plate at a temperature of 160°C or higher.

Chemicals with a high reactivity hazard number should be handled with great care, following any precautions given in the Safety Notes. Such chemicals may ignite if subjected to shock or brought in contact with metal spatulas or materials that may catalyze their decomposition.

The value of a vapor or contact hazard number may suggest the appropriate response to inhalation of or contact with a chemical. If you inhale significant amounts of a chemical's vapor or the dust of a solid chemical, you should go to a window or other area where you can breathe fresh air, unless the chemical's vapor hazard number is zero. If the chemical has a high vapor hazard number, or if inhalation was prolonged, you should notify your instructor to see if treatment is required. If a chemical with a contact hazard number other than zero contacts your skin, you should wash the area of contact with soap and water. If the chemical has a high contact hazard number, or if the exposure is extensive, you should notify your instructor to see if treatment is required. If you get chemicals in your eyes, follow the procedures described under the heading "Eye Injuries" on page 11. The higher the chemical's contact number, the more likely it is that eye damage will result without prompt medical treatment.

Chemicals designated as strong oxidants (OX) must not be allowed to contact other chemicals, except for the ones specified in the experimental directions. In fact, you should never mix chemicals together unless directed to, since mixing incompatible chemicals may result in generation of toxic gases, fire, or an explosion.

Chemicals that polymerize readily (P) may sometimes undergo spontaneous, rapid polymerization that generates heat. If this reaction occurs in a capped reagent bottle it could cause the bottle to shatter violently. Most such chemicals are stabilized by addition of a small amount of antioxidant or other stabilizer, so they are unlikely to react unless the stabilizer has been removed.

Water-sensitive chemicals (W) react violently and exothermically in contact with water, and often generate toxic fumes. Some of them generate toxic fumes when exposed to moist air. Such chemicals should be kept in tightly closed containers that are opened only for transfers and closed immediately afterward. They should be used under fume hoods at some distance from any source of water.

Carcinogens

A few chemicals used in the experiments are identified as suspected *carcinogens*—agents suspected of causing cancer (CA). Although this label should certainly not be disregarded, it is important to point out that such chemicals, if used only as directed in this book, present little risk of cancer to students handling them. The carcinogenic activity of a compound is generally established by animal tests in which high doses of the chemical are administered by various routes for prolonged periods. For example, phenacetin taken orally has been found to cause cancer in laboratory animals, but it is highly unlikely that anyone will ingest (eat) enough phenacetin during a laboratory experiment to cause cancer. Chromium(VI) compounds have been shown to cause cancers of the lungs, nasal cavity, and sinuses in humans, but the persons at risk are industrial workers who have been exposed to the dust of chromium compounds continuously in the workplace. In this course, you will use chromic acid solutions (prepared from chromium(VI) oxide) a drop or so at a time, so there will be no possibility of inhaling chromium dust. Some halogenated hydrocarbons are suspected of causing cancer if inhaled, so you should take special precautions to avoid breathing the vapors of chlorinated solvents such as chloroform and dichloromethane. Compounds such as benzene and carbon tetrachloride, which might present a significant risk of cancer under conditions likely to be encountered in an undergraduate laboratory, will not be used in this course.

It is prudent to exercise particular care when handling potential carcinogens, such as wearing protective gloves and working under a hood, but the label "carcinogen" should cause no more apprehension than other hazard warnings.

Teratogens

If you are pregnant or think you may be pregnant, inform your instructor before you enter the organic chemistry lab for the first time. Some organic chemicals are known or suspected *teratogens*, meaning that they may harm a developing fetus. As is the case for carcinogens, handling a chemical designated as a teratogen does not necessarily represent a danger to the fetus. Ethanol is a teratogen when ingested, but people are far more likely to ingest ethanol in a bar or in their homes than in a chemistry laboratory. Nevertheless, some chemicals that are routinely used in organic chemistry laboratories may represent a significant danger to a developing fetus. Therefore it is important to contact your instructor, the laboratory coordinator, or a designated Chemical Hygiene Officer to discuss your options. The best option may be to take the laboratory course after your child is delivered. If you prefer to remain in the lab course, you should make arrangements with your instructor to minimize your exposure to potential teratogens. Such arrangements might include substituting less hazardous chemicals for teratogenic ones; performing alternative experiments that do not require the use of teratogens; or using gloves, protective clothing, and a hood when handling any teratogenic chemical.

Sources of Information about Chemical Hazards

Because of space limitations, the chemical hazard information provided in this book is of necessity incomplete. The most complete source of information about the hazards associated with any chemical is its *Material Safety Data Sheet (MSDS)*, which is provided by the manufacturer or vendor of the chemical. The MSDS's for chemicals used in your organic chemistry laboratory should be available from the chemistry department office, the Chemical Hygiene Officer, or some other designated source at your institution. Other useful sources of chemical hazard information include *The Sigma-Aldrich Library of Chemical Safety Data* [Bibliography, C4], *Sax's Dangerous Properties of Industrial Materials* [C5], *Hazards in the Chemical Laboratory* [C6], and *Bretherick's Handbook of Reactive Chemical Hazards* [C8].

Disposal of Hazardous Wastes

Many of the chemicals and solutions generated in organic chemical labs are regarded as hazardous wastes, which are regulated by various state and federal agencies. Hazardous waste generators are required to collect and dispose of hazardous wastes in ways that pose minimum potential harm to health and the environment. Some hazardous wastes can be purified and reused, or converted to nonhazardous materials. Other hazardous wastes are placed in landfills, incinerated, or otherwise disposed of by commercial waste disposal firms. During your organic chemistry lab course, you will be required to place certain designated waste solvents, solutions, and other chemicals in labeled containers, as directed by your instructor.

Except when precluded by local regulations, moderate quantities of many common chemicals can be safely and acceptably disposed of down a laboratory drain, if certain procedures are followed. Small quantities (less than 100 g) of most water-soluble organic compounds, including water-soluble alcohols, aldehydes, amides, amines, carboxylic acids, esters, ethers, and ketones, can be disposed of in this way. The organic compounds should be mixed or flushed with at least 100 volumes of excess water. Dilute aqueous solutions of inorganic compounds can be disposed of down the drain if they contain cations and anions from the following list, and no others.

Cations of these metals: Al, Ca, Cu, Fe, Li, Mg, Mo, Pd, K, Na, Sn, Zn
Anions: HSO_3^-, BO_3^{3-}, $B_4O_7^{2-}$, Br^-, CO_3^{2-}, Cl^-, OCN^-, OH^-, I^-, O^{2-}, PO_4^{3-}, SO_4^{2-}, SO_3^{2-}

Some less common cations and anions, such as Cs^+ and SCN^-, are also permissable.

Water-soluble organic compounds that are highly flammable or boil below 50°C should *not* be poured down the drain. Other compounds that shouldn't go down the drain include water-insoluble organic compounds (including all hydrocarbons and halogenated hydrocarbons); flammable or explosive solids, liquids, or gases; phenols and other taste- or odor-producing substances; wastes containing poisons in toxic concentrations; and corrosive wastes capable of damaging the sewer system.

PART I

Mastering the Operations

The experiments in Part I will help you learn the basic laboratory operations of organic chemistry and become proficient in their use. The first time an operation is used in Part I, its number is highlighted in boldface type in the **Operations** list at the beginning of the experiment. The section **Before You Begin** in each experiment lists some assignments that you must complete before you arrive at the laboratory to start the experiment.

EXPERIMENT 1

Learning Basic Operations. The Effect of pH on a Food Preservative

Laboratory Orientation. Acid-Base Reactions. Reaction Stoichiometry.

Operations:

OP-1 Cleaning and Drying Glassware
OP-4 Weighing
OP-5 Measuring Volume
OP-12 Vacuum Filtration
OP-21 Drying Solids

Before You Begin

1 Read the Introduction and Laboratory Safety sections of this book.
2 Read the experiment carefully, particularly the section "Understanding the Experiment" and the directions.
3 Read the sections in Part V describing operations OP-1, OP-4, OP-5, OP-12, and OP-21. Most of the descriptions are quite brief.
4 If you are required to write up your experiments in a formal laboratory notebook, prepare the notebook as requested by your instructor after reading Appendix II.

Scenario

For this and the other experiments in this book, you will play the role of a consultant in a Consulting Chemists Institute operated by your college or university, with your instructor as the laboratory supervisor and other students in your lab section as members of your project group. Most of the Scenarios in this book describe purely hypothetical situations. Except where otherwise indicated, any similarity between a person or organization named herein and an actual person or organization is coincidental.

Fresh Foods Incorporated (FFI) is a small, family-owned chemical company that manufactures sodium benzoate and other food preservatives for sale to food processors. One of its competitors has launched an ad campaign claiming that FFI's sodium benzoate changes to a different chemical in the stomach, implying that the chemical may be harmful or ineffective. D. K. Little, FFI's marketing director, has asked your institute to find out whether the competitor's claim is valid. Your assignment is to place sodium benzoate into a medium simulating stomach acid (an aqueous solution containing hydrochloric acid at a pH of 1–3) and see whether or not a new substance forms in that environment. Since this is your first day on the job, your supervisor will show you around the laboratory and also help you develop some basic lab skills involving measurements of mass and volume.

Applying Scientific Methodology

Read Scientific Methodology (pp. 2-5) *before you read this section.*

As in the other experiments in this book, the scientific *problem* to be solved is described in the Scenario. In this experiment the fundamental problem, stated as a question, is: "Does a new substance form when sodium benzoate is placed into a medium simulating stomach acid?" After reading the experiment thoroughly you should be ready to formulate a *working hypothesis* such as "A new substance will (or will not) form when sodium benzoate is placed into a medium simulating stomach acid." Your *course of action*, described in detail in the Directions, is then fairly obvious; you will have to place some sodium benzoate in an aqueous solution of hydrochloric acid having a pH of 1–3. You can do this by dissolving the sodium benzoate in water and adding enough dilute HCl to lower the pH to 2, for example. During this process you will have to *gather evidence* suggesting whether or not a chemical reaction has occurred. Evidence for a chemical reaction may include a color change, formation of a gas, formation of a precipitate, evolution of heat, or any combination of these. As you *evaluate the evidence*, (i.e., think about what you've observed) you may decide to stay with your working hypothesis (if the evidence supports it), modify it, or discard it and formulate a new one. Then you should be able to *reach a conclusion* based on the experimental evidence and any clues you came across while reading the experiment. You will then *report your results* after referring to the Report section that follows the Directions.

Sodium Benzoate as a Food Preservative

When a resin from the Sumatran tree *Styrax benzoin* is heated to 100°C, white vapors rise and condense to form needlelike crystals of benzoic acid, which got its name from the tree and its resin, gum benzoin. Benzoic acid can easily be converted to its salt, sodium benzoate, which is used widely as a food preservative, particularly in acidic foods such as fruits and fruit juices. Benzoic acid is so widely distributed in plants that the urine of all plant-eating animals (including humans) contains hippuric acid, a compound synthesized in the kidneys by the combination of benzoic acid with the amino acid glycine. Significant amounts of benzoic acid can be isolated from such diverse natural sources as anise seed, cranberries, prunes, cherry bark, cloves, and the scent glands of the beaver. In cranberries and other plant products, benzoic acid is a natural preservative that inhibits the growth of bacteria, yeasts, and molds, and thus retards spoilage.

If you read the labels on cans and bottles in a grocery store, you will find sodium benzoate (sometimes called benzoate of soda) listed on many of them. Sodium benzoate is used as a preservative in such food products as jams and jellies, soft drinks, fruit juices, pickles, condiments, margarine, and canned and frozen seafoods, and even in such nonfood items as toothpaste and tobacco. Benzoic acid is also an effective food preservative, but sodium benzoate is a more popular food additive because it is much more soluble in

The hexagon with a circle in it represents a benzene ring, whose molecular formula in these compounds is C_6H_5.

benzoic acid   sodium benzoate

hippuric acid   glycine

water than benzoic acid, making it easier to blend into water-containing food products.

Understanding the Experiment

This experiment will help you become familiar with the laboratory environment and with some fundamental laboratory operations such as measuring mass and volume, separating solids from liquids by vacuum filtration, and drying solids.

In part **A** you will add hydrochloric acid to a solution of sodium benzoate and, if a different substance forms, recover it and measure its mass. To understand what is happening, you need to know something about the properties of the substances involved and remember what you have previously learned about acid/base chemistry and stoichiometry. When sodium benzoate dissolves in water it dissociates into benzoate ions, which are weakly basic, and sodium ions.

$$C_6H_5COONa \rightarrow C_6H_5COO^- + Na^+$$
$$\text{benzoate ion}$$

Hydrochloric acid is a solution of hydrogen chloride (HCl) in water. The strongest acid present in this solution is the hydronium ion, formed by the transfer of a proton from an HCl molecule to a water molecule.

$$HCl + H_2O \rightarrow Cl^- + H_3O^+$$
$$\text{hydronium ion}$$

Key Concept: In a Lowry-Brønsted acid-base reaction under standard conditions, the stronger acid transfers protons to a stronger base to yield a weaker acid and a weaker base.

If the hydronium ion concentration is high enough, protons will be transferred from the strong acid H_3O^+ to the basic benzoate ions. This will yield benzoic acid which, being quite insoluble in water (sodium benzoate is about 200 times more soluble), should precipitate from solution.

$$C_6H_5COO^- + H_3O^+ \rightarrow C_6H_5COOH + H_2O \qquad (1)$$
$$\text{benzoic acid}$$

The net reaction is the sum of these three reaction steps:

$$C_6H_5COONa + HCl \rightarrow C_6H_5COOH + NaCl \qquad (2)$$

Stop and Think: What is the relationship between hydronium ion concentration and pH?

The hydronium ion concentration depends on the pH of the solution; the lower the pH, the higher $[H_3O^+]$ will be. By carrying out the experiment you will discover whether or not the pH of stomach acid is low enough to convert sodium benzoate to benzoic acid.

In any experiment involving the conversion of one substance to another, it is important to understand the stoichiometry of the reaction. (Read Appendix IV if you need a review of stoichiometric calculations.) Equation 2 shows that if the reaction were complete at the pH used, a mole of benzoic acid would be formed for each mole of sodium benzoate in the reaction mixture. You will start with about 2.00 g of sodium benzoate, which is 13.9 mmol (0.0139 mol), so the theoretical yield of benzoic acid (if any forms) should be 13.9 mmol, which corresponds to a mass of 1.70 g. Keep in mind that you may not measure out exactly 2.00 g of sodium benzoate, in which case you will need to calculate the theoretical yield (to the appropriate number of significant figures) based on your measured mass of sodium benzoate.

Whether or not your yield approaches the theoretical value will depend on a number of factors, including the following. (1) You may make errors in measuring mass or volume. (2) The proposed reaction may not occur, or if it does, it may not be complete. (3) Some of the product may remain dissolved in the product mixture and may not be recovered. (4) You may lose some of the product while transferring it from one vessel to another. You should try to minimize losses by taking great pains to perform accurate measurements and to make *quantitative transfers*; that is, to scrape or rinse the last traces of solid from the glassware. Get into the habit of doing the kind of careful, meticulous work that will increase your yields and save you time and effort in the long run.

The product of a chemical preparation should always be dried to *constant mass*, meaning that its mass after drying should not change significantly between two successive weighings. For most preparations in this book you can assume that a product is sufficiently dry if—following the initial drying period—its mass does not decrease by more than 0.5% after an additional five minutes or so of oven drying. If the product is being dried at room temperature in a desiccator, the interval between weighings should be longer.

In part **B** you will determine the density of an unknown liquid by using a graphical method. The validity of your results will depend on the care you take in measuring the mass and volume of the liquid and graphing your results.

Laboratory Safety and Orientation

Before coming to lab you should have read the **Laboratory Safety** section of this book. Before you begin to work in the laboratory, your instructor will review the safety rules and tell you what safety supplies you must have such as safety goggles and protective gloves and aprons. During the first laboratory period, the instructor will show you where safety equipment is located and tell you how to use it. As you locate each item, check it off the following list and make a note of its location. (Your instructor may suggest additions or changes to the list.) You should also learn the locations of chemicals, consumable supplies (such as filter paper and boiling chips), waste containers, and various items of equipment such as balances and drying ovens.

Safety Equipment and Supplies

- Fire extinguishers
- Fire blanket
- Safety shower
-

- Eyewash fountain
- First aid supplies
- Spill clean-up supplies
-

Checking In

Obtain a locker and supply list and check into the laboratory as directed by your instructor. You will find illustrations of typical locker supplies in Appendix I at the back of this book. Pieces of glassware with chips, cracks, or star fractures (see Figure 1.1) should be replaced; they may cause cuts,

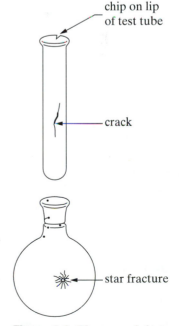

Figure 1.1 Glassware defects

break on heating, or shatter under stress. If necessary, clean up any dirty glassware in your locker [see OP-1] at this time.

Reactions and Properties

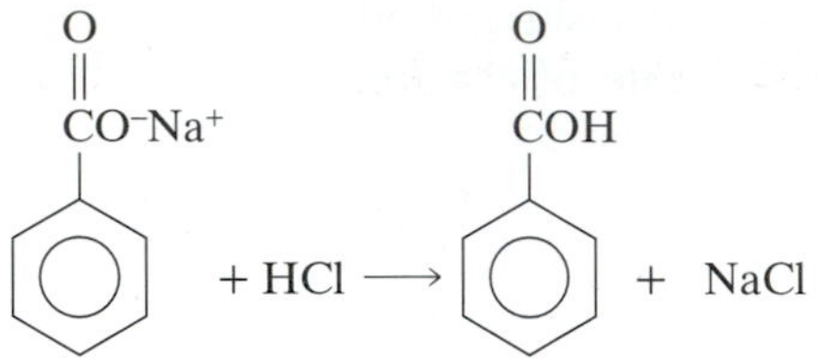

Table 1.1 Physical properties

	M.W.	m.p.	water solubility
benzoic acid	122.1	122	0.34
sodium benzoate	144.1	-	61.2

Note: M.W. = molecular weight; melting points (m.p.) are in °C; solubilities are in grams of solute per 100 mL of water at room temperature (usually 25°C).

Directions

Your instructor will demonstrate the operation of balances, automatic pipets, and any other special equipment you will use during this experiment.

A. *The Effect of pH on Sodium Benzoate*

Safety Note

Sodium benzoate and benzoic acid are mild irritants. Do not get them in your eyes or on your skin or clothing.

Reaction. Weigh [OP-4] approximately 2.00 g of sodium benzoate to the maximum accuracy of your balance and transfer it quantitatively to a 30-mL (or larger) beaker. Measure 10 mL of water (use distilled or deionized water if available) with a graduated cylinder [OP-5], pour it into the beaker, and stir until the sodium benzoate dissolves. Measure 5.0 mL of 3 *M* hydrochloric acid and add it slowly, while stirring, to the sodium benzoate solution until you have added about 4.0 mL. Then add more 3 *M* HCl drop by drop, with stirring, testing the solution frequently with pH paper, until its pH is 2. To test the pH, use your stirring rod to transfer a drop of the supernatant liquid (liquid at the surface) to a strip of pH paper. Adding a little excess HCl will do no harm. Cool the solution to 10°C or below by setting the 30-mL beaker in a larger beaker containing cracked ice and a little water, and stir occasionally.

Observe and Note: Can you detect any evidence for a chemical reaction? If so, describe it in your lab notebook.

Take Care! Do not use your thermometer as a stirring rod. The bulb may break and spill toxic mercury.

Stop and Think: What is the purpose of cooling the solution?

Stop and Think: What is the substance?

Waste Disposal: The liquid in the filter flask may be poured down the drain.

Separation. If a new substance has formed, separate it from the reaction mixture by vacuum filtration [OP-12] and wash it on the filter with about 5 mL of ice-cold water. (Be sure to transfer the solid quantitatively to and from the filter.) Use a spatula or the flat end of a flat-bottomed stirring rod to mix the crystals well with the wash water, being careful not to displace the filter paper. Let the solid air dry on the filter for a few minutes with the aspirator turned on, then dry it [OP-21] to constant mass. If an oven is used for drying, its temperature should be 90°C or lower because the solid may sublime just above that temperature. Weigh the dry product [OP-4] in a tared (preweighed) vial, and label the vial with the experiment number, the name of the product, its mass, your name, the current date, and any other information required by your instructor.

B. *Measuring the Density of an Unknown Liquid*

Safety Note

At your instructor's request, first measure the density of room-temperature distilled water by the procedure described here. If your measured density is not within 2% of the expected density of water (~0.997 g/mL at room temperature), repeat the measurement.

If one is available, use an automatic pipet set at 500 μL (0.500 mL) to measure the volume of the liquid [OP-5]. (Do not try to adjust the automatic pipet! If it does not read 500 μL, see your instructor.) If not, you can use a 1-mL graduated pipet. Have ready two clean, dry, 2-dram (or larger) screwcap vials. Use a graduated cylinder or dispensing pump to measure about 4 mL of the unknown liquid into one of the vials and take a pipet and both vials to a balance. Weigh the empty vial and its cap to the nearest milligram (0.001 g) and record the mass [OP-4]. Use the pipet to accurately measure 0.500 mL of the liquid into the empty vial, cap it immediately, and record the mass of the vial, cap, and liquid. Without delay, measure additional 0.500 mL portions of the liquid into the vial, capping it and recording the mass after each addition, until you have added a total of six 0.500 mL portions. Use the same balance for all readings.

Take Care! Avoid contact with the liquid, do not breathe its vapors.

Determine the mass of the liquid after each addition by subtracting the mass of the vial and cap. Prepare a graph of your data, plotting cumulative volume on the *x*-axis and mass on the *y*-axis. Draw the best straight line through the data points. From the slope of the graph calculate the density of the liquid to three decimal places. Show your graph and your calculated density to your instructor; if either is unacceptable you may be asked to repeat your measurements or redraw your graph.

Waste Disposal: Put the liquid in a solvent recovery container as directed by your instructor.
The cumulative volume is the total volume added up to a particular time; for example, after three additions the cumulative volume is 3 × 0.500 mL or 1.500 mL.

Stop and Think: What is the relationship between the slope of your graph and the density? Does your calculated density make sense physically?

Cleanup Routine. After completing this and all subsequent experiments you should:

1 Clean up the glassware you used during the experiment [OP-1].
2 Clear off your work area, wipe the bench top with a sponge or wet towel, and remove any refuse or equipment from the sink.
3 Turn in any items you may have checked out of the stock room and return any community supplies to their proper locations.
4 Turn in your labeled product to the instructor, when applicable.
5 See that all the items on the locker list are safely inside your locker (you may have to pay for missing supplies), then lock it.

Report. Read Appendix III before your write your report. (Your instructor may ask you to write up your results as if you were submitting them to D. K. Little of Fresh Foods Incorporated.) For part **A**, state your initial hypothesis, describe and interpret the relevant evidence, then tell whether your hypothesis changed when it was tested and if so, why. State your final conclusion and tell how it is supported by the evidence. Calculate the theoretical yield of the product based on your measured mass, and the percent yield of your preparation (show your calculations). For part **B** include the density of your unknown liquid, a table of your data, and your graph. Follow your instructor's directions regarding any additional items to be included in your report.

Exercises

1 You can confirm that a substance has been converted to a different substance by showing that the substances have different properties. Your observations should have revealed a difference in at least one property of sodium benzoate and benzoic acid. What property was that, and what observation revealed the difference?

2 Label the stronger acid, stronger base, weaker acid, and weaker base in Equation **1** for the proton transfer reaction in part **A**.

3 When you lower the pH of a solution containing aqueous sodium benzoate, a precipitate eventually forms. What will happen if you then raise the pH by adding aqueous NaOH? Why will this happen? Write a balanced equation for the proton transfer reaction involved and label the stronger acid, stronger base, weaker acid, and weaker base.

4 The water solubilities of oxalic acid and sodium oxalate at room temperature are 10 g/100 mL and 3.7 g/100 mL, respectively. Could you prepare oxalic acid by adding HCl to a solution of sodium oxalate, cooling it to room temperature, and filtering the resulting mixture? Explain why or why not.

5 The equation for a straight line can be written in the form $y = mx + b$ where x and y are variables, m is the slope, and b is the intercept. Write an equation for the straight line you obtained in your graph for part **B** and show how it yields the equation for density $d = m/v$. (Remember that m in this equation stands for mass, not the slope.)

6 (a) Calculate the ratio of dissolved benzoic acid to benzoate ion that will exist in solution at equilibrium at a pH of 2.00. The acid equilibrium constant (K_a) for benzoic acid is 6.46×10^{-5}. Note that this calculation does not account for the benzoic acid that has precipitated from solution, but you can assume that the higher the ratio, the more benzoic acid will precipitate. (b) Carry out the same calculation for a pH of 4.00 and explain why it was important to reduce the pH below 4 in this experiment.

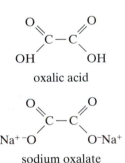

oxalic acid

sodium oxalate

Other Things You Can Do

(Starred items require your instructor's permission.)

*1 Make a flat-bottomed stirring rod and other useful laboratory items as described in Minilab 1 (p. 474).

*2 Try converting some other salts of organic acids, such as sodium salicylate and sodium oxalate, to the corresponding acids. You may need to look up the solubilities of the substances involved to explain your results.

3 Write a short research paper about food additives after consulting such sources as the *Kirk-Othmer Encyclopedia of Chemical Technology* [Bibliography, A7] and appropriate sources listed in section L of the Bibliography.

Extraction and Evaporation. Separating the Components of "Panacetin"

Separation Methods.

Operations:

OP-6 Heating
OP-11 Gravity Filtration
OP-13 Extraction
OP-14 Evaporation
OP-4   Weighing
OP-12 Vacuum Filtration
OP-21 Drying Solids

Before You Begin

1 Read the experiment carefully.
2 Read operations OP-11, OP-13, OP-14, and the section "Heat Sources" in OP-6; review the other operations listed as necessary.

Scenario

Your supervisor has been e-mailed the following message from a drug watchdog agency, the Association for Safe Pharmaceuticals (ASP).

See Experiment 1, p. 20, if you don't understand the purpose of the Scenarios.

Greetings:

Our roving agent in Southern California, Sam Surf, recently purchased some Panacetin—an analgesic drug preparation—at a drugstore in San Diego. According to the label on the bottle, the Panacetin tablets were manufactured in the United States by a legitimate pharmaceutical company, but Sam noticed some discrepancies on the label and flaws in the tablets themselves that made him suspect they might be counterfeit. Such knockoffs of a domestic drug can be manufactured cheaply elsewhere and smuggled into the United States, where they are sold at a big profit margin.

The label on the bottle lists the ingredients per tablet as aspirin (200 mg), acetaminophen (250 mg), and sucrose (50 mg). The sucrose is an inactive ingredient used to make the tablets more palatable to children. We have reason to believe that Panacetin does contain aspirin, sucrose, and another active component, but we're not sure what that component is or whether the amounts listed on the label are accurate. The unknown component is probably a chemical relative of acetaminophen, either acetanilide or phenacetin. Both of these kill pain as effectively as acetaminophen, so their presence in an analgesic drug wouldn't be detected by the consumer. But acetanilide and phenacetin are banned in the United States because of their toxicity, and we would like to keep them off the market.

We want your Consulting Chemists Institute to analyze this drug preparation to find out what percentages of aspirin, sucrose, and the unknown component it contains, and whether the unknown is acetanilide or phenacetin. You have two weeks to complete your investigation.

Les Payne, director of operations, ASP

Applying Scientific Methodology

The Scenario presents two problems for you to solve in this experiment and the next: (1) Is the composition of Panacetin as stated on the label accurate? (2) What is the identity of the unknown component in Panacetin? You will concentrate on the first problem in this experiment, following the course of action described in the Directions. Because of potential material losses, you must allow some margin for error in deciding whether the percent composition derived from the label (10% sucrose, 40% aspirin, 50% unknown component) is accurate. For the purposes of the experiment, a range of 8–12% sucrose, 35–45% aspirin, and 45–55% of the unknown component are close enough to indicate that the label is reasonably accurate. As in the previous experiment, you should start with a working hypothesis, gather and interpret evidence, change your hypothesis if the evidence does not support it, arrive at a conclusion, and report your results.

Painkilling Drugs, from Antifebrin to Tylenol

Analgesic drugs reduce pain and *antipyretic* drugs reduce fever. Some drugs, including aspirin and acetaminophen, do both. Many of the common over-the-counter analgesic/antipyretic drug preparations contain aspirin, acetaminophen, or combinations of these substances with other ingredients. For example, acetaminophen is the active ingredient of Tylenol, and Extra Strength Excedrin contains aspirin, acetaminophen, and caffeine. From their molecular structures you can see that acetaminophen is chemically related to both phenacetin and acetanilide, whose painkilling effects were discovered late in the nineteenth century.

In 1886 a pair of clinical assistants, Arnold Cahn and Paul Hepp, were looking for something that would rid their patients of a particularly unpleasant intestinal worm. The trick was to find a drug that would kill the worm but not the patient, and their method—not a very scientific one—was to test the chemicals in their stock room until they found one that worked. When someone came across an ancient bottle labeled NAPHTHALENE, they tried it out on a patient who had every malady in the book, including worms. It didn't faze the worms, but it reduced the patient's fever dramatically. Before Cahn and Hepp went out on a limb and endorsed naphthalene as a cure-all for fevers, someone noticed that the white substance in the bottle was nearly odorless. Since naphthalene has a strong mothball odor, Hepp

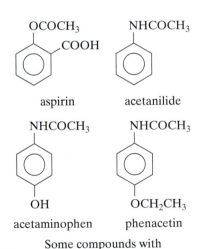

Some compounds with analgesic/antipyretic properties.

suspected that the bottle was mislabeled and sent it to his cousin, a chemist at a nearby dye factory, for analysis. The tests showed that the new drug was not naphthalene at all but acetanilide!

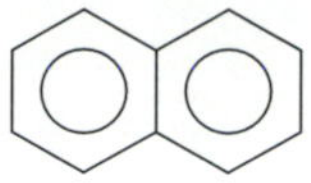

Acetanilide proved to have painkilling as well as fever-reducing properties and was soon being marketed under the proprietary name antifebrin. Unfortunately, some patients using antifebrin developed a serious form of anemia called methemoglobinemia, in which hemoglobin molecules are altered in a way that reduces their ability to transport oxygen through the bloodstream. Even though antifebrin is now considered too toxic for medicinal use, its discovery did much to stimulate the development of safer and more effective analgesic/antipyretic drugs.

About six months after the discovery of antifebrin, a similar drug was developed as the result of a storage problem. Carl Duisberg, director of research for the Friedrich Bayer Company, had to get rid of 50 tons of *para*-aminophenol—a seemingly useless yellow powder that was a by-product of dye manufacturing. Rather than pay a teamster to haul the stuff away, Duisberg decided to change it into something Bayer could sell. After reading about antifebrin, he reasoned that a compound with a similar molecular structure might have similar therapeutic uses. Duisberg knew that a hydroxyl (OH) group attached to a benzene ring is characteristic of many toxic substances (for example, phenol), so he decided to "mask" the hydroxyl group in *para*-aminophenol with an ethyl (CH_3CH_2-) group, as shown in the figure.

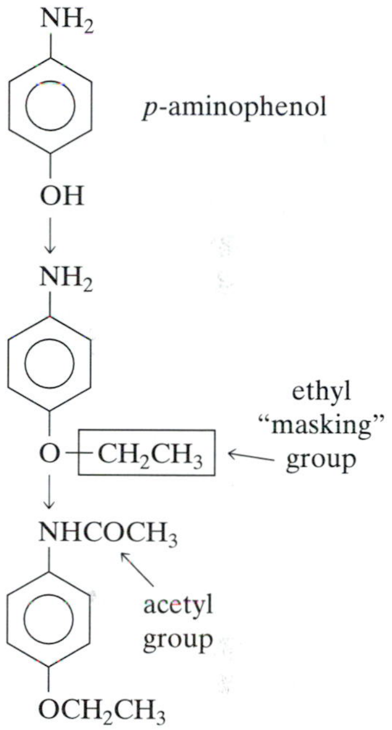

Incorporation of an acetyl (CH_3CO-) group then yielded phenacetin, which proved to be a remarkably effective and inexpensive analgesic/antipyretic drug. Until recently phenacetin was used in APC tablets (which contained aspirin, phenacetin, and caffeine) and other analgesic/antipyretic drug preparations. It is no longer approved for medicinal use in the United States because it may cause kidney damage, hemolytic anemia, or even cancer in some patients.

Ironically, the substance Duisberg would have obtained had he not masked the hydroxy group is acetaminophen, which has proven to be a safer drug than either acetanilide or phenacetin. In the body, acetanilide and phenacetin are both converted to acetaminophen, which is believed to be the active form of all three drugs.

Understanding the Experiment

Most natural products and many commercial preparations are mixtures containing a number of different substances. To obtain a pure compound from such a mixture, you must separate the desired compound from the other components of the mixture by taking advantage of differences in their physical and chemical properties. For example, substances having very different solubilities in a given solvent may be separated by extraction or filtration, and liquids with different boiling points can be separated by distillation. Acidic or basic substances are often converted to water-soluble salts, which can then be separated from the water-insoluble components of a mixture.

In this experiment you will separate the components of a simulated pharmaceutical preparation, Panacetin, making use of their solubilities and

naphthalene

p-aminophenol

ethyl "masking" group ←

acetyl group

phenacetin

Synthesis of phenacetin

Key Concept: The separation of substances from one another is based on differences in their physical and chemical properties.

acid-base properties. Panacetin contains aspirin, sucrose, and an unknown component that may be either acetanilide or phenacetin. These substances have the following solubility characteristics:

1   Sucrose is insoluble in the organic solvent dichloromethane (CH_2Cl_2, also called methylene chloride).
2   Aspirin is soluble in dichloromethane but relatively insoluble in water. Sodium hydroxide converts aspirin to a salt, sodium acetylsalicylate, which is insoluble in dichloromethane but soluble in water.
3   Acetanilide and phenacetin, like aspirin, are soluble in dichloromethane and insoluble in water. They are not converted to salts by sodium hydroxide.

Minilab 2 (page 475) will help you understand how the extraction operation works; you can carry out this minilab at your instructor's request.

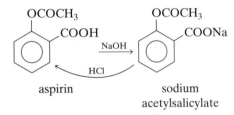

aspirin   sodium
                  acetylsalicylate

Mixing the Panacetin with dichloromethane should therefore dissolve the aspirin and the unknown component, but leave the sucrose behind as an insoluble solid that can be removed by filtration. Aspirin can then be removed from the dichloromethane solution by extraction with an aqueous solution of sodium hydroxide, which converts the aspirin to its salt as shown in the margin. This salt, being much more soluble in water than in dichloromethane, will dissolve in the aqueous layer while the unknown component will remain behind in the dichloromethane layer. Aspirin can be precipitated from the aqueous layer with hydrochloric acid and filtered from the solution by the same method you used in Experiment 1 to prepare benzoic acid from its salt. The unknown component can then be isolated by evaporating the solvent from the dichloromethane solution.

This separation process is summarized by the following flow diagram:

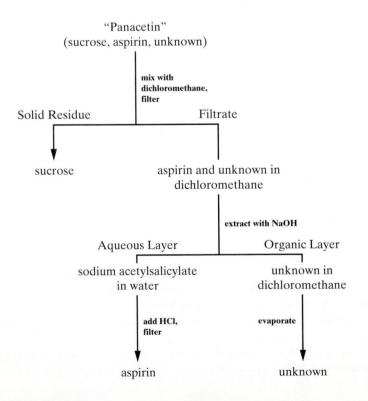

You can estimate the percentage composition of Panacetin from the masses of the dried components. Note that the actual composition may or may not be the same as the composition given in the Scenario. Careful work is required to obtain accurate results in this experiment; errors can arise from incomplete mixing with dichloromethane, incomplete extraction or precipitation of aspirin, incomplete drying of the recovered components, and losses in transferring substances from one container to another.

Directions

Dichloromethane may be harmful if ingested, inhaled, or absorbed through the skin. There is a possibility that prolonged inhalation of dichloromethane may cause cancer. Minimize contact with the liquid and do not breathe its vapors.

See Chemical Hazards, pages 14-17, for an explanation of the hazard symbols.

Separation of Sucrose. Accurately weigh [OP-4] about 3 g of Panacetin and transfer it to a clean, dry 125-mL Erlenmeyer flask. Add 50 mL of dichloromethane to the flask. Stir the mixture thoroughly to dissolve as much solid as possible, using your stirring rod to break up any lumps or granules. Using a preweighed fluted filter paper, filter the mixture by gravity [OP-11] into a small flask, saving the filtrate (the liquid that goes through the filter paper) for the next step. Set the filter paper aside, being careful not to lose any of the sucrose, and reweigh it when it is completely dry. Record the mass of the sucrose in your laboratory notebook. If requested, submit the sucrose to your instructor in a tared and labeled vial.

Separation of Aspirin. Transfer the filtrate to a separatory funnel and extract it [OP-13] with two 25-mL portions of aqueous 1 *M* sodium hydroxide. Because the dichloromethane layer will be on the bottom, you will have to transfer each layer to a different container (label the containers) and return the dichloromethane layer to the separatory funnel before the second extraction. Combine the two aqueous extracts in the same container and save the dichloromethane layer for the following step, *Isolation of the Unknown Component.*

Add 10 mL of 6 *M* hydrochloric acid slowly, with stirring, to the combined aqueous extracts. Test the pH of the solution as described in Experiment 1 and add more acid, if necessary, to bring the pH to 2 or lower. Cool the mixture in an ice/water bath, collect the aspirin by vacuum filtration [OP-12], and wash it on the filter with cold distilled water. Let the aspirin dry on the filter for a few minutes with the aspirator running, then dry it to constant mass [OP-21]. Weigh the aspirin and record its mass in your lab notebook. Submit the aspirin to your instructor in a properly labeled vial.

Isolation of the Unknown Component. Use a filter flask attached to a trap and aspirator to evaporate the solvent [OP-14] from the dichloromethane solution. Cool the trap in a beaker of ice water to recover the dichloromethane. Heating and swirling the solution over a steam bath or in

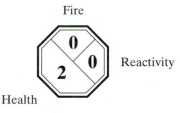

If it is quite impure, the unknown may remain liquid after all of the solvent is removed. It should solidify upon cooling.

Waste Disposal: Put the recovered dichloromethane in a designated chlorinated solvent recovery container.

a hot water bath [OP-6] will increase the evaporation rate. Discontinue evaporation when only a solid residue remains in the flask or when no more solvent evaporates. Transfer the unknown component to a tared vial and dry it to constant mass. Weigh it and save it for Experiment 3.

Report. State the problem and your initial hypothesis. Calculate your percent recovery by dividing the sum of the masses of all components by the mass of Panacetin you started with. Calculate the approximate percentage composition of Panacetin, based on the total mass of components recovered. (These percentages should add up to 100%.) Summarize and interpret the evidence. Tell whether your hypothesis changed when it was tested and if so, why. State your final conclusion and tell how it is supported by the evidence.

Exercises

1 (a) Describe any evidence that a chemical reaction occurred when you added 6 *M* HCl to the solution of sodium acetylsalicylate. (b) Explain why the changes that you observed took place.

2 Describe and explain the possible effect on your results of the following experimental errors or variations, in each case specifying the component(s) whose percentages would be too high or too low if: (a) After adding dichloromethane to Panacetin, you didn't stir the mixture long enough. (b) During the NaOH extraction you failed to mix the aqueous and organic layers thoroughly. (c) You mistakenly extracted the dichloromethane solution with 1 *M* HCl rather than 1 *M* NaOH. (d) Instead of using pH paper, you neutralized the NaOH solution to pH 7 using litmus paper.

3 Although acetanilide and phenacetin are not appreciably acidic, acetaminophen (like aspirin) is a stronger acid than water. What problem would you encounter if the unknown component were acetaminophen rather than acetanilide or phenacetin?

4 Acetaminophen is a weaker acid than carbonic acid (H_2CO_3), but aspirin is a stronger acid than carbonic acid. Prepare a flow diagram like the one in this experiment, showing a procedure for separating a mixture of sucrose, aspirin, and acetaminophen.

5 Write balanced reaction equations for the reactions involved (a) when aspirin dissolves in aqueous NaOH and (b) when aspirin is precipitated from a sodium acetylsalicylate solution by HCl. Assuming that both reactions are spontaneous under standard conditions, label the stronger acid, stronger base, weaker acid, and weaker base in each equation.

Other Things You Can Do

(Starred items require your instructor's permission.)

*1 Carry out Minilab 2 in Part III to help you visualize what happens during an extraction.

2 Write a short research paper about the chemistry and physiological effects of analgesic/antipyretic drugs using such sources as the *Kirk-Othmer Encyclopedia of Chemical Technology* [Bibliography, A7] and titles from section L of the Bibliography.

Recrystallization and Melting-Point Measurement. Identifying a Component of "Panacetin"

Purification Methods.

Operations:

OP-23 Recrystallization
OP-28 Melting Point
OP-4   Weighing
OP-6a Heat Sources
OP-21 Drying Solids

Before You Begin

1 Read the experiment and operations OP-23 and OP-28. Review the other operations as necessary.
2 Read "Using the Bibliography" in Appendix VII and familiarize yourself with the layout of the Bibliography.
3 Calculate the approximate volume of boiling water needed to dissolve all of your unknown compound if it is acetanilide *and* if it is phenacetin (see "Understanding the Experiment").

Scenario

The Scenario for this experiment is given in Experiment 2.

Applying Scientific Methodology

The problem you will be trying to solve in this experiment is "What is the identity of the unknown component of Panacetin?" Your course of action is described in the Directions; during the experiment you should be gathering and evaluating evidence that can help you solve the problem. Any working hypothesis can be no more than a guess, so you might want to wait until you carry out the *Purification* step before you formulate your initial hypothesis. You will then test your hypotheses in the *Analysis* step, which will allow you to reach a conclusion. Keep a careful record of your observations and your interpretation of the evidence so that you will be able to describe them in your report.

Using Chemical Reference Books

Knowing the physical properties of a compound—such as its melting or boiling point, refractive index, and spectral absorption bands—can help you

This and other general reference books are described under category A in Appendix VII.

identify the compound and estimate its degree of purity. Whenever a new compound is discovered or synthesized, its physical properties are measured and reported in one or more scientific journals, so that future investigators can recognize that compound when they encounter it. Physical properties and other information about the more familiar compounds are reported in chemical handbooks and other reference books. For instance, from its entry in *The Merck Index*, we learn that acetaminophen is known by at least 8 chemical and 45 proprietary drug names including Alpiny, Bickiemol, Cetadol, Dial-a-gesic, Enelfa, Finimal, Gelocatil, Homoolan, and so on down the alphabet. The same source lists the compound's important physical properties and uses, gives references describing its preparation from various starting materials, and tells where to find more information about it. For example, it refers to an evaluation of acetaminophen's effect on the kidneys on page 1238 of volume 320 (published in 1989) of the *New England Journal of Medicine*, as indicated by the following entry:

Evaluation of renal effects: D. P. Sandler *et al*, *N. Engl. J. Med.* **320**, 1238 (1989).

In such citations the abbreviated journal name is followed by the volume number, the page number on which the article begins, and the year of publication. Another convention is to list the year first (in boldface type), followed by the volume number (in italics) and page number, as illustrated for the following citation to an article in the *Journal of Organic Chemistry*.

J. Org. Chem.   **1994**,   *59*,   2546.
journal abbrev.   year   volume   page

The full names of scientific periodicals can be found in the *Chemical Abstracts Service Source Index* (*CASSI*).

Figure 3.1 reproduces the entry for acetanilide from the 14th edition of *Lange's Handbook of Chemistry*. This handbook reports that acetanilide has a melting point of 114.2°C and that it is only slightly soluble in water (aq) at 25°C, dissolving to the extent of 0.56 g per 100 mL of water at that temperature. The *Merck Index* reports that a gram of acetanilide dissolves in about 185 mL of cold (near room temperature) water or 20 mL of boiling water, and that a gram of phenacetin dissolves in 1310 mL of cold water or 82 mL of boiling water. Such entries often tell you what you need to know in order to purify a compound and identify or characterize the pure substance. These and other physical properties of acetanilide and phenacetin are summarized in Table 3.1 where solubilities from *The Merck Index* are expressed in grams of solute per 100 mL of water.

No.	Name	Formula	Formula weight	Beilstein reference	Density	Refractive index	Melting point	Boiling point	Flash point	Solubility in 100 parts solvent
a18	Acetanilide	$CH_3CONHC_6H_5$	135.17	12, 237	1.219_4^{15}		114.2	304	173 (OC)	0.56 aq^{25}; 29 alc; 2 bz; 27 chl; 25 acet; 5 eth

Figure 3.1 *Lange's Handbook of Chemistry* entry for acetanilide. (Reprinted with permission from *Lange's Handbook of Chemistry*, 14th ed., by N. A. Lange, edited by J. A. Dean. Copyright McGraw-Hill, Inc., New York, 1992.)

Understanding the Experiment

In this experiment you will purify and identify the unknown component of Panacetin from Experiment 2.

No separation is perfect; traces of impurities will always remain in a substance that has been separated from a mixture. Therefore some kind of purification process is needed to remove them. Solids can be purified by such operations as recrystallization, chromatography, and sublimation; liquids are usually purified by distillation or chromatography. The solubility information from Table 3.1 indicates that both acetaminophen and phenacetin are relatively soluble in boiling water but insoluble in cold water. This suggests that the unknown component can be purified by recrystallization, in which an impure solid is dissolved in a hot (usually boiling) solvent and crystallizes when the resulting solution is cooled to room temperature or below. Based on the mass of unknown you recovered from Experiment 2, you can estimate the volume of boiling water needed to dissolve that mass of either acetanilide or phenacetin. For example, the table shows that the solubility of acetanilide in boiling water is 5.0 g per 100 mL water. If your unknown is acetanilide and you recovered 1.15 g of the crude solid, the approximate volume of boiling water you will need to dissolve it is 23 mL.

$$1.15 \text{ g acetanilide} \times \frac{100 \text{ mL water}}{5.0 \text{ g acetanilide}} = 23 \text{ mL water}$$

Phenacetin, which is less soluble in boiling water, will require more water. You should begin the recrystallization using the smaller volume of water (don't add it all at once—see OP-23) and add more only if your compound does not dissolve in that amount of water at its boiling point.

After a compound has been purified, it should be *analyzed* to establish its identity and degree of purity. Although sophisticated instruments such as NMR spectrometers and mass spectrometers are now used to determine the structures of most newly discovered organic compounds, an operation as simple as a melting-point measurement can help identify a compound whose properties have already been reported. By itself, the melting point of a compound is not sufficient proof of its identity since thousands of compounds may share the same melting point. But when an unknown compound is thought to be one of a small number of possible compounds, its identity can often be determined by mixing the unknown with an authentic sample of each known compound and measuring the melting points of the mixtures. The use of a *mixture melting point* for identification is based on the fact that the melting point of a pure compound is lowered and its melting point range broadened when it is combined with a different compound. For example, if your unknown is phenacetin, it should melt sharply near 135°C, and a mixture of the unknown with an authentic sample of phenacetin should have essentially the same melting point. But a 1:1 mixture of phenacetin with acetanilide should melt at a considerably lower temperature over a much broader range.

The melting point of a compound can also give a rough indication of its purity. If your compound melts over a narrow range (~2°C or less) at a temperature close to the literature value, it is probably quite pure. If its melting point range is broad and substantially lower than the literature value, it is probably contaminated by water or other impurities.

Key Concept: The solubilities of most substances decrease as the temperature is lowered.

Key Concept: Impurities lower the melting point of a pure substance.

acetaminophen and phenacetin are soluble in water but insoluble in cold water.

acetanilide or phenacetine

Structures and Properties

acetanilide

phenacetin

Table 3.1 Physical properties

	M.W.	m.p.	Solubility, c.w.	Solubility, b.w.
acetanilide	135.2	114	0.54	5.0
phenacetin	179.2	135	0.076	1.22

Note: Melting points are in °C; solubilities are in grams of solute per 100 mL of cold water (c.w.) or boiling water (b.w.)

Directions

Safety Notes

> **Acetanilide and phenacetin can irritate the skin and eyes, so minimize contact with your unknown compound.**

Observe and Note: How much water was needed to dissolve it? What happens as the solution cools?

Waste Disposal: The filtrate may be poured down the drain.

Purification. Recrystallize [OP-23] the unknown drug component from Experiment 2 by boiling it [OP-6] with just enough distilled water to dissolve it completely, then letting it cool slowly to room temperature. If necessary, induce crystallization by scratching the sides of the flask with a stirring rod. Then cool the flask further in ice water to increase the yield of product. Collect the solid by vacuum filtration, washing it with a small amount of cold water. Dry the product [OP-21] to constant mass and weigh it [OP-4] in a tared vial.

Analysis. Grind a small amount of the purified, dry unknown component to a fine powder on a watch glass using a spatula or a flat-bottomed stirring rod. Divide the solid into three equal portions and prepare 1:1 mixtures by grinding one portion with a nearly equal amount of acetanilide and a second portion with a nearly equal amount of phenacetin. Measure the melting point ranges [OP-28] for (1) the purified unknown, (2) the mixture with acetanilide, and (3) the mixture with phenacetin. In each case, record the temperature at which you see the first trace of liquid and the temperature at which the sample is completely liquid. Unless your instructor indicates otherwise, you should carry out at least two measurements with each of these. Turn in the remaining product to your instructor in a vial labeled as illustrated in the margin (or as directed by your instructor).

Stop and Think: Is your unknown reasonably pure? If not, why not and what should you do?

> *Exp. 3*
> *phenacetin*
> *1.04 g*
> *m.p. 112-114°C*
> *Cynthia Sizer*
> *9-20-99*

Report. Your instructor may ask you to write up your results as if you were submitting them to Les Payne from the Association for Safe Pharmaceuticals. State the problem you attempted to solve in this experiment and state your initial hypothesis. Summarize and interpret the relevant evidence. Tell whether your hypothesis changed when it was tested and if so, why. State your final conclusion and tell how it is supported by the evidence. Calculate the percentage of the starting material that you recovered after crystallization and try to account for any significant losses (see Exercise 1).

Exercises

1 (a) How much boiling water is needed to dissolve 1.15 g of phenacetin (see Table 3.1)? (b) About how much phenacetin will remain dissolved when the water is cooled to room temperature? (c) Calculate the maximum mass of solid (undissolved) phenacetin that can be recovered when the cooled solution is filtered.

2 An unknown compound **X** is one of the four compounds listed in Table 3.2. A mixture of **X** with benzoic acid melts at 89°C, a mixture of **X** with phenyl succinate melts at 120°C, and a mixture of **X** with *m*-aminophenol melts at 102°C. Give the identity of **X** and explain your reasoning.

Table 3.2 Melting points for Exercise 2

Compound	*m.p., °C*
o-toluic acid	102
benzoic acid	121
phenyl succinate	121
m-aminophenol	122

3 Tell whether each of the following experimental errors will raise or lower the amount of product you recover, and why. (a) You failed to dry the product completely. (b) You used enough water to recrystallize phenacetin but your unknown was acetanilide. (c) In Experiment 2 you didn't extract all of the aspirin from the dichloromethane solution.

4 Tell whether each of the experimental errors in Exercise 3 will affect the melting point of the unknown component. If it will, tell how it will affect the melting point and why.

5 Using one or more of the reference books listed in Appendix VII, give the following information about the analgesic drug ibuprofen: chemical names, molecular weight, molecular formula, structural formula, melting point, and a suitable recrystallization solvent.

6 Locate citations for one or more journal articles that give procedures for preparing (a) aspirin and (b) propoxyphene (Darvon). Give the full name of each journal, the volume number, the page number(s), and the year.

Other Things You Can Do

(Starred items require your instructor's permission.)

*1 Purify an unknown solid by recrystallization as described in Minilab 3.

*2 Carry out the purification and melting-point analysis of an acetanilide-salicylic acid mixture as described in *J. Chem. Educ.* **1989**, *66*, 1063 or as directed by your instructor.

3 Calibrate your thermometer by measuring the melting points of compounds listed in Table F1 of OP-28.

4 Look up the properties of a common organic compound in at least five different reference books and compare the kind of information provided by each.

EXPERIMENT 4

Heating Under Reflux. Synthesis of Salicylic Acid from Wintergreen Oil

Preparation and Purification of Solids.

Operations:

OP-2 Using Standard-Taper Glassware
OP-4   Weighing
OP-6   Heating
OP-12 Vacuum Filtration
OP-21 Drying Solids
OP-23 Recrystallization
OP-28 Melting Point

Before You Begin

1 Read the experiment and operation OP-2. Read "Smooth Boiling Devices" and "Heating Under Reflux" in OP-6 and review the other operations as necessary.

2 Read Appendix IV, *Calculations for Organic Synthesis.* Then calculate the mass and volume of 10.0 mmol of methyl salicylate, the theoretical yield of salicylic acid from that much methyl salicylate, and the volume of water needed to recrystallize that much salicylic acid.

Be sure to distinguish millimoles (mmol) from moles (mol); most Prelab Assignments will specify amounts of chemicals in millimoles.

3 Complete the experimental plan (page 42) by specifying all sizes and quantities indicated by asterisks in the Chemicals and Supplies list. Note that boiling flasks come in 25-, 50-, 100-, 250-, and 500-mL sizes; Erlenmeyer flasks in 25-, 50-, 125-, and 250-mL sizes; and beakers in 30-, 50-, 150-, 250-, and 400-mL sizes.

Scenario

The new-age pharmaceutical company Natural Nostrums manufactures drugs from "natural" starting materials. For example, the company manufactures a painkilling drug it advertises as "organic aspirin" starting with methyl salicylate, which occurs naturally in wintergreen oil. Most commercially marketed aspirin is manufactured starting with benzene, a product of petroleum refining. An intermediate in both of these syntheses is salicylic acid.

Natural Nostrums claims that its aspirin, which is supposedly more natural than asprin made from benzene, has fewer side effects than ordinary aspirin. Critics have accused the company of false and misleading advertising, asserting that salicylic acid made from methyl salicylate is no different than salicylic acid made from benzene, and that the resulting aspirin is therefore no better than any other aspirin.

OH

.COOH

salicylic
acid

OH

COOCH₃

methyl
salicylate

OCOCH₃

.COOH

aspirin

synthesis of aspirin from methyl salicylate

OH

phenol

benzene

OH

.COOH

OCOCH₃

.COOH

salicylic
acid

aspirin

synthesis of aspirin from benzene

Les Payne, director of operations for the Association for Safe Pharmaceuticals (ASP), is investigating the company. He just shipped your supervisor a sample of salicylic acid manufactured from benzene and a bottle of methyl salicylate that one of his agents obtained from the chemical stock room at Natural Nostrums. Your assignment is to prepare salicylic acid from this methyl salicylate and find out whether or not it is the same as salicylic acid made from benzene.

Applying Scientific Methodology

As in the previous experiments, you need to state the problem as a question, formulate a working hypothesis, follow the course of action described in the Directions, gather and evaluate evidence, test your hypothesis, arrive at a conclusion, and report your findings.

Wintergreen Oil

For many years commercial methyl salicylate was obtained from wintergreen oil, an aromatic liquid distilled from the leaves of the wintergreen plant (*Gaultheria procumbens*) or the bark of sweet birch trees (*Betula lenta*). When cheap raw materials became available from petroleum, a more economical commercial process was developed for synthesizing methyl salicylate from salicylic acid and methanol (CH_3OH, methyl alcohol). Regardless of its source, methyl salicylate is methyl salicylate. There is no difference between its pure natural and synthetic forms because both are composed of the same kind of molecules and must therefore have the same properties. Any differences between oil of wintergreen obtained from natural sources and synthetic methyl salicylate are mainly due to impurities in the natural oil.

Methyl salicylate, both natural and synthetic, has been used for many years as a flavoring agent because of its pleasant penetrating odor and flavor. According to Euell Gibbons, author of *Stalking the Healthful Herbs* and other books about edible and medicinal wild plants, you can prepare a good wintergreen tea by pouring boiling water into a jar filled with freshly picked

wintergreen plant

OH

.COOH
$\xrightarrow{CH_3OH}$

salicylic
acid (from
benzene)

OH

COOCH₃

methyl
salicylate

A natural substance is one obtained directly from plant, animal, or mineral sources. A synthetic substance is one prepared by chemically altering other substances, which may themselves be either natural or synthetic.

OH   COOCH$_3$ OH   COOCH$_3$

| natural methyl salicylate | synthetic methyl salicylate |

wintergreen leaves and letting it steep overnight or longer. The distinctive flavor of wintergreen is found in a variety of commercial products including candy, chewing gum, root beer, and toothpaste. Medicinally, wintergreen oil has some of the painkilling properties of other *salicylates*, a group of related organic compounds that includes salicylic acid and aspirin. When absorbed through the skin, it produces an astringent but soothing sensation, and it is frequently used in preparations that alleviate muscular aches and arthritis such as BenGay.

Understanding the Experiment

In this experiment you will be carrying out an *organic synthesis*, the preparation of salicylic acid from methyl salicylate. An organic synthesis can be described as the preparation of a desired organic compound by chemically modifying the molecules of another organic compound, the starting material. Until 1874 commercial salicylic acid was synthesized entirely from natural wintergreen oil. In this experiment you will reproduce this nineteenth-century synthesis of salicylic acid starting with methyl salicylate, the major constituent of wintergreen oil.

Organic reactions are slower than most inorganic reactions. For example, when aqueous solutions of silver nitrate and sodium chloride are mixed, the resulting reaction is almost instantaneous, taking place as quickly as silver and chloride ions can come together to form silver chloride.

$$Ag^+ + Cl^- \rightarrow \textbf{AgCl}$$

In an organic reaction the reacting molecules may come together about as frequently as the ions in a precipitation reaction, but they may not collide with enough energy or in the right orientation to react. An ion-combination reaction has such a low activation energy that ions need only "stick together" to form a product. Covalent molecules must ordinarily undergo a process of bond breaking and bond formation to be converted to product molecules, and such processes have relatively high activation energies.

Some of the bond-breaking and bond-forming steps involved in this experiment's synthesis are shown in the *mechanism* on the next page for the conversion of methyl salicylate to salicylic acid.

The *reaction*, then, is the crucial step in an organic synthesis—it may be over in a few minutes or require a few weeks, but enough time should be allowed to bring the reaction as nearly as possible to completion before proceeding to the next steps. These steps have already been described in Experiments 2 and 3: The desired product must be *separated* from the reaction mixture, *purified* to remove residual contaminants, and *analyzed* to verify its identity and purity.

Key Concept: *A mechanism is a step-by-step description of a chemical reaction, showing what bonds are broken and formed as reactant molecules are converted to product molecules.*

Completion implies that as many reactant molecules as possible have been converted to product molecules under the conditions of the reaction.

Mechanism for the reaction. Note that the OH group of methyl salicylate is ionized at the high pH of the reaction mixture.

Methyl salicylate is the limiting reactant in this synthesis; sodium
hydroxide is used in excess to ensure a reasonably fast, complete reaction. It
is best to measure limiting reactants that are liquids by mass because the
mass of a liquid can be measured more accurately than its volume. To avoid
waste and minimize spillage at the balance, you should dispense the estimated *volume* of liquid from the reagent bottle before you take it to the balance to weigh it. This is why you were asked (in "Before you Begin") to
calculate both the mass and volume of methyl salicylate required. If the
mass of the methyl salicylate is not within about 5% of your calculated
value, you can add or remove liquid with a dropper.

See Appendix IV for a discussion of limiting reactants.

You will carry out the reaction by boiling a mixture of methyl salicylate
and sodium hydroxide in a flask equipped with a reflux condenser. A heating mantle or other flameless heat source capable of boiling water (not a
steam bath) is preferred. The purpose of the reflux condenser is to condense
water and methyl salicylate vapors and return them to the reaction flask so
that they don't boil away.

The initial product of the reaction will be a salt of salicylic acid, disodium
salicylate, and not salicylic acid itself. Adding 3 M sulfuric acid will then precipitate salicylic acid. When you recover the product by vacuum filtration,
remember that the *solvent* in the reaction mixture is water, so use cold water
(not 3 M sulfuric acid) to wash the product on the filter. The solubility of salicylic acid in water is about 0.18 g/100 mL at 20°C and 6.7 g/100 mL at the
boiling point, so it can be purified by recrystallization from water. The melting point of your product should give you a good indication of its purity; the
melting range of pure salicylic acid is reported to be 158–160°C. You can then
carry out a mixed melting point with the benzene-derived salicylic acid provided by your instructor to see whether it is identical to your own.

Stop and Think: Why? (Consider the pH of the reaction mixture at this point.)

Stop and Think: Why?

To help you organize your time efficiently, an experimental plan for the
preparation of salicylic acid is provided. You should write your own experimental plans (described in Appendix V) for subsequent experiments. Note
that the 30-minute reaction period should be used to collect the supplies
and assemble the apparatus needed for upcoming operations.

Reactions and Properties

Table 4.1 Physical properties

	M.W.	m.p.	b.p.	d
methyl salicylate	152.1	−8	223	1.174
salicylic acid	138.1	159		

Note: Melting points and boiling points are in °C; density (d) is in g/mL

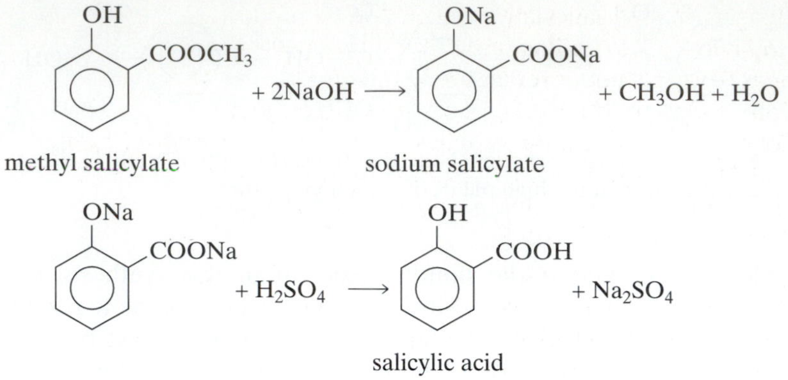

$$\text{methyl salicylate} + 2NaOH \longrightarrow \text{sodium salicylate} + CH_3OH + H_2O$$

$$\text{sodium salicylate} + H_2SO_4 \longrightarrow \text{salicylic acid} + Na_2SO_4$$

Experimental Plan

Chemicals and Supplies Needed

[An asterisk (*) indicates a number (size or quantity) to be filled in.]

Heating under reflux [OP-6]

> * g (* mL) of methyl salicylate (avoid contact, inhalation)
>
> 15 mL of 6 M NaOH (wear gloves, avoid contact)
>
> *-mL boiling flask
>
> reflux condenser, boiling chips, ring stand, and heat source (specify)

Addition of sulfuric acid

> 16 mL of 3 M H$_2$SO$_4$
>
> *-mL beaker
>
> pH paper, stirring rod, and dropper

Vacuum filtration [OP-12]

> * mL of cold water for washing
>
> Buchner funnel, filter flask, filter paper, flat-bottomed stirring rod, spatula, and watch glass

Recrystallization [OP-23]

> * mL of water for recrystallization
>
> * mL of cold water for washing
>
> two *-mL Erlenmeyer flasks, *-mL graduated cylinder
>
> funnel, fluted filter paper, and vacuum filtration apparatus (see above)

Drying [OP-21] and weighing [OP-4]

> drying dish, desiccator jar (or oven), tared vial

Melting points [OP-28]

> m.p. tubes, flat-bottomed stirring rod (or spatula), watch glass

Lab Checklist

- Collect and clean supplies for reflux
- Obtain heat source
- Assemble reflux apparatus

- Measure methyl salicylate and 6 *M* NaOH solution
- Add reactants to boiling flask, turn on condenser water
- Heat reactants under reflux for 30 minutes
- Collect supplies for H_2SO_4 addition (during reflux)
- Collect vacuum filtration supplies and assemble apparatus (during reflux)
- Collect supplies for recrystallization (during reflux)
- Measure 3 *M* H_2SO_4 solution (during reflux)
- Precipitate product with H_2SO_4
- Filter and wash product; air dry on filter
- Measure and boil water for recrystallization
- Recrystallize product from boiling water
- Filter and wash product; air dry on filter
- Dry product
- Weigh product
- Measure melting point
- Turn in product
- Clean up

Directions

Safety Notes

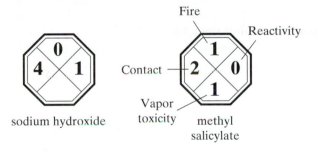

sodium hydroxide   methyl salicylate

Reaction: Obtain an appropriate heat source and position it correctly. Assemble an apparatus for reflux [OP-6], using standard-taper glassware if available [OP-2]. Make sure the apparatus is clamped securely to a ring stand. Measure 15 mL of aqueous 6 *M* (20%) sodium hydroxide into the boiling flask , followed by 10.0 mmol of methyl salicylate and a few boiling chips. Do not add both reactants through the same funnel; if you do, a solid may plug up the funnel stem. Start water flowing in the condenser jacket, then have the instructor check your apparatus before you begin heating. Heat the reaction mixture under reflux for 30 minutes, measuring from the time the solution starts to boil.

Take Care: Wear gloves, avoid contact with the NaOH.
Observe and Note: What did you observe when you combined methyl salicylate with 6 *M* NaOH? When you heated the reaction mixture to boiling?
Stop and Think: What do you think caused the changes you observed?

Observe and Note: What happened?

Waste Disposal: The filtrate may be poured down the drain.

Observe and Note: How much water was required?

Waste Disposal: The filtrate may be poured down the drain.

Stop and Think: Do the results support your initial hypothesis?

When the reaction time is up, let the reaction mixture cool down and transfer it to a beaker large enough to contain the reaction mixture and added sulfuric acid. Slowly add 16 mL of aqueous 3 *M* sulfuric acid with stirring, and test the pH of the supernatant liquid using a strip of pH paper. If the pH is above 2, add enough additional sulfuric acid drop by drop until the pH is 2 or below and no more precipitate forms with additional H_2SO_4. Cool the mixture in an ice/water bath for about 10 minutes.

Separation. Collect the salicylic acid by vacuum filtration using a small Buchner funnel [OP-12], washing it on the filter with ice-cold water.

Purification and Analysis. Purify the salicylic acid by recrystallization from boiling water [OP-23]. There should be no need to filter the hot solution. Collect the product by vacuum filtration, washing it with a little cold water. Dry the product [OP-21] to constant mass and measure its mass [OP-4]. Obtain a melting point [OP-28] of the dry product *and* of a 1:1 mixture of the product with salicylic acid synthesized from benzene. Turn in the product in a properly labeled vial.

Report. State the problem you attempted to solve in this experiment and state your initial hypothesis. Summarize and interpret the relevant evidence. Tell whether your hypothesis changed when it was tested and if so, why. State your final conclusion and tell how it is supported by the evidence. Recalculate the theoretical yield of salicylic acid based on the actual mass of methyl salicylate you used, and calculate your percentage yield. Try to account for any significant material losses (see Exercise 3).

Exercises

1 (a) Calculate the volume of 6 *M* NaOH required to react completely with 10.0 mmol of methyl salicylate. How much of the 6 *M* NaOH you used was in excess of the theoretical amount? (b) What volume of 3 *M* H_2SO_4 is needed to neutralize all of the disodium salicylate and the excess NaOH present after the initial reaction? How much sulfuric acid was in excess?

2 Refer to the reactions on page 42 as you answer the following questions. (a) Which hydrogen atom(s) of salicylic acid are acidic? Which hydrogen atom(s) of methyl salicylate would you expect to be acidic? (b) Draw the structure of the white solid that forms immediately after NaOH and methyl salicylate are combined, and write an equation for its formation.

3 Based on the amount of water you used during the recrystallization, estimate the amount of salicylic acid that was lost as a result of being dissolved in the filtrate. Assume that the recrystallization solution was cooled to 10°C; the solubility of salicylic acid at that temperature is 0.14 g per 100 mL of water.

4 Describe and explain the possible effect on your results of the following experimental errors or variations. (a) You forget to turn on the condenser water for the first ten minutes of the reaction. (b) You add only enough 3 *M* H_2SO_4 to bring the pH down to 4. (c) You stop

the reaction after 15 minutes rather than the 30 minutes specified. (d) The bottle labeled "salicylic acid from benzene" is mislabeled; it actually contains acetylsalicylic acid (aspirin).

5 From the equations in the Reactions and Properties section, you can see that methanol and sodium sulfate are by-products of the synthesis of salicylic acid. During which step of the synthesis would these compounds have been separated from the final product? Explain your answer, based on any relevant properties of salicylic acid and of the by-products. (If necessary, you can look up the properties in a reference book.)

Other Things You Can Do

(Starred items require your instructor's permission)

*1 Develop and test a hypothesis based on your observations in Minilab 4. You can also test your product from this experiment with ferric chloride as described in Minilab 4.

*2 Convert your salicylic acid to aspirin by following the procedure in Experiment 34 (you can omit the spectral analysis).

3 Find some examples of organic compounds besides methyl salicylate that are available commercially in both natural and synthetic forms. You might start hunting at a local drugstore or health food store. Look up the compounds in *The Merck Index* and give their chemical structures.

4 Write a research paper about salicylic acid and the salicylates using references from the Bibliography.

Simple Distillation, Gas Chromatography. Preparation of Synthetic Banana Oil

EXPERIMENT 5

Preparation, Purification, and Analysis of Liquids. Gas Chromatography

Operations:

OP-19 Washing Liquids
OP-20 Drying Liquids
OP-25 Simple Distillation
OP-32 Gas Chromatography
OP-2   Using Standard-Taper Glassware
OP-4   Weighing
OP-6   Heating
OP-11   Gravity Filtration

Before You Begin

1   Read the experiment and operations OP-19, OP-20, OP-25, and OP-32. Review the other operations as necessary. (Reading OP-25 and OP-32 may be deferred until the second lab period for this experiment.)
2   Calculate the mass and volume of 150 mmol of isopentyl alcohol and the theoretical yield of isopentyl acetate from this amount of the alcohol.
3   Read Appendix V *Planning an Experiment*, then write an experimental plan comparable to the one you used in Experiment 4.

Scenario

Your supervisor has just received the following message from the Cavendish Distilling Company (CDC).

Greetings:

     We have a problem. Cavendish Distilling Company markets the popular liqueur Banana Elixir, which is flavored with a natural banana extract from fruit grown on our Caribbean banana plantations. Last June, Hurricane Floyd blew down all of our banana trees. Our stock of banana extract is running low and we have no alternative source of bananas at this time. As a temporary solution, we have decided to add a synthetic banana flavoring to our remaining stock of the natural extract until our plantations start producing again.

The banana plant (Musa cavendishii and other species) is actually a gigantic herb and not a true tree. Each year the foliage-bearing part withers away and is replaced by new growth from an underground stem.

Synthetic banana flavorings are formulated mainly from isopentyl acetate, with smaller amounts of other esters. I understand that esters can be prepared economically by a process called the Fischer esterification, which involves the combination of an acid such as acetic acid with an alcohol. According to our technical staff, one problem with this method is that Fischer esterifications do not go to completion, leaving considerable amounts of starting material in the product. We can tolerate up to 15% isopentyl alcohol in our isopentyl acetate since the alcohol is a component of natural banana extract. But we cannot tolerate more than 2% acetic acid because it would tend to degrade the flavor of the liqueur. Will you have your consulting chemists prepare some isopentyl acetate and analyze it to see if it falls within our tolerances?

Amy Lester, CEO

banana tree

Applying Scientific Methodology

By now you should be familiar with the steps involved in applying scientific methodology to the solution of a problem, so they will not all be repeated hereafter. In this experiment the problem described in the Scenario does not involve the identity of the product but its purity, so your course of action will include the use of an instrumental method, gas chromatography, to determine the composition of the product. You should be able to gather some evidence regarding the presence of impurities in your isopentyl acetate before you carry out the analysis, so record and evaluate your observations as usual.

Esters and Artificial Flavorings

The word *flavor* is used to describe the overall sensory effect of a substance taken into the mouth; flavor may involve tactile, temperature, and pain sensations as well as smell and taste. Many fruits, flowers, and spices contain esters that contribute to their characteristic flavors, where an *ester* is an organic compound that contains the functional group (characteristic combination of atoms) shown in the margin. Most volatile esters have strong, pleasant odors that can best be described as "fruity." Some esters with flavors characteristic of real and "fantasy" fruits are shown in Table 5.1. The ester you will prepare in this experiment, isopentyl acetate, has a strong banana odor when undiluted and an odor reminiscent of pears in dilute solution. It is used as an ingredient in artificial coffee, butterscotch, and honey flavorings as well as in pear and banana flavorings.

Many different esters are included in the basic repertoire of the flavor chemist, who combines natural and synthetic ingredients to prepare artificial flavorings. These ingredients may include natural products, synthetic organic compounds identical to those found in nature, and synthetic compounds not found in nature but accepted as safe for use in food. Each flavor ingredient is characterized by one or more flavor *notes* that suggest the predominant impact it makes on the senses of taste and smell. Although the flavor note of a single ingredient may seem unrelated to the overall character of a natural flavor, the combination of carefully selected ingredients in the right proportions can often yield a good approximation of that flavor. A

$$\overset{\displaystyle O}{\overset{\|}{-C-O-R}}$$

functional group
of an ester

$$\overset{\displaystyle O}{\overset{\|}{CH_3COCH_2CH_2CHCH_3}}$$
$$\underset{\displaystyle CH_3}{|}$$

isopentyl acetate

Some food products such as cola beverages and Juicy Fruit gum are characterized by fantasy flavors that have no counterparts in nature, but most artificial flavorings are meant to resemble natural flavors.

Table 5.1 Flavor notes of some esters used in artificial flavorings

Name	Structure	Flavor note
Propyl acetate	$CH_3\overset{\overset{O}{\|}}{C}\!-\!OCH_2CH_2CH_3$	pears
Octyl acetate	$CH_3\overset{\overset{O}{\|}}{C}\!-\!O(CH_2)_7CH_3$	oranges
Benzyl acetate	$CH_3\overset{\overset{O}{\|}}{C}\!-\!OCH_2\!-\!\bigcirc$	peaches, strawberries
Isopentenyl acetate	$CH_3\overset{\overset{O}{\|}}{C}\!-\!OCH_2CH\!=\!\overset{\overset{CH_3}{\|}}{C}\!-\!CH_3$	"Juicy-Fruit"
Isobutyl propionate	$CH_3CH_2\overset{\overset{O}{\|}}{C}\!-\!OCH_2\overset{\overset{CH_3}{\|}}{CH}\!-\!CH_3$	rum
Ethyl butyrate	$CH_3CH_2CH_2\overset{\overset{O}{\|}}{C}\!-\!OCH_2CH_3$	pineapples

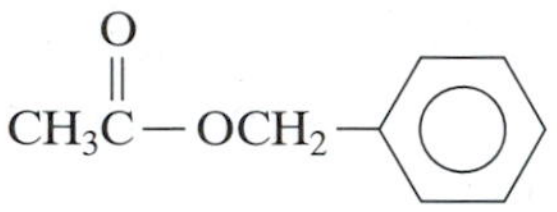

Fixatives

$$CH_2\overset{\overset{OH}{\|}}{C}H\overset{\overset{OHOH}{\|\ \ \|}}{C}H_2$$

glycerine

$$\bigcirc\!-\!\overset{\overset{O}{\|}}{C}OCH_2\!-\!\bigcirc$$

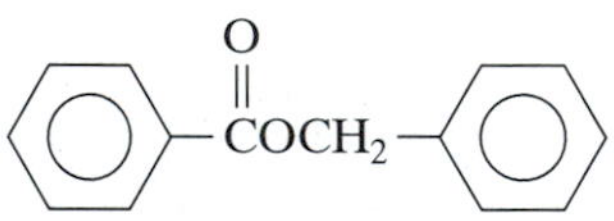

benzyl benzoate

Vehicle

$$CH_3CH_2OH$$

ethyl alcohol
(ethanol)

high-boiling *fixative* such as glycerine or benzyl benzoate is usually added to an artificial flavoring to retard vaporization of volatile components, and the flavor notes of the individual components are blended by dissolving them in a solvent called the *vehicle*. The most frequently used vehicle is ethanol (ethyl alcohol).

The formulation of artificial flavorings is perhaps as much an art as a science. The components of a strawberry flavoring, for example, may vary widely depending on the manufacturer and the specific application. Because natural flavors are usually very complex, a cheap artificial flavoring may be a poor imitation of its natural counterpart, but advances in flavor chemistry have made possible the production of superior flavorings that reproduce natural flavors very closely. Superior flavorings may contain natural oils or extracts that have been fortified with a few synthetic ingredients to enhance the overall effect and replace flavor elements lost during the distillation or extraction process. Even the most experienced flavor chemist can't hope to do as well as a strawberry plant, which may combine several hundred different flavor components in its berries. But a superior strawberry flavoring with a few dozen ingredients may be hard to distinguish from the real thing, except by the most discriminating of strawberry aficionados.

Understanding the Experiment

In this experiment you will prepare synthetic banana oil, which is known by several chemical names including isopentyl acetate, isoamyl acetate, and 3-methylbutyl ethanoate. Esters are often prepared by the Fischer esterifi-

Synthetic banana flavorings are formulated mainly from isopentyl acetate, with smaller amounts of other esters. I understand that esters can be prepared economically by a process called the Fischer esterification, which involves the combination of an acid such as acetic acid with an alcohol. According to our technical staff, one problem with this method is that Fischer esterifications do not go to completion, leaving considerable amounts of starting material in the product. We can tolerate up to 15% isopentyl alcohol in our isopentyl acetate since the alcohol is a component of natural banana extract. But we cannot tolerate more than 2% acetic acid because it would tend to degrade the flavor of the liqueur. Will you have your consulting chemists prepare some isopentyl acetate and analyze it to see if it falls within our tolerances?

Amy Lester, CEO

banana tree

Applying Scientific Methodology

By now you should be familiar with the steps involved in applying scientific methodology to the solution of a problem, so they will not all be repeated hereafter. In this experiment the problem described in the Scenario does not involve the identity of the product but its purity, so your course of action will include the use of an instrumental method, gas chromatography, to determine the composition of the product. You should be able to gather some evidence regarding the presence of impurities in your isopentyl acetate before you carry out the analysis, so record and evaluate your observations as usual.

Esters and Artificial Flavorings

The word *flavor* is used to describe the overall sensory effect of a substance taken into the mouth; flavor may involve tactile, temperature, and pain sensations as well as smell and taste. Many fruits, flowers, and spices contain esters that contribute to their characteristic flavors, where an *ester* is an organic compound that contains the functional group (characteristic combination of atoms) shown in the margin. Most volatile esters have strong, pleasant odors that can best be described as "fruity." Some esters with flavors characteristic of real and "fantasy" fruits are shown in Table 5.1. The ester you will prepare in this experiment, isopentyl acetate, has a strong banana odor when undiluted and an odor reminiscent of pears in dilute solution. It is used as an ingredient in artificial coffee, butterscotch, and honey flavorings as well as in pear and banana flavorings.

Many different esters are included in the basic repertoire of the flavor chemist, who combines natural and synthetic ingredients to prepare artificial flavorings. These ingredients may include natural products, synthetic organic compounds identical to those found in nature, and synthetic compounds not found in nature but accepted as safe for use in food. Each flavor ingredient is characterized by one or more flavor *notes* that suggest the predominant impact it makes on the senses of taste and smell. Although the flavor note of a single ingredient may seem unrelated to the overall character of a natural flavor, the combination of carefully selected ingredients in the right proportions can often yield a good approximation of that flavor. A

$$\overset{\displaystyle O}{\underset{\displaystyle }{\overset{\displaystyle \|}{-C}}} - O - R$$

functional group
of an ester

$$\overset{\displaystyle O}{\overset{\displaystyle \|}{CH_3COCH_2CH_2CHCH_3}}$$
$$\underset{\displaystyle CH_3}{\displaystyle |}$$

isopentyl acetate

Some food products such as cola beverages and Juicy Fruit gum are characterized by fantasy flavors that have no counterparts in nature, but most artificial flavorings are meant to resemble natural flavors.

Table 5.1 Flavor notes of some esters used in artificial flavorings

Name	Structure	Flavor note
Propyl acetate	O $\parallel$ $CH_3C - OCH_2CH_2CH_3$	pears
Octyl acetate	O $\parallel$ $CH_3C - O(CH_2)_7CH_3$	oranges
Benzyl acetate	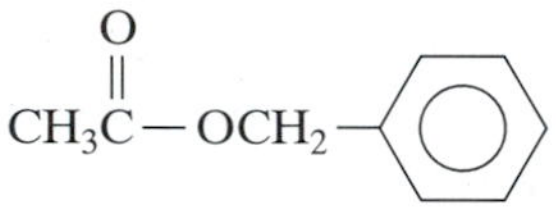 O $\parallel$ $CH_3C - OCH_2 -$ ⬡	peaches, strawberries
Isopentenyl acetate	O CH_3 $\parallel$ $\mid$ $CH_3C - OCH_2CH = C - CH_3$	"Juicy-Fruit"
Isobutyl propionate	O CH_3 $\parallel$ $\mid$ $CH_3CH_2C - OCH_2CH - CH_3$	rum
Ethyl butyrate	O $\parallel$ $CH_3CH_2CH_2C - OCH_2CH_3$	pineapples

Fixatives

$$\begin{array}{ccc} OH & OHOH \\ \mid & \mid \;\; \mid \\ CH_2 & CHCH_2 \end{array}$$

glycerine

⬡—$\overset{\overset{\textstyle O}{\parallel}}{C}OCH_2$—⬡

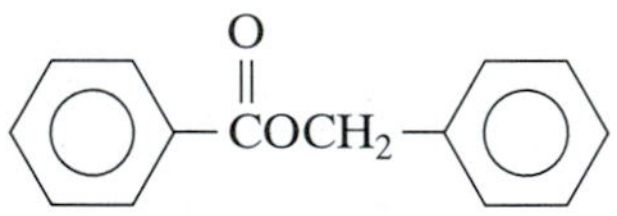

benzyl benzoate

Vehicle

$$CH_3CH_2OH$$

ethyl alcohol
(ethanol)

high-boiling *fixative* such as glycerine or benzyl benzoate is usually added to an artificial flavoring to retard vaporization of volatile components, and the flavor notes of the individual components are blended by dissolving them in a solvent called the *vehicle*. The most frequently used vehicle is ethanol (ethyl alcohol).

The formulation of artificial flavorings is perhaps as much an art as a science. The components of a strawberry flavoring, for example, may vary widely depending on the manufacturer and the specific application. Because natural flavors are usually very complex, a cheap artificial flavoring may be a poor imitation of its natural counterpart, but advances in flavor chemistry have made possible the production of superior flavorings that reproduce natural flavors very closely. Superior flavorings may contain natural oils or extracts that have been fortified with a few synthetic ingredients to enhance the overall effect and replace flavor elements lost during the distillation or extraction process. Even the most experienced flavor chemist can't hope to do as well as a strawberry plant, which may combine several hundred different flavor components in its berries. But a superior strawberry flavoring with a few dozen ingredients may be hard to distinguish from the real thing, except by the most discriminating of strawberry aficionados.

Understanding the Experiment

In this experiment you will prepare synthetic banana oil, which is known by several chemical names including isopentyl acetate, isoamyl acetate, and 3-methylbutyl ethanoate. Esters are often prepared by the Fischer esterification method, which involves heating a carboxylic acid with an alcohol in the presence of an acid catalyst, as shown by this general equation:

$$
\underset{\substack{\text{carboxylic} \\ \text{acid}}}{\overset{\overset{\displaystyle O}{\parallel}}{R C O H}} + \underset{\text{alcohol}}{H O R'} \xrightarrow{H^+} \underset{\text{ester}}{\overset{\overset{\displaystyle O}{\parallel}}{R C O R'}} + H_2O
$$

The acid catalyst is used to increase the rate of the reaction, which would otherwise require a much longer reaction time. You will synthesize isopentyl acetate by combining isopentyl alcohol (3-methyl-1-butanol) with acetic acid and sulfuric acid and then heating the reaction mixture under reflux for an hour. The alcohol is the limiting reactant, so it should be weighed; the acids can be measured by volume. The esterification reaction is reversible, having an equilibrium constant of approximately 4.2. If you were to start with equimolar amounts of acetic acid and isopentyl alcohol, only about two-thirds of each reactant would be converted to isopentyl acetate by the time equilibrium was reached. Your highest attainable yield in that case would be only 67 percent of the theoretical value. To increase the yield of isopentyl acetate you will apply Le Châtelier's principle, using a 100% excess of acetic acid—the less expensive reactant—to shift the equilibrium toward the products. Even then, the reaction will not be complete at equilibrium, so the product mixture will contain some unreacted isopentyl alcohol as well as the excess acetic acid.

    As you learned in Experiment 2, a pure component can be obtained from a mixture by separating it from all other components of the mixture, using procedures that take advantage of differences in solubility, boiling points, acid-base properties, and other characteristics. Because isopentyl acetate is a liquid, the separation and purification operations will differ from those used previously for solid products. At the end of the reflux period, the reaction mixture will contain (in addition to the ester) unreacted acetic acid, sulfuric acid, water, unreacted isopentyl alcohol, and some unwanted by-products. (Reactions carried out with acid catalysts often yield polymeric, tarlike by-products that are high boiling and insoluble in water.)

    Isopentyl acetate is quite insoluble in water, whereas both acetic acid and sulfuric acid are water soluble and acidic. This makes it easy to separate the two acids from the product by washing the reaction mixture with water and then with aqueous sodium bicarbonate. Water does not remove the acids entirely because they are somewhat soluble in the ester as well, but it removes the bulk of them and thus helps prevent a violent reaction with sodium bicarbonate in the second washing step. The aqueous sodium bicarbonate converts the acids to their salts, sodium acetate and sodium sulfate, which are insoluble in the ester but very soluble in water; these salts migrate to the aqueous layer where they can be removed. Note that the rule of thumb given in OP-19 will help you estimate the quantities of wash solvent needed.

    The water that forms during the reaction will be separated from the ester along with the wash liquids. Any traces of water that remain are then

Key Concept: For a reaction at equilibrium, adding more of a reactant or removing a product will shift the equilibrium to favor the products.

removed by a drying agent, magnesium sulfate. The rule of thumb in OP-20 will help you estimate the amount of drying agent needed.

Because isopentyl alcohol has a lower boiling point than isopentyl acetate and the by-products have higher boiling points, it should be possible—in principle—to remove the alcohol and by-products from the ester by distillation. Isopentyl alcohol should distill first, followed by the ester, and any by-products should remain behind in the boiling flask. The separation is incomplete for the reasons described in OP-25, so you will still have some isopentyl alcohol in your isopentyl acetate after the purification step.

You will determine the composition of your distillate by injecting a very small amount into an instrument called a gas chromatograph. (Your instructor will demonstrate the operation of this instrument.) Inside the gas chromatograph, the liquid will vaporize and the vapors of different components will travel through a packed column at different rates. As the vapors exit the column, their presence will be detected and recorded on a graph called a gas chromatogram, which should display several peaks of different sizes. The area under the peak for each component will be proportional to the amount of that component present, so by measuring the peak areas you can estimate the percentage of each component in the distillate. This will tell you how much (if any) isopentyl alcohol and acetic acid remain in your product.

The procedure for the synthesis of isopentyl acetate is summarized in the flow diagram in Figure 5.1, which illustrates the transformations or separations that occur during each operation.

Isopentyl acetate is known to be an *alarm pheromone* of the honeybee, where a pheromone is a "molecular messenger" that produces a response in another member of the same species such as mating behavior or aggression toward a perceived threat. When a worker honeybee stings someone it releases a tiny amount (about 1 μg) of isopentyl acetate in its stinger, attracting more honeybees to the scene. Although isopentyl acetate alone does not cause the bees to sting (other pheremones in the stinger do that), it agitates them and puts them on guard. So it might be wise to steer clear of beehives on your way home from lab!

Reactions and Properties

$$CH_3\overset{\displaystyle O}{\overset{\displaystyle \|}{C}}-OH + HOCH_2CH_2\overset{\displaystyle CH_3}{\overset{\displaystyle |}{C}}HCH_3 \rightleftharpoons CH_3\overset{\displaystyle O}{\overset{\displaystyle \|}{C}}-OCH_2CH_2\overset{\displaystyle CH_3}{\overset{\displaystyle |}{C}}HCH_3 + H_2O$$

acetic acid   isopentyl alcohol   isopentyl acetate

Table 5.2 Physical properties

	M.W.	b. p.	d	solubility
acetic acid	60.1	118	1.049	miscible
isopentyl alcohol	88.1	130	0.815	2.7
isopentyl acetate	130.2	142	0.876	0.25
sulfuric acid	98.1	290	1.84	miscible

Note: Boiling points are in °C; densities in g/mL, solubilities in g/100 mL solvent

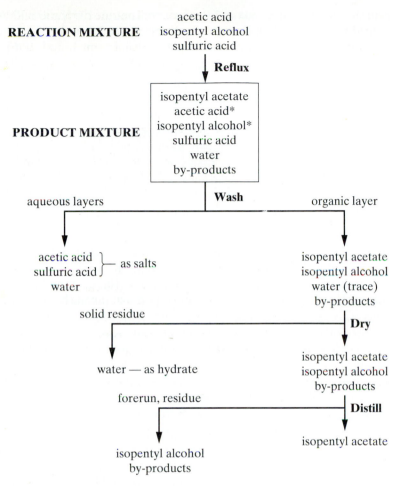

Figure 5.1   Flow diagram for the synthesis of isopentyl acetate.

Directions

If your instructor requests that you carry out a small-scale preparation, divide all quantities by three (or more) and carry out the purification using the small-scale apparatus described in OP-25b.

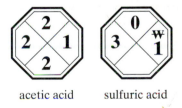

acetic acid   sulfuric acid

Safety Notes

Acetic acid causes chemical burns that can seriously damage skin and eyes; its vapors are highly irritating to the eyes and respiratory tract. Wear gloves, dispense under a hood; avoid contact and do not breathe vapors. Sulfuric acid causes chemical burns that can seriously damage skin and eyes. Wear gloves and avoid contact. Isopentyl alcohol and isopentyl acetate can irritate the skin, eyes, and respiratory tract.

Glacial acetic acid is a pure grade of acetic acid that freezes at about 17°C.

Take Care! Wear gloves, avoid contact, do not breathe vapors.

Reaction. *Under a hood,* measure 17 mL (~300 mmol) of glacial acetic acid and combine it with 150 mmol of isopentyl alcohol in a boiling flask of

Take Care! Wear gloves, avoid contact with H_2SO_4.

Observe and Note: What evidence do you observe that suggests that a chemical reaction is taking place?

Stop and Think: What is the density of isopentyl acetate? Which layer should be on top, the aqueous layer or the organic layer? Which layer should you save?

Take Care! Pressure may build up in a stoppered separatory funnel.

Stop and Think: What gas is evolved during the $NaHCO_3$ wash?

Stop and Think: What does the boiling range of the distillate tell you about the purity of your product?

Waste Disposal: Any residue (in the boiling flask) and forerun should be placed in a waste container designated by your instructor.

Stop and Think: Can you guess which peak corresponds to which component based on the peak areas?

appropriate size. Carefully mix in 1.0 mL of concentrated sulfuric acid. Add a few acid resistant boiling chips (or a stirbar if you are using a magnetic stirrer) and heat the reaction mixture under reflux for one hour [OP-6].

Separation. When the reaction time is over, allow the mixture to cool nearly to room temperature before you turn off the cooling water and remove the reflux condenser. Transfer the product mixture to a separatory funnel and wash it [OP-19] with 50 mL of water, being sure to save the right layer. Then carefully wash it with two succesive portions of 5% aqueous sodium bicarbonate, stirring the layers until gas evolution subsides before you stopper the separatory funnel, then venting it frequently thereafter. Dry the crude ester [OP-20] with anhydrous magnesium sulfate (or another suitable drying agent) and filter it by gravity [OP-11]. (**Waste Disposal:** The spent drying agent can be dissolved in the washing solvents and the mixture poured down the drain.)

Purification and Analysis. Using standard-taper glassware [OP-2], assemble an apparatus for simple distillation [OP-25] and have your instructor check it before you start. Distill the crude product (be sure the thermometer bulb is placed correctly) collecting any liquid that distills between 136°C and 143 °C. Record the actual boiling range you observe; wait until the entire thermometer bulb is moist with condensing vapors, liquid is distilling into the receiver, and the temperature is stable before you record the initial temperature. Weigh the distillate in a tared, labeled vial [OP-4]. Obtain a gas chromatogram [OP-32] of the distillate using a Carbowax column (or another suitable column) and measure the peak areas as directed by your instructor. Turn in the product to your instructor.

Report. State the problem and your initial hypothesis. Unless your instructor indicates otherwise, assume that the components of the distillate appear on the gas chromatogram in the order (1) isopentyl alcohol, (2) isopentyl acetate, (3) acetic acid. From the peak areas, calculate the percentage of each component in the distillate, then calculate your percent yield of isopentyl acetate based on its percentage in the distillate. Summarize and interpret the data and any other relevant evidence. Tell whether your hypothesis changed when it was tested and if so, why. State your conclusion and tell how it is supported by the evidence.

Exercises

Isoamyl acetate equilibrium

$i\text{-AmOH} + \text{HOAc} \rightleftharpoons$
$\quad\quad\quad\quad i\text{-AmOAc} + H_2O$

$K = \dfrac{[i\text{-AmOAc}][H_2O]}{[i\text{-AmOH}][\text{HOAc}]}$

Ac = acetyl, CH_3CO-
i-Am = isoamyl, $CH_3CHCH_2CH_2-$
$\quad\quad\quad\quad\quad\quad\quad\quad |$
$\quad\quad\quad\quad\quad\quad\quad\quad CH_3$

1 (a) Calculate the amount of isopentyl acetate (isoamyl acetate) that should be present in the product mixture at equilibrium, based on the quantities of starting materials you used and a value of 4.2 for the equilibrium constant. (Use the quadratic equation; because volumes cancel out, moles can be used in place of molar concentrations.) (b) Estimate the amount (in grams) of isopentyl acetate that was lost (1) as a result of incomplete reaction, (2) during the washings, and (3) during the distillation (see OP-25). Assume that the ester's solubility in aqueous $NaHCO_3$ is about the same as in water. Compare the sum of these estimated losses with your actual product loss and try to account for any significant differences.

2 What gas escaped during the sodium bicarbonate washing? Write balanced equations for two reactions that took place during this operation.

3 Tell how the procedure for the preparation of isopentyl acetate could be modified to increase the yield and purity of the product.

4 Describe and explain how each of the following experimental errors or variations might affect your results. (a) You fail to dry the reaction flask after it is washed with water. (b) You forget to add the sulfuric acid. (c) You use 34 mL of acetic acid instead of 17 mL. (d) You leave out the sodium bicarbonate washing step. (e) Your thermometer bulb is improperly placed in the still head, causing its readings to be 4°C too low.

5 (a) In the Methodology section it was stated that the reaction of an equimolar mixture of isopentyl alcohol and acetic acid will produce, at most, 67% of the theoretical amount of isopentyl acetate. Verify this with an equilibrium constant calculation, using $K = 4.2$. (b) Compare this with the corresponding percentage for the conditions used in this experiment (see Exercise 1). Are your results consistent with Le Châtelier's principle? Explain.

6 Based on the procedure that you used in this experiment and using the same molar quantities of reactants, develop a procedure that would be suitable for the preparation of isobutyl propionate. Specify the amounts of all materials required and a distillation range for the product. Obtain the necessary physical properties from one of the reference books listed in the Bibliography.

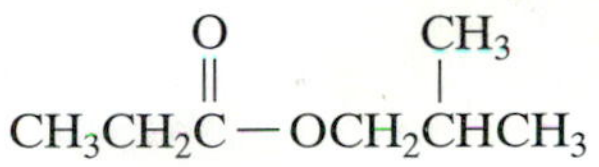

$$CH_3CH_2\overset{\overset{\displaystyle O}{\|}}{C} - OCH_2\overset{\overset{\displaystyle CH_3}{|}}{C}HCH_3$$

isobutyl propionate

Other Things You Can Do

(Starred items require your instructor's permission.)

*1 You and your coworkers can prepare a series of esters and compare their odors as described in Minilab 5.

*2 Prepare another ester of a primary alcohol, such as butyl acetate or isobutyl propionate, by the general method described for isopentyl acetate. Work out a procedure for the synthesis (see Exercise 6) and have it approved by your instructor. In some cases, a longer reflux time may be necessary for satisfactory results.

3 Read about the isolation of isopentyl (isoamyl) acetate in the alarm pheromone of the honeybee in *Nature* **1962**, *195*, 1018.

4 Read about various types of pheromones in *Chemical Communication: The Language of Pheromones* by William C. Agosta (W. H. Freeman & Co., New York, 1992).

Fractional Distillation. Separation of Petroleum Hydrocarbons

EXPERIMENT 6

Separation Methods.

Operations:

OP-27 Fractional Distillation
OP-4   Weighing
OP-6   Heating
OP-32 Gas Chromatography

Before You Begin

1 Read the experiment and operation OP-27. Review the other operations as necessary.
2 Prepare a brief experimental plan for the experiment.

Scenario

See OP-27 for information about fractional distillation and distilling columns.

cyclohexane

toluene

The Northern Pines Chemical Company specializes in manufacturing chemicals from wood products such as turpentine. For example, they use a major component of turpentine, α-pinene, to prepare organic compounds such as isobornyl acetate and terpineol, which impart a "piney" note to perfumes. To obtain pure α-pinene they must separate it from the other major component of turpentine, β-pinene. They have been carrying out this separation by fractional distillation using an expensive column packing material that has to be replaced frequently. They would like to switch to a cheaper and longer lasting packing material, but because the new packing material will be less efficient than their current type, they will need to purchase a longer column. From the difference between the boiling points of the pinenes (10°C), they estimate that such a column will need to have at least 20 theoretical plates (hypothetical column segments) to provide adequate separation. But the height of each theoretical plate, and thus the total length of the column, depends on the efficiency of the packing material.

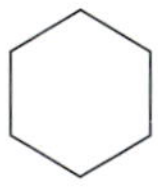

Forrest Greenwood, the operations engineer at Northern Pines, has sent your supervisor some of the new packing material for testing. Your assignment is to determine its HETP (a measure of column efficiency) and to estimate how long their new column must be to separate α-pinene from β-pinene. To do this you will fractionally distill a mixture of two petroleum hydrocarbons, toluene and cyclohexane, and measure the composition of the fractions by gas chromatography.

Applying Scientific Methodology

This experiment differs from the previous ones in that no hypothesis is being tested; your job is simply to determine the value of a physical quantity, the HETP (height equivalent to a theoretical plate) of the column and report the value as your conclusion. The accuracy of your results—and thus the validity of your conclusion—will depend on the care you take in performing the distillation.

Distillation in Petroleum Refining

Distillation has been used since antiquity to separate the components of mixtures—the ancient Egyptians were distilling wood to make an embalming fluid more than 3500 years ago. Today, distillation is used to manufacture perfumes, flavor ingredients, liquors, charcoal, coke, and a host of organic chemicals, but its most important application is probably in the refining of petroleum to produce fuels, lubricants, and petrochemicals.

The first step in petroleum refining is the separation of petroleum into different hydrocarbon fractions by distilling it through huge fractionating columns, called distillation towers, which may be up to 200 feet high. Since components having different numbers of carbon atoms usually have significantly different boiling points, this process separates the petroleum into fractions containing hydrocarbons of similar carbon content. Thus a lower-boiling fraction might contain hydrocarbons with five or six carbon atoms while a higher-boiling fraction contains hydrocarbons having from 12 to 18 carbon atoms. Vapors from the lower-boiling hydrocarbons rise to the top of the tower where they are condensed and collected as "top fractions." Hydrocarbons in top fractions, such as the one called straight-run gasoline, can be chemically modified and used in gasoline. The less volatile middle fractions are collected partway down the tower; these include kerosene and a gas-oil fraction used to produce diesel fuel, jet fuel, and home heating oil. The bottom fraction is usually subjected to vacuum distillation to produce vacuum gas oils, which can be used as fuel oils or converted to hydrocarbons suitable for gasoline. Such petroleum products as paraffin wax, lubricating grease, and asphalt also come from this fraction.

With an overall octane number range of 30–50, straight-run gasoline is not a suitable motor fuel, but its octane number can be increased to nearly 100 by a process known as catalytic reforming. In this process, the straight-run gasoline is heated to about 500°C at high pressure in the presence of a suitable catalyst. Catalytic reforming converts alkanes to cycloalkanes and cycloalkanes to aromatic compounds; the result is a mixture rich in high-octane aromatics. For example, hexane may be *cyclized* to yield cyclohexane, which may then lose hydrogen atoms in a process called *dehydrogenation* to yield the aromatic compound benzene. The cyclohexane in petroleum cannot easily be separated by distillation alone, so pure cyclohexane is generally obtained by catalytic hydrogenation of benzene, the reverse of the second step shown here.

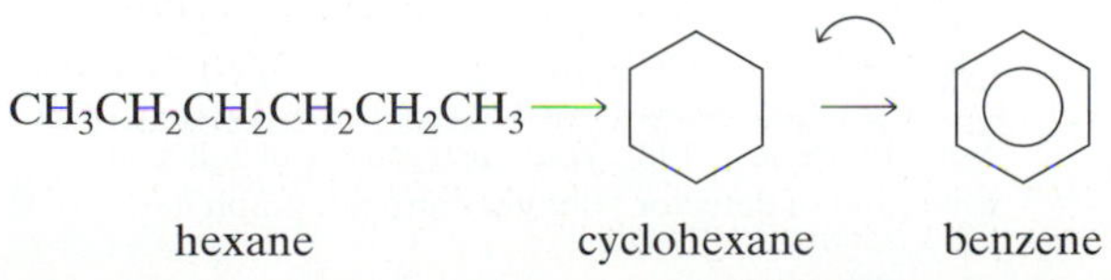

CH₃CH₂CH₂CH₂CH₂CH₃ $\longrightarrow$ $\longrightarrow$

hexane   cyclohexane   benzene

Toluene is obtained by dehydrogenation of methylcyclohexane, which can be formed by cyclization of heptane and other alkanes.

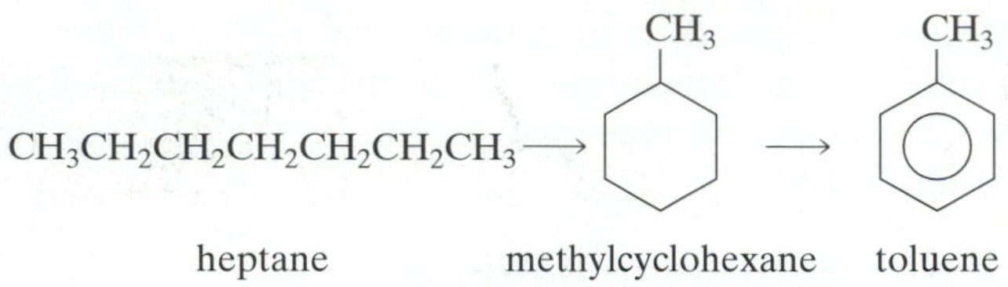

CH₃CH₂CH₂CH₂CH₂CH₂CH₃ ⟶   [cyclohexane with CH₃] ⟶ [benzene ring with CH₃]

heptane   methylcyclohexane   toluene

Understanding the Experiment

In this experiment you will separate the components of an equimolar mixture of cyclohexane and toluene by fractional distillation and assess the efficiency of the separation by measuring the composition of the fractions. The more completely the cyclohexane and toluene are separated from one another, the more efficient the separation. The degree of separation depends not only on the column packing used, but also on such factors as the stability of the heat source, the rate of distillation, and the way the column is packed. Good separation requires a low rate of distillation to maintain a high *reflux ratio*—the ratio of liquid returned to the boiling flask to liquid that distills into the receiving flask—so patience is required if you are to get good results. The efficiency of the column will be reduced if it is not packed uniformly, so it is important to distribute the packing material as evenly as possible.

You will measure the composition of each fraction you collect by gas chromatography. Your gas chromatograms should display two peaks, the cyclohexane peak being the first to appear. The areas of the peaks can be converted to relative masses by multiplying them by the appropriate correction factors from Table 6.1. You will then determine the HETP of the column packing from the composition of the first drops of liquid you collect, the *HETP sample*. The number of theoretical plates provided by your distillation apparatus can be calculated by using the Fenske equation, expressed as follows for an equimolar mixture of two components, A and B:

See OP-27 for definitions of theoretical plate and HETP and a discussion of the Fenske equation.

$$\text{total number of theoretical plates} = \frac{\log \dfrac{n_A}{n_B}}{\log \alpha} \qquad \textbf{(1)}$$

In this equation n_A/n_B is the ratio of the number of moles of cyclohexane and toluene in the HETP sample. The volatility factor, α, for the cyclohexane-

Table 6.1 Gas chromatography correction factors for cyclohexane and toluene

	TC Detector	FI Detector
cyclohexane	1.11	0.942
toluene	1.05	1.02

Note: benzene = 1.00. Your instructor will tell you what kind of detector your gas chromatograph has.

toluene mixture is 2.33. The boiling flask furnishes one theoretical plate, so you will have to subtract 1 from the total number of theoretical plates to obtain the number of plates provided by the column itself. From this number and the length of your column packing you can calculate the HETP of the column packing; the lower the HETP, the more efficient the packing.

Properties

Table 6.2 Physical properties of cyclo-hexane and toluene

	M.W.	b.p.	d
cyclohexane	84.2	81	0.774
toluene	92.2	111	0.867

Note: Boiling points are in °C and densities in g/mL

Directions

Cyclohexane is very flammable and may irritate the skin, eyes, and respiratory tract . Do not use open flames during the experiment.
Toluene is flammable, and inhalation, ingestion, or skin absorption may be harmful. Avoid contact with the liquid and do not breathe its vapors.

Safety Notes

cyclohexane

toluene

Take Care! Avoid contact with the liquid mixture and do not breathe its vapors.

Separation. All components of the fractional distillation apparatus must be clean and dry! Measure 40 mL of an equimolar mixture of cyclohexane and toluene into a 100-mL round bottomed boiling flask (keep the flask stoppered to prevent evaporation) and add some boiling chips (or a stirbar). Pack a distilling column with the column packing provided and measure the height of the packing in centimeters to the nearest 0.1 cm; the column need not be insulated. Clamp the boiling flask to a ring stand over an appropriate heat source [OP-6] and assemble an apparatus for fractional distillation [OP-27], using a small beaker as a receiver for the forerun. Clean and dry a small screwcap vial to collect the HETP sample. Clean, dry, weigh, and number four fraction collectors, preferably 5-dram (or larger) screwcap vials; remember to weigh the caps with their containers.

Heat the cyclohexane solution to a gentle boil (with the stirrer turned on, if you are using one) and adjust the heating rate so that the vapors rise slowly up the column. Reduce the heating rate if the column begins to flood (fill with liquid); the packing should be moistened by condensing vapors but should contain no flowing liquid. When the vapors reach the still head, reduce the heating rate slightly to keep the ring of condensing vapors between the top of the column packing and the sidearm for a minute or more, allowing the vapor composition to stabilize before any distillate is collected.

Increase the heating rate just enough to distill the liquid slowly into the beaker, and collect no more than five drops of distillate (the distillate

Stop and Think: Will the thermometer record the boiling temperature of the liquid as soon as it begins to boil? Why or why not?

Observe and Note: When does the thermometer begin to record the true boiling temperature? Describe what you observe at that point.

should be clear, not cloudy). Quickly replace the beaker by the HETP vial, collect the next 5–10 drops that distill, and cap the vial tightly. Replace the HETP vial with the first fraction collector, record the distillation temperature, and distill at a rate of not more than 20 drops per minute. You may have to gradually increase the heating rate to keep the distillation rate more or less uniform. When the distillation temperature reaches 85°C, replace the first fraction collector with the second one and cap the first one tightly. When the distillation temperature reaches 97°C, switch to the third collector. When the temperature reaches 107°C, remove the heat source and let all the liquid that remains in the column drain into the boiling flask. After the apparatus has cooled, transfer the contents of the boiling flask to the fourth collector. At this point you should have fractions covering the following boiling ranges:

1. 81–84.9°C
2. 85–96.9°C
3. 97–106.9°C
4. 107–111°C

Stop and Think: Why does the boiling temperature rise during the distillation?

Waste Disposal: Combine all fractions and place them in the hydrocarbon solvent recovery container.

Analysis. Weigh the four capped fraction collectors [OP-4] and analyze these fractions and the HETP sample by gas chromatography [OP-32].

Report. State the problem you attempted to solve in this experiment. Measure the peak areas on each gas chromatogram and use the appropriate correction factors to convert the areas to relative masses. Calculate the percentage (by mass) and the actual mass of cyclohexane and toluene in each fraction. Plot the component masses for each fraction (on the y-axis) as a function of boiling temperature, using the midpoints of the appropriate boiling ranges on the x-axis (use different symbols, such as $\times$ and •, for different components). Draw a smooth curve connecting the data points for cyclohexane and another one (overlapping the first) connecting the data points for toluene. (Good separation is suggested by a graph that is low in the middle and high on each end.) Calculate the number of theoretical plates provided by your fractional distillation apparatus; then calculate the HETP of your column and the length of a column (in cm) that will provide 20 theoretical plates. Summarize and interpret the data, discussing the efficiency of the separation, and state your final conclusion. Turn in your gas chromatograms with your report.

Exercises

1 A certain fractional distillation apparatus consists of a boiling flask surmounted by a 24-cm Vigreux column and a 30-cm glass tube half filled with 4×4 mm porcelain saddles. How many theoretical plates does the entire apparatus provide if the HETP of the saddles is 5 cm, the HETP of the Vigreux column is 8 cm, and the HETP of the empty glass tube is 15 cm?

2 Describe and explain how each of the following experimental errors or variations would affect your HETP value and the efficiency of

your separation. (a) You don't collect the HETP sample until midway through the distillation. (b) Your distillation rate is too high. (c) You use a column that is twice as long as the one you would ordinarily use.

3 Using the data in Table 6.3, construct a temperature-composition diagram like that in Figure E15 of OP-27. (a) From your diagram, estimate the initial composition of the distillate obtained by simple distillation of a mixture containing 20 mole percent cyclohexane and 80 mole percent toluene. (b) Estimate the initial composition if the same mixture is distilled through a three-plate column.

4 If you were to return your 97–107°C fraction to the empty boiling flask and redistill it, you might expect it all to distill between 97°C and 107°C as it did the first time. It actually yields some distillate in all four boiling ranges. Explain.

5 Derive equation **1** for an equimolar mixture of two components, from the Fenske equation in OP-27.

6 Suggest a chemical method that could be used to remove small amounts of toluene from cyclohexane.

7 Show how nylon-66 can be synthesized using cyclohexane as the starting material.

Other Things You Can Do

(Starred items require your instructor's permission.)

*1 Analyze some or all of your fractions with a refractometer rather than a gas chromatograph, using a calibration graph plotting refractive index *vs.* mole fraction (assume a linear releationship between these variables). The refractive indexes at 20°C of pure cyclohexane and toluene are 1.4260 and 1.4968, respectively.

*2 Carry out a gas chromatographic analysis of commercial xylene, as described in Minilab 6.

3 Write a research paper about distillation and its uses based on information from the *Kirk-Othmer Encyclopedia of Chemical Technology* and other sources from the Bibliography.

Table 6.3. Temperature-composition data for cyclohexane-toluene

T, °C	Mol % Cyclohexane	
	Liquid	Vapor
110.7	0	0
108.3	4.1	10.2
105.5	9.1	21.2
103.9	11.8	26.4
101.8	16.4	34.8
99.5	21.7	42.2
97.4	27.3	49.2
95.5	32.3	54.7
93.8	37.9	59.9
91.9	45.2	66.2
89.8	53.3	72.4
88.0	59.9	77.4
86.6	67.2	81.1
84.8	76.3	86.4
83.8	81.4	89.5
82.7	87.4	92.6
81.1	96.4	97.3
80.7	100.0	100.0

EXPERIMENT 7

Addition, Mixing, Sublimation.
Preparation of Camphor

Preparation and Purification of Solids. Oxidation.

Operations:

OP-7   Cooling
OP-8   Temperature Monitoring
OP-9   Mixing
OP-10 Addition
OP-24 Sublimation
OP-4   Weighing
OP-12 Vacuum Filtration
OP-21 Drying Solids
OP-28 Melting Point

Before You Begin

1   Read the experiment and read OP-7, OP-8, OP-9, OP-10, and OP-24. Review the other operations as necessary.
2   Calculate the mass of 25.0 mmol of isoborneol and the theoretical yield of camphor from that amount of isoborneol.
3   Prepare an experimental plan for the experiment.

Scenario

The Northern Pines Chemical Company was pleased with your analysis of their column packing material (see Experiment 6) so they have requested your services for another project. One of the many chemicals they manufacture from α-pinene is camphor. The last step in the preparation of camphor from α-pinene is the oxidation of isoborneol, for which Northern Pines currently uses the powerful oxidizing agent chromic acid. But the chromium-containing by-products of this reaction are classified as hazardous wastes and disposing of them properly is very costly, so Northern Pines wants your Consulting Chemists Institute to develop a more environmentally acceptable oxidation process for them to use. In searching the chemical literature your supervisor came across an article in the *Journal of Chemical Education* that describes the use of common laundry bleach (aqueous sodium hypochlorite) to oxidize a secondary alcohol, cyclohexanol, to a ketone, cyclohexanone.

See J. Chem. Educ. **1985**, *62*, 519.

The only significant by-products of this process are water and sodium chloride. The oxidation of isoborneol also involves the conversion of a secondary alcohol to a ketone, so your supervisor thinks the laundry bleach

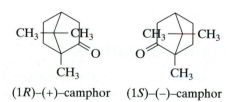

$$\text{cyclohexanol} + \text{NaOCl} \xrightarrow{\text{HOAc}} \text{cyclohexanone} + \text{NaCl} + \text{H}_2\text{O}$$

cyclohexanol   cyclohexanone

method may be just the kind of environmentally friendly process Northern Pines is looking for. The camphor that the company currently manufactures is about 95% pure, the remaining 5% being unreacted isoborneol. Your assignment is to carry out the oxidation of isoborneol with laundry bleach to find out if the reaction does in fact produce any camphor, and if so to determine whether you can obtain camphor that is at least 95% pure.

Applying Scientific Methodology

You should be able to identify the problem and develop one or more working hypotheses after reading the Scenario. You must make careful observations to detect evidence that a reaction is taking place. You will test your hypothesis by measuring the melting point of the product, which is very sensitive to the presence of impurities.

Camphor and the Camphoraceous Odor

The Chinese camphor tree, *Cinnamomum camphora*, is a tall, striking evergreen tree with dark shiny leaves.

When steam is forced through the chopped up wood of a camphor tree, camphor distills with the steam and crystallizes as a translucent white solid. Just as there are right-handed and left-handed gloves, scissors, and corkscrews, there are right-handed and left-handed camphor molecules that are mirror images of one another.

camphor tree

(1R)–(+)–camphor   (1S)–(–)–camphor

Isomers whose molecules differ only in their "handedness" are called *enantiomers*. (1R)-(+)-Camphor, the enantiomer obtained from the camphor tree, is composed of the "right-handed" molecules shown here. The less common "left-handed" enantiomer, (1S)-(−)-camphor, has been isolated from feverfew (*Chrysanthemum parthenium*), a daisylike plant unrelated to the camphor tree. Camphor is synthesized commercially from α-pinene by the pathway outlined here

α-pinene → pinene hydrochloride → camphene
→ isobornyl acetate → isoborneol → camphor

Most synthetic camphor is an equimolar mixture of left-handed and right-handed enantiomers.

The story of Bredt's quest is described in the Journal of Chemical Education **1983**, *60*, 341.

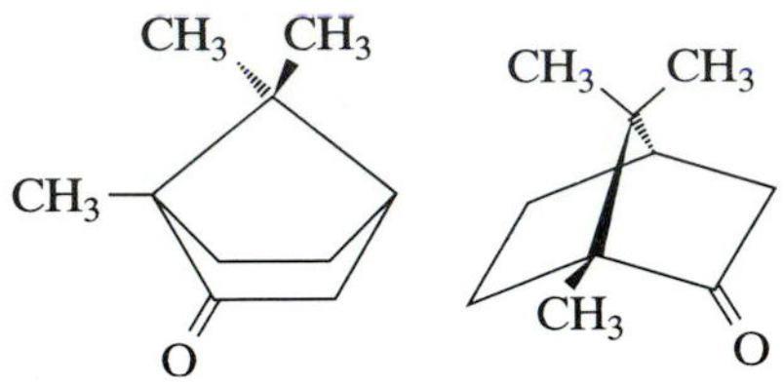

The history of camphor is longer and more involved than that of perhaps any other natural product. Scientific speculations about camphor have appeared in print since the time of Libavius (*Alchymia*, 1595). Although its molecular formula ($C_{10}H_{16}O$) was determined in 1833, its complicated bicyclic structure baffled nineteenth century scientists. Over the next 60 years they proposed more than 30 different structures for camphor, all of them wrong, before Julius Bredt finally came up with the correct structure in 1893.

The penetrating "camphoraceous" odor of camphor is shared by many compounds of similar molecular shape and size. Compounds as diverse in structure as the ones shown here all have roughly spherical molecules and similar camphoraceous odors.

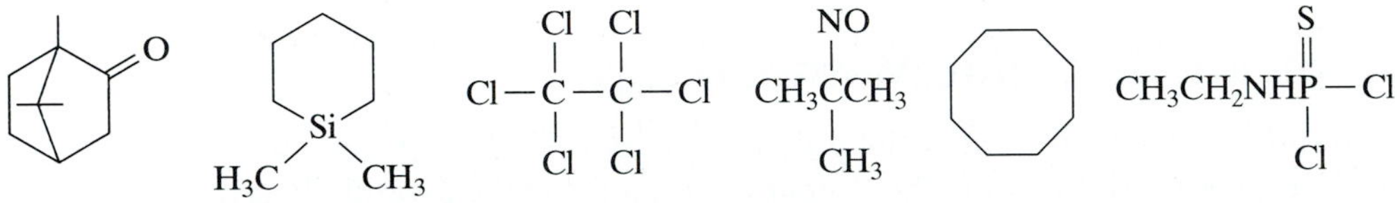

Perspective drawings that show the structure of camphor

Most scientists believe that the sense of smell is based on the presence in the nasal passageways of a large number of odor receptors, each of which is programmed to detect a specific kind of odor. This idea gained support recently (December 1996) when scientists who had induced bacteria to grow an odor receptor normally found in rats discovered that molecules of two compounds with floral odors, lilial and lyral, became strongly attached to the receptors. The similarities among molecules having camphoraceous odors suggest that the odor of a substance may, at least in part, depend on the size and shape of its molecules. A spherical molecule, for example, might fit nicely inside a hemispherically convex odor receptor, causing it to transmit a neural message that the brain interprets as a camphorlike odor. But scientists still do not fully understand how humans and other animals can detect and recognize a multitude of different odors.

Understanding the Experiment

In this experiment you will oxidize a secondary alcohol, isoborneol, to a ketone, camphor. Secondary alcohols can be converted to ketones by powerful oxidizing agents such as chromic acid, but many chromium compounds are highly toxic and corrosive and some are known to cause cancer. They also present a difficult disposal problem because they cannot legally be discharged into waterways or other places where they might harm the environment. For these reasons, you will use a safer and more environmentally friendly oxidizing agent, the familiar laundry bleach that is sold under such trade names as Clorox and Javex. Most chlorine bleaches contain about 5.25% sodium hypochlorite (NaOCl) in an aqueous solution. Adding a little acetic acid facilitates an oxidation reaction by converting most of the sodium hypochlorite to hypochlorous acid (HOCl), which is probably the active oxidizing agent.

In some previous experiments you heated the reaction mixture to speed up the reaction. In this experiment you may have to slow it down a little, because the oxidation of isoborneol is exothermic and the heat evolved may lead to the formation of unwanted by-products such as camphoric acid. An ice bath can be used to reduce the reaction temperature, if necessary. You will also control the reaction rate by adding the sodium hypochlorite a little at a time from a separatory/addition funnel. When a reaction takes place under reflux, the boiling action helps mix the reactants. In this experiment mixing must be effected by other means, such as by using a magnetic stirrer or by shaking and swirling the flask after each addition. You will have to monitor the reaction temperature by holding a thermometer with the bulb immersed in the reactants as you mix them. The directions in OP-8 tell how to do this without breaking the thermometer.

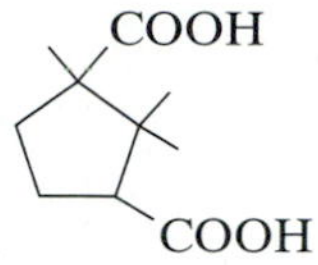

To ensure a complete reaction, you must add enough sodium hypochlorite solution to keep the oxidizing agent in excess throughout the reaction. When HOCl is present in excess, a drop of the acidic reaction mixture placed on an indicator paper impregnated with starch and potassium iodide will oxidize iodide ions to iodine, which turns the starch a deep blue-black color. Any excess HOCl that remains after the reaction is over can be destroyed by treatment with the reducing agent sodium bisulfite, according to the following equation:

$$HOCl + HSO_3^- \rightarrow HCl + HSO_4^-$$

Because of its compact and symmetrical molecular structure, camphor changes directly from a solid to a vapor when heated, which allows it to be purified by sublimation. Isoborneol also sublimes at elevated temperatures, so the sublimed camphor will probably contain some unreacted isoborneol. Assuming that isoborneol is the only significant impurity in the product, its purity can be estimated with good accuracy from its melting point because camphor has an unusually large freezing-point depression constant (remember that the melting point of a solid equals the freezing point of the corresponding liquid). The product can be regarded as a solid solution with camphor as the solvent and isoborneol as the solute, so you can use the following equation to calculate the molal concentration (m) of isoborneol in the product.

$\Delta T = K_f \times m$
ΔT = melting-point depression (reported m.p. − observed m.p.)
K_f = freezing-point depression constant for camphor = 40 °C kg mol^{-1}
m = molal concentration of isoborneol (mol isoborneol/kg camphor)

Knowing m, the number of moles of isoborneol per kilogram of camphor, you can then calculate the mass percent of isoborneol and camphor in the product.

camphoric acid

Reactions and Properties

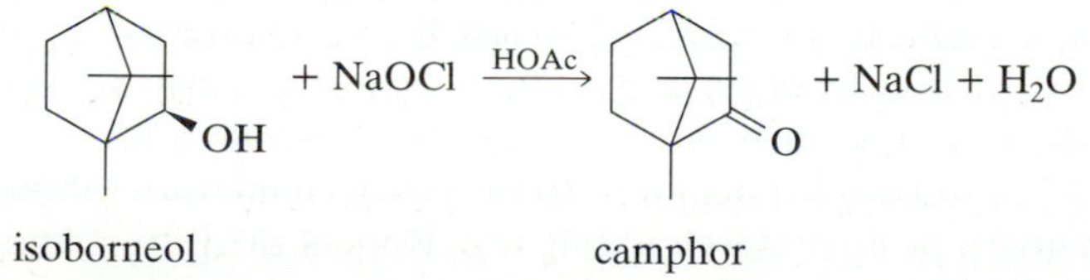

isoborneol   camphor

Table 7.1 Physical properties

	M.W.	m.p.	b.p.	d
isoborneol	154.3	212		
camphor	152.2	179	204	
sodium hypochlorite	74.4			
acetic acid	60.1	17	118	1.049

Note: m.p. and b.p. are in °C, *d* is in g/mL

Directions

Safety Notes

> Acetic acid causes chemical burns that can seriously damage skin and eyes; its vapors are highly irritating to the eyes and respiratory tract. Wear gloves, dispense under a hood; avoid contact and do not breathe vapors. The reaction mixture may evolve some chlorine gas, which can irritate the eyes and respiratory tract, so carry out the reaction under a fume hood.

acetic acid   camphor   chlorine

Take Care! Wear gloves, avoid contact with acetic acid and the NaOCl solution, do not breathe their vapors.

Reaction. *Under the hood* combine 25.0 mmol of isoborneol with 2.0 mL of glacial acetic acid in a 125-mL Erlenmeyer flask. Mix in 5.0 mL of 5.25% sodium hypochlorite solution (Clorox or another hypochlorite laundry bleach). Measure another 40 mL of 5.25% sodium hypochlorite (NaOCl) solution into a separatory/addition funnel, stopper the funnel, and support it over the flask. Place a strip of filter paper between the addition funnel's neck and the stopper to keep a vacuum from forming. Add the NaOCl solution [OP-10] to the reaction mixture in small portions, with vigorous swirling or magnetic stirring [OP-9], over a period of 10 minutes or more. Use a thermometer to monitor the temperature of the reaction mixture [OP-8], and control the rate of addition so that the temperature remains below 50°C. Have an ice bath handy to cool the reaction mixture [OP-7] if its temperature reaches 50°C.

Observe and Note: What evidence can you detect that suggests that a reaction is taking place?

When the addition is complete, stir the reactants or swirl the flask frequently [OP-9] at room temperature for 30 minutes or more. The reaction mixture should have a light greenish-yellow color while excess hypochlorous acid is present. If the color fades, test the reaction mixture for excess hypochlorite by transferring a drop of the solution to a strip of starch-iodide paper. If the test paper does *not* turn blue-black within a few seconds, add enough 5.25% sodium hypochlorite to the reaction mixture (about 1 mL at a time) to give a positive starch-iodide test. Repeat the testing and addition of NaOCl as necessary during the reaction period. When the reaction period is over, let the mixture cool to room temperature. Test it with starch-iodide paper, and if the test is positive add enough saturated sodium bisulfite solution dropwise to give a negative test.

Separation. Cool the reaction mixture to 5°C or below in an ice/water bath [OP-7]. Collect the product by vacuum filtration [OP-12], washing it on the filter with several portions of cold water. Dry the crude product [OP-21] at room temperature. (At your instructor's request, weigh the crude product and save a little for a melting point.)

Purification and Analysis. Purify the crude product by sublimation [OP-24], taking care not to char the solid by overheating. Weigh the sublimate [OP-4], dry it if necessary [OP-21], and measure its melting point [OP-28], preferably using a sealed capillary tube. Turn in the product in a labeled vial.

Report. State the problem and your initial hypothesis. Calculate your theoretical yield and percent yield of camphor. Calculate the mass percentages of isoborneol and camphor in your final product by taking as its melting point the temperature at which your camphor was completely liquefied. Summarize and interpret the data and all other relevant evidence. Tell whether your hypothesis changed when it was tested and if so, why. State your conclusion and tell how it is supported by the evidence.

Stop and Think: What is the purpose of the sodium bisulfite addition?

Waste Disposal: The filtrate may be poured down the drain.

Your instructor may request that you purify only a gram or less of the crude camphor.

Exercises

1   In this experiment you started with a white, strong-smelling solid and ended up with a white, strong-smelling solid. What evidence leads you to conclude that these two substances are in fact different compounds and that you did not just isolate the unreacted starting material?

2   Following the directions in Appendix V, construct a flow diagram for the synthesis of camphor.

3   The equation in the "Reactions and Properties" section shows only one isoborneol enantiomer and the camphor enantiomer it forms. Find out from your instructor if the isoborneol you used was the right-handed (1*R*) enantiomer shown, the left-handed (1*S*) enantiomer, or an equal mixture of both. Then rewrite the equation to show the actual reactants and products.

4   Describe how each of the following experimental errors or variations might affect your results. (a) You omit the 30-minute reaction period after the addition step. (b) You add the 20 mL of bleach all at once. (c) You mistake a negative starch-iodide test for a positive one and fail to add any bleach after the first 20 mL.

5   Write a balanced net ionic equation for the reaction of the acidified sodium hypochlorite solution with iodide ion from the starch-iodide paper, assuming that HOCl is reduced to HCl.

6   Show which carbon-carbon bond of camphor must be broken to form camphoric acid. Use molecular models if necessary.

Other Things You Can Do

(Starred projects require your instructor's permission.)

*1 Isolate caffeine from No-Doz tablets and purify it by sublimation as described in Minilab 7.

*2 Oxidize cyclohexanol to cyclohexanone as described in *J. Chem. Educ.* **1985**, *62*, 519.

3 Write a research paper about camphor and its applications, using sources listed in the Bibliography.

Boiling Point, Refractive Index. Identification of a Petroleum Hydrocarbon

Physical Properties of Liquids. Alkanes and Cycloalkanes.

Operations:

OP-29 Boiling Point
OP-30 Refractive Index
OP-4   Weighing
OP-5   Measuring Volume
OP-25 Simple Distillation

Before You Begin

1 Read the experiment and the descriptions for OP-29 and OP-30, read OP-25b about small-scale distillation, and review the other operations as needed.
2 Prepare a brief experimental plan for this experiment.

Scenario

An investigative organization known as The Consumer's Advocate (TCA) publishes a monthly magazine, *Caveat Emptor*, which evaluates consumer products and exposes scams. TCA is currently investigating an auto supply company that sells Thrust, a so-called miracle gasoline additive claimed to improve engine performance. TCA's preliminary tests show that the additive has no measurable effect on either power or mileage, and they suspect that the additive is nothing more than a hydrocarbon that burns along with the gasoline. To support its case against the company, TCA needs to know the identity and octane number of the hydrocarbon. If its octane number is higher than that of a typical no-lead gasoline (about 87) then the company's claim that the additive improves engine performance might have some merit, even though the amount of improvement is negligible when the additive is used in the quantity recommended on the can.

Because their own chemists are busy with other projects, TCA's technical director, Patsy Haven, has farmed out the job to your Institute. Your assignment is to determine the identity of the hydrocarbon in Thrust and find out whether or not its octane number is greater than 87. Your supervisor purchased a can of Thrust at a local service station, but that is only enough to provide each member of your project group with a small sample.

Applying Scientific Methodology

You should evaluate the evidence and formulate tentative hypotheses as you go along. For example, if you measure a boiling point of 79°C your tentative hypothesis might be "The alkane in Thrust is 2,4-dimethylpentane" (see Table 8.2). If you then measure its density as 0.79 g/mL you might have to change your hypothesis to "The alkane in Thrust is cyclohexane." Measuring a refractive index of 1.4262 would then confirm your second hypothesis and lead you to the conclusion that the alkane is indeed cyclohexane, whose octane number you can look up in Table 8.1.

Gasoline—A Chemical Soup

Gasoline is a kind of "chemical soup" that contains an incredibly large number of ingredients that are carefully selected and blended to produce a fuel with the desired properties. Virtually all of the main fuel components of gasoline are derived either directly or indirectly from petroleum, which must be refined before a usable fuel is obtained. The word "refine" suggests a simple separation and purification process, but the refining of petroleum is a more complex operation involving chemical as well as physical changes. As you learned in Experiment 6, petroleum is first fractionated in a distillation tower, which separates its components according to their boiling ranges. The fraction boiling between about 50°C and 150°C, straight-run gasoline, is not a good motor fuel by itself because it contains a large proportion of unbranched hydrocarbons such as heptane and hexane, in addition to branched alkanes and cycloalkanes. Straight-chain alkanes burn very rapidly, generating a shock wave in the combustion chamber that reduces power and can damage the engine. This "knocking" does not occur with highly branched alkanes, which burn more slowly and uniformly.

The octane number of a motor fuel is a measure of its antiknock qualities. The highly branched alkane 2,2,4-trimethylpentane (sometimes called "isooctane") is a very good motor fuel and has arbitrarily been assigned an

Table 8.1 Octane numbers of some petroleum hydrocarbons

Hydrocarbon	Octane No.	Hydrocarbon	Octane No.
nonane	−45	2,4-dimethylpentane	82
octane	−17	methylcyclopentane	82
heptane	0	cyclopentane	83
2-methylheptane	24	2,3-dimethylpentane	89
hexane	26	2-methylbutane	89
3-methylheptane	35	butane	92
2-methylhexane	45	2,3-dimethylbutane	95
pentane	61	2,2-dimethylbutane	96
3-methylhexane	66	2,2,3-trimethylbutane	100
methylcyclohexane	71	2,2,4-trimethylpentane	100
2-methylpentane	73	2,2,3-trimethylpentane	102
3-methylpentane	75	toluene	104
cyclohexane	77	benzene	106

octane number of 100. Heptane, with no branching, is assigned an octane number of zero. The performance of a particular motor fuel is measured relative to these two alkanes; for example, a fuel performing as well as a mixture containing 70% 2,2,4-trimethylpentane and 30% heptane is assigned an octane number of 70. Table 8.1 lists the octane numbers of selected hydrocarbons found in gasoline.

A major objective of petroleum refining is to convert the low-octane components of petroleum into higher-octane compounds. This can be accomplished by chemical processes such as *isomerization*, which converts straight-chain alkanes to branched alkanes; *cracking*, which breaks down large molecules into smaller ones; *alkylation*, which combines short-chain alkane and alkene molecules to form longer, branched molecules; and *catalytic reforming,* which converts alkanes to cycloalkanes and aromatic compounds. Aromatic hydrocarbons such as toluene have particularly high octane numbers and are used to increase the octane rating of no-lead fuels.

The chemistry involved in these reactions is described in many textbooks of organic chemistry.

Gasoline for use in automobile engines is prepared by combining varying amounts of straight-run gasoline, cracked gasoline, alkylated gasoline, reformate, and other hydrocarbon mixtures in the right proportions to give the desired boiling range and octane number. The properties of the fuel are then further adjusted with a variety of additives. Antiknock additives such as methyl *t*-butyl ether (MTBE) may be included to boost the octane rating further. Quick-start additives such as butane facilitate cold weather starting, and antifreeze additives such as isopropyl alcohol reduce icing. Antioxidants such as BHT help improve fuel stability and reduce gum formation,

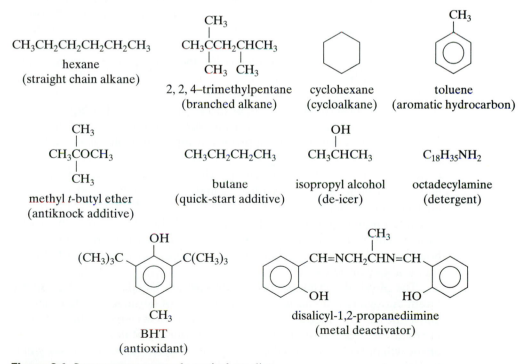

Figure 8.1 Some components of a typical gasoline

particularly in fuels that contain appreciable amounts of alkenes. Certain metals, such as copper and iron, can catalyze gum-forming reactions, so chelating compounds such as disalicyl-1,2-propanediimine may be added to deactivate these metals. Newer cars with fuel injectors require detergent additives such as octadecylamine to keep their intake systems clean. Dyes are then added for identification and visual appeal.

Understanding the Experiment

In this experiment you will attempt to identify an unknown hydrocarbon that is one of the compounds listed in Table 8.2. Identifying an unknown organic compound is somewhat like identifying the perpetrator of a crime. The investigator compiles a list of suspects, hunts for clues that might have a bearing on the case, sifts through the evidence in an effort to eliminate most of the suspects, and searches for additional evidence to build a case against the prime suspect. Many kinds of evidence may have a bearing on the identity of an organic compound: physical evidence such as boiling point and density, chemical evidence such as the appearance of a precipitate with a test reagent, and spectral evidence such as the occurrence of an infrared band that suggests the presence of a particular functional group. Alkanes and cycloalkanes are comparatively unreactive, making it difficult to gather much chemical evidence about them, and their infrared spectra are not very revealing, so in this experiment you will identify the unknown alkane by using its physical properties alone.

Before you can measure the physical properties of a liquid accurately, the liquid must be pure. In this experiment you will purify your hydrocarbon by simple distillation using a small-scale apparatus to reduce material losses. You can estimate the boiling point of the liquid as you distill it, but you will also measure its boiling point by using a micro boiling-point method described in OP-29. Because the boiling point of a liquid varies with the barometric pressure, you may have to apply a boiling-point correction as described in the operation. The density of a liquid can be obtained by accurately weighing a measured volume of the liquid. The volume is measured with an appropriate pipet, and the mass should be measured to the nearest milligram on an accurate balance. A good refractometer can be used to determine the refractive index of a pure liquid with great accuracy. With reasonable care, you should be able to measure the refractive index of your unknown to within 0.05% or better, so this value may be the most important clue to the identity of your hydrocarbon. The refractive index of a liquid is very sensitive to temperature, however, so you will need to correct your observed value if the temperature is above or below 20°C.

The physical constants of the hydrocarbons listed in Table 8.2 are different enough that an accurate determination of all three constants should allow the certain identification of an unknown as one of the 12. Your instructor may add more hydrocarbons to your list of possibilities. If so, he or she will provide you with the appropriate physical constants or ask you to look them up.

Structures and Properties

Table 8.2 List of possible hydrocarbons

Name	b.p.	n_D^{20}	d^{20}
cyclopentane	49	1.4065	0.746
2,2-dimethylbutane	50	1.3688	0.649
2,3-dimethylbutane	58	1.3750	0.662
3-methylpentane	63	1.3765	0.664
hexane	69	1.3749	0.659
methylcyclopentane	72	1.4097	0.749
2,4-dimethylpentane	80	1.3815	0.673
cyclohexane	81	1.4266	0.779
2,3-dimethylpentane	90	1.3919	0.695
heptane	98	1.3877	0.684
2,2,4-trimethylpentane	99	1.3915	0.692
methylcyclohexane	101	1.4231	0.769

Note: n_D^{20} = refractive index at 20°C using sodium D line; d^{20} = density at 20° in g/ml

Directions

Your instructor may suggest additional tests to carry out on your hydrocarbon.

> **Your unknown hydrocarbon is flammable; keep it away from flames and hot surfaces.**

Purification and Boiling-Point Determination. Obtain about 5 mL of the unknown hydrocarbon from your instructor and record its identification number and the barometric pressure in your laboratory notebook. Select a suitable heat source and assemble an apparatus for small-scale simple distillation [OP-25], using boiling chips or a stirbar. Be sure the thermometer bulb is positioned correctly in the still head. Distill the liquid slowly, setting aside any low-boiling forerun or high-boiling fraction for later disposal, and record its boiling range and the temperature when about half of it has distilled (the median boiling point). Then carry out a micro boiling-point measurement [OP-29] on the hydrocarbon. The median and micro boiling points should agree within a degree or two; if they don't, repeat the micro boiling-point measurement or redistill the hydrocarbon.

Density Measurement. The temperature of the purified hydrocarbon should be as close as possible to 20°C. Accurately measure 1 mL of the liquid into a clean, dry, tared vial using a volumetric pipet or an automatic pipet [OP-5]. (**Take Care!** Do not pipet by mouth.) Stopper the vial immediately and weigh it to the nearest milligram on an accurate balance [OP-4]. Calculate the density of your hydrocarbon from your results.

Refractive Index Measurement. Measure the refractive index of the purified hydrocarbon [OP-30] as directed by your instructor and record the temperature of the measurement.

alkanes and cycloalkanes of five to eight carbons

Safety Notes

If the hydrocarbon boils over a broad range, it should be redistilled and a pure fraction (collected over a range of 1–2°C) should be used for analysis.

Take Care! Keep the liquid away from flames or hot surfaces.

Stop and Think: What should happen if you put a few drops of your hydrocarbon in a test tube containing a little water and shake the test tube? Do it. Was your prediction correct?

Waste Disposal: Put your hydrocarbon and any liquid saved from the distillation in a designated hydrocarbon solvent recovery container.

Report. State the problem you attempted to solve in this experiment. Apply a correction to the refractive index if the temperature of the measurement was not 20°C, and a correction to the boiling point if the atmospheric pressure was below 750 torr. Summarize the relevant evidence in a table, state your initial hypothesis, and describe and explain any revisions of your hypothesis. Report your final conclusion regarding the identity of the hydrocarbon and its value as an additive and show how the evidence supports it.

Exercises

1 The following properties were measured for an unknown hydrocarbon in a laboratory with an ambient temperature of 28°C and a barometric pressure of 28.9 inches of mercury (1 inch Hg = 25.4 torr):

> boiling point: 78.2°C
> refractive index: 1.3780
> mass of 5 mL: 3.346 g

Correct the refractive index and boiling point to 20°C and 1 atmosphere, and calculate the density of the unknown. If the unknown is one of the hydrocarbons listed in Table 8.2, what is its probable identity?

2 Give names and structural formulas for all structural isomers of your unknown hydrocarbon that are alkanes or cycloalkanes.

3 Describe and explain the possible effect on your results of the following experimental errors or variations. In each case tell whether the resulting physical property (b.p., density, refractive index) will be too high or too low. (a) You record the temperature at the still head when the unknown liquid begins to boil as its boiling point. (b) To measure 1.00 mL of the liquid, you fill a Mohr pipet to the 1-mL mark and drain it completely into the weighing vial (you can ask your instructor to show you a Mohr pipet). (c) The liquid is at a temperature of 25° when you measure its mass and volume. (d) The liquid is at a temperature of 25° when you measure its refractive index, but you forget to correct it.

4 A large oil spill that resulted from the 1989 grounding of the Exxon Valdez in Prince William Sound did considerable harm to Alaska's wildlife, especially to waterfowl and aquatic mammals. Do you think a similar spill of a water-insoluble liquid having a density of 1.10 g/mL would have done as much harm, assuming the toxicity of the liquid was comparable to that of petroleum? Explain your answer.

5 (a) Write a balanced equation for the complete combustion of 2,2,4-trimethylpentane in an engine's combustion chamber. (b) Show how butane can be converted to 2,2,4-trimethylpentane using petroleum-refining processes mentioned in this experiment.

6 (a) Dioxane has a boiling point of 101°C. Could you separate dioxane from methylcyclohexane by distillation? Explain why or why not, based on the liquid and vapor compositions during the distillation. (b) How could these liquids be separated?

dioxane

Other Things You Can Do

(Starred projects require your instructor's permission.)

*1 Solve the "missing label puzzle" described in Minilab 8.

*2 With your coworkers, obtain and compare gas chromatograms of different grades of gasoline such as leaded, unleaded, and "gasahol." Refer to *J. Chem. Educ.* **1972,** *49*, 764 and *J. Chem. Educ.* **1976,** *53,* 51 for information and references that will help you interpret the chromatograms and identify some of the components.

*3 Test for lead in gasoline as follows: Saturate a piece of filter paper with gasoline and expose it to strong sunlight for several hours. Moisten the paper with 3 *M* acetic acid followed by a few drops of aqueous potassium iodide solution (16.5 g/100 mL). A yellow color after several minutes indicates the presence of lead.

 4 Write a research paper about gasoline and petroleum refining using references cited in the Bibliography.

Column Chromatography, UV-VIS Spectrometry. Isolation and Isomerization of Lycopene from Tomato Paste

EXPERIMENT 9

Isolation of Natural Products. Ultraviolet-Visible Spectrometry. Geometric Isomers.

Operations:

OP-16 Column Chromatography
OP-36 Ultraviolet-Visible Spectrometry
OP-11 Gravity Filtration
OP-13 Extraction
OP-14 Evaporation
OP-19 Washing Liquids
OP-20 Drying Liquids

Before You Begin

1 Read the experiment and read OP-16, OP-36, and the section "Liquid-Solid Extraction" in OP-13. Review the other operations as necessary.
2 Prepare an experimental plan for this experiment.

Scenario

The Consumer's Advocate (see Experiment 8) is now investigating the quality of processed foods such as tomato paste. Ideally, such processed foods would contain all the nutrients and flavor components present in the fresh fruits or vegetables, but all too often the processing methods used tend to degrade the color, flavor, and nutritional value of a food. The red pigment that colors ripe tomatoes is an all-*trans* form of lycopene, an antioxidant that is known to fight many kinds of cancer including cancers of the digestive tract, cervical cancer in women, and prostate cancer in men.

Heat, light, and certain chemicals may convert some of this pigment to its 13-*cis* isomer. Thus the presence of 13-*cis*-lycopene in a canned tomato product suggests that the fresh fruit may have been subjected to excessive heat or light during processing. TCA's technical director wants your organization to assess the quality of different brands of tomato paste on the basis of their *trans*-lycopene content. Your supervisor has found a method for doing this using ultraviolet-visible spectrometry in the *Journal of Chemical*

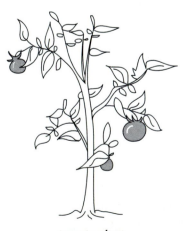

tomato plant

Education. To improve the validity of your results, you and your coworkers will work in small research teams, with each team assigned a specific brand of tomato paste.

J. Chem. Educ. **1989**, *66*, 258.

Applying Scientific Methodology

With your instructor's permission, you or another member of your research team may bring a sample of a commercial tomato paste for testing. Your opinion about the quality of the tomato paste may then be the basis for a working hypothesis, which will be tested when you analyze your lycopene by ultraviolet-visible spectrometry. During the experiment you should try to avoid conditions that might, by promoting isomerization or oxidation of lycopene, invalidate your results.

Carotenoids, Vitamin A, and Vision

Lycopene, with its 13 carbon-carbon double bonds, is one of the most unsaturated compounds in nature. Because most of its double bonds are conjugated, lycopene absorbs radiation at long wavelengths in the 400–500-nm region of the visible spectrum. Its resulting deep orange-red color is responsible for the redness of ripe tomatoes, rose hips, and many other fruits. An even more important plant pigment is the yellow-orange substance β-carotene, which is present not only in carrots but in all green leaves and many flowers as well. Both lycopene and β-carotene, along with most of the other natural *carotenoids*—compounds related to β-carotene—occur naturally in the all-*trans* forms shown in Figure 9.1.

Key Concept: *The wavelength of the UV-VIS radiation absorbed by a conjugated substance increases with the length of its conjugated system.*

Although the main function of carotenoids in plants remains somewhat of a mystery, the importance of carotenes to animals is clear—β-carotene (and, to a lesser extent, α-and γ-carotene) is converted in the intestinal wall to Vitamin A, which is then stored in the liver. Generations of children have grown up with the mealtime refrain "Eat your carrots—they're good for

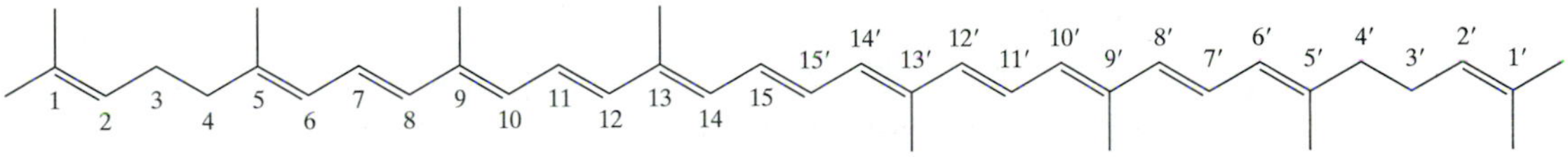

lycopene

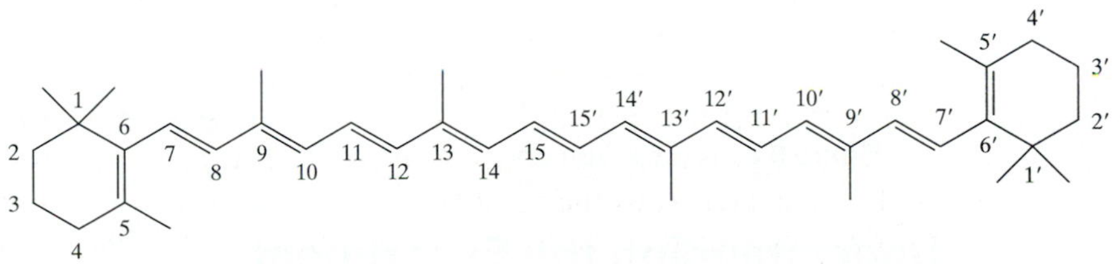

β-carotene

Figure 9.1 Structures of carotenoids. A single straight line branching off from a chain or ring stands for a methyl group in these and similar formulas. A carbon atom with the requisite number of hydrogens is at each bend of the chain.

your eyes!" In fact, Vitamin A from carotenes and other sources is an essential participant in the very complex process by which light entering your eyes causes your brain to construct a visual picture of your surroundings.

Vitamin A
(all-*trans*)

The process of vision, although extremely complex in its entirety, is based on the isomerization of an oxidized form of Vitamin A called retinal. In the rods of the retina, which are responsible for night vision, retinal occurs in combination with the complex protein opsin to form rhodopsin (visual purple). The retinal in rhodopsin assumes the shape shown in Figure 9.2A, with an 11-*cis* double bond and probably a *cisoid* conformation between the #12 and #13 carbon atoms as well. This allows a retinal molecule to fit comfortably into a cavity in an opsin molecule—much like a joey (a baby kangaroo) curled up in its mother's pouch. When a photon of light strikes a rhodopsin molecule, its 11-*cis*-retinal passenger suddenly straightens out and becomes all-*trans*-retinal. This process is incredibly fast—much faster than the blink of an eye—taking only about 0.2 trillionths of a second (200 femtoseconds).

A similar process occurs in the cones of the retina, which are responsible for color vision.

The isomerized retinal molecule no longer fits into its niche on the opsin molecule, which responds much as a mother kangaroo might when her joey creates a disturbance in her pouch—it ejects its unruly passenger, triggering the transmission of a visual message to the brain. Subsequently the *trans*-retinal is enzymatically reduced to all-*trans*-Vitamin A, which isomerizes to 11-*cis*-Vitamin A, which is oxidized back to 11-*cis*-retinal, which promptly combines with another molecule of opsin to regenerate more rhodopsin. At this point, another photon of light can start the cycle all over again.

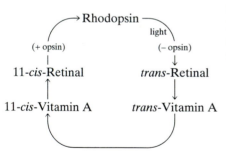

A. 11-*cis*-12-*s*-*cis*-retinal   **B.** all-*trans*-retinal

Figure 9.2. Retinal isomers

Understanding the Experiment

In this experiment you will extract the carotenold pigments (lycopene, carotenes, and xanthophylls) from canned tomato paste and separate them by column chromatography to obtain a solution containing lycopene. Then you will record the ultraviolet-visible spectrum of this solution and analyze it for evidence of isomerization. Although lycopene can be obtained directly from ripe tomatoes, it is easier to extract it from commercial tomato paste, in which the lycopene is more concentrated—a tablespoon of tomato paste yields as much lycopene as a medium ripe tomato, about 10 mg. Lycopene will isomerize if allowed to stand in solution too long, particularly in the presence of heat, light, or acids. For this reason, it is important to avoid unnecessary delays and exposure of the pigment to heat or bright light. Acids that occur naturally in tomatoes can be removed by washing the extract with potassium carbonate. Lycopene also oxidizes slowly in the presence of atmospheric oxgen, so you should try to record its UV-VIS spectrum on the same day as you isolate the lycopene solution, if possible; otherwise oxidation products may alter the spectrum.

You will extract the carotenold pigments with a solvent mixture that contains equal volumes of acetone and petroleum ether. Acetone is very soluble in water; thus, washing the extracts with water removes the acetone and leaves the pigments dissolved in the petroleum ether layer. Low-boiling petroleum ether (b.p. ~35–60°C) is preferred for the extraction because it can be evaporated readily at room temperature to yield a concentrated solution of the pigments. This concentrate is then transferred to the top of a chromatography column, which should be packed with neutral alumina (Brockmann grade II-III) by the slurry method described in OP-16. If you use a 25-mL buret as a column, it should be filled to a depth of about 15 cm, which should require about 15–20 g of alumina. You will elute the pigments with high-boiling petroleum ether (b.p. ~60–80°C) or hexanes, followed by a more polar eluent containing a little acetone with the hydrocarbon solvent. Lycopene, with its 13 double bonds, is attracted to alumina somewhat more strongly than the carotenes, which have 11–12 double bonds. Therefore the yellow-orange carotene band will move down the column faster than the orange-red lycopene band. A lycopene sample for spectrometric analysis should be collected from the more concentrated center of the orange-red band. Yellow xanthophyll pigments will trail behind the lycopene band because they contain polar hydroxyl groups that are strongly attracted to alumina.

Although lycopene is not involved in the vision cycle, it can go through a configurational change analogous to that undergone by retinal. Adding a small amount of iodine to a solution of all-*trans*-lycopene induces its partial conversion to 13-*cis*-lycopene, causing observable changes in its properties. The conversion can be followed by ultraviolet-visible spectrometry, since the UV-VIS spectrum of 13-*cis*-lycopene is significantly different from that of its all-*trans* counterpart. To determine whether your lycopene has undergone any isomerization to 13-*cis*-lycopene, you will record a spectrum of your initial lycopene solution and another spectrum of the same solution after the addition of iodine. Because 13-*cis*-lycopene absorbs at slightly lower wavelengths than all-*trans*-lycopene does, the absorption bands in the visible region of the spectrum will shift slightly to the left after isomerization. In the ultraviolet region, a characteristic *cis* peak associated with the bent geometry of the 13-*cis*-lycopene molecules should appear at 360 nm, a wavelength where all-*trans*-lycopene shows little absorption (see Figure 9.3).

If the spectrum of your un-isomerized lycopene contains a definite *cis*-peak, or if its spectrum does not change significantly as a result of the

Petroleum ether is a general term used to describe volatile petroleum distillates having varying compositions and boiling ranges. Do not confuse it with ethyl ether!

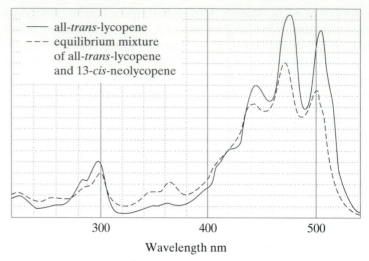

Figure 9.3 Ultraviolet-visible spectra of lycopene stereoisomers

iodine addition, the lycopene in your tomato paste may have isomerized either while the tomatoes were being processed into tomato paste *or* during your isolation of the lycopene. You can estimate the percentage of all-*trans*-lycopene in your solution before and after isomerization by the following empirical method: (1) Draw a straight, horizontal line tangent to the bottom of the valley between the last (highest wavelength) two peaks on the spectrum. (2) Measure the height of both peaks from that line, divide the smaller height by the larger, and multiply by 100%. (3) Subtract 40% from the result and then divide by 0.40. Before drawing any definite conclusion about the quality of your tomato paste, you should compare your results with those of other members of your team.

This method is derived from the Journal of Chemical Education *article referred to in the Scenario.*

Directions

Students may work together in small groups (research teams) with each group working on a specific brand, agreeing upon a conclusion, and comparing its results with those of other groups working with different brands.

Safety Notes

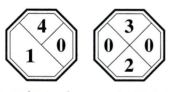

petroleum ether   acetone

Take Care! Keep the extraction mixture away from flames and hot surfaces.

> **Acetone and petroleum ether are very flammable and their vapors can irritate the eyes and upper respiratory tract. Keep petroleum ether and the petroleum ether-acetone mixture away from flames and hot surfaces, and do not breathe their vapors.**

Extraction of Pigments from Tomato Paste. Weigh about 4.0 g of tomato paste into a small beaker. Extract the solid material [OP-13d] three times with 10-mL portions of a 50% (by volume) mixture of acetone and low-boiling petroleum ether, each time filtering the extract through fluted filter paper [OP-11] into a small Erlenmeyer flask. After each extraction, decant the liquid extract onto the filter, pressing the residue in the beaker with a flat-bladed spatula to squeeze out as much liquid as possible. After the third extraction, transfer the residue to the filter and wash it with 5 mL of the extraction solvent, combining the wash liquid with the extracts. Wash the combined extracts [OP-19] with 25-mL of saturated sodium chloride solution followed by 25 mL of 10% aqueous potassium carbonate and then 25 mL of water. (**Waste Disposal:** The wash sovents can be poured down the drain.) Dry the petroleum ether layer [OP-20] with magnesium sulfate and concentrate the pigment solution to a volume of 1–2 mL by evaporating most of the petroleum ether under vacuum *without* heating [OP-14]. If you inadvertently evaporate the solution to dryness, dissolve the residue in 1 mL of petroleum ether.

Separation and Isolation of Lycopene. Prepare an alumina column for chromatography [OP-16] using high-boiling petroleum ether or hexanes as the column-packing solvent. Protect the column from strong light by wrapping it with aluminum foil. Transfer the pigment solution to the column and elute it with high-boiling petroleum ether or hexanes until the yellow-orange carotene band begins to drain out of the column. Open the foil occasionally to observe the location of the bands. Change to a 10% solution of acetone in high-boiling petroleum ether (or in hexanes) and elute the orange-red lycopene band, collecting a 5.0-mL sample of eluate from the center of this band in a small vial. (You can save the rest of the lycopene eluate and evaporate it to obtain the pigment as described on p. 80.) Use this midband sample for the spectral analysis as soon as possible. If you cannot record its spectrum on the same day, store your sample in a tightly closed container in a cool, dark place.

Spectral Analysis and Isomerization of Lycopene. Record an ultraviolet-visible spectrum [OP-36] for the midband sample of lycopene over a wavelength range extending from 600 nm in the visible region to 250 nm in the ultraviolet region. If necessary, dilute the lycopene solution with more petroleum ether to keep the strongest peak (at ~475 nm) on scale. Mix a drop of a 0.025% solution of iodine in hexane into the lycopene solution and leave the solution in the sample beam at 475 nm, monitoring its absorbance, until its absorbance remains constant (about 2 minutes). (Alternatively, leave it in bright sunlight for 15 minutes or more.) Then record another ultraviolet-visible spectrum over the same wavelength range as before. Turn in your lycopene solution in a labeled vial.

Report. State the problem you attempted to solve in this experiment and state your initial hypothesis. Estimate the percentage of all-*trans*-lycopene in your midband sample before and after isomerization. After consulting with other students who analyzed the same brand of tomato paste, decide whether the lycopene in that brand of tomato paste was isomerized significantly as a result of processing (read Exercise 1 before you do this). Tell whether your hypothesis changed when it was tested and if so, why. State your final conclusion and tell how it is supported by the evidence. Turn in your spectra with your report.

Stop and Think: What is the purpose of the K_2CO_3 wash?

Take Care! Keep the solvent away from flames and hot surfaces.

Observe and Note: Describe what you see as the bands pass down the column.

Stop and Think: Why do the bands move down the column at different rates?

Waste Disposal: Place all eluates except the lycopene eluate in a designated solvent recovery container.

Observe and Note: Do you see any differences in the spectra? If so, describe them.

Exercises

1   To estimate the percentage of all-*trans*-lycopene present in the original tomato paste, should you average the percentages obtained by all the members of your team? If not, what should you do, and why?

2 (a) Calculate the concentration of your lycopene solution before isomerization, given that the molar absorptivity of lycopene at 471 nm is 1.86×10^4. (b) Calculate the mass of lycopene in that 5.0-mL sample.

3 (a) Draw the structures of the 7-*cis*, 11-*cis*, and 13-*cis* isomers of lycopene. (b) Linus Pauling predicted that 13-*cis*-lycopene should be considerably more stable than the other two isomers. Explain.

4 Describe and explain the possible effect on your results of the following experimental errors or variations. (a) You use a can of tomato paste that has been left open in a refrigerator for several days. (b) You record the second UV-VIS spectrum immediately after adding the iodine solution. (c) You use acid-washed alumina for the chromatographic separation.

5 (a) Explain why some hydrocarbons such as lycopene and β-carotene are colored, whereas most other hydrocarbons are not. (b) The color of a lycopene solution fades and may disappear entirely if it is treated with a larger amount of iodine than you used in this experiment. Explain and give an equation for a possible reaction.

6 (a) Write an equation for the reaction that occurred during the addition of iodine to all-*trans*-lycopene. (b) Write a feasible mechanism for this reaction.

Other Things You Can Do

(Starred projects require your instructor's permission.)

*1 Concentrate your lycopene-containing eluate to a small volume and cool it to obtain crystalline lycopene. Its melting point should be about 175°C.

*2 Separate the dyes in a commerical drink mix by paper chromatography as described in Minilab 9.

*3 Make a "tomato juice rainbow" showing the effect of conjugation on color, as described in *J. Chem. Educ.* **1986**, *63*, 1092.

4 Write a research paper about Vitamin A and vision, using sources from the Bibliography.

Steam Distillation, Infrared Spectrometry. Isolation and Identification of the Major Constituent of Clove Oil

EXPERIMENT 10

Isolation of Natural Products. Infrared Spectrometry.

Operations:

OP-15 Steam Distillation
OP-34 Infrared Spectrometry
OP-4   Weighing
OP-13 Extraction
OP-14 Evaporation
OP-20 Drying Liquids

Before You Begin

1   Read the experiment and read OP-15 and OP-34. Review the other operations as necessary.
2   Write an experimental plan for this experiment.

Scenario

A professional aromatherapist, Rose Otto, uses the essential oil from cloves as a treatment for toothache, muscle pain, ringworm, warts, flatulence, and general exhaustion. But the latest batch of clove oil from her current supplier is darker than usual, has a harsh odor, and appears to be less effective than the oil she received previously. She suspects that the supplier has substituted some clove leaf oil for true clove oil, which is distilled from clove buds, the dried calyxes left after the flowers of the clove tree have fallen off.

Ms. Otto has asked you to provide her with an authentic sample of freshly distilled clove oil and tell her what's in it so that she can compare the authentic clove oil with the product she has on hand. Your assignment is to isolate clove oil from ground cloves and identify its major constituent, which is known to have the molecular formula $C_{10}H_{12}O_2$. Your supervisor believes you can identify that constituent by using infrared spectrometry.

clove bud

Applying Scientific Methodology

Reading the experiment should yield a clue about the identity of the unknown, which is one of those illustrated in Figure 10.1 (p. 85). Then you can develop a tentative hypothesis that will be tested when you obtain and interpret the infrared spectrum.

Plants and Healing

As people seek alternatives to traditional medical practices, which emphasize the use of drugs and surgery to treat illness, various fields of alternative medicine are gaining adherents around the world. These include *aromatherapy*, the use of essential oils to maintain health and treat illness; *naturopathy*, a system of treating diseases by using special diets, herbs, vitamins, and other natural healing methods; and *homeopathy*, which originally relied on the use of minute doses of drugs to cure illness but now utilizes carefully formulated mixtures of herbal medicines. Although some alternative medical practices may be associated with scientifically questionable theories, such as the idea (still accepted by some homeopathic practitioners) that the potency of a drug increases with dilution, many fields of alternative medicine utilize plant-based medicines with a long history of healing efficacy.

In our age of modern medical miracles, people often associate herbal medicine—the use of plants to treat and prevent illness—with witch doctors, shamans, or far-out medical cults. But herbal remedies have gained popularity in recent years as more and more people turn to echinaceae, goldenseal, and even garlic to help them stay healthy and cope with illness. Europe is well ahead of the United States in conducting scientific research on herbal medicines. In fact, the popularity of six of the ten top-selling herbs in the United States has resulted mainly from European research. For example, a scientific team at the University of Dusseldorf, Germany, recently studied the active principals and biological effects of the purple coneflower, *Echinacea purpurea*, which is used to treat colds and flu by stimulating the immune system. One objective of their research was to improve the standardization of echinaceae extracts, helping to ensure that each dose provides the same physiological activity.

After echinaceae, the most widely used herbal remedy in America is garlic—one of the few remedies you are more likely to find in a grocery store than a drugstore. Although researchers disagree on the virtues of garlic, there is evidence that it lowers cholesterol and triglyceride levels in blood, helps prevent blood clots that could lead to heart attacks or strokes, and lowers blood pressure. The main active ingredient in garlic is a sulfur compound called allicin, which also gives garlic its distinctive smell. Unfortunately, cooked garlic contains little if any allicin, so many of the alleged benefits of garlic are obtained only from the raw cloves or garlic capsules. Eating lots of raw garlic could have a negative impact on your social life, but that may be a small price to pay for good health!

Other popular herbal medicines include goldenseal root, used for treating peptic ulcers, infected gums, sore throats, skin infections, and various other conditions; saw palmetto berries for treating nonmalignant prostate disease; gingko leaf to improve blood flow in capillaries and arteries; aloe vera gel to heal burns, cuts, and wounds; ephedra stems to treat asthma and hay fever; and ginseng root to enhance one's general well-being and revitalize those weakened by old age or illness.

It should not surprise us that natural medicines can be effective. There are, after all, far more molecules in the world's natural life forms than have been synthesized in all the world's pharmaceutical laboratories. Many natural molecules are already known to have medicinal properties, and there must be at least as many more whose properties are yet to be discovered.

$$\overset{\displaystyle O}{\overset{\|}{CH_2=CHCH_2S-SCH_2CH=CH_2}}$$

allicin

Most of the drugs now prescribed by physicians were either derived from natural sources or developed by modifying the molecular structures of natural substances. For example, the heart stimulant digitalis is extracted from the foxglove plant, and the molecular structure of aspirin (acetylsalicylic acid) is based on natural salicylates such as salicin from willow bark, which has been used for centuries by Native Americans to treat fevers.

The six-carbon substituent in salicin is a glucose unit.

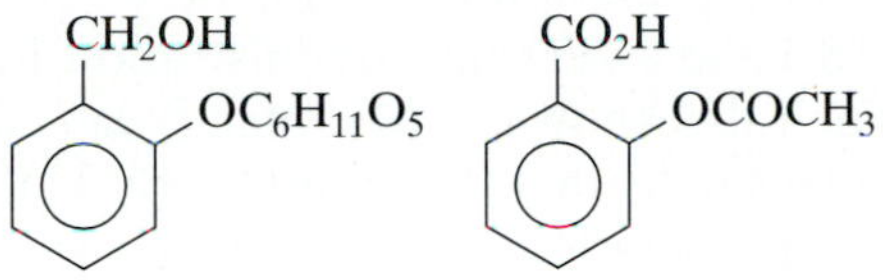

salicin   aspirin

Another herbal medicine that was known to native healers long before its discovery by modern medicine is rauwolfia, a tranquilizer and antihypertensive drug derived from the East Indian snakeroot, *Rauwolfia serpentina*. The snakeroot plant has been used for at least 3000 years as an antidote for snakebite, a treatment for insanity and stomach ache, and to soothe grumpy babies. Rauwolfia came to the attention of medical science when it was found to reduce blood pressure and cure some kinds of insomnia. When physicians reported that heart patients treated with rauwolfia acted as if they didn't have a worry in the world, psychiatrists began using it to treat mentally disturbed patients. Analysis of the plant led to the discovery and characterization of its most active component, reserpine. No commercially feasible process for synthesizing reserpine has yet been developed, so this drug is still derived from snakeroot extracts.

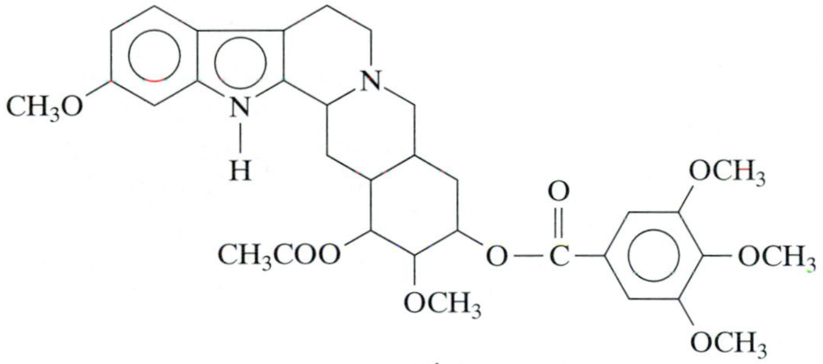

reserpine

A number of plants used for centuries by native healers and ancient physicians have not yet been discovered by modern pharmaceutical companies. Legend has it that Achilles, during the seige of Troy, used yarrow to treat the wounded Greeks. Its botanical name, *Achillea millefolium*, recognizes that tradition. The bruised leaves of yarrow help to stop bleeding, heal cuts, and relieve the pain of a wound, so the plant has been used in medical emergencies by backpackers and other outdoor adventurers. St. Johnswort (*Hypericum perforatum*) was supposedly used by the ancients to drive away evil spirits; today it is touted as an alternative to Prozac and similar drugs for treating mild to moderate depression. The indigineous North American weed boneset (*Eupatorium perfoliatum*) provides a bitter tea that was a favorite Native American remedy for fevers and other ailments. The closely related Joe Pye weed (*Eupatorium purpureum*) was named after a Native American who gained fame for using it to cure typhus. It is an effective diuretic for the treatment of kidney and bladder ailments.

Although herbal medicines are generally milder and have fewer side effects than traditional prescription drugs, they are not all harmless. According to the U.S. Food and Drug Administration, herbal preparations containing ephedrine (also known as Ma huang) can cause heart attacks, strokes, seizures, and even death if used improperly. Unlike most prescription drugs, different preparations containing the same herb may vary widely in potency and physiological effect. But when used responsibly by well-informed individuals, herbal medicine may provide a viable alternative to the use of traditional drugs for treating illness and maintaining good health.

Understanding the Experiment

The *essential oil* of a plant is a mixture of volatile, water-insoluble components that exhibits the odor and other characteristics of the plant. In this experiment you will isolate the essential oil from cloves, which are obtained from a small evergreen tree (*Syzygium aromaticum*) that grows in places such as Indonesia, Madagascar, and Zanzibar. Cloves contain about 16 percent by mass of the essential oil, a pale yellow liquid with a sweet, spicy aroma. Clove oil is unusual among essential oils in having only one major component, which comprises about 85 percent of the oil.

Because ground cloves lose their volatile components rapidly if left standing, it is best to grind fresh whole cloves just before using. Cloves can be ground quickly in an electric spice grinder, but a mortar and pestle will suffice. Essential oils are nearly always isolated by steam distillation, in which steam forced through the plant material vaporizes the essential oil, which is then condensed into a receiver along with water from the condensed steam. This process is preferable to ordinary distillation because the volatile components distill at temperatures below their normal boiling points, reducing or preventing decomposition due to overheating. During the steam distillation, the presence of clove oil in the distillate will be indicated by oily droplets or cloudiness. When all the clove oil has distilled, the emerging distillate should be as clear as water and essentially odorless. Clove oil is separated from the distillate by extraction with dichloromethane (methylene chloride), and its major component is then separated from minor components by extraction with aqueous sodium hydroxide.

The major component of clove oil is a strong-smelling liquid that has the molecular formula $C_{10}H_{12}O_2$. The structures of some natural compounds that have this formula are shown in Figure 10.1. Because these compounds have different sets of functional groups, it is possible to distinguish them by using infrared spectrometry. By observing the presence or absence of IR absorption bands corresponding to specific functional groups, you should be able to arrive at the correct structure for the major component. The section "Interpretation of Infrared Spectra" in OP-34 describes the characteristic bands of organic compounds that have these functional groups.

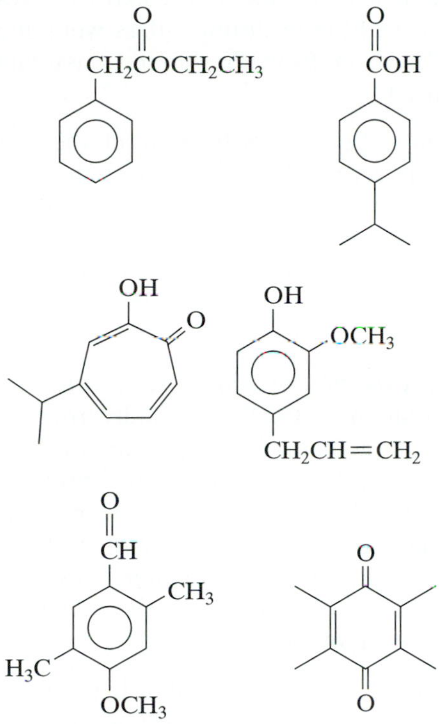

Figure 10.1 Compounds with the molecular formula $C_{10}H_{12}O_2$

Directions

Live steam can cause serious thermal burns. Make sure all steam hoses are connected tightly and turn off the steam before you withdraw the steam inlet tube from the apparatus. Dichloromethane may be harmful if ingested, inhaled, or absorbed through the skin. There is a possibility that prolonged inhalation of dichloromethane may cause cancer. Minimize contact with the liquid and do not breathe its vapors.

Safety Notes

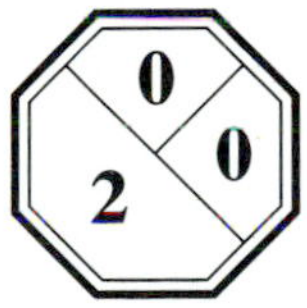

dichloromethane

Isolation of Clove Oil. Weigh about 5.0 g of fresh whole cloves and use a spice grinder or a mortar and pestle to grind them to a fine powder. Set up an apparatus for steam distillation [OP-15] using a large (250–500 mL) boiling flask and a steam trap, and have your instructor check your apparatus. Combine the ground cloves with 50 mL of water in the boiling flask, then steam distill the mixture to extract the clove oil. Continue the distillation until a drop or two of the emerging distillate, collected on a watch glass, is odorless and water-clear, with no oily droplets. You may need to distill

Observe and Note: How does the appearance of the distillate change during the distillation?

Take Care! Turn off the steam before you remove the inlet tube from the apparatus.

Take Care! Avoid contact with dichloromethane and do not breathe its vapors.

Stop and Think: What does this separation procedure suggest about the nature of the major component?

150 mL of liquid or more before the distillate becomes completely clear. Be sure to vent the steam line or raise the steam inlet tube above the liquid level in the boiling flask before you turn off the steam.

Extract the clove oil from the distillate [OP-13] with two 20-mL portions of dichloromethane and combine the extracts. (**Waste Disposal:** The clove residue should be filtered through glass wool and placed in a solid waste container. The water from the boiling flask and distillate can be poured down the sink.)

Separation of Minor Components. Extract the active component of clove oil from the dichloromethane solution with two 15-mL portions of 1 *M* aqueous sodium hydroxide and combine the aqueous layers. (**Waste Disposal:** Place the dichloromethane in a chlorinated solvent recovery container.) Acidify the aqueous solution to blue litmus paper with 10 mL or more of 3 *M* hydrochloric acid. Extract this aqueous solution with two 15-mL portions of dichloromethane and combine the extracts.

Isolation and Analysis of the Major Component. Dry the resulting dichloromethane solution [OP-20] with anhydrous magnesium sulfate. Evaporate the dichloromethane [OP-14] under vacuum using a cold trap until the boiling stops and the volume of the residue remains constant. (**Waste Disposal:** Place the recovered dichloromethane in a chlorinated solvent recovery container.) Weigh the liquid residue [OP-4] in an open container, then reweigh it a minute or so later. If its mass decreases significantly between weighings, the evaporation should be resumed. Record the infrared spectrum [OP-34] of the liquid.

Report. State the problem and your initial hypothesis. Identify as many bands in the IR spectrum as you can, and summarize and interpret this and other evidence. Deduce the structure of the active component of clove oil. Tell whether your hypothesis changed when it was tested and if so, why. State your final conclusion and tell how it is supported by the evidence. Calculate the percentage of the active component based on the mass of cloves you started with. Turn in your product and IR spectrum with your report.

Exercises

1 Derive a systematic name for the active component of clove oil, and use this to find its common name in *The Merck Index* or another reference book.

2 (a) What property of the active component of clove oil made possible its separation from the other components by the extraction process you used? Is this consistent with the structure you chose for it? Explain. (b) Write equations for the chemical reactions involved in the extraction and the subsequent acidification of the extract.

3 (a) Clove oil contains about 10% of a minor component that has the formula $C_{12}H_{14}O_3$, which can be hydrolyzed to yield the major component and acetic acid. Deduce the structure of the minor component. (b) The percentage of the major component of clove oil

actually increases as the cloves are steam distilled. Explain why and give an equation for the reaction involved.

4 Describe and explain the possible effect on your results of the following experimental errors or variations. (a) You stop the steam distillation after collecting 75 mL of distillate. (b) You omit the extraction of minor components (see Exercise 3). (c) You don't evaporate the dichloromethane long enough.

5 (a) Clove oil also contains a small amount of a substance whose systematic name is (*E*)-4,11,11-trimethyl-8-methylenebicyclo[7.2.0]undec-4-ene. Write the structure of this compound and find its common name in *The Merck Index* or another reference book.

6 Construct a flow diagram for this experiment, showing how the active component is separated from the solid part of the cloves and the component mentioned in Exercise 3.

Other Things You Can Do

(Starred projects require your instructor's permission.)

*1 Omit the *Separation of Minor Components* step in this experiment and analyze the clove oil by gas chromatography to determine the approximate percentage of each component. You can also compare the infrared spectrum of clove oil with that of its active component.

*2 Obtain and analyze an essential oil from orange peel as described in Minilab 10.

*3 Steam distill the essential oils from anise seed, caraway seed, or cumin seed following the procedure in this experiment but omitting the *Separation of Minor Components* step. Each of these essential oils contains a single major component that can be characterized by infrared spectrometry.

4 Write a research paper about herbal medicine after referring to sources cited in the Bibliography.

Thin-Layer Chromatography, NMR Spectrometry. Identification of Unknown Ketones

EXPERIMENT 11

Qualitative Analysis. Thin-Layer Chromatography. NMR Spectrometry. Ketones.

Operations:

OP-17 Thin-Layer Chromatography
OP-35 Nuclear Magnetic Resonance
OP-4   Weighing
OP-5   Measuring Volume
OP-12 Vacuum Filtration
OP-21 Drying Solids
OP-23 Recrystallization
OP-28 Melting Point

Before You Begin

1 Read the experiment and read OP-17, OP-35, and "Recrystallization from Mixed Solvents" in OP-23. Review the other operations as necessary.
2 Prepare a brief experimental plan for this experiment.

Scenario

A machine shop in your city was recently destroyed by fire under conditions that strongly suggest arson. A residue recovered at the fire's point of origin was found to contain traces of MEK (methyl ethyl ketone), one of several commercial degreasing solvents that shop employees use to clean the machinery. The primary suspect is a disgruntled ex-employee who was recently fired for sleeping on the job. While searching his garage, police discovered two unlabeled cans containing flammable liquids, which the suspect claims are charcoal starters for his grill. However, most charcoal starting fluids are mixtures of petroleum hydrocarbons, and preliminary tests indicate that the liquids found in the suspect's possession are both ketones.

The local chief of police, Spike Burns, has asked your Institute to help them solve the crime. Captain Burns provided your supervisor with samples of the two ketones. Matching one of them with the solvent recovered at the fire's point of origin will help the police discredit the suspect's claim, and may result in his conviction. Your assignment is to identify the ketones and see if one of them matches the solvent found at the scene of the crime. Your supervisor has requested that you use two different methods to identify your ketones.

Applying Scientific Methodology

You will, in effect, be performing two separate experiments. Your initial hypothesis for the first experiment might be, for example, "The first unknown solvent is (or is not) methyl ethyl ketone." You will then gather evidence to prove or disprove the hypothesis. In part **A** the evidence used to test the hypothesis will be obtained by TLC analysis and a melting-point determination; the result of either of these may lead you to revise your initial hypothesis before you arrive at a conclusion. In part **B** the only evidence will be provided by the NMR spectrum of the second solvent, which should lead you to its structure. There is, of course, no guarantee that either of the two solvents will be methyl ethyl ketone.

Crime and Chemistry

Forensic chemistry is chemistry applied to the solution of crimes. It deals with the analysis of materials that were used in committing a crime or that were inadvertently left at the scene of a crime. Materials used in committing a crime might include the ink on a forged document, an explosive used in a terrorist bombing, a toxic substance used in a fatal poisoning, or a flammable liquid used to start a fire. Such materials can be identified and sometimes traced to a particular source. Materials found at the scene of a crime might include paint chips, pieces of fiber from clothing, and particles of dust or soil, as well as any materials that were used in committing the crime. Chips of paint or glass found at the scene of a hit-and-run accident can be analyzed both chemically and under a microscope to determine the make and model of the car involved. Clothing can be traced by the dyes contained in fibers, and dust or soil particles may link a criminal to a particular occupation or location.

Bringing a suspect to trial requires that evidence be presented to establish, first, that a crime has actually been committed, and second, that the suspect is connected with the crime. In an arson case, for example, this requires proof that the fire was deliberately set as well as evidence implicating the suspect. One way of establishing that a fire was deliberately set is to prove that an *accelerant* (a flammable substance causing a fire to intensify and spread rapidly) was used to start and spread the fire. Because fires burn upward from the point of origin, some accelerant may soak downward into flooring, rags, paper, or other porous materials. When an investigator traces a fire to its point of origin, he or she can often collect samples of materials containing the accelerant, which are placed in airtight containers and sent to a forensic laboratory for analysis.

In the laboratory, a forensic chemist can separate an accelerant from debris collected at the crime scene by steam distillation or extraction. Once the accelerant has been isolated, it is usually classified according to chemical type (gasoline, turpentine, etc.) by an instrumental method such as gas chromatography or spectrometry. The forensic chemist may then try to match the accelerant sample with a control material, such as a liquid in the suspect's possession or a commercial material. Some flammable liquids, such as the industrial solvent methyl ethyl ketone (whose IUPAC name is 2-butanone), contain only one major component and can be compared to the control material by using chemical or spectrometric methods. If the original accelerant was a mixture of different components, such as gasoline, its more volatile components will evaporate and burn more rapidly in a fire. For this reason a sample of accelerant taken from the scene will probably not have the same composition as the original accelerant, or a control material that matches the original accelerant. The control material is therefore evaporated slowly and analyzed repeatedly by gas chromatography to see whether its composition at any stage of evaporation duplicates that of the recovered material. With this procedure it is often possible to determine the brand and grade of gasoline or other accelerant used.

Thin-layer chromatography (TLC) is another important technique used by forensic chemists to solve crimes. TLC provides a rapid, sensitive means of analyzing many of the materials associated with various crimes. The dyes used to color gasoline can be characterized by the pattern of spots they produce on a TLC plate, making it possible in some cases to trace an arson accelerant to its source. The U.S. Treasury Department maintains a library of pen inks catalogued according to their TLC dye patterns, allowing an investigator to match the ink on a document with one on file. In a few cases, TLC analysis has proven that the ink used to fraudulently back date a document did not even exist on the date in question! Substances suspected of being illicit drugs are frequently screened by TLC; most drugs that are mixtures of several substances, such as marijuana, produce telltale patterns that are easily recognized. A thin-layer chromatogram alone may not be sufficient to establish the identity of a suspect material, but it can narrow down the list of possibilities and thus lead to positive identification of the material by other means.

Understanding the Experiment

In this experiment you will attempt to identify one of the unknown ketones by using thin-layer chromatography and the other by using NMR spectrometry. Thin-layer chromatography can be used in the identification of pure compounds as well as complex mixtures such as drugs and dyes. When a TLC plate spotted with structurally similar organic compounds is developed with an appropriate solvent, the R_f values obtained vary more or less regularly with chain length. This correlation is illustrated in Table 11.1 for a homologous series of carboxylic acids. When an unknown compound is

Table 11.1 TLC R_f values for carboxylic acids

carboxylic acid	# carbon atoms	R_f
methanoic (formic) acid	1	0.07
ethanoic (acetic) acid	2	0.13
propanoic acid	3	0.30
butanoic acid	4	0.40
pentanoic acid	5	0.50
hexanoic acid	6	0.57
heptanoic acid	7	0.60
octanoic acid	8	0.66

Note: on silica gel, developed with a 19:1 mixture of methyl acetate and 2.5% ammonia

known to be one of a limited number of compounds, a TLC plate can be spotted with samples of the unknown and the most likely known compounds. When the plate is developed, it may be possible to match the R_f value of the unknown compound's spot with that of a known compound.

Your first unknown will be a member of a homologous series of methyl ketones represented by the formula $CH_3CO(CH_2)_nCH_3$. A ketone is often converted to its 2,4-dinitrophenylhydrazone or another colored derivative for TLC analysis because the colored spots are easily located after the plate is developed, making a visualizing reagent unnecessary. A derivative preparation is actually a small-scale organic synthesis that may produce only a fraction of a gram of product. The 2,4-Dinitrophenylhydrazone derivative of a ketone is prepared by combining the ketone with 2,4-dinitrophenylhydrazine (DNPH) reagent, which contains DNPH and sulfuric acid in aqueous ethanol. Traces of sulfuric acid may remain on the derivative and catalyze isomerization reactions that will lower its melting point, so the acid is removed by washing the derivative with sodium bicarbonate solution after vacuum filtration. The derivative is then purified by recrystallization, using a small-scale apparatus. Some 2,4-dinitrophenylhydrazones dissolve too readily in ethanol and too sparingly in water for either of these liquids to be a good recrystallization solvent. Thus you may have to use a mixture of the two, as described in the section "Recrystallization from Mixed Solvents" in OP-23. To obtain a solvent mixture of the right composition, you should first dissolve the derivative in the better solvent (the one in which it is most soluble), ethanol, and then add just enough of the poorer solvent, water, to saturate the hot solution. After measuring the melting point of your purified derivative and comparing its R_f value with those of known derivatives, you should be able to identify your unknown as one of the ketones in Table 11.2.

Modern instruments such as infrared and nuclear magnetic resonance (NMR) spectrometers can greatly reduce the time and effort required for the positive identification of an unknown. In part **B** of this experiment, you will attempt to identify an unknown saturated ketone that has 4–6 carbon atoms from its proton NMR (^{1}H NMR) spectrum. If an NMR spectrometer (or simulator) is available, you can record the spectrum yourself, with help from your instructor. Otherwise, your instructor will provide the NMR spectrum of your unknown. The section "Interpretation of ^{1}H NMR Spectra" in OP-35 contains an introduction to NMR spectral analysis that will help you deduce the structure of the second unknown.

Reactions and Properties

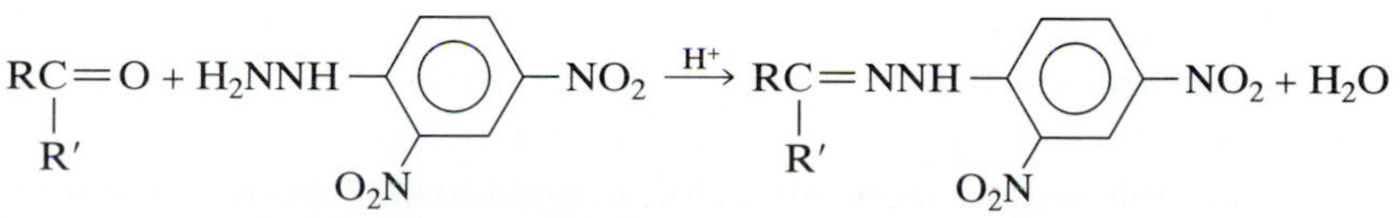

ketone   2,4-dinitrophenylhydrazine   2,4-dinitrophenylhydrazone

Table 11.2 Physical properties and derivative melting points for homologous methyl ketones

Ketone	M.W.	b.p.	*d*	derivative m.p.
2-propanone (acetone)	58.1	56	0.791	126
2-butanone	72.1	80	0.805	117
2-pentanone	86.1	102	0.809	143
2-hexanone	100.2	128	0.811	106
2-heptanone	114.2	151	0.811	89
2-octanone	128.2	173	0.819	58

Directions

With the instructor's permission, teams of 2–3 students can work together on some parts of this experiment, such as the preparation of derivatives of the known ketones.

Safety Notes

ketones

ethyl acetate

TLC solvent

> The ketones are flammable and may be harmful if inhaled or absorbed through the skin. Avoid contact, do not breathe vapors, and keep away from flames.
> 2,4-Dinitrophenylhydrazine is harmful if absorbed through the skin and will dye your hands yellow. Wear gloves, avoid contact with the DNPH reagent, and wash your hands after using it.
> Ethyl acetate and the TLC solvent are flammable and may be harmful if inhaled or absorbed through the skin. Avoid contact, do not breathe vapors, and keep away from flames.
> Deuterochloroform is toxic and may be carcinogenic; avoid contact and inhalation.

Take Care! Do not pipet by mouth.

Take Care! Wear gloves, avoid contact with the reagent.

Waste Disposal: Place the filtrate in a designated DNPH waste container or dispose of it as directed by your instructor.

Waste Disposal: Pour the filtrate down the drain.

A. *Identification of an Unknown Methyl Ketone by TLC*

Preparation of 2,4-Dinitrophenylhydrazones. Obtain an unknown methyl ketone from your instructor and record its identification number in your laboratory notebook. Using a graduated pipet [OP-5], measure 0.20 mL of the unknown into a 15-cm test tube. Dissolve it in 3 mL of 95% ethanol and stir in 7.0 mL of the 2,4-dinitrophenylhydrazine reagent, then set the test tube aside for 15 minutes. Collect the DNPH derivative by vacuum filtration [OP-12] using a Hirsch funnel. After you have removed the acidic filtrate, wash the derivative on the filter with 5 mL of cold 5% aqueous sodium bicarbonate, then with cold water. Recrystallize the derivative from 95% ethanol or ethanol-water mixed solvent [OP-23] using a test tube or a small Erlenmeyer flask. Collect it by vacuum filtration using a Hirsch funnel, washing it on the filter with a cold 3:1 mixture of ethanol and water. Dry the derivative [OP-21], weigh it [OP-4], and measure its melting point [OP-28]. Save enough of the derivative for the TLC separation and turn in the rest in a labeled vial.

TLC Separation of 2,4-Dinitrophenylhydrazones. Clean and label as many small test tubes as there are known methyl ketones available (see Table 11.2). Measure 1 mL of the 2,4-dinitrophenylhydrazine reagent into each test tube and add a drop of the appropriate methyl ketone. Set the test tubes aside until crystallization is complete, then collect the crystalline derivatives by vacuum filtration [OP-12] using a Hirsch funnel. Dissolve approximately 10 mg (0.01 g) of the unknown ketone's derivative in 0.5 mL of ethyl acetate, using a spot plate or small labeled test tube. Do the same for each known derivative. Use each solution to spot a silica gel TLC plate [OP-17] and develop the plate using a 3:1 mixture of toluene and petroleum ether as the developing solvent.

B. *Identification of an Unknown Ketone by NMR*
Obtain a second unknown ketone (or its proton NMR spectrum) and record its identification number in your laboratory notebook. If an NMR spectrometer is available for your use, make up a solution of the unknown in deuterochloroform using a TMS standard, and record and integrate its proton NMR spectrum [OP-35].

Report. For each part of the experiment, state the problem and your initial hypothesis, and tabulate and interpret the evidence (R_f values, melting points, NMR parameters). In each case tell whether your initial hypothesis changed (and if so, why), then state and support your conclusion. Turn in your NMR spectrum, TLC plate, and derivative with your report.

Waste Disposal: Place the filtrate in a designated DNPH waste container or dispose of it as directed by your instructor.

Take Care! Keep the solvent away from flames and do not breathe its vapors.

Waste Disposal: Put the developing solvent in a recovered hydrocarbon solvent recovery container.

Take Care! Avoid contact with $CDCl_3$ and do not breathe its vapors.

Waste Disposal: Put the deuterochloroform solution in a designated solvent recovery container.

Exercises

1 (a) Write a balanced equation for the reaction of your unknown methyl ketone with 2,4-dinitrophenylhydrazine. (b) The DNPH reagent contains 2.9 g of 2,4-dinitrophenylhydrazine in 100 mL of solution. What was the limiting reactant for the preparation of your 2,4-dinitrophenylhydrazone? Calculate the theoretical yield and percentage yield of the reaction.

2 Describe and explain any relationship between chain length and R_f value that you observed from your TLC separation.

3 (a) The purpose of the sodium bicarbonate washing was to prevent isomerization of the DNPH derivative during the melting-point determination. Write structures for two stereoisomers of the DNPH derivative of 2-pentanone. (b) Which one would you expect to be more stable, and why?

4 Describe and explain the possible effect on your results of the following experimental errors or variations. (a) You didn't wash the DNPH derivative of your unknown with aqueous sodium bicarbonate. (b) You let the TLC plate develop too long and you can't find the solvent front. (c) You left the TMS out of your ¹H NMR sample.

5 Draw structures for all ketones that have the molecular formula $C_5H_{10}O$, and sketch the ¹H NMR spectrum you would expect to obtain from each one. Your spectra should show the relative area, multiplicity, and approximate chemical shift of each signal.

6 Account for the fact that the DNPH derivative of 2-octanone melts at a lower temperature than the DNPH derivative of acetone, even though the molecular weight of the 2-octanone derivative is much higher.

Other Things You Can Do

(Starred items require your instructor's permission.)

*1 Use TLC to analyze some felt tip pen inks as described in Minilab 11.
*2 Use NMR spectrometry to identify an alkyl chloride that has the formula $C_4H_{10}Cl$ or another compound whose molecular formula will be provided by your instructor.
 3 Write a research paper about forensic chemistry starting with sources cited in the Bibliography.

Vacuum Distillation, Optical Rotation.
Optical Activity of α-Pinene

Separation of Liquids. Optical Activity. Stereoisomerism. Terpenes.

Operations:

OP-26 Vacuum Distillation
OP-31 Optical Rotation
OP-4   Weighing

Before You Begin

1 Read the experiment and read OP-26 and OP-31.
2 Prepare a brief experimental plan for this experiment.

Scenario

Dick Hawkshaw, private investigator, has called upon you to help him solve a mystery. A wealthy American entrepreneur, Aldo Hyde, was recently found murdered in his chalet in Cannes, France. Although the house was set on fire in an apparent attempt to cover up the murder, firefighters were able to extinguish it before the evidence was destroyed. In the house was found a half-filled can of paint thinner with no label, but the word *turpentine* was written on one side with a felt tip pen. The housekeeper informed Hawkshaw that there was no turpentine in the house before the night of the fire, so it must have been purchased by the murderer. Gas chromatographic analysis of a flammable residue found at the fire's point of origin revealed that liquid from this can was used to start the blaze.

The only suspects in the case are Aldo Hyde's widow, Dr. Jacqueline Hyde, a talented but unpredictable talk show psychologist; and Guy Framboise, a hot-tempered French businessman. Aldo and Jacqueline Hyde had lived separately for more than a year, she in New York and he in Cannes. His intention to leave her out of his will was thwarted by his untimely death, so she stands to inherit most of his fortune. Dr. Hyde arrived in Cannes from New York on the night of the murder and checked into a hotel at 10 p.m., approximately two hours before the murder took place, but there are no witnesses who can place her at the scene of the crime. Guy Framboise made threats against Aldo's life after a joint business operation failed, and a reliable witness saw the Frenchman's Peugeot parked near Hyde's chalet on the night of the murder.

Dick Hawkshaw just shipped your supervisor a sample of the "turpentine" used in the crime. Now it is up to you to discover the crucial evidence that will identify the murderer of Aldo Hyde.

Applying Scientific Methodology

The problem, of course, is "Who murdered Aldo Hyde?" As you read the experiment, you should find some clues that will help you solve the mystery—after you have gathered the experimental evidence.

Turpentine and the Terpenoids

Terpenes are among the most widely distributed natural products, occurring in nearly all plants. Terpenes are compounds that can, in principle, be broken down into two or more isopentane (2-methylbutane) units. For example, the carbon skeleton of geraniol can be separated in the middle to yield two isopentane units connected head-to-tail; that is, with the "head" end of one unit—the end nearest the side chain—connected to the "tail" end of the next.

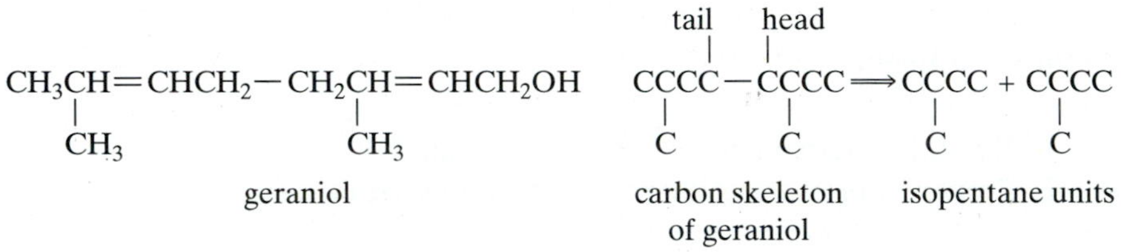

These isopentane units are also called isoprene units, after a diene having the same carbon skeleton.

$$CH_3CH=CHCH_2-CH_2CH=CHCH_2OH$$
$$\underset{CH_3}{|}\qquad\qquad\underset{CH_3}{|}$$
geraniol

tail   head
CCCC$-$CCCC $\Longrightarrow$ CCCC + CCCC
| | | |
C   C   C   C
carbon skeleton   isopentane units
of geraniol

Terpenes having oxygen-containing functional groups are sometimes called terpenoids.

Geraniol, with its rose blossom aroma, is an important constituent of the essential oil known as rose otto. It also plays an important role in the biosynthesis of terpenes. This process involves the enzymatic isomerization of isopentenyl pyrophosphate and a subsequent reaction with its isomer to yield geranyl pyrophosphate, from which the other terpenes are produced.

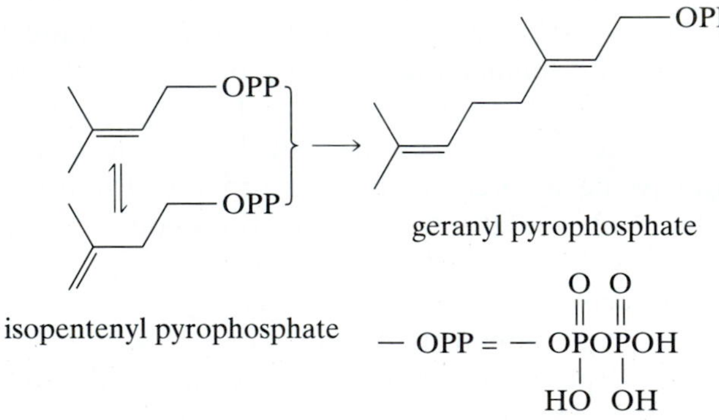

geranyl pyrophosphate

isopentenyl pyrophosphate

$$-OPP = -\underset{\underset{HO}{|}}{\overset{\overset{O}{||}}{O}}P\underset{\underset{OH}{|}}{\overset{\overset{O}{||}}{O}}POH$$

Many terpenes, especially the ones that contain oxygen, have pleasant odors and flavors and are therefore important flavoring and perfume ingredients.

An important natural source of terpenes is *turpentine,* the sticky oleoresin ("pitch") obtained from conifer trees such as the Southern longleaf pine, *Pinus palustris.* The "turpentine" sold as a paint thinner, more accurately called oil of turpentine, is distilled from this oleoresin. It is the world's most abundant essential oil, being obtained as a by-product of paper production as well as from pine pitch. The major components of American oil of turpentine are (+)-α-pinene and (−)-β-pinene, with the former predominating.

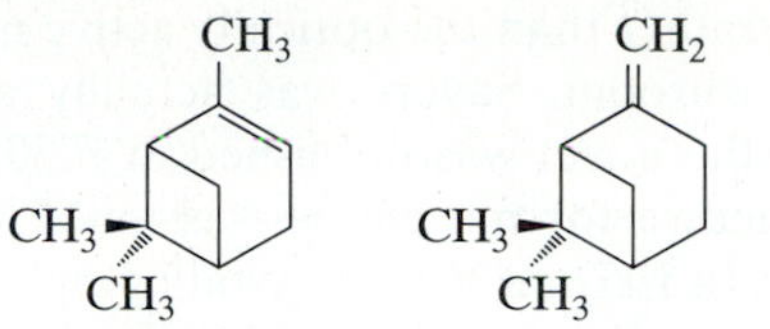

(+)-α-pinene   (−)-β-pinene

Constituents of American oil of turpentine

European oil of turpentine contains mostly the enantiomeric (−)-α-pinene and very little β-pinene. Both α- and β-pinene are used to synthesize other terpenes and their derivatives, which are used in perfumes, flavorings, and other commercial products. For example, α-pinene is a starting material in the synthesis of isobornyl acetate, which contributes a "pine needle" note to perfumes; linalool, which has a sweet woody-floral odor; and camphor, which is used in the manufacture of cosmetics, plastics, and pharmaceuticals.

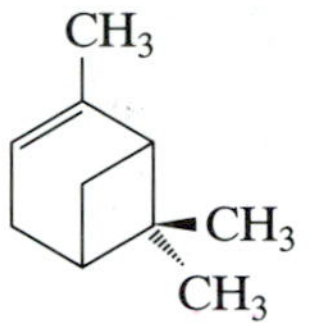

(−)-α-pinene

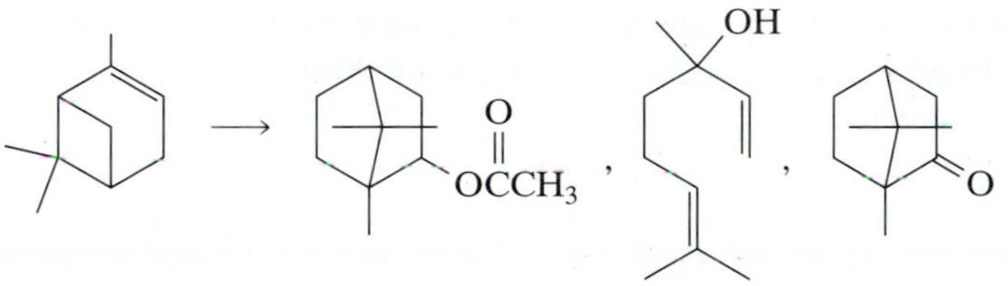

α-pinene   isobornyl acetate   linalool   camphor

Large quantities of β-pinene are converted to β-myrcene, which is then used to make perfume ingredients that have floral notes, including linalool and the aldehyde shown.

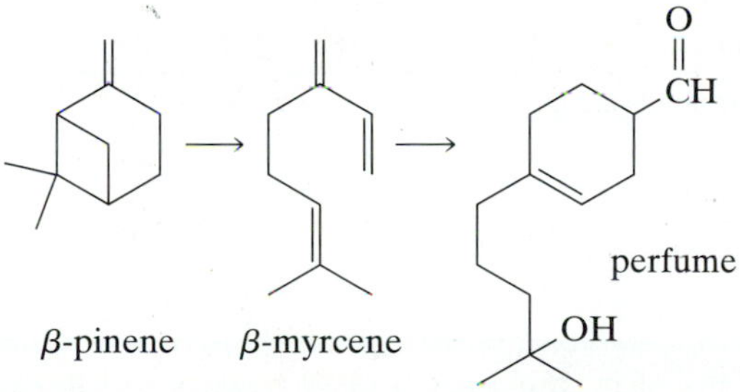

perfume aldehyde

β-pinene   β-myrcene

Understanding the Experiment

The application of chemistry to crime solving is not only the province of the forensic chemist. Mystery novels by writers from Arthur Conan Doyle to Dorothy L. Sayers contain many references to forensic chemistry. In Sayers' novel *The Documents in the Case,* a mushroom collector dies after eating a stew containing mushrooms he picked himself. The death is believed to be accidental, caused by the poisonous mushroom *Amanita muscaria,* until a chemist discovers that the stew contains optically inactive (and therefore

Dr. Watson himself claimed that Sherlock Holmes had a profound knowledge of chemistry.

Key Concept: *As a rule, chiral compounds from natural sources are optically active. Most synthetic chiral compounds are racemic mixtures and therefore optically inactive.*

CH₃ ⟍ O ⟍ CH₂N⁺(CH₃)₃

HO

(+)-muscarine

synthetic) muscarine rather than the optically active muscarine that occurs in the poisonous mushroom. Sayers was actually a step ahead of the chemists of her day (the novel was published in 1930), who—based on an incorrect achiral structure for muscarine—assumed that its natural form was optically inactive. In 1957 an X-ray crystallographer proved that Sayers had guessed right—muscarine is indeed chiral, so the premise of her novel was sound.

In this experiment you, like the chemist in Sayers' novel, will be measuring the optical rotation of a substance to help solve a mystery. First you must obtain α-pinene in a reasonably pure form. You will purify the "turpentine" sample you are issued by distilling it, performing the distillation under vacuum to reduce the likelihood of decomposition. α-Pinene distills at about 156°C at normal atmospheric pressure, but its boiling point is reduced to 52°C at a pressure of 20 torr. Any operation carried out under vacuum carries with it some risk of an implosion, which can cause injury from flying glass fragments, so it is essential to inspect all parts of the apparatus carefully to see that they are free from cracks or other defects, assemble the apparatus properly, and have it checked by your instructor.

According to *The Merck Index* (+)-α-pinene has a specific rotation of +51° and (−)-α-pinene has a specific rotation of −51° at 20°C. Because your distilled α-pinene will still contain some impurities, its specific rotation may be somewhat lower.

Properties

Table 12.1 Physical properties

	M.W.	b.p.^{760}	b.p.^{20}	$[\alpha]_D$
(+)-α-pinene	136.2	156	52	+51°
(−)-α-pinene	136.2	156	52	-51°
(−)-β-pinene	136.2	163		-22°

Note: Boiling points are in °C

Directions

Safety Notes

α-pinene

α-Pinene irritates the skin and eyes and its vapors are harmful. Avoid contact and do not breathe its vapors.
A vacuum distillation apparatus may implode if any of its components are cracked or otherwise damaged. Inspect the parts for damage and have your instructor check your apparatus. If possible, it is best to work behind a hood sash or safety shield while the apparatus is under vacuum.

Purification of α-Pinene. If you are using an aspirator and do not have a manometer, measure the temperature of the aspirator water after the aspirator has run for a while and estimate the minimum pressure the aspirator can attain. Otherwise, determine the pressure as directed by your instructor. Use the nomograph in Figure E11 of OP-26 to estimate the boiling point of α-pinene at that pressure (the actual boiling temperature may be somewhat higher). After inspecting the glassware carefully, assemble an apparatus for vacuum distillation [OP-26] using a 25-mL boiling flask and silicone-based stopcock grease and have it approved by your instructor. Add 10 mL of "turpentine" and either microporous boiling chips or a smooth boiling device such as a magnetic stirrer or capillary bubbler. Turn on the vacuum and carry out the distillation until at least two-thirds of the pinene has distilled (some undistilled residue should remain). When the distillation is finished, release the vacuum and transfer the distillate to a screwcap vial.

Analysis. Using a 25-mL volumetric flask, prepare a solution containing about 2.5 g of your α-pinene, weighed to the maximum accuracy of your balance [OP-4], in absolute ethanol. Transfer the solution to a 2-dm polarimeter cell and measure its optical rotation [OP-31], then measure the optical rotation of a blank consisting of pure absolute ethanol. Turn in the remainder of your α-pinene.

Report. State the problem and your initial hypothesis. Calculate the specific rotation of your α-pinene and give its complete name and structure. Summarize and interpret the relevant evidence. Tell whether your hypothesis changed when it was tested and if so, why. State your final conclusion and describe how it is supported by the evidence.

Stop and Think: Why should the water temperature affect the pressure an aspirator can attain?

Take Care! Be sure none of the components are cracked or otherwise damaged.

Stop and Think: Was the initial boiling point different than you had estimated? If so, why?

Stop and Think: Which α-pinene enantiomer do you have? How do you know?

Waste Disposal: Place the solution in a designated solvent recovery container. Leave the blank in its polarimeter cell.

Exercises

1 From its specific rotation estimate the purity of your α-pinene, as the mass percent of the predominant α-pinene enantiomer: (a) if the impurity is (−)-β-pinene; (b) if the impurity is optically inactive.

2 The Aldrich Chemical Company manufactures a technical (low purity) grade of pinene containing (+)-α-pinene and having a specific rotation of +43°C. Is the 15% (by mass) impurity in this product more likely to be (−)-α-pinene, (−)-β-pinene, or some optically inactive substance? Justify your answer with calculations.

3 Describe and explain the possible effect on your results of the following experimental errors or variations. (a) Your vacuum distillation apparatus has a leaky joint. (b) The "turpentine" contains synthetic α-pinene, which is racemic. (c) You use a 1-dm tube rather than a 2-dm tube to measure the optical rotation.

4 (a) Show how α-pinene can be divided into isopentane units. (b) Do the same for linalool and camphor.

5 (a) Determine the configuration (R or S) at each stereocenter of (+)-muscarine. (b) Determine the configuration at each stereocenter of (−)-β-pinene.

6 The pyrophosphate group is an excellent leaving group that can easily be lost to yield a carbocation. (a) Propose a mechanism for the synthesis of geraniol from isopentenyl pyrophosphate and its isomer

(page 96) in an acidic aqueous environment. (b) Propose a mechanism for the synthesis of α-pinene from geranyl pyrophosphate.

7 Propose a synthesis from β-myrcene of the perfume aldehyde shown on page 97.

Other Things You Can Do

(Starred projects require your instructor's permission.)

*1 Determine the percentage composition of turpentine by measuring its optical rotation, as described in Minilab 12.

*2 Purify technical grade (85–90%) α-terpineol, whose normal boiling point is 220°C, by vacuum distillation. Cool the purified product in ice, if necessary, to see if it will solidify (pure α-terpineol is said to solidify around 30°C).

3 Write a research paper about terpenes starting with sources cited in the Bibliography.

PART

II

Correlated Laboratory Experiments

The experiments in Part II are correlated with topics discussed in most introductory textbooks of organic chemistry. Correlations are indicated by the list of topics beginning each experiment. In Part II, frequently used elementary operations such as heating and weighing will no longer be noted at the beginning of an experiment or in the directions.

Investigation of a Chemical Bond by Infrared Spectrometry

EXPERIMENT 13

Resonance. Infrared Spectrometry. Molecular Mechanics. Carbonyl Compounds. Acyl Compounds.

Operation:

OP-34 Infrared Spectrometry

Before You Begin

1 Read the experiment, read or review the operations as necessary, and write an experimental plan.
2 Be prepared to predict the *relative* $C=O$ vibrational frequencies (not their actual numerical values) of compounds your instructor assigns in the lab.

Scenario

The Olfactory Factory manufactures perfumes by mixing essential oils, resins, and other ingredients as specified by secret fragrance formulas that have been kept in the Pomander family for generations. The company's quality control officer, Hyacinth Pomander, is responsible for measuring the relative amounts of certain key aroma chemicals in the ingredients to ensure that they remain within the tolerances specified by the formulas. Most of these substances are aldehydes, ketones, esters, and other compounds that contain a carbonyl $(C=O)$ group. The Olfactory Factory recently purchased a number of infrared detectors to analyze the key chemicals by measuring the intensity of the carbonyl band in each substance's infrared spectrum. Unfortunately this method has not been working well because of interference by other substances whose molecules also possess carbonyl groups. Dr. Pomander believes that by tuning the infrared detectors to the vibrational frequency of the carbonyl group in a key substance of each perfume ingredient, she should be able to minimize such interferences, but first she needs to know their vibrational frequencies. She has sent samples of several representative carbonyl and acyl compounds to your supervisor. Your assignment is to find a way to estimate in advance their $C=O$ vibrational frequencies, and then to check your predictions by measuring the vibrational frequency of each compound's carbonyl band as accurately as possible.

Applying Scientific Methodology

The basic problem in this experiment is to see if there is a way to estimate the vibrational frequencies of $C=O$ bonds in molecules that contain such bonds. After reading the experiment you should be able to develop a hypothesis and apply it to the representive compounds assigned by your instructor, then see whether or not your predictions are verified by the experimental results.

The Art and Science of Perfumery

Throughout history, both men and women have used substances with pleasant odors to attract attention, to cover up unpleasant odors, and simply for the pleasure they provide. Elizabeth I used large quantities of lavender oil, in part to mask the stench of Elizabethan England, where streets doubled as sewers and bathing was considered unhealthful. The ancient Egyptians had no such aversion to personal hygiene; they used perfumes to scent their baths and sweeten their breath. Egyptian tomb paintings depict feasts at which revelers piled lumps of scented animal fat on their heads; as the evening progressed the fat melted and coated their bodies with fragrant perfumes. A fat-based perfume found in an alabaster vase in the tomb of King Tutankhamen was still fragrant 3000 years later, and similar vases were found in the tombs of First Dynasty kings who reigned from about 5100 B.C. The aromatic resins frankincense and myrrh were among the gifts of the Magi in the biblical account of the birth of Jesus. The Queen of Sheba reputedly used myrrh to beguile King Solomon, and frankincense is still burned during rites of the Roman Catholic and Greek Orthodox churches.

Traditional perfume ingredients include natural resins, essential oils, concretes, and absolutes, which are obtained from plant or animal sources by such processes as distillation, solvent extraction, and expression (pressing). *Resins,* such as frankincense, are solid or semisolid materials exuded by plants. *Essential oils,* such as rose otto (also called attar of roses), are volatile liquid mixtures that are usually obtained by mixing finely divided plant materials with water and then distilling the mixture with steam. *Concretes* are waxy residues obtained by extracting plant components with volatile solvents and then evaporating the solvent. *Absolutes* are concretes that have been processed to remove insoluble materials. Jasmine, one of the most highly prized perfume ingredients, is made by extracting jasmine flowers with hydrocarbons, then treating the resulting concrete with alcohol to produce the absolute.

Because natural materials such as jasmine are often costly and may be subject to considerable variation in composition and availability, most modern perfumes contain synthetic aroma chemicals as well as these natural ingredients. Aroma chemicals include synthetic versions of natural compounds and even some compounds that have never existed in nature. Most aroma chemicals have an oxygen containing functional group; aldehydes, ketones, alcohols, esters, and ethers are well represented in perfumes. A few hydrocarbons are also used, and smelly sulfur and nitrogen compounds may be added in very small amounts. For example, the nitrogen containing compound indole has the repulsive odor of feces when pure, but in minute amounts it imparts a fine jasmine scent to perfumes. Some representative aroma chemicals and their odors are illustrated in Figure 13.1.

A perfumer creates a new perfume in much the same way that a composer creates a symphony. Inspired by perhaps a single aroma "note," such as a recently discovered natural product or a new synthetic aroma chemical, the perfumer creates a "theme," called an *accord,* that may consist of several related aroma chemicals. For example, *Chanel No. 5* is based on an accord that consists of a blend of aliphatic aldehydes. Additional ingredients that

isocyclocitral (carnation)   muscone (musk)   phenylethyl alcohol (rose)

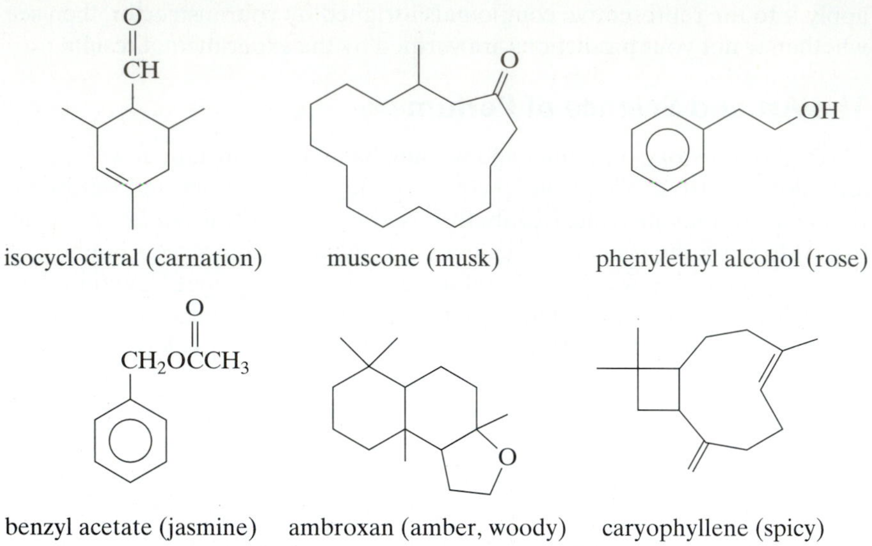

benzyl acetate (jasmine)   ambroxan (amber, woody)   caryophyllene (spicy)

Figure 13.1 Some aroma chemicals and their aroma notes

have different volatilities and aroma intensities are combined with the accord to produce a "composition" consisting of top, middle, and end notes. The top note contains the most volatile and odoriferous components, whose odors predominate immediately after a perfume is applied to the skin but fade as components evaporate. A fruity top note, for example, might consist of low molecular weight esters and some lactones (cyclic esters). The middle note provides the basic character of a fragrance. Mixtures of floral essential oils such as neroli oil with flowery aroma chemicals are often used as middle notes. The end note contains less volatile materials that persist longer on the skin, such as resins and certain oils. Musky aroma chemicals and woody materials such as sandalwood are among the ingredients of end notes. All perfume ingredients must be carefully selected and combined in just the right proportions to give the perfume a smooth odor profile so that there are no abrupt changes in odor as the ingredients evaporate. For a typical perfume or cologne, the ingredients are then dissolved in ethyl alcohol or some other alcohol-containing solvent.

There is some scientific evidence that perfumes do more than just make people smell better. Pleasant odors may improve your mood, bring to mind happy memories, enhance your creativity, and make you feel more cooperative and less confrontational toward the people you work with. On the other hand, a perfume that smells good to one person may be objectionable to someone else, and not everyone wants to be subjected to olfactory assaults from such flagrantly fragrant consumer products as scented toilet paper and fabric softeners. But an appreciation of pleasant and stimulating fragrances is rooted deeply in the human psyche, and any drawbacks of commercial scents must be outweighed by the benefits they provide.

Understanding the Experiment

In this experiment your instructor will provide you (or your research team) with three or more compounds that contain a carbonyl group. Based on the discussion to follow, you will try to arrange the compounds in order of their

C=O vibrational frequencies. At your instructor's option, you can also use a molecular mechanics computer program to estimate specific values for the frequencies. Then you will record the infrared spectra of your compounds to determine their actual vibrational frequencies and see how accurate your predictions were.

A chemical bond is similar in some ways to a coiled spring. Just as it takes energy to stretch a spring, energy is required to stretch a chemical bond. The stronger the bond, the more energy is required. A bond stretches and contracts as it vibrates, and the frequency at which the bond vibrates (ν) is proportional to the energy associated with the vibration, as given by the equation $E = h\nu$. Thus the stronger the bond, the higher its vibrational frequency. The C—O single bond in ethanol (CH_3CH_2OH) vibrates about 31 trillion (3.1×10^{13}) times a second, while the stronger C=O double bond in ethanal ($CH_3CH{=}O$) vibrates about 52 trillion times a second. A bond vibrating at a certain frequency can absorb a photon of infrared radiation having exactly the same frequency. For example, a C=O bond vibrating 5.2×10^{13} times a second can absorb a photon whose frequency is 5.2×10^{13} Hz, where 1 Hz = 1 s^{-1}.

It is convenient to express the vibrational frequency of infrared radiation in wave numbers ($\bar{\nu}$), where the wave number of a vibration expressed in cm^{-1} is the number of waves of electromagnetic radiation in one centimeter. The wave number corresponding to a vibrational frequency of $5.2 \times 10^{13} \text{ s}^{-1}$ is about 1740 cm^{-1}, so the infrared spectrum of ethanal contains an absorption band at 1740 cm^{-1} that is generated when its carbonyl bonds absorb infrared radiation.

You can convert frequency to wave number by dividing by the speed of light, expressed as 3.00×10^{10} cm/s.

To determine the relative vibrational frequencies for C=O bonds in different compounds, you can estimate their relative strengths using resonance theory. For example, suppose the carbonyl carbon in a compound is attached to some atom or group, Z, giving the compound the general formula RCOZ. Such a compound will have at least two resonance structures. We write resonance structure **B** by moving a pair of pi electrons onto the oxygen atom.

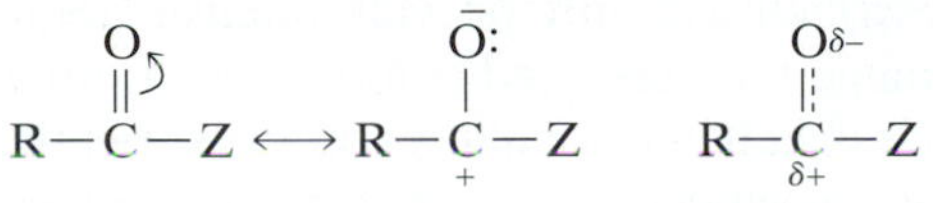

| A | B | composite structure |

The actual structure of the molecule can then be depicted by the composite structure shown, which represents the superposition of **A** and **B** in a single structure. The strength of the carbon-oxygen bond depends on the relative importance of the two resonance structures. If **A** and **B** were of equal importance, the carbon-oxygen bond would be halfway between a double bond and a single bond. This "one-and-a-half bond" would be considerably weaker than a full C=O double bond and would therefore vibrate at a lower frequency. As a rule, the more electronegative Z is, the less stable (and therefore less important) resonance structure **B** is, because an electronegative substituent withdraws electrons from the positively charged carbon, increasing the concentration of charge at that point. The less important

Key Concept: Dispersion of charge stabilizes a molecule; concentration of charge destabilizes it.

structure **B** is, the smaller contribution it will make to the structure of the compound, and the nearer the carbon-oxygen bond will be to a full double bond.

If Z is unsaturated, additional resonance structures may be drawn. For example, if Z is a vinyl (CH_2=CH—) group there will be three significant resonance structures, two having carbon-oxygen single bonds.

$$R-\overset{\overset{\displaystyle O}{\|}}{C}-CH=CH_2 \longleftrightarrow R-\underset{+}{\overset{\overset{\displaystyle \bar{O}:}{|}}{C}}-CH=CH_2 \longleftrightarrow R-\overset{\overset{\displaystyle \bar{O}:}{|}}{C}-CH=\overset{+}{C}H_2$$

$$\textbf{A} \qquad\qquad\qquad \textbf{B} \qquad\qquad\qquad \textbf{C}$$

This will increase the contribution of single bonded resonance structures, giving the carbon-oxygen bond less double bond character.

Additional resonance structures are possible if Z contains unpaired electrons on an atom next to the carbonyl carbon, as illustrated for Z = NH_2.

$$R-\overset{\overset{\displaystyle O}{\|}}{C}-\ddot{N}H_2 \longleftrightarrow R-\underset{+}{\overset{\overset{\displaystyle \bar{O}:}{|}}{C}}-\ddot{N}H_2 \longleftrightarrow R-\overset{\overset{\displaystyle \bar{O}:}{|}}{C}=\overset{+}{N}H_2$$

$$\textbf{A} \qquad\qquad\qquad \textbf{B} \qquad\qquad\qquad \textbf{C}$$

Stop and Think: Why?

If resonance structure **C** is important, it will decrease the double bond character of the carbon-oxygen bond. As a rule, structures like **C** are less important when the atom next to the carbonyl carbon is more electronegative than nitrogen. For the purposes of this experiment, you can disregard the effect of such resonance structures when the Z group has an oxygen atom at its point of attachment to the carbonyl carbon.

The method described will only provide relative frequencies; that is, it should help you arrange your assigned compounds in order of their C=O vibrational frequencies but it will not allow you to predict the actual frequencies. With a molecular mechanics program you can use a computer to carry out complicated quantum mechanical calculations to determine electron distributions and other features of molecules. To do so, you will first draw the structure of each molecule on the computer screen. Then you will have the program calculate the electron distribution about the C=O bond (it may provide a pictorial representation of the electron distribution), from which it can estimate the bond's vibrational frequency expressed in Hz or cm^{-1}.

When you record the infrared spectra of your compounds, you will need to identify the carbonyl band in each spectrum and determine its wave number as accurately as possible. Generally the C=O band will be the only strong band within a region extending from about 1650 cm^{-1} to 1800 cm^{-1}. If you are using a Fourier transform infrared (FT-IR) spectrometer, the wave numbers of the important bands may be printed directly on the spectrum or you may have to select the carbonyl band on a screen display of the spectrum and read its wave number off the monitor. With a dispersive infrared spectrometer, it is essential that the chart paper be properly aligned because you will have to estimate the wave number of the carbonyl band from its position on the chart paper.

Directions

This procedure is adapted from an article published in the *Journal of Chemical Education* **1996**, *73*, 188.

Your instructor will provide three or more organic compounds and may assign you to a research team of three or more students.

> **The compounds you are assigned may be hazardous if inhaled or absorbed through the skin. Avoid contact and inhalation.**

Safety Note

Estimating C═O Frequencies. If the structures of your compounds are not given, write their structures from their systematic names. Write as many significant resonance structures as you can for each compound. Based on your assessment of the relative importance of the resonance structures, predict the relative strength of the carbon-oxygen bond in each compound. Then list the compounds in order of their carbonyl vibrational frequencies, from higher to lower frequency.

At your instructor's request and following his or her instructions, use a molecular modeling program to estimate the vibrational frequency of the C═O bond in each compound.

Obtaining C═O Frequencies from the Infrared Spectrum: Record the infrared spectra [OP-34] of the assigned compounds. If you are working in a group, each member should record the spectrum of one compound. Accurately determine the wave number of each compound's carbonyl band from its spectrum.

Stop and Think: Were the results as you predicted? If not, why not?

Waste Disposal: Place any unused liquids in a designated waste container.

Report. Calculate the vibrational frequency of each compound's carbonyl group from the wave number of its carbonyl band and compare the results with your predictions. Your report should include a statement of the problem and an account of how you applied scientific methodology to solve it.

Exercises

1 Predict the relative C═O vibrational frequencies of isocyclocitral, muscone, and benzyl acetate (see Figure 13.1), listing them in order from lower to higher frequency.

2 Describe and explain the possible effect on your results of the following experimental or conceptual errors. (a) The infrared chart paper on a dispersive IR spectrometer is not aligned correctly when you record your spectrum. (b) One of your compounds is 3,7-dimethyl-2,6-octadienal. In writing its structure you inadvertently write the #2 double bond

between the #3 and #4 carbon atoms. (c) In comparing an aldhyde and a ketone, you assume that an alkyl group is electron-withdrawing relative to hydrogen because in general chemistry you learned that carbon is more electronegative than hydrogen.

3 Calculate the wavelength in μm of the infrared radiation absorbed by the carbonyl group of each compound whose wave number you recorded.

4 Write resonance structures for the following compounds and predict their relative C=O vibrational frequencies, listing them in order from lower to higher frequency.

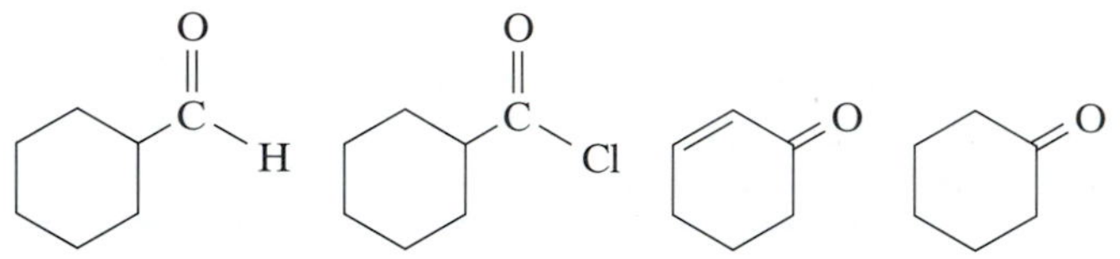

5 (a) Outline a synthesis of phenylethyl alcohol (2-phenylethanol) starting with benzene and ethylene oxide. (b) Outline a synthesis of benzyl acetate starting with toluene and ethanol.

Other Things You Can Do

(Starred items require your instructor's permission.)

1 Construct molecular models of some representative organic molecules, as described in Minilab 13.

2 Interpret one or more of your infrared spectra, indicating what kind of bond is responsible for each significant IR band.

*3 Use the molecular mechanics program to determine the preferred geometry for one or more or your compounds, or for another compound suggested by your instructor.

4 Write a research paper about perfumes, starting with sources listed in the Bibliography.

Properties of Common Functional Groups

Functional Group Chemistry. Qualitative Analysis. Infrared Spectrometry.

Operations:

OP-5   Measuring Volume
OP-25   Simple Distillation
OP-29   Boiling Point
OP-34   Infrared Spectrometry

Before You Begin

1   Read the experiment, read or review the operations, and write a brief experimental plan.
2   Read or review the section "Interpretation of Infrared Spectra" in OP-34.

Scenario

Marvelous Molecules Incorporated (MMI) manufactures fine chemicals for use by research chemists in educational institutions and industry. One of their sales representatives, Al Keene, just flew in from Milwaukee to meet an important client. He brought along samples of some representative chemicals to demonstrate the quality of MMI's products. Federal regulations prevented him from transporting the chemicals on the airliner so he had them sent ahead by freight carrier. On arrival he discovered, to his dismay, that the bottles of chemicals had been exposed to high heat and humidity during transit, causing all of their labels to fall off. The labels were salvaged, but he has no idea which label belongs to which bottle, so he placed a desperate phone call to your supervisor asking for help. Your assignment is to match each chemical with the correct label. Fortunately Al Keene selected chemicals from different families of organic compounds, so you will only have to identify the functional group present in each compound to find out what it is.

Applying Scientific Methodology

You will be working in teams, with each member of the team responsible for the identification of one compound. After you observe some of your compound's physical and chemical properties, you should be able to formulate a tentative hypothesis about the identity of its functional group. You will then record the infrared spectrum of your compound to test your hypothesis and arrive at a conclusion.

Chemical Taxonomy

A botanist attempting to identify an unknown flowering plant may first examine the flowering parts to detect features that suggest the family of plants it belongs to, then study the whole plant systematically to determine its genus and species. For example, a botanist coming across a plant with four symmetrical flower petals and six stamens (four long and two short) might tentatively classify it into the mustard family (*Cruciferae*). Further observation of white flower petals, lyre-shaped leaves, and a red, globular root might then lead the botanist to the conclusion that the plant is a specimen of *Raphanus sativus*. On the other hand, an experienced gardener might immediately recognize the same specimen as a radish plant based only on its general appearance.

Similarly, people experienced in handling chemicals may learn how to recognize a familiar organic compound from its odor and general characteristics, but a systematic approach is needed for positive identification of a range of organic compounds. Qualitative organic analysis, the identification of organic compounds based on their physical and chemical properties, is analogous in some ways to the identification of plants according to their *taxonomy*—their structural features and presumed natural relationships. To classify an organic compound into a given family requires first detecting a specific functional group (characteristic set of atoms) in the molecules of the compound. Table 14.1 lists some functional groups found frequently in organic compounds. Because functional groups influence the physical, chemical, and spectral properties of an organic compound, a chemist can identify a compound's functional groups by measuring certain physical properties, observing its chemical behavior with different classification

Table 14.1 Common functional groups and their families

Functional group	Functional group name	Family
$-C=C-$	carbon-carbon double bond	alkene
$-Cl$	chlorine atom	alkyl chloride
$-OH$	hydroxyl group	alcohol (also phenol)
$\overset{O}{\overset{\|}{-C}}-H\ (-CHO)$	carbonyl group, with H on carbonyl carbon	aldehyde
$\overset{O}{\overset{\|}{-C}}-\ (-CO-)$	carbonyl group, no H on carbonyl carbon	ketone
$\overset{O}{\overset{\|}{-C}}-OH\ (-COOH)$	carboxyl group (*carb*onyl + hydr*oxyl*)	carboxylic acid
$-NH_2$	amino group	amine (primary)

Note: Condensed representations of some of the functional groups are given in parentheses.

reagents, and studying its infrared spectrum. The chemical name and structure of the compound can then be determined by methods described in Part IV of this book.

Understanding the Experiment

In this experiment you will study a compound that belongs to one of the families listed in Table 14.1 and then share your results with your coworkers, who will be studying the remaining compounds. All of the compounds have molecules of about the same size and mass, so differences in their physical, chemical, and spectral properties will depend primarily on their functional groups.

Physical properties that are affected by functional groups include boiling point, density, and water solubility.

Molecular Structure and Boiling Points

Boiling occurs in a liquid when the kinetic energy of its component molecules becomes high enough to overcome the forces between them, allowing them to leave the surface of the liquid and enter the gaseous state (see Figure 14.1). Since the kinetic energy of molecules increases with temperature, and the energy required to separate molecules depends on the strength of the forces between them, we can expect liquids that have strong intermolecular forces to have high boiling points as well.

The important kinds of intermolecular forces occurring between organic molecules are, in order of increasing strength: (1) dispersion forces (sometimes called "van der Waals forces"), (2) dipole-dipole interactions, and (3) hydrogen bonding. Dispersion forces are caused by alternating transient charge separations on the surfaces of molecules. They occur among all kinds of molecules, polar or nonpolar, causing them to "stick together" on contact, somewhat like the styrofoam peanuts used as packing materials that cling to one another by static electricity. Dipole-dipole interactions result when molecules that have permanent bond dipoles line up so that the negative end of one molecule's dipole is opposite the positive end of another's, and vice versa. Hydrogen bonding is a special kind of dipole-dipole interaction involving the attraction of a highly polarized hydrogen atom for an electron donating atom (such as oxygen or nitrogen) on another molecule.

Among compounds of similar molecular shape and molecular weight, those capable of forming hydrogen bonds tend to have the highest boiling points, followed by compounds with polar groups capable of dipole-dipole interactions. Hydrogen bonding makes an important contribution to the boiling point only for organic compounds containing O—H and N—H bonds, with OH groups forming the strongest hydrogen bonds. Compounds with no polar groups are held together only by dispersion forces and tend to have the lowest boiling points.

You will distill your liquid to remove impurities that might affect your results. You can estimate its boiling point during the distillation if it is reasonably pure; otherwise you may need to redistill it or carry out a micro boiling-point determination.

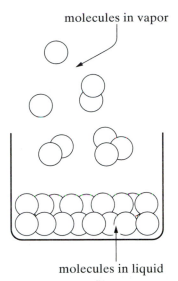

Figure 14.1 Boiling of a liquid

molecules in vapor

molecules in liquid

dipole-dipole interaction (formaldehyde)

hydrogen bonding (water)

Molecular Structure and Density

The density of an organic compound depends, to some extent, on the mass/volume ratio of its constituent atoms. Atoms that have a high nuclear mass confined within a small atomic volume, such as those in the top right-hand corner of the periodic table, have a high atomic density (atomic mass/atomic volume), so compounds whose molecules contain such atoms have comparatively high densities. Among the atoms encountered in this experiment (other than hydrogen), oxygen has the highest atomic density, followed by chlorine, nitrogen, and carbon.

Atomic density of some "heavy" atoms, relative to carbon = 1.00		
O: 1.50	Cl: 1.35	N: 1.26

The density of an organic liquid will thus depend on the number and kind of these "heavy" atoms it contains and on the fraction of its molecular weight they contribute. For example, chlorobenzene (C_6H_5Cl) has a higher density than phenol (C_6H_5OH) even though chlorine has a lower atomic density than oxygen, because the chlorine atom contributes about one-third of chlorobenzene's molecular weight while phenol's oxygen atom makes up only one-sixth of its molecular weight. You will determine the density of your liquid by weighing a precisely measured volume of the liquid.

Molecular Structure and Solubility

A compound will generally be soluble in a given solvent if the forces holding its own molecules together are similar to the forces holding the molecules of the solvent together, *or* if the compound can form hydrogen bonds with the solvent. Thus hexane dissolves readily in benzene because both compounds are hydrocarbons whose molecules are held together by dispersion forces. Ethanol dissolves in water because both compounds contain OH groups capable of forming hydrogen bonds. Formaldehyde, although it cannot form hydrogen bonds among its own molecules, can hydrogen bond to solvents that contain OH groups and is therefore soluble in water.

Key Concept: *"Like dissolves like." Polar compounds tend to dissolve in polar solvents and nonpolar compounds in nonpolar solvents.*

$$CH_3CH_2CH_2CH_2CH_2CH_3$$

hexane

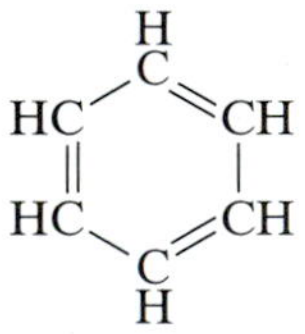

benzene

Hydrogen bonding interactions

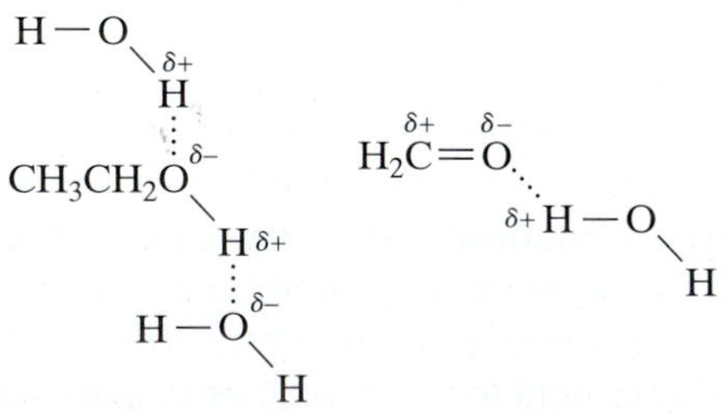

ethanol
and water

formaldehyde
and water

Keep in mind that "solubility" is a relative term; there are varying degrees of solubility. Terms commonly used to indicate the extent to which one compound dissolves in another are, in order of decreasing solubility, miscible (∞), very soluble (v), soluble (s), sparingly soluble (δ) and insoluble (i).

You will measure the solubility of your liquid by shaking it with water in a small test tube. For the purposes of this experiment, a liquid will be classified as *miscible* if it dissolves in an equal volume of water and as *soluble* if 0.2 mL of the liquid dissolves in about 6 mL of water. If the liquid dissolves, the resulting solution should be water-clear. If it does not dissolve you should observe a second liquid layer, cloudiness, or liquid droplets (not air bubbles) in the water. Since low-boiling liquids may evaporate rapidly, you should keep the test tube stoppered while you make your observations.

Classification Tests for Functional Groups

Chemists have developed a number of simple chemical tests that are positive only for compounds having certain kinds of functional groups. A litmus test, for example, can be used to detect acidic and basic functional groups. When dissolved in water, carboxylic acids turn blue litmus paper red and amines turn red litmus paper blue.

Compounds that are easily oxidized react with a solution of chromium(VI) oxide in sulfuric acid, commonly referred to as "chromic acid." Primary and secondary alcohols react within 2–3 seconds to form an opaque blue-green suspension. Aldehydes give the same result but usually take 10 seconds or more to react.

Aldehydes and ketones both react with 2,4-dinitrophenylhydrazine (DNPH) reagent to yield yellow or orange precipitates within a few minutes. Alkenes react readily with dilute aqueous potassium permanganate to form a brown precipitate as the purple color of the permanganate disappears. Alkyl halides give a green flame in the Beilstein test, which involves heating a copper wire moistened with the unknown in a burner flame.

Conflicting or ambiguous results may be produced with some tests because of impurities in the unknown or because the tests themselves are open to misinterpretation. For example, an aldehyde often contains traces of the corresponding carboxylic acid as an impurity, leading to a false positive litmus test. In such cases it may be necessary to repeat a test or redistill the unknown. For additional information and general equations for the reactions involved, see the Classification Tests section of Part IV.

Infrared Spectra

Each kind of functional group is associated with one or more characteristic infrared bands that can reveal its presence. A carboxyl group (COOH), for example, contains $C\!=\!O$, $C\!-\!O$, and $O\!-\!H$ bonds, all of which give rise to strong infrared absorption bands. Infrared bands that may help you classify your unknown include the two-pronged $N\!-\!H$ band characteristic of a primary amine, which is around 3400–3300 cm^{-1}; the broad $O\!-\!H$ band centered near 3300 cm^{-1} for an alcohol and 3000 cm^{-1} for a carboxylic acid; the strong $C\!=\!O$ band near 1700 cm^{-1}; and the $C\!-\!Cl$ band between 600 and 800 cm^{-1}. Alkenes are often characterized by vinylic $C\!-\!H$ bending vibrations in the 650–1000 cm^{-1} region and by a $C\!=\!C$ band (sometimes weak or absent) near 1650 cm^{-1}. See the section "Interpretation of Infrared Spectra" in OP-34 for more detailed information about these and other infrared bands associated with common functional groups.

Directions

Your instructor will give you the names that were on the labels described in the Scenario. The experiment can be performed in teams of as many as seven students, each team working with a full set of unknown liquids. Alternatively, students can work individually and report their data to the instructor for posting.

Safety Notes

> All of the unknown liquids are flammable, and some of them are caustic or have hazardous vapors. Avoid contact with the liquids, do not breathe their vapors, and keep them away from flames or hot surfaces.
> 2,4-Dinitrophenylhydrazine is harmful if absorbed through the skin and it will dye your hands yellow. Avoid contact with the DNPH reagent and wash your hands after using it.
> The chromic acid reagent is corrosive, very toxic, and carcinogenic. Wear gloves and avoid contact with the reagent.

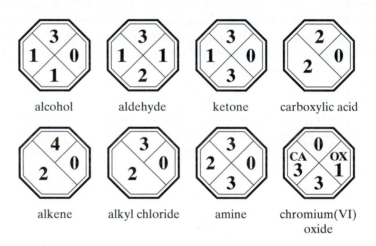

alcohol   aldehyde   ketone   carboxylic acid

alkene   alkyl chloride   amine   chromium(VI) oxide

A. *Physical Properties of the Unknowns*

Obtain a 5-mL sample of an unknown organic liquid and purify it by simple distillation [OP-25] using a small-scale apparatus. (**Take Care!** Avoid contact and do not breathe vapors.) If the liquid boils over a range of 3°C or less, you can use the median distillation temperature as its boiling point; otherwise, determine the boiling point of the purified liquid by a micro method [OP-27].

Using a volumetric pipet or automatic pipet [OP-5], accurately measure one mL of the purified unknown into a tared vial, stopper the vial, and weigh it to the nearest milligram on an accurate balance. (**Take Care!** Do not pipet by mouth!)

Using a measuring pipet measure 0.20 mL of the liquid into a 10-cm test tube. Add 0.20 mL of water, stopper the tube, and shake it vigorously for a few seconds. If two liquid layers separate on standing, or if the mixture is cloudy or contains undissolved droplets, add another 6 mL of water and shake it again. If the unknown dissolves, save the solution for the litmus test in part **B**.

Stop and Think: Compare your boiling point with those obtained by other students in your group. What can you say about your compound's intermolecular forces?

Stop and Think: Compare your density with those obtained by other students in your group. Does your compound contain any "heavy" atoms such as O, N, or Cl?

Stop and Think: If your compound forms a separate layer, does it float or sink? Why?

B. *Chemical Tests*

Except for the litmus test, all of the following tests should be carried out with the purified unknown liquid. Keep a careful record of your observations in your lab notebook.

Litmus test: If the unknown is miscible or soluble in water, test its *aqueous solution* with red and blue litmus paper.

2,4-Dinitrophenylhydrazine test: Add 1 drop of the unknown to 1 mL of the DNPH reagent in a test tube. Stopper and shake the test tube, and let the mixture stand for 15 minutes.

Chromic acid test: Dissolve 1 drop of the unknown in 1 mL of reagent grade acetone in a test tube. Add 1 drop of the chromic acid reagent, shake the mixture, and observe it for at least one minute, noting the time required for any positive test.

Potassium permanganate test: Dissolve 1 drop of the unknown in 2 mL of 95% ethanol, and add 10 drops of 0.1 *M* potassium permanganate with shaking.

Beilstein's test: Make a small loop in the end of a copper wire. Heat the loop to redness in a burner flame. (Do this under a hood or in an area away from the unknowns or other flammable liquids.) Dip the loop into a little of your unknown liquid and hold it in the lower outside part of the flame.

C. *Infrared Spectrum*

Record an infrared spectrum [OP-34] of your unknown, using the neat liquid, or obtain the spectrum from your instructor.

Decide which functional group is present in your compound and give its name and structure. If your instructor assigns Exercise 1, report the chemical family, boiling point, density, and solubility of your unknown to your coworkers (or instructor) and obtain the same information for the other liquids. If your instructor assigns Exercise 2, make copies of your IR spectrum for the other members of your group and obtain copies of their spectra in return.

Report. Your report should include a statement of the problem and an account of how you applied scientific methodology to solve it, describing clearly how you deduced the chemical family of your unknown and telling how you eliminated the other possibilities. Tabulate the physical properties of all the liquids and the results of your classification tests. Interpret the infrared spectrum of your unknown as completely as you can, and turn in the spectrum to your instructor.

Waste Disposal: Put wastes for the DNPH, chromic acid, and potassium permanganate tests in the appropriate containers.

Take Care! Wear gloves, avoid contact.

Stop and Think: At this point, which functional group do you think your compound contains?

Waste Disposal: Place the rest of your unknown in a waste container designated by your instructor.

Exercises

1 Using the data provided by your coworkers, group together compounds with similar physical properties. You should have one set of groups arranged according to boiling point, another set for density, and a third set for water solubility. Then explain the differences between the groups, specifying the type of intermolecular force involved where appropriate. For example, you can group together compounds that have boiling points within 10°C of one another, listed from higher to lower boiling point, and explain why the compounds in the first group have higher boiling points than those in the second group, etc.

2   Compare the infrared spectra obtained by the members of your group, looking for significant similarities and differences. In each spectrum, point out any IR bands that are associated with a functional group and indicate what kind of chemical bond is responsible for each band.

3   Describe and explain the possible effect on your results of the following experimental errors or variations. (a) You inadvertently use a 2-mL volumetric pipet to measure your liquid for the density determination. (b) Most of your liquid (an aldehyde) distills around 75°C, but you continue to collect distillate as the temperature rises above 100°C. (c) You clean a test tube for the DNPH test by rinsing it with acetone.

4   Draw a flowchart that could be used to efficiently classify compounds from the families in Table 14.1, using classification tests from this experiment. There should be two branches for each classification test, one leading to compounds that give a positive result and the other leading to compounds that give a negative result.

5   Draw structures for all possible constitutional isomers of your unknown that have the same functional group (if any), and give their systematic names.

6   Compound X, having the molecular formula $C_4H_6O_2$, is believed to be one of the compounds in the margin. (a) State the chemical family or families to which each compound belongs. (b) Draw a flowchart diagramming a procedure that could be used to determine the identity of compound X, using chemical tests described in Part IV.

1. $CH_3CH_2\overset{\displaystyle O}{\overset{\|}{C}}-\overset{\displaystyle O}{\overset{\|}{C}}H$

2. $CH_3\overset{\displaystyle O}{\overset{\|}{C}}-\overset{\displaystyle O}{\overset{\|}{C}}CH_3$

3. $CH_2{=}CH\overset{OH}{\overset{|}{C}}H-\overset{\displaystyle O}{\overset{\|}{C}}H$

4. $CH_3CH{=}CH\overset{\displaystyle O}{\overset{\|}{C}}-OH$

Other Things You Can Do

(Starred items require your instructor's permission.)

*1   If the reagents are available, you can use additional or alternative classification tests such as Tollens' test (C-23) for aldehydes, sodium iodide in acetone (C-22) for alkyl chlorides, and bromine (C-7) for alkenes.

*2   In Minilab 14, find out who else has the same compound that you have.

3   Write a research paper describing how hydrogen bonding can be detected and studied by spectrometric methods, using sources listed in the Bibliography or elsewhere.

Thin-Layer Chromatographic Analysis of Drug Components

Qualitative Analysis. Thin-Layer Chromatography.

Operations:

OP-11 Gravity Filtration
OP-17 Thin-Layer Chromatography

Before You Begin

Read the experiment, read or review the operations as necessary, and write a brief experimental plan.

Scenario

A patient who swallowed a large number of drug tablets was just wheeled into the emergency room of a local hospital. The bottle that contained the drug is missing, but some unused tablets were left on a counter in the bathroom. The patient's spouse recalls seeing a bottle of an over-the-counter analgesic drug in the medicine chest, but doesn't remember what was on the label. The hospital has rushed the tablets to your supervisor for analysis. Your assignment is to determine the identity of the drug so that emergency room physicians can apply the appropriate treatement.

Applying Scientific Methodology

This experiment requires you to identify the components of an analgesic drug tablet and then identify the tablet as one of the commercial drug preparations listed in Table 15.1. You will probably not be able to formulate a meaningful hypothesis until after you examine your TLC plate under ultraviolet light. Then you can use another visualization method to test your hypothesis.

Drugstore Chemicals

Whenever you go to the drugstore to buy a bottle of aspirin, Advil, Tylenol, or another of the dozens of different analgesic (painkilling) drug preparations available, you are purchasing an organic chemical or a mixture of several chemicals. That shouldn't be surprising since all matter—including yourself—is made up of chemicals, where *chemical* is just another name for a substance (element or compound). But most of the materials we encounter on a day-to-day basis consist of complex mixtures of substances, while many drugstore chemicals are reasonably pure substances or mixtures of only a few substances. In fact, the majority of analgesic drug preparations contain only one or two active ingredients, either aspirin, acetaminophen, ibuprofen, or some combination of these. The most popular combination is aspirin and acetaminophen, but salicylamide (a chemical relative of aspirin) is combined with aspirin in a few drug preparations. Caffeine is sometimes added to an analgesic preparation for its stimulant effect, but it is not classified as an analgesic.

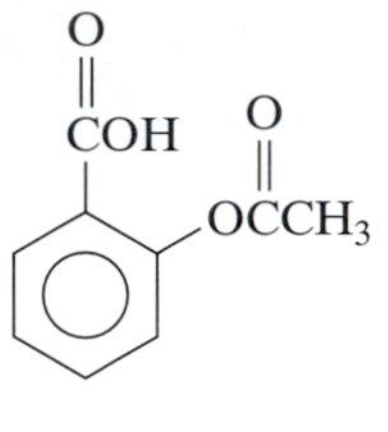

aspirin 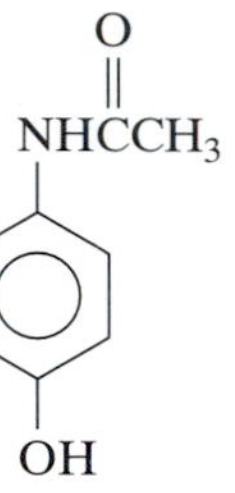   acetaminophen 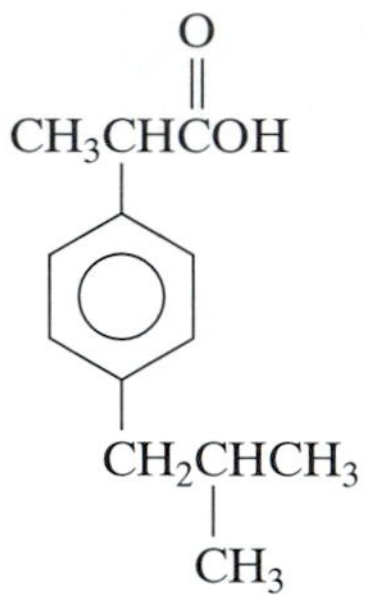   ibuprofen

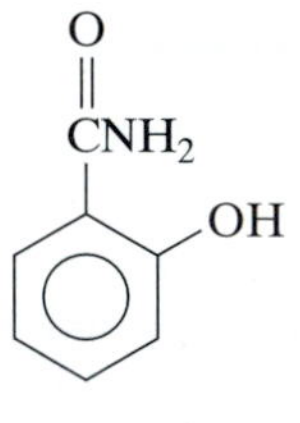

salicylamide 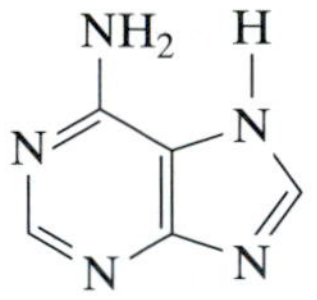   caffeine

A little starch is generally added as a binder to hold the tablets together. Table 15.1 shows the composition (excluding the starch) of some representative analgesic preparations.

Most analgesic drugs do more than just kill pain. Aspirin, acetaminophen, and ibuprofen are also antipyretics, meaning that they reduce fever, while aspirin and ibuprofen are both non-steroidal anti-inflammatory drugs (NSAIDs), meaning that they help reduce swelling and other symptoms of inflammation. Aspirin, in fact, has such a surprisingly wide range of benefits that it can be regarded as a true "wonder drug." In recent years it has

Certain steroids such as cortisone are used as anti-inflammatory drugs.

Table 15.1   Composition of some analgesic drug preparations (in milligrams per tablet)

Drug name	aspirin	acetaminophen	ibuprofen	salicylamide	caffeine
Advil			200		
Anacin	400				32
Aspirin*	325				
B.C. Tablets	325			95	16
Excedrin	250	250			65
Tylenol		325			

*5 grain tablet (1 grain = 64.8 mg)

been shown to reduce the incidence of heart disease, strokes, and certain cancers. It may also improve brain function in people who have suffered small strokes, help prevent cataracts, and reduce the occurrence of gallstones.

Although aspirin is the most widely used drug in the world, nobody understood how it worked until 1971, when the British pharmacologist John Vane showed that aspirin blocks the overproduction of natural biological substances called *prostaglandins* by deactivating a key enzyme required to manufacture them, prostaglandin synthase. Although prostaglandins are essential biological regulators, an oversupply of certain prostaglandins can promote the formation of blood clots that lead to heart attacks or strokes while others trigger pain, fever, and inflammation.

Vane shared the 1982 Nobel Prize in medicine as a result of his discovery.

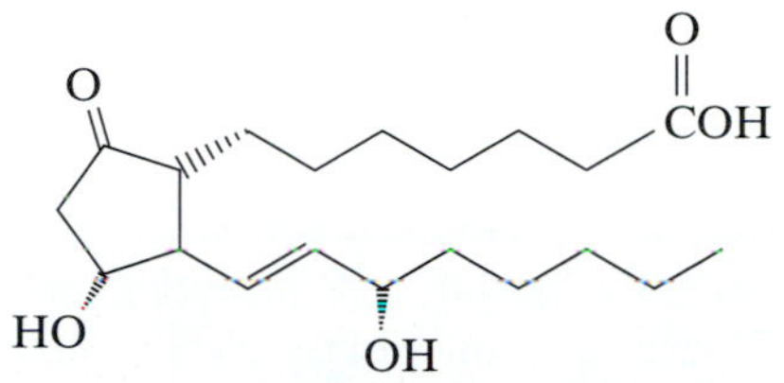

prostaglandin E$_2$

Recent studies have shown that prostaglandin synthase functions like a factory assembly line; raw materials enter one end of a channel that passes through the enzyme and leave the other end as fully assembled prostaglandin molecules. Aspirin molecules sabotage this operation by blocking the channel, thereby preventing raw materials from getting by. Aspirin appears to inhibit cancers of the digestive system by a different mechanism, stimulating the production of cancer fighting substances utilized by the body's immune system.

If the other analgesics can't compete with aspirin in versatility, they do have certain advantages over aspirin. Aspirin tends to promote bleeding, especially in the stomach, and it has been implicated in Reye's syndrome, a rare but often fatal disease that affects children. Acetaminophen is just as effective as aspirin at reducing pain but has fewer side effects at normal dosages. Ibuprofen is a powerful analgesic with about the same painkilling effect as aspirin at one-third the dosage. It is especially effective in treating arthritis and menstrual cramps, but it can cause problems in people with high blood pressure or kidney, liver, or heart disease.

Understanding the Experiment

In this experiment you will use thin-layer chromatography to identify the components of the drug tablet your instructor assigns. Your tablet may contain from one to three of these substances: acetaminophen, aspirin, caffeine, ibuprofen, and salicylamide.

You will first prepare a solution of the crushed tablet in ethanol/dichloromethane, then spot a TLC plate with the unknown solution along with standard solutions of all the active substances the drug tablet is likely to contain. The TLC plate can be developed with a solvent mixture such as ethyl acetate/acetic acid (200:1). By using several different methods to visualize the spots and comparing the R_f values of your unknown's spot(s) with those of the standards, you should be able to match each unknown spot with the spot of a known substance. Then you can refer to Table 15.1 to find out which commercial drug preparation contains the components you have identified.

Caffeine is only a minor component of a drug such as Excedrin, so its spot might be too faint to see clearly in some cases. In this event you can usually arrive at the correct commercial drug by comparing the remaining components with those in the table. For example, Excedrin is the only drug listed that contains both aspirin and acetaminophen, and B.C. is the only one that contains salicylamide.

Directions

Dichloromethane may be harmful if ingested, inhaled, or absorbed through the skin. There is a possibility that prolonged inhalation of dichloromethane may cause cancer. Minimize contact with the drug and standard solutions and do not breathe their vapors.
Avoid contact with the developing solvent and do not breathe its vapors.

Safety Notes

dichloromethane

Take Care! Avoid contact with the developing solvent and do not breathe its vapors.

Take Care! Avoid contact, do not breathe vapors.

Use a scrap piece of TLC plate to practice your spotting technique if one is available.

Preparation of the Developing Chamber. Obtain a suitable developing chamber [see OP-17] such as a wide mouth screwcap jar or a large beaker, and add enough developing solvent to cover its bottom to a level of about 5 mm. Prepare a paper wick by submerging the middle of a long strip of chromatography paper (about 3 cm wide) in the solvent so that its ends extend up both sides of the developing chamber. Seal the developing chamber with a screwcap lid or plastic wrap. Swirl the chamber to moisten all of the wick and let it sit to equilibrate.

Preparation of the TLC Plate. Obtain a quarter tablet of the unknown analgesic drug and grind it to a powder with a flat-bottomed stirring rod or spatula blade. (If the tablet is coated, remove as much of the coating as you can before grinding.) Transfer the powder to a test tube and add 3.0 mL of 1:1 ethanol/dichloromethane. Use a stirring rod to thoroughly mix and crush the solid in the solvent to dissolve as much of it as possible, then filter the solution [OP-11] through a Pasteur pipet that has a plug of cotton at the constriction, and collect it in a small vial.

Obtain a silica gel TLC sheet (containing a fluorescent indicator) of appropriate size and lightly pencil a starting line about 1.5 cm from the bottom, taking care not to touch the surface of the adsorbent with your fingers. Using micropipets (such as Drummond Micro Caps) or other spotting devices, carefully spot your unknown solution and the standard solutions provided on the TLC plate. To avoid cross-contamination, use a different micropipet for each solution. If possible, the unknown should be spotted in different concentrations (1–3 applications) at two or more locations. With the standard solutions, use 2–3 successive applications for each spot. Allow the solution to dry after each application and try to produce spots that are no more than 2 mm in diameter. The spots should be about 1 cm apart and the outer spots about 1.5 cm from the edge of the plate.

Development of the TLC Plate. Place the TLC plate in the pre-equilibrated developing chamber so that it straddles the wick but does not touch it. Develop the plate until the solvent front is 1 cm or less from the top. Mark the solvent front with a pencil and let the TLC plate dry under the hood.

Visualization and Analysis. Observe the spots under short-wavelength (254 nm) ultraviolet light and outline them with a pencil, marking the center of greatest intensity for each spot. Then visualize the spots by using iodine vapor or another visualizing reagent selected by your instructor. Calculate the R_f value of each spot and identify the active ingredients of your unknown by comparison with the R_f values and visualization results for the known spots. Use Table 15.1 to determine the commercial name of your analgesic drug preparation. (Your tablet may be a generic equivalent of the one listed.)

Report: Your report should include a statement of the problem and an account of how you applied scientific methodology to solve it. Show how you calculated your R_f values. Tabulate your results neatly and turn in your TLC plate with your report.

Waste Disposal: Place the developing solvent in the designated waste container.

Take care: Do not look directly at the light source.

Observe and Note: Record in your lab notebook any characteristics of the spots that might help you identify them.

Waste Disposal: Place the solution of your unknown in an appropriate waste container.

Exercises

1 Suppose you carry out a TLC separation of acetaminophen and phenacetin (see page 28) on silica gel using a nonpolar developing solvent. Which should have the higher R_f value and why?

2 Assuming that the drug components whose spots you identified were completely dissolved when you extracted the crushed tablet with methanol, what was the solid that you filtered from your methanol solution?

3 Describe and explain the possible effect on your results of the following experimental errors. (a) You allowed your chromatogram to develop too long and you can't find the solvent front. (b) The lab assistant who prepared the developing solvent mistakenly used aqueous ammonia in place of acetic acid. (c) You used a melting-point capillary tube to apply the spots. (d) You marked the starting line with a ballpoint pen.

4 Describe one or more simple chemical tests that would distinguish acetaminophen from ibuprofen. Refer to Part IV of this book, if necessary.

5 A new analgesic drug, Aleve, contains the sodium salt of naproxen as its active ingredient. (a) Look up the systematic name and structure of naproxen. (b) The physiological effects of naproxen are very similar to those of a drug component you studied in this experiment. Which one do you think it is, and why?

Other Things You Can Do

(Starred items require your instructor's permission.)

*1 Carry out the TLC analysis of felt tip pen inks as described in Minilab 11.

*2 Carry out a quantitative or qualitative analysis of an analgesic drug mixture using high-performance liquid chromatography. See OP-33 and *J. Chem. Educ.* **1983**, *60*, pages 163 and 1000.

 3 Write a research paper describing some of the methods used to test for drugs and their metabolites in body fluids, starting with sources listed in the Bibliography. For example, you might tell how the presence of proscribed drugs can be detected in professional athletes or race horses.

Separation of an Alkane Clathrate

Reactions of Alkanes. Clathrates. Infrared Spectrometry.

Operations:

OP-9   Mixing
OP-12 Vacuum Filtration
OP-13 Extraction
OP-14 Evaporation
OP-20 Drying Liquids
OP-21 Drying Solids
OP-34 Infrared Spectrometry

Before You Begin

1 Read the experiment, read or review the operations as necessary, and write a brief experimental plan.
2 Calculate the mass of 10.0 mmol of hexadecane and the mass of 200 mmol of urea.

Scenario

The Petit Prix Racing Group mixes its own auto racing fuel by combining clean-burning, high-octane hydrocarbons such as 2,2,4-trimethylpentane (also called "isooctane") with other racing fuel components such as methanol and nitromethane. The last batch of fuel was being mixed by an inexperienced employee, Rusty Tappet, when he accidentally added a can of diesel fuel intended for his employer's Mercedes-Benz to a barrel containing 2,2,4-trimethylpentane and methanol. Diesel fuel is not a clean-burning fuel and the hydrocarbons it contains have extremely low octane numbers, so the contents of the barrel are now worthless as racing fuel. But the racing group can't afford to waste expensive fuel, so it has contacted your supervisor to see whether your Institute's consulting chemists can develop a method of separating the diesel fuel components from the mixture.

Your supervisor thinks it should be possible to trap the straight-chain hydrocarbon molecules of diesel fuel inside molecular "cavities" that are too small to hold the highly branched 2,2,4-trimethylpentane molecules. Your assignment is to carry out preliminary tests using a mixture of hexadecane (a major component of diesel fuel), 2,2,4-trimethylpentane, and methanol to see if this method can be used to remove the hexadecane from the mixture.

$$CH_3CH_2CH_2CH_2CH_2CH_2CH_2CH_2CH_2CH_2CH_2CH_2CH_2CH_2CH_2CH_3$$

hexadecane

$$\begin{array}{c} CH_3 \\ | \\ CH_3CCH_2CHCH_3 \\ | \quad\quad | \\ CH_3 \quad CH_3 \end{array}$$

2,2,4-trimethylpentane

Applying Scientific Methodology

As in the previous experiments, you need to state the problem as a question, formulate a working hypothesis, follow the course of action described in the Directions, gather and evaluate evidence, test your hypothesis, arrive at a conclusion, and report your findings. Your working hypothesis will be tested when you record the infrared spectrum of the hydrocarbon you isolate.

Clathrates and Fuel Quality

Figure 16.1 Structure of a methane hydrate

In 1976 scientists drilling off the coast of Central America brought up a core sample that contained softball sized lumps of an icelike solid that sizzled like frying bacon and emitted liquid droplets and a flammable gas, leaving nothing behind but a puddle of water. The icelike solid was a gas hydrate, consisting mostly of single methane molecules trapped inside a cage made up of water molecules. Methane hydrate consists of 20 water molecules arranged to form a nearly spherical 12-sided geometric figure (a dodecahedron), with a free methane molecule floating inside its watery cage.

Recent discoveries of enormous methane hydrate deposits beneath ocean floors have attracted attention because they represent both an opportunity and a potential threat. It is estimated that about 20 quadrillion (2×10^{16}) cubic meters of methane are trapped in hydrates, which is energetically equivalent to approximately twice the earth's coal, oil, and gas reserves combined. If the oceans could somehow be "mined" for methane hydrates they would represent a huge energy reserve. But methane is also an efficient greenhouse gas, and some scientists believe that global warming or geological disturbances could cause hydrate deposits to decompose, releasing huge quantities of methane into the atmosphere. This could cause a runaway global warming effect that would drastically alter the earth's weather patterns.

Methane hydrate is an example of a *clathrate,* a complex formed when molecules of one compound are enclosed in cavities within other molecules or crystal lattices. In 1941 a German chemist discovered that urea combines with straight-chain alkanes having seven or more carbon atoms to form a crystalline clathrate, but does not combine with most branched alkanes.

O
||
H₂NCNH₂
urea

The quality of a motor fuel is measured by its octane number, which compares its antiknock properties to those of the highly branched alkane 2,2,4-trimethylpentane (octane number = 100). Long straight-chain alkanes have very low octane numbers, so treating a gasoline-grade petroleum fraction with urea to remove such alkanes increases its octane number. Jet fuels are also improved by treatment with urea because straight-chain alkanes tend to freeze sooner than branched alkanes of the same carbon number. Unless the straight-chain alkanes are removed, wax crystals could form in jet fuel at high altitude, where the temperature can drop to $-60°C$.

The quality of a diesel fuel, on the other hand, is measured by its *cetane number,* which compares its ignition properties to those of hexadecane (also called cetane). The cetane number of a fuel is established by matching the fuel's performance to that of a mixture containing hexadecane, whose cetane number is 100, and 2,2,4,4,6,8,8-heptamethylnonane (HMN), whose cetane number is 15. For example, a diesel fuel that has the same ignition properties as a mixture containing 40% hexadecane and 60% HMN is assigned a cetane number of

$$100(0.40) + 15(0.60) = 49$$

Most diesel fuels marketed in the United States have cetane numbers between 40 and 65.

$$
\begin{array}{cccc}
CH_3 & CH_3 & & CH_3 \\
| & | & & | \\
CH_3CCH_2CCH_2CHCH_2CCH_3 \\
| & | & | & | \\
CH_3 & CH_3 & CH_3 & CH_3 \\
\end{array}
$$
HMN

Understanding the Experiment

In this experiment, you will combine a solution of urea in methanol with a mixture of hexadecane and 2,2,4-trimethylpentane to see if the urea will form a clathrate with one of the alkanes. Many organic compounds—not just alkanes—combine with urea in solution to form crystalline clathrates in which six or more molecules of urea, the *host* compound, form a tubular channel large enough to hold a molecule of the *guest* compound. The channel's walls are formed by interpenetrating spirals of urea molecules that are hydrogen bonded to one another. The guest molecule is not covalently bonded to the host molecules, but is held inside the channel by relatively weak intermolecular forces. Whether a compound qualifies as a guest depends mainly on the size and shape of its molecules. The diameter of the channel formed by urea molecules is about 0.52 nm, which is large enough to hold straight-chain molecules but not most branched molecules. In addition to alkanes, guest compounds can include straight-chain primary alcohols, carboxylic acids, and esters that have terminal functional groups and at least seven carbon atoms.

The longer the guest molecule, the more urea molecules are needed to surround it. The number of urea molecules per guest molecule (the host/guest ratio) can be estimated using equation **1**, where n is the number of carbon atoms in the guest molecule.

$$\text{host/guest ratio for urea clathrates} \cong 1.5 + 0.65n \qquad \textbf{(1)}$$

For example, about eight urea molecules $[1.5 + (0.65 \times 10)]$ are sufficient to confine a molecule of decane.

Adding water to a urea clathrate causes it to decompose, releasing the guest compound. If urea forms a clathrate with a hydrocarbon in your mixture of hexadecane and 2,2,4-trimethylpentane, the identity of the guest alkane can be determined from its infrared spectrum. *Stretching* vibrations of C—H bonds in methyl (CH_3) groups produce infrared bands around 2960 cm^{-1} and 2870 cm^{-1}. The corresponding methylene (CH_2) bands are near 2925 cm^{-1} and 2850 cm^{-1}. Often the methyl and methylene bands overlap so that only two or three peaks are observed. Methyl groups also show asymmetrical and symmetrical *bending* vibrations near 1450 cm^{-1} and 1375 cm^{-1}. If there are two or three methyl groups on the same carbon atom, the symmetrical bending band is split into two or more closely spaced peaks near 1385 cm^{-1} and 1370 cm^{-1}. Bending vibrations of methylene groups may also give rise to bands near 1465 cm^{-1}, between 1350 cm^{-1} and 1150 cm^{-1} (often weak), and around 720 cm^{-1}. The intensity of the 720 cm^{-1} band (called the methylene rocking band) increases in proportion to the number of adjacent methylene groups, so a band in this region is characteristic of unbranched long-chain alkanes. The terms used to describe some of these bond vibrations—scissoring, twisting, wagging, rocking—testify to the fact that molecules are dynamic entities and not the static particles that molecular models might suggest.

Hexadecane	$CH_3(CH_2)_{14}CH_3$	2924.1	720.9
		1467.1	
		1378.0	

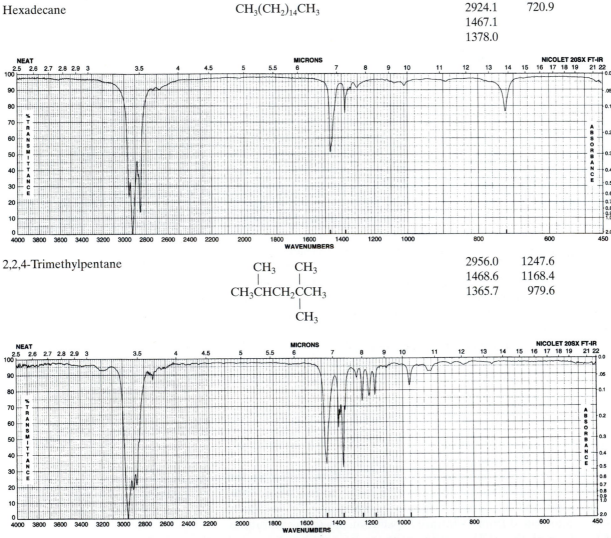

2,2,4-Trimethylpentane		2956.0	1247.6
		1468.6	1168.4
		1365.7	979.6

$$CH_3CHCH_2CCH_3$$

Figure 16.2 IR spectra of two alkanes (These and other IR spectra in this book are reproduced from *The Aldrich Library of FT-IR Spectra, Edition II,* with the permission of the Aldrich Chemical Company.)
Note: The exact wave numbers of certain absorption bands (designated by "tick" markers at the base of the spectrum) will be listed above most of the IR spectra reproduced in this book.

From the masses of the clathrate and the recovered guest alkane, you can determine the host/guest ratio—the number of urea molecules that surround each alkane molecule. For this result to be valid, material losses must be kept to a minimum and the masses of the clathrate and recovered guest alkane must be measured as accurately as possible. It is also essential that the clathrate be completely dry and the alkane be entirely free of solvent.

Reactions and Properties

Table 16.1 Physical properties

	M.W.	b.p.	m.p.	d	n_D^{20}
hexadecane	226.4	287	18	0.773	1.4345
2,2,4-trimethylpentane	114.2	99	−107	0.692	1.3915
urea	60.06		135	1.323	

Note: m.p. and b.p. are in °C, density in g/mL; refractive index is for 20°C

$$C_nH_{(2n+2)} + n\ H_2NCNH_2 \longrightarrow C_nH_{(2n+2)} \cdot [H_2NCNH_2]_n$$

$$\underset{\text{alkane}}{} \quad \underset{\text{urea}}{} \quad \underset{\text{alkane-urea clathrate}}{}$$

alkane   urea   alkane-urea clathrate

Directions

Safety Notes

> **Hexadecane and 2,2,4-trimethylpentane are flammable; keep them away from flames and hot surfaces.**
> **Methanol is flammable and harmful if ingested, inhaled, or absorbed through the skin. Avoid contact with the liquid and do not breathe its vapors.**
> **Dichloromethane may be harmful if ingested, inhaled, or absorbed through the skin. There is a possibility that prolonged inhalation of dichloromethane may cause cancer. Minimize contact with the liquid and do not breathe its vapors.**

2,2,4-trimethyl-
pentane

methanol   dichloromethane

Reaction. Accurately weigh 10.0 mmol of hexadecane and combine it with 5.0 mL of 2,2,4-trimethylpentane. In a 125-mL Erlenmeyer flask, combine 200 mmol of urea with 50 mL of methanol. Warm the mixture to 50°C, swirling or stirring magnetically [OP-9] until all the urea has dissolved; do

Take Care! Avoid contact with methanol, do not breathe its vapors.

Stop and Think: What do you think is in the solid?

Waste Disposal: Place the filtrate in an appropriate solvent recovery container.

Observe and Note: What happens as the clathrate dissolves?

Take Care! Avoid contact with dichloromethane, do not breathe its vapors.

Waste Disposal: Place the recovered dichloromethane in an appropriate solvent recovery container.

not heat the methanol to boiling. While the solution is still warm, add the mixture of alkanes and stir or swirl until a white solid begins to separate. Set the solution aside and let it cool slowly to 30°C or below; then cool it in an ice bath for 10 minutes or more until crystallization is complete.

Separation. Collect the clathrate (the solid product) by vacuum filtration [OP-12] and wash it on the filter with ice-cold methanol. Let the product dry [OP-21] to constant mass at room temperature (heating will cause it to decompose). If the product cannot be left overnight or longer to dry, it should be air dried thoroughly on the filter, blotted between filter papers, and left to dry in a well-ventilated location such as a hood. Weigh the dry product accurately.

Mix the clathrate with 25 mL of warm water and stir the mixture on a steam bath or boiling water bath for several minutes until the crystals dissolve. Cool the mixture in an ice bath and extract [OP-13] the guest alkane with two portions of dichloromethane. (If the alkane solidifies on cooling, stir in the first portion of dichloromethane and transfer the solution to the separatory funnel after the alkane dissolves.) Dry the combined extracts over anhydrous calcium chloride [OP-20]. Evaporate the solvent *completely* [OP-14] using a cold trap and weigh the guest alkane accurately.

Analysis. Record the infrared spectrum [OP-34] of the guest alkane and use it to identify the alkane. Turn in the infrared spectrum and the alkane to your instructor. Give its correct name on the label.

Report. Your report should include a statement of the problem and an account of how you applied scientific methodology to solve it. Explain how you identified the guest alkane and interpret its infrared spectrum as completely as you can. Calculate the percentage recovery and the host/guest ratio from your experimental results.

Exercises

1 Estimate the host/guest ratio for hexadecane using equation 1 and compare that value with your experimental value. Try to account for any significant differences.

2 Compare the infrared spectrum of hexadecane with that of 2,2,4-trimethylpentane and point out some bands that might help you distinguish straight-chain alkanes from branched ones.

3 Describe and explain the possible effect on your results (including your host/guest ratio) of the following experimental errors. (a) The lab assistant accidentally put octane in the "isooctane" (2,2,4-trimethylpentane) bottle. (b) The lab assistant put hexane in the hexadecane bottle. (c) You dried the clathrate in a 90°C oven. (d) The clathrate crystals were not dry when you weighed them.

4 Following the format in Appendix V, construct a flow diagram for the procedure used in this experiment.

5 What is the cetane number of a fuel that has the same ignition properties as a mixture containing 20% hexadecane and 80% HMN? Would it make a good diesel fuel?

6   Thiourea forms tubular clathrates that are similar in structure to those formed by urea. Combining a mixture of hexadecane and 2,2,4-trimethylpentane with thiourea (H_2NCSNH_2) produces a crystalline solid that decomposes in water to yield 2,2,4-trimethylpentane, but no hexadecane. Propose an explanation for this result.

7   Outline a possible synthesis of 2,2,4,4,6,8,8-heptamethylnonane (HMN) starting with isobutylene.

Other Things You Can Do

(Starred items require your instructor's permission.)

*1   Identify the guest alkane by measuring its refractive index instead of (or as well as) its infrared spectrum.

*2   Recover the other alkane by adding a large amount of water to the filtrate and separating the alkane layer. After it has been dried with calcium chloride, the alkane can be characterized by obtaining its infrared spectrum and refractive index.

3   Write a research paper about the use of urea and thiourea in the manufacture of barbiturates, starting with sources listed in the Bibliography. Describe some of the therapeutic uses and side effects of these drugs.

Isomers and Isomerization Reactions

Reactions of Alkenes. Isomerization Reactions. Isomerism.

Operations:

    OP-9   Mixing
    OP-11   Gravity Filtration
    OP-12   Vacuum Filtration
    OP-21   Drying Solids
    OP-28   Melting Point

Before You Begin

1   Read the experiment, read or review the operations as necessary, and write an experimental plan.
2   Calculate the mass of 11.0 mmol of potassium cyanate and the mass and volume of 8.0 mmol of dimethyl maleate.

Scenario

A roving science historian, Dr. Perry Celsus, travels around the country delivering lectures on the history of chemistry at scientific meetings and educational institutions. Dr. Celsus would like to enliven his lectures by demonstrating some of the pivotal experiments in chemistry that led to major advances in the field. These experiments include: (1) Friedrich Wöhler's isomerization of ammonium cyanate to urea, which helped to demolish the vitalistic theory of organic chemistry; (2) the conversion of maleic acid to its geometric isomer, fumaric acid, from which J. H. van't Hoff developed the concept of geometric isomerism.

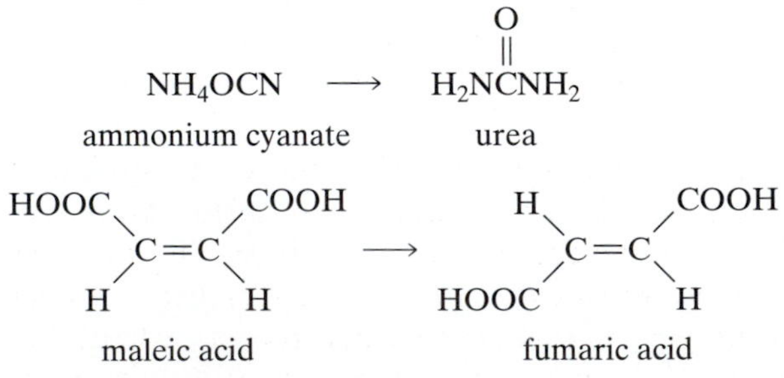

$$NH_4OCN \longrightarrow H_2NCNH_2$$
$$\text{ammonium cyanate} \qquad \text{urea}$$

maleic acid   fumaric acid

These experiments are not very suitable for demonstrations because of the time and conditions required, so Dr. Celsus has asked your Institute for help in developing updated versions of the ammonium cyanate and fumaric acid experiments that still demonstrate the scientific principles involved. Your supervisor thinks it may be possible to re-enact these nineteenth century experiments in modern form by using phenylammonium cyanate in place of ammonium cyanate and dimethyl maleate in place of maleic acid. Your assignment is to find out whether these two substances do, in fact, yield constitutional and geometric isomers analogous to the ones in the original experiments.

Applying Scientific Methodology

Since you will be performing two separate experiments you will need to develop two working hypotheses, which will be tested when you obtain the melting points of any products you obtain.

Isomerism in the History of Chemistry

Early in the nineteenth century, most scientists believed that plants and animals possessed some "vital force" that made it possible for them to convert inorganic substances into organic compounds. Since this vital force was presumably absent in inanimate objects, it was assumed that organic compounds could not possibly be synthesized from inorganic substances in the laboratory. Vitalism received a serious blow in 1828, when Friedrich Wöhler mixed cyanic acid (HOCN) with ammonia expecting to obtain ammonium cyanate and came up instead with urea, a product of protein metabolism that is excreted in urine. This "lucky accident" was the first known synthesis of an organic compound from an inorganic one to take place outside a living organism. As Wöhler put it in a letter to another eminent chemist, J. J. Berzelius, "I can make urea without the aid of kidneys, either man or dog!" Wöhler's synthesis of urea suggested that there is no essential difference between inorganic and organic compounds, and it paved the way for the flowering of organic chemistry in the last half of the nineteenth century.

Ammonium cyanate and urea share the same molecular formula, CH_4N_2O, and are therefore *isomers.* At one time scientists believed that no two compounds could be formed out of the same set of atoms. There is, after all, only one kind of H_2SO_4 and one kind of NaCl. So when two independent investigators reported the same elemental composition for cyanic acid (HOCN) and fulminic acid (HONC), Berzelius thought one of them had made a mistake. After further investigation he became convinced that a given set of atoms can indeed combine in different ways to form compounds that have different properties, and he coined the term *isomerism* to describe this phenomenon.

Isomers such as ammonium cyanate and urea, which differ in the way their atoms are attached to one another, are called *constitutional isomers* or *structural isomers.* Isomers that differ only in the way their atoms are arranged in space are called *stereoisomers. Geometric isomers* are stereoisomers that differ from one another because of restricted rotation about the bonds connecting two or more atoms. Maleic acid and fumaric acid are geometric isomers because their carbon-carbon double bonds prevent their interconversion under ordinary conditions. A geometric isomer that has certain groups (other than hydrogen) on the same side of the double bond is called a *cis* isomer, and one with those groups on opposite sides is a *trans*

Key Concept: The properties of a substance arise from its chemical structure, not from its source.

The term isomer *is derived from Greek words meaning "equal parts," referring to the fact that isomers have the same number and kind of atoms.*

isomer. Maleic acid is thus *cis*-2-butenedioic acid and fumaric acid is *trans*-2-butenedioic acid.

Understanding the Experiment

Wöhler prepared urea by several different methods. One of them involved mixing ammonium chloride and silver cyanate and then evaporating the resulting ammonium cyanate solution to dryness. When heated, ammonium cyanate decomposes to ammonia and cyanic acid, which combine to form urea.

$$NH_4Cl + AgOCN \longrightarrow NH_4OCN + AgCl$$
ammonium
cyanate

$$NH_4OCN \xrightarrow{\text{heat}} NH_3 + HOC \equiv N \longrightarrow \overset{\overset{\textstyle O}{\|}}{H_2NCNH_2}$$
cyanic   urea
acid

You will attempt to isomerize the phenyl derivative of ammonium cyanate by a similar reaction. As shown in the "Reactions and Properties" section, aniline reacts with hydrochloric acid to form the organic salt phenylammonium chloride. By combining an aqueous solution of phenylammonium chloride with potassium cyanate, you can prepare a solution containing phenylammonium cyanate. Then you will see whether heating this solution yields a product. If it does, a melting-point determination should tell you whether you have synthesized a constitutional isomer of phenylammonium cyanate, phenylurea.

Sometimes one of a pair of geometric isomers can be converted to the other under conditions that cause temporary cleavage of a pi bond. When maleic acid is heated with a trace of bromine in the presence of light, some of the bromine molecules break apart into bromine atoms. A bromine atom then adds to one end of the double bond of a maleic acid molecule, breaking the pi bond. This allows the molecule to rotate freely about the remaining sigma bond. If the bromine atom is ejected when the COOH groups are opposite one another, the pi bond reforms to yield a molecule of fumaric acid.

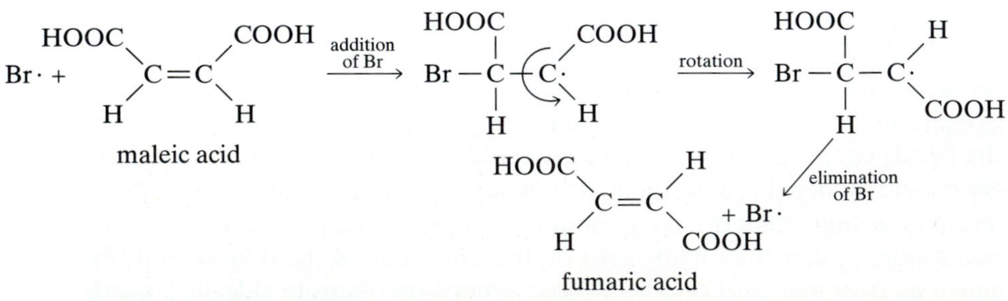

Maleic acid and fumaric acid are both solids, so this reaction must be run in solution and the fumaric acid separated by filtration. But dimethyl maleate (the methyl ester of maleic acid) is a liquid while dimethyl fumarate is a solid, so if dimethyl maleate can be induced to isomerize to dimethyl fumarate, the liquid *cis* isomer should change to a solid *trans* isomer. If molecular model sets are available, you can simulate the isomerization of dimethyl maleate after building a molecular model of it.

Reactions and Properties

A. $PhNH_2 + HCl \longrightarrow PhNH_3Cl$
     aniline   phenylammonium
                                chloride

$PhNH_3Cl + KOCN \longrightarrow PhNH_3OCN + KCl$
                              phenylammonium
                                   cyanate

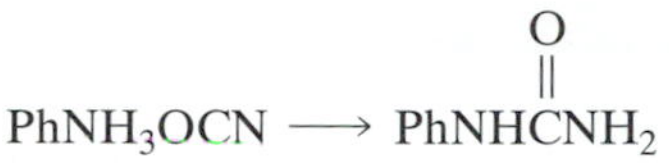

$$PhNH_3OCN \longrightarrow PhNHC\overset{\displaystyle O}{\overset{\displaystyle \|}{}}NH_2$$
                        phenylurea

B.

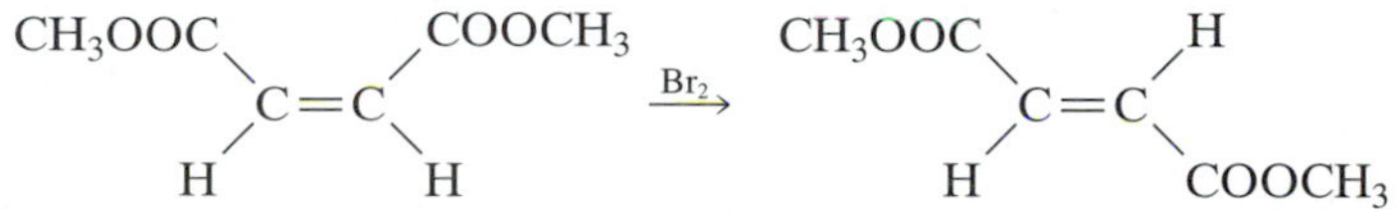

$$CH_3OOC\!\diagdown \quad \diagup COOCH_3 \xrightarrow{\;Br_2\;} CH_3OOC\!\diagdown \quad \diagup H$$
$$C{=}C C{=}C$$
$$\diagup H H\diagdown \diagup H COOCH_3\diagdown$$
dimethyl maleate   dimethyl fumarate

Table 17.1 Physical properties

	M.W.	m.p.	b.p.	*d*
aniline	93.1	−6	184	1.022
potassium cyanate	81.1	800 d		
phenylurea	136.2	147	238	
dimethyl maleate	144.1	8	205	1.151
dimethyl fumarate	144.1	104	193	

Note: m.p. and b.p. are in °C, *d* is in g/mL

Directions

aniline

Safety Note

> Aniline is poisonous and may be carcinogenic. It can cause serious injury or death if swallowed, inhaled, or absorbed through the skin. Wear protective gloves and dispense under a fume hood; avoid contact and do not breathe vapors.

Safety Notes

bromine

A. *Isomerization of Phenylammonium Cyanate*

Under the hood, prepare a solution of phenylammonium chloride by combining 5 mL of 3 *M* hydrochloric acid with 5 mL of water in a 50-mL beaker and adding 1.0 mL (~ 11 mmol) of recently distilled aniline. Dissolve 11.0 mmol of potassium cyanate in 10 mL of water in another small beaker, and mix this solution with the phenylammonium chloride solution to form phenylammonium cyanate. Stir for 2–3 minutes [OP-9], then add 7 mL of water and heat the solution to 80°C to dissolve any solid. Add about 0.05 g of decolorizing carbon (preferably pelletized Norit) and stir for 2 minutes at 80°C. Filter the solution by gravity [OP-11] while it is still hot.

Let the solution cool slowly to room temperature while you proceed with part **B**, then cool it further in ice. Collect any product by vacuum filtration [OP-12] and dry it thoroughly [OP-21]. Weigh the dry product and measure its melting point [OP-28].

B. *Isomerization of Dimethyl Maleate*

> **Bromine is toxic and corrosive, and its vapors are very harmful. Dichloromethane may be harmful if ingested, inhaled, or absorbed through the skin. Avoid contact with the bromine/dichloromethane solution and do not breathe its vapors.**

Measure 8.0 mmol of dimethyl maleate into a small test tube and mix in a drop of 1 *M* bromine in dichloromethane. Set the test tube in a 150-mL beaker that is half full of boiling water and place the beaker about 15 cm away from an unfrosted 75- or 100-watt light bulb. Keep the water boiling for 10 minutes or more with the light switched on, then remove the test tube and cool it in ice water until crystallization is complete. Collect the product by vacuum filtration in a Hirsch funnel [OP-12], wash it on the filter with a little cold 95% ethanol, and dry it. Measure the mass and melting point [OP-28] of the product.

If molecular model kits are available, construct a molecular model of dimethyl maleate and show that it cannot be converted to dimethyl fumarate without breaking bonds. Using a model of Br_2, simulate the course of the reaction you carried out in this experiment, referring to the mechanism for the isomerization of maleic acid.

Report. Your report should include a statement of the problem for each part of the experiment and an account of how you applied scientific methodology to solve it. Calculate the theoretical and percent yields for each part of the experiment.

Exercises

1 (a) Estimate the percentage of your dimethyl maleate that was converted to dimethyl fumarate. (b) Would it be possible to make dimethyl maleate in good yield from dimethyl fumarate (using bromine and light) under experimental conditions that permitted recovery of the dimethyl maleate? Why or why not?

2 Methylenemalonic acid, $CH_2\!=\!C(COOH)_2$, is an isomer of maleic acid. (a) What kind of isomers are methylenemalonic acid and maleic acid, constitutional or geometric? (b) Would methylenemalonic acid isomerize under the conditions of this experiment (part **B**)? If so, give the structure of the product; if not, explain why.

3 Describe and explain the possible effect on your results of the following experimental errors or variations. (a) Your instructor was out of aniline so *p*-toluidine (*p*-methylaniline) was substituted for it. (b) It was a dark and stormy day and the lights were out in the laboratory (but the gas was on). (c) You misread the label on a bottle of dimethyl malonate and used it instead of dimethyl maleate.

4 (a) Explain why dimethyl maleate has a higher boiling point than dimethyl fumarate. (b) Explain why dimethyl fumarate has a higher melting point than dimethyl maleate.

5 Propose a mechanism for the isomerization of phenylammonium cyanate to phenylurea.

Other Things You Can Do

(Starred items require your instuctor's permission.)

*1 Investigate the structures and properties of isomers as described in Minilab 15.

*2 Make models for as many isomers having the molecular formula C_3H_6O as you can, and identify any pairs of geometric isomers. You should also be able to construct models for pairs of molecules that, like your two hands, are nonsuperposable mirror images of one another. These are called mirror image isomers, or enantiomers.

*3 Record the ^1H NMR spectra of dimethyl maleate and dimethyl fumarate. Measure and compare the coupling constants for the vinylic protons in the two compounds. Interpret the spectra as completely as you can.

4 Starting with sources listed in the Bibliography, write a research paper about the manufacture of urea and its use in the production of such commercial products as urethane plastics, urea-formaldehyde residues, barbiturates, and jet fuel.

EXPERIMENT 18

Structures and Properties of Stereoisomers

Isomerization Reactions. Stereoisomerism. Infrared Spectrometry.

Operations:

OP-6   Heating
OP-13 Extraction
OP-14 Evaporation
OP-20 Drying Liquids
OP-29 Boiling Point
OP-30 Refractive Index
OP-31 Optical Rotation
OP-34 Infrared Spectrometry

Before You Begin

1 Read the experiment, read or review the operations as necessary, and write an experimental plan.
2 Before beginning this experiment, you should know how to draw stereochemical structural formulas, how to designate configurations at stereocenters, and how to classify stereoisomers as enantiomers and diastereomers. Review the sections of your lecture textbook on stereochemistry if you need help in any of these areas.

Scenario

The new-age herbalist Basil Wormwood explores the composition and uses of natural products from plants. For example, he isolates and analyzes essential oils from various plants using techniques he learned in his college organic chemistry course. While pursuing his studies he observed several phenomena that puzzled him, so he contacted the Institute to see if its consulting chemists could explain them.

1 After isolating the major component of caraway oil (from caraway seeds) and the major component of spearmint oil (from spearmint leaves) he discovered that these substances, whose odors are very different, have virtually identical physical and chemical properties.
2 His assistant accidentally spilled some vinegar in a batch of peppermint oil while Basil was away. When he returned he removed the vinegar by extraction, but the aroma of the oil had changed markedly.
3 When examined under a microscope, the tartaric acid crystals he received from a supplier looked different than the crystals of natural tartaric acid that he had used before.

caraway plant   spearmint plant

Your assignment is threefold: (1) To find out whether the major components of caraway and spearmint oils are identical or if they differ in some way; (2) To find out what happens when menthone from peppermint oil is treated with acid; (3) To find out whether the tartaric acid is natural L-(+)-tartaric acid or an "unnatural" tartaric acid stereoisomer.

Applying Scientific Methodology

You should find enough information relating to the first two problems in the experiment to formulate a working hypothesis for each. You will need to gather some experimental evidence before you can develop a meaningful hypothesis for the third problem.

Alice Through the Looking Glass: Dissymmetry and Life

In *Through the Looking Glass,* Alice, contemplating the world she views in her mirror, wonders aloud to her cat whether looking-glass milk would be good to drink. Such an inquiry might seem naive, even to a cat. If you hold a glass of milk in front of a mirror, the milk and its reflection look exactly alike, so why should "mirror image milk" be any different than the milk we ordinarily drink? You may be suprised to learn that mirror image milk would *not* be very good to drink; it would be quite indigestible and its taste would be bitter and unpleasant. To understand why, let's take an imaginary trip to a looking-glass world.

Imagine yourself on board the spaceship *Icarus* bound for the planetary system of Barnard's star. Arriving on the surface of its earthlike third planet, called *Arret* by the inhabitants, your landing party is invited to a feast by some friendly Arretians. You are served their standard banquet fare: roast *krop,* overdone *iloc'corb,* crusty *yawarac*-seed rolls, and an aromatic *t'nim* tea. You had been looking forward to a change from the monotonous space diet, but you soon lose your appetite. Most of the food tastes flat or bitter, the seeds on the roll have a minty flavor, and the tea smells like caraway seeds! After politely declining a second helping of *krop,* you return to your shuttlecraft with the rest of the landing party. Before long you and the rest of the crew are suffering from indigestion, causing a serious breach in interplanetary relations.

Earthly organisms are composed mostly of *chiral* molecules, such as the D-monosaccharides in complex carbohydrates and the L-amino acids that make up proteins and enzymes. It is conceivable that somewhere in the universe there exists a mirror image planet, otherwise similar to Earth, where carbohydrates are composed of L-monosaccharides and proteins of D-amino acids, and the configurations of other chiral compounds are reversed as well. On such a planet many foods would taste different than their earthly counterparts, and we Earthlings would find them indigestible. The chiral compounds on that planet would still be optically active; that is, they would rotate plane-polarized light, as ours do. But they would rotate the light in the opposite direction.

One of the great unsolved mysteries of life concerns the origin of optically active compounds. Living systems are composed of dissymmetric (chiral) molecules, and molecular dissymmetry is necessary for life as we know it, but how did such dissymmetry come about? Was there some "molecular

Key Concepts: A chiral compound is one exhibiting "handedness." Like a glove, a chiral molecule lacks a plane of symmetry and cannot be superposed on its mirror image.

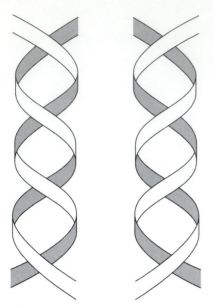

Figure 18.1 Left-handed and right-handed double helixes

Adam," a single dissymmetric molecule that gave rise to all molecular dissymmetry on Earth? Or is molecular dissymmetry a phenomenon that appeared independently at many different locations as a consequence of some kind of fundamental dissymmetry in the universe? The discovery (in 1997) that some amino acids in an Australian meteorite contain an excess of the L-enantiomer appears to support the hypothesis that some natural process favors one enantiomer over another. But convincing proof of such a hypothesis is hard to come by, and the answer may never be fully known.

The basic molecules of life—proteins, carbohydrates, nucleic acids, and enzymes—are chiral and are built up of smaller units that are also chiral. A strand of DNA, for example, consists of two long chains, each having a backbone of linked D-2-deoxyribose molecules twisted into a right-handed double helix. DNA and RNA regulate the synthesis of proteins from L-amino acids, which are combined in specific sequences inside cellular structures called ribosomes. Some of these proteins make up the enzymes that assist in the digestion of carbohydrates, yielding D-glucose to be used by the body for fuel. If life somewhere else in the universe were based on a mirror-image DNA made up of L-2-deoxyribose and twisted into a left-handed double helix, then protein synthesis could utilize only D-amino acids and the corresponding enzymes could digest only L-carbohydrates. In other words, if the configuration of one link in the chain of life (as we know it) is reversed, all the rest must be reversed as well.

If we could, by some magical contrivance, pass through the looking glass as Alice did, all the people, plants, and other organic matter in the looking glass world would presumably be constructed of these mirror image molecules. We could not survive in such a world (though Alice did, in Lewis Carroll's imagination) because digestion, metabolism, reproduction, and other life processes involving chiral molecules would be inhibited or prevented entirely.

Understanding the Experiment

Peppermint oil is a mixture of several optically active components, including (−)-menthol and the corresponding ketone, (−)-menthone. When a ketone has a stereocenter next to its carbonyl group, the configuration at the stereocenter may change under acidic or basic conditions, as illustrated by the following mechanism for the isomerization of (R)-3-methyl-2-butanone to its enantiomer.

Mechanism of the acid-catalyzed isomerization of (R)-3-methyl-2-butanone to (S)-3-methyl-2-butanone

In part **A** of this experiment you will see whether treating (−)-menthone from peppermint oil with acid can bring about a similar isomerization. Note that (−)-menthone has two stereocenters (see p. 140), so changing the configuration at just one of them will yield a diastereomer of (−)-menthone, not its enantiomer. If the configuration at the carbon next to the carbonyl group changes, the product will be (+)-isomenthone. Since the specific rotation of (−)-menthone is −30° and the specific rotation of (+)-isomenthone is +92°, it should be easy to detect the occurrence of isomerization by measuring the optical rotation of the product, even if the isomerization is not complete.

Enantiomers have identical physical properties except the direction in which they rotate plane-polarized light. If one enantiomer of menthone, for example, rotates polarized light in a clockwise (+) direction, the other enantiomer will rotate polarized light by the same angle in a counterclockwise (−) direction. Enantiomers may also differ in some chemical properties, such as the way they interact with chiral substrates. These differences are particularly important in biochemical systems. For example, some kinds of olfactory receptors are apparently chiral, so the (+)- and (−)-enantiomers of a compound that interacts with such receptors may smell different. Carvone, a ketone found in the essential oils of both caraway seeds and the spearmint plant, is an example of a chiral compound whose enantiomers have decidedly different odors.

Experiments that proved the odor differences between carvone enantiomers are described in Science **1971,** *172,* 1043.

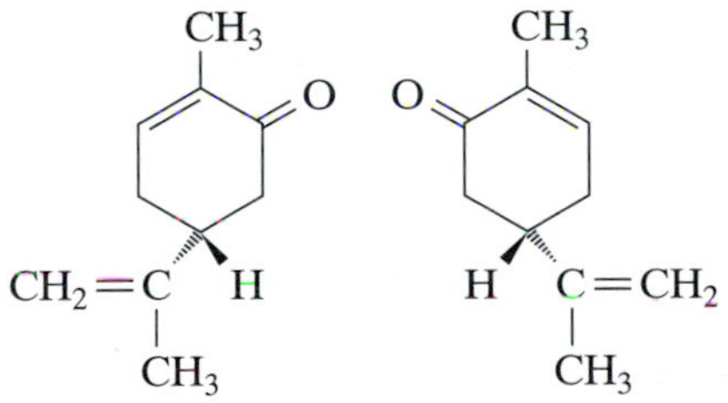

(R)-(−)-carvone   (S)-(+)-carvone

In part **B** of this experiment you and your coworkers will compare some physical properties and the infrared spectra of the carvones from both spearmint and caraway oils, as well as their odors. You will also determine their specific rotations to find out whether the carvones are the same or different, and if they are different which enantiomer is present in each oil.

Tartaric acid can be produced from potassium hydrogen tartrate (cream of tartar), a by-product of winemaking. This natural tartaric acid is the L-(+)-form illustrated here. Since tartaric acid has two stereocenters, several other stereoisomers are possible. In part **C** of this experiment you will measure the optical rotation of the tartaric acid provided to determine whether or not it is the natural form and, if not, what its structure is.

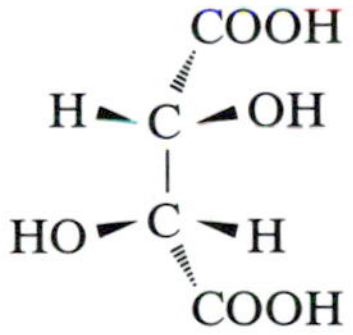

If molecular models are available, you can use them to study the stereochemical relationships among the compounds you will be working with in this experiment.

(L)-(+)-tartaric acid

Reactions and Properties

CH$_3$ ... $\xrightarrow{\text{acid}}$ ... CH$_3$ ... O

(–)-menthone   (+)-isomenthone

Table 18.1 Physical properties

	M.W.	d	$[\alpha]$
(R)-carvone	150.2	0.96	−62°
(S)-carvone	150.2	0.96	+62°
(−)-menthone	154.2	0.895	−30°
(+)-isomenthone	154.2	0.900	+92°

Note: Densities are in g/mL

Directions

A. *Isomerization of (−)-Menthone*

Safety Notes

acetic acid   ethyl ether

Take Care! Wear gloves, do not breathe vapors of acetic acid.

Take Care! Keep ether away from flames and hot surfaces.

Stop and Think: Has any of the menthone isomerized? How do you know?

Waste Disposal: Polarimeter solutions from this and other parts of the experiment should be put in a designated waste container unless your instructor indicates otherwise.

Stop and Think: Which carvone enantiomer do you have? How do you know?

> **Acetic acid causes chemical burns that can seriously damage skin and eyes; its vapors are highly irritating to the eyes and respiratory tract. Wear gloves, dispense under a hood; avoid contact and do not breathe vapors. Diethyl ether is extremely flammable and may be harmful if inhaled. Do not breathe the vapors, and keep it away from flames and hot surfaces.**

Under the hood, mix 1.0 mL of (−)-menthone with 5.0 mL of glacial acetic acid and 5.0 mL of 1 *M* HCl. Heat the mixture under reflux for 30 minutes [OP-6], then let it cool to room temperature and transfer it to a beaker. Add enough 3 *M* NaOH to raise the pH of the solution to 10 (30 mL or so may be needed), then extract it [OP-13] with two portions of diethyl ether. Dry the combined ether extracts with magnesium sulfate [OP-20] and evaporate the ether [OP-14]. Using 0.5–1.0 g of solute per 10 mL of solution (measure accurately), prepare a solution of the product in absolute ethanol, then measure its optical rotation [OP-31] and the optical rotation of a blank. (If your instructor requests, measure the optical rotation of the (−)-menthone as well.) Turn in your remaining product in a labeled vial.

Construct a molecular model of (−)-menthone in its most stable chair conformation. Show how it can be converted to a model of (+)-isomenthone by interchanging atoms or groups. Use the models to confirm that the two compounds are diastereomers, not enantiomers, and decide which one is the most stable. Determine the configuration (*R* or *S*) at each stereocenter of each isomer.

B. *Properties of Carvones from Spearmint and Caraway Oils*
Students should work in pairs, with each student using 1–2 mL of carvone from a different oil. Measure the boiling point [OP-29] and refractive index [OP-30] of your carvone. Using 0.5–1.0 g of solute per 10 mL of solution (measure accurately), prepare a solution of your carvone in 95% ethanol, then measure its optical rotation [OP-31] and the optical rotation of a blank. Record its infrared spectrum [OP-34], or obtain the spectrum from your instructor. Compare the odors of the two carvones and compare the

properties you measured with those of your lab partner's carvone. Compare their infrared spectra as well.

After you have decided which carvone you have, construct a molecular model for your carvone and compare it with your lab partner's molecular model. Describe the stereochemical relationship between the two models (identical, enantiomers, or diastereomers). Determine the configuration (*R* or *S*) at the stereocenter of each model.

C. *Identification of a Tartaric Acid Stereoisomer*
Using about 1.0 g of solute per 10 mL of solution (measure accurately), prepare a solution of the tartaric acid stereoisomer in water, then measure its optical rotation [OP-31] and the optical rotation of a blank. Using the fact that the specific rotation of L-(+)-tartaric acid is +12°, predict the specific rotations of the other stereoisomers and deduce the structure of your tartaric acid (assume that it is not a racemic mixture). Construct a molecular model of your tartaric acid stereoisomer and determine the configuration at each stereocenter.

Report. Your report should include statements of the problems and an account of how you applied scientific methodology to solve them. Include stereochemical drawings representing the structures of all the models you constructed, giving the configurations of the stereocenters, and the following calculations: (1) the specific rotations of all compounds whose optical rotations you measured; (2) the percent composition of the product mixture from the reaction of menthone; (3) the optical purity of your carvone.

Waste Disposal: Place any unused carvone in a designated container for recovery and re-use.

Exercises

1 Explain any similarities or differences in the properties of the carvone enantiomers and interpret your infrared spectrum as completely as you can.
2 (a) Draw chair-form structures for the most stable conformations of (−)-menthone and (+)-isomenthone, and decide which diastereomer should be more stable. (b) Is the composition of your isomerization mixture consistent with this conclusion? Explain.
3 Describe and explain the possible effect on your results of the following experimental errors or variations. (a) You try to isomerize (−)-menthol rather than (−)-menthone in part **A**. (b) In part **B**, you mix your carvone with an equal amount of your partner's carvone before measuring the optical rotation. (c) Your tartaric acid is the (−)-stereoisomer, but in measuring its optical rotation you read the scale when the analyzer is rotated 180° from the correct setting.
4 Propose a mechanism for the isomerization of (−)-menthone to (+)-isomenthone in the presence of acid.
5 A synthetic form of tartaric acid has a specific rotation of 0°, but its melting point is very different from that of *meso*-tartaric acid. Explain.

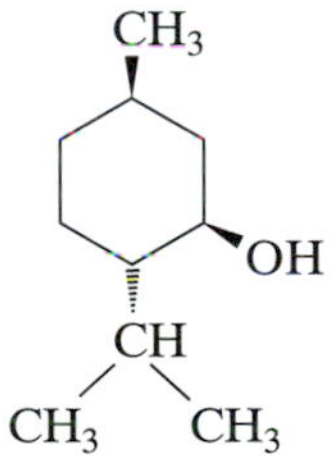

(−)-menthol

6 The steroid cholic acid is said to have 2047 possible stereoisomers. Indicate each stereogenic carbon atom on the cholic acid molecule with an asterisk, and perform a calculation to confirm this isomer number.

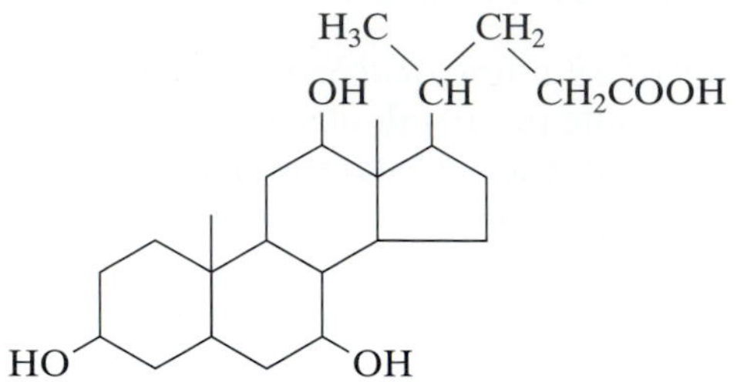

cholic acid

Other Things You Can Do

(Starred items require your instructor's permission.)

*1 Isolate your carvone enantiomer for part **B** from spearmint or caraway oil by column chromatography. Slurry-pack a chromatography column [OP-16] with silica gel, using high-boiling petroleum ether (fire hazard!). Introduce about 2 g of the essential oil onto the top of the column and elute it with the following solvents: 25 mL of high-boiling petroleum ether; 50 mL of 10% dichloromethane/petroleum ether; 25 mL of 20% dichloromethane/petroleum ether; and 125 mL of 50% dichloromethane/petroleum ether. Put the first 100 mL of eluate in a designated solvent recovery container and collect the rest in 25-mL fraction collectors. Evaporate the fractions and measure the refractive indexes of the residues. Use the purest carvone samples ($n_D \cong 1.499$ at 20°C) in part **B**.

*2 Obtain gas chromatograms of (−)-menthone and the product from part **A** using a 10% Carbowax-10M column packing at 125°. Identify the two major peaks, calculate the percentage of isomenthone in the mixture (assuming equal detector response factors), and compare your result with the percent composition calculated from the specific rotation of the mixture.

3 Starting with sources listed in the Bibliography, write a research paper about chiral drugs, describing differences in the physiological properties of enantiomers of the same drugs.

Bridgehead Reactivity in an S_N1 Solvolysis Reaction

Reactions of Alkyl Halides. Nucleophilic Substitution. Carbocations.
Reaction Kinetics

Operations:

OP-5   Measuring Volume
OP-6   Heating
OP-9   Mixing

Before You Begin

Read the experiment, read or review the operations, and write a brief experimental plan.

Scenario

Bridgehead Balms, Inc., is a small pharmaceutical company seeking to develop new drugs and more economical routes to existing drugs based on the adamantane ring system. One such drug, amantadine, is an antiviral agent that is effective against rubella (German measles) and some influenza viruses.

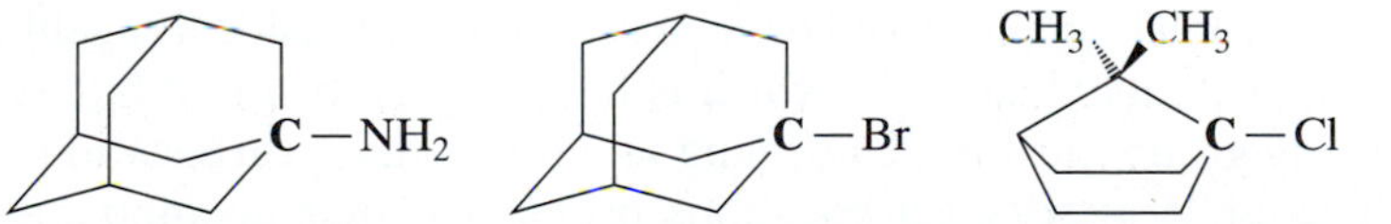

amantadine	1-bromoadamantane	apocamphyl chloride

C = bridgehead carbon

One promising route to such compounds might involve nucleophilic substitution reactions of 1-bromoadamantane, but similar cage compounds having a halogen atom at the *bridgehead* (the point at which fused rings are joined), such as apocamphyl chloride, are quite resistant to such reactions. Bridgehead Balms has commissioned your Institute to study the reactivity of 1-bromoadamantane to determine (1) whether it will undergo nucleophilic substitution reactions with hydroxylic solvents and if so, (2) how its reactivity compares with that of a comparable open-chain tertiary halide, 2-bromo-2-methylpropane (*t*-butyl bromide). In order to compare the reactivities of these compounds, you must determine the rate constants for their reactions.

Applying Scientific Methodology

After reading the experiment you should be able to formulate working hypotheses based on each of the problems outlined in the Scenario. You will need to complete the rate measurements to obtain a quantitative result relating to the second problem, but you should be able to decide which tertiary halide should react faster before you begin.

A Gem Among Molecules

Adamantane, whose name is derived from a Greek word for diamond, has molecules of elegant symmetry that can be regarded as fragments of a tetrahedral diamond lattice. Molecular models of this unique molecule reveal that it consists of four interlocking chair-form cyclohexane rings, arranged somewhat like the four planes of a tetrahedron. A space filling model is nearly spherical, and this molecular shape results in a particularly stable crystal lattice that is responsible for adamantane's unusually high melting point of 268°C.

Like the spherical "soccer ball" structure of buckminsterfullerene (an exciting new form of carbon), the extraordinary structure of adamantane is intriguing to chemists because adamantyl systems have properties that make them almost ideal for the study of certain chemical phenomena. The rigid adamantane skeleton results in a system of known geometry with unstrained, tetrahedral bond angles; the cyclohexane rings making up the adamantane molecule come together at four points, forming a bridgehead at each junction; and the cagelike structure prevents certain kinds of interactions and reaction mechanisms, thereby simplifying the analysis of reaction parameters.

In 1939, P. D. Bartlett and L. H. Knox discovered that another bridged compound, apocamphyl chloride, is surprisingly inert to reagents that usually bring about nucleophilic substitution reactions. For a nucleophile to attack from the back side of its bridgehead carbon, as required for an S_N2 reaction, the nucleophile would have to somehow get inside a cagelike molecule of apocamphyl chloride, which is highly unlikely. In an S_N1 reaction the leaving group must leave, forming an intermediate carbocation, before the nucleophile attacks. Carbocations prefer a planar geometry, as illustrated here for the *t*-butyl carbocation—the intermediate in the S_N1 reactions of 2-bromo-2-methylpropane. But attaining a planar geometry at the bridgehead carbon is impossible in the rigid apocamphyl system.

Unlike the highly strained cage of apocamphyl chloride, a molecule of 1-bromoadamantane has normal tetrahedral bond angles with no bond-angle strain. In a 1-adamantyl carbocation, the bridgehead carbon can flatten out a bit to attain a bond angle of 113°, which is somewhere between the tetrahedral angle of the adamantane ring and the 120° angle of the *t*-butyl carbocation.

adamantane

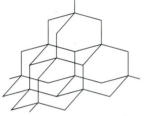

fragment of
diamond lattice

You can read the paper by Bartlett and Knox in J. Amer. Chem. Soc. **1939,** *61,* 3184.

t-butyl carbocation

1-adamantyl carbocation

In an S_N1 reaction, a more stable carbocation forms faster than one that is less stable. Therefore any factor that decreases the stability of a carbocation would tend to decrease the reactivity of the substrate from which the carbocation forms. You will be testing this principle as you compare the reactivities of 1-bromoadamantane and 2-bromo-2-methylpropane.

Key Concept: The more stable a reactive intermediate is, the faster it will form.

Understanding the Experiment

When Bartlett and Knox showed that bridgehead compounds such as apocamphyl chloride are quite unreactive with nucleophiles, they suggested that studies of bridgehead reactivity might yield valuable information about reaction mechanisms and the geometry of transition states. That suggestion was taken up by numerous investigators; particularly fruitful results have been obtained in the study of *solvolysis reactions*—nucleophilic substitution reactions in which the solvent acts as the nucleophile. In this experiment you will measure the first-order rate constants for the solvolysis reactions of 1-bromoadamantane and 2-bromo-2-methylpropane with a mixed ethanol-water solvent.

The solvolysis of 2-bromo-2-methylpropane and similar halides by hydroxylic solvents such as water and ethanol is believed to proceed by an S_N1 mechanism involving the formation of a carbocation intermediate.

t-Butyl bromide solvolysis mechanism

1. t-BuBr $\longrightarrow$ t-Bu$^+$ + Br$^-$ (slow)

2. t-Bu$^+$ + SOH $\longrightarrow$ t-Bu$\overset{H}{\underset{\oplus}{\text{OS}}}$ (fast)

(SOH = hydroxylic solvent)

3. t-Bu$\overset{H}{\underset{\oplus}{\text{OS}}}$ + Base: $\longrightarrow$ t-BuOS + Base: H$^+$ (fast)

Dissociation of the alkyl bromide is the rate determining step, so the reaction is first order with the rate equation

$$\frac{-d[\text{RBr}]}{dt} = k\,[\text{RBr}]$$

Although the solvent does not appear in the rate equation, it can affect the reaction rate by assisting in the formation of the carbocation. Some solvent molecules may help push the leaving group off from the rear while others pull it off from the front. Thus the rate of an S_N1 solvolysis reaction should depend on both the polarity of the solvent and the stability of the carbocation formed in the first step; the more polar the solvent and the more stable the carbocation, the faster the reaction.

Role of solvent in displacement reactions

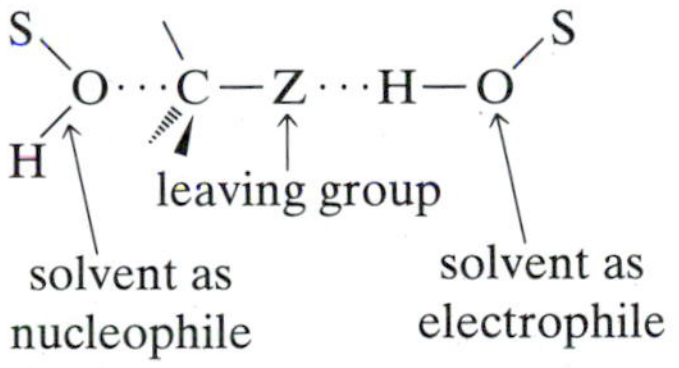

solvent as nucleophile   solvent as electrophile

leaving group

The rate constant for a first-order reaction can be calculated from the following integrated rate law, where c_0 is the initial concentration of the substrate and c is its concentration at time t.

$$\ln \frac{c}{c_0} = -kt \tag{1}$$

A solvolysis reaction of an alkyl bromide with a hydroxylic solvent (SOH, where S = H or R) will produce hydrogen bromide according to the general equation

$$R-Br + SOH \rightarrow R-OS + HBr$$

Therefore the concentration (c) in equation **1** can be measured indirectly by measuring the amount of HBr that forms during the reaction. You will do this by adding measured portions of a KOH solution from a buret and recording the time it takes for the HBr evolved to neutralize the added KOH, as shown by the color change of an indicator. At any given time during the reaction, the concentration of the alkyl bromide (c) will equal its initial concentration (c_o) minus the concentration of the HBr that has formed by then (c_{HBr}). The concentration of HBr is proportional to the volume of KOH needed to neutralize it (V), and the initial concentration of the alkyl bromide is proportional to the total volume of KOH solution needed to neutralize all the HBr produced (V_∞). Thus we can derive an expression for the concentration factor (c/c_o) from equation **1** in terms of the volume of KOH solution added.

$$\frac{c}{c_o} = \frac{c_o - c_{HBr}}{c_o} = 1 - \frac{c_{HBr}}{c_o} = 1 - \frac{V}{V_\infty}$$

The solvolysis of 1-bromoadamantane will be studied in 40% aqueous ethanol, in which both water and ethanol act as nucleophiles. Because the reaction of 2-bromo-2-methylpropane in 40% ethanol is difficult to measure accurately, its solvolysis reaction will be conducted in 80% ethanol instead. In order to make a meaningful comparison between the two reactions, you will have to estimate the rate constant for 2-bromo-2-methylpropane in 40% ethanol using equation **2** (the Winstein–Grunwald equation), which relates the rate of a solvolysis reaction to the ionizing power of the solvent.

$$\ln \frac{k}{k_o} = mY \tag{2}$$

In this equation k_o is the rate constant for a reaction in the reference solvent, 80% ethanol, and k is the rate constant in the actual reaction solvent (40% ethanol in this experiment) at the same temperature. Y is a measure of the reaction solvent's ionizing power and m measures the sensitivity of the substrate to changes in ionizing power. The value of m for 2-bromo-2-methylpropane is 0.94, and Y for 40% ethanol is 2.20.

For each kinetic run, you will prepare the appropriate reaction solvent by combining 95% ethanol and water in such proportions that, after you have added the alkyl bromide, the solvent will be 40% or 80% aqueous ethanol. It is important to measure the solvents accurately because an error in solvent composition can markedly affect the solvolysis rate. You will then add some bromthymol blue indicator and the alkyl bromide to the reaction solvent, followed by a measured portion of potassium hydroxide in the appropriate solvent. The indicator should turn blue as each 1-mL portion of KOH solution is added, changing to green when enough HBr is produced by the solvolysis reaction to neutralize the added KOH. As the solution becomes more acidic, its color fades to yellow. Each portion of KOH solution will consume the HBr produced by the reaction of approximately 5 percent of the alkyl halide; to get sufficient data for the rate calculations, you should continue the run until the reaction is at least 50 percent complete, which requires 10 portions or more of the KOH solution. After the last portion of KOH has been added, you will heat the reaction flask gently to bring the reaction to completion, then titrate the solution with more KOH to determine V_∞.

Reactions and Properties

$$RBr + H_2O \rightarrow ROH + HBr \quad and$$

$$RBr + CH_3CH_2OH \rightarrow ROCH_2CH_3 + HBr$$
(R = 1-adamantyl or *t*-butyl)

$$HBr + KOH \rightarrow H_2O + KBr$$

Table 19.1 Physical properties

	M.W.	m.p.	b.p.	d
1-bromoadamantane	215.1	118		
2-bromo-2-methylpropane	137.0	−16	73	1.221

Note: m.p. and b.p. are in °C, density is in g/mL

Directions

The alkyl halides must be protected from moisture; be sure your glassware is clean and dry. With your instructor's permission, work in pairs, with one student recording the times and the other adding the KOH solution.

2-Bromo-2-methylpropane is harmful if inhaled or absorbed through the skin, and it may be carcinogenic. Avoid contact and inhalation.

Safety Note

A. *Solvolysis of 1-Bromoadamantane in 40% Ethanol*

Preparation of Solutions. Prepare an indicator blank by measuring 25 mL of a pH 6.9 buffer into a 125-mL Erlenmeyer flask and adding 4 drops of bromthymol blue indicator solution (the blank should be green). Set this flask aside while you prepare the reaction mixture.

Using 50-mL burets and measuring as accurately as possible [OP-5], combine 20.0 mL of 95% ethanol and 29.0 mL of distilled water in another 125-mL Erlenmeyer flask. Add 8 drops of bromthymol blue solution and swirl to mix. If you have a magnetic stirrer, add a stirbar to the flask. Clamp this reaction flask to a ring stand and lower it into a water bath containing room temperature (20–25°C) water. The water temperature should remain nearly constant throughout a kinetic run. Adjust the water level so that the bath is about two-thirds full and measure the water temperature. Now fill a clean, dry 25-mL buret with a 0.005 *M* solution of KOH in <u>40%</u> ethanol, and record the initial buret reading accurately. Obtain a timer or a watch that measures seconds, if you don't have one.

Take Care! Be sure to use the right KOH solution.

Kinetic Run. Start the stirrer if you have one; otherwise swirl the flask after each addition and at intervals between additions [OP-9]. Add 1.0 mL of a

Observe and Note: Take note of the color changes throughout the kinetic runs and try to explain them.

The color change will be easier to see if you set the flasks on a sheet of white paper.

Observe and Note: Does the time between color changes increase, decrease, or stay the same as the reaction proceeds?

Waste Disposal: Unless your instructor directs otherwise, the reaction mixture and unused KOH solution can be poured down the drain.

Take Care! Be sure to use the right KOH solution.

Take Care! Avoid contact, do not breathe vapors, do not pipet by mouth.

Stop and Think: Why does the time between buret readings change as the reaction proceeds?

Waste Disposal: Unless your instructor directs otherwise, the reaction mixture and unused KOH solution can be poured down the drain.

freshly prepared 0.1 *M* solution of 1-bromoadamantane in absolute ethanol to the reaction flask and *immediately* record the time of addition to the nearest second (or start the timer). Without delay, add 1.0 mL of the KOH/40% ethanol solution to the reaction mixture and record the buret reading. (If the solution does not turn blue, add more KOH until it does, and then record the buret reading.) Place the indicator blank beside the reaction mixture and record the time (to the nearest second) when the solution changes to the same shade of green as the blank. Within a minute of the color change, add another 1.0-mL portion of the KOH solution, and record the time when the solution again turns from blue to green. Repeat the addition of 1.0-mL portions of KOH, recording the buret reading and the time of the color change after each addition, until at least 10 portions have been added.

Determination of V_∞. After the last color change, seal the flask with a square of Parafilm and heat the reaction mixture [OP-6] in a 60°C water bath for 30 minutes (alternatively, let it stand overnight or longer at room temperature). Cool the solution to room temperature, then titrate it with the 0.005 *M* solution of KOH in 40% ethanol to the green end point. If the green color fades to yellow after standing a few minutes, heat the flask in the water bath about 10 minutes longer. After cooling, again titrate the solution to the green end point. Subtract your initial buret reading (recorded before you started the kinetic run) from the final reading to get V_∞.

B. *Solvolysis of 2-bromo-2-methylpropane in 80% Ethanol*

Preparation of Solutions. Fill a clean, dry 25-mL buret with the 0.05 *M* solution of KOH in 80% ethanol, and record the initial buret reading. Accurately measure 21.0 mL of 95% ethanol and 4.0 mL of distilled water into a 125-mL Erlenmeyer flask. Add 4 drops of bromthymol blue solution, swirl to mix, and support the flask in the room temperature water bath.

Kinetic Run. Start the stirrer if you have one; otherwise swirl after each addition and at intervals between additions [OP-9]. Use an automatic pipet or a *dry* graduated pipet with a pipet pump [OP-5] to measure 0.10 mL of 2-bromo-2-methylpropane into the reaction flask. Immediately record the time of addition (or start your timer). Carry out the kinetic run by the same procedure you followed in part **A**, using at least ten 1.0-mL portions of the KOH solution in 80% ethanol.

Determination of V_∞. After the last color change, seal the flask with Parafilm and heat the reaction mixture in a water bath at 60°C for about 10 minutes. Then titrate it with the 0.05 *M* solution of KOH in 80% ethanol to the green end point and calculate V_∞ as for part **A**.

Report. Your report should include a statement of the problem and an account of how you applied scientific methodology to solve it. For each run, compute the time (*t*) of each color change in seconds, measured from the time of addition of the alkyl bromide ($t = 0$). Calculate $\ln(1 - V/V_\infty)$ for each *t* value, where *V* is the total volume of KOH that has been added up to that time. Using good graph paper, plot $\ln(1 - V/V_\infty)$ versus *t* and determine the value of *k* (in s^{-1}) for the alkyl bromide from the slope of the line. (Alternatively, you may use a calculator or computer that has a linear regression program to determine the least-squares slope from your data.) From the rate constant for 2-bromo-2-methylpropane in 80% ethanol (k_o), use equation **2** to estimate its rate constant in 40% ethanol. Then calculate the relative solvolysis rate for 1-bromoadamantane by dividing its rate constant by the rate constant for 2-bromo-2-methylpropane in 40% ethanol.

Exercises

1 (a) Which tertiary carbocation is more stable, 1-adamantyl or *t*-butyl? Explain why it is more stable and tell how your experimental results support this conclusion. (b) Which solvent promotes the formation of a carbocation more effectively, 40% ethanol or 80% ethanol? Expain.

2 Write a mechanism for the solvolysis reaction of 1-bromoadamantane with ethanol.

3 Describe and explain the possible effect on your results of the following experimental errors or variations. (a) The pH of thc buffer for the blank is 3.9 rather than 6.9. (b) You run both reactions in absolute ethanol. (c) You use norbornyl bromide (1-bromobicyclo[2.2.1]heptane) in place of 1-bromoadamantane. (d) Your pipet is wet when you use it to measure the 2-bromo-2-methylpropane.

4 Hydroxide ion is a stronger nucleophile than either water or ethanol, yet the addition of KOH during the kinetic runs in this experiment has virtually no effect on the reaction rates. Explain.

5 (a) During its solvolysis reaction, some 2-bromo-2-methylpropane molecules lose HBr by an E1 reaction to form an alkene. Would you expect this to affect the measured reaction rate? Why or why not? (b) None of the 1-bromoadamantane undergoes elimination during its solvolysis reaction. Explain.

6 (a) Outline a synthesis of amantadine from 1-bromoadamantane. (b) Outline a synthesis of the antiviral agent rimantadine [RCH(NH$_2$)CH$_3$, where R = 1-adamantyl] from 1-bromoadamantane using an organocuprate.

7 (a) Using your experimental rate constant, calculate the time it should take for 90 percent of your 1-bromoadamantane to react at room temperature. (b) How long should it take for 90 percent of the 2-bromo-2-methylpropane in 40% ethanol to react under the same conditions?

Other Things You Can Do

(Starred items require your instructor's permission.)

*1 Collect data for this experiment using a pH probe with a computer interface as described in *J. Chem. Educ.* **1991,** *68*, 609.

*2 Measure the relative reactivities of different alkyl halides as described in Minilab 16.

*3 Carry out an S$_N$1 reaction of trityl bromide with ethanol and isolate the product as described in Minilab 17.

4 Write a research paper about medical uses of adamantane derivatives, starting with the article in *J. Chem. Educ.* **1973**, *50*, 780 and using sources listed in the Bibliography.

EXPERIMENT 20

Reaction of Iodoethane with Sodium Saccharin, an Ambident Nucleophile

Nucleophilic Substitution. NMR Spectrometry. Carboxylic Acid Derivatives. Heterocyclic Compounds.

Operations:

OP-33 (optional) High-Performance Liquid Chromatography
OP-12 Vacuum Filtration
OP-28 Melting Point
OP-35 Nuclear Magnetic Resonance Spectrometry

Before You Begin

1 If you will be doing the optional HPLC analysis, read OP-33. Read the experiment, read or review the other operations, and write a brief experimental plan.
2 Calculate the mass of 10.0 mmol of sodium saccharin.

Scenario

A nutritional Calorie is 1000 times as large as a scientific calorie.

Saccharin is a nonnutritive sweetener, meaning that it is not metabolized by the body to produce energy. But saccharin is usually mixed with fructose or other Calorie laden sweeteners to mask its bitter aftertaste, giving the mixture about half as many Calories as sucrose and thus making it less attractive as a sugar substitute. Dulcinea Petty IV directs a product development team at Sweet Nothings Ltd., which manufactures saccharin. She has learned that substances with N—H bonds often have bitter tastes, so she wonders if converting the N—H bond of saccharin to an N—C bond by alkylating it will mask the bitter taste and thus yield a better sweetener. Saccharin is converted to its more nucleophilic sodium salt prior to alkylation, but resonance structures of the salt reveal that it is an ambident nucleophile; that is, it has two potentially nucleophilic atoms, the nitrogen atom and an oxygen atom.

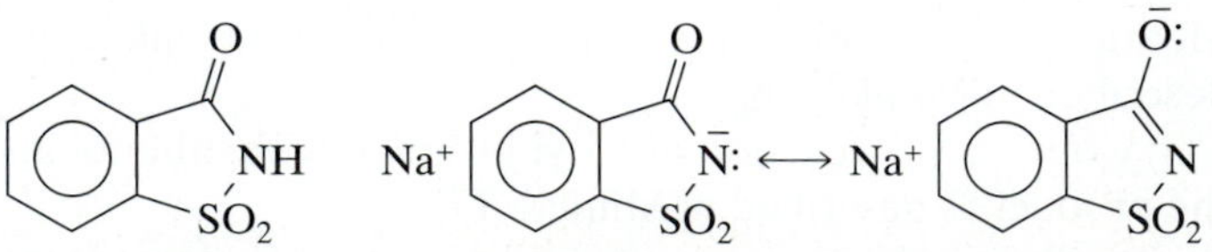

saccharin   resonance structures of sodium saccharin

Before their quest for a better sweetener can be pursued, Sweet Nothings needs to know whether or not alkylation will occur mainly on the nitrogen atom. Your assignment is to carry out the alkylation of sodium saccharin with iodoethane and analyze the product mixture to determine the structure of the major product.

Applying Scientific Methodology

The scientific problem can be stated as "Will the alkylation of sodium saccharin with iodoethane produce *N*-ethylsaccharin as the major product?" You will not taste the product, but you can ask your instructor about its taste after the experiment.

Saccharin, an Accidental Sweetener

One rule that most chemists follow scrupulously is to never, *ever,* taste anything they make in the laboratory. A chemist should not even eat or drink anything else while working in the lab because of possible contamination by toxic chemicals. During the nineteenth century, however, chemists were not so fastidious. It was a common practice to perform a "taste test" on any new chemical, sometimes with unfortunate results; but occasionally an accidental or deliberate tasting paid off with a new discovery.

Ira Remsen, a Johns Hopkins University chemistry professor, studied chemistry under a student of the "father of organic chemistry," Friedrich Wöhler, and became the most famous American chemist of the nineteenth century. In 1878 a German student working in Remsen's research group, Constantin Fahlberg, prepared some white crystals of a previously unknown compound from *o*-toluenesulfonamide. He later ate a piece of bread and was astonished to find that it tasted intensely sweet. It didn't take Fahlberg long to trace the sweet taste to the new compound he had just handled, which he named saccharin after the Latin word for sugar, *saccharum.*

Saccharin is about 500 times sweeter than sucrose, common table sugar. Its sweetness came as a surprise because no one was looking for a synthetic sweetener at the time—most scientists believed that only natural compounds could be sweet. Fahlberg recognized the commercial possibilities of a nonfattening sweetener, so he applied for a patent and began to manufacture saccharin. Despite its somewhat bitter aftertaste, saccharin was the most popular artificial sweetener during most of the twentieth century, outselling other synthetic sweeteners such as dulcin (from the Latin *dulcis,* meaning sweet), which was discovered just six years after saccharin.

Concerns about the safety of saccharin cropped up from time to time, inspiring Theodore Roosevelt to proclaim, "anyone who says saccharin is injurious to health is an idiot." Roosevelt, who liked to sweeten his chewing tobacco with saccharin, was no authority on the safety of commercial products, but his words must have reassured many Americans about saccharin. Then in 1977 a Canadian study showed that some rats developed bladder tumors when they were fed a diet containing 5% saccharin. Although the rats' diet was equivalent to a human consuming about 1000 cans of diet soda per day, saccharin was promptly removed from the GRAS (generally recognized as safe) list and later banned in the United States. Reacting to

A review of the status of saccharin may eventually lead to removal of the warning label.

protests by diabetics and overweight Americans, for whom consuming sugar was a far greater health risk than the remote possibility of saccharin induced cancer, Congress suspended the ban in 1979. As a result, you can still buy Sweet 'n Low at your local grocery store, but the packets carry a mandatory warning label.

Because of the cancer scare and competition from aspartame (Nutra-Sweet), saccharin use has declined sharply in recent years. Lacking the bitter aftertaste of saccharin, aspartame has become our most popular artificial sweetener, but it may also face some tough competition before long. A French sweetener called superaspartame is 300 times sweeter than aspartame and—unlike aspartame—can be used in baking and frying. The natural sweetener thaumatin, which is extracted from the west African plant ketemfe, is reported to be nearly 100,000 times sweeter than sucrose, making it the sweetest natural substance ever discovered. It is also (like aspartame) a flavor enhancer, so it has been used to persuade farm animals to eat more—pigs gain up to 10 percent more weight when thaumatin is added to their feed!

Understanding the Experiment

In this experiment you will carry out the reaction of sodium saccharin with iodoethane in the solvent N,N-dimethylformamide (DMF). This is a nucleophilic substitution reaction in which the nucleophilic atom can be either nitrogen or oxygen and the leaving group is iodide ion (I^-). The rate of a nucleophilic substitution reaction can be very sensitive to the solvent used. Polar protic solvents (solvents capable of hydrogen bonding) such as water and ethanol form bulky solvation shells around a charged nucleophile, reducing its nucleophilic strength. Polar aprotic solvents such as DMF do not solvate the nucleophile strongly, leaving it free to attack the substrate. Thus they accelerate the rates of many substitution reactions, particularly S_N2 reactions in which the strength of the nucleophile has a large effect on the reaction rate.

The composition of the product will depend on whether nitrogen or oxygen acts as the nucleophilic atom most of the time. As shown in the "Reactions and Properties" section, nucleophilic attack by nitrogen will yield N-ethylsaccharin, while nucleophilic attack by oxygen will yield O-ethylsaccharin. Predicting the major product is not easy because a number of competing factors may come into play. N-Ethylsaccharin is more stable than O-ethylsaccharin so it should be the major product if the reaction reaches thermal equilibrium. But the oxygen atom of sodium saccharin has a higher partial negative charge than the nitrogen atom (oxygen is more electronegative than nitrogen), so a reaction involving oxygen as the nucleophile might occur faster than one involving nitrogen. For example, the reaction of potassium saccharin with 2-bromopropane in N,N-dimethylformamide yields mainly O-isopropylsaccharin.

You will determine the composition of your product by using proton nuclear magnetic resonance (^{1}H NMR) spectrometry. An oxygen atom has a stronger deshielding effect on nearby protons than a nitrogen atom, so the signal for the methylene protons (highlighted) of an —OCH_2CH_3 group will appear farther downfield ($\delta = 4.7$ ppm) than the corresponding signal for an —NCH_2CH_3 group ($\delta = 3.9$ ppm). Because the methylene protons have three methyl protons as neighbors, their signal in either case will be a quartet. By measuring the integrated signal areas for both quartets (assuming the product is a mixture) you will be able to determine the percentages of *N*-ethylsaccharin and *O*-ethylsaccharin in your product.

At your instructor's option, you can analyze your product using high-performance liquid chromatography (HPLC) in addition to or instead of NMR spectrometry.

Reactions and Properties

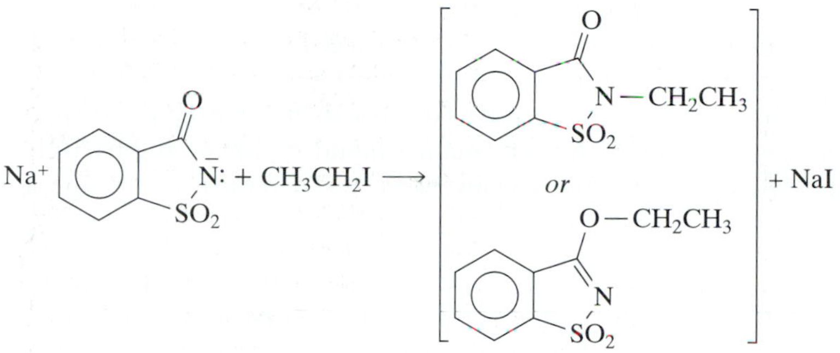

$$\text{Na}^+ \; \overset{O}{\underset{SO_2}{\bigcirc\!\!\!\!\Vert}}\!\!\bar{N}\!: \; + \; CH_3CH_2I \; \longrightarrow \; \left[\; \underset{SO_2}{\overset{O}{\bigcirc\!\!\!\!\Vert}}\!N\!-\!CH_2CH_3 \quad or \quad \underset{SO_2}{\overset{O-CH_2CH_3}{\bigcirc\!\!\!\!N}} \; \right] \; + \; NaI$$

Table 20.1 Physical properties

	M. W.	b. p.	m. p.	*d*
sodium saccharin	205.2			
iodoethane	156.0	72		1.950
N,N-dimethylformamide	73.1	153		0.945
N-ethylsaccharin	211.2		95	
O-ethylsaccharin	211.2		211	

Note: m.p. and b.p. are in °C, density is in g/mL

Directions

This procedure is adapted from an article published in the *Journal of Chemical Education* **1990,** *67,* 611.

Safety Notes

Iodoethane severely irritates the eyes, skin, and respiratory tract and it may be carcinogenic. Wear gloves, avoid contact, and do not breathe vapors.

N,N-Dimethylformamide is harmful by inhalation, ingestion, and absorption through the skin. Avoid contact, do not breathe vapors.

Deuterochloroform is harmful if inhaled, ingested, or absorbed through the skin, and it may be carcinogenic. Avoid contact and do not breathe vapors.

N,N-dimethylformamide

Take Care! Avoid contact with DMF, do not breathe its vapors.

Take Care! Wear gloves, avoid contact, do not breathe vapors.

Take Care! Do *not* taste the product.

Waste Disposal: Put the filtrate in a designated solvent recovery container.

Stop and Think: Is the product a single compound or a mixture? How can you tell?

Waste Disposal: Put the deuterochloroform solution in a designated solvent recovery container.

Reaction. Carry out the reaction under the hood. Weigh out 10.0 mmol of sodium saccharin and add it to 5.0 mL of *N,N*-dimethyl formamide in a 125-mL Erlenmeyer flask. Heat the mixture in an 80°C water bath with swirling until the solid dissolves, then add 0.80 mL (~10 mmol) of iodoethane using a dispenser or an automatic pipet. Seal the flask with Parafilm and heat the mixture in the water bath for 10 minutes, keeping the temperature around 80°C.

Separation. Let the reaction mixture cool to room temperature, add 75 mL of water, and shake the stoppered flask until any liquid residue that forms has solidified. Cool the flask in an ice water bath, break up the solid with a stirring rod if necessary, and collect the product by vacuum filtration [OP-12], washing it twice with 5-mL portions of cold water.

Analysis. Dry the product and measure its mass and melting point range [OP-28]. Obtain an integrated ^{1}H NMR spectrum [OP-35] of the product in deuterochloroform. (**Take Care!** Avoid contact, do not breathe vapors.) Turn in the rest of your product to your instructor. You can also analyze the product mixture by high-performance liquid chromatography [OP-33] as directed by your instructor. The instructor will demonstrate the operation of the instrument.

Report. Your report should include a statement of the problem and an account of how you applied scientific methodology to solve it. Calculate the product yield, decide whether *N*-ethylsaccharin or *O*-ethylsaccharin is the major product, and calculate their percentages in the mixture. Interpret your NMR spectrum as completely as you can.

Exercises

1 (a) Assuming that the reaction was S_N2 and that the major product was the one that formed faster, which atom appears to be more nucleophilic, N or O? (b) Write a mechanism showing the transition state of the reaction that led to your major product.

2 Describe and explain the possible effect on your results of the following experimental errors or variations. (a) The reagent bottle labeled "sodium saccharin" contained saccharin instead. (b) You used water as the reaction solvent rather than DMF. (c) You heated the reaction mixture for 3 hours under reflux.

3 Following the format in Appendix V, write a flow diagram for this experiment.

4 Most compounds containing N—H bonds are basic but saccharin is acidic. Explain why, using resonance structures.

5 Outline a synthesis of saccharin from *o*-toluenesulfonamide.

6 One objection raised to the use of aspartame is that it decomposes in the presence of moisture to produce phenylalanine, which must be avoided by people who have the genetic condition phenylketonuria, and methanol, which can have an effect on mental behavior. Write an equation for the hydrolysis of aspartame yielding both of these products.

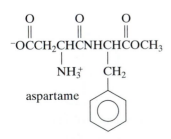

aspartame

Other Things You Can Do

(Starred items require your instructor's permission.)

*1 Add some aqueous sodium bicarbonate to a solution of saccharin (not sodium saccharin) and explain the result, writing an equation for the reaction.

*2 Carry out an S_N1 reaction of trityl bromide with ethanol as described in Minilab 17.

3 Write a research paper about artificial sweeteners, starting with sources listed in the Bibliography.

EXPERIMENT 21   # Dehydration of Methylcyclohexanols

Reactions of Alcohols. Preparation of Alkenes. Elimination Reactions. Carbocations. Regioselectivity.

Operations:

OP-19 Washing Liquids
OP-20 Drying Liquids
OP-25 Simple Distillation
OP-32 Gas Chromatography

Before You Begin

1 Read the experiment, read or review the operations as necessary, and write an experimental plan.
2 Calculate the mass and volume of 150 mmol of 2- and 4-methylcyclohexanol and the theoretical yield of methylcyclohexenes from either alcohol.

Scenario

(Unlike most other Scenarios in this book, this one describes a real chemical mystery, which is documented in the *Journal of Chemical Education,* **1994**, *71*, 440. Not even the names of the characters have been changed.)

For many years the dehydration of 2-methylcyclohexanol to a mixture of alkenes has been carried out in college organic chemistry labs to demonstrate the application of Zaitzev's rule and the occurrence of the E1 mechanism in alcohol dehydration reactions. In 1994 David Todd, then a chemistry professor at Pomona College, was distilling the product alkenes out of the reaction mixture when he was interrupted by an urgent summons to lunch with the chemistry department secretary, Evelyn Jacoby. Upon returning from lunch he decided, out of curiosity, to replace the distillation receiver with a new one and collect a second fraction. He then worked up both fractions and analyzed them by gas chromatography. Much to his surprise, the second fraction contained a markedly lower percentage of the expected product, 1-methylcyclohexene, than the first. Because his decision to replace the receiver with a new one was a direct result of the secretary's request, Professor Todd named this unexpected result the "Evelyn effect."

Although several mechanistic hypotheses have been proposed to explain the Evelyn effect, it is by no means certain that any of them are correct. Your project group's assignment is to verify the existence of the Evelyn effect for the dehydration of 2-methylcyclohexanol and to see if a similar effect exists for 4-methylcyclohexanol. You may then want to speculate about some possible causes of the Evelyn effect.

Applying Scientific Methodology

As in the previous experiments, you need to state the problem as a question, formulate a working hypothesis, follow the course of action described in the Directions, gather and evaluate evidence, test your hypothesis, arrive at a conclusion, and report your findings.

Zaitzev's Rule and the Evelyn Effect

More than a century ago at the University of Kazan, Vladimir Vasilevich Markovnikov and Alexander Zaitzev were investigating a chemical reaction both backward and forward: Markovnikov was adding hydrogen iodide to alkenes to prepare alkyl iodides, and Zaitzev was removing hydrogen iodide from alkyl iodides to prepare alkenes. Markovnikov discovered that hydrogen iodide adds to propene to form mainly 2-iodopropane. From this and other results, Markovnikov formulated his well-known rule, which can be expressed as follows for a hydrogen-containing species represented by HZ:

$$CH_3CH{=}CH_2 + HI \longrightarrow CH_3\overset{\displaystyle I}{\underset{\displaystyle |}{C}}HCH_3$$

> *Markovnikov's rule:* When HZ adds to the carbon-carbon double bond of an unsymmetrical alkene, hydrogen adds preferentially to the carbon atom that already has more hydrogens.

In the meantime, Zaitzev learned that dehydrohalogenation of 2-iodobutane by alcoholic potassium hydroxide yields mainly 2-butene. He proposed an analogous rule for elimination reactions.

$$CH_3CH_2\overset{\displaystyle I}{\underset{\displaystyle |}{C}}HCH_3 \xrightarrow{\text{KOH}} CH_3CH{=}CHCH_3$$

> *Zaitzev's rule:* When HZ is removed from a species to form an alkene, hydrogen is lost preferentially from the carbon atom that has fewer hydrogens.

Markovnikov's and Zaitzev's rules together can be paraphrased by the well-known maxim "The rich get richer and the poor get poorer."

These examples show that organic reactants often react selectively, favoring some products and not others—Zaitzev's reaction *might* have yielded as much 1-butene as 2-butene, but it did not. When a reaction could produce two or more different structural isomers but in fact yields mainly one of them, the reaction is said to be *regioselective.* Zaitzev's rule works because, in most cases, it predicts the formation of the most stable alkene. 2-Butene was the major product of Zaitzev's reaction not because hydrogen-poor carbon atoms have some innate tendency to lose the hydrogens they have, but because 2-butene is more stable than 1-butene.

Although generalizations like Zaitzev's rule can help us predict the products of many organic reactions, organic chemistry remains an empirical science—we cannot be certain that a rule that is valid for one system under a given set of conditions will apply equally well under different circumstances. Chemists must study each system experimentally to see if it behaves in the expected manner, and if it doesn't, try to find out why. For example, neomenthyl chloride undergoes dehydrohalogenation to yield a product mixture that consists of mostly alkene **A**, the one predicted by Zaitzev's rule. But menthyl chloride, which differs only in the geometry of the C—Cl bond, yields 100% of alkene **B**, and none of the Zaitzev product. It also reacts much more slowly than neomenthyl chloride.

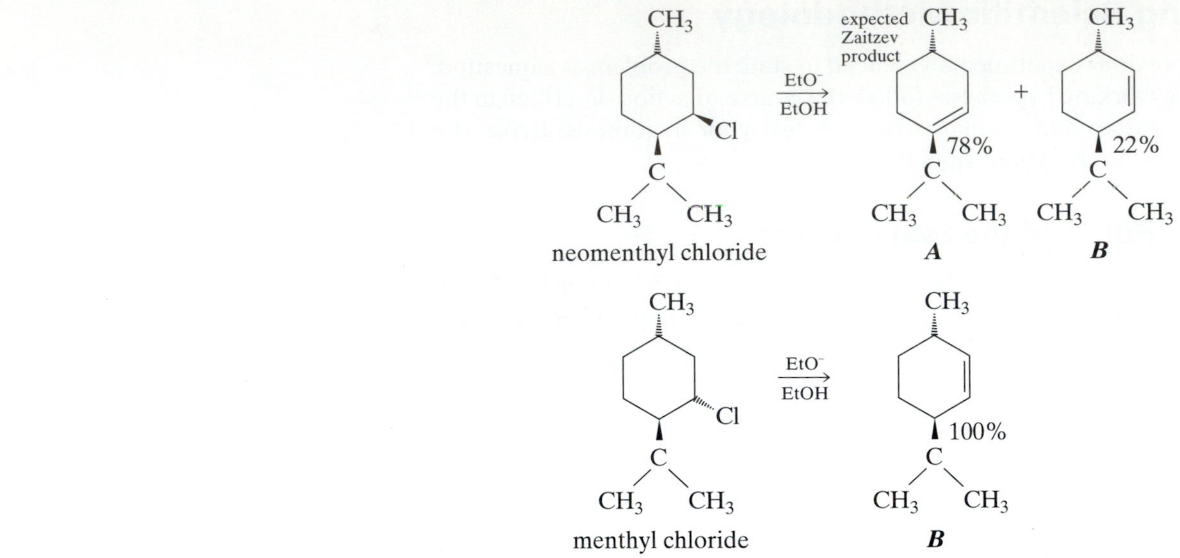

This result can be explained by assuming that the reaction occurs by an E2 (elimination, bimolecular) mechanism, requiring that the H and Cl atoms being eliminated lie in the same plane and on opposite sides of the C—C bond separating them; this is called *anti*-periplanar geometry.

E2 mechanism for elimination of HCl

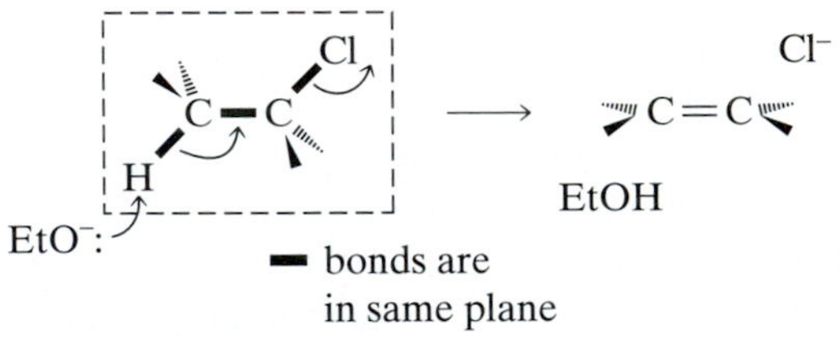

Neomenthyl chloride, in its most stable ring conformation, has the desired *anti*-periplanar geometry for formation of either **A** or **B**. Since **A** is the more stable alkene, it is the major product.

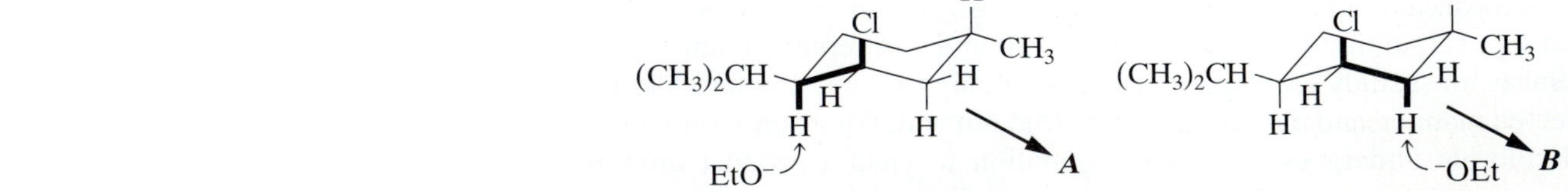

In its more stable conformation (with all large groups equatorial), menthyl chloride does not have the geometry necessary to form either product. In its less stable conformation, the *anti*-periplanar geometry needed to form product **A** cannot be attained because the isopropyl group rather than a hydrogen atom is *anti* to the chlorine atom. This conformation is suitable for the formation of product **B**, so it is the only product isolated.

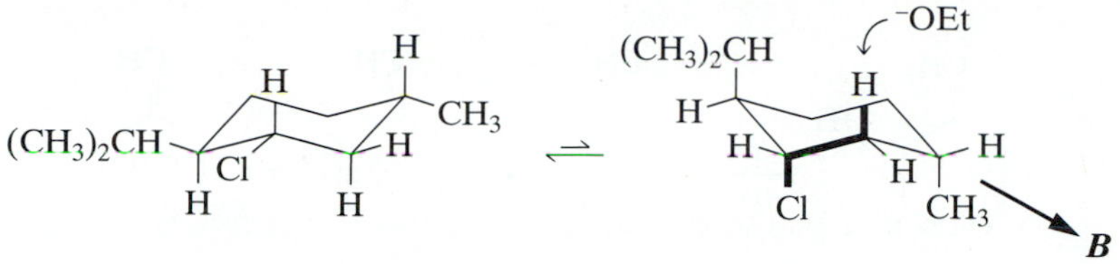

Because only a very small percentage of the menthyl chloride molecules are in the less stable conformation at any time, the reaction is much slower than the reaction of neomenthyl chloride. This example shows that whenever a substrate yields the less stable alkene as a major product of an elimination reaction, there may be some stereochemical constraints inhibiting the formation of the Zaitzev product.

The acid catalyzed dehydration of alcohols, which usually follows Zaitzev's rule, is generally believed to occur by an E1 (elimination, unimolecular) mechanism involving protonation of the hydroxyl group, loss of water to form a carbocation intermediate, and loss of a proton.

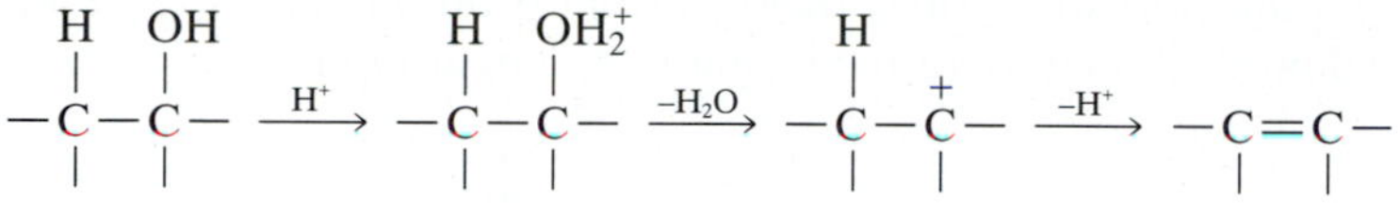

Note that there are *no* stereochemical constraints in an E1 reaction because the leaving group leaves before the proton is lost. Thus in the E1 dehydration of an alcohol, H and OH do not need to be *anti*-periplanar or in any other particular orientation in order for elimination to occur.

Unlike E2 reactions, E1 reactions may involve rearrangements in which the initial carbocation rearranges to a more stable carbocation before it loses H^+. A carbocation rearrangement may involve a *hydride shift* during which a hydrogen next to the positively charged carbon moves to that carbon, taking its bonding electron pair along with it. Such rearrangements lead to alkenes whose double bond connects carbon atoms that were not originally bonded to the hydroxyl group. Postulating such a rearrangement can explain the formation of 2-methyl-2-butene in the dehydration of 2-methyl-1-butanol, for example.

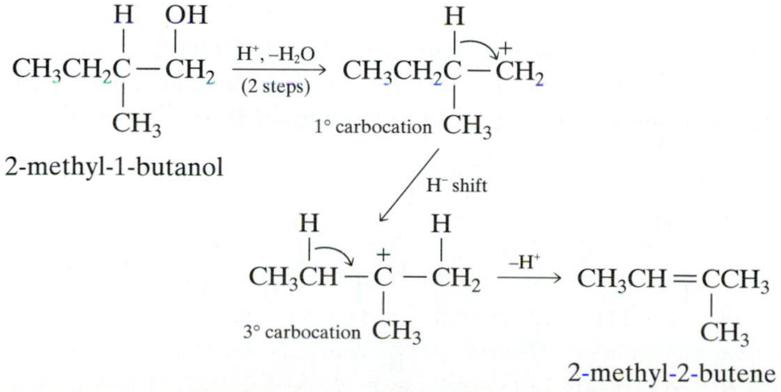

This brings us to the Evelyn effect. When Professor Todd carried out the dehydration of 2-methylcyclohexanol he obtained the following mixture of alkenes.

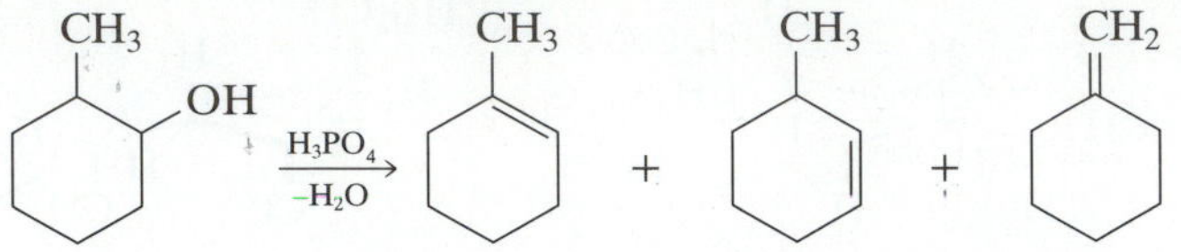

The reaction is performed by distilling the alkenes as they are formed, and the distillate typically contains 75–80 percent of product **A**, the product predicted by Zaitzev's rule. When the distillate is collected in separate fractions and the fractions are analyzed separately, the first 10 percent of the distillate contains about 93% **A** while the final distillate contains as little as 55% **A**. There is a clue to the origin of the Evelyn effect in the catalog of the Aldrich Chemical Company, where the 2-methylcyclohexanol used by Professor Todd is described as a mixture of *cis* and *trans* isomers. In fact, it is a nearly equimolar mixture of the two isomers. Previous researchers had reported that the *cis*-isomer reacted much faster than the *trans*-isomer, so Todd reasoned that the initial product mixture formed mainly by dehydration of the *cis*-isomer while the final product mixture formed mainly by dehydration of the *trans*-isomer.

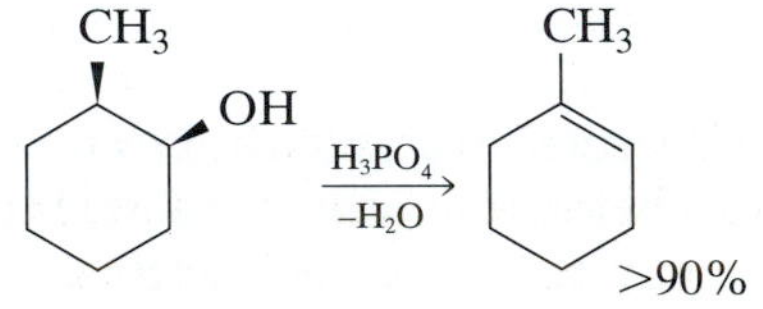

cis-2-methylcyclohexanol   **A**

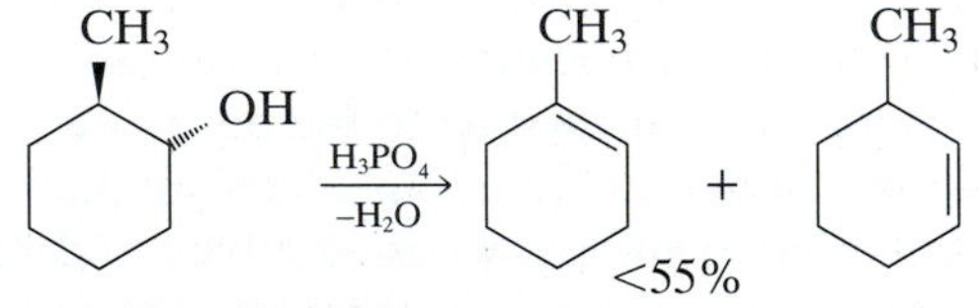

trans-2-methylcyclohexanol   **A**   **B**

Since the *trans*-isomer yields an unexpectedly large percentage of the less stable alkene, it appears that this reaction, like the E2 dehydrohalogenation of menthyl chloride, has some stereochemical constraints. The occurrence of E2 elimination from a protonated alcohol could explain a reduction in the amount of the expected product **A**, since elimination via an *anti*-periplanar geometry can yield only product **B** and not product **A**.

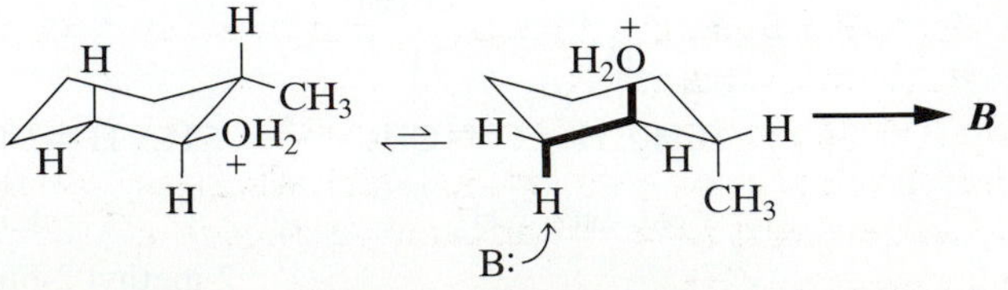

It could also explain the lower reactivity of the *trans*-alcohol, which can achieve the *anti*-periplanar geometry only in its less stable *diaxial* confor- mation. However it does *not* explain why the *trans*-alcohol yields any **A** at all, nor does it explain the existence of a small amount of methylenecyclo- hexane (product **C**) in the product mixture. Product **C** might be obtained via an E1 mechanism involving a carbocation rearrangement, but not by an ordinary E2 mechanism.

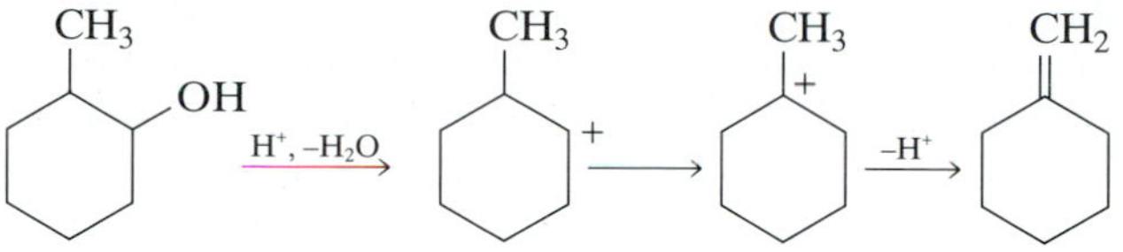

Does the reaction proceed by both E1 and E2 mechanisms? That possi- bility, raised by Todd, has been questioned by two other researchers, John J. Cawley and Patrick E. Lindner. Cawley and Lindner proposed an "E2-like" mechanism involving bridged ions (*J. Chem. Educ.* **1997,** *74,* 102), but it remains to be seen whether their mechanism will gain general acceptance. A mechanism is, after all, a scientific hypothesis about processes that we can't observe directly—the things that molecules do as they redistribute their atoms and are thereby transformed into new molecules—and is therefore subject to revision.

The Evelyn effect illustrates how science often works. For decades the results of alcohol dehydration reactions are adequately explained by the E1 hypothesis; no other explanation seems necessary. Then a chance obser- vation shows the inadequacy of the accepted hypothesis. A different hypothesis—that both E1 and E2 mechanisms are involved—is proposed and contested, followed by another hypothesis, and so on. The road to sci- entific discovery is a rocky one and there may be many detours along the way, but every failed hypothesis yields new information, new ideas, and often new applications. Science is not simply a body of established facts and theories; the facts and theories of science are always subject to further inquiry that may disprove or modify them. Science is a dynamic *process* by which knowledge is acquired, ideas are debated, theories are proposed, and new ways of doing things are discovered.

Understanding the Experiment

In this experiment, you and your coworkers will carry out the dehydration of 2-methylcyclohexanol and 4-methylcyclohexanol by heating the alcohols in the presence of phosphoric acid. Unless your instructor indicates other- wise, both alcohols will be mixtures of *cis* and *trans* isomers, so either or both may exhibit an Evelyn effect.

Dehydration of a secondary alcohol proceeds readily with about half a mole of phosphoric acid for every mole of the alcohol. By protonating an alcohol, the acid catalyst converts the poor leaving group —OH to a much better leaving group, $-OH_2^+$.

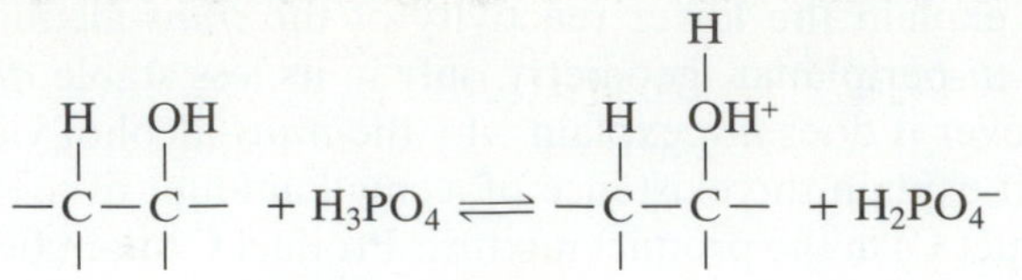

$$\begin{array}{cc} H\ \ OH \\ |\ \ \ | \\ -C-C- \\ |\ \ \ | \end{array} + H_3PO_4 \rightleftharpoons \begin{array}{cc} \qquad H \\ \qquad | \\ H\ \ OH^+ \\ |\ \ \ \ | \\ -C-C- \\ |\ \ \ \ | \end{array} + H_2PO_4^-$$

Elimination of H^+ and H_2O from the protonated alcohol yields an alkene, with the unprotonated alcohol serving as the reaction solvent.

According to LeChâtelier's principle, removing a product from a chemical system at equilibrium shifts the equilibrium in the direction favoring the formation of the products. You will carry out the dehydration reaction in a distillation apparatus so that the products (water and alkene) will continuously distill out of the reaction mixture as they are formed. Their removal will shift the equilibrium to the right and thus increase the yield of alkene.

The upward-pointing arrows in the equation indicate that the products are vaporized under the reaction conditions, not that they are gases at room temperature.

$$alcohol \rightleftharpoons alkene\uparrow + water\uparrow$$

If the reaction mixture is heated to a temperature above the boiling points of the product alkenes but below that of the alcohol, most of the unreacted alcohol will remain in the reaction flask while the alkenes and water distill into the receiving flask. Interposing a vertical column between the boiling flask and the still head will further reduce the amounts of unwanted alcohol, acid, and by-products in the distillate.

You will follow the progress of the reaction by measuring the volume of alkene in the distillate, collecting two fractions of approximately equal volume. When the reaction is over, the residue in the reaction flask may begin to foam and emit white vapors. You should remove the heat source at this time because overheating the residue may form a black tar and generate toxic fumes.

After washing and drying the organic layer of each fraction, you will analyze the fractions by gas chromatography. If you started with 2-methylcyclohexanol, your gas chromatograms may show peaks for both 1- and 3-methylcyclohexene (the methylenecyclohexene peak may be resolved only if you use a capillary column). If you started with 4-methylcyclohexanol you may obtain only 4-methylcyclohexene or a mixture of products including 3-methylcyclohexene and 1-methylcyclohexene (3- and 4-methylcyclohexene may not be separated on a packed column). From the relative areas of your peaks, you can estimate the percentage composition of the product mixture in each fraction.

Reactions and Properties

Table 21.1 Physical properties

	M.W.	b.p.	d
2-methylcyclohexanol*	114.2	166	0.930
4-methylcyclohexanol*	114.2	173	0.914
1-methylcyclohexene	96.2	110	0.813
3-methylcyclohexene	96.2	104	0.801
4-methylcyclohexene	96.2	102	0.799
phosphoric acid (85%)	98.0		1.70

*Mixture of *cis-* and *trans*-isomers.

Note: The molecular weight given for phosphoric acid is for the pure acid; 85% phosphoric acid is about 14.7 *M*; b.p. is in °C, density is in g/mL.

Directions

Students can work in pairs, each student using one of the two methylcyclohexanols.

The methylcyclohexanols and alkenes are flammable, and inhalation, ingestion, or skin absorption may be harmful. Avoid contact, do not breathe their vapors, and keep them away from flames or hot surfaces. Phosphoric acid can cause serious burns, particularly to the eyes; avoid skin or eye contact.

Safety Notes

Reaction and Separation. Measure 150 mmol of 2-methylcyclohexanol *or* 4-methylcyclohexanol (*cis-trans* mixtures) into a 50-mL round-bottomed flask. Mix in 5 mL of 85% phosphoric acid and drop in a stirbar or a few acid resistant boiling chips. Clamp the flask to a ring stand over a suitable heat source and assemble an apparatus for simple distillation [OP-25], inserting an unpacked column between the flask and the still head as illustrated in Figure E18, OP-27. Use a 10-mL graduated cylinder or another graduated container as the receiver.

Take Care! Avoid contact with the acid and alcohol, do not breathe vapors.

   Start the stirrer if you are using one, and boil the reactants gently so that vapors ascend slowly up the column and begin to condense into the receiver. When the vapors reach the still head and the temperature has stabilized, record the still-head temperature. Observe it at intervals throughout the reaction. Control the rate of heating so that the distillation rate is 1 drop per second or less and the temperature stays below 120°C. Have ready two clean, numbered 15-mL conical centrifuge tubes. Monitor the volume of the alkene (top) layer in the distillate. When the alkene volume is about 8 mL, quickly pour the distillate into the first centrifuge tube, cap the tube, and replace the graduated receiver. After the alkene volume reaches ~6 mL, observe the still-head temperature continually. Lower the heat source and turn it off when you observe a marked temperature change at the still head *or* foaming and dense white fumes in the reaction flask. Pour the distillate into the second centrifuge tube and cap the tube. For each

The water that codistills with the alkene reduces its boiling temperature, so the still-head temperature may be lower than the expected boiling point of the product.

Stop and Think: Why does the temperature change as the reaction proceeds?

Waste Disposal: Place the residue from the reaction flask into a designated waste container.

Waste Disposal: The aqueous layers can be poured down the drain.

separate fraction, wash [OP-19] the distillate by shaking it with two 5-mL portions of saturated aqueous sodium bicarbonate and carefully removing the aqueous layer with a Pasteur pipet. Dry [OP-20] each alkene mixture separately by adding anhydrous calcium chloride or another suitable drying agent to the centrifuge tube, centrifuging it if necessary, and decanting the liquid into a labeled, tared screwcap vial. Weigh each alkene mixture and calculate the total mass of alkenes.

Analysis. Analyze both fractions by gas chromatography [OP-32] as directed by your instructor. Measure the area and retention time of each peak on your gas chromatograms. Identify the peaks by comparison with a chromatogram provided by your instructor or by spiking your product mixture with an authentic sample of 1-methylcyclohexene and obtaining a chromatogram of the resulting mixture. Alkene peaks should appear on the gas chromatogram in order of the alkene boiling points. Turn in the remaining products.

Stop and Think: Were the results as you expected?

Report. Your report should include a statement of the problem and an account of how you applied scientific methodology to solve it. Assuming that the detector response factors for the alkenes are equal, calculate the percentage composition of each fraction and obtain the same data from a coworker who started with the other alcohol. Calculate the percent yield of alkenes based on their combined mass and the total mass of each alkene in your product. Try to explain your results as clearly and completely as you can.

If peaks for 3- and 4-methylcyclohexene are not separated on the gas chromatogram, calculate their combined percentage.

Exercises

1 (a) Which kind of mechanism can better account for the product mixture obtained from the dehydration of *cis* and *trans* 4-methylcyclohexanol: E1, E2, or a combination of the two? (Keep in mind that the actual mechanism may be none of these.) (b) Based on your answer, write detailed mechanisms explaining the formation of all the observed products.

2 Construct a flow diagram for this experiment following the format outlined in Appendix V.

3 Describe and explain the possible effect on your results of the following experimental errors or variations. (a) You forgot to add the phosphoric acid. (b) Throughout the reaction, you ran cold water through the jacket of the distilling column. (c) The 2-methylcyclohexanol you used was actually the pure *trans*-isomer rather than a mixture of isomers.

4 In "Understanding the Experiment," 1-methylcyclohexene and 3-methylcyclohexene were mentioned as possible products of the dehydration of 2-methylcyclohexanol. Why was 2-methylcyclohexene not mentioned as a possible product?

5 In an E2 reaction a base removes H^+ as the leaving group leaves the substrate. Draw structures of at least three possible bases (they may be weak bases) present in the reaction mixture for the reaction of 2-methylcyclohexanol.

6 (a) Predict the major alkene product that would result from dehydrating each of the following alcohols, with no carbocation rearrangements.

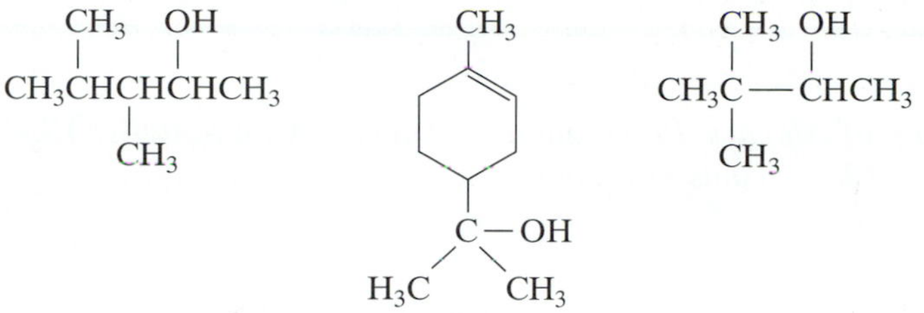

CH₃ OH
| |
CH₃CHCHCHCH₃
|
CH₃

(b) In each case, predict the most stable dehydration product that could result after a single carbocation rearrangement.

Other Things You Can Do

(Starred items require your instructor's permission.)

*1 Prepare and test a gaseous alkene as described in Minilab 18.
*2 Test your product mixtures with bromine or potassium permanganate solution, and interpret the results. (See classification tests C-7 and C-19 in Part IV.)
*3 Dehydrate another alcohol, such as cyclohexanol, 3-methylcyclohexanol, or 4-methyl-2-pentanol, by the same procedure, adjusting the distillation temperature for the alkene or alkene mixture anticipated. Analyze the product mixtures by gas chromatography and interpret the results.
4 The methylcyclohexanols used in this experiment are synthesized by catalytic hydrogenation of the corresponding cresols (methylphenols). Write a research paper about the production and uses of cresols, starting with sources listed in the Bibliography.

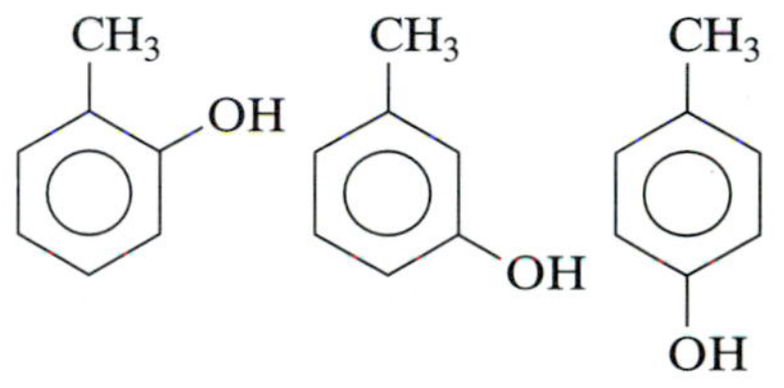

o-cresol   *m*-cresol   *p*-cresol

The Synthesis of 7,7-Dichloronorcarane Using a Phase-Transfer Catalyst

EXPERIMENT 22

Reactions of Alkenes. Preparation of Alicyclic Compounds. Alkyl Halides. Carbenes. Phase-Transfer Catalysis.

Operations:

OP-8   Temperature Monitoring
OP-9   Mixing
OP-13   Extraction
OP-14   Evaporation
OP-19   Washing Liquids
OP-20   Drying Liquids
OP-25   Simple Distillation
OP-34   Infrared Spectrometry

Before You Begin:

1 Read the experiment, read or review the operations as necessary, and write an experimental plan.
2 Calculate the mass and volume of 100 mmol of cyclohexene and the theoretical yield of 7,7-dichloronorcarane.

Scenario

Erewhon is an imaginary kingdom described in the 1872 novel of the same name by Samuel Butler.

The somewhat backward kingdom of Erewhon has just modernized its air force, which previously consisted of 24 supercharged Sopwith Camels. But the F-14 Tomcats the Erewhonians just purchased don't perform well on aviation fuel designed for Camels, so they have commissioned your Institute to help them develop a high energy jet fuel.

Hydrocarbons with a high degree of bond-angle strain have unusually high energies, so when burned they release more energy per kilogram than unstrained hydrocarbons. Cyclopropane has the highest bond-angle strain of any monocyclic hydrocarbon, and the strain increases when cyclopropane rings are fused onto other rings, as illustrated by the following examples.

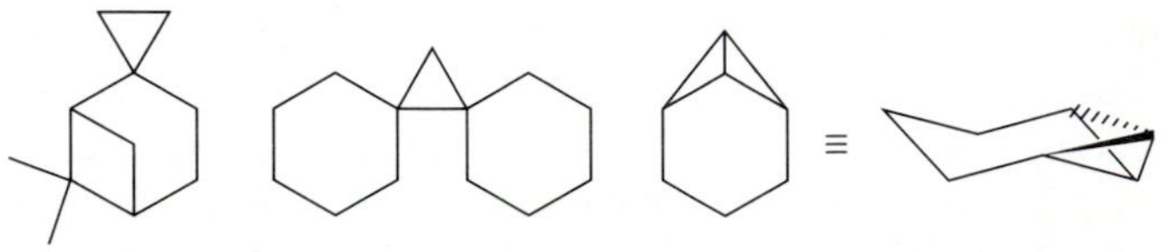

1   *2*   *3*

Compounds **1** and **2** were developed by the Monsanto Research Corporation as part of a U.S. Air Force program to develop new jet fuels that have high heats of combustion. Compound **3** (tricyclo[4.1.0.0^{2,7}]heptane), with two cyclopropane rings fused side by side onto a cyclohexane ring, has an extremely high bond-angle strain, which should make it a very energetic aircraft fuel. Your supervisor thinks it should be possible to prepare compound **3** by the following reaction of 7,7-dichloronorcarane, which proceeds via an intermediate carbene.

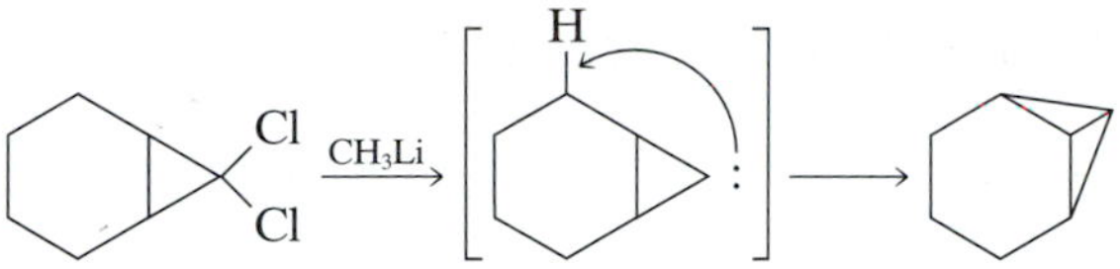

7,7-dichloronorcarane　　　carbene　　　**3**
intermediate

Before this idea can be tested, you will need to prepare some 7,7-dichloronorcarane, which can be synthesized by the addition of dichlorocarbene (:CCl$_2$) to cyclohexene. The synthesis involves a two-phase reaction mixture consisting of aqueous NaOH and an organic phase containing chloroform and cyclohexene, but the reaction tends to be very slow because some of the reactants are separated by the phase boundary. Two-phase reactions can often be accelerated by phase-transfer catalysts, so your assignment is to see whether 7,7-dichloronorcarane can be prepared from cyclohexene using a phase-transfer catalyst.

Applying Scientific Methodology

The problem is essentially to see whether following the course of action described in the Directions will lead to the preparation of 7,7-dichloronorcarane. You will test your hypothesis by obtaining and analyzing an infrared spectrum of the product.

The Ubiquitous Triangle

It often appears that nature loves the hexagon, since so many natural compounds contain six-membered rings in their molecules. This is not surprising since both the aromatic benzene ring and the unstrained cyclohexane ring are unusually stable compared to other possible ring structures. But it is surprising to find numerous examples of the triangle, in the form of the highly strained cyclopropane ring, in everything from arborvitae to water molds.

Henry David Thoreau reported that northwoods lumbermen of the last century were accustomed to drink "a quart of arborvitae, to make (them) strong and mighty." An extract of the leaves of the arborvitae tree (white cedar) was thought to impart strength and prevent illness, particularly rheumatism. One of the main constituents of the oil from arborvitae and other *Thuja* species is the bicyclic terpenoid thujone, which contains a three-membered ring fused onto a cyclopentane ring. Thujone is also found in wormwood (*Artemisia absinthium*), an ingredient in the addictive nineteenth century drink absinthe. It has been speculated that thujone was

hexagon

triangle

Euell Gibbons, an authority on edible wild plants, tasted arborvitae tea and declared that he "would almost prefer rheumatism."

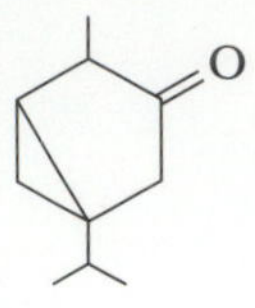

thujone

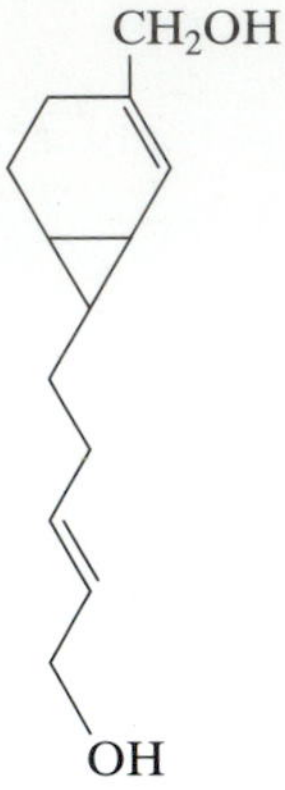

sirenin

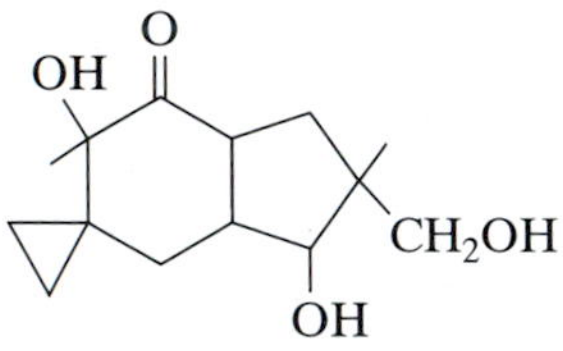

illudin-S

responsible for the artist Vincent van Gogh's mental illness. He was an absinthe drinker who once cut off part of his left ear and later committed suicide.

The sirens of Greek mythology, beautiful female creatures who lured sailors to destruction with their singing, inspired the name for sirenin, a sperm attractant produced by the female gametes of a water mold, *Allomyces javanicus.* Like the compound you will synthesize in this experiment, sirenin has a three-membered ring fused onto a six-membered ring.

One of the more startling sights in nature is a pumpkin colored mushroom that glows in the dark around Halloween. This is the poisonous jack-o'-lantern fungus, *Clitocybe illudens,* which is sometimes mistaken for the edible chanterelle mushroom with unpleasant consequences. It contains an antibiotic substance, illudin-S, which has a cyclopropane ring attached at its apex to a six-membered ring.

Pyrethrin is a natural biodegradable insecticide, nontoxic to humans, obtained from the flowers of a daisylike plant, *Chrysanthemum cineariae-folium.* It is made up of a mixture of esters such as cinerin I that contain a cyclopropane ring in the carboxylic acid portion. In recent years a number of *pyrethroids,* synthetic analogs of the natural pyrethrins, have been developed in an effort to find relatively safe but effective insecticides to replace environmentally unsound "hard" pesticides such as DDT. One of the most powerful of these is decamethrin, which is more than 60 times more lethal to houseflies than parathion and over 600 times more effective against certain mosquitoes than DDT.

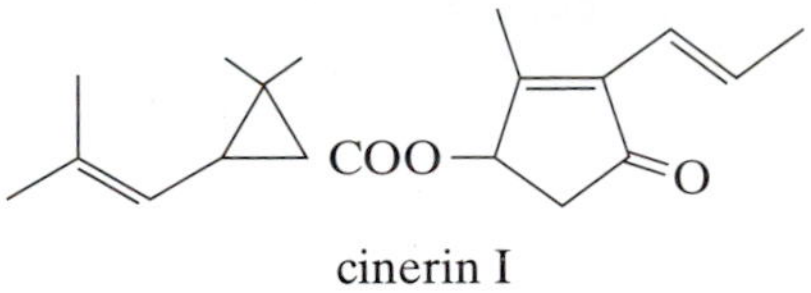

cinerin I

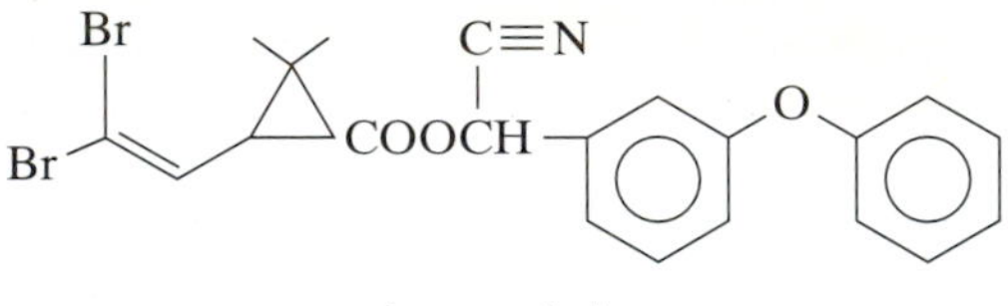

decamethrin

Understanding the Experiment

In this experiment you will generate the highly reactive intermediate dichlorocarbene by the reaction of chloroform with a strong aqueous solution of sodium hydroxide. Chloroform is toxic and is suspected of causing cancer in humans, so you should wear gloves and use a hood while you are working with it. The carbene should then combine with cyclohexene to form 7,7-dichloronorcarane. But sodium hydroxide is insoluble in the organic phase and both chloroform and cyclohexene are insoluble in water, so the reactants would ordinarily have a hard time getting together. To circumvent this difficulty you will use a phase-transfer catalyst to "escort" reactant molecules across the phase boundary.

The principles of phase-transfer catalysis (PTC) can be explained by reference to the nucleophilic substitution reaction of an alkyl halide with sodium cyanide to form a nitrile. If a high-molecular-weight halide such as 1-chlorooctane is heated with aqueous sodium cyanide, the reaction is exceedingly slow. Cyanide ions stay in the aqueous layer and alkyl halide molecules stay in the organic layer, so the reactants can only meet at the phase boundary. If, instead of sodium cyanide, a quaternary ammonium (Q) salt such as tetrabutylammonium cyanide is used, the reaction proceeds quite readily and gives a high yield of product.

Reaction of 1-chlorooctane with sodium cyanide (R = n-C$_8$H$_{17}$):

$$RCl + Na^+CN^- \longrightarrow RCN + Na^+Cl^-$$

Improved reaction with quaternary ammonium cyanide:

$$RCl + Q^+CN^- \longrightarrow Q^+Cl^- + RCN$$

$$[Q^+ = (CH_3CH_2CH_2CH_2)_4N^+]$$

There are several reasons for this enhanced reactivity of the cyanide. First (and most important), the 16 carbon atoms of the cation make it soluble in the organic phase, and where the cation goes, the anion must follow. Second, the cyanide ion is more reactive in the organic phase than it would have been in the aqueous phase since it is not solvated by water molecules that would otherwise shield it from the alkyl halide and decrease its reactivity. Finally, the bulky alkyl groups around the positive nitrogen of the cation decrease the attractive forces between cation and anion and allow the cyanide ion more freedom to attack the alkyl halide. So substituting a quaternary ammonium ion for the sodium ion in the cyanide salt allows the desired reaction to proceed at a much higher rate. Its main drawback is that the quaternary ammonium cyanide is much more expensive than sodium cyanide.

The PTC technique gets around the high cost of quaternary ammonium salts by recycling them after each reaction step. If the quaternary ammonium cation in the product (Q$^+$Cl$^-$ = tetrabutylammonium chloride) can be made to pick up some more cyanide to react with the alkyl halide, it will function as a true catalyst, accelerating the reaction without being used up. All that is needed is a reservoir of cyanide ion and a small amount of the quaternary ammonium salt to keep the reaction going. The reservoir can be provided by an aqueous layer that contains sodium cyanide.

Figure 22.1 diagrams the process, which occurs as follows: A catalytic amount of Q$^+$Cl$^-$ combines with cyanide ion in the aqueous phase, and the Q$^+$CN$^-$ that forms crosses over to the organic layer. There it reacts with the alkyl halide to produce the nitrile (RCN) and form more Q$^+$Cl$^-$,

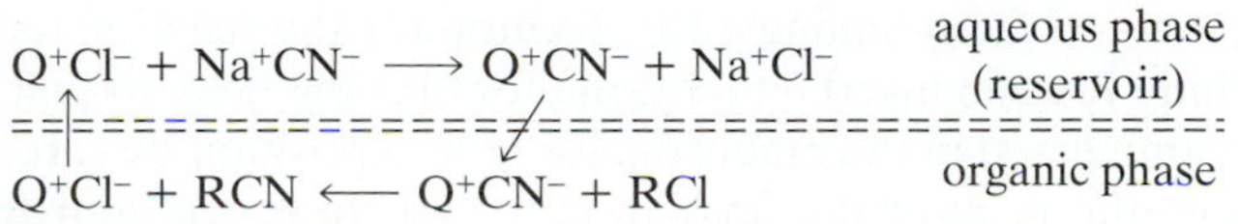

$$Q^+Cl^- + Na^+CN^- \longrightarrow Q^+CN^- + Na^+Cl^-$$

aqueous phase (reservoir)

$$Q^+Cl^- + RCN \longleftarrow Q^+CN^- + RCl$$

organic phase

Figure 22.1 Phase-transfer process for nucleophilic substitution reaction

$[CH_3(CH_2)_n]_3NCH_3^+Cl^-$
tricaprylmethylammonium
chloride (Aliquat 336)
$n = 7$ or 9

which migrates across the interface, picks up some more cyanide, shuttles it back into the organic layer to react with the alkyl halide and form more product and Q^+Cl^-, and so on, until the alkyl halide or the cyanide ion is used up.

You will use the viscous liquid tricaprylmethylammonium chloride (also known as Aliquat 336) as the phase-transfer catalyst for the synthesis of 7,7-dichloronorcarane. In this catalyst, the quaternary cation contains a methyl group and three alkyl chains of 8 or 10 carbons attached to a nitrogen atom. The initial reaction mixture will consist of an organic phase containing cyclohexene and chloroform and an aqueous phase containing sodium hydroxide, with the catalyst distributed between both phases. The sequence of events in the subsequent reaction is not entirely clear, but the following scenario seems reasonable. A hydroxide ion associated with the quaternary cation can remove a proton from chloroform at the phase boundary to produce CCl_3^-. This ionic species would ordinarily linger in the aqueous phase; but when paired with the quaternary cation, it can invade the organic phase, where it loses a chloride ion to form dichlorocarbene ($:CCl_2$). The highly reactive carbene then attacks a cyclohexene molecule in the organic phase to form the product, and the quaternary cation can return to the aqueous phase to repeat the process.

In the procedure used for this experiment, the phase-transfer catalyst promotes the formation of an emulsion, in which the organic phase is dispersed throughout the aqueous medium in minute spherical clusters called *micelles*. This greatly increases the area of contact between the phases, which speeds up the reaction by increasing the rate at which reactant molecules cross the phase boundary. The formation of a thick emulsion, with color and texture similar to thick cream, is essential to the success of the reaction. If it does not form, the reaction temperature will not rise much above 40°C and you will recover little, if any, product. A little cyclohexanol will help stabilize the emulsion, and cleaning your glassware thoroughly should remove impurities that might break up the emulsion or prevent its formation. The emulsion can be produced by either manual mixing or magnetic stirring. The manual method requires swirling the reaction flask vigorously enough to "whip" the reactants into a frothy liquid that eventually thickens and becomes completely opaque. With a little practice, you should be able to hold the neck between your thumb and fingers and use the middle finger to rock the flask back and forth with just enough twisting motion to keep the liquid moving around the flask in a circular path. Your hand and arm should move hardly at all; only the bottom of the flask should be in rapid motion. Shaking the flask with too much enthusiasm may actually break up the emulsion, so try to keep the motion vigorous but not violent. Once the emulsion has stabilized, the swirling need not be as vigorous as before.

The reaction is exothermic so it may be necessary to cool the flask to keep the reactants from boiling away. To monitor the reaction temperature while swirling, you can insert a thermometer into the flask so that its bulb is completely immersed in the emulsion. Keep it in position by carefully holding it against the neck of the flask in the "vee" between your thumb and forefinger. When the reaction is nearly over, the temperature will start to drop spontaneously.

The product is separated from the reaction mixture by extraction and purified by simple distillation. A forerun that distills below 180°C should contain cyclohexanol and unreacted starting materials. You will characterize your product by recording its infrared spectrum. The cyclopropane ring of 7,7-dichloronorcarane is highly strained, and bonds to strained ring carbons tend to vibrate at higher frequencies than normal. The carbon-chlorine stretching band for an alkyl halide can occur anywhere between 550 and 850 cm^{-1}, but it tends to be at the higher end of the range when two or more chlorine atoms are bonded to the same carbon; thus chloroform (CHCl$_3$) has a strong C—Cl band near 750 cm^{-1} (see Figure 22.2).

Choloroform

Cl
|
Cl—C—H
|
Cl

3019.0   669.2
1215.5
758.7

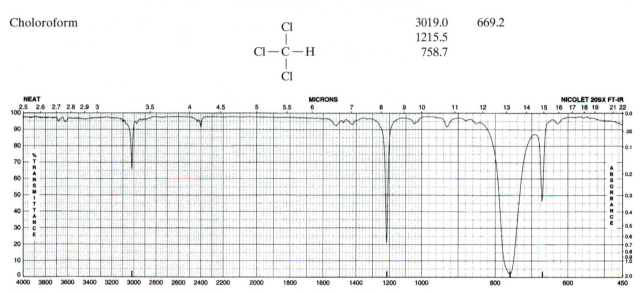

Figure 22.2 IR spectrum of chloroform (Reproduced from *The Aldrich Library of FT-IR Spectra, Edition II*, with the permission of the Aldrich Chemical Company.)

Reactions and Properties

cyclohexene   + CHCl$_3$ + NaOH $\xrightarrow{\text{PTC}}$ 7,7-dichloronorcarane   + H$_2$O + NaCl

Table 22.1 Physical properties

	M.W.	b.p.	d
cyclohexene	82.15	83	0.810
chloroform	119.4	62	1.483
tricaprylmethylammonium chloride	404.2		0.884
7,7-dichloronorcarane	165.1	198	

Note: b.p. is in °C, density in g/mL

Directions

Chloroform is harmful if inhaled, swallowed, or absorbed through the skin. It is known to cause kidney and liver tumors when ingested by rats or mice and is a suspected carcinogen in humans. Wear gloves and work under a hood; avoid contact with the liquid, and do not breathe its vapors. Cyclohexene is flammable and inhalation or skin absorption may be harmful. Avoid contact, do not breathe vapors.
Sodium hydroxide is toxic and corrosive and will cause severe damage to skin, eyes, and mucous membranes. Wear gloves and avoid contact with the NaOH solution.
Ethyl ether is extremely flammable and may be harmful if inhaled. Do not breathe the vapors, and keep it away from flames and hot surfaces.

chloroform   cyclohexene   sodium hydroxide   ethyl ether

The 50% sodium hydroxide solution can be prepared by measuring 15 g of NaOH pellets [wear gloves!] into an Erlenmeyer flask, adding 15 mL of water under a hood, swirling the flask until the solid dissolves, and cooling the hot solution.

Take Care! Keep your gloves on.

Stop and Think: Why is it important to maintain the emulsion?

Reaction. Carry out this reaction under an efficient fume hood and wear gloves throughout. If magnetic stirrers are available, you can use the alternative procedure described. In a small Erlenmeyer flask combine 100 mmol of cyclohexene with 9.0 mL (~0.11 mol) of chloroform (**Take Care!** Wear gloves, avoid contact, do not breathe vapors.) and 20 drops (0.4–0.5g) of tricaprylmethylammonium chloride, then swirl to mix. Measure 20 mL of 50% aqueous sodium hydroxide (**Take Care!** Wear gloves, avoid contact.) into a clean 250-mL Erlenmeyer flask, and add the cyclohexene-chloroform mixture. Have ready an ice/water bath large enough to accomodate the reaction flask. Add 1 mL of cyclohexanol and swirl the flask vigorously [OP-9] to whip the liquid into an opaque, creamy emulsion. (See "Understanding the Experiment" for a description of the technique.) Insert a thermometer into the flask [OP-8] and hold it securely so that its bulb is covered by the liquid while you continue to swirl the flask vigorously enough to maintain the emulsion. If the temperature rises to 60°C, swirl the flask in the ice/water bath just long enough to bring it down to 55°C. Repeat this operation as necessary to keep the temperature below 60°C. When the temperature drops spontaneously without external cooling, continue to swirl until it reaches 45°C; then let the reaction mixture stand until it reaches 35°C or below. (Go on to the *Separation* step.)

Alternative Reaction with Magnetic Stirring. Combine the reactants as described previously (hood, gloves!), using a 125-mL Erlenmeyer flask as the reaction flask. Add a stirbar (40 mm or longer, if possible) and adjust the stirring motor [OP-9] to a speed sufficient to whip the mixture into a frothy, cloudy liquid, then add 1 mL of cyclohexanol. When the solution forms an opaque, creamy emulsion, monitor the temperature continually [OP-8]. If the temperature reaches 60°C, remove the flask momentarily (turn off the stirrer first) and swirl it in an ice/water bath to bring the temperature down to 55°C, then resume stirring. Repeat as necessary to keep the temperature below 60°C. When the temperature drops spontaneously without external cooling, continue stirring until it reaches 50°C. Then remove the reaction mixture from the stirrer, and let it stand until its temperature is 35°C or below.

Separation. Add 50 mL of saturated aqueous sodium chloride to the reaction mixture, and extract it [OP-15] with two 25-mL portions of ethyl ether. Use the first portion to rinse out the reaction flask. Be sure to save the right layer. (**Take Care!** Remember that the aqueous layer contains caustic NaOH.) Wash the combined extracts [OP-19] with two portions of saturated aqueous sodium chloride. Dry the ether solution [OP-20] over anhydrous calcium chloride for five minutes or more. Evaporate the solvent [OP-14] under aspirator vacuum using a cold trap.

Waste Disposal: Carefully pour the aqueous layer down the drain and wash it down with plenty of water.

Purification and Analysis. Purify the residue by small-scale simple distillation [OP-25], collecting the product around 190–200°C. Weigh the 7,7-dichloronorcarane and record its infrared spectrum [OP-34]. Turn in the product to your instructor.

Waste Disposal: Pour the aqueous layers down the drain and place the recovered ether in a designated solvent recovery container.

Report. Your report should include a statement of the problem and an account of how you applied scientific methodology to solve it. Identify as many bands as you can in the infrared spectrum of 7,7-dichloronorcarane. In particular, point out the C—Cl and C—H stretching bands involving bonds to the cyclopropane ring.

Waste Disposal: Place any forerun and residue in a designated waste container.

Exercises

1 (a) Diagram the phase-transfer process for the reaction you carried out, using a format like that illustrated in Figure 22.1. (b) Write a mechanism for the reaction of cyclohexene with chloroform in the presence of sodium hydroxide and a phase-transfer catalyst, showing the role of the catalyst.

2 Give the systematic (IUPAC) name of 7,7-dichloronorcarane.

3 Describe and explain the possible effect on your results of the following experimental errors or variations. (a) You forgot to add the tricaprylmethylammonium chloride. (b) The lab assistant misplaced a decimal point and prepared 5.0% NaOH for this experiment. (c) You didn't swirl the reaction flask properly and no emulsion formed.

4 Construct a flow diagram for the synthesis of 7,7-dichloronorcarane using the format described in Appendix V.

5 The carbon-carbon sigma bonds of cyclopropane rings exhibit some of the characteristics of pi bonds in other systems. For example, the cyclopropylmethyl cation undergoes reactions suggesting that it is

best represented by the resonance structures shown. Explain the iso-merization in acidic solution of chrysanthemyl alcohol to yomogi alcohol and artemisia alcohol by writing appropriate mechanisms.

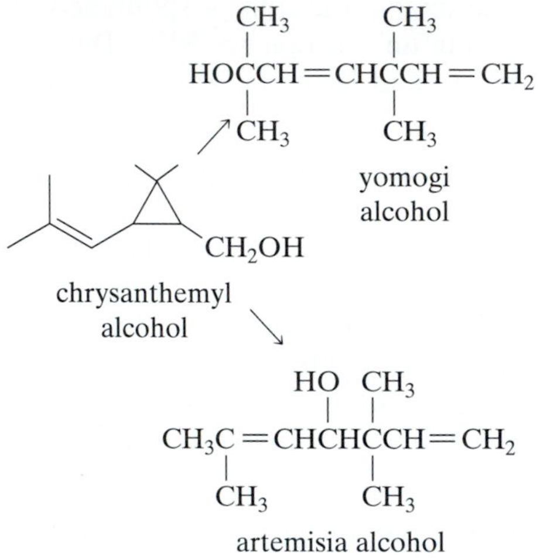

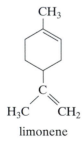

chrysanthemyl
alcohol

yomogi
alcohol

artemisia alcohol

6 Outline syntheses of compounds **1** and **2** from the Scenario, starting with an appropriate alkene in each case.

Other Things You Can Do

(Starred items require your instructor's permission.)

*1 Assess the purity of your product by gas chromatography, or record its ^{1}H NMR spectrum in deuterochloroform. A general purpose silicone oil/Chromosorb W column at ~110°C can be used for the chromatography.

*2 See what happens when you mix iodine with turpentine, whose components have strained rings, in Minilab 19.

3 Look up the article in *Tetrahedron Letters* **1975**, 3013 to find out which of the double bonds of limonene is attacked first by dichloro-carbene. Give the structure of the major product of this reaction and summarize the reaction conditions used, indicating how they differ from the conditions used in this experiment.

limonene

Stereochemistry of the Addition of Bromine to *trans*-Cinnamic Acid

EXPERIMENT 23

Reactions of Alkenes. Preparation of Alkyl Halides. Electrophilic Addition. Stereoselectivity.

Operations:

OP-9 Mixing
OP-10 Addition
OP-12 Vacuum Filtration
OP-21 Drying Solids
OP-28 Melting Point

Before You Begin

1 Read the experiment, read or review the operations as necessary, and write an experimental plan.
2 Calculate the mass of 10.0 mmol of *trans*-cinnamic acid and the theoretical yield of 2,3-dibromo-3-phenylpropanoic acid.

Scenario

The Bond Triplex is a chemical specialties company that supplies alkynes to order. Many of their alkynes are prepared by adding bromine to the corresponding alkenes and dehydrohalogenating the resulting dibromide. For example, they convert *trans*-cinnamic acid to 3-phenylpropynoic acid by way of 2,3-dibromo-3-phenylpropanoic acid.

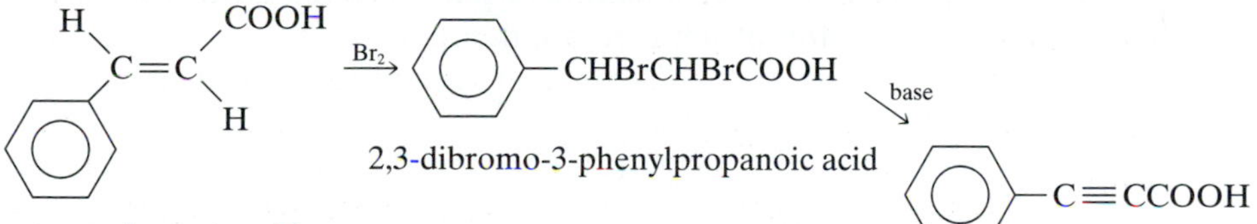

trans-cinnamic acid
2,3-dibromo-3-phenylpropanoic acid
3-phenylpropynoic acid

At a recent meeting of the three Bond partners (Sigmund, Bridget, and James) with the chemical engineers who developed their manufacturing processes, the engineers were compelled to admit that they weren't certain of the stereochemical structures of the intermediate dibromides. They had simply assumed that the bromine addition reactions proceeded by the well-established bromonium ion mechanism (see p. 178), resulting in *anti* addition of bromine, and deduced the stereochemistry of the dibromides

trans-anethole

accordingly. But changing the structure of a reactant can often alter the mechanism of a reaction, and some of the alkenes the Bond Triplex is using have electron donating substituents that might alter the mechanism of the addition reaction. For example, *trans*-anethole undergoes significant amounts of *syn* addition of bromine by an alternative mechanism. Because knowing the stereochemistry of the intermediate dibromides could help the company's chemical engineers design more efficient processes for converting them to alkynes, Sigmund Bond has contacted your Institute for help in characterizing the intermediates. Your assignment is to carry out the bromination of *trans*-cinnamic acid, determine the stereochemical structure of the dibromide, and find out whether the reaction proceeds by the usual bromonium ion mechanism or some other mechanism.

Applying Scientific Methodology

Your hypothesis should include a prediction regarding the mechanism of the reaction. You will test your hypothesis by measuring the melting point of the product, which will reveal its stereochemistry.

The Cinnamic Acid Connection

cinnamic acid

cinnamaldehyde

Cinnamic acid and its close relatives, cinnamaldehyde and cinnamyl alcohol, are naturally occurring compounds that are important as flavoring and perfume ingredients and as sources for pharmaceuticals. Cinnamaldehyde, the major component of cinnamon oil, is used to flavor many foods and beverages and to contribute a spicy, "oriental" note to perfumes. Cinnamic acid itself plays an important role in secondary plant metabolism. As an intermediate in the shikimic acid pathway of plant biosynthesis, cinnamic acid is the source of an enormous number of natural substances that (to give only a few examples) contribute structural strength to wood, give flavor to cloves, nutmeg, and sassafras, and produce many of the brilliant colors of nature—the flower pigments that attract insects for pollination, the vivid and delicate shades of a butterfly's wings, and the radiant colors of leaves in autumn.

coniferyl alcohol

In nature, cinnamic acid is formed by the enzymatic deamination (removal of ammonia) of the amino acid phenylalanine, which in turn is biosynthesized in a series of steps from shikimic acid.

myristicin

shikimic acid

many steps →

phenylalanine

$\xrightarrow{-NH_3}$

cinnamic acid

safrole

It can then be converted, by a wide variety of biosynthetic pathways, to coniferyl alcohol (a precursor of lignin) in sapwood, myristicin in nutmeg, safrole in sassafras bark, and flavonoids in a wide variety of plant structures.

The flavonoids are natural substances characterized by the 2-arylbenzopyran structure found in flavanone, which itself is biosynthesized from cinnamic acid by a process involving the linkage of three acetate units to the carboxyl group of the acid.

[handwritten: 177]

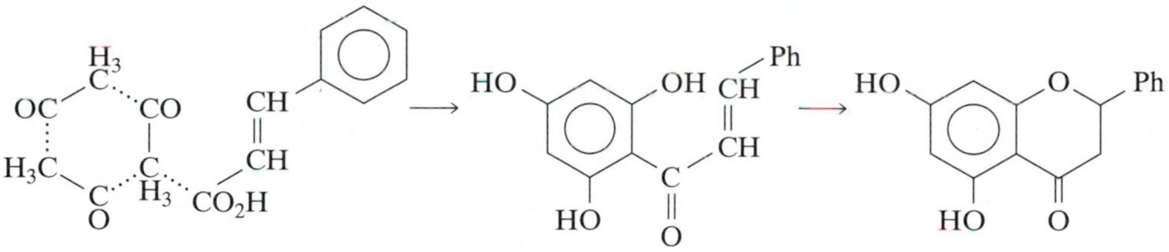

acetates   cinnamic acid   flavanone

[handwritten: COJH at the top, not CHO]

Flavonoids perform no single function in plants. Many are highly colored and attract insects for pollination or animals for seed dispersal; others help regulate seed germination and plant growth or protect plants from fungal and bacterial diseases. Certain flavonoids contribute the bitter taste to lemons and the bracing astringency of cocoa, tea, and beer. Some flavonoids in foods appear to act as antioxidants that may boost your immune system and help prevent cancer. Although flavonoids and the other derivatives of cinnamic acid may not have attained the commercial success of some other natural chemicals, they provide much to delight the eye and stimulate the senses, and the world would be a drearier place without them.

Understanding the Experiment

In this experiment you will carry out the addition of bromine to *trans*-cinnamic acid and identify the product from its melting point. The product, actually a mixture of enantiomers, could be either *erythro*-2,3-dibromo-3-phenylpropanoic acid [whose enantiomers have the (2*R*,3*S*) and (2*S*,3*R*) configurations], the *threo*-dibromide [(2*R*,3*R*) and (2*S*,3*S*)], or a mixture of the *erythro* and *threo* dibromides. The *erythro-threo* nomenclature is used to describe the configurations of compounds having two chiral centers but no plane of symmetry. It is based on the structures of the two simple sugars erythrose and threose.

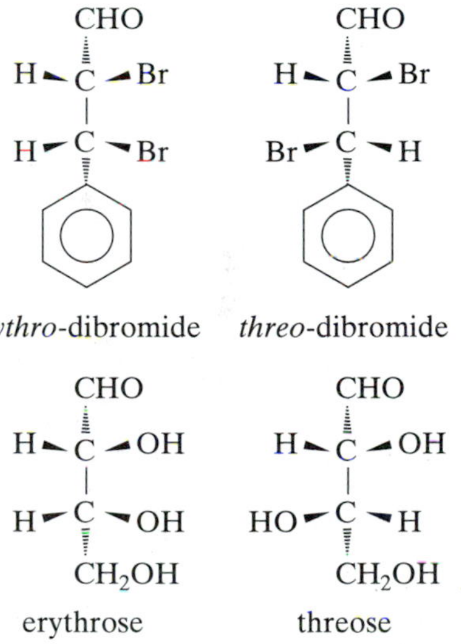

The product you obtain will depend on the stereochemical course of the reaction. As explained in Experiment 21, a reaction is said to be regioselective if it might produce two or more structural isomers but in fact yields one of them preferentially. Similarly, a reaction is said to be *stereoselective* if it might produce two or more stereoisomers but in fact yields mainly (or entirely) one of them. For example, the electrophilic addition of bromine to cyclopentene is stereoselective because it yields *trans*-dibromocyclopentane and no *cis*-dibromocyclopentane, indicating that the components of Br$_2$ must add to opposite sides of the carbon-carbon double bond. This mode of addition is called *anti* addition, while addition of the components of a reagent to the same side of a double bond is called *syn* addition.

The following scenario has been devised to explain the *anti* addition of bromine to cyclopentene. As a bromine molecule approaches perpendicular to the negatively charged pi cloud of the carbon-carbon double bond, its

erythro-dibromide   threo-dibromide

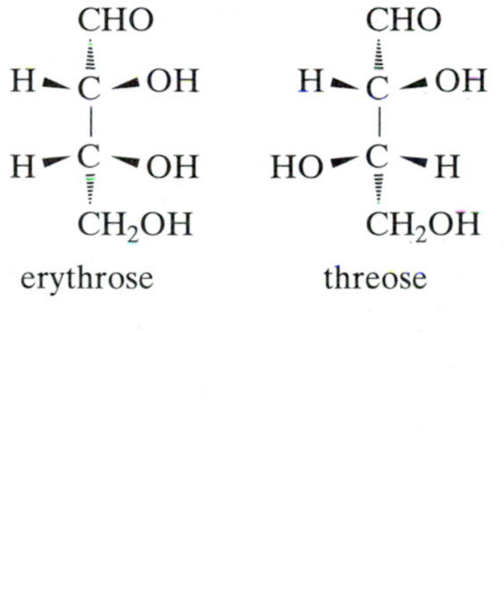

erythrose   threose

bonding electrons are repelled away from the bromine atom nearer the double bond, leaving it with a partial positive charge. As the positively charged bromine penetrates the pi cloud, a negative bromide ion breaks away from it, leaving a cyclic *bromonium ion* in which the positive bromine is bonded to two carbon atoms. Backside attack on the bromonium ion by a bromide ion results in the observed *trans* product.

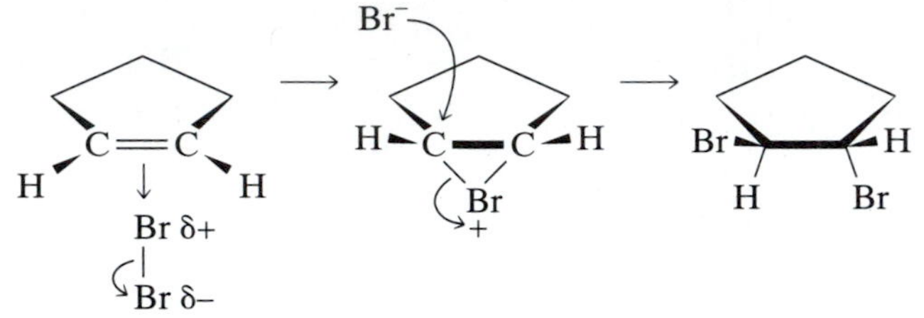

Other electrophilic addition mechanisms may lead to different stereochemical outcomes. In the addition of bromine to *trans*-anethole, conjugation with the ring stabilizes a carbocation intermediate that can be attacked on either side, leading to *syn* as well as *anti* addition.

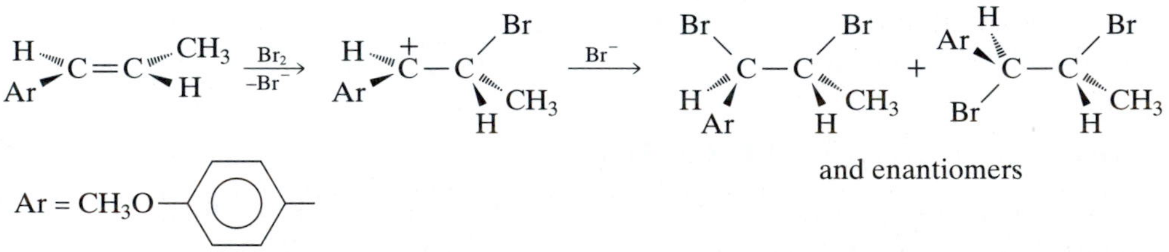

In this reaction about 35 percent of the product results from *syn*-addition and 65 percent from *anti*-addition. Another possible mechanism might involve a concerted addition to the carbon-carbon double bond that, like the catalytic hydrogenation of alkenes, leads exclusively to *syn*-addition.

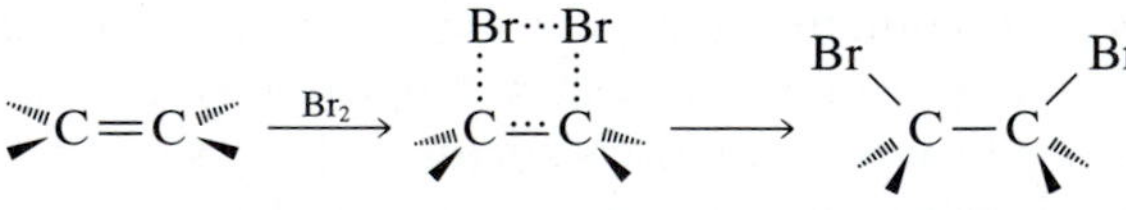

You will carry out the reaction by slowly adding a solution of bromine in acetic acid to a solution of *trans*-cinnamic acid in the same solvent. The dibromide begins to precipitate from solution during the reaction and is separated by vacuum filtration. Because the melting points of the *erythro* and *threo* dibromides are separated by more than 100°C, the product can be easily identified from its melting point. A mixture of both products would melt over a broad range that should not coincide with the melting point of either pure dibromide. From the identity of your product, you should be able to deduce whether the addition of bromine to *trans*-cinnamic acid involves *syn* or *anti* addition or a mixture of the two. Molecular models will help you relate the configurations of the *syn* and *anti* addition products to the stereochemical structures shown for the *erythro* and *threo* dibromides.

Reactions and Properties

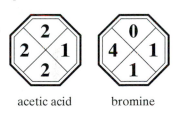

trans-cinnamic acid $+ Br_2 \xrightarrow{HOAc}$ CHBrCHBrCOOH

2,3-dibromo-3-phenylpropanoic acid

Table 23.1 Physical properties

	M.W.	m.p.	b.p.	d
trans-cinnamic acid	148.2	136		
bromine	159.8	−7	59	3.12
acetic acid	60.1	17	118	1.049
erythro-2,3-dibromo-3-phenylpropanoic acid	308.0	204		
threo-2,3-dibromo-3-phenylpropanoic acid	308.0	95		

Note: m.p. and b.p. are in °C, *d* is in g/mL

Directions

Bromine is highly toxic and corrosive, and its vapors can damage the eyes and respiratory tract. Wear gloves when handling the bromine solution, and dispense it under a hood; avoid contact and do not inhale the vapors. Acetic acid causes chemical burns that can seriously damage skin and eyes; its vapors are highly irritating to the eyes and respiratory tract. Wear gloves, dispense under a hood; avoid contact and do not breathe vapors.

Safety Notes

acetic acid   bromine

Take Care! Wear gloves, avoid contact with acetic acid and the bromine solution, do not breathe their vapors.

Reaction. Carry out the reaction under the hood and wear gloves throughout. In a 50-mL Erlenmeyer flask combine 10.0 mmol of *trans*-cinnamic acid with 6.0 mL of glacial acetic acid and add a stirbar if you have a magnetic stirrer. Securely support a separatory/addition funnel over the Erlenmeyer flask and be sure the stopcock is closed [OP-10]. Pour 8.2 mL of a 1.25 *M* solution of bromine in acetic acid into the separatory funnel, and stopper it immediately. Start the stirrer (or swirl the flask after each addition) [OP-9] and add the bromine/acetic acid solution in five or more portions, waiting until the color has faded to light orange before adding the next portion. The cinnamic acid should dissolve shortly after addition of the first portion, and the entire addition should take about a half hour. After the last addition, let the reaction mixture stand at room temperature for 15 minutes with stirring or occasional swirling. If the mixture becomes colorless during this period, add more of the bromine/acetic acid solution dropwise until the color just persists. If the mixture has a distinct color

Observe and Note: What evidence suggests that a reaction has occurred?

Stop and Think: What is the purpose of the cyclohexene? What reaction is involved?

Waste Disposal: Unless your instructor indicates otherwise, flush the filtrate down the drain with water.

Stop and Think: Was the result what you expected? If not, why not?

(yellow or orange) at the end of the reaction period, add cyclohexene dropwise until it is colorless or nearly so.

Separation and Analysis. Cool the mixture in ice water until crystallization is complete, then collect the product by vacuum filtration [OP-12], washing it on the filter with cold water until the acetic acid odor is hardly noticeable. (You can obtain additional product of lower purity by adding about 10 mL of water to the filtrate, cooling it in ice water, and collecting the solid by vacuum filtration.) Dry the 2,3-dibromo-3-phenylpropanoic acid [OP-21] and measure its melting point [OP-28].

Stereochemistry of Bromine Addition. Construct a molecular model of *trans*-cinnamic acid. Simulate the *syn* addition of bromine by removing one of the C=C connectors and inserting two orange bromine atoms, with connectors, into the vacant holes. (You may want to replace the remaining flexible connector by a rigid one.) Rotate around the carbon-carbon single bond that remains until the model corresponds to the stereochemical projection for either *threo-* or *erythro*-2,3-dibromo-3-phenylpropanoic acid. Simulate *anti* addition of bromine by removing the upper connector of the carbon-carbon double bond and moving one end of the lower connector from the hole it occupies to the vacant hole in the same carbon atom. Be careful not to rotate either carbon atom as you do so. Insert two bromine atoms, with connectors, into the vacant holes; then rotate the model as before until it matches one of the stereochemical projections.

Report. Your report should include a statement of the problem and an account of how you applied scientific methodology to solve it. Give the name and structure of your product and tell whether it was formed by *syn* addition, *anti* addition, or a mixture of both. Tell whether or not the reaction is stereoselective, and write a mechanism that explains its stereochemistry.

Exercises

1 (a) Write resonance structures showing how the aryl group of *trans*-anethole stabilizes the intermediate carbocation shown in "Understanding the Experiment." (b) Based on your results, explain any differences or similarities in the stereochemistry of the bromine addition reactions of *trans*-cinnamic acid and *trans*-anethole.

2 What product would you expect to obtain by the addition of bromine to *cis*-cinnamic acid, assuming that it reacts by the same mechanism the *trans* acid?

3 (a) Write a mechanism showing why the bromination of *trans*-cinnamic acid produces a racemic mixture of enantiomers. (b) Draw a stereochemical projection for each enantiomer and specify the configuration (R or S) at each stereocenter.

4 Describe and explain the possible effect on your results of the following experimental errors or variations. (a) The cinnamic acid you used was actually a mixture of *cis* and *trans* isomers. (b) You stopped the addition after you had added just 4 mL of the bromine solution. (c) You misread the label on a bottle of cyclohexane and used it in place of cyclohexene.

5 Following the format in Appendix V, construct a flow diagram for the synthesis you carried out in this experiment.

6 Draw stereochemical projections for the products of bromine addition to maleic acid and fumaric acid (*cis-* and *trans-*HOOCCH=CHCOOH), assuming that bromine adds in the same way to these compounds as it does to cinnamic acid.

7 (a) Would you expect the product from this experiment to be optically active? Could it be resolved into optically active constituents? Explain. (b) Would the product of the bromination of fumaric acid (see Exercise 5) be optically active? Could it be resolved into optically active constituents? Explain.

Other Things You Can Do

(Starred projects require your instructor's permission.)

*1 Test some commercial products for unsaturation as described in Minilab 20.

*2 Synthesize 3-phenylpropynoic acid from your product by scaling down the procedure given in *J. Am. Chem. Soc.* **1942**, *64*, 2510. Find a suitable recrystallization solvent to use in place of carbon tetrachloride.

3 Write a research paper about the industrial preparation and commercial uses of cinnamic acid and its derivatives, starting with sources listed in the Bibliography.

EXPERIMENT 24

Hydration of a Difunctional Alkyne

Reactions of Alkynes. Preparation of Carbonyl Compounds. Electrophilic Addition. Infrared Spectrometry.

Operations:

OP-6   Heating
OP-10 Addition
OP-13 Extraction
OP-14 Evaporation
OP-15 Steam Distillation
OP-19 Washing Liquids
OP-20 Drying Liquids
OP-25 Simple Distillation
OP-34 Infrared Spectrometry

Before You Begin

1 Read the experiment, read or review the operations, and write an experimental plan.
2 Calculate the mass and volume of 100 mmol of 2-methyl-3-butyn-2-ol and the theoretical yield of the product.

Scenario

Malthusian Lozenges were introduced in Aldous Huxley's futuristic novel Brave New World.

Malthusian Solutions manufactures a line of birth control pills, called Malthusian Lozenges, which contain synthetic sex hormones as their active ingredients. Megestrol acetate, a sex hormone the company would like to incorporate in their pills, is one of the most powerful ovulation inhibitors known. Megestrol acetate is classified as an *acetoxyketo steroid* because it is a ketone with an acetoxy (CH_3COO) group on its five-membered ring (the D ring). A number of synthetic sex hormones such as norethynodrel are *alkynol steroids* having an ethynyl group ($HC \equiv C-$) and a hydroxyl function on the D ring.

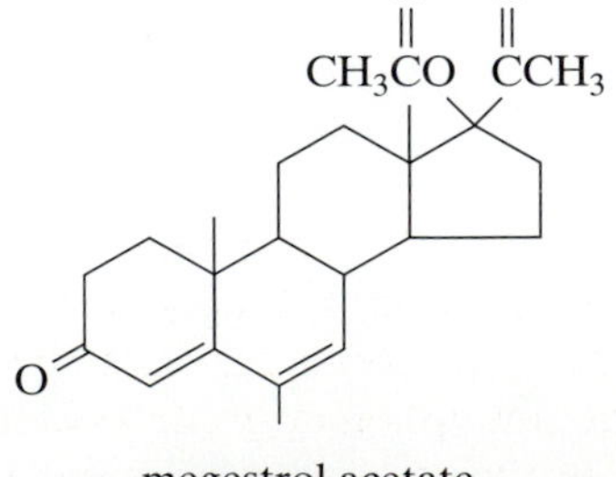

megestrol acetate 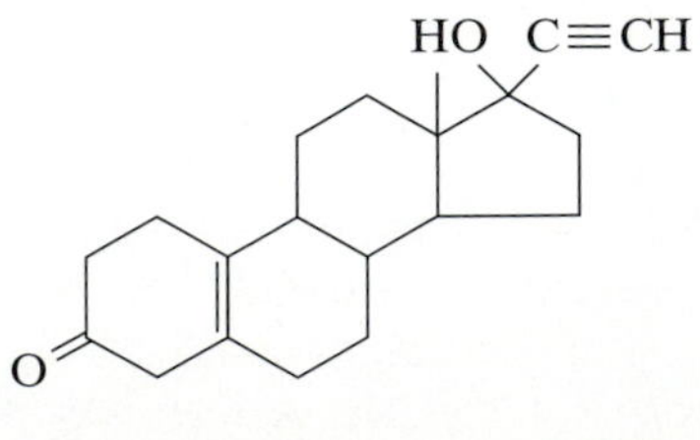   norethynodrel

Harry D. Stork, the product development director at Malthusian Solutions, thinks it should be possible to convert an alkynol steroid such as norethynodrel to an acetoxyketo steroid by hydrating the carbon-carbon triple bond and acetylating the hydroxyl group.

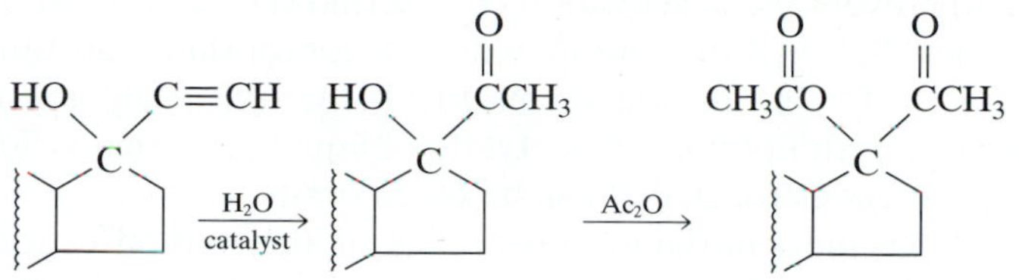

acetoxyketo
steroid

If this reaction pathway is feasible, Malthusian solutions might use it to produce megestrol acetate and related acetoxyketo steroids. But first they must establish that the alkyne hydration follows Markovnikov's rule and yields the desired carbonyl compound. Norethynodrel and other synthetic sex hormones are quite expensive, so you will test Stork's idea by carrying out the hydration reaction on a simpler "model compound," 2-methyl-3-butyn-2-ol.

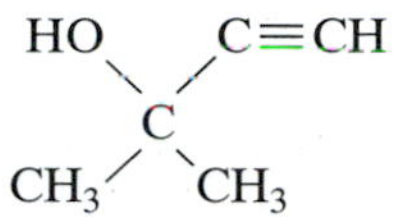

2-methyl-3-butyn-2-ol

Applying Scientific Methodology

In developing your hypothesis, consider not only whether the hydration of the model compound will occur, but also whether it will yield a product analogous to an acetoxyketo steroid. You will use infrared spectroscopy to test your hypothesis .

From "The Pill" to Oblivon

Natural *estrogens* such as estrone and 17β-estradiol are responsible for promoting the development of secondary sex characteristics in females during puberty. Estrogens are also frequently prescribed to alleviate the mental and physical discomfort associated with menopause and to prevent osteoporosis (weakening of bone structure) in older women. Natural *progestins* such as progesterone are necessary to maintain pregnancy in mammals.

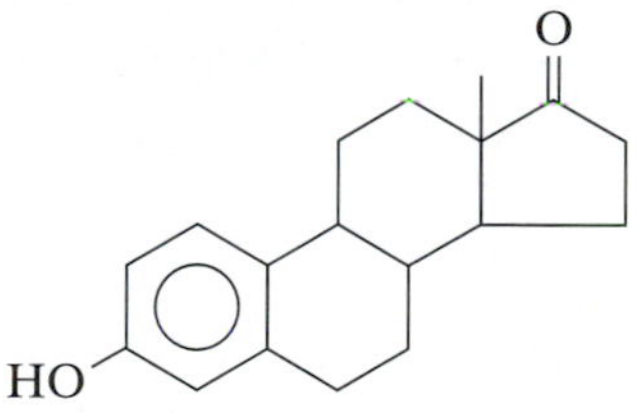

estrone

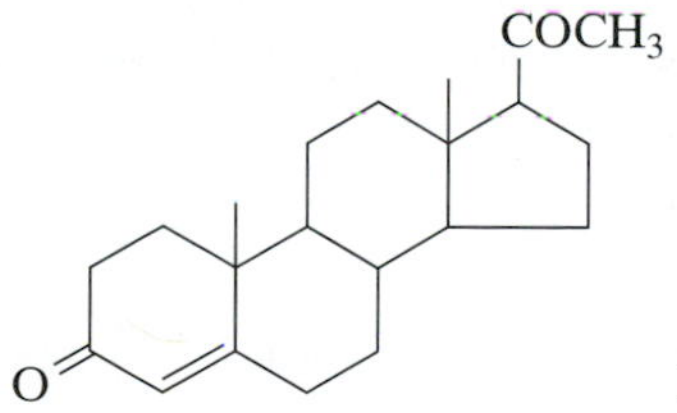

progesterone

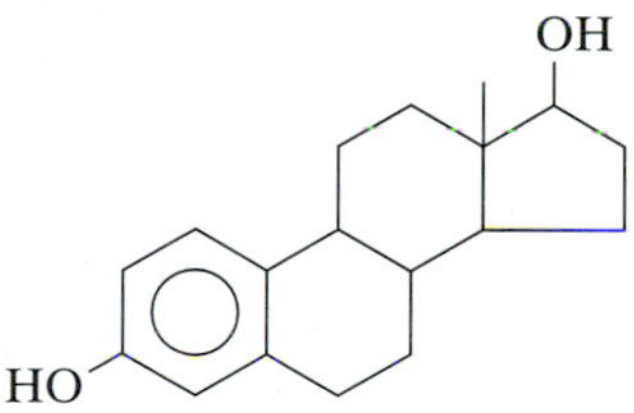

17β-estradiol

Progestins have been used to treat menstrual disorders and uterine bleeding and to prevent miscarriages. Both types of hormones belong to an important class of natural products called the *steroids*, which are characterized by a basic four-ring skeleton, the perhydrocyclopentanophenanthrene nucleus. The most widely known steroid, cholesterol, is often regarded as an undesirable

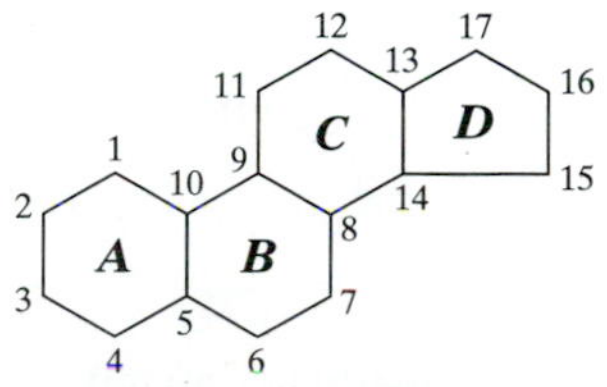

perhydrocyclopentanophenanthrene

cholesterol

ethinyl estradiol

ingredient in our food because of its role in cardiovascular disease, but it is also an essential precursor of sex hormones and other body regulators.

Natural hormones cannot be taken orally because they are rapidly deactivated in the liver, so the search for synthetic hormones with similar activity began soon after the natural ones were isolated and characterized. One way to prevent the deactivation of a steroid such as 17β-estradiol is to stabilize the C-17 hydroxy group with an appropriate substituent. The ethynyl (HC$\equiv$C—) function seems to fill the bill nicely. For example, treating estrone with potassium acetylide in liquid ammonia yields ethinyl estradiol, a potent estrogen that can be taken orally.

One of the most intriguing properties of the natural estrogens and progestins is their ability to inhibit ovulation in females. During the 1940s it was a common practice to treat certain menstrual disorders with these hormones because preventing ovulation stops menstruation. Of course, preventing ovulation also prevents pregnancy, so the idea of using hormones for birth control undoubtedly occurred to some scientists. At that time there were no suitable sex hormones that could be taken orally, but by the early 1950s the picture had changed. The discovery that certain Mexican wild yams of the species *Dioscorea* contain the natural steroid diosgenin provided chemists and pharmaceutical manufacturers with a cheap, abundant starting material for the preparation of synthetic steroids. The first synthesis of a steroid oral contraceptive, 19-norprogesterone, was accomplished in 1951 by a research team at Syntex led by Carl Djerassi. Early in 1953 this and other synthetic steroids were evaluated for anti-ovulatory activity by Gregory Pincus and his colleagues. They soon discovered that the combination of a synthetic progestin with a small amount of synthetic estrogen provided the highest anti-ovulatory activity, and "the pill" was born soon afterward. The G. D. Searle Company marketed the first birth control pill, Enovid, which contained 9.85 mg of the progestin norethynodrel and 0.15 mg of the estrogen mestranol. Modern oral contraceptives of the "minipill" type contain considerably smaller amounts of a progestin and no estrogen; although they are slightly less effective than the combination pills, they have fewer side effects. When used properly, most oral contraceptives are 99–100% effective in preventing pregnancy, and their impact on society has been enormous.

Alkynols having a hydroxyl group adjacent to the triple bond (as in norethynodrel) are comparatively easy to synthesize commercially and they are used in a number of pharmaceuticals. The 2-methyl-3-butyn-2-ol used in this experiment is prepared by combining acetone and acetylene with an alkali metal in liquid ammonia.

$$CH_3-\overset{\overset{\displaystyle O}{\|}}{\underset{\underset{\displaystyle CH_3}{|}}{C}} + HC\equiv CH \xrightarrow{\text{Li, NH}_3} CH_3-\overset{\overset{\displaystyle OH}{|}}{\underset{\underset{\displaystyle CH_3}{|}}{C}}-C\equiv CH$$

Preparation of 2-methyl-3-butyn-2-ol

methylparafynol
(Oblivon)

A similar alkynol named methylparafynol has been used in sleeping pills under the trade name Oblivon. Apparently the tertiary alcohol portion of the molecule causes methylparafynol to act as sedative or depressant while the acetylenic group gives it hypnotic (sleep inducing) properties.

Understanding the Experiment

In this experiment you will carry out the hydration of 2-methyl-3-butyn-2-ol and identify the product by infrared spectrometry. Alkyne hydration, the electrophilic addition of water to a carbon-carbon triple bond, is generally accomplished by heating the alkyne with water in the presence of an acid and a mercury(II) salt. As shown in the proposed reaction mechanism here, mercuric ion catalyzes the reaction by a process that involves its addition to the triple bond to form a pi complex (**1**). The pi complex is then attacked by water to give intermediate **2**, which loses a proton and is hydrolyzed to an enol (**3**). Isomerization of the enol to the corresponding keto tautomer yields the final product.

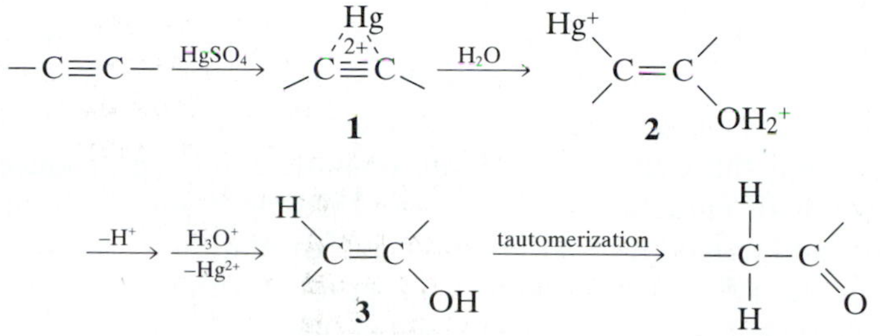

The hydration of alkenes is regioselective because the intermediate is a carbocation that can be stabilized by electron donating alkyl substituents, resulting in Markovnikov addition. The illustrated mechanism for hydration of alkynes does not involve a free carbocation, so it is not obvious that such a reaction should follow Markovnikov's rule, especially in the presence of an electron withdrawing group such as the OH in 2-methyl-3-butyn-2-ol. If it does, the product from hydration of a terminal alkyne (one having a hydrogen on a triple-bonded carbon) should be a methyl ketone. Anti-Markovnikov addition of water to a terminal alkyne should yield an aldehyde.

See Experiment 21 for a discussion of regioselective reactions.

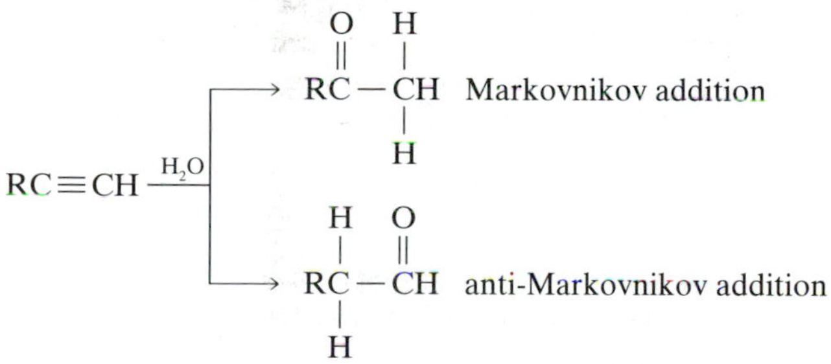

You will heat the reactant under reflux with water containing sulfuric acid and mercury(II) sulfate. Because the reaction is strongly exothermic, the alkyne is added slowly to the reaction mixture from an addition funnel. The product can be separated from any nonvolatile impurities in the reaction mixture by codistillation with water. During the reaction, some mercury(II) ion may be reduced to metallic mercury, which may codistill with the product. Because mercury vapor is very toxic, it is important to recover and dispose of this substance properly.

Any acid that distills with the product will be neutralized by adding some potassium carbonate to the distillate. The potassium carbonate, along with some sodium chloride, will be used to "salt out" the product so that it can be extracted more completely. Mercury and other heavy metals can react with terminal alkynes to form organometallic salts that may explode if heated to dryness. Because such salts do not form at a low pH, it is unlikely that the product will contain any, but you will wash the dichloromethane extract with dilute acid as a precautionary measure. After evaporating the solvent, you will purify the crude product by simple distillation.

Infrared spectrometry is particularly useful for detecting functional groups in molecules, so it can readily confirm the conversion of one functional group to another in this synthesis. As you can see in Figure 24.1, the infrared spectrum of the reactant is characterized by a strong, narrow ≡CH stretching band near 3300 cm^{-1} and a weak C≡C stretching band near 2100 cm^{-1}. Because the broad O—H band of the reactant occurs in the 3300–3400 cm^{-1} region, it partially obscures the sharper ≡C—H band. Aldehydes and ketones both give rise to strong carbonyl bands near 1700 cm^{-1}, but the CHO group of an aliphatic aldehyde produces sharp C—H bands of moderate strength near 2720 cm^{-1} and 2840 cm^{-1}. (See OP-34 for additional information about interpretation of IR spectra.) By comparing the infrared spectrum of your product with that of the reactant, you should be able to decide whether or not the expected functional group conversion has taken place and whether or not the reaction obeyed Markovnikov's rule.

Reactions and Properties

2-methyl-3-butyn-2-ol	OH │ CH$_3$C—C≡CH │ CH$_3$	3299.1 2985.1 1365.4	1168.2 962.2 888.6	707.2 649.0 558.3

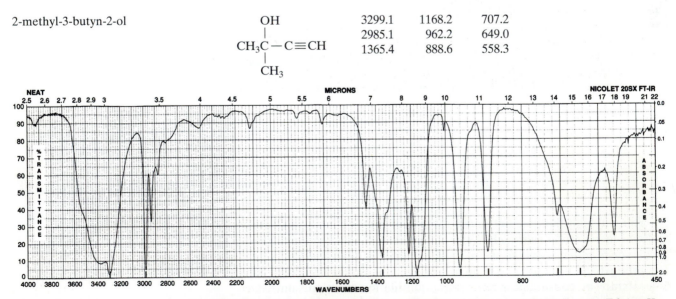

Figure 24.1 Infrared spectrum of 2-methyl-3-butyn-2-ol (Reproduced from *The Aldrich Library of FT-IR Spectra, Edition II,* with the permission of the Aldrich Chemical Company.)

$$\underset{\substack{\text{2-methyl-3-butyn-2-ol}}}{\underset{\displaystyle CH_3}{\overset{\displaystyle OH}{\underset{|}{\overset{|}{CH_3C}}}} - C \equiv CH} + H_2O \xrightarrow[\text{HgSO}_4]{\text{H}_2\text{SO}_4} \underset{\substack{\text{3-hydroxy-3-methyl-2-butanone}}}{\underset{\displaystyle CH_3}{\overset{\displaystyle OH\ \ O}{\underset{|}{\overset{|\ \ \ ||}{CH_3C}}}} - CCH_3} \quad or \quad \underset{\substack{\text{3-hydroxy-3-methylbutanal}}}{\underset{\displaystyle CH_3}{\overset{\displaystyle OH\ \ \ \ O}{\underset{|}{\overset{|\ \ \ \ \ \ ||}{CH_3C}}}} - CH_2CH}$$

Table 24.1 Physical properties

	M.W.	b. p.	d
2-methyl-3-butyn-2-ol	84.1	104	0.868
mercuric sulfate	296.7		
3-hydroxy-3-methyl-2-butanone	102.1	141	0.953
3-hydroxy-3-methylbutanal	102.1	67^{13}	

Directions

Safety Notes

Mercury(II) sulfate is *very poisonous* if inhaled or ingested. Do not breathe its dust or allow it to contact skin or eyes. Wash hands thoroughly after handling the compound.

2-Methyl-3-butyn-2-ol and 3-hydroxy-3-methyl-2-butanone may be harmful by inhalation, ingestion, or skin absorption. Avoid contact with the reactant and product and do not breathe their vapors.

Mercury (formed during the reaction) emits toxic vapors whose concentration can build up to dangerous levels in poorly ventilated areas. Dispose of mercury-containing residues properly, and clean up any spills immediately.

Dichloromethane may be harmful if ingested, inhaled, or absorbed through the skin. There is a possibility that prolonged inhalation of dichloromethane may cause cancer. Minimize contact with the liquid and do not breathe its vapors.

If the reaction mixture is not strongly acidic, it might contain an unstable organomercury compound that could explode if heated to dryness. For that reason, monitor the distillations carefully and do not distill to dryness.

Reaction. Assemble an apparatus for addition under reflux [OP-10]. Put 100 mmol of 2-methyl-3-butyn-2-ol in the addition funnel, and put boiling chips or a stirbar in the reaction flask. Carefully weigh out 0.5 g of mercury(II) sulfate. Dissolve it in 50 mL of 3 *M* sulfuric acid, stirring with gentle heating if necessary, and transfer this solution to the boiling flask. Start the stirrer (if you have one) and heat the reaction mixture to boiling; then continue heating under gentle reflux [OP-6] as you add the 2-methyl-3-butyn-2-ol drop by drop over about a 10-minute period. If you have no stirrer, carefully shake the apparatus after each addition. Heat the reaction mixture under reflux for 30 minutes after all the alkyne has been added.

Separation. Assemble the internal steam distillation apparatus illustrated in Figure C11 of OP-15, using a graduated cylinder as the receiver. Put 50 mL of water in the addition funnel and begin distilling the product mixture. Add

Take Care! Poison! Avoid contact with or ingestion of mercury(II) sulfate.

Take Care! Foaming may occur.

Stop and Think: What is the purpose of the water?

Waste Disposal: If the residue in the boiling flask contains any droplets of mercury metal, carefully decant the supernatant liquid and place the mercury in a mercury wastes container. Wash the liquid down the drain with water.

Take Care! Avoid contact with dichloromethane, do not breathe its vapors.

Waste Disposal: Place the recovered dichloromethane in a chlorinated solvents waste container.

Take Care! Do not distill to dryness.

water during the distillation to maintain the water level in the boiling flask, until all the water has been added. Continue distilling until about 75 mL of distillate has been collected; there should be about 25 mL of liquid left in the boiling flask at this time. (**Take Care!** Do not distill to dryness.)

Shake the distillate with 15 g of potassium carbonate sesquihydrate (or dihydrate), followed by enough solid sodium chloride to saturate the solution. (A separate organic layer may form after the salts are added. If so, decant both layers into the separatory funnel.) When most of the sodium chloride has dissolved, carefully decant the liquid into a separatory funnel, leaving behind any droplets of metallic mercury and undissolved sodium chloride. (**Waste Disposal:** Place this residue in a mercury wastes container.) Extract the liquid mixture in the separatory funnel [OP-13] with two 25-mL portions of dichloromethane and wash the combined extracts [OP-19] with 25 mL of 1 M sulfuric acid, followed by 25 mL of saturated aqueous sodium chloride. Dry the organic layer [OP-20] with about 4 g of *anhydrous* potassium carbonate and evaporate the solvent [OP-14] from the filtered solution using a cold trap.

Purification and Analysis. Purify your crude product by simple distillation using a small-scale apparatus [OP-25], collecting the 3-hydroxy-3-methyl-2-butanone that distills around 138–141°C. Weigh the purified product, then record its infrared spectrum [OP-34] or obtain the spectrum from your instructor. Turn in the product to your instructor.

Report. Your report should include a statement of the problem and an account of how you applied scientific methodology to solve it. Calculate the percent yield of the synthesis. Interpret the infrared spectrum of the product as completely as you can, and explain clearly how it shows the functional group conversion that has taken place.

Exercises

1   Write a mechanism for the hydration of 2-methyl-3-butyn-2-ol that explains your results, assuming that the bridged ion **1** in the alkyne hydration mechanism (p. 185) is asymmetric, with a larger partial positive charge on the alkyl-substituted carbon.

2   Describe and explain the possible effect on your results of the following experimental errors or variations. (a) You add potassium carbonate to the reaction flask (after the reflux period) rather than to the initial distillate, forget to wash the dichloromethane solution with 1 M sulfuric acid, and distill the final product to dryness. (b) You add concentrated sulfuric acid instead of 3 M sulfuric acid to the reaction mixture. (c) You forget to add potassium carbonate and sodium chloride to the initial distillate.

3   (a) Identify each signal in the ^{1}H NMR spectrum of the Markovnikov addition product 3-hydroxy-3-methyl-2-butanone (Figure 24.2) by indicating the proton set responsible for it. (b) Sketch the ^{1}H NMR spectrum you would expect to obtain for the anti-Markovnikov addition product 3-hydroxy-3-methylbutanal, showing approximate signal areas, multiplicities, and chemical shifts.

4   Following the format in Appendix V, construct a flow diagram for this synthesis.

3-hydroxy-3-methyl-2-butanone

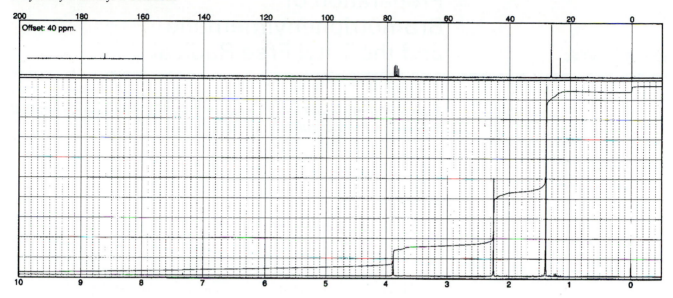

Figure 24.2 ^{13}C and ^{1}H NMR spectra of 3-hydroxy-3-methyl-2-butanone. (This and all other nuclear magnetic resonance spectra in this book are reproduced from *The Aldrich Library of ^{13}C and ^{1}H FT-NMR Spectra,* by C. J. Pouchert and J. Behnke, with the permission of the Aldrich Chemical Company.)
Note: All ^{1}*H NMR* spectra in this book were run in $CDCl_3$ unless otherwise indicated.

5 Draw structures for the products that you would expect to obtain from the hydration of acetylene (ethyne), 1-butyne, 2-butyne, and norethynodrel.

6 Using reactions discussed in this experiment, outline a synthetic pathway for the conversion of estrone methyl ether to the acetoxyketo steroid shown.

Other Things You Can Do

(Starred projects require your instructor's permission.)

*1 Test your reactant and product with bromine (test C-7 in Part IV) and potassium permanganate (test C-19). Explain the results.

*2 To verify the direction of addition, prepare a semicarbazone derivative of the product (Procedure D-4 in Part IV) and obtain its melting point. The semicarbazone of the Markovnikov product melts at 163°C; that of the anti-Markovnikov product melts at 222–223°C.

3 Oral contraceptives have been the subject of scientific and ethical controversy since Enovid was first marketed in 1960. Starting with sources listed in the Bibliography, write a research paper about recent developments in the field of contraception, discussing the impact of modern birth control methods on society.

estrone methyl ether

an acetoxyketo steroid

Preparation of Bromotriphenylmethane and the Trityl Free Radical

EXPERIMENT 25

Reactions of Hydrocarbons. Preparation of Alkyl Halides. Free Radical Substitution. Free Radicals.

Operations

> **OP-22** Drying and Trapping Gases
> OP-6   Heating
> OP-9   Mixing
> OP-10 Addition
> OP-11 Gravity Filtration
> OP-12 Vacuum Filtration
> OP-14 Evaporation
> OP-21 Drying Solids
> OP-23 Recrystallization
> OP-28 Melting Point

Before You Begin

1 Read the experiment and operation OP-22, read or review the other operations as necessary, and write an experimental plan.
2 Calculate the mass of 5.00 mmol of triphenylmethane and the theoretical yield of bromotriphenylmethane.

Scenario

J. Am. Chem. Soc. **1900,** *22,* 757.

Dr. Perry Celsus, the science historian who asked you to help develop a modern version of Wöhler's urea synthesis (Exp. 17), now wants you to re-create the experiment that led to the discovery of the first stable free radical by Moses Gomberg. In a landmark paper that was published in the 1900 *Journal of the American Chemical Society,* Gomberg described his preparation of the triphenylmethyl (trityl) free radical by the reaction of bromotriphenylmethane (trityl bromide) with various metals. Dr. Celsus has requested that you prepare some bromotriphenylmethane and find out whether it can, in fact, be converted by metallic zinc to the free radical that Gomberg described.

Applying Scientific Methodology

Your working hypothesis should deal with the likelihood that the free radical can be prepared from your product by the procedure described in part **B** of the Directions. You will test the hypothesis by observing and interpreting the behavior of the reaction mixture obtained by treating bromitriphenyl-methane with zinc.

The Case of the Disappearing Dimer

In the mid-nineteenth century, many chemists were convinced that carbon could exist in a trivalent state as a free *radical,* where a radical is a group of atoms such as methyl (CH_3) that generally exists only in combination with other atoms or groups, as in methyl bromide (CH_3Br). For example, it seemed reasonable to believe that if magnesium chloride could react with sodium metal to yield magnesium, then methyl halides should react with sodium to form methyl. But when Charles Wurtz added sodium to methyl iodide, he obtained not methyl but ethane in a reaction known today as the Wurtz reaction.

> Reaction of methyl iodide and sodium:
> $CH_3I + Na \rightarrow CH_3 + NaI$ (expected reaction)
> $2CH_3I + Na \rightarrow CH_3CH_3 + 2NaI$ (actual reaction)

After many similar attempts to make free radicals ended in failure, chemists began to doubt that they could exist at all. Then in 1900, Moses Gomberg, a young chemistry instructor at the University of Michigan, published a remarkable paper describing his discovery of the world's first stable free radical, triphenylmethyl.

Gomberg never meant to make a free radical. He was trying to synthesize hexaphenylethane to prove a point that, had he been successful, would be remembered today by only a handful of scientists. His initial attempts to prepare this compound using the Wurtz reaction were not successful, so he tried different metals such as silver and zinc.

> Gomberg's attempted synthesis of hexaphenylethane:
> $2Ph_3CBr + Zn \rightarrow Ph_3C—CPh_3 + ZnBr_2$ (expected reaction)

Each time he obtained a snow white solid that melted at 185°C and gave the wrong analysis for hexaphenylethane. After repeated attempts to prepare this compound, it finally occurred to him that the product was reacting with oxygen in the air and forming triphenylmethyl peroxide, $Ph_3COOCPh_3$. When Gomberg next ran the reaction, he was careful to exclude air from the reaction mixture, and he obtained a white solid that melted at 147°C and gave the correct analysis for hexaphenylethane. But this compound behaved very strangely for a hydrocarbon. It reacted in solution with air to form triphenylmethyl peroxide and it rapidly decolorized dilute halogen solutions—something no ordinary hydrocarbon would do. Gomberg eventually concluded that he had synthesized the world's first stable free radical, triphenylmethyl.

$$2Ph_3CBr + Zn \rightarrow 2Ph_3C \cdot + ZnBr_2$$

He had, in a way, reversed Wurtz's experience; Wurtz tried to make a radical and obtained its dimer, while Gomberg tried to make the dimer of a radical and ended up with the radical instead.

Gomberg believed the colored trityl radical was in equilibrium with hexaphenylethane in solution.

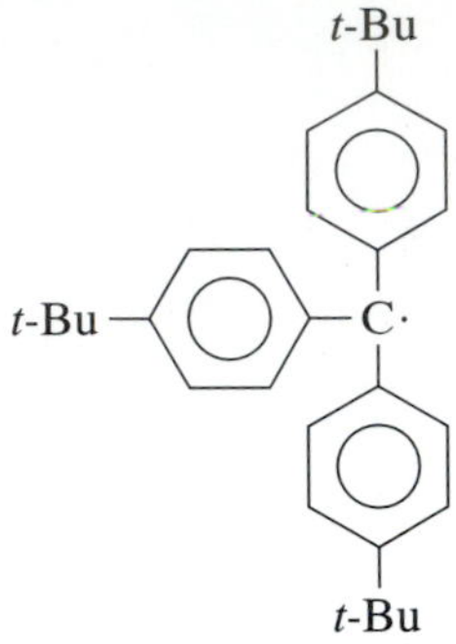

t-Bu

t-Bu—⟨⟩—C·

t-Bu

tris(4-*t*-butylphenyl)methyl

CH₃
|
t-Bu = CH₃C —
|
CH₃

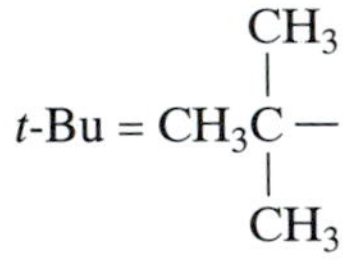

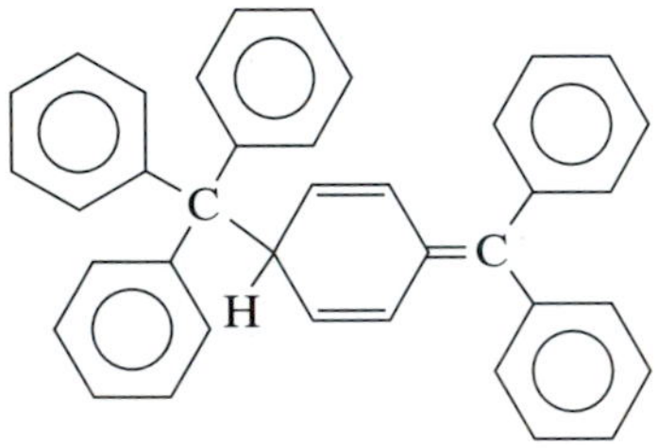

dimer of triphenylmethyl

Gomberg's proposed equilibrium

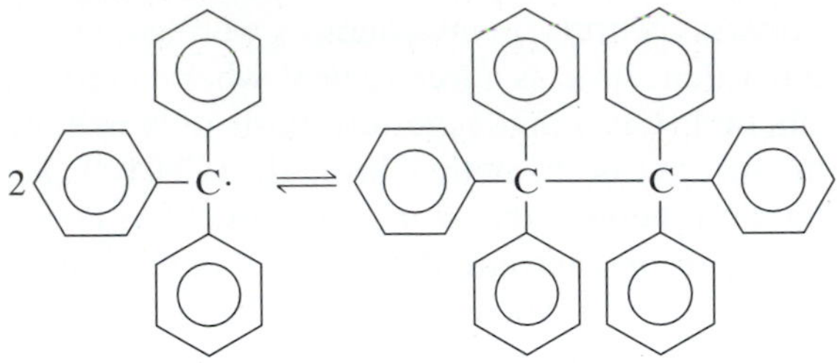

triphenylmethyl   hexaphenylethane (assumed)
(trityl) free
radical

But in 1968 a team of chemists from the Netherlands reported that they had prepared a similar free radical that would not dimerize. This new radical, *tris*(4-*t*-butylphenyl)methyl, had bulky *t*-butyl groups at each *para* position. Because the Dutch chemists could not explain why such substituents would prevent the formation of a hexaphenylethane-type dimer, they decided to prepare Gomberg's dimer and find out whether it had the structure he proposed. It didn't—NMR analysis showed that it has the structure shown in the margin. So Gomberg's dimer was not hexaphenylethane after all; that elusive hydrocarbon has probably never existed!

Understanding the Experiment

In this experiment you will attempt to brominate triphenylmethane with elemental bromine using light to initiate the reaction. The bromination reaction proceeds by a chain mechanism similar to that for the chlorination of methane and other hydrocarbons. A source of atomic bromine is needed to initiate the chain reaction. In a typical alkane bromination, bromine atoms are produced by irradiating molecular bromine in solution.

$$Br_2 \xrightarrow{h\nu} 2Br\cdot$$

In the chain propagating stage of the reaction, triphenylmethane should react with bromine atoms to produce trityl radicals. (Thus you will be making this radical twice during the experiment; once as an intermediate in the synthesis of bromotriphenylmethane, and later as a result of its reaction with zinc.) Each trityl radical should then react with molecular bromine, yielding a molecule of the product and another bromine atom that starts another cycle of chain propagating steps.

Chain propagating steps

$$Ph_3C-H + Br\cdot \longrightarrow Ph_3C\cdot + HBr$$
$$Ph_3C\cdot + Br_2 \longrightarrow Ph_3C-Br + Br\cdot$$

You will carry out the reaction by adding a solution of bromine in dichloromethane to a solution of triphenylmethane in the same solvent while heating the reactants under reflux. The reaction mixture will be irradiated using an ordinary unfrosted light bulb. The reaction evolves hydrogen bromide which, along with any escaping bromine vapors, can be captured using a sodium hydroxide gas trap. When the reaction is complete, some excess bromine may remain that will color the reaction mixture. It can be removed by adding a drop or two of cyclohexene.

Removal of excess bromine

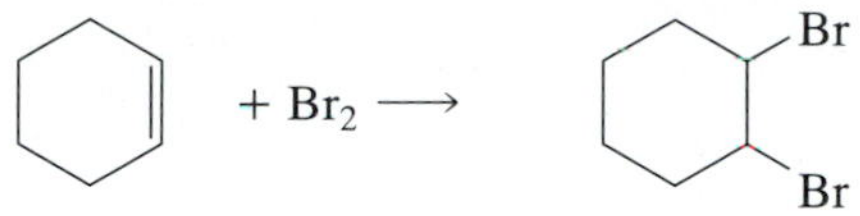

cyclohexene   1,2-dibromocyclohexane
                                    (a liquid)

Bromotriphenylmethane will be isolated by evaporating the solvent and purified by recrystallization from hexane or high-boiling petroleum ether. Since you are starting out with just over a gram of the reactant, it will be important to minimize material losses by making quantitative transfers and using the minimal amounts of recrystallization and washing solvents.

You will then attempt to prepare the triphenylmethyl radical by treating bromotriphenylmethane with metallic zinc. Exposing the solution to air should shift the radical-dimer equilibrium and thus provide evidence for the existence of the dimer.

Reactions and Properties

A.

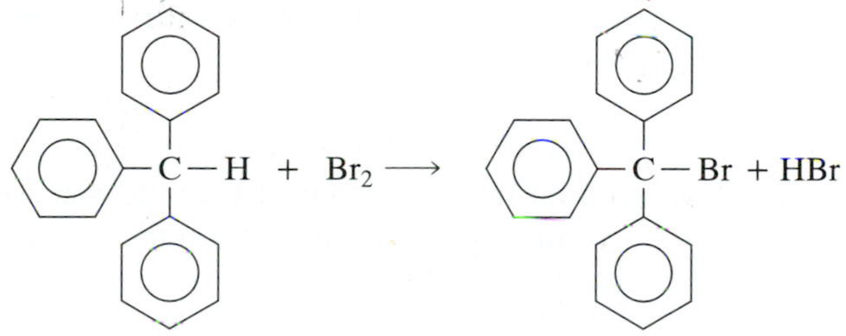

triphenylmethane   triphenylmethyl bromide

B. 2

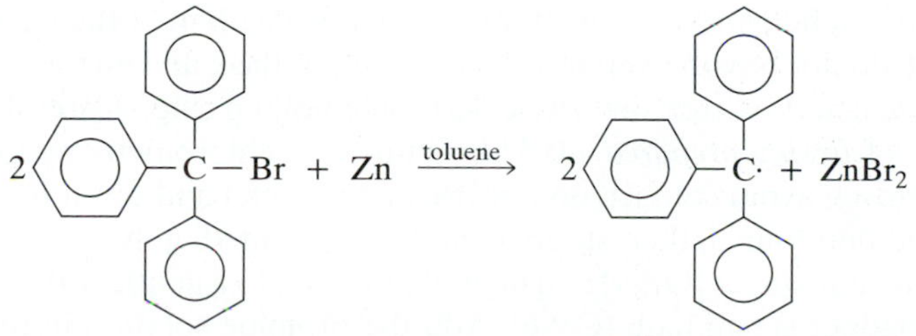

triphenylmethyl radical
(dimerizes in solution)

Table 25.1 Physical properties

	M.W.	m.p.	b.p.	*d*
triphenylmethane	244.3	94	359	
bromine	159.8	−7	59	3.12
dichloromethane	84.9	−97	40.5	1.326
bromotriphenylmethane	323.2	154		

Note: m.p. and b.p. are in °C, density is in g/mL

Directions

Safety Notes

> Bromine is toxic and corrosive, and its vapors are very harmful. Avoid contact with the bromine solution and do not breathe its vapors.
> Dichloromethane may be harmful if ingested, inhaled, or absorbed through the skin. There is a possibility that prolonged inhalation of dichloromethane may cause cancer. Avoid contact with the liquid and do not breathe its vapors.
> Bromotriphenylmethane is harmful if inhaled or absorbed through the skin. Avoid contact with the product.
> Petroleum ether is extremely flammable, so keep it away from flames and hot surfaces.

bromine   dichloromethane   petroleum ether

A. *Preparation of Bromotriphenylmethane*

Stop and Think: Why?

Reaction. All glassware that will be in contact with the reaction mixture or product must be dry. Assemble an apparatus for addition under reflux [OP-10] using a 50-mL round-bottomed flask and attaching a gas trap [OP-22] containing dilute NaOH to the top of the reflux condenser. (The gas trap can be omitted if the reaction is carried out under a hood.) Clamp an electrical socket containing an unfrosted 75- or 100-watt light bulb about 5 cm from the reaction flask (two or more students can use the same bulb). Dissolve 5.00 mmol of triphenymethane in 10 ml of dichloromethane (**Take Care!** Avoid contact, do not breathe vapors.) in the reaction flask and add a stirbar if you have a magnetic stirrer; otherwise add some boiling chips. Carefully measure 5.2 mL of *freshly prepared* 1.0 *M* bromine in dichloromethane (**Take Care!** Wear gloves, avoid contact, do not breathe vapors.) and put it in the separatory/addition funnel, then stopper the funnel immediately. Turn on the light and magnetic stirrer [OP-9] and heat the mixture to a gentle reflux using a hot water bath or steam bath [OP-6]. Add the bromine solution in five or more portions, waiting until the color of the solution has faded to light yellow-orange before adding the next portion. (If you don't have a stirrer, shake the

This solution can be prepared by carefully pipetting 0.25 mL of bromine (use of an autopipet is recommended) into 5.6 mL of dichloromethane. This should only be done under a hood with appropriate protective equipment (safety goggles, gloves, apron) and an instructor present.

apparatus gently after each bromine addition to mix the reactants.) Continue heating for 30 minutes after the last addition (or until the solution becomes light yellow) then let the reaction mixture cool to room temperature and dis-assemble the apparatus, washing the addition funnel with dilute sodium thio-sulfate under the hood. If the reaction mixture has a distinct orange color, add cyclohexene drop by drop until the color disappears or fades to light yellow.

Separation. Evaporate the dichloromethane [OP-14] under vacuum using a cold trap, with gentle heating if necessary. You can evaporate it directly from the reaction flask using the apparatus shown in Figure C10C of OP-14.

Waste Disposal: Place the recovered dichloromethane in a chlorinated solvent recovery container.

Purification and Analysis. Recrystallize the crude bromotriphenyl-methane [OP-23] from hexanes or high-boiling petroleum ether. To avoid transfers, you can recrystallize it in the reaction flask and filter the hot solution by gravity when most of the solid has dissolved. Collect the product with a Hirsch funnel [OP-12] and wash it on the filter with a little cold petroleum ether (preferably low-boiling). Dry the bromotriphenylmethane [OP-21] and measure its mass and melting point [OP-28]. Save enough product for Part **B** and turn in the rest to your instructor.

Take Care! Keep away from flames or hot surfaces.

B. *Preparation and Reactions of the Trityl Free Radical*
At your instructor's option, this part can be carried out in small groups using either commercial trityl bromide or the purest product from each group.

toluene

Safety Note

> Toluene is flammable, and inhalation, ingestion, or skin absorption may be harmful. Avoid contact with the liquid and do not breathe its vapors.

Dissolve about 0.2 g of purified bromotriphenylmethane in 4 mL of toluene in a small test tube. (**Take Care!** Avoid contact and inhalation.) Add 0.5 g of 30- or 40-mesh zinc, stopper the tube immediately, and shake vigorously for 10 minutes. Quickly filter the mixture [OP-11] through a thin layer of glass wool into another test tube, and stopper this test tube immediately. Shake the solution, and let it stand (tightly stoppered) for about 5 minutes. Bubble dry air through it until the color fades, stopper and shake it, and let it stand for a few more minutes. Repeat this process until the solution remains colorless for a minute or so after stoppering.

Observe and Note: During this procedure, take careful notes and describe your observations in detail.

Stop and Think: What evidence do you see for a dynamic equilibrium involving the radical and its dimer?

Waste Disposal: Place the solution in a designated waste container.

Report. Your report should include a statement of the problem and an account of how you applied scientific methodology to solve it. Calculate the percentage yield of bromotriphenylmethane. Explain what was happening in part **B**, beginning with the addition of zinc to the bromotriphenylmethane solution. Give names and structures of all species and equations for all reactions for which you saw evidence in part **B**.

Exercises

1 (a) Show how Gomberg's dimer, whose true structure was given previously, can form by the combination of two trityl radicals. (b) Explain why *tris*(4-*t*-butylphenyl)methyl does not form an analogous dimer.

2 Describe and explain the possible effect on your results of the following experimental errors or variations. (a) You forgot to switch on the light. (b) The reaction flask for part **A** was wet. (c) In part **B**, you bubbled nitrogen into the reaction mixture rather than air.

3 Following the format in Appendix V, construct a flow diagram for the synthesis of bromotriphenylmethane.

4 Assuming a free radical mechanism for the bromination of the following compounds, arrange them in order of their expected reactivity, starting with the most reactive.

a. 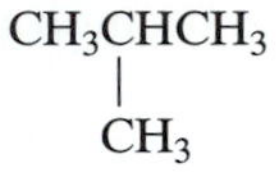 CH_3CHCH_3 with CH_3 substituent   **b.** 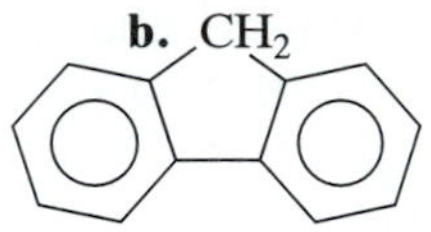 CH_2 (fluorene structure)   **c.** 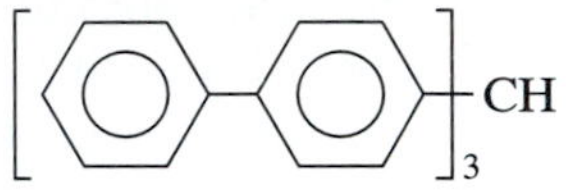 (biphenyl)$_3$CH structure

d. 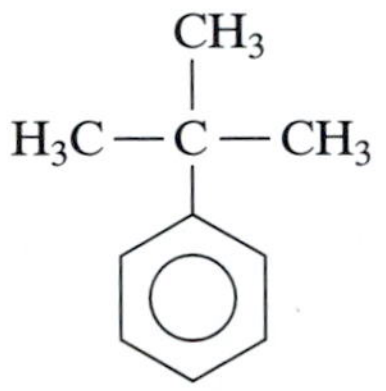 $H_3C{-}\underset{\underset{\text{(phenyl)}}{|}}{\overset{\overset{CH_3}{|}}{C}}{-}CH_3$ **e.** 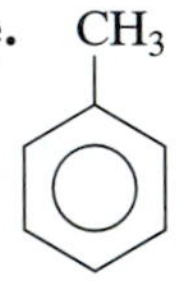 CH_3 (toluene)   **f.**  (phenyl)$_3$CH structure

5 Explain why Gomberg's dimer will remove the purple color of a solution of iodine in dichloromethane, and write equations for any relevant reactions.

6 Write equations for three chain terminating steps that can occur during the synthesis of bromotriphenylmethane.

Other Things You Can Do

(Starred projects require your instructor's permission.)

*1 Convert your bromotriphenylmethane to an ether in Minilab 17.

*2 Measure the bromination rates of some aromatic hydrocarbons in Minilab 21.

*3 Collect the precipitate that formed in part **B** by vacuum filtration and wash it with petroleum ether. Measure the melting point of the dry solid and give its structure.

4 Read the selections from Gomberg's work in *A Sourcebook in Chemistry, 1400–1900,* edited by H. M. Leicester and H. S. Klickstein (Boston: Harvard University Press, 1952). Describe the reasoning that led him to the conclusion that he had prepared the trityl free radical, and tell why he tried to prepare hexaphenylethane in the first place.

Chain-Growth Polymerization of Styrene and Methyl Methacrylate

Reactions of Alkenes. Preparation of Vinyl Polymers. Addition Polymerization. Free Radicals.

Operations:

OP-6   Heating
OP-9   Mixing
OP-12   Vacuum Filtration
OP-21   Drying Solids
OP-25   Simple Distillation
OP-34   Infrared Spectrometry

Before You Begin

Read the experiment, read or review the operations as necessary, and write an experimental plan.

Scenario

The Consulting Chemists Institute has two clients this week, both seeking information about polymers and polymerization reactions. Polly von Ilek, an artisan who specializes in nature crafts, wants to embed natural objects such as leaves, insects, and flowers in a clear, recycled plastic. She wants you to develop a procedure for depolymerizing granulated poly(methylmethacrylate) and repolymerizing the monomer to clear Lucite, under conditions that would allow objects to be embedded in the matrix as it hardens.

Mega Molecules, Inc., which manufactures styrofoam packing "peanuts" and other polymer-based products, recently had a serious fire in the plant where they manufacture polystyrene. Some benzoyl peroxide (a polymerization initiator) exploded, setting fire to a batch of styrene monomer in toluene. As a result of this accident they want to develop a safer alternative to their current solution polymerization process for making polystyrene. Your supervisor has worked out an emulsion polymerization procedure that uses water as the polymerization medium and potassium peroxydisulfate, a more stable initiator.

Your assignments are (1) to see whether granulated poly(methylmethacrylate) can be depolymerized and then repolymerized to a clear plastic in which objects can be embedded, and (2) to find out whether the emulsion polymerization procedure can be used to synthesize polystyrene.

Applying Scientific Methodology

In each part of this experiment the scientific problem is, in effect, to try something and see whether it works. If you can convert liquid styrene to a solid that has the properties of a plastic, that is evidence that a polymerization reaction has occurred, but you can confirm the identity of the product by obtaining its infrared spectrum.

Chain-Growth Polymers

The giant molecules we call polymers are immensely important to humankind—the meat, fruit, and vegetables we eat, the clothing we wear, and the wood we use for housing and furniture all consist partly or entirely of organic polymers. In fact, we are *made* of polymers—proteins in muscles, organs, blood cells, enzymes, and protoplasm; lipids in nerve sheaths, cell walls, and energy-storing fat tissues; and nucleic acids in the chromosomes that control our heredity. Compared to these natural polymers, synthetic polymers are newcomers on the scene. Polystyrene was first synthesized in 1839 (the same year Charles Goodyear learned how to vulcanize another polymer, natural rubber) but its properties were not appreciated at the time. The first commercially useful synthetic polymer did not appear until 1907 when Leo Baekeland synthesized Bakelite from phenol and formaldehyde. Bakelite is a good electrical insulator that is still used to make electrical plugs and switches. The synthesis of Nylon 66 and polyethylene in 1939 and the development of synthetic rubbers by German chemists during World War II, when Germany's supply of natural rubber was cut off, gave added impetus to the search for useful synthetic polymers. Today it is possible to design "tailor-made" polymers having almost any desired combination of properties by (1) using special catalysts (called Ziegler–Natta catalysts) to regulate the stereochemistry of a polymer, and (2) programming the sequence of monomers in copolymers, in which two or more different monomers are combined in various ways.

Synthetic polymers can be divided into the two broad categories of addition and condensation polymers. *Addition polymers* are built up by combining monomer units without eliminating any by-product molecules. The repeating unit of the polymer, therefore, has the same chemical constitution as the monomer, as illustrated for polyethylene. *Condensation polymers* are built up from monomer units containing two or more reactive functional groups that lose a small molecule such as water or HCl as they combine. Thus the repeating unit of a condensation polymer does not have the same formula as the monomer (or monomers), as illustrated for Nylon 6 synthesized from 6-aminohexanoic acid. But Nylon 6 can also be formed by an addition polymerization reaction from ω-caprolactam, so it is somewhat arbitrary to classify such polymers as either addition or condensation polymers.

The terms *chain-growth* and *step-growth* polymerization are used to describe two basic polymerization processes. The former refers to a process that starts when some monomer molecules are activated by a polymerization initiator; each polymer chain then grows very rapidly by adding more monomer molecules to its reactive end. The latter refers to a process by which each polymer chain grows from both ends via a series of individual reaction steps.

In a sense, polystyrene is a "natural" polymer since it is found in styrax from the sweet gum tree.

Formation of polyethylene by addition polymerization

$$n\text{CH}_2\!=\!\text{CH}_2 \longrightarrow \overset{\text{repeating unit}}{-\!(\text{CH}_2\text{CH}_2)_n\!-}$$

ethylene   polyethylene

(*n* is a large but indeterminate number.)

Formation of Nylon 6 by condensation polymerization

$$n\text{H}_2\text{N}(\text{CH}_2)_5\overset{\overset{\displaystyle O}{\|}}{\text{C}}\text{OH} \longrightarrow$$

6-aminohexanoic acid

$$-(\text{NH}(\text{CH}_2)_5\overset{\overset{\displaystyle O}{\|}}{\text{C}})_n\!- + \text{H}_2\text{O}$$

Nylon 6

Formation of Nylon 6 by addition polymerization

$$n \qquad \longrightarrow -(\text{NH}(\text{CH}_2)_5\overset{\overset{\displaystyle O}{\|}}{\text{C}})_n\!-$$

Nylon 6

ω-caprolactam

The polymerization of styrene with a free radical initiator is a typical example of chain-growth polymerization. The initiator (usually an organic peroxide) decomposes under the influence of heat or light to form free radicals, which add to styrene molecules according to Markovnikov's rule. Each activated styrene molecule then adds in similar fashion to another styrene molecule, leaving an unpaired electron at the end of the chain after each step. This process continues indefinitely until a reaction such as radical coupling or disproportionation occurs between chain ends (or with impurities) and deactivates the chain ends by forming stable products.

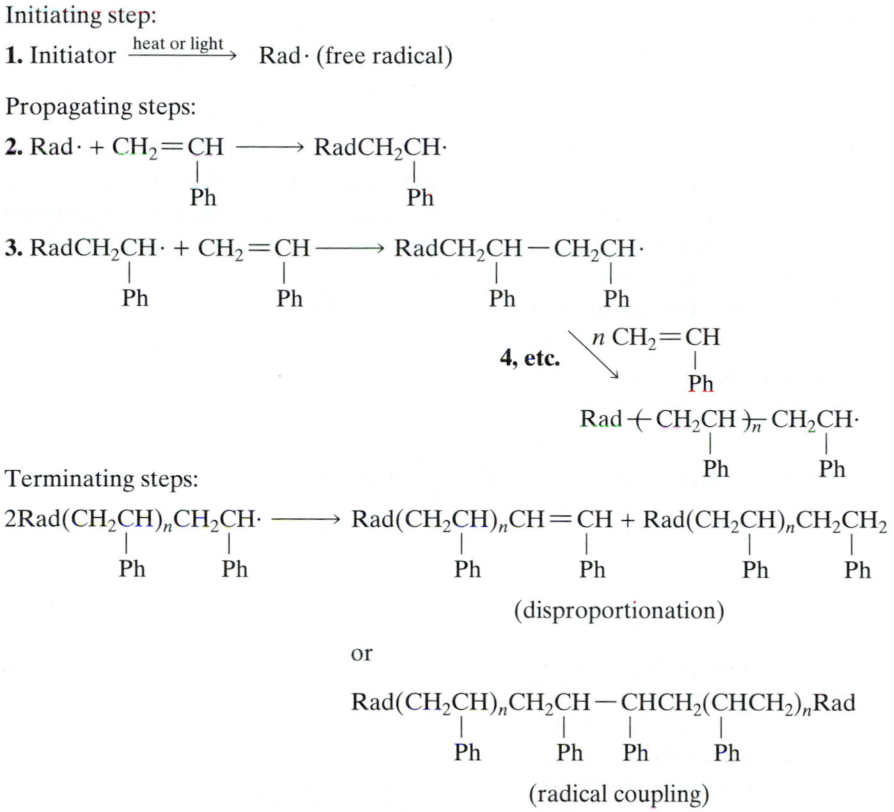

Initiating step:

1. Initiator $\xrightarrow{\text{heat or light}}$ Rad· (free radical)

Propagating steps:

2. Rad· + CH$_2$=CH $\longrightarrow$ RadCH$_2$CH·
　　　　　　｜　　　　　　　　　｜
　　　　　　Ph　　　　　　　　Ph

3. RadCH$_2$CH· + CH$_2$=CH $\longrightarrow$ RadCH$_2$CH—CH$_2$CH·
　　　　　｜　　　　　｜　　　　　　　　｜　　　　｜
　　　　　Ph　　　　Ph　　　　　　　Ph　　　Ph

4, etc. $\searrow$ n CH$_2$=CH
　　　　　　　　　　　　　　｜
　　　　　　　　　　　　　Ph

Rad$\left(\text{CH}_2\text{CH}\right)_nCH_2$CH·
　　　　　｜　　　　　　　｜
　　　　　Ph　　　　　　Ph

Terminating steps:

2Rad(CH$_2$CH)$_n$CH$_2$CH· $\longrightarrow$ Rad(CH$_2$CH)$_n$CH=CH + Rad(CH$_2$CH)$_n$CH$_2$CH$_2$
　　｜　　　　｜　　　　　　　　　｜　　　　｜　　　　　　｜　　　　｜
　　Ph　　　Ph　　　　　　　　Ph　　Ph　　　　　Ph　　　Ph

(disproportionation)

or

Rad(CH$_2$CH)$_n$CH$_2$CH—CHCH$_2$(CHCH$_2$)$_n$Rad
　　｜　　　　｜　　　｜　　　｜
　　Ph　　　Ph　　Ph　　Ph

(radical coupling)

Mechanism of chain-growth polymerization of polystyrene

Throughout a chain-growth polymerization, the bulk of the reaction mixture will consist of finished polymer molecules and unreacted monomers waiting to meet up with a reactive chain end. Because chain growth is so rapid once activation occurs, only one among many millions of molecules may be involved in the growth process at any given instant.

A number of different experimental methods are used for chain-growth polymerization, each with its advantages and disadvantages. *Bulk polymerization* is the simplest; it is carried out by adding a suitable initiator to the pure monomer and using heat or light to promote the reaction. The high heat of reaction makes bulk polymerization of vinyl monomers hard to control, and the method is seldom used commercially except for some polystyrene and poly(methyl methacrylate) products. Bulk polymerization can be used to preserve a botanical or zoological specimen (such as a desiccated flower or beetle) by suspending the specimen in a liquid monomer, which is then allowed to polymerize around it. Poly(methyl methacrylate), known by the trade names Lucite and Plexiglas, is often used for that purpose.

In *solution polymerization* the monomer is dissolved in an organic solvent, which dilutes it enough to reduce the problem of heat generation. Because it is often difficult to remove the solvent entirely, solution polymerization works best for polymers that are commonly used in solution, such as acrylic finishes. *Suspension polymerization* is carried out by mechanically dispersing the monomer in water or a similar solvent, so that the polymer is obtained in the form of granular beads that can be easily isolated. *Emulsion polymerization* is similar to suspension polymerization in that the monomer is dispersed in a solvent, usually water. In this method, however, the initiator is dissolved in the aqueous phase, and the monomer is emulsified (broken up into tiny droplets) by a detergent or some other surfactant (surface-active agent). The surfactant forms *micelles,* spherical aggregates of surfactant molecules, at the interface between monomer droplets and the aqueous solution. Polymerization starts in the micelles rather than in the monomer droplets, which provide a reservoir of monomer molecules that are continually fed into the growing polymer. The resulting dispersion resembles a rubber latex and can be used in that form, but it is more often coagulated and isolated as a finely divided powder.

Understanding the Experiment

In the first part of this experiment you will attempt to carry out the emulsion polymerization of styrene by dispersing the monomer in an aqueous detergent solution and heating the emulsion in the presence of a water soluble initiator, potassium peroxydisulfate. Oxygen inhibits this reaction by deactivating free radical intermediates, so you will bubble some nitrogen through the reaction mixture to remove it. The rubbery latex that forms is broken up by adding an alum solution to precipitate finely divided polystyrene.

Polystyrene is essentially an aromatic hydrocarbon with a very long alkyl side chain, so its infrared spectrum should resemble that of an ordinary arene. The spectrum of a thin polymer film will often display a number of extraneous small peaks called *interference fringes,* which are caused by interference between beams of infrared radiation reflected from the film's surfaces. As illustrated in Figure 26.1, the thickness of such a film can be estimated by counting the number of fringes in a given wave number interval, $\Delta\bar{\nu}$.

$$\text{Film thickness} = \frac{\text{number of fringes}}{2n(\Delta\bar{\nu})}$$
$$n = \text{refractive index} \ (\sim 1.60 \text{ for polystyrene})$$

In part **B** you will heat some poly(methyl methacrylate) in a distillation apparatus to break it down into the monomer, methyl methacrylate. This monomer will be free of the polymerization inhibitors (usually aromatic

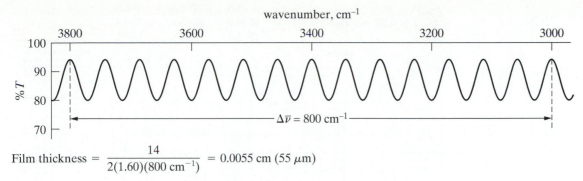

$$\text{Film thickness} = \frac{14}{2(1.60)(800 \text{ cm}^{-1})} = 0.0055 \text{ cm } (55 \text{ } \mu\text{m})$$

Figure 26.1 Determination of film thickness by counting interference fringes

phenols) that are added to commercial monomers to keep them from poly-merizing in storage. You will add a small amount of a polymerization initia-tor, *t*-butyl peroxybenzoate, to the monomer and (if you like) suspend a small object of your choice in it. During the next lab period you will find out whether or not it has polymerized with the object imbedded inside. It is dif-ficult (if not impossible) to get the polymer out without breaking its con-tainer; if you use a disposable glass container your instructor may allow you to break the glass and take the polymer with you.

Reactions and Properties

A. n styrene $\text{CH}=\text{CH}_2 \longrightarrow -(\text{CH}-\text{CH}_2)_n-$

polystyrene

B. $n\text{CH}_2=\underset{\underset{\text{CH}_3}{|}}{\text{CCOOCH}_3} \longrightarrow -(\text{CH}_2-\underset{\underset{\text{CH}_3}{|}}{\overset{\overset{\text{COOCH}_3}{|}}{\text{C}}})_n-$

methyl methacrylate   poly(methyl methacrylate)

Table 26.1 Physical properties

	M.W.	m.p.	b.p.	d
styrene	104.2	−31	146	0.909
potassium peroxydisulfate	270.3	100d		
methyl methacrylate	100.1	−48	100	0.944

Note: b.p. and m.p. are in °C, density is in g/mL; d = decomposes on melting

Styrene

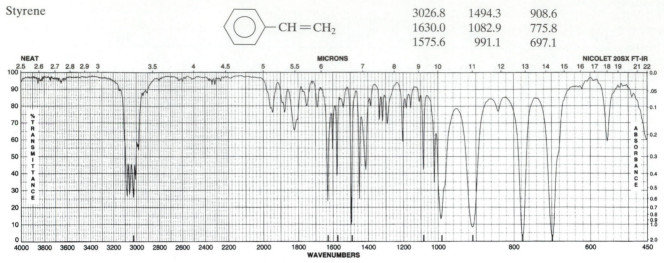

3026.8	1494.3	908.6
1630.0	1082.9	775.8
1575.6	991.1	697.1

Figure 26.2 IR spectrum of styrene (Reproduced from *The Aldrich Library of FT-IR Spectra, Edition II*, with the permission of the Aldrich Chemical Company.)

Directions

A. *Emulsion Polymerization of Styrene*
If this reaction cannot be carried out under a hood, it should be done under reflux, using a boiling flask in place of the 125-mL Erlenmeyer flask.

Safety Notes

> **Styrene has irritating vapors and inhalation, ingestion, or skin absorption may be harmful. Avoid contact with the liquid, do not breathe its vapors, and keep it away from flames.**
> **Potassium peroxydisulfate can react violently with oxidizable materials and alkalis; keep it away from other chemicals.**
> **Tetrahydrofuran is extremely flammable and can be harmful if inhaled. Do not breathe its vapors and keep it away from flames and hot surfaces.**

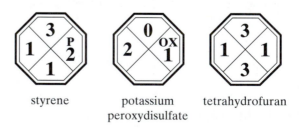

       styrene   potassium   tetrahydrofuran
                      peroxydisulfate

Take Care! Avoid contact with styrene, do not breathe its vapors.

Stop and Think: What is the purpose of the potassium peroxydisulfate? Of the sodium lauryl sulfate?

Reaction. Under the hood, combine 5.0 g of styrene with 50 mL of water in a 125-mL Erlenmeyer flask. Gently bubble nitrogen through the solution for about 5 minutes. Add 0.10 g of potassium peroxydisulfate, 0.35 g of sodium lauryl sulfate, and a magnetic stirbar. Stretch a square of Parafilm over the mouth of the flask, or insert a cork loosely. Stir the reaction mixture

[OP-9] at a rate sufficient to maintain a stable emulsion while heating it in a 40–50°C water bath for two hours [OP-6]. You may work on part **B** during the reaction period.

Observe and Note: What evidence do you see that a reaction is occurring?

Separation. Precipitate the product by adding 10 mL of a saturated alum solution and boiling for a few minutes. Collect the product by vacuum filtration [OP-12], transfer it to a small beaker, and stir it with small portions of methanol (decant the methanol after each washing) until it is no longer sticky. Again collect the solid by vacuum filtration. Dry the product in a 110°C oven [OP-21] or leave it to dry in a desiccator, and weigh it.

Waste Disposal: Unless your instructor indicates otherwise, pour the filtrates down the drain.

Analysis. [At your instructor's option, you can record the IR spectrum of a prepared polystyrene film rather than your product.] Dissolve about 0.05 g of the product in 10 mL of tetrahydrofuran. This solution can be shared by a number of students. Use a disposable pipet to transfer about 10 drops of the solution to a clean microscope slide and tilt the slide to coat it evenly. Let the solvent evaporate completely under the hood (a heat lamp will speed up the process), then use a razor blade to carefully strip off the polymer film and mount it in an IR cell holder. Record the infrared spectrum [OP-34] of the film. Turn in the remaining product to your instructor.

Take Care! Do not breathe vapors, keep away from flames.

Take Care! Don't cut yourself.

B. *Bulk Polymerization of Methyl Methacrylate*
Your instructor may opt to omit the depolymerization procedure, in which case you can start at the polymerization step.

> Methyl methacrylate is flammable and lachrymatory (tear producing) and can irritate the skin and eyes. Avoid contact, do not breathe the vapors, and keep it away from flames. Use protective gloves and a fume hood.
> *t*-Butyl peroxybenzoate may react violently with strong oxidizing or reducing agents, or explode when strongly heated. Keep it away from combustible materials, heat, and flames.

Safety Notes

methyl   *t*-butyl
methacrylate   peroxybenzoate

Depolymerization of Poly(methyl methacrylate). Position your heating mantle so that it can be lowered quickly if necessary, then assemble a simple distillation apparatus [OP-25] using a 25-mL boiling flask and a tared shell vial (or other disposable glass container) as the receiver. Place 10 g of granulated low-molecular weight poly(methyl methacrylate) and a boiling chip or two in the reaction flask and heat it at a medium-high power setting [OP-6] until the solid softens and starts to liquefy. Then adjust the heating rate so that the monomer distills slowly. Stop the distillation when the residue in the flask begins to discolor and become viscous. Clean the reaction apparatus with acetone as soon as it has cooled. If any residue solidifies in the reaction flask, cover it with ethyl acetate and let it stand until the residue dissolves (this sometimes takes several hours or days).

Waste Disposal: Place the residue and the acetone wash liquid in a designated waste container.

Polymerization of Methyl Methacrylate. Add 5 drops of *t*-butyl peroxybenzoate (**Take Care!** Keep it away from combustibles and heat sources.) to the distillate in the shell vial and shake gently to dissolve it and mix the solution. Heat the vial in a boiling water bath for 20 minutes or more. If you like, suspend a small object in the liquid by imbedding a hook (made from a piece of copper wire or part of a paper clip) in a cork of the appropriate size and hanging the object from a thread. Metal objects should first be coated with clear enamel. Stopper the vial loosely with the cork (or seal it with Parafilm) and set it in direct sunlight until the next laboratory period or until the clear polymer has hardened. Measure the mass of the polymer. With your instructor's permission, wrap the glass container in a towel and carefully break it inside a cardboard box using a pestle or other hard object, then remove the polymer and place the glass in a sharps collector.

Report. Your report should include statements of the problems and an account of how you applied scientific methodology to solve them. Calculate the percentage yield of polystyrene. Interpret the infrared spectrum as completely as you can, and calculate the thickness of the film.

Exercises

1 Propose a mechanism for the polymerization of methyl methacrylate in the presence of *t*-butyl peroxybenzoate, which decomposes as shown here.

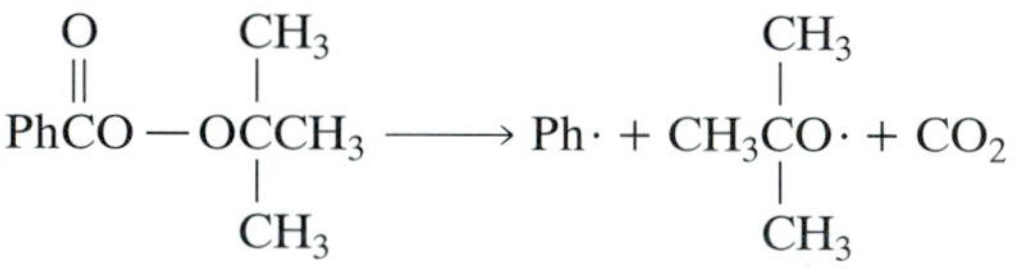

$$
\underset{\substack{|\\ CH_3}}{\overset{\substack{O\quad\quad CH_3\\ \parallel\quad\quad |}}{PhCO-OCCH_3}} \longrightarrow Ph\cdot + \underset{\substack{|\\ CH_3}}{\overset{CH_3}{CH_3CO\cdot}} + CO_2
$$

2 Compare your polystyrene spectrum with the infrared spectrum of styrene in Figure 26.2, and account for any significant similarities and differences.

3 Describe and explain the possible effect on your results of the following experimental errors or variations. (a) You forgot to add the sodium lauryl sulfate in part **A**. (b) In part **A**, you bubbled air instead of nitrogen into the reaction mixture. (c) For part **B**, you used commercial methyl methacrylate containing 50 ppm of 4-methoxyphenol.

4 Draw the structures of the monomers needed to prepare polymers having the following repeating units.

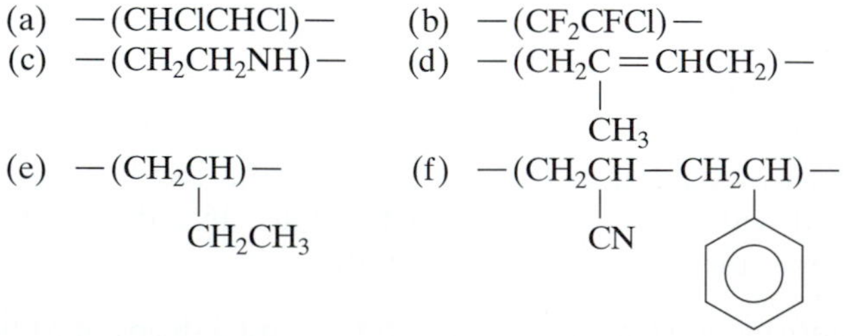

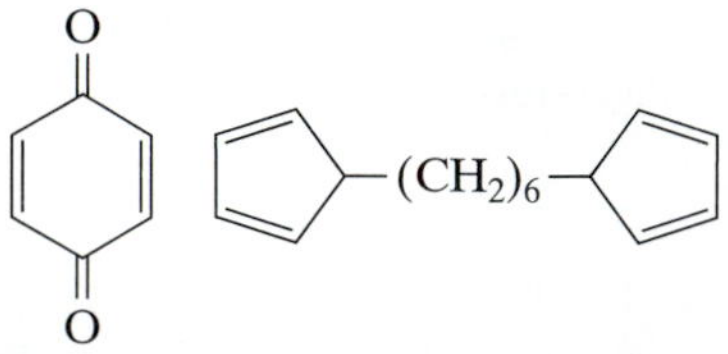

(a) $-(CHClCHCl)-$ (b) $-(CF_2CFCl)-$
(c) $-(CH_2CH_2NH)-$ (d) $-(CH_2C=CHCH_2)-$
$\qquad\qquad\qquad\qquad\qquad\qquad\qquad\quad |$
$\qquad\qquad\qquad\qquad\qquad\qquad\qquad\ CH_3$

(e) $-(CH_2CH)-$ (f) $-(CH_2CH-CH_2CH)-$
$\qquad\quad |$ $\qquad\qquad\qquad\qquad |$
$\qquad\ CH_2CH_3$ $\qquad\qquad\quad CN$

5 Show how the two monomers illustrated could combine to form a Diels–Alder addition polymer, and give the structure of the polymeric repeating unit.

6 Poly(ethylene glycol) can be prepared from either ethylene glycol or ethylene oxide. Write a balanced equation for each reaction, and classify each as addition or condensation polymerization.

7 Methyl methacrylate can be prepared commercially from acetone, hydrogen cyanide, and methanol. Propose a synthesis of methyl methacrylate from these starting materials, using any necessary inorganic reagents or solvents.

$-(CH_2CH_2O)-$
poly(ethylene glycol)

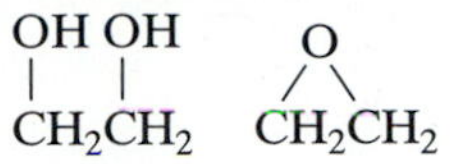

$\underset{\text{ethylene}}{\underset{\text{glycol}}{\overset{\text{OH OH}}{\underset{|\quad|}{CH_2CH_2}}}}$ $\underset{\text{ethylene}}{\underset{\text{oxide}}{\overset{O}{\overset{/\backslash}{CH_2CH_2}}}}$

Other Things You Can Do

(Starred projects require your instructor's permission.)

*1 Three students can work together on part **B**, one using the standard procedure, the second using no *t*-butyl peroxybenzoate, and the third using 0.1 g of hydroquinone in place of the *t*-butyl peroxybenzoate. The results should be compared and explained.

*2 Perform a "nylon rope trick" in Minilab 22.

3 Beginning with sources from the Bibliography, write a research paper on the use of Ziegler–Natta catalysts to synthesize stereoregular polymers and describe some properties and uses of these polymers.

Synthesis of Ethanol by Fermentation

Reactions of Carbohydrates. Preparation of Alcohols. Biosynthesis of Organic Compounds.

Operations:

> OP-12 Vacuum Filtration
> OP-27 Fractional Distillation

Before You Begin

1 Read the experiment, read or review the operations as necessary, and write an experimental plan.

2 Calculate the mass of 0.100 mol of sucrose and the theoretical mass and volume of ethanol that can be produced from that much sucrose.

Scenario

The Great Plains Farm Cooperative helps farmers market their crops, to help ensure that they receive a fair price. Unfortunately, a bumper crop of sugar beets has resulted in a surplus of beet sugar and lowered its price drastically. The Co-op has advised farmers to store their beets until the price rises, and in the meantime it is exploring alternative uses for the surplus beets. One alternative would be to convert the beet sugar to ethanol for use as an energy source in gasahol and other commercial fuels.

The Co-op has sent you a quantity of beet sugar to see whether an ethanol conversion process would be feasible. To price the beet sugar ethanol low enough to make it competitive with ethanol from corn and other sources, they require a product that is at least 180 proof. Your assignment is to ferment the sugar, distill it to concentrate the ethanol, and find out whether your product meets their specifications.

Applying Scientific Methodology

Whether or not you solve the "problem" in this experiment—that is, prepare ethanol that is at least 180 proof—depends in part on your skill in carrying out the experiment.

The Chemistry of Brewing and Winemaking

The single-celled fungi called yeasts obtain the chemical energy they need to grow and reproduce by breaking down molecules of monosaccharides (simple sugars) such as glucose and fructose, and disaccharides (two-unit sugars) such as sucrose and maltose. To the yeast cells, ethyl alcohol is a use-less by-product of this energy generating process, but humans have developed a liking for that by-product as it exists in alcoholic beverages.

The fermentation of plant carbohydrates to produce alcohol is as old as civilization, and probably began in prehistoric times. The first alcoholic beverage may have been a kind of mead (honey wine) produced when some honey-sweetened water was forgotten and allowed to ferment. Early civilizations discovered that starchy grains such as barley, wheat, and millet could be treated to make them fermentable, leading to the brewing of beer and similar grain based beverages. A poem dedicated to the Sumerian goddess of brewing contains a recipe for beer that dates back to about 2800 B.C. Since grains contain starch but no sugars that yeast cells can ferment, they are first allowed to germinate for several days, forming enzymes that break down the stored starch into glucose. This process is known as *malting*. Ancient Egyptians made beer by baking malted grain into bread, then soaking the bread in water and allowing it to ferment. Modern brewers soak the dried, malted barley in warm water to extract its sugars and flavor components and then separate the liquid, called *wort,* from the solid residue. Additional cereals (rice, corn, or wheat) may be added to boost the wort's carbohydrate content, hops are included for their bitter, aromatic flavor, and the wort is boiled to sterilize and concentrate it. Brewing yeast is then added and the wort is allowed to ferment, during which the yeast cells transform the sugars into alcohol.

In 1989 an American brewer reproduced this honey-sweetened beer by following the ancient Sumerian recipe.

Grape vines grew wild in Central Asia and Western Europe long before humans appeared on Earth. By the sixth millenium B.C., grapes had been domesticated in Asia Minor, and the cultivation and use of grapes for food and wine spread from there to Egypt and Mesopotamia. Noah planted the first vineyard mentioned in the Bible; he was also the first biblical character reported to have become intoxicated by drinking wine. Legend has it that Dionysus, the Greek god of wine, introduced wine drinking in Crete, and by Homer's time (ca. 700 B.C.) the ancient Greeks were making strong, sweet, thick wine that was always watered down before drinking. During Roman times the Gauls, like the French who succeeded them, were the best vine growers in Europe. Their wine was popular in Rome until the emperor Domitian, to protect Italian winemakers from Gallic competition, ordered that half the grape vines in Gaul be uprooted. Wine may have been partially responsible for the decline and fall of Rome, because a syrup used to preserve wines was boiled in lead-lined pots, and one of the symptoms of lead poisoning is mental confusion. After Rome fell to the barbarians, the art of winemaking was kept alive in European monasteries, which cultivated grapes to supply their sacramental wine. A Benedictine monk, Dom Pérignon, is usually credited with the invention of champagne, and many of the best wines of France are still grown around former monasteries.

Even today, winemaking may be as much an art as a science. The first step is crushing ripe grapes into juice, called the *must,* which contains glucose and fructose along with smaller quantities of fruit acids (such as tartaric acid), tannins, and other components. To make white wine, the skins are removed before fermentation, so the wine contains little pigmentation or tannic material. Red wine musts are fermented in contact with red grape skins for most of the fermentation period. The longer the must is in contact with the skins, the deeper is its color and the more astringent or "tannic" its taste. Sulfur dioxide is added to prevent oxidation of flavor components and inhibit the growth of bacteria and wild yeasts that would convert the must to vinegar. To balance their sweet and sour components, sugar is added to overly acidic musts and tartaric acid to musts having insufficient acid.

Some grape skins are covered with a dusting of yeasts of the type needed for fermentation (saccharomycetes); with other grapes a yeast culture must be added. The yeast not only converts grape sugars to alcohol but also biosynthesizes flavor components—mainly esters and long-chain alcohols— that don't exist in the grapes themselves. A long, slow fermentation process at low temperature produces the most flavor components, giving the wine a more pronounced "nose" (aroma). Fermentation stops when nearly all of the sugar has been transformed. The yeast is then allowed to settle and the new wine is drawn off; this *racking* process is repeated several times as the wine ages. The wine may also be *fined,* treated with a substance that coagulates suspended fine particles and carries them to the bottom. New wine has a harsh, raw flavor and a simple aroma, so it is aged in wooden casks to soften the flavor and make the aroma more complex. A large number of chemical reactions take place during this stage, forming more flavor components and removing some components with disagreeable characteristics. Many of these are oxidation reactions, made possible by pores in the wooden casks that allow air to enter the wine. The cask itself may contribute flavor, as vanillin and other components of the wood leach into the wine. If oxidation proceeds too long it begins to degrade the wine, so after six months to two years of aging the wine is bottled to protect it from excessive oxidation. The chemical composition of a bottled wine continues to change, and a good wine may improve in flavor when aged in the bottle for several years, or even several decades.

Distilled beverages such as whiskey, brandy, and vodka are made by distilling beer, wine, or another fermented beverage to increase the alcohol content and concentrate the volatile flavor components, then aging the distillate to make the flavor richer and mellower. Thus brandy is distilled from wine, scotch from a special distiller's beer, rum from fermented molasses or sugar cane juice, and vodka from fermented potatoes or grain.

Drinking alcoholic beverages of any kind can be both beneficial and harmful. Wine may enhance our enjoyment of a good meal, and a drink or two can help us relax and stimulate social interactions. In moderation, wine and other alcoholic beverages confer certain health benefits. Beer, wine, and liquor all raise the concentration of "good cholesterol" (HDL) in the blood, thereby reducing the risk of heart attacks. Chemicals in wine may also boost the immune system, inhibit cancer, and cut the risk of Alzheimer's disease.

Balanced against these beneficial effects of moderate drinking are the unquestionably harmful effects of excessive drinking, which include drunkenness, hangovers, alcoholism, cirrhosis of the liver, birth abnormalities, antisocial behavior, broken homes, wasted lives, and death. Alcohol is a drug, often a highly addictive one, that acts as a central nervous system depressant. Its apparent stimulating influence, which is associated with a loss of inhibitions, is actually a result of alcohol's depressant effect on the brain centers that normally keep our social behavior in check. Because the destructive potential of beverage alcohol is so much greater than its apparent health benefit, most medical professionals agree that nondrinkers should not begin drinking alcohol for their health, and drinkers should limit their alcohol intake to one or two drinks a day.

Understanding the Experiment

In this experiment you will ferment an aqueous solution of beet sugar (sucrose) with baker's yeast (*Saccharomyces cerevisiae*) to obtain a dilute ethanol solution, which you will concentrate by fractional distillation. The fermentation mixture will contain Pasteur's salts, a mixture of nutrients that are necessary for the growth and reproduction of yeast. The fermentation takes time, especially at low temperatures, so you should be prepared to get it started a week or two in advance.

Sucrose is a disaccharide, its molecules containing units of the two simple sugars glucose and fructose. The yeast you will use contains many *enzymes* (biochemical catalysts) that are necessary for the conversion of sucrose to ethanol. Invertase catalyzes the hydrolysis of sucrose to glucose and fructose. Then zymase, which is actually a group of at least 22 separate enzymes, catalyzes a series of steps in the very complex conversion of these simple sugars to ethanol and carbon dioxide.

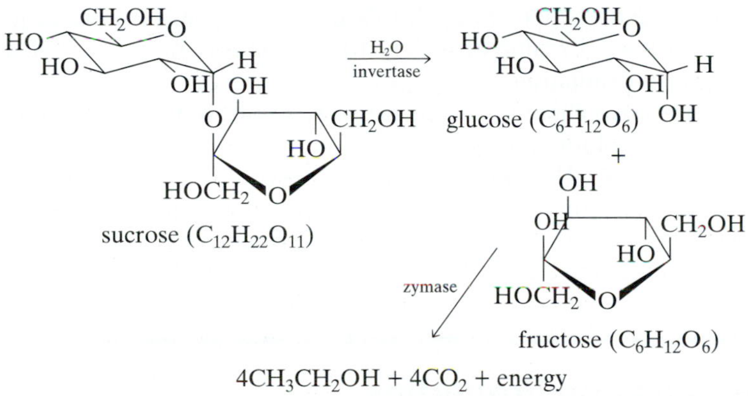

$$4CH_3CH_2OH + 4CO_2 + energy$$

From a yeast cell's point of view, the important product of this process is energy. Ethanol is a waste product that is actually harmful to the yeast cells, which are deactivated after the alcohol concentration reaches 15 percent (v/v) or so. Fermentation also produces a number of other by-products by different pathways, such as glycerol, which improves a wine's "legs" (its ability to coat the inside of a wine glass) by increasing its surface tension. Another fermentation by-product is a mixture of alcohols called *fusel oils,* which are responsible for the headaches that cheap wines often cause. Fusel oils include 1-propanol, 2-methyl-1-propanol, 2-methyl-2-butanol, and 3-methyl-1-butanol. Ethanal (acetaldehyde) is also produced along the biosynthetic pathway that leads to ethanol, and some of it remains at the end of the fermentation.

Some by-products of fermentation

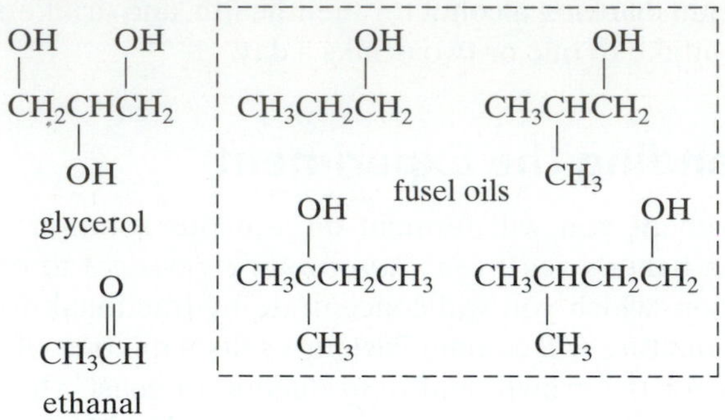

$$\underset{\text{glycerol}}{\overset{\displaystyle \overset{\text{OH}}{|} \quad \overset{\text{OH}}{|}}{\underset{\overset{|}{\text{OH}}}{\text{CH}_2\text{CHCH}_2}}}$$

fusel oils

$$\overset{\overset{\text{OH}}{|}}{\text{CH}_3\text{CH}_2\text{CH}_2} \qquad \underset{\overset{|}{\text{CH}_3}}{\overset{\overset{\text{OH}}{|}}{\text{CH}_3\text{CHCH}_2}}$$

$$\underset{\overset{|}{\text{CH}_3}}{\overset{\overset{\text{OH}}{|}}{\text{CH}_3\text{CCH}_2\text{CH}_3}} \qquad \underset{\overset{|}{\text{CH}_3}}{\overset{\overset{\text{OH}}{|}}{\text{CH}_3\text{CHCH}_2\text{CH}_2}}$$

$$\underset{\text{ethanal}}{\overset{\displaystyle \overset{\text{O}}{\underset{\displaystyle \|}{}}}{\text{CH}_3\text{CH}}}$$

To prepare alcohol for nonbeverage purposes, all of these by-products must be removed, along with excess water and spent yeast cells. The yeast cells, which collect as a sediment at the bottom of the fermentation flask, can be removed by decanting and filtration. The remaining components, including most of the water, are removed by fractional distillation. Ethanal boils at 21°C, ethanol at 78°C, water at 100°C, the fusel oil alcohols at temperatures ranging from 97°C to 132°C and higher, and glycerol at 290°C. Thus any ethanal present should distill in the forerun, while most of the water and the other components should remain in the boiling flask. Water and ethanol form a constant-boiling mixture (called an *azeotrope*) with a composition of 95.6% ethanol/4.4% water (by mass), so it is impossible to obtain ethanol purer than 95.6 mass percent by distillation alone. Your percentage may be substantially lower because of such factors as the column packing efficiency and the rate of distillation.

The *proof* of a commercial alcohol or alcoholic beverage is determined by doubling its volume percentage; that is, 1 degree of proof is equivalent to 0.5 volume percent. Thus an 80 proof bourbon contains 40% ethanol by volume. You can determine the volume and mass percentage of ethanol in your product, and from that its proof and your percent yield, by measuring its density and using the data in Table 27.2.

Reactions and Properties

$$\underset{\text{sucrose}}{C_{12}H_{22}O_{11}} + H_2O \longrightarrow 4\underset{\text{ethanol}}{CH_3CH_2OH} + 4CO_2$$

Table 27.1 Physical properties

	M.W.	m.p.	b.p.
sucrose	342.3	186d	
ethanol	46.1	−115	78.3

Note: m.p. and b.p. are in °C; d = decomposes at melting point

Directions

Fermentation. In a 500-mL Erlenmeyer flask, dissolve 0.100 mol of sucrose in 250 mL of water that has been warmed to about 30°C. Then add 30 mL of Pasteur's salts and 3.5 g (about one-half packet) of dry baker's

Table 27.2 Density and composition of ethanol-water solutions at 20°C

Density at 20°C	Mass % ethanol	Vol. % ethanol
0.914	50.0	57.8
0.903	55.0	62.8
0.891	60.0	67.7
0.879	65.0	72.4
0.868	70.0	76.9
0.856	75.0	81.3
0.844	80.0	85.5
0.831	85.0	89.5
0.818	90.0	93.3
0.804	95.0	96.8
0.789	100.0	100.0

yeast. Mix the contents by shaking the flask vigorously. Insert a one-hole rubber stopper fitted with a bent glass tube, connect a length of rubber tube and a 6-inch length of straight glass tube, and insert the straight tube in a bottle or beaker containing limewater (saturated aqueous calcium hydroxide) with a thin layer of mineral oil on top (see Figure 27.1). Label the flask with your name and let the mixture stand at room temperature (preferably 25°C or higher) for a week or more until no more gas is evolved.

Pasteur's salts contains 1.0 g potassium dihyrogen phosphate, 0.10 g calcium dihyrogen phosphate, 0.10 g magnesium sulfate, and 5.0 g ammonium tartrate in about 500 mL of water.

Stop and Think: What is the purpose of the limewater? The mineral oil?

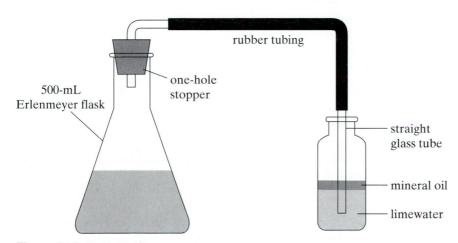

Figure 27.1 Fermentation apparatus.

Separation. At the end of the fermentation period, remove the stopper from the Erlenmeyer flask. Trying not to disturb the sediment, which contains tiny yeast cells, decant (pour off) the fermentation mixture into a large beaker. (You can also siphon it out with a length of plastic tubing.) Stop decanting, leaving some liquid behind, when the sediment begins to rise and is about to pass into the beaker. If the liquid in the beaker is not clear, filter it by the following procedure: Assemble a vacuum filtration apparatus [OP-12], using the largest available filter flask and including a trap if you are using a water aspirator, and turn on the vacuum. Measure about 10 g of Celite filtering aid into a large Erlenmeyer flask and shake the Celite with 200 mL of water until it is well dispersed. Swirling to keep the Celite suspended, pour this mixture onto the filter paper to form a thin layer of Celite on top of it. Discard the water in the filter flask, then filter the decanted liquid. You may also be able to filter some of the liquid that remains with the sediment and combine it with the decanted liquid, but stop filtering if the filtrate becomes cloudy or when the filtration rate becomes very slow.

Waste Disposal: Flush the sediment and supernatant liquid down the drain with water.

Purification. Assemble an apparatus for fractional distillation using a 500-mL round-bottomed flask and an appropriate column packing [OP-27]. Distill the ethanol/water mixture and collect all distillate that boils between 78°C and 82°C. Stop the distillation when the temperature rises above 82°C.

Waste Disposal: Pour the contents of the boiling flask down the drain.

Analysis. Measure the mass of the distillate. With the temperature of this liquid as close to 20°C as possible, use an automatic pipet, volumetric pipet, or other appropriate measuring device to accurately measure 1.00 mL of distillate into a tared screwcap vial. Then weigh it and determine its density.

Report. Your report should include a statement of the problem and an account of how you applied scientific methodology to solve it. If your ethanol was *not* 180 proof or higher, describe what you could have done differently or what else you could have done to increase its purity. Prepare a graph of mass percent and volume percent ethanol (on the y axis) vs. density. Use the graph to determine the percent composition and proof of your distillate, then calculate the mass of ethanol it contains and your percent yield.

Exercises

1 A strong vodka has a density of 0.909 g/mL. What is its proof? (Vodka is essentially a solution of ethanol and water.)

2 Describe and explain the possible effect on your results of the following experimental errors or variations. (a) You forgot to add the Pasteur's salts. (b) During the fractional distillation you continued to collect distillate until the temperature reached 98°C. (c) Someone had reset the automatic pipet you used for the density determination to 0.950 mL.

3 Following the format in Appendix V, construct a flow diagram for the biosynthesis of ethanol.

4 Calculate the volume of carbon dioxide at 25°C and 1 atm that would be produced by complete fermentation of 0.100 mol of sucrose.

5 (a) Write a balanced equation for the reaction of carbon dioxide with $Ca(OH)_2$ (b) Calculate the theoretical mass of calcium carbonate that could be produced by the CO_2 generated from 0.100 mol of sucrose.

6 One of the by-products of fermentation is diethylacetal, $CH_3CH(OCH_2CH_3)_2$. Write a balanced equation for its formation from components of the fermentation mixture.

Other Things You Can Do

(Starred projects require your instructor's permission.)

*1 Estimate the ethanol concentration of a commercial vodka by the method used in this experiment.

2 Find out how absolute (100%) ethanol is manufactured, and how ethanol is denatured to make it undrinkable.

3 Write a research paper about the chemistry and technology of wine-making, starting with sources listed in the Bibliography.

EXPERIMENT 28

The Reaction of Butanols with Hydrobromic Acid

Reactions of Alcohols. Preparation of Alkyl Halides. Nucleophilic Aliphatic Substitution. Infrared Spectrometry.

Operations:

OP-6   Heating
OP-15 Steam Distillation
OP-19 Washing Liquids
OP-20 Drying Liquids
OP-22 Drying and Trapping Gases
OP-25 Simple Distillation
OP-34 Infrared Spectrometry

Before You Begin

1 Read the experiment, read or review the operations as necessary, and write an experimental plan.
2 Calculate the mass and volume of 72.0 mmol of 1-butanol and the theoretical yield of 1-bromobutane (both mass and volume). Repeat these calculations for 2-butanol and 2-bromobutane.

Scenario

The Bond Triplex (see Experiment 23) manufactures not only alkynes but a variety of chemicals that can be prepared from ethyne (acetylene) and other alkynes. For example, they manufacture 1-butanol from ethyne by the following process.

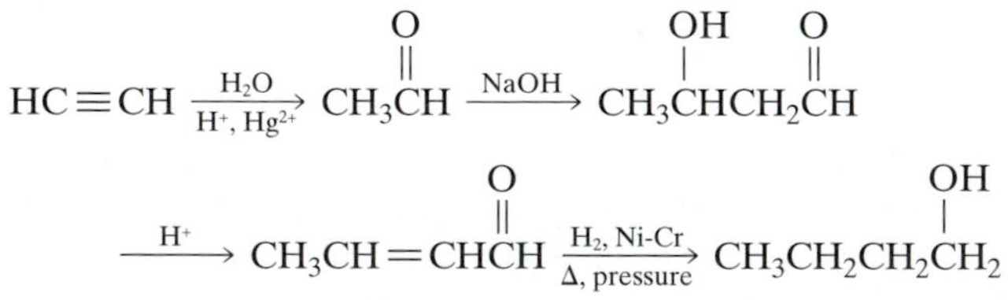

$$HC\equiv CH \xrightarrow[H^+, Hg^{2+}]{H_2O} CH_3\overset{O}{\overset{\|}{CH}} \xrightarrow{NaOH} CH_3\overset{OH}{\overset{|}{CH}}CH_2\overset{O}{\overset{\|}{CH}}$$

$$\xrightarrow{H^+} CH_3CH=CH\overset{O}{\overset{\|}{CH}} \xrightarrow[\Delta, \text{pressure}]{H_2, \text{Ni-Cr}} CH_3CH_2CH_2\overset{OH}{\overset{|}{CH_2}}$$

Now they are preparing to manufacture 1-bromobutane and 2-bromobutane by treating the corresponding butanols with hydrogen bromide in the presence of a sulfuric acid catalyst. Economic considerations require that they use the amount of catalyst that will produce each alkyl bromide at the lowest cost per kilogram. They want your project group to carry out the syntheses of 1- and 2-bromobutane using different catalyst/substrate ratios, then do a cost analysis to find out which ratio produces each product most economically.

Applying Scientific Methodology

Your working hypothesis should include a prediction of the effect on each reaction of increasing the catalyst/substrate ratio.

The Controversial Halides

The familiar expression "You can't get along with them and you can't get along without them" applies to many organic halogen compounds. Organic halides tend to be very stable. However, the stability that makes a particular halide a useful commercial chemical can be a liability if the halide is released into the environment, where it may present a serious and persistent health hazard. Most people first became aware that chlorinated organics might be a problem with the publication of Rachel Carson's book *Silent Spring,* in which she documented the effect of DDT on wildlife and the environment.

The polybrominated and polychlorinated biphenyls (PBBs and PCBs) have also caused serious environmental problems because they are toxic and believed to be carcinogenic and because they persist in the environment. PBBs are mixtures of various bromine derivatives of biphenyl, such as 2,2',4,4',5,5'-hexabromobiphenyl, a major component of the fire retardant mixture Firemaster BP-6. In 1973, Firemaster BP-6 was accidentally added to cattle feed in Michigan, resulting in the illness of some dairy farmers and the death of many farm animals; it has since been withdrawn from the market. Polychlorinated biphenyls have been used in a great number of applications, from cooling electrical transformers to manufacturing plastic liners for baby bottles. In the United States the manufacture, processing, and distribution of PCBs has been prohibited since 1979. Nevertheless, PCBs have persisted in the environment and may now be the most widespread pollutants on earth. They occur in dangerously high concentrations in some fish from the Great Lakes, and they have been detected in the tissues of many other organisms, even polar bears from the high Arctic.

Because of these and other instances of pollution, some environmental groups have called for a total ban on the manufacture of chlorine and chlorinated compounds, and the International Joint Commission, a U.S./Canadian government-appointed group whose function is to preserve water quality in the Great Lakes, supports a gradual phasing out of these chemicals. People on opposite sides of this issue could hardly be farther apart. According to a spokesman for Dow Chemical Company, one of the largest producers and users of chlorine, "It is the single most important ingredient in [industrial] chemistry." Chlorine is used to manufacture an enormous number of consumer products, including plastics, pharmaceuticals, perfume bases, cosmetics, paper, medical devices, household adhesives, anesthetics, nonstick cookware, dry cleaners, refrigerants, photographic film, magnetic recording tape, floor coverings, wallpaper, food wraps, PVC pipe and fittings, electrical insulation, seat covers, baby strollers, compact discs, shoes, paints, solvents, brake fluid, food additives, and crop protection chemicals. Reliable estimates are hard to come by, but chlorine probably accounts for tens of billions of dollars worth of commerce and at least a million jobs. Yet a Greenpeace research analyst has stated, "There are no uses of chlorine that we regard as safe."

DDT

2,2',4,4',5,5'-hexabromobiphenyl

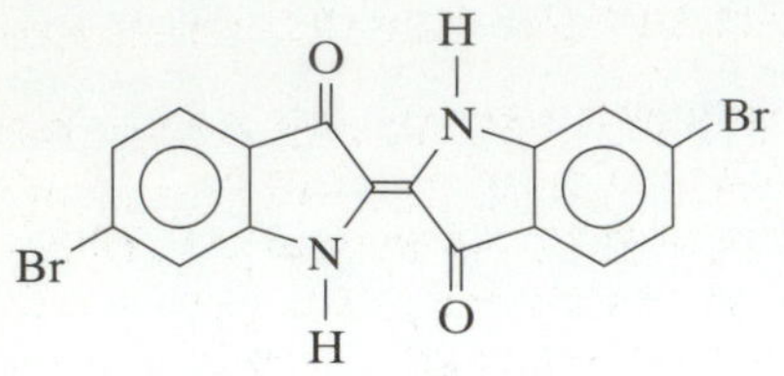

Tyrian purple (dibromoindigo)

Media reports on the issue tend to give the impression that all organic halides are synthetic compounds that have never occurred naturally. In fact, there are a number of natural organic halides. Tyrian purple is a bromine-containing dye derived from spiny carnivorous snails of the murex family. Because of its rarity—about 9000 snails are needed to produce a gram of the dye—Tyrian purple was a costly status symbol in the ancient world. Julius Caesar wore a purple toga as a sign of office, but a Roman senator was allowed only a purple stripe on his white toga. Some seaweeds produce trichloromethane (chloroform, $CHCl_3$) and tetrachloromethane (carbon tetrachloride, CCl_4), compounds that are also produced as by-products of water chlorination. Bromomethane (methyl bromide, CH_3Br), an important insect fumigant, is produced in large quantities by ocean algae. The sharp smell of seawater is probably generated by these and other organohalogen compounds.

As various forest fungi break down dead trees and other plant materials, they produce about 200 times as much natural chloromethane (methyl chloride, CH_3Cl) as humans produce synthetically. Volcanoes also produce chloromethane, along with dichloromethane (methylene chloride) and chlorofluorocarbons such as dichlorodifluoromethane (Freon 12). Our own bodies utilize organic iodides in the form of thyroid hormones, and our white blood cells oxidize chloride ions in the blood to molecular chlorine, which kills invading bacteria and other microorganisms.

Polychlorodibenzodioxins (PCDDs) and polychlorodibenzofurans (PCDFs) are among the most fear inspiring organohalogen compounds.

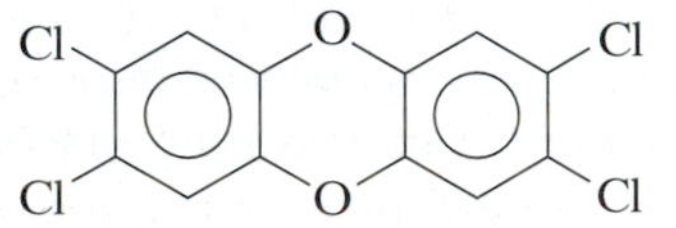

2,3,6,7-tetrachlorodibenzodioxin   2,3,6,7-tetrachlorodibenzofuran
(a PCDD)   (a PCDF)

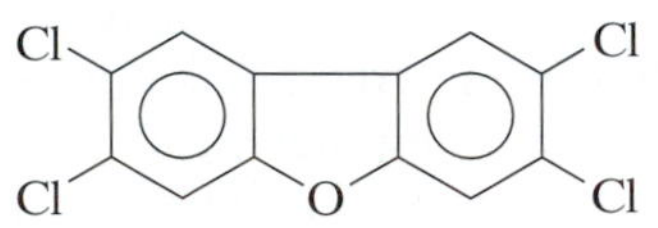

2,3,6,7-Tetrachlorodibenzodioxin, better known as "dioxin," occurs as a by-product when the herbicide 2,4,5-trichlorophenoxyacetic acid (2,4,5-T) is manufactured from 2,4,5-trichlorophenol.

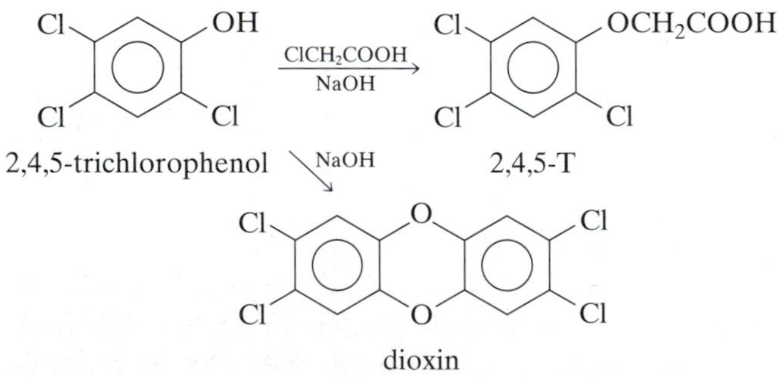

dioxin

2,4,5-T was the main constituent of Agent Orange, the Vietnam War defoliant believed to have caused health problems for many Vietnam veterans.

Thus 2,4,5-T is almost invariably contaminated with dioxin. Dioxin is regarded as one of the most toxic substances known, and some PCDFs may be even more toxic. Yet both PCDDs and PCDFs are produced by burning wood, which contains natural chlorides, and some researchers believe that forest fires and brush fires are the major source of dioxins in the environment. In addition, natural enzymes act on humic acids—phenolic acids formed by the decomposition of organic matter in the soil—to yield a number of chlorinated compounds, including chlorophenols that, like 2,4,5-trichlorophenol, could give rise to TCDDs. Dioxins have been identified in ancient marine sediments formed as early as 6000 B.C., and they are even produced in garden compost piles.

This is not to say that synthetic organohalogen compounds are not harmful because many of them are also produced naturally—many natural substances are hazardous to our health. And the levels of certain organohalogen compounds in humans, including PCDDs and PCDFs, are considerably higher now than they were in preindustrial times, when the only sources of these substances were natural. Environmental groups and health organizations are concerned that they may already be causing reproductive problems, such as declining sperm counts, and other health problems in humans.

Most scientists agree that we should weigh the hazards and benefits of each compound individually rather than banning a whole family of compounds simply because it contains some bad actors, but those who support a chlorine ban believe we face a potential crisis that can't wait for a chemical-by-chemical study of all 15,000 or so chlorine-containing products that are now on the market. Meanwhile, measures are being taken to reduce the production of the most hazardous organohalogen compounds. For example, the Environmental Protection Agency has proposed a regulation that would force pulp and paper mills—an important source of dioxin—to curtail their use of chlorine for bleaching. Nevertheless, the demand for many organohalogen compounds has continued to grow, and there is little evidence to date of a slowdown in the production of industrial chlorine and its compounds.

Understanding the Experiment

In this experiment you will convert either 1-butanol or 2-butanol to the corresponding alkyl bromide with HBr, using sulfuric acid as a catalyst. Catalysts are substances that accelerate chemical reactions without themselves being consumed in the process. For example, an alcohol cannot be converted to an alkyl bromide by sodium bromide because —OH is too poor a leaving group to be displaced by bromide ion. A strong acid is needed to convert this poor leaving group to a better one, $-OH_2^+$, by protonating it.

$$R-OH + H^+ \rightarrow R-OH_2^+$$

Sulfuric acid catalyzes the overall reaction by increasing the concentration of the protonated alcohol, which can then react with bromide ion by either an S_N1 or S_N2 mechanism to form an alkyl bromide. Primary alcohols tend to react by the direct substitution (S_N2) mechanism; secondary and tertiary alcohols are more likely to form an intermediate carbocation, which then combines with halide ion.

See your lecture-course textbook for a more detailed discussion of nucleophilic substitution reactions.

S$_N$2 reaction of 1-butanol

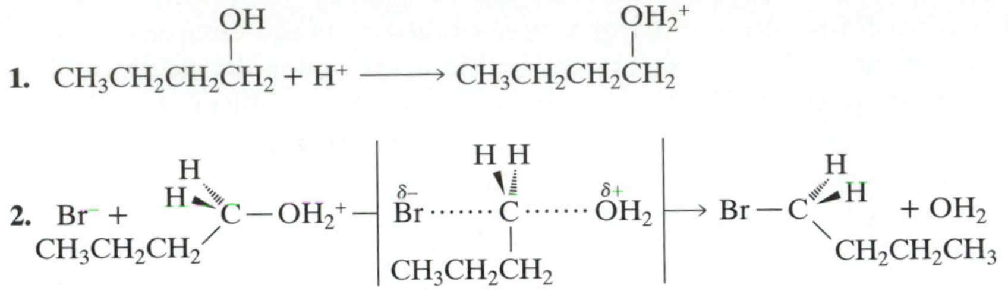

1. $\underset{\displaystyle CH_3CH_2CH_2CH_2}{\overset{\displaystyle OH}{|}}$ + H$^+$ $\longrightarrow$ $\underset{\displaystyle CH_3CH_2CH_2CH_2}{\overset{\displaystyle OH_2{}^+}{|}}$

2. Br$^-$ + $\underset{\displaystyle CH_3CH_2CH_2}{\overset{\displaystyle H}{\underset{\displaystyle}{H}}}$C—OH$_2{}^+$ $\rightarrow$ $\left[\,\overset{\delta^-}{Br}\cdots\cdots\underset{\displaystyle CH_3CH_2CH_2}{\overset{\displaystyle H\ H}{C}}\cdots\cdots\overset{\delta^+}{OH_2}\,\right]$ $\longmapsto$ Br—$\underset{\displaystyle CH_2CH_2CH_3}{\overset{\displaystyle H}{C}}$H + OH$_2$

S$_N$1 reaction of 2-butanol

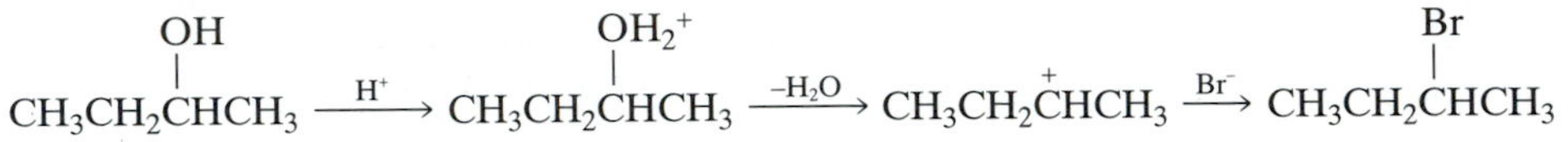

$\underset{\displaystyle CH_3CH_2CHCH_3}{\overset{\displaystyle OH}{|}}$ $\overset{H^+}{\longrightarrow}$ $\underset{\displaystyle CH_3CH_2CHCH_3}{\overset{\displaystyle OH_2{}^+}{|}}$ $\overset{-H_2O}{\longrightarrow}$ $CH_3CH_2\overset{+}{C}HCH_3$ $\overset{Br^-}{\longrightarrow}$ $\underset{\displaystyle CH_3CH_2CHCH_3}{\overset{\displaystyle Br}{|}}$

Since the reaction rate for both mechanisms depends on the concentration of the protonated alcohol, increasing the catalyst concentration should increase the rate. We might expect this to in turn increase the amount of alkyl bromide produced in a given reaction time, but it could increase the rates of side reactions as well. Accelerating such side reactions as E1 elimination, polymerization, and ether formation could actually reduce the product yield by converting the substrate into unwanted by-products.

By using from 1 to 4 mL of sulfuric acid with each of the two alcohols, your project group will vary the ratio of catalyst to substrate from 0.25 to 1.00 (see Table 28.1). Then you will carry out a cost analysis to determine which catalyst : substrate ratio will produce each product at the lowest cost per kilogram. Because variable product losses during the purification step may invalidate the results, the masses of the crude products will be used for the cost analyses.

You will carry out the reaction by refluxing the alcohol with about a 25% excess of 48% hydrobromic acid and the assigned volume of sulfuric acid. Since sulfur dioxide and HBr fumes are evolved during the reaction, a hood or gas trap must be used to keep them from escaping into the laboratory. The alkyl bromide will be separated from the reaction mixture by internal steam distillation using the water present in the reaction flask. The product ordinarily forms the lower layer in the distillate and can be separated from the aqueous layer in a separatory funnel. However with a low catalyst : substrate ratio, there may be enough unreacted alcohol in the distillate to make the organic layer less dense than water. The organic layer may contain varying amounts of HBr, sulfuric acid, unreacted alcohol, and by-products; these impurities can be removed by washing with sulfuric acid, water, and aqueous sodium bicarbonate, followed by distillation. It is important to be aware that the alkyl bromides are less dense than sulfuric acid but more dense than water, so the alkyl bromide layer will be on the top after the first washing but on the bottom in the next two.

You can confirm the conversion of reactant to product by using infrared spectrometry. The infrared spectra of alkyl bromides are characterized by a strong C—Br stretching band at low frequency, from 690 to 515 cm^{-1}. Since the range of some infrared spectrometers does not extend below 600 cm^{-1}, the entire C—Br band is not observed in all cases.

Reactions and Properties

$$CH_3CH_2CH_2CH_2OH + HBr \xrightarrow{H_2SO_4} CH_3CH_2CH_2CH_2Br + H_2O$$

or

$$CH_3CH_2\overset{\overset{\displaystyle OH}{|}}{C}HCH_3 + HBr \xrightarrow{H_2SO_4} CH_3CH_2\overset{\overset{\displaystyle Br}{|}}{C}HCH_3 + H_2O$$

Table 28.1 Ratio of catalyst to alcohol in the eight runs

Run	alcohol (72 mmol)	mmol catalyst	catalyst:substrate mole ratio
1	1-butanol	18	0.25
2	1-butanol	36	0.50
3	1-butanol	54	0.75
4	1-butanol	72	1.00
5	2-butanol	18	0.25
6	2-butanol	36	0.50
7	2-butanol	54	0.75
8	2-butanol	72	1.00

Table 28.2 Physical properties

	M.W.	b.p.	d
1-butanol	74.1	117	0.810
2-butanol	74.1	99.5	0.808
hydrobromic acid (48%)	80.9		1.49
1-bromobutane	137.0	102	1.276
2-bromobutane	137.0	91	1.259

Note: m.p. and b.p. are in °C, density is in g/mL

1-butanol $CH_3CH_2CH_2CH_2OH$

3335.4	1378.6	952.2
2959.7	1072.6	846.7
1466.0	1010.6	737.5

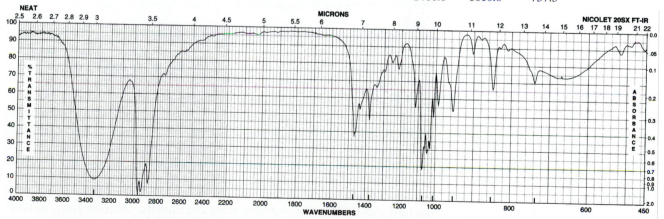

2-butanol

OH
|
CH₃CHCH₂CH₃

3343.1	1374.6	991.0
2967.2	1299.9	912.6
1457.7	1108.9	668.4

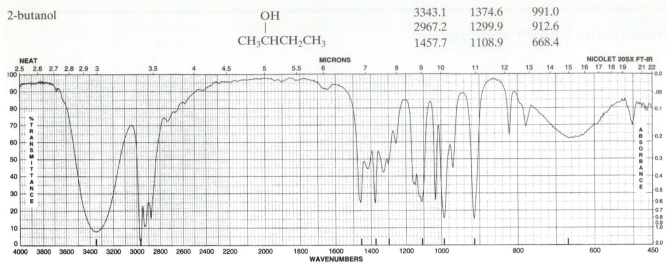

Figure 28.1 IR spectra of the starting materials (Reproduced from *The Aldrich Library of FT-IR Spectra, Edition II*, with the permission of the Aldrich Chemical Company.)

Directions

You will be assigned one of the two alcohols and either 1, 2, 3, or 4 mL of sulfuric acid. Measure the quantities of all chemicals carefully for the cost analysis.

Hydrobromic acid is toxic and very corrosive and can cause very serious damage to the skin, eyes, and respiratory tract. Wear gloves and dispense under a hood. Avoid contact with the acid, and do not inhale its vapors. Sulfuric acid causes chemical burns that can seriously damage the skin and eyes. Wear gloves and avoid contact.
1-Butanol and 2-butanol are flammable and may cause eye or skin irritation. Avoid contact.
1-Bromobutane and 2-bromobutane are flammable and may be harmful if inhaled or absorbed through the skin. 2-Bromobutane is a suspected carcinogen. Avoid contact with the products and do not breathe their vapors.

Safety Notes

sulfuric acid

butanols

bromobutanes

Take Care! Wear gloves, avoid contact with HBr, do not breathe its vapors.

Reaction. Assemble an apparatus for heating under reflux [OP-6] using a 50-mL round-bottomed flask, a water cooled reflux condenser, and a gas trap [OP-22] containing 1 *M* NaOH. The gas trap can be omitted if you are performing the reaction under an efficient fume hood. Weigh 72.0 mmol of your assigned alcohol (1-butanol or 2-butanol) into the round-bottomed flask. Cool the flask in ice water and *under the hood* cautiously add 10.0 mL (89 mmol) of 48% aqueous hydrobromic acid. Then add your assigned volume of concentrated sulfuric acid. (**Take Care!** Wear gloves, avoid contact.) Drop in some acid resistant boiling chips or a stirbar and heat the mixture under reflux for one hour from the time the solution starts to boil.

Separation. Codistill the alkyl bromide with the water present in the reaction mixture [OP-15], using a simple distillation apparatus (no addition funnel) with a graduated cylinder as the receiver. Stop the distillation when the distillate is no longer cloudy and the organic layer no longer increases in volume over a period of 5–10 minutes. Add 10 mL of water to the distillate, shake the mixture in a separatory funnel, and separate the layers cleanly. If you're not sure which is the organic layer, test a drop of each layer with a drop of water. Cautiously wash the organic layer [OP-19] with 5 mL of cold concentrated sulfuric acid and save the top layer. (**Take Care!** Possible violent reaction! Wear gloves, avoid contact.) Then wash it with a 10-mL portion of water followed by a 10-mL portion of saturated aqueous sodium bicarbonate, saving the *bottom* layer each time. Dry the alkyl bromide [OP-20] with a little anhydrous calcium chloride. Decant and weigh the product at this point, and report the crude yield to your instructor or coworkers.

Purification and Analysis. Purify the alkyl bromide by simple distillation [OP-25] from a 25-mL boiling flask (you can use a small-scale distillation apparatus). Collect the product over a 3–4° range centered around the expected boiling point. Measure the mass of the purified product. At your instructor's option, obtain its infrared spectrum [OP-34]. Turn in the product to your instructor.

Cost Analysis. Using prices provided by your instructor (preferably from the *Chemical Marketing Reporter*), calculate the total cost of all the chemicals (except water) that you used in the "Reaction" phase of the experiment. Since sulfuric acid can be recycled (at some cost), divide its cost by 3. Divide the cost in dollars by the mass of your product in kilograms.

Report. Your report should include a statement of the problem and an account of how you applied scientific methodology in this experiment. Calculate the percent yield of your crude and purified products. Using the cost/mass values reported by other members of your project group, determine the catalyst:substrate ratio that yields each product at the lowest cost per kilogram.

Waste Disposal: The residue is corrosive and produces very irritating fumes. Let it cool to room temperature, then dispose of it in the acid wastes container under the hood.

Stop and Think: Under what conditions might the organic layer be on top?

Waste Disposal: Put the acid into the acid wastes container.

Waste Disposal: Pour all aqueous layers down the drain.

Exercises

1   Based on their structures, attempt to explain any difference in the catalyst:substrate ratio for the two alcohols.

2   (a) Write structures for at least three by-products that might have formed during the reaction of 2-butanol. (b) Propose a mechanism for the formation of each by-product.

3   If you recorded the infrared spectrum of your product, interpret the spectrum as completely as you can, and use the appropriate starting-material spectrum in Figure 28.1 to show that the expected functional group conversion has taken place.

4   Construct a flow diagram for this synthesis following the format described in Appendix V.

5   Describe and explain the possible effect on your results of the following experimental errors or variations. (a) You used 1 mL of sulfuric acid in the reaction of 2-butanol and saved the bottom layer from

the distillate. (b) You omitted the sulfuric acid wash. (c) The fresh distillate was still cloudy when you stopped the initial distillation.

6 Write balanced equations showing how HBr and SO_2 are consumed in a gas trap containing NaOH.

7 Write a mechanism showing how 2,4,5-trichlorophenol is converted to 2,4,5-T.

Other Things You Can Do

(Starred projects require your instructor's permission.)

*1 Obtain an NMR spectrum of your product in deuterochloroform, and interpret it as completely as you can.

*2 Compare the nucleophilic substitution rates of some alcohols in Minilab 23.

3 Starting with sources from the Bibliography, write a research paper describing the manufacture, uses, and environmental effects of some chlorinated pesticides.

Borohydride Reduction of Vanillin to Vanillyl Alcohol

Preparation of Alcohols. Reactions of Carbonyl Compounds. Reduction Reactions. Nucleophilic Addition.

Operations:

The operations you use will depend on the procedure you develop.

Before You Begin

After reading the experiment, develop a procedure and experimental plan for the sodium borohydride reduction of 25.0 mmol of vanillin to vanillyl alcohol. Calculate or estimate the quantities you will need, describe the reaction conditions, and tell how you intend to separate and purify the product. Your procedure should be clear and detailed enough so that anyone with sufficient background could carry it out successfully. (Alternatively, a group of students can work out a set of procedures in which some experimental parameters are varied, to see how such variations alter the yield and purity of the product.)

Scenario

Pulpchem Inc., a subsidiary of a large paper company, produces useful chemicals from lignin and other by-products of paper production. Chemical treatment of lignin yields large quantities of vanillin, a white solid that is also responsible for the characteristic aroma of vanilla beans. Woody Aspin, Pulpchem's product development manager, has asked your Institute to develop a method for reducing vanillin to vanillyl alcohol, which shows promise as a starting material for the synthesis of synthetic drugs and flavoring ingredients. There is a procedure for the preparation of vanillyl alcohol in an obscure Swedish journal, *Acta Universitatis Lundensis,* which ceased publication in the 1960s. Unless you understand Swedish and can get your hands on a copy of the journal article, you will have to develop your own procedure for the synthesis.

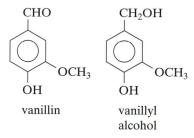

vanillin   vanillyl alcohol

Applying Scientific Methodology

The problem, whether vanillin can be converted to vanillyl alcohol by a procedure you have developed yourself, will be solved when (and if) you obtain and characterize the product. You will test your working hypothesis by measuring the product's melting point, and you can record its infrared spectrum for confirmation.

vanilla plant

Fragrant and Fiery Aromatics

Chemists recognized at an early date that certain compounds obtained from natural sources showed a higher ratio of carbon to hydrogen than did typical aliphatic compounds. These compounds also had distinctly different chemical properties. Because many of them came from such pleasant-smelling sources as the essential oils of cloves, sassafras, cinnamon, anise, bitter almonds, and vanilla, they were called *aromatic* compounds. The name stuck, although it is no longer associated with the odors of such compounds but with their structures and properties. Many aromatic compounds do live up to the original meaning of the term, however. Among the most interesting and important of these are vanillin and the *vanilloids*, structurally related compounds that may exhibit vanillin's characteristic ring-substitution pattern.

In 1520 the Spanish conquistador Hernando Cortez was served an exotic drink by Montezuma II, emperor of the Aztecs, at his capital of Tenochtitlán on the site of modern Mexico City. Cortez enjoyed the drink, a combination of chocolate and vanilla, and it soon found its way back to Europe. The vanilla plant, a climbing orchid, *Vanilla planifolia,* was also shipped back to the Old World in the hope that it could be cultivated there. The transported plants grew well, but mysteriously would not fruit. This mystery remained unsolved for more than 300 years, until someone discovered that the plant was pollinated by a native Mexican bee with an exceptionally long proboscis. A method of hand pollination was soon developed, allowing the cultivation of vanilla outside of Mexico.

Vanilla flavoring comes from the fruit of the vanilla plant, a long, narrow pod that, after curing, looks somewhat like a dark brown string bean. The principal component of vanilla flavoring, vanillin (3-methoxy-4-hydroxybenzaldehyde), does not exist as such in the fresh vanilla bean but is formed by the enzymatic breakdown of a glucoside during the curing process. Although the finest vanilla flavoring is still obtained from natural vanilla, synthetic vanillin is far less costly. It is used as a component of flavorings and perfumes and as a starting material for the synthesis of such drugs as L-dopa, which is used for treating Parkinson's disease.

At one time synthetic vanillin was made mostly from isoeugenol, a naturally occurring and widely used perfume ingredient.

isoeugenol   vanillin

Most vanillin is now synthesized using lignin derived from wood pulp. Lignin is a complex polymer that gives rigidity to trees and other woody plants; after cellulose, it is the second most abundant organic material on earth. Many of the basic structural units of lignin are guiacylpropane units, and such units can be broken down chemically to yield vanillin.

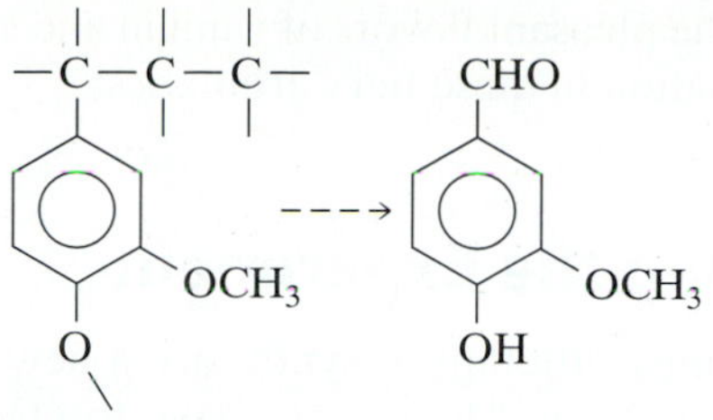

$$
\begin{array}{ccc}
-\text{C}-\text{C}-\text{C}- & & \text{CHO} \\
\end{array}
$$

guiacylpropane unit

Safrole, a fragrant liquid derived from the roots and bark of the sassafras tree (*Sassafras albidum*), is structurally related to vanillin. Once widely used as a flavoring in root beer, toothpaste, and chewing gum, safrole can no longer be added to such products because of its toxic, irritant qualities and because it produces liver tumors in rats and mice. It has also been used by illicit drug manufacturers to synthesize the dangerous designer drug "Ecstasy" (methylenedioxymethamphetamine), whose use can cause mental confusion, kidney failure, and death. As a result, the sale of safrole is now strictly controlled. But black pepper, anise, nutmeg, and other spices contain minute amounts of safrole, so you probably consume a little of it every day.

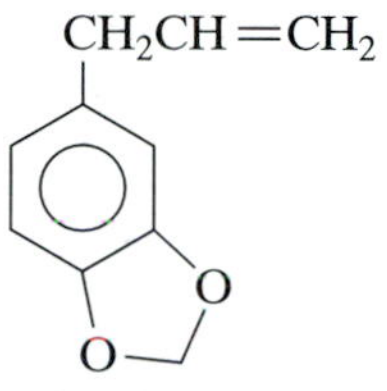

Black pepper gets its "bite" from piperine, which contains the same methylenedioxy ($-\text{OCH}_2\text{O}-$) unit found in safrole.

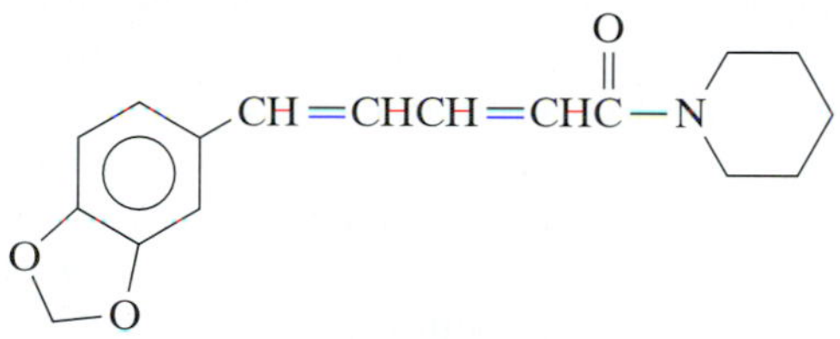

safrole

piperine (*trans* double bonds)

Capsaicin, piperine's much hotter cousin, has the 3-methoxy-4-hydroxy grouping characteristic of vanillin.

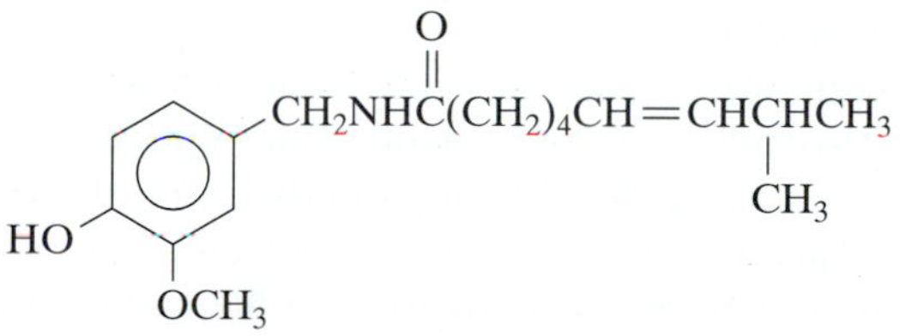

capsaicin

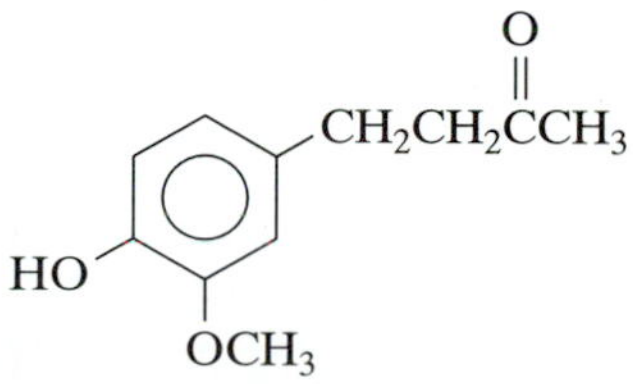

Capsaicin is a fiery component of many *capsicums*, pungent peppers such as cayennes, jalapeños, and the ultrahot habaneros. Capsicum peppers are used to make Tabasco Sauce, Jamaica Hell Fire and other hot sauces, as well as most commercial salsas. Another hot compound with the vanillin structural unit is zingerone, the pungent principle of ginger. The structural units that

zingerone

we associate with the pleasant flavors of vanillin and safrole contribute a far different taste sensation in these fiery aromatics!

Understanding the Experiment

When organic chemists attempt to synthesize a new compound, they have no set procedure to follow. They either have to (1) adapt and modify an existing procedure for a similar synthesis; (2) devise a procedure based on their knowledge of the substrate, reagent, product, and reaction involved; or (3) invent a new way of synthesizing the compound they wish to make. In this experiment you will follow the second option. You will develop your own procedure for reducing vanillin to vanillyl alcohol based on information that you will find in this section about the reducing agent, sodium borohydride, and the reaction, reduction of a carbonyl compound to an alcohol, as well as information in the "Reaction and Properties" section about the substrate and product.

Reduction Reactions. In organic chemistry *reduction* usually refers to a reaction that is accompanied by a gain of hydrogen atoms or a loss of oxygen atoms, or both. For example, a carbonyl compound is reduced to an alcohol when its carbonyl group gains two hydrogen atoms. The hydrogen is provided by an appropriate reducing agent.

Reducing Agents. When lithium aluminum hydride (LiAlH$_4$) was introduced as a reducing agent in the late 1940s, it brought about a revolution in the preparation of alcohols by reduction. At that time the two most popular reducing agents for carbonyl compounds were sodium metal and gaseous hydrogen under pressure. The greater simplicity and convenience of hydride reduction soon made it the preferred method for a broad spectrum of chemical reductions. Lithium aluminum hydride is a powerful reducing agent whose high reactivity is a disadvantage in some applications. Because it reacts violently with water and other hydroxylic solvents to release hydrogen gas, it can only be used in aprotic solvents, such as ethyl ether, under strictly anhydrous conditions. It is also expensive and somewhat hazardous to use— even grinding it in a mortar may cause a fire. By contrast, sodium borohydride (NaBH$_4$) is a much milder reducing agent that is comparatively safe to handle in the solid form. Unlike lithium aluminum hydride, it can even be used in aqueous or alcoholic solutions. NaBH$_4$ does decompose slowly in moist air, so its container should be tightly capped when it is not in use.

Sodium borohydride reductions involve nucleophilic addition of hydride ion (H:$^-$) to the carbonyl carbon, but apparently no free hydride ions are generated. Kinetic evidence suggests that one solvent molecule bonds to the boron atom while it is transferring hydride to the carbonyl compound, and

Reduction of a carbonyl group

$$\underset{\displaystyle |}{\overset{\displaystyle O}{\underset{\displaystyle |}{\overset{\displaystyle \|}{-C-}}}} \xrightarrow{\text{(2H)}} \underset{\displaystyle |}{\overset{\displaystyle O-H}{\underset{\displaystyle |}{\overset{\displaystyle |}{-C-H}}}}$$

$$R-O-H + O=C \overset{\diagup}{\underset{\diagdown}{}} + H-B \overset{\ominus}{} + O-R \longrightarrow R-\overset{\ominus}{O} + H-O-\overset{\displaystyle |}{\underset{\displaystyle |}{C}}-H + B\overset{\ominus}{-}OR + H^+$$

(R = alkyl or H)

another solvent molecule provides a proton to the carbonyl oxygen. This process can continue until all hydride ions on BH_4^- have been used up.

Reaction Stoichiometry. The overall stoichiometry of the borohydride reduction of a carbonyl compound ($R' = $ alkyl or H) is given by the following general equation.

$$4RCR' + NaBH_4 + 4H_2O \longrightarrow 4RCHR' + H_3BO_3 + 4NaOH$$
$$\overset{O}{\overset{\|}{}} \qquad\qquad \overset{OH}{\overset{|}{}}$$

In practice, it is best to use a 50–100% excess of sodium borohydride to compensate for any that reacts with the solvent or decomposes from other causes. Since the reaction is first order in sodium borohydride (as well as the carbonyl compound), using an excess will also increase the reaction rate.

Reaction Solvents. Sodium borohydride reductions are usually carried out in a dilute (~1 M) aqueous NaOH solution or in an alcohol such as methanol, ethanol, or 2-propanol. The reagent is not stable at low pH, and even in a neutral aqueous solution it decomposes to the extent of about 4.5% per hour at 25°C. Acidic functional groups such as COOH and the OH group of a phenol may cause rapid decomposition of sodium borohydride. When carbonyl compounds having such functional groups are being reduced, enough 1 M NaOH should be used to neutralize the functional groups *and* maintain a pH of 10 or higher. Sodium borohydride reacts slowly with alcohols, but ethanol and methanol are usually suitable solvents when there are no acidic functional groups and the reaction time is no more than 30 minutes at 25°C. For longer reaction times or reactions at higher temperatures, isopropyl alcohol is a better solvent, but it is more difficult to remove after the reaction is over.

Reaction Conditions. In most reactions with sodium borohydride, the aldehyde or ketone is dissolved in the reaction solvent and a solution of sodium borohydride is added, with external cooling if necessary, at a rate slow enough to keep the reaction temperature below 25°C. Higher temperatures may decompose the hydride, especially in methanol or ethanol. The amount of solvent is not crucial, but enough should be used to completely dissolve the reactants and facilitate the workup of the reaction mixture. The solubility of sodium borohydride per 100 g of solvent is reported to be 55 g in water at 25°C, 16.4 g in methanol at 20°C, and 4.0 g in ethanol at 20°C.

The time required to complete the reaction depends on the reaction temperature and the reactivity of the substrate. Kinetic studies of borohydride reduction in isopropyl alcohol have shown that aldehydes are considerably more reactive than ketones and that aliphatic carbonyl compounds are more reactive than aromatic ones. Most reactions of aldehydes and aliphatic ketones are complete in 30 minutes at room temperature, but those of aromatic ketones or particularly hindered ketones may require more time or higher reaction temperatures. For example, the comparatively reactive ketone 4-*t*-butylcyclohexanone is completely reduced at room temperature in 20 minutes, but benzophenone is reduced only by heating it at the boiling point of isopropyl alcohol for 30 minutes.

Et   = Ethyl
i-Pr = isopropyl
t-Bu = *t*-butyl
r.t.   = room temperature

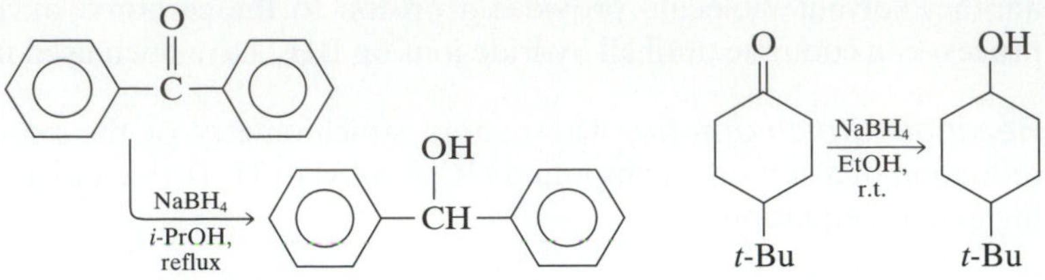

Reaction conditions in borohydride reductions

Workup of the Reaction Mixture. After the reaction is complete, the excess sodium borohydride is decomposed by acidifying the reaction mixture to pH 6 or below (slowly and with stirring) using dilute (~3 *M*) hydrochloric acid. Hydrogen gas is evolved during this process as the excess sodium borohydride decomposes, so there must be no flames in the vicinity; working under a hood is advisable. Addition of acid may also generate some diborane (B_2H_6), which can cause side reactions if other reducible groups such as COOH, COOR, and C=C are present.

Depending on the properties of the product and the reaction solvent used, the product can be separated from the reaction mixture by filtration, extraction, or partial evaporation of the solvent followed by extraction. If the product is a solid that crystallizes from the reaction mixture, it can be collected by vacuum filtration. The yield of the solid product can usually be increased by extracting the filtrate with ethyl ether or another suitable solvent, then drying and evaporating the ether. Liquids or water soluble products are generally separated from an aqueous reaction mixture by extraction with ethyl ether and recovered by evaporating the dried ether. If the reaction solvent is an alcohol, the reaction mixture is usually concentrated by evaporating most of the alcohol. Water is then added and the product is extracted with a suitable solvent.

Purification. The product can be purified by any appropriate method, based on its physical state and properties.

Reactions and Properties

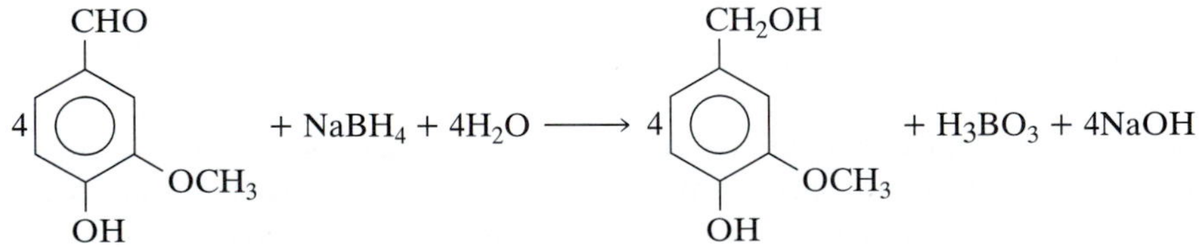

Table 29.1 Physical properties

	M.W.	m.p.	b.p.
vanillin	152.2	79	285
sodium borohydride	37.83		400d
vanillyl alcohol	154.2	115	d

Note: m.p. and b.p. are in °C, density is in g/mL; d = decomposes

Vanillyl alcohol is reported to be soluble *cold* in alcohol and ether and soluble *hot* in water, alcohol, ether, and benzene. It is relatively insoluble in cold water and cold benzene. Vanillyl alcohol tends to form supersaturated solutions in water.

Vanillin

3171.4	1509.6	859.6
1665.7	1266.6	733.5
1588.0	1154.9	633.2

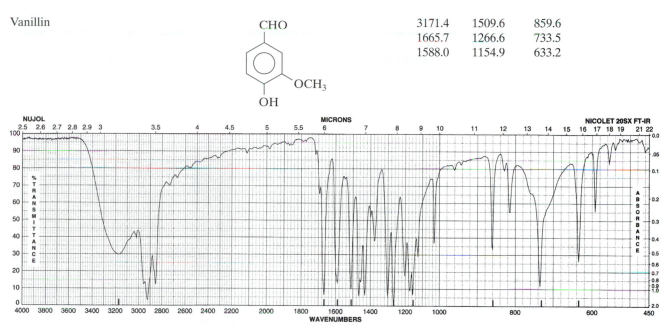

Figure 29.1 IR spectrum of vanillin (Reproduced from *The Aldrich Library of FT-IR Spectra, Edition II*, with the permission of the Aldrich Chemical Company.)

Directions

> Sodium borohydride is toxic and corrosive, and it can react violently with concentrated acids, oxidizing agents, and other substances. Aqueous borohydride solutions with pH values below 10.5 have been known to decompose violently, so be sure your reaction mixture (if aqueous) is sufficiently alkaline. Avoid contact, do not breathe dust, and keep $NaBH_4$ away from other chemicals.
>
> Hydrogen is generated when the reaction mixture is acidified, so be sure there are no flames nearby during this step.

Safety Notes

sodium borohydride

Develop your own procedure for this experiment and submit it to your instructor for approval. Carry out the synthesis in the laboratory, measure the yield and melting point of the purified vanillyl alcohol, and turn it in. At your instructor's option, you can obtain the infrared spectrum of your product and compare it with the spectrum of the starting material to confirm that the expected functional group conversion has occurred. Dispose of any waste solvents, etc., as directed by your instructor.

Report. Your report should include a statement of the problem and an account of how you applied scientific methodology to solve it. Calculate your percent yield and compare your results with those of other students in your laboratory section. Suggest ways of improving the yield and/or purity of your product.

Exercises

1 Write a mechanism for the reduction of vanillin by sodium borohydride under the reaction conditions you used.

2 Write a balanced equation for the decomposition of sodium borohydride in water to which HCl has been added.

3 Describe and explain the possible effect on your results of the following experimental errors or variations. (a) You used pure water as the reaction solvent. (b) You used 5 mL of 1 M NaOH for the $NaBH_4$ solution and 10 mL of 1 M NaOH for the vanillin. (c) Since you started with 25 mmol of vanillin, you decided that a 100% excess of $NaBH_4$ was 50 mmol.

4 Construct a flow diagram for the procedure you used in preparing vanillyl alcohol, based on the format described in Appendix V.

5 Sodium borohydride is a strong base as well as a reducing agent. In the reduction of base-sensitive compounds, should the substrate be added to sodium borohydride (both in solution) or should sodium borohydride be added to the substrate? Explain.

6 (a) Draw the structure of the product you would have isolated if you had used $NaBD_4$ in D_2O in this experiment, and write a mechanism explaining the result. (b) Give the product structure for the same reaction using $NaBD_4$ in H_2O. Assume that there is no hydrogen exchange between $NaBD_4$ and the solvent.

7 A student dissolved vanillin in 10 M sodium hydroxide and let the mixture stand overnight before adding sodium borohydride. Although the product was a white solid, it melted over a broad temperature range that was much lower than the melting point of vanillyl alcohol. The instructor suggested washing the product with dilute sodium bicarbonate; when the student did so, about half of the product dissolved and the remainder melted at 115°C. Explain what happened and write an equation for the reaction.

Other Things You Can Do

(Starred projects require your instructor's permission.)

*1 Carry out the following tests from Part III on vanillin and on your product: 2,4-dinitrophenylhydrazine (Test C-11), ferric chloride (Test C-13), and Tollens' test (Test C-23). Interpret the results and comment on the purity of your product.

*2 Carry out a photoreduction reaction that uses isopropyl alcohol as the reducing agent in Minilab 24.

*3 Reduce another carbonyl compound (such as benzaldehyde or camphor) with sodium borohydride after developing a detailed procedure, and turn in a pure sample of the resulting alcohol. Your instructor must approve the procedure before you start.

4 Starting with sources listed in the Bibliography, write a research paper on the sources, nature, and uses of lignin, including a description of a commercial process for producing vanillin from the by-products of papermaking.

Multistep Synthesis of Cyclopentanol from Cyclohexanol

Reactions of Alcohols. Preparation of Alcohols. Oxidation. Ring Cleavage. Ring Closure. Reduction. Multistep Synthesis.

Operations:

OP-6	Heating
OP-9	Mixing
OP-12	Vacuum Filtration
OP-13	Extraction
OP-14	Evaporation
OP-19	Washing Liquids
OP-20	Drying Liquids
OP-21	Drying Solids
OP-25	Simple Distillation
OP-28	Melting Point
OP-29	Boiling Point
OP-34	Infrared Spectrometry

Before You Begin

1 Read the experiment, read or review the operations as necessary, and write an experimental plan.

2 Calculate the mass and volume of 75.0 mmol of cyclohexanol and the theoretical yields of adipic acid, cyclopentanone, and cyclopentanol from that much cyclohexanol.

Scenario

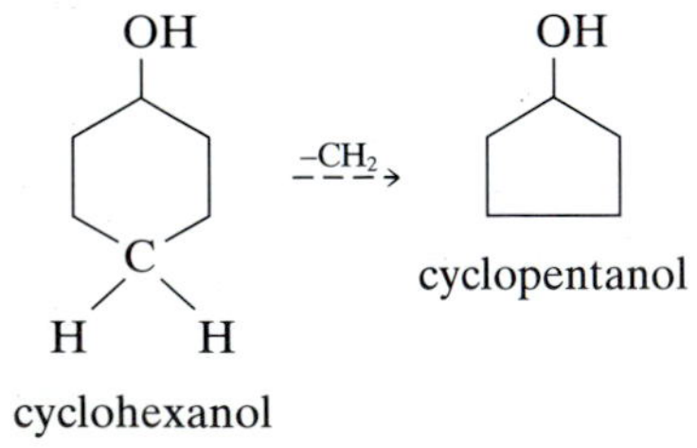

cyclohexanol

cyclopentanol

A client who had placed a large order for cyclohexanol from Marvelous Molecules Inc. just went bankrupt, leaving the company with a surplus of that chemical. Marketing executive Penny Wise noticed that MMI's selling price for cyclopentanol in their chemicals catalog is about five times the price of cyclohexanol. She thinks it should be possible to convert the surplus cyclohexanol to cyclopentanol by simply removing one methylene (CH_2) group from each cyclohexanol molecule. She has engaged your Institute to study the problem and see if there is any economically feasible way to convert cyclohexanol to cyclopentanol. The synthesis is not as simple as she imagines, but your supervisor has come up with a synthetic pathway that might work. It involves an oxidation reaction accompanied by ring cleavage, a ring closure reaction accompanied by decarboxylation, and a reduction reaction. Your assignment is to test this multistep synthesis in the lab and assess its practicality by comparing the cost of the cyclohexanol with the price MMI can get for the amount of cyclopentanol you recover.

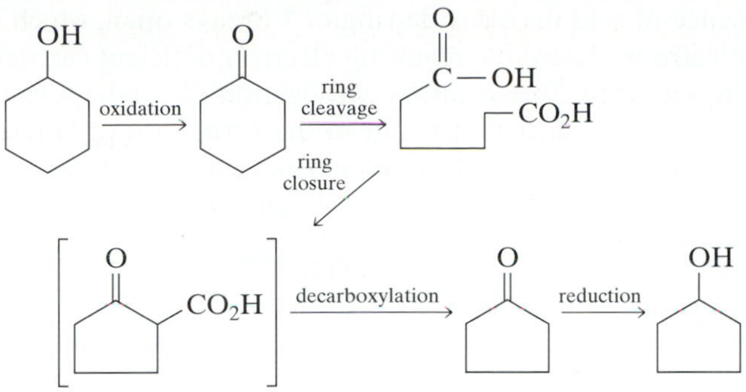

Applying Scientific Methodology

One way to state the problem might be "Can cyclopentanol be synthesized from cyclohexanol in sufficient yield so that the value of the cyclopentanol will exceed the cost of the cyclohexanol used?" Any working hypothesis based on the problem will be tested when you measure the mass of the product and confirm its identity.

Making Carbon Rings

Some of the most interesting and useful organic molecules contain one or more rings of carbon atoms. For example, lanosterol, a component of lanolin and an intermediate in the biosynthesis of cholesterol and other hormones, contains the four-ring system characteristic of steroids. In your body, lanosterol is biosyntheized from an open-chain precursor, squalene oxide (**1**), by a process whose first step resembles the closing of a zipper.

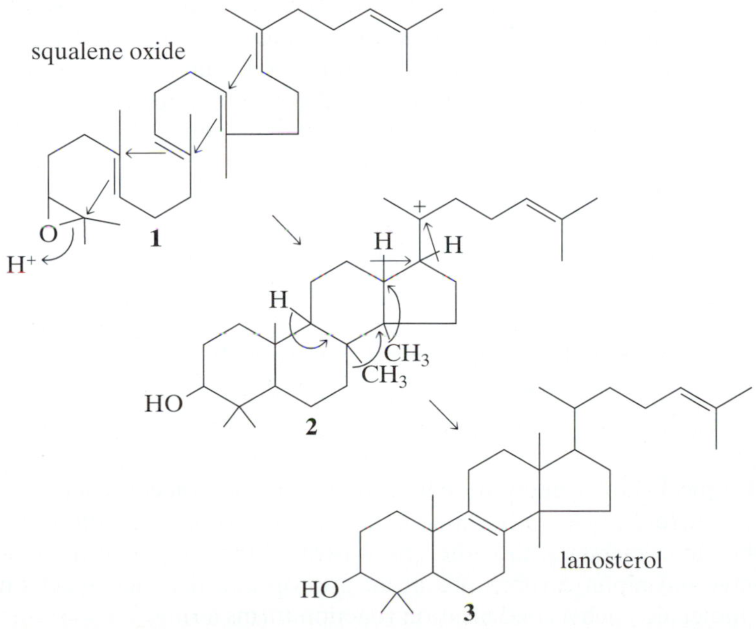

In the presence of acid the epoxide ring of **1** breaks open, which initiates a flow of pi electrons toward the resulting electron-deficient carbon, "zipping up" four rings at once. The resulting carbocation (**2**) undergoes a series of methide and hydride shifts that result in the formation of lanosterol. This remarkable process is mediated by an enzyme, which is a kind of biochemical catalyst. Stretched out, a molecule of squalene oxide would never be able to form even one ring.

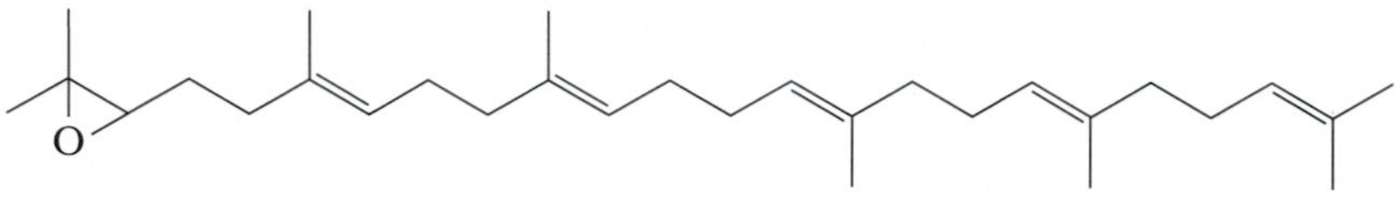

1

Like a cabinetmaker's bench clamps that hold pieces of wood in place while they are being glued together, an enzyme is capable is holding a reacting molecule in just the right orientation for electron shifts to "glue" a new molecule together.

Organic chemists deal with matter in bulk; they don't have the enzyme's ability to arrange the individual molecules making up a reaction mixture in the optimum orientation for a reaction to occur (at least not yet). But molecules are constantly in motion, undergoing conformational changes that— given enough time—allow them to adopt almost any conceivable orientation. Relying on a molecule's ability to eventually orient itself in the right conformation for a reaction, organic chemists have developed a number of useful ring-forming strategies. Some chemists have even rivaled the enzyme's accomplishment by closing as many as four rings in one operation, but more often they form just one ring at a time.

Many carbon-ring forming reactions are just variations of well-known carbon-carbon bond forming reactions. For example, eliminating two bromine atoms from a 1,2-dibromide with zinc yields a carbon-carbon pi bond, forming a double bond, so eliminating two bromine atoms from a 1,3-dibromide yields a sigma bond, forming a three-membered ring.

A typical aldol condensation links together two molecules of a carbonyl compound, forming a double bond between the carbonyl carbon of the one and the alpha carbon of the other, as illustrated for acetone. If the reacting carbonyl and alpha carbons are at an appropriate distance apart on the same molecule, such a condensation reaction forms a ring.

Aldol condensation

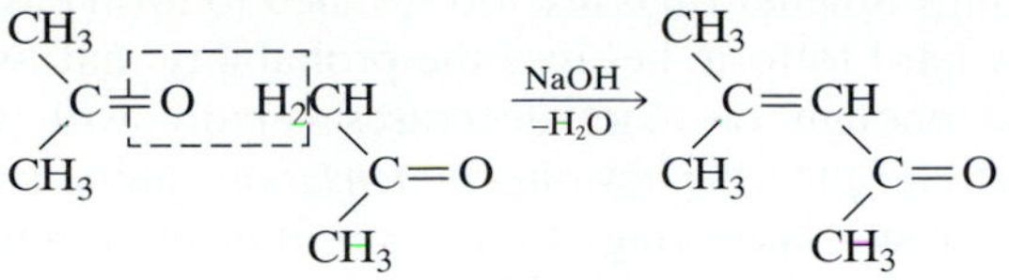

Ring-forming aldol condensation

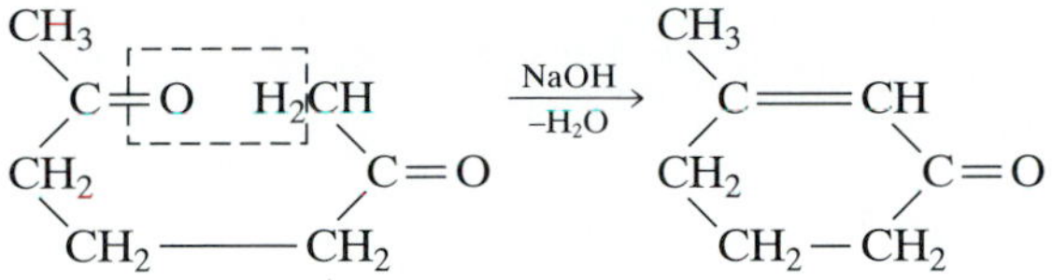

An aldol condensation can be combined with a Michael addition to form a ring from two previously unconnected molecules. This ring forming sequence, called the Robinson annulation, is illustrated for the reaction of ethanal (acetaldehyde) with 3-buten-2-one (methyl vinyl ketone).

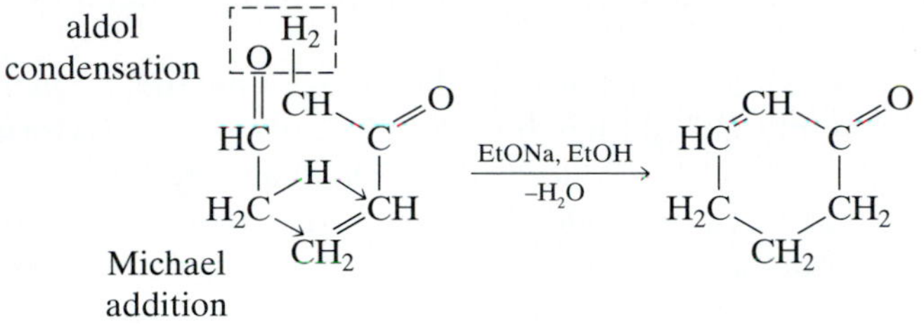

Because the product is a carbonyl compound that can react with 3-buten-2-one in the same way as ethanal, this sequence can be repeated indefinitely to build up large polycyclic systems, as shown here for two repetitions.

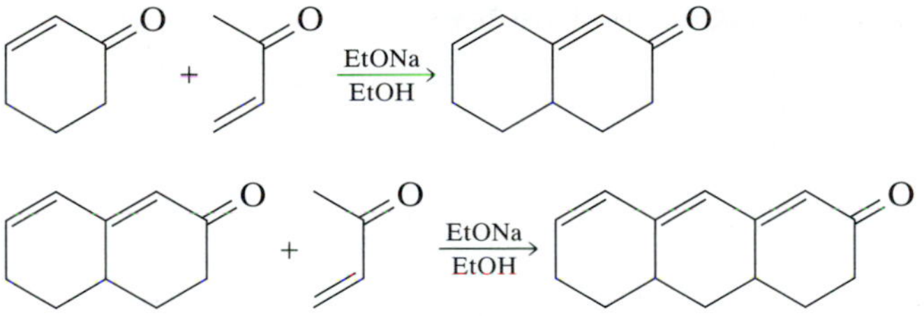

Another ring forming condensation reaction, the Dieckmann condensation, is simply an intramolecular Claisen condensation.

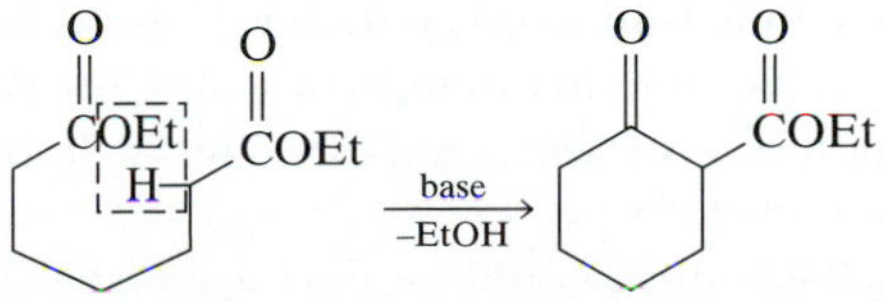

Condensation reactions such as these are most effective for making five- and six-membered rings. Smaller rings are too strained to form easily, and larger rings are usually hard to form because the probability that two chain ends will come close enough to react decreases rapidly with chain length. However, a reaction called the *acyloin condensation* can be used to synthesize medium-sized and large rings by the reaction of sodium metal with open-chain diesters, as illustrated for the preparation of 2-hydroxycyclodecanone from dimethyl decanedioate.

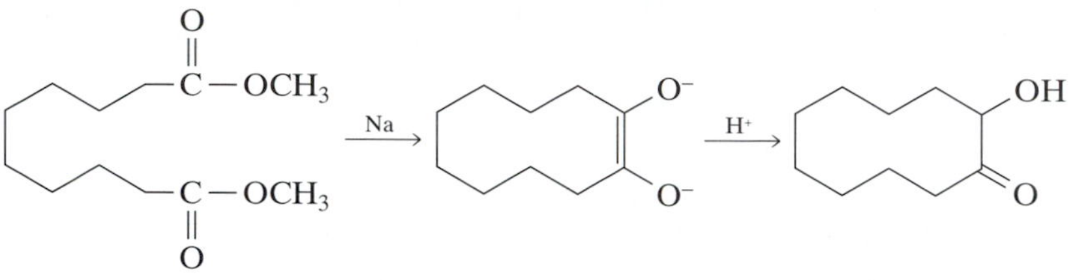

It is speculated that the ends of the diester become attached at nearby sites on the surface of the sodium, increasing the probability that they will "find" one another.

The cycloalkane was labeled with deuterium atoms to help establish that its ring was, in fact, interlinked.

The acyloin condensation was used to prepare the first *catenane,* a compound that has interlinked rings. Initially a 34-carbon ring was formed by an acyloin condensation and reduced to a cycloalkane, cyclotetratriacontane. When the same acyloin condensation was repeated in the presence of cyclotetratriacontane, some 1–2 percent of the product that was isolated consisted of two interlocked rings, formed when molecules of the open-chain diester threaded through cyclotetratriacontane rings before they closed.

Other ring-forming reactions include the Diels–Alder reaction, which is discussed in Experiment 32.

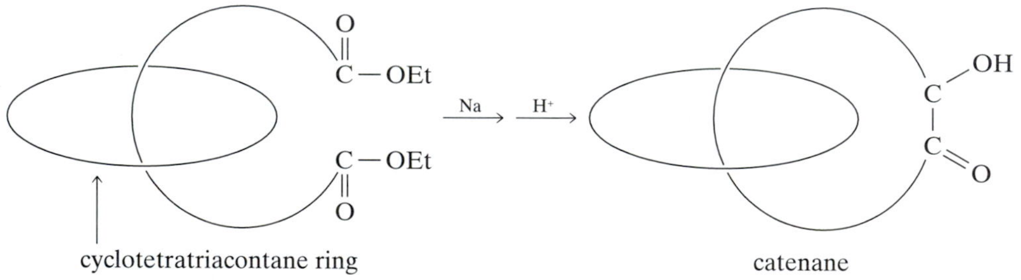

cyclotetratriacontane ring   catenane

Understanding the Experiment

This experiment involves three separate synthetic steps beginning with cyclohexanol. The product from each step becomes the reactant for the next one, so material losses are cumulative. Unless you are careful to minimize your losses you may find yourself with very little product after the last step. For example, if you obtain 60 percent yields for each step of a three-step synthesis, the yield of the final product will be only 22 percent $(0.60^3 \times 100\%)$.

The first step, the oxidation of cyclohexanol with nitric acid, opens the six-membered ring to yield adipic acid (hexanedioic acid). The oxidizing agent, nitric acid, may generate hazardous fumes of nitrogen oxides, so you must work under a hood. Adipic acid precipitates from the reaction mixture and is separated by vacuum filtration.

The conversion of adipic acid to cyclopentanone involves a base-catalyzed cyclization step similar to a Dieckmann condensation, followed by loss of carbon dioxide from the remaining carboxyl group. The reaction is carried out by heating a melted mixture of adipic acid and barium hyroxide (or barium oxide) in a simple distillation apparatus, so that the cyclopentanone distills from the reaction mixture as it is formed. The reaction requires a higher temperature than can be attained with a heating mantle or most other common heat sources. At your instructor's option a burner flame can be used, but you must be very careful to keep the flammable cyclopentanone (or any other flammable liquid) well away from the flame. A hot plate with an aluminum block machined to accomodate a 25-mL boiling flask may also be suitable. Cyclopentanone should begin to distill when the temperature of the molten acid-barium hydroxide mixture reaches 285°C, but it is advisable not to measure the temperature of the melt with a glass thermometer since it may break as it is being inserted in or withdrawn from the hot melt.

Cyclopentanone is reduced to cyclopentanol using sodium borohydride in dilute aqeous sodium hydroxide, as in Experiment 29 for the reduction of vanillin to vanillyl alcohol. Although cyclopentanone does not dissolve completely in the reaction mixture, it is soluble enough to react fairly rapidly. The product is separated by extraction with ether and purified by simple distillation. You will confirm the identity of the product by obtaining its boiling point and infrared spectrum. Infrared spectra of the other products may be recorded as well.

Reactions and Properties

A.

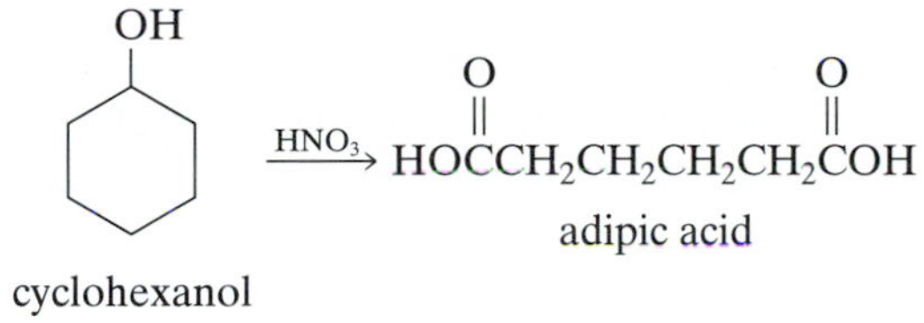

cyclohexanol $\xrightarrow{HNO_3}$ adipic acid

$$\text{HOCCH}_2\text{CH}_2\text{CH}_2\text{CH}_2\text{COH}$$

B.

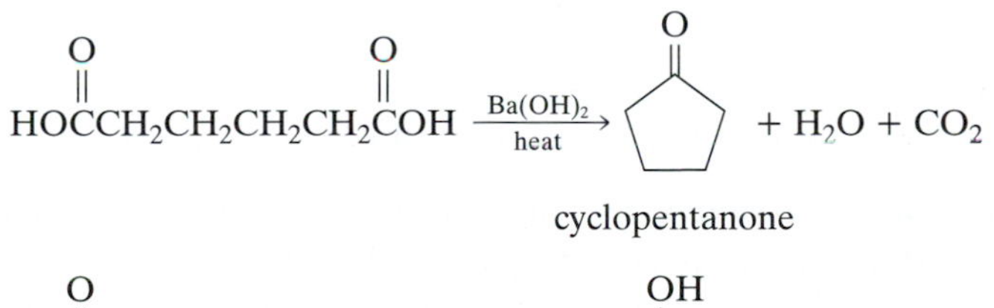

$$\text{HOCCH}_2\text{CH}_2\text{CH}_2\text{CH}_2\text{COH} \xrightarrow[\text{heat}]{Ba(OH)_2}$$

cyclopentanone $+ \text{H}_2\text{O} + \text{CO}_2$

C.

$$4 \;(\text{cyclopentanone}) + \text{NaBH}_4 + 4\text{H}_2\text{O} \longrightarrow 4 \;(\text{cyclopentanol}) + \text{H}_3\text{BO}_3 + 4\text{NaOH}$$

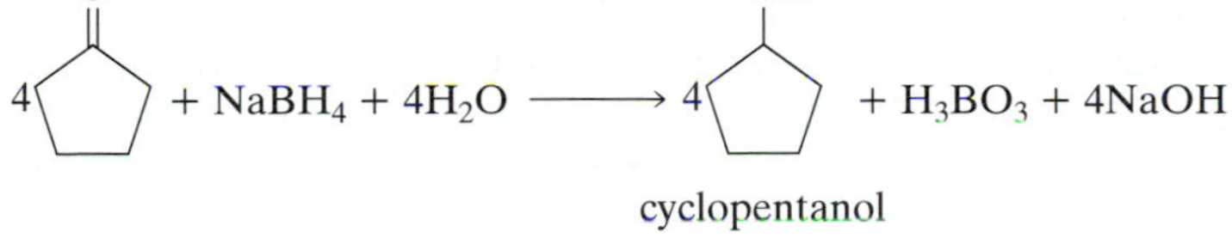

Table 30.1 Physical properties

	M.W.	m.p.	b.p.	d
cyclohexanol	100.2	25	161	0.942
adipic acid	146.1	152	338	
cyclopentanone	84.1	−58	131	0.951
cyclopentanol	86.1	−19	141	0.949

Note: m.p. and b.p. are in °C, density is in g/mL

Directions

A. *Oxidation of Cyclohexanol to Adipic Acid (Hexanedioic Acid)*

Safety Notes

> **Cyclohexanol may be harmful by inhalation, ingestion, and skin contact. Avoid contact and do not breathe vapors.**
> **Nitric acid is corrosive to eyes, skin, and mucous membranes and it can cause severe upper respiratory irritation. Wear gloves, use a hood, avoid contact, and do not breathe vapors.**
> **Adipic acid is a severe eye irritant; avoid contact with eyes.**

cyclohexanol   adipic acid   nitric acid

10.6 M *HNO₃ is prepared by adding 2 volumes of concentrated HNO₃ to 1 volume of water, with cooling.*

Take Care! Wear gloves, avoid contact with HNO_3, do not breathe its vapors.

Take Care! Violent reaction!

Waste Disposal: Dispose of the filtrate as directed by your instructor.

The adipic acid can be recrystallized from ethanol/water if it seems impure.

Reaction. All operations in the "Reaction" step should be performed under an efficient fume hood. Accurately weigh 75 mmol of cyclohexanol. Measure 75 mL of 10.6 *M* aqueous nitric acid into 250-mL Erlenmeyer flask. Add a stirbar [OP-9] or some boiling chips, insert a thermometer, and heat the mixture to 85°C under the hood. Start the stirrer (or swirl after each addition), then discontinue heating and add *no more than* 0.5 mL (1/2 mL) of cyclohexanol to the solution. Moniter the temperature carefully and add another small (~0.5 mL) portion of cyclohexanol each time the temperature drops to 80°C, until all of the cyclohexanol has been added. Be sure none of the toxic red-brown nitrogen oxides generated escape into the laboratory; if any do, *notify your instructor and do not breathe the vapors.* When the addition is complete, heat the mixture at about 90° for 5 minutes.

Separation and Analysis. Let the solution cool to room temperature, inducing crystallization by scratching the inside of the flask with a stirring rod if necessary. Then cool the reaction mixture in an ice/water bath for 30 minutes until crystallization appears complete. Collect the adipic acid by vacuum filtration [OP-12] and wash it on the filter with several small portions of ice-cold water. Dry the adipic acid to constant mass [OP-21], measure its mass and melting point [OP-28], and calculate the percent yield. At your instructor's request, obtain its infrared spectrum [OP-34].

B. *Cyclization of Adipic Acid to Cyclopentanone*

Cyclopentanone is a skin and eye irritant; avoid contact.

Reaction. Using a mortar and pestle, grind together your *dry,* preweighed adipic acid with 1/10 its mass of barium hydroxide until the mixture is a fine powder. Transfer the powder to a 25 mL boiling flask and assemble an apparatus for small-scale simple distillation [OP-25], placing the receiving flask in an ice/water bath. Use heat resistant boiling chips (*not* teflon). Heat the flask using a suitable high temperature heat source [OP-6] until the powder is completely melted, then heat it more strongly until cyclopentanone begins to distill. Continue to distill slowly, keeping the still-head temperature below 150°C, until only a solid residue is left in the flask. To keep the standard taper joints from freezing, loosen them carefully as soon as possible(use heat-protective gloves if available), but wait until the glassware has cooled down to dismantle the apparatus completely. Protect the thermometer from sudden temperature changes that might break it by leaving it in place until the apparatus has cooled down. Break up the residue in the boiling flask and wash out the apparatus as soon as possible after use, being especially careful to clean the joints thoroughly.

Lubricate the ground-glass joints to prevent freezing.

If any black, carbonized residue remains in the boiling flask, dissolve it by covering the residue with a strong NaOH solution [Contact hazard, wear gloves] for 30 minutes or more.

Waste Disposal: Place the residue in a designated waste container.

Separation, Purification, and Analysis. Transfer the distillate to a conical centrifuge tube using 3 mL of saturated aqueous sodium chloride in the transfer, shake the mixture to "salt out" the organic layer, and let it stand. When the layers are well separated, remove the aqueous layer completely with a Pasteur pipet and dry the organic layer over anhydrous potassium carbonate [OP-20]. Purify the cyclopentanone by small-scale simple distillation [OP-25] over a range of about 128–135°C. Measure the mass of the product and calculate the percent yield of this step and the percent yield from cyclohexanol. At your instructor's request, obtain the micro boiling point [OP-29] and infrared spectrum [OP-34] of the cyclopentanone.

Waste Disposal: The aqueous layer can be poured down the drain.

C. *Reduction of Cyclopentanone to Cyclopentanol*

Sodium borohydride is toxic and corrosive, and it can react violently with concentrated acids, oxidizing agents, and other substances. Aqueous borohydride solutions with pH values below 10.5 have been known to decompose violently, so be sure your reaction mixture (if aqueous) is sufficiently alkaline. Avoid contact, do not breathe the dust, and keep NaBH$_4$ away from other chemicals.
Hydrogen is generated when the reaction mixture is acidified, so be sure there are no flames nearby during this step.

Because hydrogen gas is released, this experiment should be carried out under a hood or in a well-ventilated laboratory.

Reaction. For every gram of cyclopentanone, measure out 0.18 g of sodium borohydride, 3.0 mL of 1 *M* aqueous sodium hydroxide, and 1 g of sodium chloride. Put the aqueous NaOH in a 50 mL Erlenmeyer flask and

Take Care! Avoid contact with NaBH$_4$, keep it away from other chemicals.

add the sodium borohydride, stirring magnetically or swirling until it dissolves [OP-9]. With continued swirling or stirring add cyclopentanone drop by drop from a Pasteur pipet, taking about 3–5 minutes for the addition. When the addition is complete, stir the mixture at room temperature for 20 minutes or more.

Take Care! Hydrogen gas is evolved, so keep away from ignition sources.

Separation. Add 6 *M* hydrochloric acid drop by drop with stirring or swirling until the reaction (fizzing) has stopped and the solution is neutral to blue litmus paper Add the previously weighed sodium chloride and stir or swirl until it dissolves, or nearly so. Transfer the solution to a separatory funnel, using 5 mL of solvent grade ethyl ether in the transfer, and use the ether to extract the product from the aqueous layer [OP-13]. Extract twice more with separate 5-mL portions of solvent grade ether and combine the extracts. Dry the extracts with anhydrous magnesium sulfate or sodium sulfate [OP-20], then evaporate the ether [OP-14].

Waste Disposal: Place the recovered ether in a designated solvent recovery container.

Purification and Analysis. Purify the product by small-scale simple distillation [OP-25], collecting the liquid that distills between 135°C and 145°C. Measure the mass of the cyclopentanol. Obtain its micro boiling point [OP-29] and infrared spectrum [OP-34]. Calculate the percent yield of this step and the percent yield from cyclohexanol, then turn in the remaining product to your instructor.

Report. Your report should include a statement of the problem and an account of how you applied scientific methodology to solve it. Tabulate your yield data and try to account for significant material losses.

Exercises

1 (a) Write a mechanism for the ring closure reaction in part **B** that shows the role of the basic catalyst. (b) Write a mechanism for the subsequent decarboxylation reaction. This reaction involves a six-center transition state that incorporates the ring carbonyl group.

2 Using the format in Appendix V, write a flow diagram for all parts of this experiment.

3 Describe and explain the possible effect on your results of the following experimental errors or variations. (a) The bottle labeled "cyclohexanol" actually contained 1-hexanol. (b) In part **B** you used a steam bath as the heat source for the reaction. (c) In part **C** you used 1 *M* HCl rather than 1 *M* NaOH as the reaction solvent.

4 (a) Write a mechanism for the Robinson annulation reaction of ethanal and 3-buten-2-one. (b) Write a mechanism for the reaction of the product from (a) with another molecule of 3-buten-2-one.

5 (a) When condensation reactions are used to form large rings, as in the aldol condensation of 7-oxooctanal shown here, the reactions are usually run in very dilute solutions. Explain why, writing an equation for at least one competing reaction that might occur in a more concentrated solution.

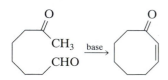

6 The dimethyl ester of adipic acid reacts with 1,6-diaminohexane to form the polymer Nylon 66. Draw the structure of a repeating unit of Nylon 66.

7 Suppose that in the catenane synthesis described on page 236, the 2-ring catenane made up 2 percent of the product isolated. (a) What compound apparently made up the other 98 percent of the product? Give its systematic name. (b) What is the probability that (under the reaction conditions) a molecule of the 34-carbon diester will be threaded through a cyclotetratriacontane ring before it closes?

Other Things You Can Do

(Starred projects require your instructor's permission.)

*1 The experiment can be performed as a four-step synthesis by (1) oxidizing cyclohexanol to cyclohexanone with laundry bleach (see *J. Chem. Educ.* **1981**, *58*, 824) and (2) oxidizing cyclohexanone to adipic acid by the procedure in part **A**, first adding 0.5 g of sodium nitrite to the 75 mL of 10.6 *M* nitric acid.

*2 Investigate the oxidation of alcohols by $KMnO_4$ in Minilab 25.

*3 Carry out another kind of ring-forming reaction by completing Minilab 27.

4 Read the original paper describing the synthesis of the first catenane in *J. Am. Chem. Soc.* **1960**, *82*, 4433.

EXPERIMENT 31

Synthesis of Triphenylmethanol and the Trityl Carbocation

Reactions of Carbonyl Compounds. Reactions of Organic Halides. Preparation of Alcohols. Nucleophilic Addition. Organometallic Compounds. Carbocations. Infrared Spectrometry.

Operations:

OP-6   Heating
OP-9   Mixing
OP-10   Addition
OP-11   Gravity Filtration
OP-12   Vacuum Filtration
OP-13   Extraction
OP-14   Evaporation
OP-19   Washing Liquids
OP-20   Drying Liquids
OP-21   Drying Solids
OP-22   Drying and Trapping Gases
OP-23   Recrystallization
OP-28   Melting Point
OP-34   Infrared Spectrometry

Before You Begin

1 Read the experiment, read or review the operations as necessary, and write an experimental plan.
2 Calculate the mass of 22.0 mmol of magnesium, the mass of 20.0 mmol of benzophenone, the mass and volume of 22.0 mmol of bromobenzene, and the theoretical yield of triphenylmethanol. Calculate the theoretical yield of trityl fluoborate from 1.00 g of triphenylmethanol.

Scenario

The Complimentary Colors Company manufactures synthetic dyes, including a number of triphenylmethane dyes. Gilda Lillie, product development director for the company, would like to develop some new colors to improve their market share in the dye industry. She has learned, for example, that having two *para*-dimethylamino groups on two of the three benzene rings in the parent structure yields a green dye (malachite green), while having three of them yields a violet dye (crystal violet). The company's techicians

carbocation structure   iminium ion structure

Two resonance structures for malachite green, a triphenylmethane dye

need to know how the kinds and positions of substituents on the parent
triphenylmethyl ring structure affect the color of triphenylmethane dyes.
For that they need a sample of the unsubstituted parent compound of these
dyes, the triphenylmethyl (trityl) carbocation. Your assignment is to prepare
triphenylmethanol by a Grignard reaction, convert it to trityl fluoborate (a
salt that contains the trityl carbocation), and find out what color it is.

triphenylmethyl (trityl)
fluoborate

Applying Scientific Methodology

The main scientific problem, determining the color of trityl fluoborate, will
be solved when you prepare the salt. At your instructor's option, you can
obtain a more quantitative measure of its color by recording its UV-VIS
spectrum.

The Colorful Career of the Triphenylmethanes

During his Easter vacation from the Royal College of Chemistry, 18-year-
old William Perkin was trying to synthesize quinine when he came up with
an unpromising black solid that had none of the properties of quinine.
When Perkin dissolved it in water or alcohol, the solid dissolved to yield a
purplish solution that could be used to dye cloth. Later named mauve, this
was the first dye to be produced synthetically, and its preparation marked
the birth of the synthetic dye industry. The first triphenylmethane dye,
fuchsin, was synthesized a few years later, and it was followed soon after by
malachite green, crystal violet, and other triphenylmethane dyes. The race
to develop commercially marketable dyes also stimulated research into the
molecular basis of color. Why, for example, is malachite green "green" and
its reduced form colorless?

    The *chromophore* of a compound is the part of its molecule over which
electrons can be delocalized and which is responsible for its absorption of
ultraviolet or visible light. Triphenylmethane dyes come in all colors of the
rainbow, from the red of rosaniline through malachite green and Victoria
blue to crystal violet. The chromophores responsible for these colors appear
to be nitrogen-substituted triphenylmethyl (trityl) cations, but they are not
true carbocations because most of their positive charge is distributed to the

*The synthesis of quinine was not accom-
plished until 1944—88 years after Perkin
attempted it.*

Reduced form of malachite green

NMe$_2$

Me$_2$N — C — H

A resonance structure of the triphenylmethyl cation

nitrogen atoms, as in the iminium ion form of malachite green shown. According to resonance theory, the iminium ion and carbocation forms are regarded as contributing structures of a resonance hybrid that has some characteristics of each. The electron delocalization suggested by such structures is responsible for their colors. As a rule, compounds with long chromophores that allow extensive electron delocalization tend to be colored; the longer the chromophore, the higher the wavelength of light they absorb. The reduced form of malachite green is not colored because the saturated carbon that connects the rings prevents delocalization over the three-ring system.

Although it is nearly a trillion times less stable than crystal violet, the trityl cation is unusually stable for a carbocation. When protected from atmospheric moisture, trityl salts will keep almost indefinitely. The cation owes this unusual stability to delocalization of the positive charge about its three benzene rings. The cation is apparently shaped somewhat like a propeller with the "blades" (benzene rings) pitched at a 32° angle, because steric interference between the ortho hydrogen atoms makes a planar configuration impossible.

Shape of triphenylmethyl cation

The year 1900 marked two significant milestones in organic chemistry. That was the year Moses Gomberg announced his discovery of the trityl free radical (see Experiment 25) and Victor Grignard reported the development of Grignard reagents. Just a year later, trityl carbocations were being prepared from triphenylmethanol, which is most easily synthesized using a Grignard reagent.

Understanding the Experiment

Victor Grignard's original procedure for preparing a Grignard reagent was as follows: About a mole of magnesium metal was placed in a dry two-necked round-bottomed flask fitted with a reflux condenser and dropping funnel. A mole of the organic halide was dissolved in ethyl ether and 50 mL of this solution was added to the magnesium. When a white turbidity appeared at the metal surface and effervescence began, more ethyl ether was added in portions, with cooling, followed by drop-by-drop addition of the remainder of the halide-ether solution. The reaction was brought to completion by refluxing in a water bath, resulting in a nearly colorless liquid. Essentially the same method is used today for making many Grignard reagents, although extensive studies of the reaction have led to some modifications of the reaction conditions.

It is extremely important that the reagents and apparatus be as dry as possible since water not only reacts with Grignard reagents but also inhibits their formation. In a study using butyl bromide, it was found that the *induction period* (the time between the combination of reactants and the start of a noticeable reaction) for forming the Grignard reagent was $7\frac{1}{2}$ minutes using sodium dried ethyl ether, 20 minutes using commercial absolute ethyl ether, and 2 hours using ethyl ether half saturated with water. It is apparent that careful drying of the reaction apparatus and reagents saves time in the long run; it can increase the yield of Grignard reagent as well.

The type and quantity of reagents and solvents used are also important. For most purposes, an ordinary grade of well-dried magnesium turnings is suitable, if they are rubbed and crushed with a glass rod to remove some of the oxide coating and to provide a fresh surface for reaction. Alternatively, magnesium ribbon can be scraped to remove oxide and cut into small pieces. Commercial anhydrous ethyl ether is suitable for most routine preparations, but the optimum quantity of ethyl ether depends on the kind of Grignard reagent. One study showed that the highest yields of phenylmagnesium bromide were obtained with 5 moles of ethyl ether per mole of bromobenzene.

In this experiment, you will prepare a Grignard reagent, phenylmagnesium bromide, by adding a solution of bromobenzene in dry ethyl ether to magnesium metal. To help get the reaction started, the magnesium can be activated by crushing it with a crystal or two of iodine and using a heat gun to sublime the iodine onto the magnesium turnings. The onset of the reaction is signaled by the formation of small bubbles at the surface of the magnesium, usually followed by the appearance of a cloudy precipitate. Before long the ether should begin to boil spontaneously. If no reaction is observed after 15 minutes or so, you can sometimes "jump-start" the reaction by adding a little previously prepared Grignard reagent to the reaction mixture.

Using a high concentration of bromobenzene helps to get the reaction started, but it can promote the formation of an undesirable by-product, biphenyl, through a side reaction at the metal's surface. Thus the bromobenzene solution is diluted with ethyl ether as soon as the reaction gets underway. Phenylmagnesium bromide reacts rapidly with water to form benzene and more slowly with oxygen to form a magnesium salt of phenol, so the reaction apparatus must be protected from moisture and the Grignard reagent should be used promptly after it is prepared.

When benzophenone is added to the Grignard reagent, a magnesium salt of triphenylmethanol precipitates from the reaction mixture, which usually turns pink during the addition. This salt is converted to triphenylmethanol by adding water. Then dilute hydrochloric acid is added to dissolve the basic magnesium salts that form along with the triphenylmethanol. After extraction from the reaction mixture with ethyl ether, the crude triphenylmethanol is treated with petroleum ether to remove biphenyl, then purified by recrystallization.

Carbocations can be prepared by mixing alcohols with a strong acid such as fluoboric acid (tetrafluoroboric acid); the fluoborate anion is a very weak nucleophile that doesn't react with the resulting carbocation. You will prepare trityl fluoborate by the reaction of triphenylmethanol with 48% fluoboric acid. The water in the aqueous fluoboric acid solution, as well as that produced during the carbocation-forming reaction, could prevent or reverse the reaction. Acetic anhydride is added to consume the water via the reaction in the margin.

Preparation of carbocations using fluoboric acid

$$ROH + HBF_4 \rightleftharpoons ROH_2^+ + BF_4^-$$
$$\longrightarrow R^+BF_4^- + H_2O$$

$$\underset{\substack{\text{acetic} \\ \text{anhydride}}}{CH_3\overset{\overset{\displaystyle O}{\|}}{C}O\overset{\overset{\displaystyle O}{\|}}{C}CH_3} + H_2O \longrightarrow \underset{\substack{\text{acetic} \\ \text{acid}}}{2CH_3\overset{\overset{\displaystyle O}{\|}}{C}OH}$$

Reactions and Properties

A.

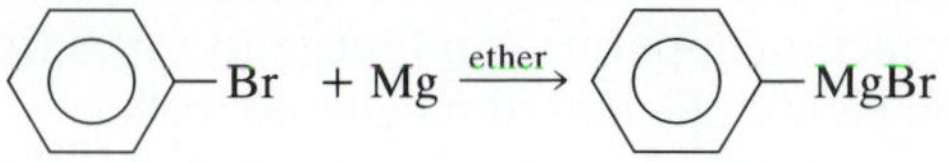

bromobenzene   phenylmagnesium bromide

B.

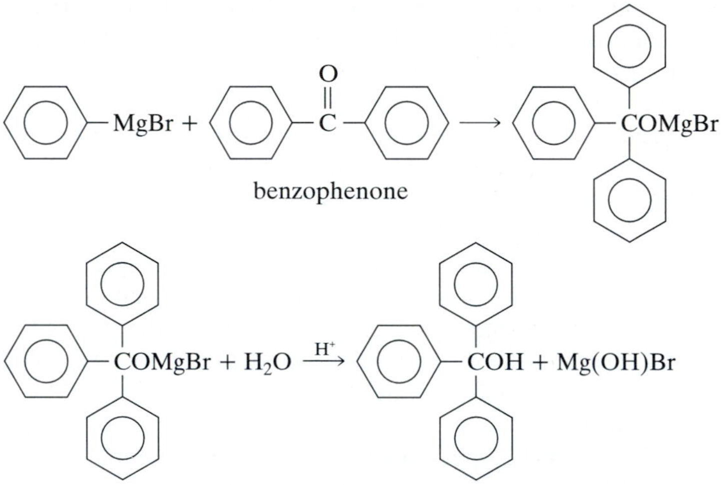

benzophenone

triphenylmethanol

C.

trityl fluoborate

Table 31.1 Physical properties

	M.W.	m.p.	b.p.	d
bromobenzene	157.0	−31	156	1.495
magnesium	24.3			
ethyl ether	74.1	−116	34.5	0.714
benzophenone	182.2	48	306	
triphenylmethanol	260.3	164	380	
fluoboric acid (48%)	87.8			1.41
acetic anhydride	102.1	−73	140	1.082
trityl fluoborate	330.1			

Note: m.p. and b.p. are in °C, density is in g/mL

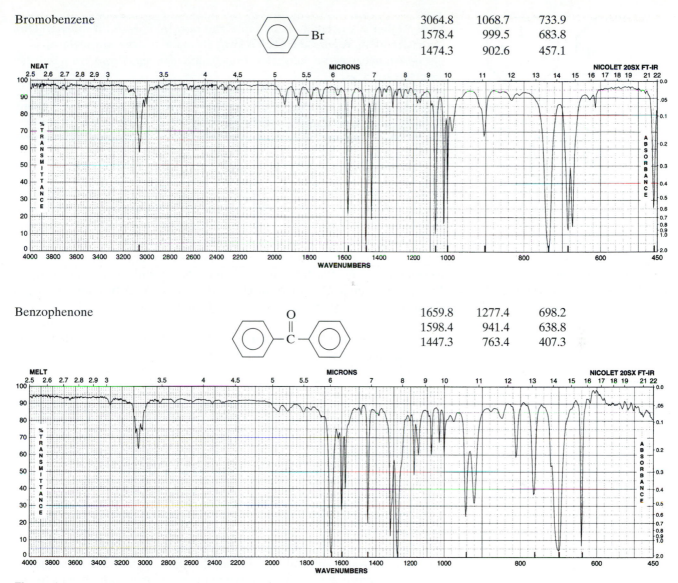

Bromobenzene

3064.8	1068.7	733.9
1578.4	999.5	683.8
1474.3	902.6	457.1

Benzophenone

1659.8	1277.4	698.2
1598.4	941.4	638.8
1447.3	763.4	407.3

Figure 31.1 IR spectra of the starting materials (Reproduced from *The Aldrich Library of FT-IR Spectra, Edition II*, with the permission of the Aldrich Chemical Company.)

Directions

If possible, clean and predry the glassware needed for the reaction before the lab period to reduce the drying time.

Bromobenzene causes eye and skin irritation, and inhalation, ingestion, or skin absorption may be harmful. Avoid contact with the liquid and do not breathe its vapors.

Safety Notes

bromobenzene   ethyl ether

magnesium

Stop and Think: What might happen if the apparatus isn't completely dry?

Stop and Think: What is the purpose of this procedure?

Take Care! Keep ether away from flames and hot surfaces.

Stop and Think: What is the purpose of the filter paper?

Observe and Note: What happens?

Stop and Think: What is the limiting reactant in this synthesis?

> **Ethyl ether is extremely flammable and may be harmful if inhaled. Do not breathe its vapors and keep it away from flames and hot surfaces. Magnesium can cause dangerous fires if ignited; keep it away from flames and hot surfaces.**
> **Petroleum ether is extremely flammable and can be harmful if inhaled or absorbed through the skin. Avoid inhalation and prolonged contact. Keep it away from flames and hot surfaces.**

A. *Preparation of Phenylmagnesium Bromide*

Reaction. It is essential that all apparatus used during this reaction step be clean and scrupulously dried. Dry the following glassware in a 110°C oven for at least 30 minutes: 100-mL round-bottomed flask, Claisen connecting tube, reflux condenser, separatory/addition funnel, glass stopper, thermometer adapter (remove the rubber connector), drying tube filled with calcium chloride, flat-bottomed stirring rod, 50-mL Erlenmeyer flask. Meanwhile, weigh 22.0 mmol of clean, dry magnesium turnings and keep them dry. As soon as the glassware is cool enough to handle, add one or two *small* iodine crystals with the magnesium to the reaction flask and *carefully* crush the magnesium turnings using the flat end of a flat-bottomed stirring rod (don't punch a hole in the flask!). Without delay, assemble an apparatus for addition under reflux [OP-10], inserting the drying tube [OP-22] in the top of the reflux condenser. Using a heat gun (a hair dryer will work) or heating mantle, gently heat the bottom of the reaction flask until purple vapors rise from the iodine and condense onto the magnesium turnings.

Weigh 22.0 mmol of dry bromobenzene into the dried Erlenmeyer flask and dissolve it in 5.0 mL of *anhydrous* ethyl ether, then transfer the solution to the dry separatory/addition funnel and stopper it, placing a strip of filter paper between the stopper and the neck of the flask. Add the bromobenzene solution all at once to the reaction flask, and replace it in the addition funnel by 6.5 mL of anhydrous ethyl ether. Observe the reaction mixture closely for evidence of a reaction. If the ethyl ether does not start to boil after a few minutes, warm the flask in the palm of your hand until it does begin to boil, then remove your hand and see whether the reaction proceeds unassisted; if it does not, consult your instructor. When the reaction mixture begins to boil (without external heating) quite vigorously, add the ethyl ether drop by drop over about a one-minute period, shaking the flask occasionally to keep the reactants mixed. Let the reaction continue until the boiling has nearly stopped, then use a heating mantle or warm water bath to heat the reaction mixture at a gentle reflux for another 10–15 minutes [OP-6]. The reflux ring of condensing ether should be in the lower third of the condenser. If any ethyl ether evaporates, reducing its volume in the reaction flask, replace it with fresh anhydrous ethyl ether. Do not stop at this point because the phenylmagnesium bromide solution will not keep for long.

B. *Preparation of Triphenylmethanol*

Reaction. While the reaction mixture is cooling to room temperature, dissolve 20.0 mmol of benzophenone in 10 ml of anhydrous ethyl ether in a dry Erlenmeyer flask and place it in the separatory/addition funnel. Add this solution drop by drop to the cooled reaction mixture with shaking or magnetic stirring [OP-9]. The addition rate should be sufficient to keep the ethyl ether boiling gently without external heating. When the addition is complete, use a heating mantle or warm water bath to heat the reaction mixture under gentle reflux for another 15 minutes. (If you stop after the addition and allow the reaction mixture to stand overnight or longer, this heating period can be omitted.) Let the reaction mixture cool to room temperature, during which time the magnesium salt of triphenylmethanol should solidify.

With shaking or magnetic stirring, add 5 mL of water drop by drop through the separatory/addition funnel, followed by 25 mL of 5% (1.4 *M*) hydrochloric acid. The white solid should eventually dissolve, possibly leaving a residue of unreacted magnesium. If any white solid remains undissolved, add some solvent grade (not anhydrous) ethyl ether with shaking or stirring, followed by more 5% HCl if the ether doesn't dissolve it.

Separation. Add solvent grade ethyl ether as needed to replace any that was lost by evaporation. If there is undissolved magnesium present, remove it by gravity filtration through glass wool [OP-11], washing the magnesium and glass wool with a little solvent grade ethyl ether. Transfer the solution to a separatory funnel and separate the layers, saving them both. Extract the aqueous layer [OP-13] with 10 mL of solvent grade ethyl ether and combine this extract with the original ethyl ether layer. Carefully wash the combined ether solution [OP-19] with 15 mL of 5% aqueous sodium bicarbonate. (**Take Care!** A gas may be evolved; vent as necessary.) Then wash it with 15 mL of saturated aqueous sodium chloride. Dry the ether solution over anhydrous magnesium sulfate [OP-20], removing the drying agent by gravity filtration and rinsing it with a little solvent grade ethyl ether. Evaporate the ethyl ether under vacuum [OP-14] using a cold trap, with gentle heating if necessary. Add 10 mL of hexanes (or high-boiling petroleum ether) to the solid residue, then stir and crush the solid in this solvent for several minutes to remove biphenyl. Collect the product by vacuum filtration [OP-12].

Purification and Analysis. Recrystallize the crude triphenylmethanol [OP-23] from a 2:1 mixture of hexanes (or high-boiling petroleum ether) with absolute ethanol. Triphenylmethanol crystals form slowly, so allow plenty of time for complete crystallization. Dry the purified triphenylmethanol [OP-21], weigh it, and measure its melting point [OP-28]. If requested, obtain the infrared spectrum of triphenylmethanol [OP-34].

C.   *Preparation of Trityl Fluoborate*

> **Acetic anhydride can cause severe damage to skin and eyes, its vapors are very harmful if inhaled, and it reacts violently with water. Use gloves and a hood; avoid contact with the liquid, do not breathe its vapors, and keep it away from water.**
>
> **Fluoboric acid is poisonous, its solutions can cause severe damage to skin and eyes, and its vapors irritate the respiratory system. Use gloves and a hood; avoid contact with the acid solution, and do not breathe its vapors.**

Under the hood, mix 1.00 g of triphenylmethanol with 7.0 mL of acetic anhydride in a small Erlenmeyer flask. Carefully add 1.0 mL of 48% fluoboric

Observe and Note: What happens?

Waste Disposal: The aqueous layers can be poured down the drain.

Waste Disposal: Place the ethyl ether in a designated solvent recovery container.

Waste Disposal: Place the filtrate in a designated solvent recovery container.

Waste Disposal: Place the filtrate in a designated solvent recovery container.

Safety Notes

acetic anhydride

fluoboric acid

Take Care! Wear gloves, avoid contact, do not breathe vapors, keep away from water.

Waste Disposal: Place the filtrate in a designated waste container.

Observe and Note: What color is the product?

acid and swirl to dissolve the solid. Stopper the flask and let the mixture stand for about 15 minutes; then cool it in ice until crystallization is complete. Collect the product by vacuum filtration [OP-10] using a Hirsch funnel and wash it on the filter with cold anhydrous ethyl ether. Weigh the trityl fluoborate in a *dry* tared vial and turn it in, along with the rest of your triphenylmethanol.

Report. Your report should include a statement of the problem and an account of how you applied scientific methodology to solve it. Calculate your percent yield of triphenylmethanol and try to account for significant losses. Calculate your percent yield of trityl fluoborate.

Exercises

1 (a) Write a balanced equation for the coupling reaction of bromobenzene at the metal surface to form biphenyl. (b) Write balanced equations for the reactions of phenylmagnesium bromide and trityl fluoborate with water.

2 If you obtained the IR spectrum of triphenylmethanol, compare it with the spectra in Figure 31.1 and describe the evidence indicating that the expected reaction has taken place. Interpret your spectrum as completely as you can.

3 Describe and explain the possible effect on your results of the following experimental errors or variations. (a) You use a steam bath to sublime the iodine crystals before the apparatus is assembled. (b) You use solvent grade (not anhydrous) ethyl ether for the reaction in part **A**. (c) You forget to add the 6.5 mL portion of anhydrous ethyl ether to the reaction mixture. (d) You use ethyl ether, rather than petroleum ether, to remove biphenyl from the crude triphenylmethanol.

4 Following the format in Appendix V, construct a flow diagram for the synthesis of triphenylmethanol (parts **A** and **B**).

5 The reaction of phenylmagnesium bromide with benzophenone to form the salt of triphenylmethanol is an example of nucleophilic addition; its reaction with ethyl benzoate yielding the same product involves nucleophilic substitution followed by a nucleophilic addition step. Write reasonable mechanisms for both reactions.

6 Outline a synthetic pathway for preparing each of the following compounds, using the Grignard reaction and starting with benzene or toluene: (a) 1,1-diphenylethanol; (b) 1,2-diphenylethanol; (c) 2,2-diphenylethanol; and (d) 2,3-diphenyl-2-butanol.

7 Excluding alternative Kekulè structures for the benzene rings: (a) Draw all possible resonance structures for the trityl cation; (b) Draw all possible resonance structures for malachite green.

8 One possible by-product from the triphenylmethanol synthesis is ethoxytriphenylmethane. Explain how and when it might form, giving an equation and a mechanism for the reaction.

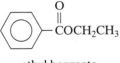

ethyl benzoate

Other Things You Can Do

(Starred projects require your instructor's permission.)

*1 Record the ultraviolet-visible spectrum (200–600 nm) of trityl fluoborate in dry acetone.

*2 Dissolve a small amount of trityl fluoborate in dry methanol, and record your observations. Dissolve about 0.1 g of trityl fluoborate in 1 mL of dry acetone; then add a solution of sodium iodide in dry acetone (0.1 g NaI in 1 mL acetone) drop by drop until no more changes are observed. Write balanced equations to explain your observations.

*3 Prepare the fluorescent dye fluorescein in Minilab 26.

4 Write a research paper about the structures, properties, and applications of Grignard reagents, starting with sources listed in the Bibliography.

EXPERIMENT 32

Identification of a Conjugated Diene from Eucalyptus Oil

Reactions of Dienes. Preparation of Bicyclic Compounds. Cycloaddition. Infrared Spectrometry. Qualitative Analysis.

Operations:

OP-6   Heating
OP-12 Vacuum Filtration
OP-21 Drying Solids
OP-23 Recrystallization
OP-28 Melting Point
OP-32 Gas Chromatography
OP-34 Infrared Spectrometry

Before You Begin

1 Read the experiment, read or review the operations as necessary, and write an experimental plan.
2 Be prepared to carry out the calculations described in part **A** of the Directions.
3 Record the gas chromatogram of the "eucalyptus oil" during the previous experiment, if possible.

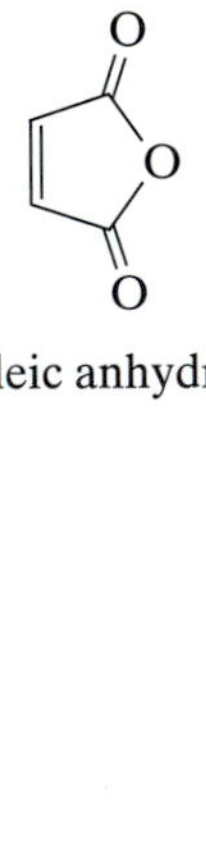

maleic anhydride

Scenario

In 1927, Otto Diels and Kurt Alder treated a constituent of one kind of eucalyptus oil with maleic anhydride and obtained a new compound that they described as forming *grosse glasglänzende Krystalle von ungewöhnlicher Schönheit* (large lustrous crystals of unusual beauty). The reaction Diels and Alder used for this preparation was eventually named for them, and the lustrous crystals belonged to the Diels–Alder adduct of a natural diene. Gondwana Natural Products Ltd., which acquires and markets useful products from Australian flora, is investigating the commercial possibilities of an essential oil from *Eucalyptus dives* that has been used to treat colds as well as malaria and other fevers. The ultraviolet-visible spectrum of the eucalyptus oil suggests that it contains the same diene that Diels and Alder studied —one of the natural dienes described next. They want your Institute to identify the diene from a sample of the eucalyptus oil they have provided. Since only conjugated dienes undergo the Diels–Alder reaction, this reaction can be used to separate the diene from the eucalyptus oil as well as identify it.

Applying Scientific Methodology

Since the problem involves the identity of an unknown diene, your working hypotheses can only be a guess, but you should be able to eliminate some of the possibilities from the list in the next section before you begin. Your hypothesis will be tested when you obtain the melting point of the adduct. Note that a triene having two conjugated double bonds can also qualify as the "diene" for a Diels–Alder reaction.

Dienes and Trienes in Nature

Dienes and trienes occur in the essential oils of a number of plants and contribute to the flavors and aromas of such plants. For example, limonene has a pleasant lemony odor that enhances the flavor of lemons, oranges, and other citrus fruits, while β-myrcene is responsible for much of the fragrance and flavor of bay leaves (*Myrcia acris*). β-Myrcene is also present in hops, verbena, and lemongrass oil. β-Ocimene was first isolated from the Javanese oil of basil (*Ocimium basilicum*) and is usually found in combination with *allo*-ocimene, which is also synthesized from α-pinene, the most abundant component of oil of turpentine. Both of the phellandrenes derive their name from the water fennel *Phellandrium aquaticum*, but α-phellandrene apparently doesn't occur in that plant; it was mistaken for its isomer, β-phellandrene, which does. α-Phellandrene *is* found in the oils of bitter fennel, ginger grass, cinnamon, and star anise, while β-phellandrene also occurs in lemon oil and Japanese peppermint oil. Another cyclic diene, α-terpinene, is obtained from the essential oils of cardamom, marjoram, and coriander.

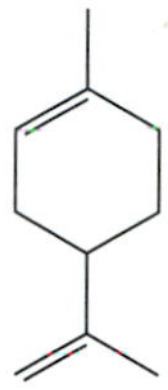

limonene

$$CH_3C=CHCH_2CH_2CCH=CH_2$$
$$\quad|\qquad\qquad\qquad\quad||$$
$$\quad CH_3\qquad\qquad\qquad CH_2$$

β-myrcene

$$CH_3C=CHCH_2CH=CCH=CH_2$$
$$\quad|\qquad\qquad\qquad\quad|$$
$$\quad CH_3\qquad\qquad\qquad CH_3$$

β-ocimene

$$CH_3C=CHCH=CHC=CHCH_3$$
$$\quad|\qquad\qquad\qquad|$$
$$\quad CH_3\qquad\qquad CH_3$$

allo-ocimene

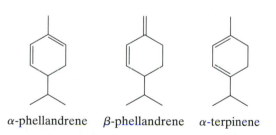

α-phellandrene   β-phellandrene   α-terpinene

Understanding the Experiment

Most conjugated dienes have the the ability to form Diels–Alder adducts with maleic anhydride. Trienes such as β-myrcene can also form such adducts if at least two of their double bonds are conjugated. The adducts are usually crystalline solids that can be separated from the other components of an essential oil and used to identify the diene.

The Diels–Alder reaction is classified as a 4 + 2 cycloaddition reaction because one reactant (the *diene*) contributes four carbons and the other reactant (the *dienophile*) contributes two carbons to the six-membered ring of the resulting cyclic compound (the *adduct*). As illustrated for the following reaction of 1,3-butadiene and ethene, the diene must be able to exist in an *s-cis* conformation, in which the carbon atoms bonding to the dienophile are on the same side of the C—C single bond.

A 4 + 2 cycloaddition reaction

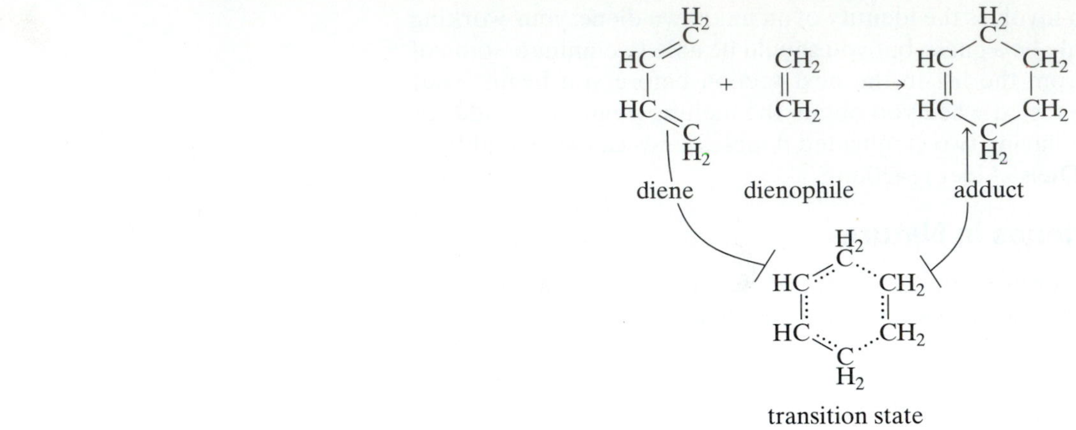

transition state

The dienophile must have either a double or a triple bond, often connected to one or more carbonyl groups or other electron-withdrawing groups.

The Diels–Alder reaction is stereoselective, usually yielding only one of several possible stereoisomers. For example, maleic acid could react with cyclopentadiene to yield either of two adducts, designated *exo* and *endo*. In fact, it yields entirely the *endo* adduct, in which the bulkier parts of the dienophile are closer to the developing carbon-carbon double bond. This orientation results from the fact that overlap between the pi electrons of the diene and those of the dienophile stabilizes the transition state leading to the adduct. Such overlap is possible only when the carbon-carbon double bonds of the diene are in close proximity to the carbonyl groups of the dienophile.

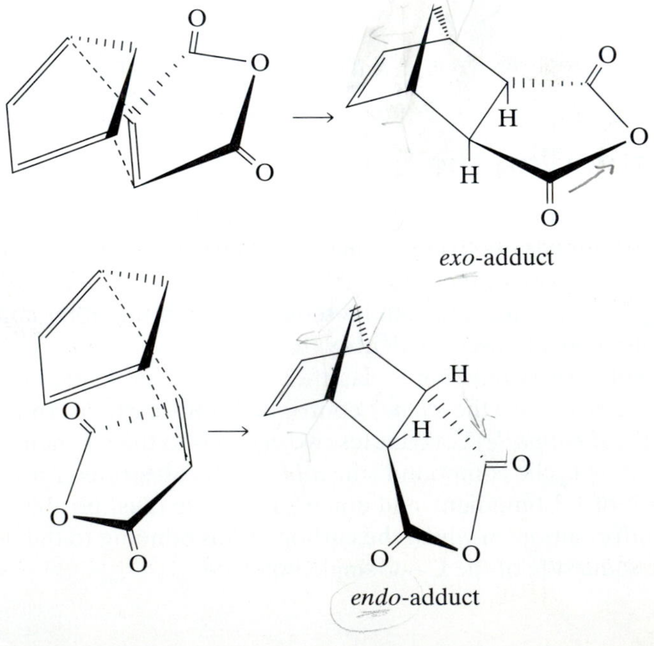

exo-adduct

endo-adduct

Sometimes more than one *endo* adduct is possible; in that event, the dienophile will tend to approach the diene from its less hindered side to give the more stable adduct.

In this experiment you will prepare the Diels–Alder adduct of the unknown conjugated diene in "eucalyptus oil," separate the adduct, and identify the diene from the melting point of its adduct. The unknown (whose molecular formula is $C_{10}H_{16}$) will be one of four conjugated dienes that were discussed in the previous section. Their names and the melting points of their adducts are listed in Table 32.2. Using gas chromatography, you will determine the approximate percentage of diene in the eucalyptus oil so that you can estimate the amount of maleic anhydride needed to react with the diene. It is best to avoid using too much maleic anhydride because the excess can be difficult to remove from the product. Both maleic anhydride and the adduct can be hydrolyzed by water, so it is important to use dry glassware and to keep out moisture during the reaction and workup. Because of its moisture sensitivity, maleic anhydride is usually sold in the form of briquettes, which must be pulverized with a mortar and pestle before being used.

A Diels–Alder reaction is sometimes carried out by simply mixing and heating the reactants. The reaction can be very exothermic, however, so a solvent is often used to keep it from proceeding too rapidly. You will prepare the adduct by heating the reactants in ethyl ether, which serves as the reaction solvent. The adduct should precipitate from the reaction mixture as beautiful rectangular crystals; slow cooling may yield crystals several centimeters long. The adduct can then be separated by vacuum filtration and purified by recrystallization from methanol. Because the adduct may react with methanol to form a solvolysis product, you should avoid prolonged boiling during recrystallization. For the same reason, it is not a good idea to leave the crystallized adduct in methanol for more than a few hours.

Once you have identified the adduct, you should be able to deduce its structure with the help of molecular models. You can also characterize the adduct by recording its infrared spectrum. The infrared spectra of anhydrides show two carbonyl stretching bands that arise from symmetric and asymmetric stretching modes; maleic anhydride itself has $C{=}O$ bands near 1780 and 1850 cm^{-1}, as shown in Figure 32.1. The C—CO—O—CO—C grouping can vibrate as a unit, causing additional bands that occur near 900 cm^{-1} and 1250 cm^{-1} for cyclic anhydrides.

Hydrolysis of maleic anhydride

maleic acid

Reaction and Properties

The actual structure of the adduct depends on the structure of the unknown diene.

Table 32.1 Physical properties

	M.W.	m.p.	b.p.	d
diene from eucalyptus oil	136.2		172	0.841
maleic anhydride	98.1	53	202	
ethyl ether	74.1	−116	34.5	0.714

Note: m.p. and b.p. are in °C, density is in g/mL

Table 32.2 Melting points of maleic anhydride adducts of the dienes

diene	m.p. of adduct
β-myrcene	33–34
allo-ocimene	83–84
α-phellandrene	126–127
α-terpinene	60–61

Maleic anhydride

3118.7	1240.1	840.6
1851.1	1058.1	695.9
1778.5	890.8	561.7

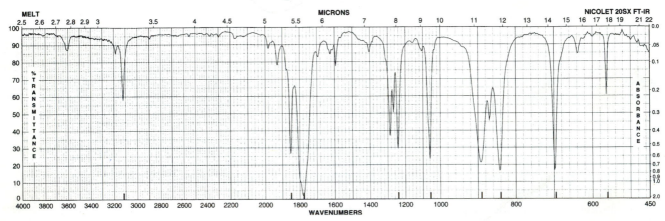

Figure 32.1 IR spectrum of maleic anhydride (Reproduced from *The Aldrich Library of FT-IR Spectra, Edition II*, with the permission of the Aldrich Chemical Company.)

Directions

Safety Notes

Maleic anhydride is corrosive and can cause severe damage to the eyes, skin, and upper respiratory tract. Avoid contact with skin, eyes, or clothing, and do not breathe the dust. If you must pulverize maleic anhydride briquettes, wear gloves and work under the hood.

Ethyl ether and petroleum ether are extremely flammable and may be harmful if inhaled. Do not breathe vapors and keep away from ignition sources.

Methanol is harmful if inhaled or absorbed through the skin. Avoid contact and do not breathe vapors.

maleic anhydride

ethyl ether

petroleum ether

methanol

A. *Preliminary Analysis and Calculations*

Obtain a gas chromatogram [OP-32] of the eucalyptus oil provided, using the column and conditions recommended by your instructor. Assuming that the unknown diene is responsible for the largest peak on the chromatogram and that peak areas are proportional to component masses, estimate the mass of the unknown diene in 5.00 g of the oil and the mass of maleic anhydride needed to react with that much diene.

B.   Preparation of the Adduct

Reaction.   *All glassware must be dry.* In a clean, dry, round-bottomed flask dissolve 5.00 g of the eucalyptus oil in 10.0 mL of anhydrous ethyl ether and add the calculated mass of powdered maleic anhydride. Heat the reaction mixture gently under reflux [OP-6] on a steam bath or hot water bath for 45 minutes or more. While it is still warm, transfer the mixture to a small Erlenmeyer flask, cover the flask with a watch glass, and let it cool slowly to room temperature. If crystallization does not begin by the time the reaction mixture reaches room temperature, scratch the inside of the flask with a glass stirring rod. Cool it further in an ice bath to allow complete crystallization.

Separation. Collect the adduct by vacuum filtration [OP-12], washing the crystals on the filter with 10 mL of cold, low-boiling petroleum ether.

Purification and Analysis. Recrystallize the adduct [OP-23] from dry methanol, avoiding prolonged boiling, and collect it by vacuum filtration. Dry the adduct [OP-21] and measure its mass and melting point [OP-28]. Deduce the identity of the diene from the melting point of its adduct. Record the infrared spectrum [OP-34] of the adduct or obtain a spectrum from your instructor. Turn in your product to the instructor.

Stereochemistry of the Adduct (Optional). Construct molecular models for maleic anhydride and the diene. By moving their bonds around, find a way to connect them to make a model representing one form of the adduct. Disconnect and reconnect the diene and dienophile units until you have made models representing all possible structures for the adduct. Based on the description of the Diels–Alder reaction in "Understanding the Experiment," decide which is the most likely structure.

Report. Your report should include a statement of the problem and an account of how you applied scientific methodology to solve it. Give the name and structure of the diene in eucalyptus oil, and justify your conclusion. Calculate the theoretical yield of the adduct based on the mass of maleic anhydride you used, and calculate your percentage yield. Write all possible structural formulas for the adduct, showing their stereochemistry clearly; designate the structure you believe is the correct one.

Take Care! Keep ether away from flames and hot surfaces.

Take Care! Avoid contact with maleic anhydride, do not breathe its dust.

Observe and Note: What happens?

Take Care! Keep petroleum ether away from flames and hot surfaces.

Waste Disposal: Place all filtrates in designated solvent recovery containers.

4.0g

Exercises

1 If you obtained an infrared spectrum of the adduct, interpret it as completely as you can. Compare the adduct's spectrum with that of maleic anhydride; point out and try to explain any significant similarities or differences.

2 Write a balanced equation for the reaction. Show the stereochemistry of the adduct and explain why it has that stereochemistry.

3 Which two dienes whose structures are shown in this experiment will not form Diels–Alder adducts with maleic anhydride? Explain why in each case.

4 Describe and explain the possible effect on your results of the following experimental errors or variations. (a) You calculated the mass of maleic anhydride needed based on the total mass of the eucalyptus oil. (b) Your reaction flask was wet. (c) You dissolved the adduct in hot methanol and then stored the recrystallization solution until the next lab period.

5 Write the structure of the compound that would result if the adduct were heated too long in the recrystallization solvent; also write a balanced equation for its formation.

6 Following the format in Appendix V, construct a flow diagram for this experiment.

7 The side reaction most often encountered in Diels–Alder syntheses is dimerization or polymerization in which the diene also acts as a dienophile. For example, butadiene can react with itself to yield 4-vinylcyclohexene as shown. Draw the structures of four possible Diels–Alder dimers of your diene.

Dimerization of 1,3-butadiene

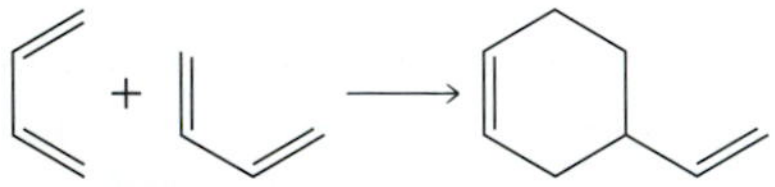

Other Things You Can Do

(Starred projects require your instructor's permission.)

*1 Prepare the Diels–Alder adduct of an aromatic "diene" in Minilab 27.

2 A number of polychlorinated insecticides, such as dieldrin, aldrin, and chlordane, are synthesized using one or more Diels–Alder reactions. Starting with sources listed in the Bibliography, write a research paper about such insecticides, giving equations for their manufacture from cyclopentadiene and reporting on their uses and environmental effects.

Spectral Identification of Terpenoids   **EXPERIMENT 33**

Infrared Spectrometry. Ultraviolet-Visible Spectrometry. Qualitative Analysis.

Operations:

OP-34 Infrared Spectrometry
OP-36 Ultraviolet-Visible Spectrometry
OP-35 (optional) Nuclear Magnetic Resonance Spectrometry

Before You Begin

Read the experiment, read or review the operations as necessary, and write an experimental plan.

Scenario

Uncommon Scents, Inc., extracts essential oils from various plants and ships them to buyers around the world. Some of their buyers are aromatherapists from countries that classify aromatherapy oils as drugs and strictly regulate their contents. These countries require that the suppliers of essential oils report their major ingredients and list them on the label. Uncommon Scents has air freighted your Institute samples of the major components of one of their products, the essential oil of the motley marigold, *Calendula salma-gundi.* Your project group's assignment is to identify the components using appropriate spectral methods and determine what names should appear on the label.

Applying Scientific Methodology

This experiment can be performed as a group project in which each individual obtains the spectra for one of the components and all members of the group, working together, arrive at the identity of each component. You should be able to formulate a working hypothesis regarding the identity of your component after you record its spectra, but be prepared to have your hypothesis tested and either accepted or rejected by the other members of your group.

Monoterpenoids from Plants

Terpenes are hydrocarbons that can, in principle, be broken down into two or more isopentane units. Monoterpenes are composed of two such units, sesquiterpenes of three, diterpenes of four, etc. Analogous compounds that contain oxygen or other atoms in addition to carbon and hydrogen are called *terpenoids.* Many terpenes and terpenoids are important constituents of the essential oils of plants, which are usually obtained by steam-distilling the roots, bark, leaves, flowers, or other parts of the plant. The components of essential oils are volatile, water-insoluble substances

See Experiment 12 for more information about terpenes and terpenoids.

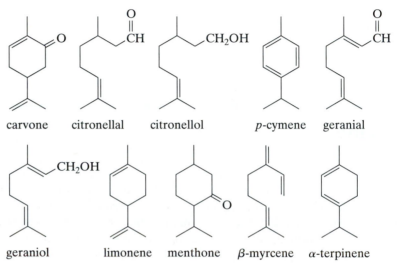

Isopentane units in terpenes and terpenoids

Figure 33.1 Structures of monoterpenes and monoterpenoids from essential oils

that often have pronounced aromas. Some monoterpenes and monoterpenoids found in essential oils are shown in Figure 33.1.

Many of these compounds are found in a wide variety of plants. For example, limonene occurs in the oils of bergamot, black pepper, cardamon, caraway, coriander, cypress, dill, eucalyptus, grapefruit, lemon, lime, neroli, and orange, among others. Carvone, which is obtained from caraway, coriander, dill, and peppermint oils, is used to flavor liqueurs and perfume soaps. Citronellal, the active ingredient of insect-repelling citronella candles, occurs in lemon and lemongrass oils as well as oil of citronella. Citronellol is also a constituent of citronella oil, but it is found in rose and geranium oils as well. The aromatic hydrocarbon *p*-cymene is a component of marjoram and oregano oils.

Geranial and its geometric isomer neral often occur together in essential oils. The isomer mixture, called citral, is the major component of lemongrass oil and is also found in oils of lemon, orange, and verbena. Geraniol occurs in the oils of citronella, lemongrass, and roses, and is used widely in perfumery. Menthone, like the corresponding alcohol menthol, is a constituent of peppermint oil, and it also occurs in the oils of pennyroyal and geranium. β-Myrcene is found in the oils of bay, juniper, and hops, and is an important intermediate in the manufacture of perfumes. α-Terpinene is a constituent of the oils from cardamom and marjoram.

    You might think that the aromas of plant constituents should resemble the aromas of the plants in which they occur. In a few cases this is so; cinnamaldehyde, the major component of cinnamon bark, has an odor much like that of the spice. But most plants contain many different volatile components, and the odor of a single component may bear little or no resemblance to the overall odor of the plant or its essential oil. That aroma is due to the combined effect on your olfactory receptors of many different kinds of molecules. The aroma of a given compound may also vary with its stereochemisty. Thus one carvone enantiomer has an odor of spearmint while the other smells like caraway seeds.

cinnamaldehyde

Understanding the Experiment

In this experiment you will use infrared and ultraviolet-visible spectrometry to identify the unknown compounds, which will be chosen from those in Figure 33.1 or from a list provided by your instructor. With your instructor's permission, you may use ^{1}H NMR spectrometry as well.

    The spectral interpretation section of OP-34 will help you interpret the infrared spectra of the compounds assigned. The four functional groups represented, as well as the aromatic ring of p-cymene, are relatively easy to identify. Both aldehydes and ketones give rise to a strong carbonyl (C=O) band near 1700 cm^{-1}, and aliphatic aldehydes also show medium-intensity C—H stretching bands near 2720 and 2840 cm^{-1}. Primary alcohols are characterized mainly by their O—H and C—O stretching bands around 3300 and 1050 cm^{-1}. The carbon-carbon double bonds of alkenes usually give rise to =C—H stretching bands just above 3000 cm^{-1}, moderate to weak C=C stretching bands in the 1670–1640 cm^{-1} region (sometimes these are quite weak), and =C—H bending bands that are sensitive to substitution patterns in the 1000–650 cm^{-1} region. Aromatic rings are characterized by Ar—H stretching bands just above 3000 cm^{-1} and Ar—H bending in the 900—690 cm^{-1} region. The Ar—H bending patterns vary with the number and location of substituents on the benzene ring.

    Other structural features may affect the positions of certain absorption bands. For example, carbonyl groups that are conjugated with carbon-carbon double bonds or aromatic rings appear at lower wavenumbers than usual. Thus while most saturated ketones have C=O bands near 1715 cm^{-1}, α, β-unsaturated ketones have C=O bands closer to 1670 cm^{-1}. Conjugation with a carbonyl group also affects both the position and intensity of the C=C stretching band; its intensity increases while its wavenumber is lowered by about 30 cm^{-1}. Unsymmetrical conjugated dienes may have two C=C stretching bands near 1650 cm^{-1} and 1600 cm^{-1}.

    Ultraviolet-visible spectra generally contain only a few relatively broad absorption bands. Compounds that appear colored to the human eye absorb radiation in the visible range ($\sim$400–800 nm) as well as the ultraviolet range ($\sim$200–400 nm). Such compounds have extensive chromophores, usually with many conjugated double bonds or aromatic rings. Colorless compounds with less extensive conjugated systems absorb ultraviolet radiation in the 200–400 nm range. Compounds with unconjugated double bonds and no nonbonded electrons do not have absorption bands above 200 nm.

When a compound absorbs radiation in the UV-VIS region of the electromagnetic spectrum, its electrons undergo transitions from lower to higher energy levels. In a π-π^* transition, pi electrons in their ground state energy levels (π molecular orbitals) jump to unoccupied higher energy levels (π^* molecular orbitals). Often it is possible to estimate the λ_{max} value (wavelength of maximum absorption) of a conjugated compound's π-π^* absorption band using a set of rules developed by chemistry Nobel laureate Robert B. Woodward and modified by Louis Fieser. Woodward–Fieser rules for $C{=}C{-}C{=}C$ conjugated systems are given in Table 33.1.

Table 33.1 Woodward–Fieser rules for $C{=}C{-}C{=}C$ systems

Structural feature	Wavelength or increment
base value for conjugated diene	214 nm
homoannular diene	+39 nm
double bond extending conjugation	+30 nm
alkyl group	+5 nm
exocyclic double bond	+5 nm

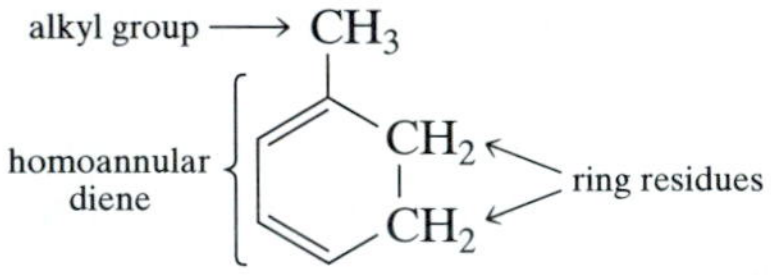

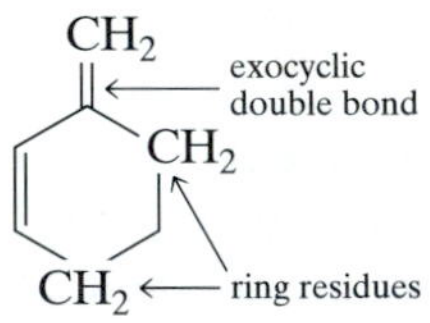

base value	214 nm	base value	214 nm
homoannular diene	+39 nm	exocyclic double bond	+5 nm
3 alkyl groups	+15 nm	2 alkyl groups	+10 nm
total	268 nm	total	229 nm

An alkyl group includes either an open-chain group such as methyl or a ring residue—a carbon-containing group that is part of a ring. To be counted, the alkyl group must be attached directly to one of the carbons of the conjugated system. A homoannular diene is one in which both double bonds of the diene are in the same ring. An exocyclic double bond is one attached to a ring carbon from outside of the ring. To apply the rules, start with the base value for

Table 33.2 Woodward–Fieser rules for $C{=}C{-}C{=}O$ systems

Structural feature	Wavelength or increment
base value for conjugated ketone	215 nm
base value for conjugated aldehyde	210 nm
homoannular double bond extending conjugation	+69 nm
α-alkyl group	+10 nm
β-alkyl group	+12 nm
exocyclic double bond	+5 nm

the conjugated system and add wavelength increments for each of the designated structural features, as shown by the examples following Table 33.1.

Rules for $C{=}C{-}C{=}O$ conjugated systems are given in Table 33.2, where an α-alkyl group is on the carbon atom adjacent to the carbonyl group and a β-alkyl group is on the second carbon from the carbonyl group. The value of λ_{max} for a conjugated carbonyl compound may also depend on the solvent used. The values in the table are for ethanol; using a nonpolar solvent such as hexane can increase λ_{max} by as much as 11 nm.

The following examples illustrate the application of these rules.

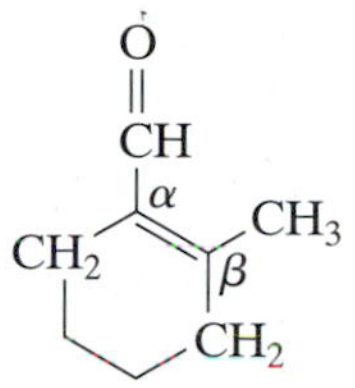

base value, aldehyde	210 nm	base value, ketone	215 nm
α-alkyl group	+10 nm	exocyclic double bond	+5 nm
2 β-alkyl groups	+24 nm	α-alkyl group	+10 nm
total	244 nm	β-alkyl group	+12 nm
		total	243 nm

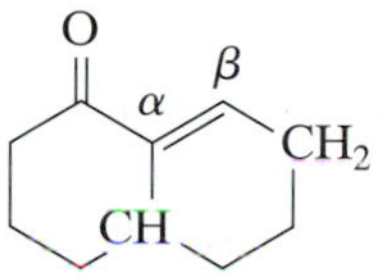

Directions

Unless your instructor indicates otherwise, you can work in groups. Each group can be provided with a selection of unknowns to be apportioned among its members, or the instructor can assign individual unknowns. If necessary, your instructor will show you how to operate the instruments.

Citral, a mixture of geranial and its (Z) isomer neral, may be substituted for geranial.

The terpenes and terpenoids may irritate the eyes and skin, and some are quite flammable. Minimize contact and keep them away from ignition sources.

Deuterochloroform is harmful if inhaled or absorbed through the skin, and it is a suspected human carcinogen. Avoid contact with the liquid and do not breathe its vapors.

Safety Notes

Infrared Spectrum. Record the identifying number of the unknown in your lab notebook as soon as you receive it. Obtain an infrared (preferably FTIR) spectrum of your unknown monoterpenoid as the neat liquid [OP-34]. Record accurate wavenumbers for all significant absorption bands in the spectrum.

Ultraviolet Spectrum. Using a microliter syringe measure 1 μL of the unknown monoterpenoid and dissolve it in 25 mL of 95% ethanol in a small Erlenmeyer flask. Using a 1 cm quartz cell and a scanning ultraviolet-visible spectrometer, scan the spectrum of the solution over the 200–400 nm range [OP-36]. By adjusting the instrument's absorbance range or diluting the

Waste Disposal: Turn in your unknown when you are finished with it.

Take Care! Avoid contact with CDCl₃, do not breathe its vapors.

Weak UV bands near 200 nm may arise from transitions of non-bonded electrons rather than conjugated systems.

sample with more 95% ethanol, obtain a spectrum in which the top of the highest absorption band is between the midpoint and top of the absorbance scale. Record the λ_{max} value for the strongest band and for any other significant bands.

¹H NMR Spectrum (optional). Obtain an ¹H NMR spectrum [OP-35] of your unknown compound in deuterochloroform.

Identification of the Terpenoids. From its infrared spectrum, decide what functional groups are present in your compound and deduce any other structural information you can, such as the existence of conjugation in a carbonyl compound. Based on this information decide which of the compounds in Figure 33.1 have structures that are consistent with your infrared spectrum. From its ultraviolet spectrum, deduce whether your compound contains conjugated double bonds or conjugated carbonyl groups and (using the Woodward–Fieser rules) decide which structure or structures are consistent with both spectra. Use your ¹H NMR spectrum, if you obtained one, to decide which is the most likely structure. Share your spectrum with the other members of your group and examine their spectra until you come to a group consensus regarding the identities of all the monoterpenes and monoterpenoids.

Report. Your report should include a statement of the problem and an account of how you applied scientific methodology to solve it. Give the numbers and identities of all the compounds your group examined. Interpret your spectra as completely as you can, turn them in with your report, and explain how you arrived at the identity of your assigned unknown.

Exercises

1 Describe and explain the possible effect on your results of the following experimental errors or variations. (a) You used a solution cell with a 0.1 mm spacer for the infrared spectrum; (b) You dissolved your unknown in petroleum ether for the UV analysis; (c) There was no deuterochloroform available for the NMR sample so you used ordinary chloroform instead.

2 Tell how infrared spectroscopy can be used to differentiate the following compounds. In each case indicate the significant IR bands that will be observed and their approximate wavenumbers. (a) 1-butanol, 2-butanol, and 2-methyl-2-butanol; (b) *ortho*-xylene, *meta*-xylene, and *para*-xylene (the xylenes are dimethylbenzenes); (c) butanal, 1-butanol, 2-butanone, butanoic acid, and butyl acetate.

3 Tell how ultraviolet-visible spectrometry can be used to differentiate the following compounds. Calculate approximate λ_{max} values for each compound, when you can. (a) 1,4-hexadiene, 2,4-hexadiene, and 1,3-cyclohexadiene; (b) cyclohexanone, 2-cyclohexenone, and 2,4-cyclohexadienone; (c) the following compounds.

    1   **2**   **3**

4 Classify each of the following (as a monoterpene, sesquiterpenoid, etc.) and show how each can be divided into isopentane units.

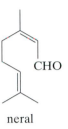

linalool   caryophyllene   β-selinene

neral

5 Tell how you could use a spectrometric method to distinguish geranial from neral.

Other Things You Can Do

(Starred projects require your instructor's permission.)

*1 Identify an unknown arene by NMR spectrometry as described in Minilab 28.

*2 Obtain and interpret the mass spectrum of a compound in Minilab 29.

3 Starting with sources listed in the Bibliography, write a research paper about cholesterol and its biological functions, including a discussion of the biosynthesis of cholesterol from the triterpene squalene.

EXPERIMENT 34

Synthesis and Spectral Analysis of Aspirin

Reactions of Phenols. Preparation of Esters. Nucleophilic Acyl Substitution. Infrared Spectrometry. NMR Spectrometry.

Operations:

OP-6   Heating
OP-9   Mixing
OP-11 Gravity Filtration
OP-12 Vacuum Filtration
OP-21 Drying Solids
OP-23 Recrystallization
OP-34 Infrared Spectrometry
OP-35 Nuclear Magnetic Resonance Spectrometry

Before You Begin

1 Read the experiment, read or review the operations as necessary, and write an experimental plan.
2 Calculate the mass of 15.0 mmol of salicylic acid and the theoretical yield of aspirin.

Scenario

The Spectrum Publishing Co., a subsidiary of The Fulcourt Press, wants to produce a multimedia CD-ROM illustrating the principles of spectral interpretation for use with their new instrumental analysis textbook. The CD-ROM would, for example, use an animated aspirin molecule to simulate bond vibrations and their relationship to IR spectral bands and to illustrate nuclear magnetic transitions and how they give rise to ^{1}H and ^{13}C NMR signals. To make the CD-ROM as accurate as possible, they have asked your Institute to perform an in-depth analysis of the IR and NMR spectra of aspirin. You can't use aspirin tablets because they contain impurities such as starch, a binder. Your assignment, therefore, is to prepare some pure aspirin, record its infrared and NMR spectra, and propose assignments for its NMR signals and its significant IR bands.

Applying Scientific Methodology

The initial scientific problem in this experiment is whether or not pure aspirin (free from unreacted starting material) can be prepared as described in the Directions. You will use a simple chemical method to test your initial hypothesis. As you study the spectra you can develop hypotheses about the origins of various IR bands and NMR signals, then test your hypotheses by consulting sources of spectral data and studying the spectra of related compounds.

The Aspirin Saga

The more we learn about aspirin, the more it appears to be a true wonder drug. The family doctor who advises you to, "Take two aspirin and call me in the morning," knows that aspirin lowers fever, reduces inflammation, and relieves pain. In recent years aspirin has been shown to reduce the incidence of heart disease, strokes, and certain cancers as well. It may also improve brain function in people who have suffered small strokes, help prevent cataracts, and reduce the occurrence of gallstones. But when it was first prepared nearly a century and a half ago, aspirin was considered so unremarkable that it was set aside and temporarily forgotten.

The aspirin saga begins with its "family tree," featuring a group of related compounds known as salicylates. Native Americans have known for centuries that willow bark can relieve pain and fight fever. The bark of the white willow and related species contains salicin, a natural salicylate chemically related to aspirin. Salicin can be broken down in the presence of water and an oxidant to form molecules of glucose (a simple sugar) and a sweet-smelling liquid named salicylaldehyde. German chemists discovered that salicylaldehyde obtained from another source, the meadowsweet plant (*Spiraea* species), reacts with strong alkali to yield a white solid on neutralization. The new compound was named *spirsäure* (<u>Spir</u>aea + <u>säure</u>, the German name for an acid) by its discoverers but was called salicylic acid by the English, who traced its lineage back to the white willow tree.

Salicin was brought to the attention of the scientific world in 1763 when a clergyman named Edward Stone read a paper to the Royal Society of London about the treatment of malarial fever with willow bark. Stone's paper eventually led scientists to look for medical applications of salicylic acid, which can be prepared synthetically from phenol and carbon dioxide using a method developed by the brilliant but irascible German chemist Hermann Kolbe. Phenol, also known as carbolic acid, has long been used as a surgical antiseptic. It is much too caustic to be taken internally, causing painful burns in the mouth and upper digestive tract. But Kolbe had an idea. What if the formation of salicylic acid from phenol and CO_2 were reversed inside the human body? Then a patient could swallow salicylic acid, which would break down inside the body to yield phenol, which would then (Kolbe hoped) kill the germs responsible for the patient's illness. He carried out tests that "proved" to his satisfaction that salicylic acid was indeed an effective germ killer, and soon recommended its use on patients suffering from a variety of bacterial diseases and infections. The first reports seemed promising—patients were still dying after salicylic acid treatments, but they felt much better while doing so! Before long doctors began to suspect that the salicylic acid "cured" only those patients who would have survived without any medication.

Kolbe's idea was wrong; salicylic acid does *not* produce phenol in the body, and his tests were found to be invalid. But like many scientific hypotheses that fail, this one led to important new discoveries. Although salicylic acid does not cure bacterial illnesses, it reduces fever and is a better

The aspirin family tree

Kolbe achieved scientific infamy with a vitriolic attack on J. H. van't Hoff, who helped develop the stereochemical theory of organic chemistry.

phenol

pain reliever than salicin, so it was soon being prescribed for rheumatism, sciatica, headaches, and other painful conditions. But salicylic acid has a serious side effect: It irritates mucous membranes that line the mouth, esophagus, and stomach. One patient who suffered intensely from this side effect was the father of Felix Hoffman, a chemist working for the Bayer division of a German pharmaceutical company. By a happy coincidence, Bayer was interested in finding a substitute for salicylic acid at about the same time Hoffman's father was experiencing its side effects, and this convergence of Hoffmann's research and personal interests provided the motivation that led to aspirin's rediscovery in 1893.

Hoffman knew that phenolic compounds were corrosive because of the presence of a free OH group on the benzene ring, and he reasoned that "masking" the OH with some easily removed substituent might provide the benefits of salicylic acid without the irritation. While studying some of the known derivatives of salicylic acid, Hoffman came across one in which the hydrogen of the OH group was replaced by an acetyl group. This was acetylsalicylic acid, which had been prepared some 40 years earlier by Charles Gerhardt. Tests showed that acetylsalicylic acid was superior to all known painkillers, both in its effectiveness against pain and fever and its freedom from serious side effects. Thus rediscovered, acetylsalicylic acid was introduced by Bayer in 1899 under the trade name "aspirin," and it soon became the world's most popular drug.

aspirin

As described in Experiment 15, aspirin functions by blocking the synthesis of prostaglandins.

acetic anhydride

Understanding the Experiment

In this experiment you will use acetic anhydride to convert salicylic acid to aspirin, and also to serve as a solvent for the reaction. Since acetic anhydride is so reactive, it will not be necessary to heat the reactants to reflux temperatures; warming them in a water bath is sufficient. When the reaction is complete, water is added to destroy the excess acetic anhydride, converting it to water-soluble acetic acid.

The most likely impurities in the crude product are salicylic acid itself and a polymeric by-product formed by intramolecular reactions of salicylic acid. The polymer can be removed by dissolving the product in aqueous sodium bicarbonate solution. Aspirin reacts with sodium bicarbonate to form a water-soluble sodium salt, whereas the polymeric by-product is insoluble and is removed by filtration. Aspirin is reprecipitated from the solution of its salt by acidifying the filtrate with hydrochloric acid, then purified by recrystallization.

Aspirin decomposes at high temperatures so its melting point is not a reliable indicator of its purity. You will assess the purity of your aspirin by testing it for the presence of salicylic acid with ferric chloride, which forms highly colored complexes with phenolic compounds. By testing the crude and partly purified product as well as the final product, you may be able to determine whether the impurity (if there is any) resulted from incomplete reaction of the starting materials or was formed during the workup of the product.

Aspirin is both an ester and a carboxylic acid, so its infrared spectrum shows characteristics of both kinds of compounds. In the liquid state, most carboxylic acids exist as dimers held together by strong hydrogen bonding.

The O—H stretching band of a monomeric carboxylic acid appears as a rather sharp band near 3520 cm^{-1}, but hydrogen bonding in the dimer moves the band to lower frequencies and causes it to spread out over much of the region between 3300 cm^{-1} and 2500 cm^{-1}. The infrared spectrum of a typical carboxylic acid also features a C=O stretching band around 1725 cm^{-1}, an acyl C—O stretching band in the 1315–1280 cm^{-1} region, and an out-of-plane O—H bending band near 920 cm^{-1}. (The "C—O" bond vibrations for carboxylic acids and esters are actually coupled vibrations involving some adjacent atoms.) Conjugation with a benzene ring or C=C bond moves a carbonyl band to a lower frequency. For example, the C=O band of benzoic acid occurs at 1688 cm^{-1}.

The carbonyl stretching band of an ester usually occurs at higher frequency than that of a carboxylic acid, and the acyl C—O stretching band of an ester ranges from about 1240 cm^{-1} to 1140 cm^{-1}. The aromatic ring of an aryl ester increases the C=O frequency and decreases the acyl C—O frequency of the ester function. For example, these bands occur at 1765 cm^{-1} and 1193 cm^{-1}, respectively, for phenyl acetate, compared to 1742 cm^{-1} and 1241 cm^{-1} for ethyl acetate.

The ^{1}H NMR spectrum of aspirin shows a complex pattern of aromatic proton signals characteristic of *ortho* substitution by groups of differing electronegativity. The —COOH substituent withdraws electrons from the ring, deshielding nearby ring protons, and the —OCOCH$_3$ substituent donates electrons by resonance, shielding the ring protons. This effect is particularly noticeable for proton H$_a$ in Figure 34.1, whose signal occurs well downfield of the rest because of its proximity to the —COOH group. Although H$_a$ has only one "nearest neighbor" proton, its ^{1}H NMR signal has four peaks because the delocalized pi cloud of the ring allows long-range coupling between nonadjacent protons. Thus the signal due to H$_a$ is split into a doublet by proton H$_b$, and each peak of that doublet is split into

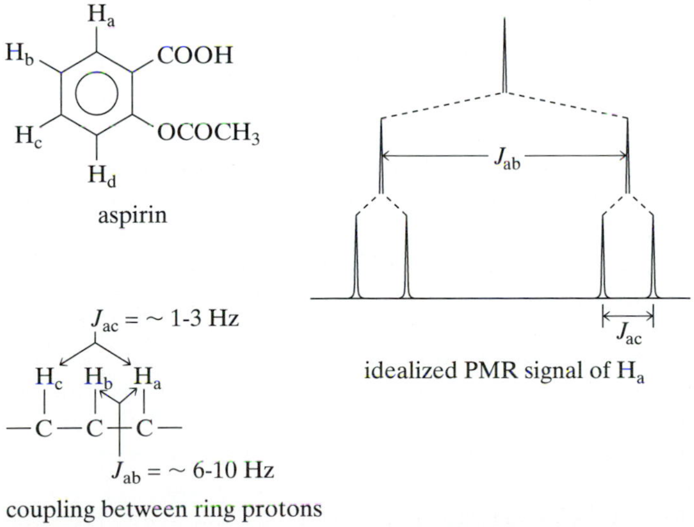

Stretching vibrations of carboxylic acids and esters

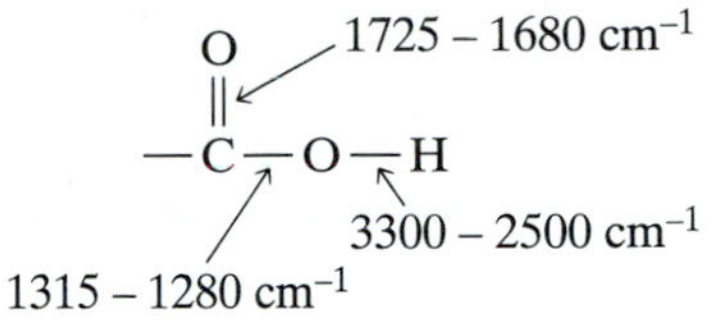

carboxylic acid

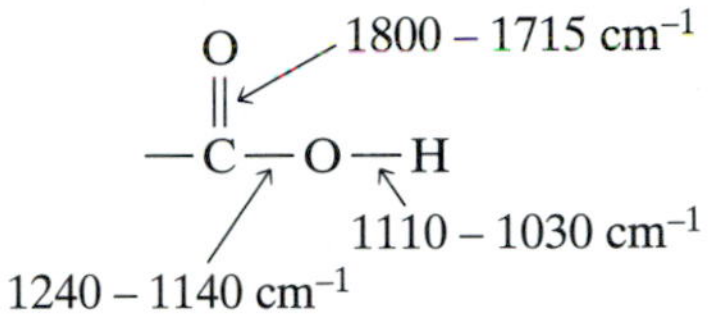

carboxylic ester

Figure 34.1 The effect of long-range coupling on an aromatic proton signal

another closely spaced doublet by proton H$_c$, as illustrated in Figure 34.1. The long-range coupling constant for protons *meta* to one another is only about 1–3 Hz, compared to typical coupling constants for *ortho* protons of 6–10 Hz.

If you obtain a ^{13}C NMR spectrum of your aspirin, refer to OP-35 and your lecture textbook or another source to help you interpret it.

Reactions and Properties

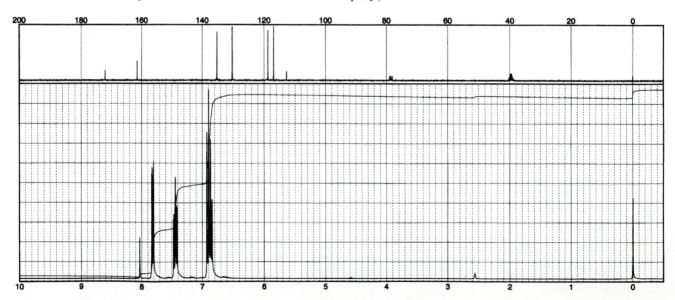

$$\text{salicylic acid} \quad + (CH_3CO)_2O \xrightarrow{H_2SO_4} \text{aspirin} \quad + CH_3COOH$$

salicylic acid   acetic anhydride   aspirin   acetic acid

Table 34.1 Physical properties

	M.W.	m.p.	b.p.	*d*
salicylic acid	138.1	159		
acetic anhydride	102.1	-73	140	1.082
aspirin	180.2	135d		

Note: m.p. and b.p. are in °C, density is in g/mL

Figure 34.2 NMR and IR spectra of salicylic acid (^{13}C and ^{1}H NMR spectrum reproduced from *The Aldrich Library of ^{13}C and ^{1}H FT-NMR Spectra* by C. J. Pouchert and J. Behnke, IR spectrum reproduced from *The Aldrich Library of FT-IR Spectra, Edition II*; both with the permission of the Aldrich Chemical Company.)

Salicylic acid

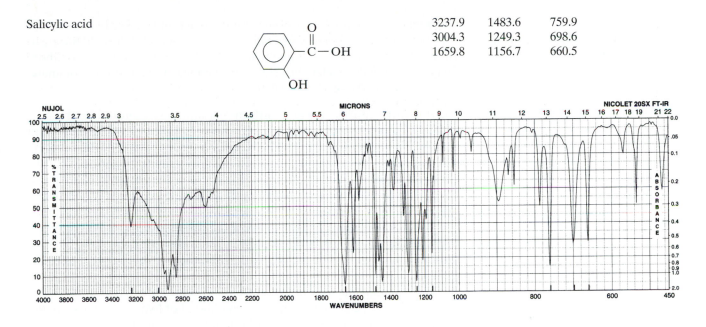

3237.9	1483.6	759.9
3004.3	1249.3	698.6
1659.8	1156.7	660.5

Directions

Acetic anhydride can cause severe damage to skin and eyes, its vapors are very harmful if inhaled, and it reacts violently with water. Use gloves and a hood; avoid contact with the liquid, do not breathe its vapors, and keep it away from water.
Sulfuric acid causes chemical burns that can seriously damage skin and eyes. Wear gloves and avoid contact.
Deuterochloroform is harmful if inhaled or absorbed through the skin, and it is a suspected human carcinogen. Avoid contact with the liquid and do not breathe its vapors.

Safety Notes

acetic anhydride   sulfuric acid

Reaction. Under the hood, add 5.0 mL (~42 mmol) of acetic anhydride to 15.0 mmol of salicylic acid in a dry 125-mL Erlenmeyer flask. Add 3–4 drops of concentrated sulfuric acid and stir or swirl the resulting mixture [OP-9]. Heat the mixture in a 45–50°C water bath [OP-6] with frequent swirling or magnetic stirring for about 6 minutes. If all of the salicylic acid has not dissolved by this time, continue heating until the solution is clear.

Let the flask stand at room temperature until crystallization begins. If no product has precipitated when the solution is near room temperature, induce crystallization by scratching the wall of the flask at the surface of the solution with a glass stirring rod, or by adding a few seed crystals of pure aspirin. When a heavy precipitate has formed, stir in 30 mL of water and break up any lumps with a flat-bottomed stirring rod. Cool the mixture in an ice bath until crystallization is complete. The reaction mixture should be a thick, semisolid slurry at this point.

Take Care! Wear gloves, avoid contact with acetic acid and sulfuric acid, do not breathe vapors.

Observe and Note: What happens during this time?

Stop and Think: Why might this help induce crystallization?

Waste Disposal: Pour the filtrate down the drain.

Stop and Think: What reaction is responsible for the sound?

Stop and Think: What products would result from hydrolysis?

Waste Disposal: Pour the filtrate down the drain.

Stop and Think: Which aspirin is the purest, and why?

Take Care! Avoid contact with $CDCl_3$, do not breathe its vapors.

Separation. Separate the aspirin from the reaction mixture by vacuum filtration [OP-12], washing it with cold water. (Use a little of the cold filtrate to rinse any aspirin adhering to the reaction flask into the funnel.) Save about 50 mg of the crude product in a clean, *dry*, labeled 10-cm test tube for analysis, and transfer the rest to a beaker.

Purification. Slowly stir in 25 mL of 5% aqueous sodium bicarbonate, using a flat-bottomed stirring rod or a spatula to break up the solid and dissolve as much of it as possible. When the fizzing sound of the reaction has stopped, remove any undissolved solid by gravity filtration [OP-11]. (**Waste Disposal:** Pour the filtrate down the drain.) Slowly add 8.0 mL of 3 *M* hydrochloric acid, with continuous stirring, to precipitate the aspirin. Test the mixture with pH paper to see that it is strongly acidic; if it is not, add enough 3 *M* HCl to bring the pH below 3. Cool the mixture in an ice bath and collect the aspirin by vacuum filtration [OP-12]. Save another small sample of the partly purified aspirin in a dry test tube for analysis.

Recrystallize the aspirin from an ethanol/water mixture [OP-23], using the following procedure to reduce the likelihood of hydrolysis. Dissolve the aspirin in the minimum volume of boiling 95% ethanol, and add another 1 mL of ethanol, measuring the total volume of ethanol used. Add twice that volume of warm ($\sim 60°C$) water to the solution while it is still at the boiling point, and swirl to mix. If some precipitate forms, heat the solution gently until it is clear, but do not boil it. Then let it cool until crystallization is complete. Collect the aspirin by vacuum filtration, washing it on the filter with ice cold water. Save another small sample of the aspirin in a dry test tube for analysis. Dry [OP-21] and weigh the remaining aspirin.

Analysis. Place a small quantity of salicylic acid in a labeled 10-cm test tube, then dissolve each reserved aspirin sample and the salicylic acid in 1 mL of 95% ethanol. The aspirin samples need not be completely dry. Add a drop of 1% ferric chloride to each test tube and record your observations.

Record the infrared spectrum [OP-34] of your aspirin in a KBr disc or Nujol mull. Record its ^1H NMR spectrum [OP-35] in deuterochloroform. If possible, use a sweep offset for the COOH signal and a scale expansion for the aromatic proton signals when you record the NMR spectrum. At your instructor's option you can record a ^{13}C NMR spectrum of aspirin or obtain its spectrum from your instructor.

Report. Your report should include a statement of the problem and an account of how you applied scientific methodology to solve it. Discuss the purity of your aspirin samples and account for the presence of salicylic acid in any of them. Identify as many IR bands as you can, indicating the kind of bond vibration responsible. Locate the ^1H NMR signal for the proton designated H_a in Figure 34.1, determine the values of J_{ab} and J_{ac}, and assign as many of the other signals as you can.

Exercises

1 Compare your IR and ^{1}H NMR spectra of aspirin with those of salicylic acid in Figure 34.2, and explain any significant similarities and
differences.

2 (a) Write an equation for a reaction that might form salicylic acid
during the workup of the product. (b) Tell how you could reduce or
prevent the contamination of your product due to this reaction.

3 Describe and explain the possible effect on your results of the following experimental errors or variations. (a) The reagent bottle
labeled "acetic anhydride" actually contained acetic acid; (b) the test
tubes used for analysis were cleaned with water and not completely
dried; (c) you boiled the recrystallization mixture after adding water.

4 Following the format in Appendix V, construct a flow diagram for the
synthesis of aspirin.

5 Write a detailed mechanism for the reaction of salicylic acid with
acetic anhydride, showing clearly the function of the catalyst, sulfuric
acid.

6 Propose a structure for the polymer (draw several repeating units)
formed during the synthesis of aspirin.

7 A small bottle of 5-grain aspirin tablets holds 100 tablets, each containing 0.325 g of aspirin. Calculate the cost of the acetic anhydride
and salicylic acid required to prepare the aspirin in such a bottle,
assuming equimolar quantities of the reactants (the sulfuric acid,
being a catalyst, can be recovered). You can look up their current
prices in the *Chemical Marketing Reporter* or another source recommended by your instructor.

8 Which would you expect to be the stronger acid, aspirin or salicylic
acid? Explain your answer.

Other Things You Can Do

(Starred projects require your instructor's permission.)

*1 Interpret the mass spectrum of a compound in Minilab 29.

*2 Test commercial aspirin tablets for salicylic acid by the ferric chloride test described. Try the test on freshly purchased aspirin and on
aspirin that has been in use for some time. (Try to find some old
aspirin that has a "vinegar" odor.) Test them for starch (often used as
a binder) by boiling 2 mg of ground up tablets in 2 mL of water and
adding a drop of a solution of iodine in potassium iodide. Starch
forms a blue-violet complex with iodine.

3 Starting with sources listed in the Bibliography, write a research
paper on some commercial uses for acetic anhydride other than in
aspirin production.

EXPERIMENT 35

Preparation of Nonbenzenoid Aromatic Compounds

Reactions of Unsaturated Hydrocarbons. Preparation of Nonbenzenoid Aromatic Compounds. Hydride-Transfer Reactions. Carbocations. Aromaticity. Infrared Spectrometry. NMR Spectrometry.

Operations:

OP-6   Heating
OP-7   Cooling
OP-9   Mixing
OP-12 Vacuum Filtration
OP-21 Drying Solids
OP-22 Drying and Trapping Gases
OP-34 Infrared Spectrometry
OP-35 Nuclear Magnetic Resonance Spectrometry

Before You Begin

1 Read the experiment, read or review the operations as necessary, and write an experimental plan.
2 Calculate the mass and volume of 3.00 mmol of benzaldehyde and the theoretical yield of *meso*-tetraphenylporphin for part **A**. Calculate the mass of 4.00 mmol of triphenylmethanol and the theoretical yield of tropylium fluoborate for part **B**.

Scenario

In 1931, the German theoretical chemist Erich Hückel predicted that a seven-membered cyclic compound with six pi electrons would show aromatic properties. Hückel was unaware that just such a compound, tropylium bromide, had already been synthesized some 40 years before! Now Dr. Perry Celsus, the science historian (see Experiments 17 and 25), wants to reenact the discovery of this substance, the first nonbenzenoid aromatic compound. Dr. Celsus first wants to establish that the method used to prepare tropylium bromide did, in fact, yield the tropylium ion and that the ion is truly aromatic. To demonstrate this he needs an authentic sample containing the tropylium ion, prepared by a different route, with which to compare it. Your supervisor has assigned you the task of preparing tropylium fluoborate and analyzing its spectra to look for evidence of aromaticity. You will also prepare a sample of another nonbenzenoid aromatic compound, tetraphenylporphin, for an analytical chemist who is interested in using this compound to prepare colored metal ion complexes that might be used for spectrometric analysis of the ions.

Applying Scientific Methodology

The scientific problem regarding the tropylium ion should be apparent from the Scenario. After reading the experiment you should be able to develop a tentative hypothesis, which you will test by spectral analysis.

Nonbenzenoid Aromatics, from Dracula to Dewar

In 1891, a team of German chemists was trying to determine the structure of atropine, an important alkaloid from the plant *Atropa belladonna*. G. Merling had just brominated a degradation product from atropine, cycloheptatriene, and was distilling the liquid dibromide when a mass of yellow crystals collected in his distilling apparatus. He faithfully reported this observation in a German chemical journal, but it didn't appear to reveal anything about the structure of atropine, so it was filed away and forgotten. About 40 years later, Erich Hückel, a pioneer in applying the theories of quantum mechanics to organic molecules, formulated his well-known *Hückel rule*, which predicts that certain cyclic, unsaturated systems having $4n + 2$ pi electrons will be aromatic. At that time aromaticity was associated exclusively with benzene and related six-membered ring compounds, but Hückel predicted, for example, that a compound containing a cyclic $C_7H_7^+$ cation (which has six pi electrons delocalized around a seven-membered ring) would show aromatic properties. When Merling's yellow salt was again prepared in 1954, it was identified as tropylium bromide, a nonbenzenoid aromatic compound of just the kind that Hückel had foretold.

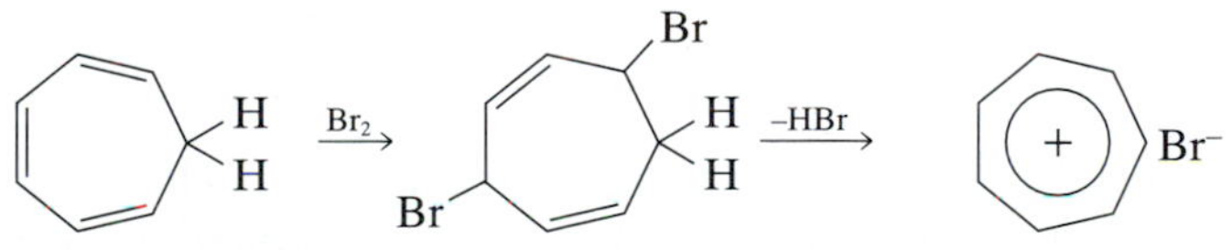

cycloheptatriene   tropylium bromide

Merling's preparation of tropylium bromide

Although their aromatic properties were unknown to scientists until a few decades ago, nonbenzenoid aromatic compounds have existed in nature for as long as green plants and animals have populated the earth. The most important of these by far are the *porphyrins*, which play an important role in both photosynthesis and respiration. Porphyrins are constructed around the system of four linked pyrrole rings that constitutes the parent compound, porphin. The heavy line in the porphin molecule on the following page traces out an aromatic ring whose pi electrons number 18—one of the numbers that can be derived from Hückel's $4n + 2$ rule [when $n = 4$, $(4n + 2) = 18$].

porphin

heme

Hemoglobin is an extremely complex protein whose structure differs slightly for different animal species; a typical empirical formula is $C_{738}H_{1166}O_{208}N_{203}S_2Fe$.

stipitatic acid

tropolone   resonance structure of
                     tropolone

Stop and Think: What other resonance structures can be drawn for tropolone?

Hemoglobin, which is a conjugated protein containing an iron-complexed porphyrin known as heme, shuttles oxygen through the bloodstream and keeps the respiratory process going. Hemoglobin itself is blue in color; when it picks up oxygen in the lungs, it is converted to bright red oxyhemoglobin, which gradually gives up its oxygen to the cells and is reduced to hemoglobin again before it is carried back to the lungs. Ordinarily, any excess porphyrins in the body are metabolized by the liver into iron-free substances such as biliverdin and bilirubin, which collectively make up the bile pigments. If this metabolic process breaks down, a disease known as porphyria results, which often causes agonizing attacks that may resemble psychotic episodes. According to Raymond McNally, a history professor from Boston College, the original Count Dracula (also known as Vlad the Impaler, for his habit of skewering people on stakes) may have suffered from porphyria. The medieval treatment for the disease was drinking blood, which Count Dracula obtained from serfs who farmed the valley below his castle. Victims of the disease may forage for food at night because they fear light, and they often have receding gums that make their teeth appear longer. All of these reported characteristics of the medieval Count Dracula were later attributed to Bram Stoker's fictional vampire of the same name.

Not long after Bela Lugosi turned Dracula into a Hollywood success story, a British chemist named Michael J. S. Dewar was puzzling over a natural substance called stipitatic acid, which had been isolated from a culture of the mold *Penicillum stipitatum*. Because stipitatic acid showed aromatic properties such as undergoing substitution rather than addition reactions with bromine, other scientists had assigned it a structure containing a benzene ring. Dewar disagreed and, with little experimental evidence to go on, he proposed a seven-membered ring structure and declared that the compound was one of a previously unknown class of aromatic compounds that he named tropolones. Dewar expected tropolone rings to show aromatic properties because one can draw resonance structures in which six pi electrons are delocalized over the seven-membered ring. His guess turned out to be correct, and before long other investigators were proposing tropolone ring structures for compounds with similar properties. For example, a sample of an essential oil from the heartwood of western red cedar (*Thuja plicata*) had been set aside and nearly forgotten for 16 years while it slowly crystallized. The crystals yielded β-thujaplicin, an effective fungicide and antibiotic, which is an iso-propyl derivative of tropolone. This compound and two other fungus-destroy-ing thujaplicins are believed to be responsible for the great durability of red cedar, which was used by Native Americans to build canoes long before set-tlers discovered its value in fence posts, shingles, and cedar chests.

β-thujaplicin

Understanding the Experiment

Besides being the first representative of the class of nonbenzenoid aromat-ics, tropylium was probably the first true carbocation synthesized. One way to prepare a stable carbocation is to remove a hydride ion from an appropri-ate hydrocarbon by a hydride-transfer reaction. Just as a strong acid can transfer a proton to a strong base, leaving behind a weaker (more stable) base, a hydrocarbon can transfer a hydride ion to a reactive carbocation, leaving behind a less reactive (more stable) carbocation. In this experiment, the reactive ion will be the trityl (triphenylmethyl) cation and the stable one will be the tropylium ion. The hydride transfer takes place when trityl fluob-orate reacts with the hydrocarbon cycloheptatriene, yielding tropylium flu-oborate and triphenylmethane. Trityl fluoborate itself is prepared as in Experiment 31, by treating the corresponding alcohol with aqueous fluo-boric acid in acetic anhydride. In the procedure used here, the trityl fluobo-rate will not be isolated but will be combined with cycloheptatriene while still in solution. Tropylium fluoborate precipitates as the dark color of the trityl carbocation disappears; adding ethyl ether completes the precipitation.

Proton transfer

$$HB_1 \;+\; B_2^- \longrightarrow$$
stronger   stronger
acid   base

$$HB_2 \;+\; B_1^-$$
weaker   weaker
acid   base

Hydride transfer

$$R_1H \;+\; R_2^+ \longrightarrow$$
less stable
carbocation

$$R_2H \;+\; R_1^+$$
more stable
carbocation

trityl fluoborate

Aromatic compounds are characterized by an uninterrupted system of delocalized pi electrons that extends around the circumference of an unsat-urated ring (see Figure 35.2). Evidence for a compound's aromaticity can be obtained from its spectra. The infrared spectra of aromatic molecules show characteristic bands: Aromatic C—H stretching vibrations generally occur between 3100 and 3000 cm^{-1}; C—H out-of-plane bending vibrations give rise to bands in the 900–650 cm^{-1} region; and skeletal vibrations involving carbon-carbon stretching within the ring result in one or more bands in the 1600–1400 cm^{-1} region. The infrared spectra of symmetrical unsubstituted aromatic compounds are particularly simple, as shown by the infrared spec-trum of benzene in Figure 35.1.

The most unequivocal evidence for aromaticity is provided by NMR spectrometry. The circulation of electrons around an aromatic ring in a mag-netic field results in a so-called ring current that deshields ring protons, moving their ^{1}H NMR signals well downfield. As a rule, the larger the ring, the stronger its ring current. The six-carbon benzene ring absorbs at about 7.3 δ, whereas the 18-carbon aromatic compound [18]annulene has protons that absorb near 8.9 δ. The protons of a tropylium ion are also deshielded because of its ion charge, which reduces the electron density at each ring atom by one-seventh. The protons on most unsubstituted aromatic rings are equivalent, so the ^{1}H NMR spectra of unsubstituted aromatic compounds are very simple, usually consisting of one or two signals. By comparing the IR and ^{1}H NMR spectra of your product with those of the starting material (see Figure 35.3), you should be able to determine whether tropylium fluob-orate is or is not aromatic.

Benzene

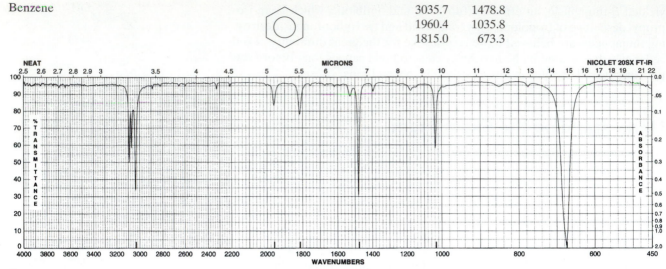

3035.7	1478.8
1960.4	1035.8
1815.0	673.3

Figure 35.1 IR spectrum of benzene (Reproduced from *The Aldrich Library of FT-IR Spectra, Edition II*, with permission of the Aldrich Chemical Company.)

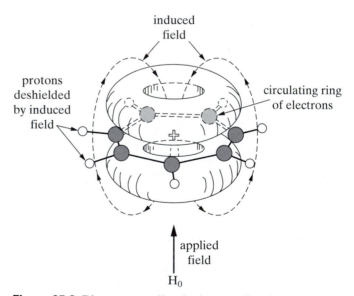

Figure 35.2 Ring current effect in the tropylium ion

You will also prepare a porphin derivative, *meso*-tetraphenylporphin by a condensation reaction between benzaldehyde and pyrrole, using boiling propionic (propanoic acid) acid as the reaction solvent. The product precipitates from the cooled reaction mixture as lustrous purple crystal that are collected by vacuum filtration.

Reactions and Properties

A. $Ph_3COH + HBF_4 \xrightarrow{Ac_2O} Ph_3C^+BF_4^- + H_2O$ $Ph_3C^+BF_4^- + $ ⬡ $\longrightarrow Ph_3CH + $ ⬡⁺ BF_4^-

B. 4 (pyrrole) $+ 4$ (benzaldehyde, CHO) $+ \frac{3}{2}O_2 \xrightarrow{CH_3CH_2COOH}$ (meso-tetraphenylporphin) $+ 7H_2O$

pyrrole   benzaldehyde   *meso*-tetraphenylporphin

Table 35.1 Physical properties

	M.W.	m.p.	b.p.	d
benzaldehyde	106.1	−26	178	1.046
propanoic acid	74.1	−21.5	141	0.993
pyrrole	67.1		130-1	0.969
meso-tetraphenylporphin	614.8			
acetic anhydride	102.1	−73	140	1.082
cycloheptatriene	92.15	−80	117	0.887
fluoboric acid (48 wt%)	87.8			1.400
triphenylmethanol	260.3	164	380	

Figure 35.3 NMR and IR spectra of 1,3,5-cycloheptatriene (NMR spectra reproduced from *The Aldrich Library of ^{13}C and ^1H FT-NMR Spectra* by C. J. Pouchert and J. Behnke, IR spectrum reproduced from *The Aldrich Library of FT-IR Spectra, Edition II*, both with the permission of the Aldrich Chemical Company.)

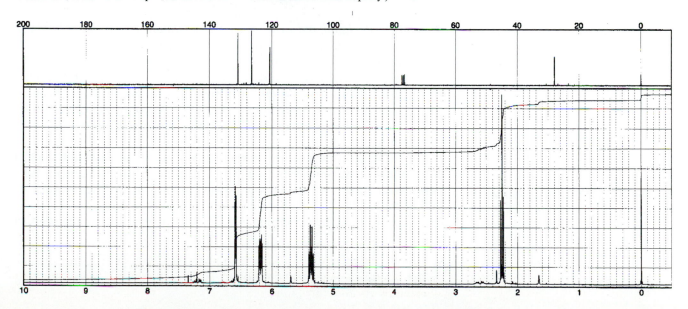

1,3,5-Cycloheptatriene

3024.0	907.2	708.9
1434.5	792.4	656.8
1294.8	741.8	588.2

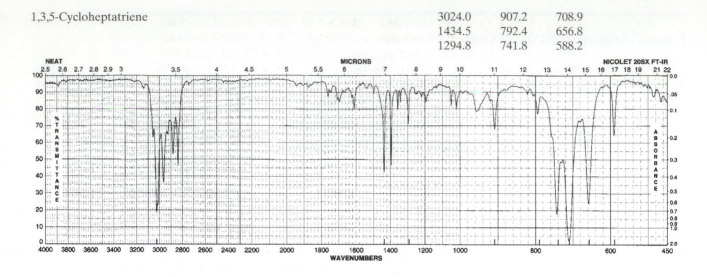

propionic acid

Safety Notes

If propionic acid is not available, you can use glacial acetic acid and increase the reflux time to 1 hour.

Take Care! Avoid contact with propionic acid and pyrrole, do not breathe their vapors.

Waste Disposal: Before you wash the product on the filter, place the filtrate in the designated waste container.

Directions

A. *Preparation of* meso-*Tetraphenylporphin*

> **Propionic acid can burn the skin and eyes severely, and its vapors irritate the eyes and respiratory system. Wear gloves, avoid contact, and do not breathe vapors.**
> **Pyrrole is poisonous and may be very harmful if inhaled, ingested, or allowed to contact the skin or eyes. Wear gloves, avoid contact, and do not breathe vapors.**

Work under the hood and wear protective gloves until the product has been filtered and washed. Measure 10.0 mL of propionic acid into a 25-mL boiling flask, and add 0.21 mL (~3 mmol) of freshly distilled pyrrole followed by 3.00 mmol of benzaldehyde. Add a condenser and boiling chips and heat the the mixture under reflux [OP-6] for 30 minutes or more. (If the reflux cannot be conducted under the hood, use a gas trap containing 1 *M* NaOH [OP-22] to keep fumes out of the lab.) During this time carry out part **B** of this experiment.

Cool the reaction mixture to room temperature and carefully collect the product by vacuum filtration [OP-12], using a Hirsch funnel. (**Take Care!** Wear gloves and use a hood.) Wash the solid on the filter thoroughly with methanol, then with hot water until the odor of propionic acid is gone. Dry the *meso*-tetraphenylporphin [OP-21] and weigh it. Turn in the product to your instructor.

B. *Preparation of Tropylium Fluoborate*

> Fluoboric acid is poisonous, its solutions can cause severe damage to skin and eyes, and its vapors irritate the respiratory system. Use gloves and a hood; avoid contact with the acid solution, and do not breathe its vapors. Acetic anhydride can cause severe damage to skin and eyes, its vapors are very harmful if inhaled, and it reacts violently with water. Use gloves and a hood; avoid contact with the liquid, do not breathe its vapors, and keep it away from water.
> Cycloheptatriene is harmful if inhaled or absorbed through the skin; avoid contact and do not breathe vapors.
> Ethyl ether is extremely flammable and may be harmful if inhaled. Do not breathe its vapors and keep it away from flames and hot surfaces.
> Tropylium fluoborate is corrosive and can damage the eyes, skin, and respiratory tract; avoid contact with the solid, and do not breathe its dust.

Safety Notes

acetic anhydride

Reaction. *Carry out the reaction under the hood and wear protective gloves.* Carefully measure 10.0 mL of acetic anhydride into a 50-mL Erlenmeyer flask and cool the flask in an ice/water bath [OP-7]. Add slowly, with swirling or magnetic stirring [OP-9], 0.65 mL (~5 mmol) of 48% aqueous fluoboric acid. With continued cooling and shaking or stirring, add 4.00 mmol of pure triphenylmethanol in small portions to yield a solution of trityl fluoborate. Add 0.50 mL (~5 mmol) of cycloheptatriene drop by drop, with cooling and shaking. If the solution remains dark, add more cycloheptatriene dropwise until the color disappears or fades to pale yellow. Then mix in 15 mL of anhydrous ethyl ether and let the mixture stand in the ice bath for at least 10 minutes until precipitation is complete.

Take Care! Avoid contact with acetic anhydride, do not breathe its vapors.

Take Care! Avoid contact with fluoboric acid, do not breathe its vapors.

Stop and Think: What evidence do you see for the formation of the trityl cation?

Take Care! Keep ether away from ignition sources, do not breathe its vapors.

Separation and Analysis. Collect the tropylium fluoborate by vacuum filtration using a Hirsch funnel [OP-12], and wash it on the filter with anhydrous ethyl ether. Let the product air dry thoroughly [OP-21], and weigh it. Record the infrared spectrum of tropylium fluoborate [OP-34] using a KBr disc or Nujol mull, and its ^{1}H NMR spectrum [OP-35] in dimethyl sulfoxide (DMSO) or dimethyl sulfoxide-d_6, or obtain the spectra from your instructor. Ordinary DMSO will give a strong NMR signal near 2.5 δ, but this will not interfere with the tropylium fluoborate signal. Turn in the remaining product to your instructor.

Stop and Think: Why is tropylium fluoborate insoluble in ether, although cycloheptatriene is soluble?

Waste Disposal: Place the filtrate in a designated waste container.

Report. Your report should include a statement of the problem and an account of how you applied scientific methodology to solve it. Interpret your spectra and describe the spectral evidence supporting (or not) an aromatic structure for tropylium fluoborate. Calculate the percentage yields of the products from parts **A** and **B**.

Exercises

1 Compare the spectra of the product with those of cycloheptatriene in Figure 35.3, and account for any significant similarities or differences.

2 Aqueous tropylium fluoborate has a K_a of 1.8×10^{-5}, making it as strong an acid as acetic acid. Write a balanced equation for an equilibrium reaction that explains the acidity of tropylium fluoborate.

3 Describe and explain the possible effect on your results of the following experimental errors or variations. (a) In part **A**, you refluxed the reactants for 30 minutes in acetic acid rather than propionic acid. (b) You used methanol in place of triphenylmethanol in part **B**. (c) The ethyl ether you used to precipitate tropylium fluoborate was not anhydrous.

4. (a) Trityl fluoborate from Experiment 31 could have been used to make tropylium fluoborate in this experiment. Write a balanced equation for a reaction that could take place if you accidentally spilled some water on the trityl fluoborate. Assume a 1:1 mole ratio of the reactants. (b) If you started with 1.30 g of trityl fluoborate, what percentage of it would be decomposed by the complete reaction of one drop of water according to your equation in (a)? (Assume that 1 drop $\cong$ 0.050 mL.)

5 Following the format in Appendix V, construct a flow diagram for the synthesis in part **A**.

6 (a) Explain why tropylium bromide (C_7H_7Br) is aromatic but the corresponding "oxide" $[(C_7H_7)_2O]$ is not. Draw structures for both compounds, showing clearly the kind of bonding between each ring and a Br or O atom. (b) What kind of compound is the "oxide?"

7 Predict which of the following compounds will show aromatic properties.

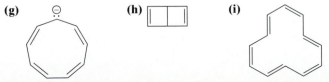

Other Things You Can Do

(Starred projects require your instructor's permission.)

   *1   Record and compare the ultraviolet spectra, from 200–400 nm, of ~10^{-4} solutions of tropylium fluoborate in 0.1 M HCl and of cyclo-heptatriene in absolute ethanol. You can also obtain and interpret a ^{13}C NMR spectra of tropylium fluoborate.

   *2   Prepare ditropyl ether by dissolving about 1 g of tropylium fluo-borate in 5 mL of water and slowly adding 4 mL of saturated aqueous sodium carbonate, with stirring. Extract the product with dichloromethane; then wash the dichloromethane solution with water, dry it, and evaporate the solvent without heating. Obtain the ^{1}H NMR spectrum of the product and decide whether or not it is aromatic.

   3.   Write a research paper about the importance of porphyrins to life on earth, starting with sources listed in the Bibliography.

Mechanism of the Nitration of Arenes by Nitronium Fluoborate

EXPERIMENT 36

Reactions of Arenes. Preparation of Nitro Compounds. Electrophilic Aromatic Substitution. Reaction Rates. Reaction Mechanisms.

Operations:

OP-13 Extraction
OP-14 Evaporation
OP-19 Washing Liquids
OP-20 Drying Liquids
OP-22 Drying and Trapping Gases
OP-32 Gas Chromatography

Before You Begin

Read the experiment, read or review the operations as necessary, and write an experimental plan.

Scenario

(*Note:* Professor Olah is a real chemist whose research group carried out the work described here.)

The research group of George A. Olah, recipient of the 1994 Nobel Prize in Chemistry, developed a new, highly reactive nitrating reagent, nitronium fluoborate (NO_2BF_4), and used it in a study of the mechanisms of aromatic nitration reactions. During their investigation they found evidence that the initial intermediate formed in nitronium fluoborate reactions is different than the one formed in traditional nitration reactions, which use a mixture of nitric and sulfuric acids. Such an intermediate could be either a sigma complex, a symmetrical pi complex, or an oriented pi complex. Your supervisor has devised two experiments that may reveal which kind of intermediate is involved in NO_2BF_4 reactions: (1) nitration of a mixture of toluene and mesitylene and measurement of their relative reaction rates; (2) nitration of *t*-butylbenzene and measurement of the product *ortho/para* ratio. Your assignment is to carry out these nitration reactions and, from their results, deduce the nature of the intermediate.

Applying Scientific Methodology

After reading the experiment you can formulate a working hypothesis about the nature of the intermediate, which you and your coworkers will test by analyzing the product mixtures of the reactions described in the Scenario.

Intermediates in Nitration Reactions

The electrophile in most aromatic nitrations is the nitronium ion (NO_2^+) which is usually generated by mixing nitric acid with sulfuric acid.

$$HNO_3 + 2H_2SO_4 \rightleftharpoons NO_2^+ + H_3O^+ + 2HSO_4^-$$

With this "mixed acid" reagent, the equilibrium concentration of NO_2^+ is so low that elevated temperatures and long reaction times are generally required for successful nitrations. Nitronium fluoborate (NO_2BF_4) is a salt that ionizes to yield NO_2^+ at a much higher concentration, allowing rapid nitration of most aromatic compounds at room temperature or below.

The nitronium ion is an electron-hungry species—an *electrophile*—that can steal a pair of pi electrons from a benzene ring's aromatic sextet. In a general mechanism proposed for electrophilic aromatic substitution, such an electrophile attacks the aromatic ring to form a *sigma complex*, in which the electrophile (symbolized here by E^+) is connected to a ring carbon by a sigma bond.

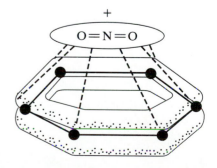

The sigma complex (also called an arenium ion) then loses a proton to a basic species in the reaction mixture to yield the corresponding substituted benzene. The ring atoms *ortho* and *para* to the incoming electrophile bear most of the positive charge in the sigma complex. Thus electron-donating groups located on those atoms stabilize the complex, favoring the formation of *ortho*- and *para*-substituted products. For example, only 4 percent of the product obtained from the nitration of toluene in mixed acid is *meta*-nitrotoluene. The other 96 percent is a mixture of *ortho*- and *para*-nitrotoluene.

In some aromatic substitution reactions the initial intermediate may be a pi complex, in which the electrophile is loosely bonded to the pi electrons of an aromatic ring. Such an intermediate would presumably rearrange to a sigma complex before giving rise to the product. A pi complex involving nitronium ion and benzene might be pictured as shown, with the electrophile sitting atop a "doughnut" of pi electrons, equidistant from all the ring atoms.

Pi complex involving nitronium ion and benzene

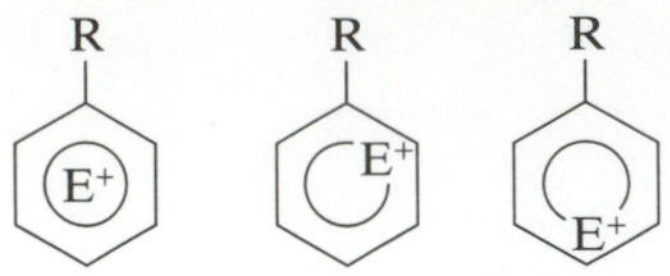

R

R

R

symmetrical
pi complex

oriented
pi complexes

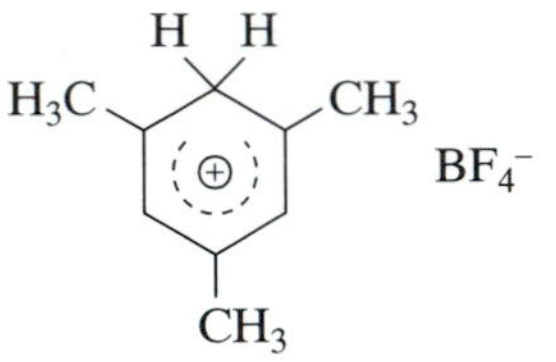

H H

H₃C — — CH₃

⊕

BF_4^-

CH₃

HBF_4 sigma complex
of mesitylene

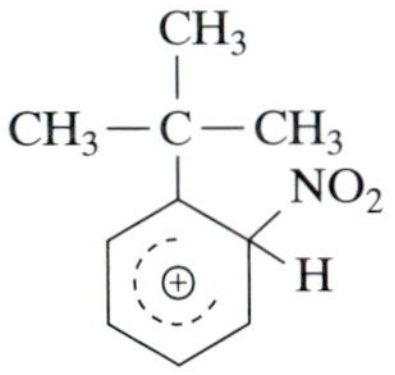

CH_3

$CH_3 — C — CH_3$

NO_2

⊕   H

sigma complex
for nitration of
t-butylbenzene

The pi cloud of an alkyl-substituted benzene is "lumpier" than that of benzene, with bulges at the positions *ortho* and *para* to the alkyl group. An electrophile that spends more time near these regions of higher electron density will form an *oriented* pi complex rather than a symmetrical benzene-type complex. Both possibilities are illustrated here.

Although alkyl groups alter the electron distribution in a pi cloud, they have only a small effect on its total electron density. Thus increasing the number of methyl groups on a benzene ring should not increase the stability of its pi complexes very much. On the other hand, the stabilities of sigma complexes are very sensitive to the electronic effects of substituents. Mesitylene, with three methyl groups, forms a sigma complex with HBF_4 that is nearly 300,000 times more stable than the corresponding sigma complex involving toluene; but the pi complex that mesitylene forms with HCl is only about twice as stable as the complex that toluene forms. As a rule, the more stable an intermediate is, the faster it forms. If the initial intermediate in nitration by nitronium fluoborate is a pi complex, the nitration rates for mesitylene and toluene should be within a factor of 10 or so of one another. But if the initial intermediate is a sigma complex, the nitration rate for mesitylene should be thousands of times greater than that for toluene.

When a substituent influences the outcome of a reaction as a result of its bulkiness (as opposed to its electronic effects), we say that a *steric effect* is operating. Steric effects in aromatic substitution reactions can be detected by measuring the ratios of *ortho* products to *para* products for the same reaction with different substituents. There are twice as many *ortho* as *para* hydrogens in a monosubstituted arene, suggesting that the ratio of *ortho*- to *para*-substituted products should be about 2:1 in the absence of steric effects. As shown in Table 36.1, *ortho:para* ratios for the nitration of arenes in mixed acid are considerably lower than 2:1, especially for very bulky substituents such as *t*-butyl. Apparently a bulky alkyl group hinders substitution at the *ortho* position by crowding the attacking electrophile in the transition state that leads to the sigma complex.

Table 36.1 *Ortho:para* ratios for the nitration of arenes in mixed acid

Arene	*o:p* ratio
toluene	1.57
ethylbenzene	0.93
isopropylbenzene	0.48
t-butylbenzene	0.22

The steric requirements of pi complexes have not been as thoroughly studied as those of sigma complexes. If an incoming electrophile approaches the ring from the top to form a symmetrical pi complex, there should be little if any steric effect even with a bulky substituent such as the *t*-butyl group. On the other hand, the formation of an oriented pi complex should be markedly influenced by steric factors, giving product ratios comparable to those observed for mixed acid nitrations.

Understanding the Experiment

Nitronium fluoborate is prepared by treating nitric acid with hydrofluoric acid and boron trifluoride:

$$HNO_3 + HF + 2BF_3 \rightarrow NO_2BF_4 + BF_3 \cdot H_2O$$

The stable crystalline salt ionizes in polar solvents to provide "ready-made" nitronium ions:

$$NO_2BF_4 \rightarrow NO_2^+ + BF_4^-$$

Nitronium fluoborate was first used for aromatic nitrations by George A. Olah and his coworkers, who discovered that it nitrates aromatic hydrocarbons rapidly at room temperature or below, producing nearly quantitative (100%) yields of the products.

In this experiment, you will nitrate a mixture of mesitylene and toluene with nitronium fluoborate and analyze the product mixture by gas chromatography to determine their relative reaction rates. You will also nitrate *t*-butylbenzene and measure the product *ortho:para* ratio. From the results, you should be able to decide whether the initial intermediate is a sigma complex, a symmetrical pi complex, or an oriented pi complex.

You will measure the relative rates of nitration for mesitylene and toluene by carrying out a competitive nitration reaction in which equimolar quantities of the two arenes compete for a limited amount of nitronium fluoborate. The arene that competes most successfully will form the most product, so the relative rates for the two arenes should be proportional to the relative amounts of nitroarene they produce:

$$\frac{\text{reaction rate for mesitylene}}{\text{reaction rate for toluene}} = \frac{\text{moles of nitromesitylene}}{\text{moles of nitrotoluenes}}$$

For a meaningful rate comparison, you should determine the relative rates per reaction site; otherwise, toluene, with five ring hydrogens, will have a statistical advantage over mesitylene, with only three. The rate per reaction site is proportional to the number of moles of product divided by the number of reaction sites, so the relative reactivity of a mesitylene site is given by this equation:

$$\frac{\text{reactivity of mesitylene site}}{\text{reactivity of toluene site}} = \frac{\text{moles of nitromesitylene}/3}{\text{moles of nitrotoluenes}/5}$$

Since the area of a peak on a gas chromatogram is proportional to the mass of the component that produces it, you can estimate the relative number of moles of each product by dividing its peak area by its molecular weight. (The relative masses of different components may not be in the exact ratio of their peak areas, but they will be close enough for the purposes of this experiment.) The peak areas for the three nitrotoluenes should be combined for this calculation.

The nitration reactions will be carried out at room temperature by adding nitronium fluoborate in sulfolane to an excess of each arene in the same solvent. The excess reactant prevents the formation of di- and trinitrated products, which would skew the results. Sulfolane is an excellent solvent for the reaction because it dissolves both the nitronium salt and the arene, thus providing a homogeneous reaction mixture. It is also miscible with water, making it easy to separate the products from the reaction

sulfolane

mixture. When water and ethyl ether are added to the reaction mixture, sulfolane and fluoboric acid (a by-product of the reaction) end up in the water layer, and the aromatic compounds are extracted into the ether layer. Evaporation of the ether leaves a mixture of unreacted arenes and nitrated products.

The reaction mixtures will be analyzed by gas chromatography on a column that separates aromatic compounds in order of their boiling points. Unreacted arenes should elute from the column first, followed by *ortho-*, *meta-*, and *para*-nitroarenes, in that order. The nitromesitylene peak should appear later than all the nitrotoluene peaks. You will see an initial ethyl ether peak if you failed to evaporate the ether completely, and there may be additional peaks due to impurities in the commercial arenes used as starting materials. Compare your chromatograms with chromatograms of those arenes if such peaks make interpretation difficult.

Reaction and Properties

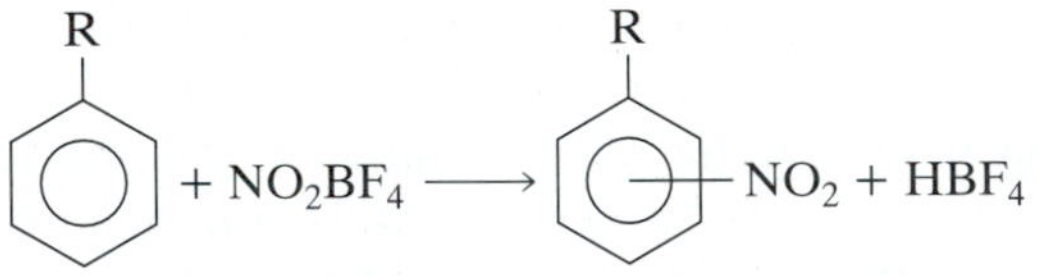

R = methyl or *t*-butyl

(NO$_2$ is predominantly *ortho* and *para*)

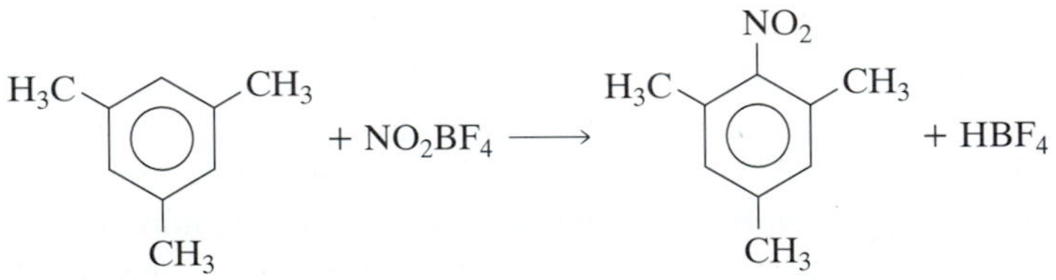

Table 36.2 Physical properties

	M.W.	m.p.	b.p.	*d*
toluene	92.1	−95	111	0.867
t-butylbenzene	134.2	−58	169	0.867
mesitylene	120.2	−45	165	0.862
nitronium fluoborate	132.8			
sulfolane	120.2	28	285	1.260
o-nitrotoluene	137.1	10	222	1.163
m-nitrotoluene	137.1	16	233	1.157
p-nitrotoluene	137.1	55	238	1.104
nitromesitylene	165.2	44	255	

Note: m.p. and b.p. are in °C, density is in g/mL

Directions

If desired, students can work in pairs, with each student being responsible for one nitration reaction.

Safety Notes

toluene

ethyl ether

p-nitrotoluene

Reactions. *Carry out the reactions under a hood, and wear protective gloves. All glassware must be clean and dry.* Measure 0.6 mL of an equimolar mixture of mesitylene and toluene into a clean, dry, labeled centrifuge tube, preferably one with a screw cap. Then mix in 1.0 mL of sulfolane, capping the tube (or sealing it with Parafilm) and shaking gently. In a second centrifuge tube, mix 0.5 mL of dry *t*-butylbenzene with 1.0 mL of sulfolane. Slowly add 1.0 mL of 0.5 M nitronium fluoborate in sulfolane to each centrifuge tube, swirling to mix the reactants. Cap or seal each tube and shake gently. Then let the tubes stand at room temperature for 10 minutes, with occasional shaking.

Take Care! Do not breathe vapors, keep away from flames.

Take Care! Wear gloves, avoid contact with nitronium fluoborate.

Separation. Carry out the following procedure with each reaction mixture. Add 2 mL of solvent grade ethyl ether and 3 mL of water to the centrifuge tube, and shake to extract [OP-13] the products and unreacted arene into the ether layer. (**Take Care!** Cap the tube, vent it often, and wear gloves.) Carefully remove the aqueous layer with a Pasteur pipet. Wash the ether layer with 2 mL of water [OP-19]. Then remove the aqueous layer and dry the ether solution [OP-20] with a little anhydrous calcium chloride, decanting it into a screwcap vial. Under the hood, carefully evaporate [OP-14] the ethyl ether using a stream of dry air or nitrogen [OP-22]. Keep the air velocity low so that the ether does not splash out.

Take Care! Don't breathe ether vapors, keep away from ignition sources.

Waste Disposal: Place the aqueous layer, which contains hydrofluoric acid and fluoboric acid, in a designated waste container.

Analysis. Obtain a gas chromatogram [OP-32] of each product mixture as directed by your instructor. Identify the peaks on the gas chromatograms and measure the peak areas for all of the nitrated products. Turn in or dispose of the products as directed by your instructor.

Stop and Think: Is there any evidence of incomplete evaporation? What gives rise to the largest peak(s) on each chromatogram?

Report. Your report should include a statement of the problem and an account of how you applied scientific methodology to solve it. Calculate the *ortho:para* ratio for the nitration of *t*-butylbenzene and the reactivity (per reaction site) of mesitylene relative to toluene. Tell whether your results suggest a significant steric effect in the nitration by nitronium fluoborate, and cite the evidence supporting your conclusion. Decide whether the initial intermediate in these nitrations by nitronium fluoborate is most likely to be a sigma complex, a symmetrical pi complex, or an oriented pi complex and justify your conclusion.

Exercises

1 Based on your results, write a detailed mechanism for the reaction of *t*-butylbenzene with nitronium fluoborate to yield *p*-nitro-*t*-butylbenzene.

2 (a) It has been estimated that the *meta* product arising from direct nitration of *t*-butylbenzene by NO_2BF_4 makes up about 2.0% of the product mixture, the remaining *meta* product arising from isomerization of the *ortho* product. Use this estimate to calculate a more accurate value of your *ortho:para* nitration ratio for *t*-butylbenzene. (b) Propose a mechanism for the isomerization reaction, which is apparently promoted by the HBF_4 formed during the reaction.

3 Describe and explain the possible effect on your results of the following experimental errors or variations. (a) Your reaction tube was rinsed with water and not dried completely. (b) You inadvertently added toluene rather than *t*-butylbenzene to the second centrifuge tube. (c) When analyzing the gas chromatogram from the *t*-butylbenzene nitration, you misidentified the *t*-butylbenzene peak as the peak for the *ortho*-product, and assumed that the next two peaks were for the *meta* and *para*-products.

4 Following the format in Appendix V, construct a flow diagram for the nitration of *t*-butylbenzene.

5 In a solution of bromine in acetic acid, mesitylene is brominated nearly 300 million times faster than benzene. Do you think a sigma complex or a pi complex is formed in the rate-determining step of this reaction? Explain.

6 Predict the major product or products of the mononitration of (a) ethyl benzoate, (b) phenyl acetate, (c) phenyl benzoate, (d) *m*-nitrotoluene, (e) *p*-methoxybenzaldehyde.

Other Things You Can Do

(Starred projects require your instructor's permission.)

Take Care! Wear gloves, avoid contact with the acids, do not breathe their vapors, keep them away from other chemicals.

*1 Carry out the same nitrations with mixed acid and compare the results with those you obtained with nitronium fluoborate. Under the hood, cool 1.0 mL of concentrated nitric acid in an ice bath, and slowly add 1.0 mL of concentrated sulfuric acid. Measure 80 mmol of *t*-butylbenzene into a flask and add the cold mixed acid drop by drop with cooling and stirring. Stir or shake for 30 minutes at room temperature. Carefully wash the organic layer with ice-cold water and saturated aqueous Na_2CO_3, then dry it with calcium chloride. Repeat using a mixture of 80 mmol of mesitylene and 80 mmol of toluene. Analyze both mixtures by gas chromatography.

*2 Carry out the mixed acid nitration of napthalene by the procedure in Minilab 30.

 3 Read the paper by George A. Olah and his coworkers that describes the nitration of arenes with nitronium fluoborate (*J. Am. Chem. Soc.* **1961**, *83*, 4571) and compare their results and conclusions with your own.

EXPERIMENT 37

Friedel–Crafts Acylation of Anisole

*Reactions of Aromatic Ethers. Preparation of Carbonyl Compounds.
Electrophilic Aromatic Substitution. Infrared Spectrometry.*

Operations:

OP-6   Heating
OP-9   Mixing
OP-10 Addition
OP-12 Vacuum Filtration
OP-19 Washing Liquids
OP-20 Drying Liquids
OP-21 Drying Solids
OP-22 Drying and Trapping Gases
OP-25 Simple Distillation
OP-28 Melting Point
OP-34 Infrared Spectrometry

Before You Begin

1 Read the experiment, read or review the operations as necessary,
   and write an experimental plan.
2 Calculate the mass of 22.0 mmol of aluminum chloride, the mass and
   volume of 10.0 mmol of anisole, and the theoretical yield of methoxy-
   acetophenone.

hawthorn blossom

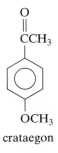

crataegon

Scenario

The purchasing agent for the Olfactory Factory mistakenly ordered
500 kg of anisole rather than 500 pounds, so the company needs to find
some way to use up the excess anisole by converting it to perfume ingredients. One possibility is *p*-methoxyacetophenone, also known as crataegon, which occurs naturally in hawthorn blossoms (*Crataegus* spp.).
Their chemical technicians think it should be possible to synthesize
p-methoxyacetophenone from anisole by a Friedel–Crafts reaction, but
they are concerned that the reaction may yield the wrong isomer or a
mixture of isomers that will be difficult to separate. According to their
business manager, the Olfactory Factory cannot sell the product at a
competitive price if they have to invest in expensive separation equipment. Your assignment is to see whether or not the Friedel–Crafts
acetylation of anisole yields mainly *p*-methoxyacetophenone, another
isomer, or a mixture of isomers.

Applying Scientific Methodology

After reading the experiment you should be able to develop a working hypothesis related to the problem, which you will test by obtaining the melting point and infrared spectrum of the product.

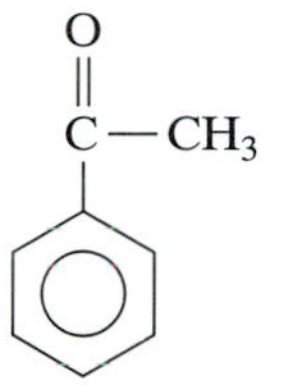

Friedel, Crafts, and Phenones

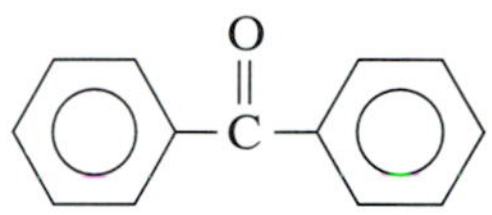

Aromatic ketones that have the carbonyl group adjacent to the benzene ring are called *phenones*. The simplest member of this group is acetophenone, a pleasant-smelling liquid that has been used to impart an odor of orange blossoms to perfumes, and is also prescribed as a sleep-producing drug under the generic name hypnone. Charles Friedel first prepared acetophenone in 1857 by distilling a mixture of calcium benzoate and calcium acetate. Another fragrant phenone, benzophenone, is a white solid with a geranium-like odor that has been used as a fixative for perfumes and as a starting material for the manufacture of drugs and insecticides. Theodor Zincke first prepared benzophenone by heating benzoyl chloride with a metal in benzene. This was essentially a Friedel–Crafts reaction, but Zincke didn't know it because that reaction had not been discovered yet!

The natural and synthetic musks are powerfully odoriferous substances that supply the long-lasting, musky "end note" characteristic of some perfumes, deodorants, and aftershave lotions. The large-ring ketone called muscone (3-methylcyclopentadecanone) is the major constituent of natural musk, which is a secretion from the musk pod of the male musk deer. Muscone is very costly and its use threatens the existence of the deer, so it has been largely replaced by synthetic musks. Among these are the phenones known as musk ketone and Celestolide.

Natural and synthetic musks

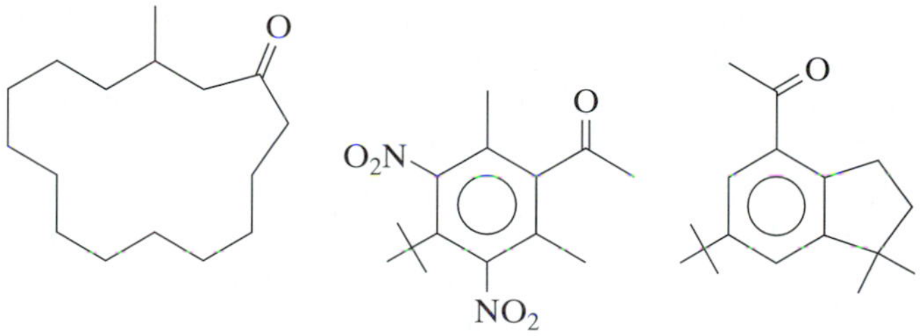

muscone   musk ketone   Celestolide

Musk ketone is prepared from *m*-xylene by two Friedel–Crafts reactions—alkylation with *t*-butyl chloride and acylation with acetyl chloride—followed by nitration of the aromatic ring.

Like benzophenone and the synthetic musks, most phenones can be prepared by a Friedel–Crafts reaction of an aromatic compound with an appropriate acylating agent. The Friedel–Crafts reaction might well have

acetophenone

benzophenone

Synthesis of musk ketone

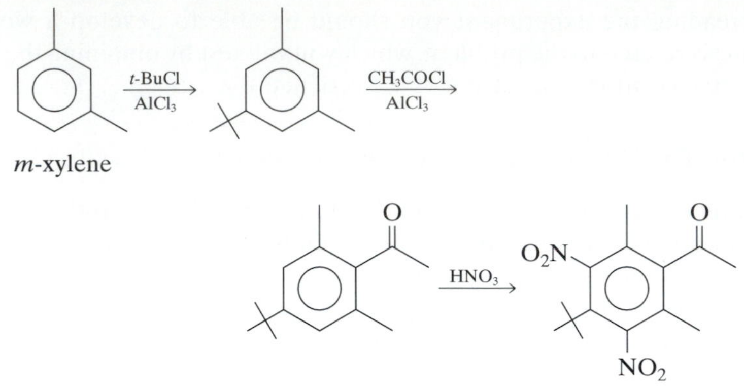

m-xylene

been named the "Zincke reaction" if Theodor Zincke had understood the significance of an experiment that failed. In 1869 Zincke tried to synthesize 3-phenylpropanoic acid by combining benzyl chloride and chloroacetic acid in the presence of metallic silver—a variation of the Wurtz reaction. While carrying out the reaction with benzene as the solvent, to his surprise, Zincke observed that a great deal of hydrogen chloride was evolved and that the major product was diphenylmethane instead of the expected carboxylic acid.

See Experiment 25 for the story of another Wurtz reaction that failed, with momentous consequences.

Zincke's attempted synthesis

$$PhCH_2Cl + ClCH_2COOH \xrightarrow[\text{benzene}]{\text{Ag}} PhCH_2CH_2COOH$$

The "Zincke reaction"

$$PhCH_2Cl + PhH \text{ (benzene)} \xrightarrow{\text{Ag}} PhCH_2Ph + HCl$$

About four years later, a Frenchman named Charles Friedel was watching a student in Wurtz's laboratory perform a "Zincke reaction" using (appropriately) powdered zinc as the catalyst. When the reaction suddenly became violent, Friedel helped the student separate the solution from the zinc powder, thinking that removing the catalyst would moderate the reaction. To the astonishment of both, the reaction was just as violent in the absence of zinc. Although there is no record of his thought processes after this event, Friedel must have recognized its significance. In 1877, he and his collaborator, an American named Charles Mason Crafts, published a paper that marked the inception of the Friedel–Crafts reaction as one of the most important synthetic procedures in the history of organic chemistry. Friedel and Crafts' basic discovery was a simple one—it was a chloride of the metal, and not the metal itself, that catalyzed the reaction of organic halides with aromatic compounds. In Zincke's experiment, traces of silver chloride had formed during the reaction as a result of the oxidation of the metal. Friedel and Crafts found that anhydrous aluminum chloride was the most effective catalyst of those then available. It is still the catalyst of choice for most Friedel–Crafts reactions.

Understanding the Experiment

The Friedel–Crafts reaction is not a single reaction type, although the term has most often been applied to alkylations and acylations of aromatic compounds using aluminum chloride (or another Lewis acid catalyst) and a suitable alkylating or acylating agent. A typical Friedel–Crafts acylation reaction uses a carboxylic acid chloride as the acylating agent and anhydrous aluminum chloride as the catalyst. The aluminum chloride, a Lewis acid, removes a leaving group from the acylating agent, forming an acylium ion, as illustrated in the following mechanism for the reaction of benzene with an acyl chloride. The acylium ion then attacks the benzene ring to form an arenium ion, which loses a proton to yield the product and regenerate the catalyst. When the acyl group is acetyl (CH_3CO), acetic anhydride is often used as the acylating agent rather than acetyl chloride. The anhydride is safer to work with and it usually provides better yields and a simpler workup. More catalyst is needed with acetic anhydride, however, because some of the aluminum chloride forms complexes with the acetic acid produced during the reaction, making it ineffective as a catalyst. As a rule, 2–3 moles of $AlCl_3$ are used per mole of acetic anhydride.

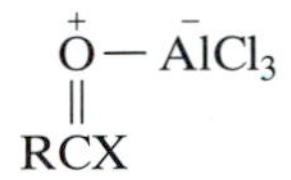

In this experiment you will use acetic anhydride as the acylating agent and dichloromethane as the reaction solvent. The reaction is highly exothermic so it will be carried out by slowly adding acetic anhydride to the other reactants, then heating under reflux to complete the reaction. Pouring the product into ice water will decompose the aluminum chloride complex of the product and transfer inorganic salts to the aqueous phase. The product can then be recovered by evaporating the organic solvent and distilling the residue. On cooling to room temperature, the product should crystallize. Its infrared spectrum can be obtained by the method for melts described in OP-34, or by some other method.

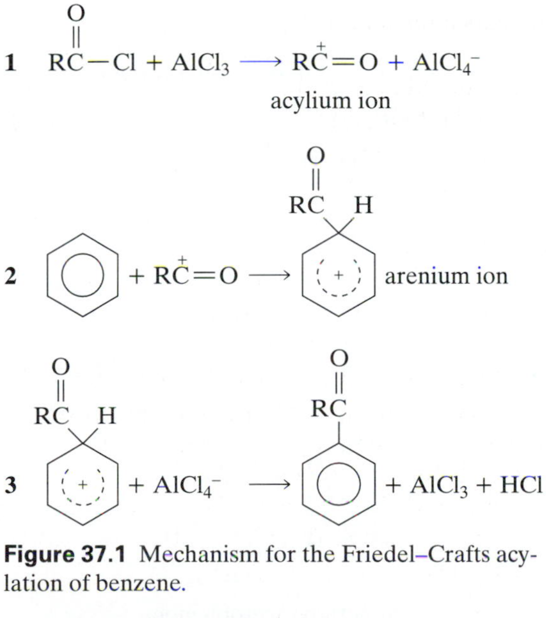

Aluminum chloride complex with acyl compound

$$\overset{+}{O} - \overset{-}{AlCl_3}$$
$$\|$$
$$RCX$$

$$\mathbf{1} \quad \overset{\overset{\textstyle O}{\|}}{RC}-Cl + AlCl_3 \longrightarrow R\overset{+}{C}{=}O + AlCl_4^-$$

acylium ion

$$\mathbf{2} \quad \bighexagon + R\overset{+}{C}{=}O \longrightarrow \text{arenium ion}$$

$$\mathbf{3} \quad \longrightarrow + AlCl_4^- \longrightarrow + AlCl_3 + HCl$$

Figure 37.1 Mechanism for the Friedel–Crafts acylation of benzene.

In principle, acylation of a monosubstituted benzene can yield any or all of three different disubstituted products. From the melting point and infrared spectrum of your product, you should be able to determine whether it is predominantly a single compound or a mixture of isomers, and if it is a single compound to establish its identity. Disubstituted benzenes can be distinguished by the location of their out-of-plane C—H bending bands, which occur at frequencies (expressed in wavenumbers) below 850 cm^{-1}. The frequency of such a band decreases with the number of adjacent hydrogens on the ring, as shown in Table 37.1. Thus a *para*-disubstituted benzene, with its two sets of two adjacent hydrogens, should show an absorption band in the 840–810 cm^{-1} region; *meta* compounds, with three adjacent ring hydrogens, absorb in the 810–750 cm^{-1} region; and *ortho* compounds, with four adjacent ring hydrogens, absorb in the 770–735 cm^{-1} region. Absorption by the isolated hydrogen of a *meta* compound is usually very weak, and its frequency may vary. Monosubstituted and *meta*-disubstituted benzenes have an additional band in the 710–680 cm^{-1} region, which arises from a vibration involving the entire benzene ring.

Your product will be an ether as well as a ketone, so its infrared spectrum will contain bands characteristic of both functional groups. The carbonyl band of a phenone generally appears in the 1685–1665 cm^{-1} region, and a weak carbonyl overtone band may be observed at twice the frequency

Table 37.1 Frequencies of C—H out-of-plane bending bands in aromatic hydrocarbons

No. of adjacent hydrogens	Frequency range, cm^{-1}
1	900–860 (weak)
2	840–810
3	810–750
4	770–735
5	770–730

Possible products from the Friedel-Crafts acylation of anisole

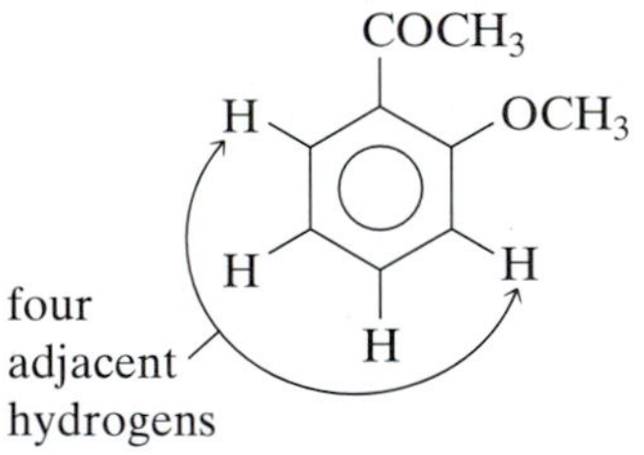

o-methoxyacetophenone

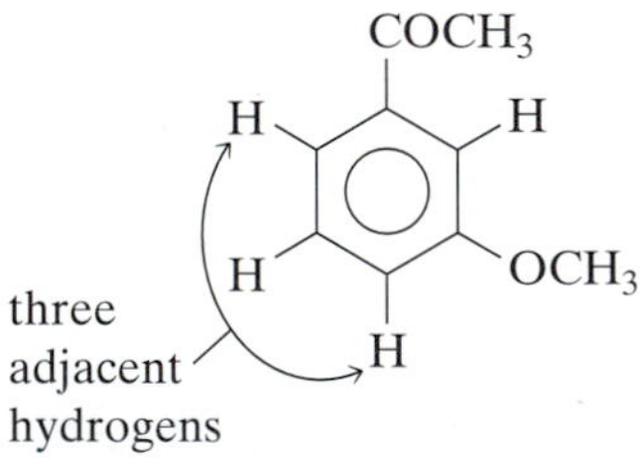

m-methoxyacetophenone

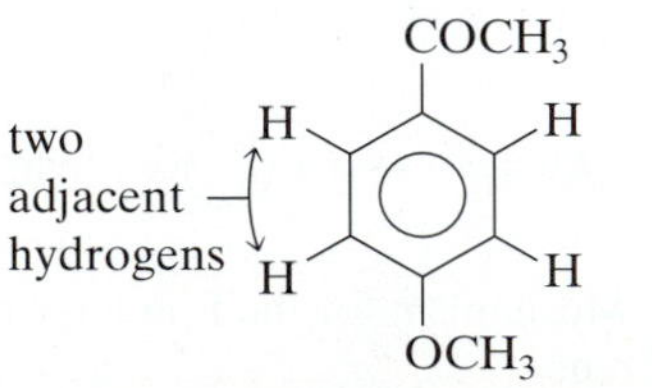

p-methoxyacetophenone

Anisole

1600.9	1247.3	784.1
1497.9	1172.6	754.4
1303.0	1040.5	692.0

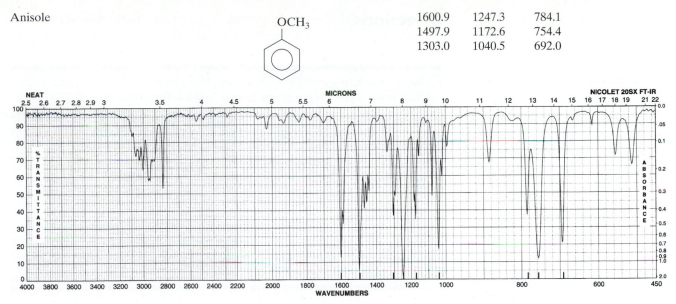

Figure 37.2 IR spectrum of anisole (Reproduced from *The Aldrich Library of FT-IR Spectra, Edition II*, with the permission of the Aldrich Chemical Company.)

of the fundamental band. Aryl alkyl ethers display an asymmetrical C—O—C stretching band at 1275–1200 cm^{-1} and a symmetrical C—O—C band near 1075–1020 cm^{-1}. These ether bands can be seen in the spectrum of anisole in Figure 37.2.

Reactions and Properties

anisole   acetic   ?-methoxy-
                 anyhdride   acetophenone

Table 37.2 Physical properties

	M.W.	m.p.	b.p.	d
anisole	108.2	238	155	0.996
acetic anhydride	102.1	−73	140	1.082
aluminum chloride	133.3	193	subl	
dichloromethane	84.9	−95	40	1.327
o-methoxyacetophenone	150.2		245	1.090
m-methoxyacetophenone	150.2		240	1.034
p-methoxyacetophenone	150.2	39	258	1.082[41]

Note: m.p. and b.p. are in °C, density is in g/mL (subl = sublimes)

Directions

Safety Notes

aluminum
chloride

acetic
anhydride

dichloromethane

anisole

Aluminum chloride reacts with atmospheric moisture and violently with water, generating HCl vapors. It can cause painful burns on moist skin and eyes, and inhaling the dust or vapors can damage the respiratory tract. Weigh it under a hood, wear gloves and safety goggles, avoid contact, do not inhale dust or vapors, and keep it away from water.

Acetic anhydride can cause severe damage to skin and eyes, its vapors are very harmful if inhaled, and it reacts violently with water. Use gloves and a hood; avoid contact with the liquid, do not breathe its vapors, and keep it away from water.

Dichloromethane may be harmful if ingested, inhaled, or absorbed through the skin. There is a possibility that prolonged inhalation of dichloromethane may cause cancer. Minimize contact with the liquid and do not breathe its vapors.

Stop and Think: How could HCl be formed during the reaction?

Take Care! Wear gloves and goggles, avoid contact with AlCl$_3$, do not breathe its vapors.

Take Care! Avoid contact with dichloromethane, do not breathe its vapors.

Take Care! Wear gloves, avoid contact with acetic anhydride, do not breathe its vapors, keep it away from water.

Take Care! Splattering may occur.

Stop and Think: Which layer is it?

Reaction. Wear gloves and eye protection! Work under a hood, if possible. Anhydrous aluminum chloride is deactivated by water, so protect it from atmospheric moisture and be sure your glassware is thoroughly dried. Assemble an apparatus for addition and reflux [OP-10] using a 25-mL boiling flask, and attach a drying tube [OP-22] to the top of the reflux condenser. If you cannot carry out the reaction under a hood, attach a gas trap as well, using dilute sodium hydroxide to react with any HCl evolved. Clamp the apparatus securely to a ring stand so that you can lower the boiling flask to add the reactants without disturbing the rest of the apparatus. *Under the hood,* carefully weigh 22.0 mmol of finely powdered anhydrous aluminum chloride into a large *dry* vial (don't let it tip over) and immediately cap the vial. Cap the aluminum chloride reagent bottle tightly, without delay. Add 10 mL of dichloromethane, 10.0 mmol of anisole, and a stirbar or boiling chips to the reaction flask. Have a beaker of cold water ready to cool the mixture if it begins to boil, then cautiously add the aluminum chloride in small portions through a dry powder funnel, stirring or shaking the flask [OP-9] after each addition. Wash any adherent aluminum chloride into the flask with a little dichloromethane, and reassemble the apparatus. *Under the hood,* measure 1.0 mL (~11 mmol) of acetic anhydride into the addition funnel. Stopper the funnel immediately and put it in place on the reaction apparatus. Add the acetic anhydride slowly (about a drop every 3 seconds) so that the reaction mixture boils gently, while stirring or shaking to mix the reactants. Have a beaker of cold water handy to moderate the reaction if necessary. When the addition is complete, heat the reaction mixture under gentle reflux [OP-6] for 30 minutes using a hot water bath or steam bath.

Separation and Purification. Wear gloves and eye protection! Under the hood, pour the warm reaction mixture *slowly* with vigorous stirring onto about 10 g of cracked ice in a large beaker. Use a little ice water to rinse any residue out of the flask into the beaker. Separate the dichloromethane layer in a separatory funnel, and wash it [OP-19] with 5 mL of 3 *M* sodium hydroxide followed by 5 mL of saturated aqueous sodium chloride. Dry the dichloromethane layer [OP-20] with anhydrous magnesium sulfate. (**Waste Disposal:** Pour the aqueous layers down the drain.)

Assemble an apparatus for small-scale simple distillation [OP-25] using a 25-mL round-bottomed flask. Cool the receiving flask in an ice/water bath. Transfer the dry dichloromethane solution to the flask and, heating gently with a water bath or steam bath, remove the dichloromethane by distillation (it should distill around 40°C). Then remove the cooling bath and use a heating mantle or other heat source to distill the product, collecting everything that distills over about 240°C. If the distillate solidifies in the receiver, melt it with a beaker of hot water. While it is still liquid, transfer the distillate to a watch glass or evaporating dish, and set it aside to crystallize. Wash the product on a Hirsch funnel [OP-12] with a little cold low-boiling petroleum ether, and air dry it [OP-21]. (**Waste Disposal:** Place the petroleum ether in a designated solvent recovery container.)

Analysis. When the product is completely dry, weigh it and measure its melting point [OP-28]. Record its infrared spectrum [OP-34] or obtain a spectrum from your instructor.

Report. Your report should include a statement of the problem and an account of how you applied scientific methodology to solve it. Interpret your infrared spectrum as completely as you can. Decide whether the product is a mixture or a single compound, and if the latter, deduce its structure. Cite all experimental evidence supporting your conclusion.

Stop and Think: What does the NaOH remove from the dichloromethane solution?

Waste Disposal: Place the dichloromethane in a halogenated solvent recovery container.

Take Care! If distillate begins to solidify in the vacuum adapter drip tube, stop the distillation and melt it with a heat gun or other heating device before you proceed.

Exercises

1 If your product was a single compound, explain why it was that compound rather than another isomer.
2 Write a mechanism for the Friedel–Crafts reaction of anisole with acetic anhydride.
3 Describe and explain the possible effect on your results of the following experimental errors or variations. (a) You used only 10 mmol of aluminum chloride for the reaction. (b) You used acetic acid as the acylating agent. (c) You used acetophenone as the substrate.
4. (a) Write an equation for the reaction of aluminum chloride with a large excess of water. (b) Write equations for one or more reactions that would account for the production of HCl during the acylation reaction.
5 Following the format in Appendix V, construct a flow diagram for the synthesis you carried out.
6 Explain why Friedel–Crafts reactions are usually carried out by adding the alkylating or acylating agent *to* the aromatic compound rather than vice versa.
7 2,5-Dichloro-2,5-dimethylhexane is an important starting material for the aroma chemicals called tetralin musks. Outline a synthesis of the tetralin musk versalide from this starting material and benzene, using any necessary inorganic or organic reagents.

versalide

Other Things You Can Do

(Starred projects require your instructor's permission.)

*1 Record the ^{1}H NMR spectrum of the product in deuterochloroform. Interpret it as completely as you can, assigning the signals due to protons on the ring as well as those due to protons on the side chains.

*2 Carry out some Friedel–Crafts reactions that yield colored products by doing Minilab 31.

 3 Starting with sources listed in the Bibliography, write a research paper about the Friedel–Crafts reaction, including specific examples and industrial applications.

Determination of the Structure of a Natural Product in Anise Oil

Reactions of Alkenylbenzenes. Preparation of Carboxylic Acids. Side Chain Oxidation. Structure Determination. Infrared Spectrometry.

Operations:

> OP-6   Heating
> OP-9   Mixing
> OP-12   Vacuum Filtration
> OP-21   Drying Solids
> OP-23   Recrystallization
> OP-28   Melting Point
> OP-34   Infrared Spectrometry

Before You Begin

Read the experiment, read or review the operations as necessary, and write an experimental plan.

Scenario

Basil Wormwood, the new-age herbalist with a chemistry degree, has another puzzle for you (see Experiment 18 for his previous puzzles). He obtained some Chinese star anise from an Oriental foods wholesaler, steam-distilled its essential oil, and isolated the major component of the oil. Not knowing its identity, he tentatively named this compound anisene. He sent it off to a chemical analyst for elemental analysis, and the results indicate that its molecular formula is $C_{10}H_{12}O$. He also carried out some experiments (described next) that show that the compound contains a methoxyl group and a three-carbon side chain on a benzene ring, but he doesn't know the identity of the side chain or where it is located with respect to the methoxyl group. He has just shipped a sample of the compound to your supervisor hoping that your Institute's consulting chemists can solve this structure puzzle. Your supervisor thinks that the position of the side chain can be determined by oxidizing it to a COOH group, and that its structure can be determined by infrared analysis. Your assignment is to carry out the experimental work and thereby determine the complete structure of anisene.

Applying Scientific Methodology

You will have to carry out some experimental work before you can propose a meaningful hypothesis about the structure of anisene.

The Structure Puzzle—Taking Molecules Apart and Putting Them Back Together

star anise seed clusters

Chinese star anise (*Illicium verum*) is a small evergreen tree of the magnolia family. When its dried, star shaped seed clusters are ground up and steam distilled, they yield an oily liquid with a strong odor of licorice. Star anise oil, or its synthetic equivalent, is widely used as a flavoring for licorice, cough drops, chewing gum, and liqueurs such as ouzo and anisette. In this experiment you will be using both classical and modern methods of structural analysis to determine the complete structure of its major component, which we will call "anisene" (not its real name).

Today, when a chemist can run an NMR or mass spectrum of an organic compound and often determine its structure in a matter of minutes, it is hard to imagine how much time and effort were once required to determine the structures of even the simpler natural products. In a "classical" structure determination, the molecular formula of a compound is first obtained by elemental analysis and molecular-weight measurement. Then the compound is degraded (broken down) into smaller structural units that are isolated and, if possible, identified. Finding how the smaller units fit together to form the original molecule is an intellectual challenge that might be compared to putting together a jigsaw puzzle with some pieces missing, others that don't belong, and still others that have been chewed up by the family dog and are no longer recognizable. Finally, when enough information has been amassed to suggest a possible structure, that structure must usually be proven by an independent synthesis in which the compound is built up again, from known compounds, by reactions whose outcome can be reliably predicted.

In many cases, classical structure determinations involved the efforts of dozens or even hundreds of chemists over many decades, with the generation of much irrelevant or misleading information and many synthetic dead ends. The advent of modern spectrometric methods has simplified the process enormously by providing detailed structural information that was not readily available to the chemists of earlier times.

Understanding the Experiment

In this experiment you will attempt to determine the structure of the major component of star anise oil, which has the molecular formula $C_{10}H_{12}O$. Most open-chain saturated organic compounds (except those containing nitrogen, phosphorus, or halogen atoms) have $2n + 2$ hydrogen atoms for every n carbon atoms. If anisene were such a compound it would have $2(10) + 2 = 22$ hydrogen atoms, but since it has only 12 it is said to be "deficient" by 10 hydrogens. Every ring or pi bond in a molecule represents a deficit of 2 hydrogens. That is, an open-chain compound must lose 2 hydrogen atoms to form a ring, and a saturated compound must lose 2 to form a pi bond (or pi-bond equivalent in the Kekulé structure of an aromatic ring). Thus its deficiency of 10 hydrogens indicates that there must be a total of 5 rings and/or pi bonds in an anisene molecule. The number 5 is its *index of hydrogen deficiency* (IHD), which can be calculated from the following formula that applies to a hydrocarbon or oxygen-containing compound with n carbon atoms and x hydrogen atoms.

$$\text{IHD} = \frac{(2n + 2) - x}{2}$$

Catalytic hydrogenation of anisene yields a saturated compound with the formula $C_{10}H_{20}O$. The gain of eight hydrogens indicates that anisene has four pi bonds, so it must contain only one ring. A high carbon:hydrogen ratio often indicates an aromatic structure, and we can account for the ring and three pi bonds by supposing that anisene contains a benzene ring.

Heating anisene with hydriodic acid yields a phenol with the molecular formula C_9H_9OH and a volatile compound identified as methyl iodide. This reaction is used to test for certain ether functions. Methyl ethers yield methyl iodide, and the formation of a phenol indicates that anisene is an aryl methyl ether, whose formula we write as $C_9H_9OCH_3$ in the following equation for the reaction.

$$C_9H_9OCH_3 + HI \rightarrow C_9H_9OH + CH_3I$$

At this point, we know that anisene contains a methoxyl $(-OCH_3)$ group and a benzene ring, which accounts for seven carbon atoms and three pi bonds. That leaves three more carbons and one pi bond to be accounted for. This remaining fragment could be a three-carbon unsaturated side chain, whose formula can be determined by subtracting the fragments already identified from the molecular formula of anisene:

molecular formula:	$C_{10}H_{12}O$
disubstituted benzene ring:	$- C_6H_4$
methoxyl group:	$- CH_3O$
side chain:	C_3H_5

Now we can write a partial structure for anisene, as shown in the margin. All that remains is to determine the structure of the unsaturated side chain and its location on the benzene ring.

Potassium permanganate is capable of oxidizing most aliphatic side chains all the way down to the benzylic carbon atom, leaving a COOH group where the side chain was originally located. Oxidizing anisene should yield one of three possible methoxybenzoic acids, whose melting points are given in Table 38.3. By identifying the oxidation product as one of these three, you will establish the position of anisene's side chain.

Although aqueous potassium permanganate is a powerful oxidizing agent, it reacts slowly with water-insoluble organic compounds because $KMnO_4$ is essentially insoluble in the organic phase. In 1974 Herriot and Picker added a quaternary ammonium salt to a stirred heterogeneous mixture of aqueous $KMnO_4$ and benzene, causing permanganate ions to dissolve in the organic layer and form "purple benzene." The quaternary salt acted as a phase-transfer catalyst, escorting the permanganate ions across the phase boundary into the organic phase (see Experiment 22 for a discussion of phase-transfer catalysis). When an oxidizable organic compound is dissolved in purple benzene, it reacts much more rapidly and under milder conditions than it would with aqueous $KMnO_4$.

Benzene is toxic and can cause leukemia in humans, so you will use a simplified procedure in which anisene and a phase-transfer catalyst are combined directly with aqueous potassium permanganate. Thus anisene itself will be the organic phase of the two-phase system, and no organic solvent is needed. Because you require only enough product for a melting point, you can start with a few drops of anisene. Excess permanganate is used because some of it may decompose during the reaction. As the reaction proceeds, permanganate ion is reduced to manganese dioxide, which forms a fine

Another possibility, that anisene has two side chains, is explored in Exercise 3.

OCH_3

C_3H_5

partial structure of anisene

OCH_3

COOH

methoxybenzoic acid

Table 38.1 Out-of-plane bending vibrations of vinylic C—H bonds

Structure type	Frequency range, cm^{-1}
RCH=CH$_2$	995–985 and 915–905
RCH=CHR (*cis*)	730–665
RCH=CHR (*trans*)	980–960
R$_2$CH=CH$_2$	895–885

R = alkyl or aryl

Table 38.2 Out-of-plane vibrations of aromatic C—H bonds

Ring substitution	Frequency range, cm^{-1}
ortho	770–735
meta	810–750 and 710–690
para	840–810

brown precipitate that is difficult to filter and wash. Fortunately, this precipitate can be dissolved during the workup by acidifying the solution and adding sodium bisulfite, which reduces manganese dioxide (and any unreacted permanganate ion) to soluble manganese(II) sulfate.

Removal of manganese dioxide

$$MnO_2 + NaHSO_3 + H^+ \rightarrow MnSO_4 + H_2O + Na^+$$

The methoxybenzoic acid can then be extracted with dichloromethane and purified by recrystallization from water.

Carbon-carbon double bonds give rise to characteristic =C—H out-of-plane bending bands in the 1000–650 cm^{-1} region of an infrared spectum. The frequencies (expressed in cm^{-1}) of these bands can reveal the number and location of substituents on the carbon-carbon double bond, as shown in Table 38.1. There are four possible structures for an unsaturated C$_3$H$_5$ side chain, corresponding to the four structure types in this table. From the wavenumber(s) of anisene's =C—H bending band(s), you should be able to deduce the structure of the side chain. But first you must locate the right absorption bands, which is more easily said than done because aromatic C—H bonds give rise to strong bands in the same region, as shown in Table 38.2. Once you learn the position of the side chain on anisene's benzene ring, you should be able to locate any bands due to ring C—H bonds in its infrared spectrum, which will help you pick out the vinylic C—H bands from the remaining strong bands in the 1000–650 cm^{-1} region.

Reactions and Properties

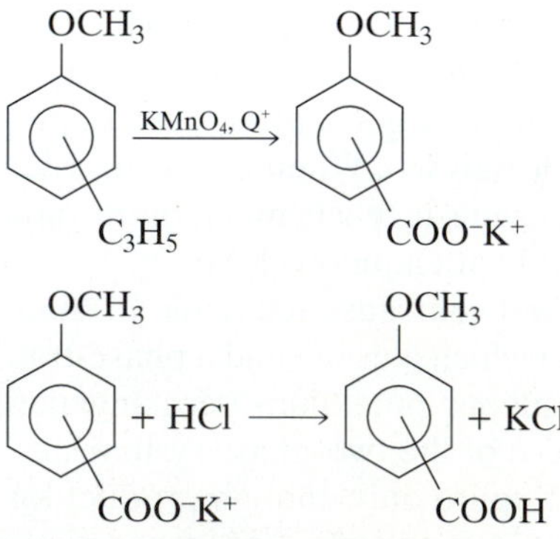

$$Q^+ \approx (CH_3(CH_2)_7)_3\overset{\oplus}{N}CH_3$$

Table 38.3 Physical properties

	M.W.	m.p.	b.p.	d
potassium permanganate	158.0			
o-methoxybenzoic acid	152.2	101		
m-methoxybenzoic acid	152.2	110		
p-methoxybenzoic acid	152.2	185		
toluene	92.2	−95	111	0.867

Directions

> **Potassium permanganate can react violently with oxidizable materials; keep it away from other chemicals and combustibles.**
> **Sodium bisulfite produces harmful vapors when it reacts with acids; do not breathe them.**

Reaction. Obtain some anisene (or anise oil) from your instructor, or isolate it from anise seeds as described in "Other Things You Can Do." Add 0.50 g of crystalline potassium permanganate and 2 drops of tricapryl-methylammonium chloride (Aliquat 336) to 10 mL of water in a 25-mL boiling flask, then drop in a stirbar. Heat with stirring [OP-9] for several minutes until the KMnO$_4$ appears to have dissolved. Add 5 drops of anisene or anise oil and heat the mixture under reflux [OP-6] with vigorous stirring for 15 minutes or more.

Cool the reaction mixture to room temperature and transfer it to a small beaker. Add 1 mL of 6 M hydrochloric acid and test the solution with blue litmus paper; if it is not acidic, add more HCl until it is. Add just enough solid sodium bisulfite, in small portions with stirring or vigorous shaking, to reduce any excess permanganate and dissolve any brown precipitate (0.5–1.0 g of NaHSO$_3$ should be sufficient). Test the solution with blue litmus paper after each bisulfite addition, and add 6 M HCl as needed to keep it acidic. When the brown color has disappeared and only a white precipitate remains, test the solution with pH paper; if necessary, reduce the pH to 2 with additional HCl.

Separation. Separate the product from the reaction mixture by vacuum filtration [OP-12] using a Hirsch funnel.

Purification and Analysis. Recrystallize the product from boiling water [OP-23] and dry it to constant mass [OP-21]. Measure the melting point [OP-28] of the methoxybenzoic acid. Record the infrared spectrum [OP-34] of anisene (*not* of the methoxybenzoic acid), or obtain a spectrum from your instructor.

Report. Your report should include a statement of the problem and an account of how you applied scientific methodology to solve it. Interpret the infrared spectrum as completely as you can. Deduce the location and structure of the side chain, and draw the structure of anisene. Justify your conclusion, citing all the experimental evidence that supports it.

potassium
permanganate

Safety Notes

m.P = 165 °C
170 °C

5 drops

Take Care! Keep KMnO$_4$ away from oxidizable materials

after adding 5 drop anisene purple turned to Brown
15 min we waited.
after adding 5 drop Hcl pH
blue paper turned pink

Take Care! Do not breathe the vapors that may be produced.

Stop and Think: What vapors may form, and how are they produced?

Stop and Think: What are the brown and white precipitates?

Waste Disposal: Place the filtrate in a designated waste container.

with adding enough NaHSO3 turned basic again blue paper showed basicity. added 3 drop Hcl turn pink again
we used Hirsh funnel
are purified the precipita

Exercises

1 Derive a systematic name for anisene and find its common name in *The Merck Index* or another reference book.

2 (a) Write a balanced equation for the reaction of anisene with potassium permanganate, assuming that acetic acid is a by-product of the reaction. (*Note:* the reaction mixture is alkaline.) (b) Assuming that 5 drops of anisene is about 1.0 mmol, calculate the mass of potassium permanganate required to oxidize that much anisene, and the percentage in excess that was actually used.

3 (a) The three carbon atoms of anisene's side chain might have formed two side chains rather than one. Give the structures of these side chains. (b) Give the structures of all carboxylic acids that could have resulted from complete side-chain oxidation of anisene had it contained these two side chains.

4 Describe and explain the possible effect on your results of the following experimental errors or variations. (a) You forgot to add the tricaprylammonium chloride; (b) The pH of the reaction mixture was 7 when you filtered it, and you obtained a brown solid. (c) You recorded the infrared spectrum of the oxidation product rather than of anisene itself.

5 Describe the probable role of the phase-transfer catalyst in this reaction, giving equations for the relevant reactions.

6 Following the format in Appendix V, construct a flow diagram for the synthesis of your methoxybenzoic acid.

7 (a) Draw the structure of the compound $C_{10}H_{20}O$ that is obtained by the catalytic hydrogenation of anisene. (b) Draw the structure of the compound C_9H_9OH that is obtained when anisene is treated with hydriodic acid.

8 You could confirm the structure of anisene by synthesizing it from known starting materials. Outline a synthesis of anisene from benzene and alcohols that have four carbon atoms or fewer.

9 The structure shown has been proposed for coniferyl alcohol, which can be obtained by the hydrolysis of coniferin, a natural product found in the sap of conifer trees. Assuming that the structure of coniferyl alcohol had not been reported in the literature, describe how you would go about proving its structure. Indicate what chemical tests and degradations might be carried out, describing the expected results and conclusions. Summarize the information that could be derived from infrared analysis. Then show how the alcohol could be synthesized from readily available starting materials.

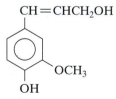

proposed structure for
coniferyl alcohol

Other Things You Can Do

(Starred projects require your instructor's permission.)

*1 Isolate anise oil from anise seeds (or star anise) as follows. Weigh out 10 g or more of fresh anise seeds and grind them finely using a spice grinder or a mortar and pestle. Isolate the anise oil by steam distillation and extraction of the distillate with dichloromethane, as described for clove oil in Experiment 10. (Do *not* extract the dichloromethane layer with NaOH.) Dry the dichloromethane solution with magnesium sulfate and evaporate the solvent completely. You can obtain a gas chromatogram of the oil and estimate the percentage of anisene it contains.

*2 Use air as an oxidizing agent to convert fluorene to fluorenone as described in Minilab 32.

 3 Starting with sources listed in the Bibliography, write a research paper on the use of chemical methods for structure determination of natural products, illustrating it with examples of actual structure determinations.

<table>
<tr><td>

EXPERIMENT 39

</td><td>

Identification of an Oxygen-Containing Organic Compound

</td></tr>
</table>

Reactions of Aldehydes and Ketones. Reactions of Alcohols. Infrared Spectrometry. Qualitative Analysis.

Operations:

OP-21 Drying Solids
OP-23 Recrystallization
OP-25 Simple Distillation
OP-28 Melting Point
OP-29 Boiling Point
OP-34 Infrared Spectrometry
OP-35 (optional) Nuclear Magnetic Resonance Spectrometry

Before You Begin

1 Read the experiment, read or review the operations as necessary, and write an experimental plan.
2 Read pp. 530–531 in Part IV, Systematic Qualitative Organic Analysis.

Scenario

You can find a description of Arkham in the Dictionary of Imaginary Places, Expanded Edition *(Harcourt Brace Jovanovich, 1987).*

Dr. Keziah Armitage, professor of medieval metaphysics at Miskatonic University in Arkham, Mass., was exploring an abandoned and nearly forgotten room in the basement of the metaphysics building when she came across a grime-encrusted bottle containing an unknown liquid. Its label had long since decomposed to dust but the liquid in the bottle appeared relatively pure. Curious about its contents, she sent the bottle and its contents to your Institute for analysis. She thinks it might be a potion used in unmentionable rites practiced by her ancestor, Keziah Mason, a witch whose trial scandalized Arkham in 1692. Your supervisor thinks it is more likely to be an alcohol or carbonyl compound. Your assignment is to find out what family the mysterious liquid belongs to and then identify it.

Applying Scientific Methodology

As you carry out the experiment you should propose provisional hypotheses about the nature and identity of your unknown, which you will test—and perhaps reject or revise—as you gather additional experimental evidence. Your conclusion should, if possible, be consistent with all of the experimental evidence you obtain. If any evidence is not consistent with your conclusion, you should attempt to explain why.

The Chemist as Detective

The process of identifying an unknown compound can be compared to the approach used by a detective in identifying the perpetrator of a crime. The detective first looks for clues that help to characterize the criminal and indicate the most productive areas of investigation. Once a list of possible suspects has been assembled, the detective has to evaluate the evidence already acquired and gather additional evidence to help narrow the list of suspects, focusing the investigation on the most likely suspects. Finally, the detective has to evaluate all of the evidence, come to a conclusion regarding the identity of the perpetrator, and organize the facts of the case in such a way as to convince a jury that the accused is, in fact, guilty of the crime.

In carrying out the identification of an organic compound you, like the detective, should be constantly on the lookout for clues to its identity. Chemical and spectral data should allow you to confine your search to a particular chemical family. Additional physical and chemical evidence will help you narrow down the list of "suspects" and focus your attention on a few of the most probable compounds. Finally, the preparation of one or more derivatives should lead you to a definite conclusion and provide you with sufficient evidence to convince the "jury" (your instructor) that your compound is, in fact, what you believe it to be.

As in the solution of any other problem, you must first ask yourself the right questions before you can arrive at the correct answer. Some important questions to be answered regarding an unknown compound are these: (1) Is it pure? (2) What functional group(s) does it contain? (3) Are there any other significant structural features that might aid in its identification? Each bit of evidence that you obtain should, if interpreted correctly, help reveal the answer to one or more of these questions. All of them combined should provide you with an answer to the ultimate question, "What is it?"

A detective trying to solve a case will almost invariably come upon clues that lead nowhere or, even worse, to false conclusions. The same is true in chemical problem solving, so it is important to keep an open mind throughout your investigation and to avoid jumping to conclusions before all the evidence is in. You may formulate tentative assumptions based on your initial observations (for example: "It turns chromic anhydride reagent green, so it must be an alcohol"), but you should be ready to revise or discard such assumptions if they are not supported by subsequent observations (for example: "Its IR spectrum has a $C{=}O$ stretching band but no $O{-}H$ band, so it may be an aldehyde instead").

Chemical and physical evidence can be misleading for a variety of reasons:

1 Some compounds of a given family may undergo an atypical reaction with a given reagent and yield either a false positive or a false negative result.
2 Some reagents give positive tests with more than one functional group.
3 Impurities may complicate or invalidate a test.
4 Spectral bands may occur outside the expected frequency ranges or may be incorrectly assigned.

Because of these and other possible sources of error, it is not wise to rely on a single piece of evidence in formulating a conclusion. For example, the classification of an unknown as a secondary alcohol can be established with some certainty by a positive chromic anhydride test, a slow reaction with Lucas's reagent, *and* an infrared band in the 1100 cm^{-1} region, but not by any one of these alone.

Understanding the Experiment

This section provides a general discussion of most of the procedures you will follow to identify your unknown. See the appropriate sections in Part IV for more detailed information about the interpretation of test results. For information about the interpretation of spectra, see the corresponding operations.

Throughout this experiment you should have your lab notebook handy to record the data you collect and all of your observations as you make them. Keeping meticulous records can often mean the difference between the successful identification of a compound and a failure that could prove very costly, insofar as it affects your lab grade.

Since an unknown liquid may be impure, it should be purified by distillation before any chemical tests or spectra are run. The median distillation temperature should also give you a good estimate of its boiling point. Solids can be purified by recrystallization but unless your instructor indicates otherwise you can assume that an unknown solid is pure enough to use without purification. It is very important to measure the boiling point or melting point of your unknown as accurately as you can since your list of possibilities will be based on the value you obtain. If your measured boiling point or melting point is inaccurate, your list of possibilities may not even contain the name of your unknown compound, and identifying it correctly may then be impossible.

A preliminary examination of your unknown may provide some clues that will help you identify it. For example, observing that a compound is a liquid at room temperature eliminates most compounds with reported melting points of 30°C or higher. The ignition behavior of a substance can provide clues about its structure; many oxygen-containing compounds burn with a blue flame, but those with a high molecular weight may exhibit a clear yellow flame and those with aromatic rings a sooty yellow flame. The solubility behavior of your unknown compound in water can also tell you a little about its structure. Most alcohols and carbonyl compounds containing up to four carbon atoms are soluble and most with six carbons or more are relatively insoluble.

To find out what family your compound belongs to, you will carry out several classification tests and record its infrared spectrum. Procedures for the classification tests and directions for their interpretation are given on pages 538–551 in Part IV. The tests you will use include: (a) the 2,4-dinitrophenylhydrazine (DNPH) test, which is positive for both aldehydes and ketones; (b) the chromic acid test, which is positive for aldehydes and for 1° and 2° alcohols but is faster with alcohols; (c) the Tollens test, which is positive only for aldehydes. Infrared bands that you should look for include the strong, broad O—H stretching band near 3300 cm^{-1}, the strong carbonyl (C=O) stretching band near 1700 cm^{-1}, and one or two weak to moderate bands in the 2700–2850 cm^{-1} region, which arise from C—H stretching vibrations involving the carbonyl carbon of an aldehyde.

Additional structural information can be obtained both from chemical tests and from your infrared spectrum. Thus the Lucas test can tell you whether an alcohol is primary, secondary, or tertiary—if the alcohol's boiling point is below 150°C. The test is invalid for most alcohols with higher boiling points. The iodoform test is positive for methyl ketones and methyl carbinols (alcohols having a CH_3 group bonded to the carbon that holds the OH). And the bromine test can show whether your compound contains any carbon-carbon double or triple bonds.

The infrared C—O band of an alcohol is sensitive to the type of alcohol. Most primary, secondary, and tertiary acyclic alcohols absorb around 1050 cm^{-1}, 1110 cm^{-1}, and 1175 cm^{-1}, respectively. The C—O band occurs about 25–50 cm^{-1} lower (to the right) for cyclic alcohols and alcohols having aromatic rings or C=C groups on the carbinol carbon. Other structural features that can be detected from infrared spectra include aromatic rings and carbon-carbon double or triple bonds. (See OP-34 for more detailed information on the interpretation of IR spectra.) NMR spectra can provide a great deal of information about the structure of a molecule. If your instructor allows you to obtain the ^{1}H or ^{13}C NMR spectrum of your unknown, see OP-35 and/or your textbook for information about NMR spectral interpretation.

After you carry out the classification tests and interpret your spectra you should have a short list of possible compounds. To narrow the list down to one compound—preferably the correct one—you will need to prepare a derivative. A derivative preparation is actually a small-scale chemical synthesis in which your unknown compound is converted to a different compound, a solid whose melting point may indicate the identity of the unknown. As for any synthesis, the formation of by-products, incomplete purification, and insufficient drying can lower the melting point of the product. This makes it important to follow the directions carefully and be certain the product is completely dry before you measure its melting point. It is also important to select the right derivative. Depending on the reagents available, you can prepare a *p*-nitrobenzoate, 3,5-dinitrobenzoate, *α*-naphthylurethane, or phenylurethane if you have an alcohol; and a 2,4-dinitrophenylhydrazone, semicarbazone, or oxime if you have an aldehyde or ketone. But a derivative may be unsuitable because its melting point is too low or is not listed for some of the compounds on your short list. Derivatives with melting points of 60° or below are often hard to purify because they tend to melt to an oil in the hot recrystallization solvent. You should also avoid derivatives whose melting points (for the compounds on your list) are too close together. For example, the semicarbazones of 3-methyl-2-butanone and 2-pentanone melt at 113°C and 112°C, while their 2,4-dinitrophenylhydrazones melt at 124°C and 143°C, making the second derivative a better choice. If you have difficulty preparing a certain derivative, or if the derivative you prepare does not eliminate all the possibilities but one, you should then prepare a second derivative.

When you think you have gathered enough evidence to identify your unknown with some certainty, you are free to write down your conclusion. But keep in mind that your evidence should be sufficient to convince your instructor—and yourself—that your conclusion is correct. Thus you should go back over the evidence and make sure that it all points to the same conclusion. If some evidence is not consistent with that conclusion—for example, if a chemical test does not give the result expected for a compound with the structure you arrived at—be prepared to either re-evaluate your conclusion or explain the inconsistency.

Reactions and Properties

General equations for classification test reactions are given in the *Classification Tests* section of Part IV (pp. 538–551). General equations for derivative preparations are given in the *Derivatives* section of Part IV (pp. 557–559).

The properties of the different classes of compounds and their derivatives are given in the following tables in Appendix VI.

class	table	page
alcohols	1	776
aldehydes	2	777
ketones	3	777

Directions

Safety Notes

> **You should consider your unknown compound to be flammable and harmful by inhalation, ingestion, and skin absorption. Minimize your contact with the unknown and do not breathe its vapors.**
> **Safety information for chemicals used in classification tests and derivative preparations are included with the corresponding procedures in Part IV.**

Take Care! Minimize contact with the unknown and do not breathe its vapors.

Premliminary Work. Obtain an unknown compound from your instructor and record its code number in your laboratory notebook. If the unknown is a liquid, purify it by simple distillation and record its distillation boiling range and median boiling temperature [OP-25], then carry out a micro boiling point determination on the pure liquid [OP-29]. If it is a solid, measure its melting point [OP-28]. Describe the physical state, general appearance, and any other notable characteristics of the purified compound in your notebook. Carry out an ignition test as described on p. 532. Test the solubility of the unknown in water by shaking 0.10 mL of the liquid with 3.0 mL of water as described on p. 535.

Functional Class Determination. Test the unknown with 2,4-dinitrophenylhydrazine reagent (DNPH, classification test C-11), chromic acid reagent (C-9), and Tollens' reagent (C-23), or carry out other classification tests suggested by your instructor.

Record the infrared spectrum of your unknown [OP-34]. Use it and the results of the classification tests to determine whether your unknown is an alcohol, aldehyde, or ketone. In your lab notebook, list all compounds from the appropriate table in Appendix VI that have melting or boiling points within ±10°C of your observed value, recording their melting or boiling points and the melting points of the derivatives listed. At your instructor's option, show him or her your list; the instructor will approve the list if it includes your unknown or suggest additional work if it doesn't.

Detection of Structural Features. Write the structure of every compound on your list and consider whether additional classification tests [such as the bromine test (C-7), iodoform test (C-16), or Lucas test (C-17)] will help you eliminate any compounds from the list. Also look for evidence from your observations and your IR spectrum that suggest specific structural features such as aromatic rings, conjugation with double bonds, or the structural class (1°, 2°, or 3°) of an alcohol. With your instructor's permission, you can obtain and interpret the ^{1}H or ^{13}C NMR spectrum [OP-35] of your unknown as well. Prepare a "short list" of compounds by eliminating the least likely possibilities. Keep in mind, however, that classification tests and infrared wavenumbers are not infallible indicators of molecular structure, so it may be necessary later to reconsider some of the compounds you eliminated to arrive at your short list.

Preparation of a Derivative. Using procedure D-1, D-2, D-3, D-4, or D-5, prepare a suitable derivative of your unknown. Purify the derivative by recrystallization [OP-23] as described in the appropriate procedure, dry it thoroughly [OP-21], and obtain its melting point [OP-28]. Deduce the identity of your unknown from the derivative melting point and all other relevant evidence.

Report. Your report should include a statement of the problem and an account of how you applied scientific methodology to solve it. Tabulate your data and results, including your observations for all tests. Justify your conclusion based on the evidence you accumulated.

Exercises

1   Interpret any spectra you obtained as completely as you can.
2   (a) Write balanced equations for the reactions involved in all of the classification tests for which you obtained a positive result. (b) Write balanced equations for the reaction(s) involved in your derivative preparation(s).
3   Describe and explain the possible effect on your results of the following experimental errors or variations. (a) The test tube you used to carry out a DNPH test had just been rinsed with acetone. (b) The watch glass you used for the ignition test had previously been used to weigh sodium sulfate. (c) You performed a Lucas test on a compound that had a boiling point of 175°C. (d) Your derivative formed an oil when you heated it in the recrystallization solvent, but the oil solidified on cooling so you used it to obtain a melting point.
4   Following the format in Appendix V, construct a flow diagram showing the process you followed to identify your unknown.
5   The unknown assigned to a student was an aldehyde, but about half of the sample distilled around 65°C and the rest of it distilled near 155°. What was the aldehyde? What else was in the sample, and why? Write a balanced equation for its formation.
6   An unknown liquid is water soluble and reacts with chromic acid within two seconds. It dissolves in the Lucas reagent but the solution remains clear for 30 minutes. The iodoform test yields a yellow precipitate. Give the name and structure of the unknown.

For Exercises 5–7, assume that the unknown is one listed in Appendix VI.

7   An unknown liquid with a boiling range of 179–181°C is insoluble in water, gives a blue-green suspension with chromic acid, and immediately forms a separate layer when shaken with the Lucas reagent. Its IR spectrum contains bands at 3060 cm^{-1}, 2805 cm^{-1}, 2730 cm^{-1}, 1705 cm^{-1}, 745 cm^{-1}, and 690 cm^{-1}. Give the name and structure of the unknown.

8   Write mechanisms for the following reactions, which are used in chemical tests and derivative preparations. (a) The reaction of butanal with 2,4-dinitrophenylhydrazine reagent. (b) The preparation of the 3,5-dinitrobenzoate of 1-butanol. (c) The iodoform reaction of 2-butanone. (d) The reaction in the Lucas test of 2-methyl-2-butanol.

Other Things You Can Do

(Starred projects require your instructor's permission.)

*1   Observe the effect of potassium permanganate on different classes of alcohols in Minilab 25.

*2   Use ^{1}H NMR to identify an unknown arene in Minilab 28.

3   Starting with sources listed in the Bibliography, write a research paper about the use of gas chromatography-mass spectrometry (GC-MS) to identify illicit drug samples and trace them back to their sources.

Wittig Synthesis of 1,4-Diphenyl-1,3-butadiene

EXPERIMENT 40

Reactions of Carbonyl Compounds. Preparation of Dienes. Nucleophilic Addition. Wittig Reaction. Ylides.

Operations:

OP-6   Heating
OP-9   Mixing
OP-12 Vacuum Filtration
OP-13 Extraction
OP-14 Evaporation
OP-21 Drying Solids
OP-23 Recrystallization
OP-28 Melting Point

Before You Begin

1 Read the experiment, read or review the operations as necessary, and write an experimental plan.
2 Calculate the mass and volume of 10.0 mmol of *trans*-cinnamaldehyde.

Scenario

Marvelous Molecules Incorporated has experienced a lower than expected demand for its product line of aldehydes, so it has a large inventory of cinnamaldehyde that it would like to reduce. MMI's peripatetic marketing executive Penny Wise has learned about a demand for novel dienes such as 1,4-diphenyl-1,3-butadiene, which can be converted to interesting products such as *p*-terphenyl via Diels–Alder reactions.

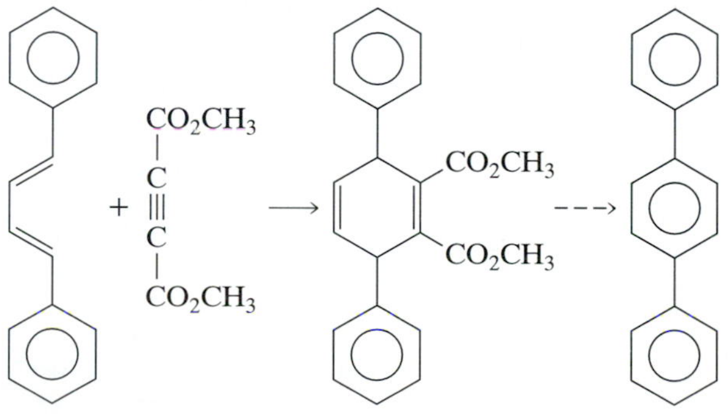

(*E,E*)-1,4-diphenyl-1,3-butadiene
(*s-cis* conformation)

p-terphenyl

A staff chemist has informed her that it should be possible to prepare 1,4-diphenyl-1,3-butadiene from cinnamaldehyde by a procedure known as the Wittig synthesis, but its value as a Diels–Alder diene will depend on its stereochemistry. The (*E,E*) diene can easily attain the *s-cis* conformation needed to form a Diels–Alder adduct, but the (*E,Z*) diene is less likely to do so. Your assignment is to find out whether or not you can prepare 1,4-diphenyl-1,3-butadiene from *trans*-cinnamaldehyde and whether or not the major product has the desired (*E,E*) stereochemistry.

Applying Scientific Methodology

The scientific problems in this experiment are implicit in the Scenario. You should be able to develop a working hypothesis, which will be tested by obtaining the melting point of the major product.

Bark Spices and Cinnamaldehyde

cinnamon
stick

Tree bark is not a common source of food products, but the spices cinnamon and cassia come from the bark of trees of the genus *Cinnamomum*. True cinnamon is obtained from *Cinnamomum zeylanicum*, a tree that grows in Sri Lanka (formerly Ceyon) and southern India. Cinnamon is obtained by peeling the bark from the cut branches of the cinnamon tree and then scraping off the wood and outer layers of bark. The thin strips of inner bark are sun dried, forming rolled up "quills" up to a meter in length. These are generally cut into shorter lengths to produce cinnamon sticks, or ground up to make powdered cinnamon. Cinnamon bark from Sri Lanka contains 1–2 percent of an aldehyde-rich essential oil of which about 70 percent is cinnamaldehyde (3-phenyl-2-propenal). Other components of cinnamon oil include benzaldehyde, *p*-isopropylbenzaldehyde, 3-phenylpropanal, nonanal, and 2-furaldehyde, along with such non-aldehyde ingredients as 2-heptanone, caryophyllene, and various other flavor components. Cinnamon leaves, surprisingly, contain no cinnamaldehyde but are rich in eugenol, the main flavor ingredient of cloves.

Nearly all of the "cinnamon" consumed in the United States is actually cassia, which is obtained from the Chinese cassia tree, *Cinnamomum cassia*, and several related species grown in southeast Asia. Chinese cassia oil is 80–95 percent cinnamaldehyde, but its strong spicy-sweet flavor is quite different from that of true cinnamon oil, which is less sweet but more complex and fragrant, with citrus overtones.

Synthetic cinnamaldehyde, which is mainly the *trans*-isomer, is prepared by an aldol condensation reaction of benzaldehyde and ethanal (acetaldehyde).

Cinnamaldehyde, as well as natural cinnamon and cassia oils, is used to flavor candies, chewing gum, and baked goods, and as an ingredient in perfumes.

Understanding the Experiment

The German chemist Georg Wittig developed the Wittig reaction in 1954, but it took 25 years before he received full recognition for originating one of the most synthetically useful reactions in organic chemistry. In 1979 Wittig shared the Nobel Prize in chemistry with another well-known synthetic organic chemist, Herbert C. Brown of the United States.

Like the aldol condensation, the Wittig reaction is used to construct larger molecules from smaller ones, connecting the components of the smaller molecules with carbon-carbon double bonds. Both reactions involve the attack of a nucleophilic carbon atom, stabilized by a neighboring electron-withdrawing group, at the carbonyl carbon atom of an aldehyde or ketone, as shown by the following examples.

aldol condensation

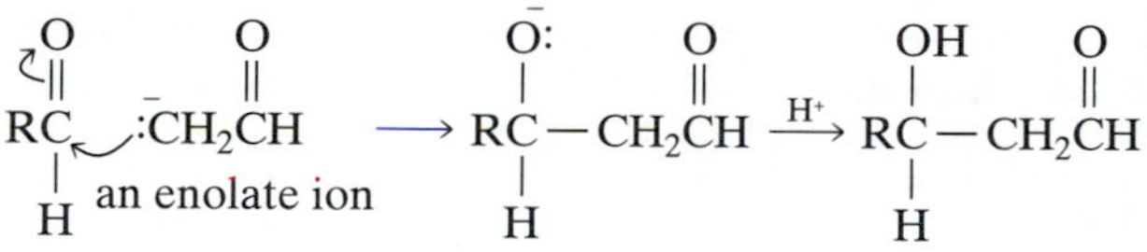

$$RC\overset{O}{\underset{H}{\|}}{:}CH_2CH\overset{O}{\|} \longrightarrow RC\overset{\overset{-}{O}:}{-}CH_2CH\overset{O}{\|} \overset{H^+}{\longrightarrow} RC\overset{OH}{-}CH_2CH\overset{O}{\|}$$

an enolate ion

Wittig reaction

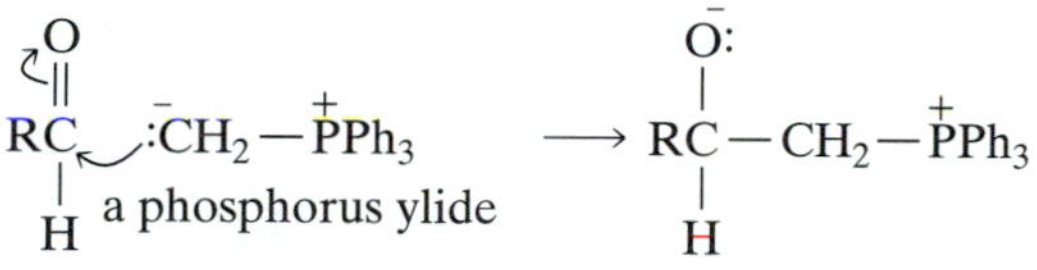

$$RC\overset{O}{\underset{H}{\|}}{:}CH_2{-}\overset{+}{P}Ph_3 \longrightarrow RC\overset{\overset{-}{O}:}{-}CH_2{-}\overset{+}{P}Ph_3$$

a phosphorus ylide

In the aldol condensation the nucleophile is an enolate ion, which is stabilized by resonance involving the carbonyl group. In the Wittig reaction the nucleophile is a phosphorus ylide, which is stabilized by resonance involving a triphenylphosphonium group. In both reactions the carbon-carbon double bond then forms by elimination (in several steps) of a molecular species; H_2O in the case of the aldol condensation and $Ph_3P{=}O$ in the case of the Wittig reaction.

aldol condensation

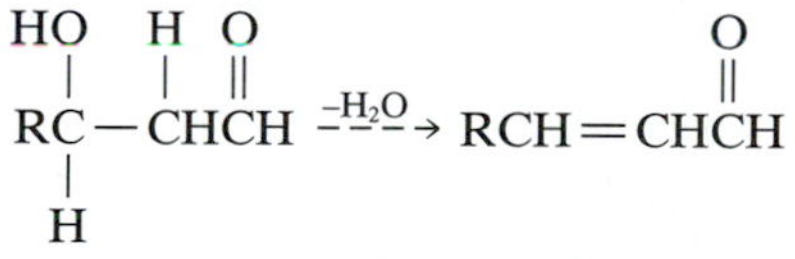

$$RC\overset{HO}{-}CHCH\overset{H}{\underset{H}{|}}\overset{O}{\|} \overset{-H_2O}{\longrightarrow} RCH{=}CHCH\overset{O}{\|}$$

Wittig reaction

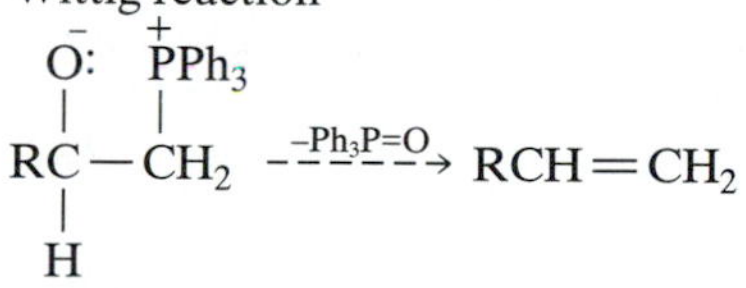

$$RC\overset{\overset{-}{O}:}{-}CH_2\overset{\overset{+}{P}Ph_3}{} \overset{-Ph_3P{=}O}{\dashrightarrow} RCH{=}CH_2$$

In the Wittig reaction the Ph_3P group needed to stabilize the nucleophilic carbon atom is lost along with the carbonyl oxygen. Thus the Wittig reaction, unlike the aldol condensation, can be used for the synthesis of unsaturated hydrocarbons that have no other functional groups.

See your lecture textbook for a more complete description and mechanism of the Wittig reaction.

The key intermediate in a Wittig synthesis is the resonance-stabilized phosphorus ylide, which is typically prepared by the reaction of triphenylphosphine with an alkyl halide to yield a phosphonium salt, followed by treatment with a strong base such as butyllithium.

$$Ph_3P + CH_3I \longrightarrow [Ph_3\overset{+}{P}-CH_3]I^- \xrightarrow{C_4H_9Li} [Ph_3\overset{+}{P}-\overset{..}{C}H_2 \longleftrightarrow Ph_3P=CH_2]$$

In this experiment the phosphonium salt will be benzyltriphenylphosphonium chloride, which is prepared by the reaction of triphenylphosphine with benzyl chloride. This is an S_N2 reaction in which triphenylphosphine (the nucleophile) displaces chloride ion (the leaving group) from benzyl chloride (the substrate).

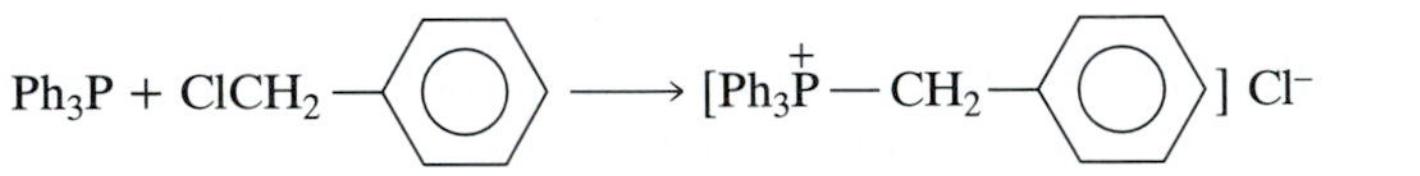

$$Ph_3P + ClCH_2 \!-\!\!\langle\bigcirc\rangle \longrightarrow [Ph_3\overset{+}{P}-CH_2\!-\!\!\langle\bigcirc\rangle]\,Cl^-$$

benzyltriphenylphosphonium chloride

The phosphonium salt may be provided or you may have to synthesize it as described in part **A** of the directions. The reaction is carried out in the high-boiling solvent *p*-cymene (1-isopropyl-4-methylbenzene) over a 2-hour reaction period.

Formation of the ylide is facilitated by the presence of a phenyl group that helps stabilize it, so concentrated sodium hydroxide is used rather than a stronger base such as butyllithium.

$$[Ph_3\overset{+}{P}-CH_2\!-\!\!\langle\bigcirc\rangle]\,Cl^- \xrightarrow{NaOH} Ph_3\overset{+}{P}-\overset{..}{\overset{-}{C}}H\!-\!\!\langle\bigcirc\rangle$$

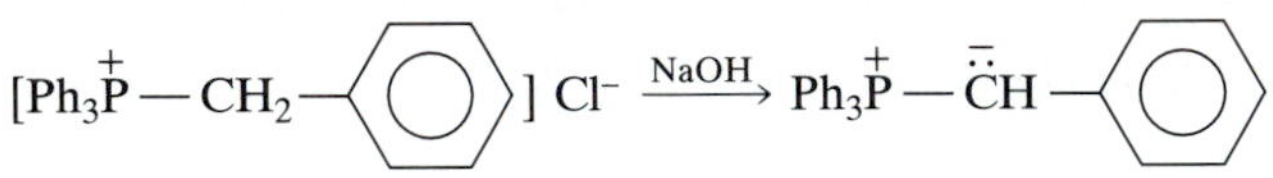

You will carry out the ylide-forming reaction in the presence of cinnamaldehyde, which reacts with the ylide to form the product, 1,4-diphenyl-1,3-butadiene.

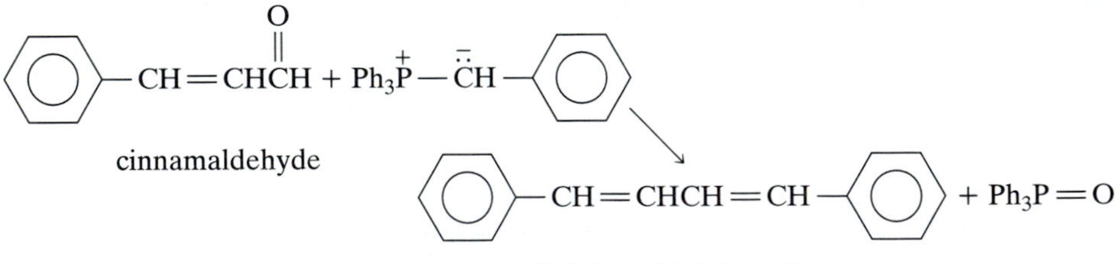

$$\langle\bigcirc\rangle\!-\!CH=CH\overset{\overset{\textstyle O}{\|}}{C}H + Ph_3\overset{+}{P}-\overset{..}{\overset{-}{C}}H\!-\!\!\langle\bigcirc\rangle$$

cinnamaldehyde

$$\langle\bigcirc\rangle\!-\!CH=CHCH=CH\!-\!\!\langle\bigcirc\rangle + Ph_3P=O$$

1,4-diphenyl-1,3-butadiene

This reaction is carried out at room temperature in a two-phase mixture with dichloromethane as the organic phase. Benzyltriphenylphosphonium chloride itself acts as a phase-transfer catalyst, apparently escorting the polar ylide from the aqueous to the organic layer, where it can encounter and react with cinnamaldehyde. Most of the major product will remain in the organic phase; the rest is extracted from the aqueous phase with additional dichloromethane. The crude product is treated with 60% aqueous ethanol to remove triphenylphosphine oxide and a minor product of the reaction (which can be isolated as described in "Other Things You Can Do"), then purified further by recrystallization from 95% ethanol. The flat crystals tend to stick to glass surfaces and are hard to scrape off, so you should avoid all unnecessary transfers.

You will be using the *trans*-isomer of cinnamaldehyde so the stereochemistry about the first double bond of the product will also be *trans*. The stereochemistry about the second double bond—the one formed in the reaction—could be either *cis* or *trans*, allowing for the formation of two possible products, (*E,E*)- and (*E,Z*)-1,4-diphenyl-1,3-butadiene.

(*E,E*)-1,4-diphenyl-1,3-butadiene   (*E,Z*)-1,4-diphenyl-1,3-butadiene

A melting point measurement should tell you whether or not the desired (*E,E*)-isomer is the major product.

Reactions and Properties

A. $Ph_3P + PhCH_2Cl \longrightarrow [Ph_3\overset{+}{P}-CH_2Ph]\ Cl^-$

       benzyl chloride   benzyltriphenylphosphonium chloride

B. $[Ph_3\overset{+}{P}-CH_2Ph]\ Cl^- + NaOH \longrightarrow Ph_3\overset{+}{P}-\overset{..}{\overset{-}{C}}HPh + H_2O + NaCl$

trans-cinnamaldehyde   (*E,E*)- or (*E,Z*)-1,4-diphenyl-1,3-butadiene

Table 40.1 Physical properties

	M.W.	m.p.	b.p.	d
benzyl chloride	126.6	−43	179	1.100
triphenylphosphine	262.3	80.5	377	
p-cymene	134.2	−68	177	0.857
trans-cinnamaldehyde	132.2	−7.5	246	1.050
dichloromethane	84.9	−97	40.5	1.326
(*E,E*)-1,4-diphenyl-1,3-butadiene	206.3	153		
(*E,Z*)-1,4-diphenyl-1,3-butadiene	206.3	88		

Note: m.p. and b.p. are in °C, density is in g/mL

Directions

A. *Preparation of Benzyltriphenylphosphonium Chloride*
This step can be omitted if previously prepared benzyltriphenylphosphonium chloride is available.

Safety Notes

benzyl chloride   *p*-cymene

> **Benzyl chloride is toxic by inhalation and highly irritating to skin and
> eyes. It is a suspected carcinogen and an experimental teratogen (a sub-
> stance that may harm a developing fetus). It may react violently with oxi-
> dants. Wear gloves and use a hood; avoid contact and do not breathe
> vapors; keep away from other chemicals.**
> **Triphenylphosphine is toxic by ingestion and may be harmful if inhaled;
> do not breathe vapors.**
> ***p*-Cymene is flammable and irritates the skin. Avoid contact and keep
> away from flames.**

Take Care! Wear gloves, avoid contact, do not breathe vapors.

Reaction. *Under the hood*, combine 2.00 mL (~17 mmol) of benzyl chloride
with 6.00 g (~23 mmol) of triphenylphosphine and 30 mL of *p*-cymene in a
100-mL round-bottomed flask. Add a stirbar [OP-9] and heat the reaction
mixture under reflux [OP-6] for two hours or more, stirring to prevent bumping as the solid product forms. If you don't have a stirrer use boiling chips, but
keep the boiling gentle enough to minimize bumping. Let the mixture cool to
room temperature, then cool it in an ice/water bath for 15 minutes and collect the product by vacuum filtration [OP-12], washing it with 10 mL of cold
low-boiling petroleum ether to remove the *p*-cymene. Dry the benzyltriphenylphosphonium chloride [OP-21] and measure its mass. Save 3.90 g for
part **B** and turn in the rest, or dispose of it as directed by your instructor.

Stop and Think: Why does the
product precipitate from this reaction mixture?
Waste Disposal: Place the *p*-cymene
in the designated solvent recovery
container.

B. *Preparation of 1,4-Diphenyl-1,3-butadiene*
If you obtained less than 3.90 g of benzyltriphenylammonium chloride from
part **A**, scale down all quantities of reactants and solvents proportionately.

Safety Notes

dichloromethane   sodium hydroxide

> ***trans*-Cinnamaldehyde is a skin irritant; avoid contact.**
> **Dichloromethane may be harmful if ingested, inhaled, or absorbed
> through the skin. There is a possibility that prolonged inhalation of
> dichloromethane may cause cancer. Minimize contact with the liquid and
> do not breathe its vapors.**
> **Sodium hydroxide is toxic and corrosive, causing severe damage to skin,
> eyes, and mucous membranes. Wear gloves and avoid contact with the
> NaOH solution.**

Take Care! Avoid contact with
dichloromethane, do not breathe
its vapors.

Take Care! Wear gloves, avoid
contact with the NaOH solution.

Stop and Think: What is the purpose of the sodium hydroxide?

Reaction. In a 50-mL Erlenmeyer flask combine 3.90 g of benzyltriphenylphosphonium chloride, 10.0 mmol of pure *trans*-cinnamaldehyde,
and 10 mL of dichloromethane. Add 5 mL of 50% (~19 *M*) aqueous sodium
hydroxide and a magnetic stirbar [OP-9] and stir the mixture vigorously at
room temperature for 30 minutes.

Separation. Transfer the reaction mixture to a separatory funnel using
20 mL of dichloromethane followed by 15 mL of water for the transfer.

Shake gently to extract the product into the dichloromethane layer [OP-13], and drain the dichloromethane layer into an Erlenmeyer flask. Dry the dichloromethane solution over anhydrous magnesium sulfate [OP-21]. Evaporate the dichloromethane under vacuum with gentle heating [OP-14], using a cold trap. Continue the evaporation until no more vapor bubbles emerge from the oily residue. The residue should solidify as it cools.

Purification and Analysis. Add 35 mL of 60% (by volume) aqueous ethanol to the solidified residue and break up the solid as finely as possible with a spatula or flat-bottomed stirring rod. This procedure removes triphenylphosphine oxide and other impurities. Collect the solid by vacuum filtration [OP-12], washing it on the filter with several small portions of ice-cold 60% ethanol. Recrystallize the crude 1,4-diphenyl-1,3-butadiene from 95% ethanol [OP-23], dry it [OP-21], and measure its mass and melting point [OP-28].

Report. Your report should include a statement of the problem and an account of how you applied scientific methodology to solve it. Calculate the percent yield of 1,4-diphenyl-1,3-butadiene, based on the mass of the limiting reactant.

Waste Disposal: Flush the aqueous layer down the drain with plenty of water.

Waste Disposal: Place the recovered dichloromethane in a chlorinated solvents recovery container.

Waste Disposal: At your instructor's option, save the filtrate for further work. Otherwise dispose of it as requested.

Exercises

1 Why do you think the product you obtained, rather than another isomer, is the major product of the Wittig reaction?

2 Of the three 1,4-diphenyl-1,3-butadiene isomers, (E,E), (E,Z), and (Z,Z), which would be most suitable as a diene in the Diels–Alder reaction? Which would be least suitable? Explain. (b) Why is (Z,Z)-1,4-diphenyl-1,3-butadiene not a likely product of this reaction?

3 Describe and explain the possible effect on your results of the following experimental errors or variations. (a) The lab assistant accidentally put chlorobenzene in the bottle labeled "benzyl chloride." (b) There was no *p*-cymene available so you used toluene instead. (c) You didn't have a magnetic stirrer, so in part **B** you shook the reaction mixture for a few seconds every minute or so.

4 Following the format in Appendix V, construct a flow diagram for the synthesis of 1,4-diphenyl-1,3-butadiene in part **B**.

5 Outline a Wittig synthesis of each of the following from an appropriate alkyl halide and carbonyl compound.

(a)   (b)   (c)

Other Things You Can Do

(Starred projects require your instructor's permission.)

*1 You can isolate another isomer of 1,4-diphenyl-1,3-butadiene from the filtrate saved from the first vacuum filtration in part **B**. Transfer the filtrate to a large test tube and withdraw any insoluble oil using a Pasteur pipet. Dissove the oil in 10 mL of dichloromethane (more if necessary), dry the dichloromethane layer with anhydrous magnesium sulfate, and evaporate the solvent. Keep this substance in the dark, as it is converted to yet another isomer by light. Dissolve a small crystal of the substance in ~10 mL of hexane and obtain its UV spectrum between 200 and 400 nm, diluting the solution with more hexane as necessary to keep the peaks on scale. Add a small crystal of iodine and illuminate the solution with a 100 watt light bulb for 10 minutes or more, then obtain its UV spectrum in the same region. Compare both spectra with the UV spectrum of your original isomer from part **B**, recorded in the same way, and explain your results.

*2 Carry out a synthesis that involves C$=$N bond-forming reactions in Minilab 33.

3 Starting with sources listed in the Bibliography, write a research paper about the mechanism, stereochemistry, and applications of the Wittig reaction.

Effect of Reaction Conditions on the Condensation of Furfural with Cyclopentanone

Reactions of Carbonyl Compounds. Preparation of α,β-Unsaturated Carbonyl Compounds. Nucleophilic Addition. Condensation Reactions. Enolate Ions. NMR Spectrometry.

Operations:

OP-7   Cooling
OP-9   Mixing
OP-12 Vacuum Filtration
OP-13 Extraction
OP-14 Evaporation
OP-19 Washing Liquids
OP-20 Drying Liquids
OP-21 Drying Solids
OP-23 Recrystallization
OP-26 Vacuum Distillation
OP-28 Melting Point
OP-35 Nuclear Magnetic Resonance Spectrometry

Before You Begin

1 Read the experiment, read or review the operations as necessary, and write an experimental plan. Read OP-26 carefully if you have not performed a vacuum distillation before.
2 For part **A**, calculate the mass and volume of 50.0 mmol of cyclopentanone and of 50.0 mmol of furfural.
3 For part **B**, calculate the mass and volume of 10.0 mmol of cyclopentanone.

Scenario

Grits 'N Groats, a breakfast cereal manufacturer, produces huge quantities of oat hulls and corncobs while processing cereal grains. In the past they have simply disposed of these "waste" products, but the rising cost of waste disposal has convinced the board of directors that Grits 'N Groats should find a way to profit from them rather than paying to get rid of them. Oat hulls, corncobs, and other agricultural wastes can be processed to yield furfural, an aldehyde with an aromatic furan ring. While searching the chemical literature for references to furfural, project leader Farina Millet came across a paper in the *Journal of Organic Chemistry* describing a Claisen–Schmidt

reaction (a type of aldol condensation) between cyclopentanone and furfural. Under one set of conditions the reaction yields a low-melting yellow
solid in approximately 60% yield, but under a different set of conditions it
produces a high-melting golden-orange solid in nearly 100% yield. Dr.
Millet needs samples of these compounds, and their structures, so that she
can explore their potential for commercial development. Your assignment is
to synthesize both compounds and identify them by using your chemical
intuition, with some help from NMR spectrometry.

Applying Scientific Methodology

After reading the experiment and comparing the reaction conditions, you
should be able to formulate a tentative hypothesis about the structures of
the products, and to predict the kind of ^{1}H NMR spectrum each product
should have. You will test your hypothesis when you obtain the actual NMR
spectra of the products.

From Oats to Furfural

furfural

Furfural, also known as 2-furaldehyde, is the most important member of the
furan series of aromatic compounds. The furan ring is aromatic because it
has six pi electrons (two from the oxygen atom) distributed about a five-
membered ring (see Figure 41.1). The aromatic sextet of a furan ring is less
stable than that of a benzene ring, so furans undergo reactions such as electrophilic addition, cycloaddition, and cleavage more readily than the corresponding benzene compounds.

Furfural can be prepared in large quantities by treating such agricultural by-products as bran, oat hulls, corncobs, and peanut shells with dilute
acids. These materials contain polysaccharides known as pentosans that are
hydrolyzed to pentoses (simple five-carbon sugars) under acidic conditions.
These in turn are converted to furfural by acid-catalyzed dehydration.

Conversion of a pentose to furfural. (This equation represents
the overall process, not the reaction mechanism.)

a pentose $\xrightarrow{-3H_2O}$ furfural

The first commercial process for manufacturing furfural was developed
in 1922 by the Quaker Oats Company, which was trying to convert oat hulls
into better cattle feed at the time. It instead came up with a valuable commercial product that can be made cheaply on a large scale. Furfural is used in
the purification of lubricating oils, the extractive distillation of 1,3-butadiene
(used in the manufacture of rubber), the synthesis of phenolic resins, and
the manufacture of a large number of chemical intermediates.

Furfural behaves like a typical aromatic aldehyde in many of its reactions; it can be oxidized and reduced to the corresponding carboxylic acid

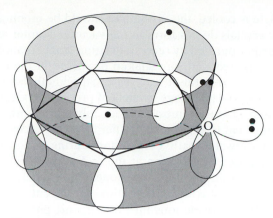

Figure 41.1 Aromatic furan ring

and alcohol and, like benzaldehyde, undergoes the Cannizzaro reaction and the benzoin condensation. It also reacts with compounds that have active methylene groups, such as aldehydes and ketones, to yield condensation products. For example, the reaction of furfural with acetone yields furfurylideneacetone by a Claisen–Schmidt reaction, and its condensation with acetic anhydride and sodium acetate forms furylacrylic acid by a Perkin reaction.

Claisen-Schmidt condensation of furfural and acetone

$$\text{furfural} - \text{CHO} + \text{CH}_3\overset{\overset{\displaystyle O}{\|}}{\text{C}}\text{CH}_3 \xrightarrow{\text{OH}^-} \text{furfural} - \text{CH}=\text{CH}\overset{\overset{\displaystyle O}{\|}}{\text{C}}\text{CH}_3$$

furfurylideneacetone

$$\text{furyl} - \text{CH}=\text{CH}\overset{\overset{\displaystyle O}{\|}}{\text{C}} - \text{OH}$$

furylacrylic acid

Understanding the Experiment

The Claisen–Schmidt reaction is a kind of crossed aldol condensation between an aromatic aldehyde and an aliphatic aldehyde or ketone that yields an α,β-unsaturated aldehyde or ketone. As illustrated for the reaction of benzaldehyde and acetone, the aliphatic carbonyl compound loses a proton to form an enolate ion, which attacks the carbonyl carbon to yield (after protonation) a β-hydroxy carbonyl compound. This intermediate is generally not isolated but undergoes base-catalyzed dehydration by an E1cb mechanism, in which a proton is removed from the α-carbon to form another enolate ion, which loses an OH^- ion to form the unsaturated product.

In one of the procedures referred to in the Scenario, equimolar amounts of cyclopentanone and furfural are dissolved in ethyl ether and shaken with an aqueous solution of dilute sodium hydroxide. Furfural tends to oxidize and turn dark brown in storage, so furfural from a previously opened bottle should be distilled before use. Because the organic reactants are not very soluble in water they tend to stay in the ether phase, making the reaction rather slow. Extraction of the reaction mixture with ethyl ether yields an impure yellow liquid that is vacuum distilled to produce a low-melting yellow solid, product **A**. Even though the product usually does not solidify

Mechanism of a Claisen-Schmidt condensation

$$\text{CH}_3\overset{\overset{\displaystyle O}{\|}}{\text{C}}\text{CH}_3 \xrightarrow{\text{OH}^-} \overset{\ominus}{\text{CH}}_2\overset{\overset{\displaystyle O}{\|}}{\text{C}}\text{CH}_3 \xrightarrow{\text{PhCHO}}$$

$$\text{PhCH}\overset{\displaystyle O^\ominus}{|} - \text{CH}_2\overset{\overset{\displaystyle O}{\|}}{\text{C}}\text{CH}_3 \xrightarrow{\text{H}_2\text{O}}$$

$$\text{PhCH}\overset{\displaystyle OH}{|}\text{CH}_2\overset{\overset{\displaystyle O}{\|}}{\text{C}}\text{CH}_3 \xrightarrow{\text{OH}^-}$$

$$\text{PhCH}\overset{\displaystyle OH}{|}\overset{\ominus}{\text{CH}}\overset{\overset{\displaystyle O}{\|}}{\text{C}}\text{CH}_3 \xrightarrow{-\text{OH}^-}$$

$$\text{PhCH}=\text{CH}\overset{\overset{\displaystyle O}{\|}}{\text{C}}\text{CH}_3$$

until the distillate is cooled, the distillation should be monitored closely to make sure that crystals do not form in the vacuum adapter and plug it up. Product **A** is further purified by recrystallization from ethanol.

When the same reactants are shaken or stirred in the presence of an effective phase-transfer catalyst such as tricaprylmethylammonium chloride (Aliquat 336), golden-orange crystals of product **B** begin to crystallize from the reaction mixture almost immediately. The exothermic reaction generates enough heat to vaporize the ether; thus it is necessary to cool the reactants in an ice/water bath before and during the reaction period. Product **B** is separated by filtration at the end of the reaction period and purified by recrystallization from 2-butanone. The role played by the phase-transfer catalyst in aldol-type condensations is not entirely clear, but the catalyst may carry hydroxide ions into the organic phase where they can generate enolate ions, which then react with furfural molecules. (See Experiment 22 for further information about phase-transfer catalysis.)

From the structures of the reactants and your knowledge of aldol-type condensation reactions, you should be able to propose likely structures for the two products. The ^{1}H NMR spectra will show clearly which product you have actually prepared in each case. You should pay particular attention to the number and multiplicity of signals produced by the methylene protons of the cyclopentanone ring. Comparing your spectra with the NMR spectra of the reactants in Figure 41.2 should help you identify the signals from particular proton sets.

Although the stereochemistry of the products has not been reported in the literature, most reactions of this type yield the (E) isomers, as in the condensation of benzaldehyde with 4,4-dimethyl-1-tetralone:

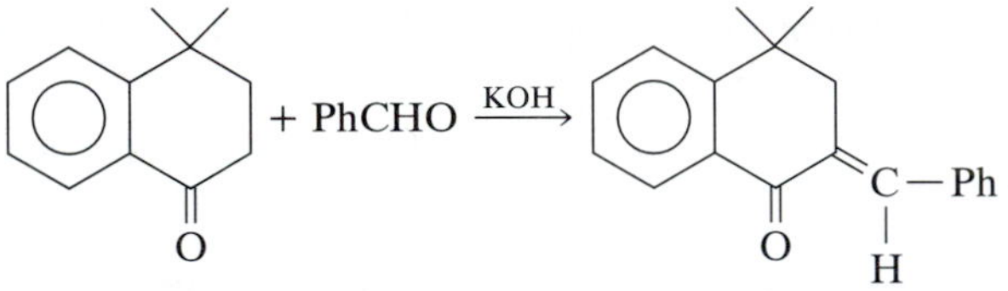

| 4,4-dimethyl- | (E)-2-benzal-4,4-dimethyl- |
| 1-tetralone | 1-tetralone |

The nearby carbonyl group has a deshielding effect on the vinylic proton of the (E) isomer, raising its chemical shift to 7.7 ppm compared to a value of 6.6 ppm in the (Z) isomer. The infrared spectra of these compounds also differ in that the C=C stretching band for the (E) isomer (at 1610 cm^{-1}) is nearly as strong as the carbonyl band, but for the (Z) isomer appears to be absent.

Reactions and Properties

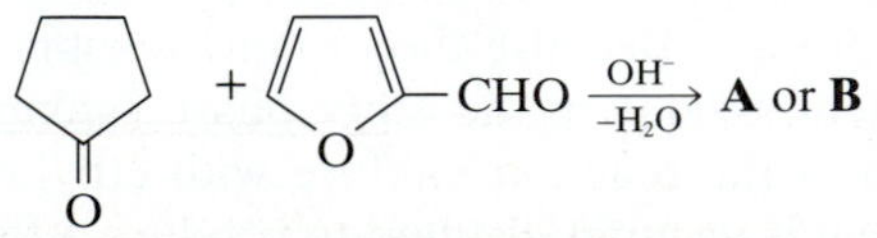

cyclopentanone   furfural   (equation not balanced)

Table 41.1 Physical properties

	M.W.	m.p.	b.p.	d
furfural	96.1	−39	162	1.159
cyclopentanone	84.1	−51	131	0.949
2-butanone	72.1	−86	80	0.805
tricaprylmethylammonium chloride	404.2			0.884
product **A**	162.2	60.5	154[15]	
product **B**	240.3	162		

Note: m.p. and b.p. are in °C, density is in g/mL

Cyclopentanone

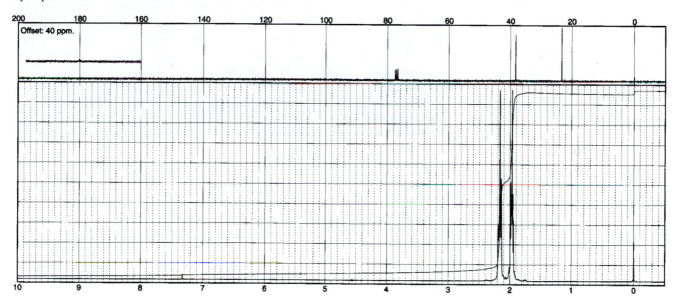

2-Furaldehyde (furfural)

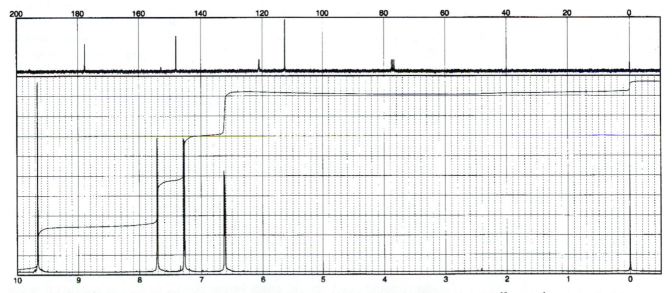

Figure 41.2 NMR spectra of the starting materials (Reproduced from *The Aldrich Library of ^{13}C and ^{1}H FT-NMR Spectra* by C. J. Pouchert and J. Behnke, with the permission of the Aldrich Chemical Company).

Directions

A. *Claisen–Schmidt Reaction of Cyclopentanone and Furfural*

Safety Notes

furfural

cyclopentanone

ethyl ether

> **Furfural irritates the skin, eyes, and respiratory tract and may cause allergic skin or respiratory reactions. Wear gloves while handling furfural and the reaction mixture (which will stain your hands yellow), and do not breathe the vapors.**
> **Cyclopentanone is a skin and severe eye irritant; avoid contact.**
> **Ethyl ether is extremely flammable and may be harmful if inhaled. Do not breathe its vapors and keep it away from flames and hot surfaces.**
> **A vacuum distillation apparatus may implode if any of its components are cracked or otherwise damaged. Inspect the parts for damage and have your instructor check your apparatus. If possible, it is best to work behind a hood sash or safety shield while the apparatus is under vacuum.**
> **Deuterochloroform is harmful if inhaled or absorbed through the skin, and it is a suspected human carcinogen. Avoid contact with the liquid and do not breathe its vapors.**

Take Care! Keep ether away from ignition sources, do not breathe its vapors.

Take Care! Wear gloves, avoid contact with furfural, do not breathe its vapors

Observe and Note: What happens during the reaction period?

Stop and Think: What could this solid be?

Waste Disposal: Place the recovered ether in a solvent recovery container. Pour the aqueous layers down the drain.

Stop and Think: At about what temperature should the product distill?

Reaction. In a 125-mL Erlenmeyer flask, dissolve 50.0 mmol of cyclopentanone in 25 mL of ethyl ether and add 45 mL of 0.10 M aqueous sodium hydroxide. Cool the mixture to 5°C in an ice bath [OP-7], and add 50.0 mmol of freshly distilled furfural and a magnetic stirbar, if you have one. Seal the flask with Parafilm and stir the reaction mixture vigorously [OP-9] in a cold water bath (10–15°C) for 45 minutes. (If you don't have a magnetic stirrer, wear gloves and shake the flask vigorously and continuously with cooling throughout the reaction period.) If necessary, replace any ether that evaporates.

Separation. Filter the reaction mixture by vacuum filtration [OP-12], saving the filtrate, and thoroughly wash any solid on the filter with 10 mL of ethyl ether, combining the wash liquid with the filtrate. Save this solid and weigh it when it is dry. If a solid continues to form in the reaction mixture after the initial filtration, remove it by filtration before you evaporate the solvent.

Shake the filtrate (both aqueous and ether layers) in a separatory funnel to extract some product from the aqueous layer, and separate and save both layers. Extract the aqueous layer [OP-13] with 15 mL of ethyl ether, and combine the ether extract with the initial ether layer. Wash this solution [OP-19] twice with saturated aqueous sodium chloride, dry it over anhydrous magnesium sulfate [OP-20], and evaporate the ether [OP-14] using a cold trap.

Purification and Analysis. Purify the residue by vacuum distillation [OP-26], following the precautions described in OP-26 and the Safety Notes, using a small-scale apparatus and *no* cooling bath. Monitor the distillation carefully. If any solid begins to form in the outlet tube, melt it with a heat gun or other appropriate heating device. Any unreacted starting materials and residual ether should distill below 100°C. After the product has distilled, cool the distillate if necessary until it completely solidifies. Purify this solid by mixed solvent recrystallization from 95% ethanol and water [OP-23]. Dry it thoroughly at room temperature [OP-21]. Measure the mass and melting point [OP-28] of product **A**. Record its ¹H NMR spectrum [OP-35] in deuterochloroform or obtain a spectrum from your instructor.

B. *Claisen–Schmidt Reaction Using a Phase-Transfer Catalyst*

> **2-Butanone is flammable and ingestion, inhalation, or skin absorption may be harmful. Avoid contact, do not breathe its vapors, and keep it away from flames or hot surfaces.**
> **See part A for additional precautions.**

Reaction. In a 125-mL Erlenmeyer flask, dissolve 10.0 mmol of cyclopentanone in 10 mL of ethyl ether. Then add 12 mL of 0.10 *M* aqueous sodium hydroxide, 6 drops of tricaprylmethylammonium chloride (or 0.2 g of another suitable phase-transfer catalyst), and a stirbar, if you have one. Cool the mixture to 5°C in an ice/water bath [OP-7], and add 2.0 mL of furfural with stirring or swirling. Seal the flask with Parafilm and stir it vigorously [OP-9] at room temperature for 15 minutes, occasionally swirling it in the ice water bath to reduce pressure buildup from vaporizing ether. (If you don't have a magnetic stirrer, wear gloves and shake/swirl the flask vigorously and continuously with occasional cooling throughout the reaction period.) Let the reaction mixture stand at room temperature, with stirring or occasional shaking, for 10 minutes more.

Separation. Collect the product by vacuum filtration [OP-12] and wash it on the filter with two portions of ethyl ether. Partially dry the crude product by blotting it between filter papers (wear gloves).

Purification and Analysis. Recrystallize product **B** [OP-23] from 2-butanone, using about 12 mL of the solvent per gram of crude product. Wash the product on the filter with ethyl ether, dry it [OP-21], and weigh it. Measure the melting point [OP-28] of product **B**. Record its ¹H NMR spectrum [OP-35] in deuterochloroform or obtain a spectrum from your instructor.

Report. Your report should include a statement of the problem and an account of how you applied scientific methodology to solve it. Interpret the ¹H NMR spectra as completely as you can. Draw the structures of products **A** and **B** and name them.

Waste Disposal: Place the low-boiling forerun in an appropriate waste container.

Take Care! Avoid contact with $CDCl_3$, do not breathe its vapors.

Safety Notes

2-butanone

Take Care! Keep ether away from ignition sources, do not breathe its vapors.

Take Care! Wear gloves, avoid contact with furfural, do not breathe its vapors.

Observe and Note: Compare your observations during this reaction with your observations during the first reaction.

Waste Disposal: Place the filtrate in a designated solvent recovery container.

Take Care! Do not breathe vapors of 2-butanone, keep it away from ignition sources.

Waste Disposal: Place the filtrate in a designated solvent recovery container.

Take Care! Avoid contact with $CDCl_3$, do not breathe its vapors.

Exercises

1   Discuss the effect of reaction conditions on the outcome of the Claisen–Schmidt reaction, and tell what conditions promote the formation of each product and why.

2   (a) What is the probable identity of the solid that was filtered from the reaction mixture in part **A**? How could you have confirmed its

identity? (b) What percentage of the cyclopentanone you started with in part **A** was converted to condensation products? (This is not the same as the percent yield you calculated.)

3   Write balanced equations and detailed mechanisms for the formation of both products, **A** and **B**.

4   Describe and explain the possible effect on your results of the following experimental errors or variations. (a) You used 10 *M* NaOH rather than 0.10 *M* NaOH in part **A**. (b) You left out the tricaprylmethylammonium chloride in part **B**. (c) You used 50 mmol of cyclopentanone in both parts, **A** and **B**.

5   Diagram a possible phase-transfer process for the formation of product **B**, using the format illustrated in Experiment 22.

6   (a) Following the format in Appendix V, construct a flow diagram for the synthesis in part **A**. (b) Construct a flow diagram for the synthesis in part **B**.

7   From your ^{1}H NMR spectra, is it more likely that your products are (*Z*) or (*E*) stereoisomers? Explain your answer.

8   A student, Mel A. Droyt, forgot to add the furfural in part **A** but recovered a small amount of liquid that distilled at 139–142°C at 20 torr and did not solidify on cooling. The ^{1}H NMR spectrum of the liquid showed no signals from vinylic or hydroxylic protons. Propose a structure for this product and write a mechanism for its formation.

Other Things You Can Do

(Starred projects require your instructor's permission.)

*1   As a group project, carry out part **B** using different phase-transfer catalysts and compare the crude yields to find out which catalysts are most effective. Suggested catalysts are tetrabutylammonium bromide, tetrabutylphosphonium bromide, cetyltrimethylammonium bromide, and 1-hexadecylpyridinium chloride.

*2   Prepare some other aldol condensation products by completing Minilab 34.

3   Starting with sources listed in the Bibliography, write a research paper about the production and uses of furan, furfural, and some derivatives of these compounds.

Haloform Oxidation of 4'-Methoxyacetophenone

Reactions of Methyl Ketones. Preparation of Carboxylic Acids. Oxidation. Haloform Reaction. Enolate Ions. Infrared Spectrometry.

Operations:

OP-6   Heating
OP-7   Cooling
OP-9   Mixing
OP-12 Vacuum Filtration
OP-19 Washing Liquids
OP-22 Drying and Trapping Gases
OP-23 Recrystallization
OP-28 Melting Point
OP-34 Infrared Spectrometry

Before You Begin

1 Read the experiment, read or review the operations as necessary, and write an experimental plan.
2 Calculate the mass of 10.0 mmol of 4'-methoxyacetophenone.

Scenario

The Olfactory Factory, which previously commissioned your Institute to prepare 4'-methoxyacetophenone from anisole (see Experiment 37), has begun to produce this compound, but the demand for it has not met their expectations. Accordingly they would like to explore the possibility of producing other useful aroma chemicals from 4'-methoxyacetophenone. One possibility would be to convert it to 4-methoxybenzoic acid (*p*-anisic acid), which could be used to make pleasant-smelling esters. The Olfactory Factory has now asked the Institute to come up with a safe and inexpensive way of carrying out the conversion of 4'-methoxyacetophenone to 4-methoxybenzoic acid. The conversion of the $-COCH_3$ group of a methyl ketone to a $-COOH$ group can be accomplished by the haloform reaction. In a typical procedure for the haloform reaction, the methyl ketone is stirred with bromine in aqueous sodium hydroxide for several hours, but your supervisor believes the reaction can be carried out much more quickly and safely using ordinary laundry bleach and a phase-transfer catalyst. Your assignment is to see whether or not 4'-methoxyacetophenone can be converted to 4-methoxybenzoic acid by laundry bleach.

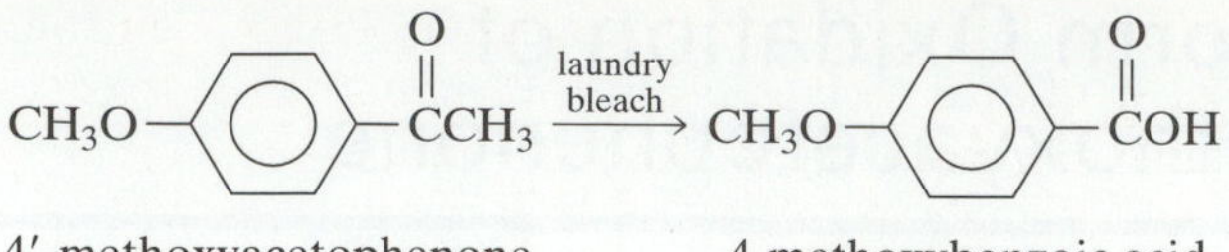

$$\text{CH}_3\text{O}-\!\!\langle\bigcirc\rangle\!-\!\overset{\overset{\displaystyle O}{\|}}{\text{CCH}_3} \xrightarrow{\text{laundry bleach}} \text{CH}_3\text{O}-\!\!\langle\bigcirc\rangle\!-\!\overset{\overset{\displaystyle O}{\|}}{\text{COH}}$$

4′-methoxyacetophenone   4-methoxybenzoic acid

Proposed synthesis of 4-methoxybenzoic acid

Applying Scientific Methodology

You should develop a working hypothesis related to the problem posed in the Scenario. You will test your hypothesis by obtaining a melting point and IR spectrum of the product, if you obtain a product.

Haloforms

The haloforms—chloroform, bromoform, and iodoform—are trihalomethanes that received their trivial names from the fact that they are <u>halo</u>gen compounds that can be hydrolyzed to <u>form</u>ic acid, HCO_2H.

$$\text{CHCl}_3 \qquad \text{CHBr}_3 \qquad \text{CHI}_3$$
chloroform   bromoform   iodoform

They can all be prepared by a haloform reaction in which acetone is treated with an appropriate hypohalite salt in basic solution.

$$\text{CH}_3\overset{\overset{\displaystyle O}{\|}}{\text{C}}\text{CH}_3 \xrightarrow[\text{OH}]{\text{OX}^-} \text{CH}_3\overset{\overset{\displaystyle O}{\|}}{\text{C}}\text{O}^- + \text{CHX}_3 \quad (\text{X} = \text{Cl, Br, I})$$

Chloroform is also prepared commercially by the controlled chlorination of methane.

$$\text{CH}_4 + 3\text{Cl}_2 \xrightarrow{h\nu} \text{CHCl}_3 + 3\text{HCl}$$

Chloroform was the first substance to be used as a general anesthetic in childbirth and (with ethyl ether) was also one of the first synthetic chemicals used by physicians. The administration of chloroform in childbirth, originated by the British obstetrician James Young Simpson, was initially criticized as "unnatural" because, as God cast Adam and Eve out of the Garden of Eden He said to Eve "… in pain will you bring forth children" (Genesis 3:13). The criticism diminished, however, after Simpson used chloroform to deliver Queen Victoria's seventh child. Chloroform was subsequently replaced by other general anesthetics because it is quite toxic, providing the anesthesiologist little margin for error. An inhaled chloroform concentration of about 1.5 percent is necessary to maintain anesthesia for surgery, but a concentration of 2 percent causes respiratory arrest. Chloroform has also been used as an industrial solvent, a dry-cleaning agent, a fire extinguisher, and a starting material for the synthesis of chlorofluorocarbons (CFCs), but its toxicity and suspected carcinogenic properties, as well as the current phase-out of ozone-depleting CFCs, have reduced its use considerably in recent years.

When surface water is purified by treatment with chlorine, very small amounts of chloroform (averaging about 20 parts per billion) and even smaller amounts of other trihalomethanes are produced. These trihalomethanes result from the reaction of chlorine with natural organic compounds such as humic acids, which are produced from decaying vegetable matter. No one knows for certain whether the small amounts of trihalomethanes in chlorinated water represent a serious health threat, but the U.S. Environmental Protection Agency has established a Maximum Contaminant Level (MCL) of 100 ppb for total trihalomethanes in drinking water.

The MCL for a substance is a legally enforceable national standard set by the EPA to protect the public from hazardous materials in water.

Bromoform is a very dense liquid ($d = 2.90$ g/mL) that has been used to separate low-density minerals from mineral mixtures. For example, quartz, calcite, halite, and feldspar all float on bromoform because they have densities lower than 2.9 g/mL.

Iodoform, a yellow solid with a disagreeable "medicinal" odor, was once used as an antiseptic to keep wounds from becoming infected, but it has been replaced for that purpose by safer and less odorous materials. It is the product of the well-known iodoform test for carbonyl compounds that contain the $-COCH_3$ group and alcohols that can be oxidized to such carbonyl compounds. The test is carried out by heating the unknown compound in an alkaline iodine solution.

$$\underset{\text{O}}{\overset{\text{O}}{R\overset{\parallel}{C}CH_3}} \xrightarrow{I_2,\ NaOH} R\overset{\overset{\text{O}}{\parallel}}{C}O^- + CHI_3$$

$$\underset{\text{OH}}{R\overset{|}{C}HCH_3} \xrightarrow{I_2,\ NaOH} R\overset{\overset{\text{O}}{\parallel}}{C}O^- + CHI_3$$

Iodoform test reactions

Understanding the Experiment

In this experiment you will attempt to convert 4'-methoxyacteophenone to 4-methoxybenzoic acid using laundry bleach, which is about 5.25% aqueous NaOCl. Under the conditions of the reaction, molecular chlorine and hydroxide ion are produced according to the following equation.

$$OCl^- + Cl^- + H_2O \rightarrow Cl_2 + 2OH^-$$

The haloform reaction is, in effect, two separate reactions that occur under the same conditions. The first is a base-catalyzed halogenation reaction, proceeding vial enolate ions, that replaces all three hydrogen atoms on the methyl group by chlorine atoms.

$$Ar\overset{\overset{\text{O}}{\parallel}}{C}-CH_3 \xrightarrow{OH^-} [Ar\overset{\overset{\text{O}}{\parallel}}{C}-CH_2^- \longleftrightarrow Ar\overset{\overset{\text{O}^-}{|}}{C}=CH_2] \quad (Ar = p\text{-}CH_3OC_6H_5\text{-})$$

enolate ion

$$\xrightarrow{Cl_2} Ar\overset{\overset{\text{O}}{\parallel}}{C}-CH_2Cl \xrightarrow{OH^-,\ Cl_2} Ar\overset{\overset{\text{O}}{\parallel}}{C}-CCl_3$$

The second involves a nucleophilic acyl substitution reaction in which hydroxide ion is the nucleophile and CCl_3^- is the leaving group.

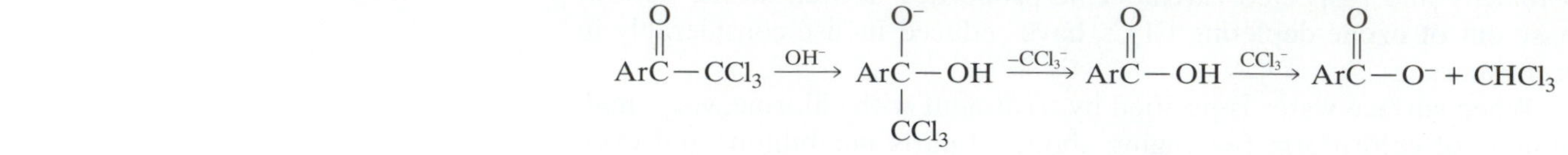

$$ArC\!-\!CCl_3 \overset{OH^-}{\longrightarrow} \underset{\underset{CCl_3}{|}}{ArC\!-\!OH} \overset{-CCl_3^-}{\longrightarrow} ArC\!-\!OH \overset{CCl_3^-}{\longrightarrow} ArC\!-\!O^- + CHCl_3$$

(with C=O double bonds shown on the ArC carbonyls)

Carbanions are ordinarily very poor leaving groups, but the three chlorine atoms on CCl_3^- stabilize its negative charge, allowing its displacement by OH^-. The last step of this reaction sequence is an acid-base reaction that yields the haloform, in this case chloroform, and the salt of the carboxylic acid.

Chloroform is toxic and some of its vapors may escape from the reaction mixture, so you should carry out the reaction under a hood. 4'-Methoxyactophenone is a low-melting solid that should melt to an oily liquid shortly after the reaction begins. Since the liquid is insoluble in the aqueous hypochlorite solution, it is important to keep the reaction mixture well stirred. A phase-transfer catalyst, tricaprylmethylammonium chloride, is used to help ionic species involved in the reaction cross the phase boundary. During the reaction, the reaction mixture is tested occasionally with starch-iodide paper to make sure that sodium hypochlorite is still present. NaOCl tends to bleach out the blue-black color that signifies a positive test, but you should see a fringe of color around the bleached spot. Any excess NaOCl left at the end of the reaction period is consumed by treatment with acetone, and the aqueous solution is washed with diethyl ether to remove chloroform and any unreacted starting material. The product is then precipitated with dilute hydrochloric acid and recrystallized from aqueous ethanol.

If your product is 4-methoxybenzoic acid, you should be able to identify the C=O, acyl C—O, and O—H stretching bands and an O—H bending band arising from the carboxyl group in its IR spectrum. Additional C—O stretching and Ar—H bending bands from the ether function and benzene ring should be observed as well. See OP-34 for additional information about the interpretation of IR spectra.

Reactions and Properties

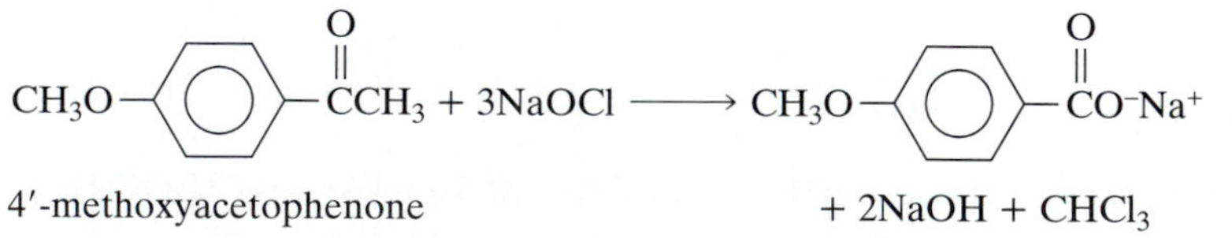

$$CH_3O\!-\!\langle\bigcirc\rangle\!-\!\overset{O}{\overset{\|}{C}}CH_3 + 3NaOCl \longrightarrow CH_3O\!-\!\langle\bigcirc\rangle\!-\!\overset{O}{\overset{\|}{C}}O^-Na^+$$

4'-methoxyacetophenone

$$+ 2NaOH + CHCl_3$$

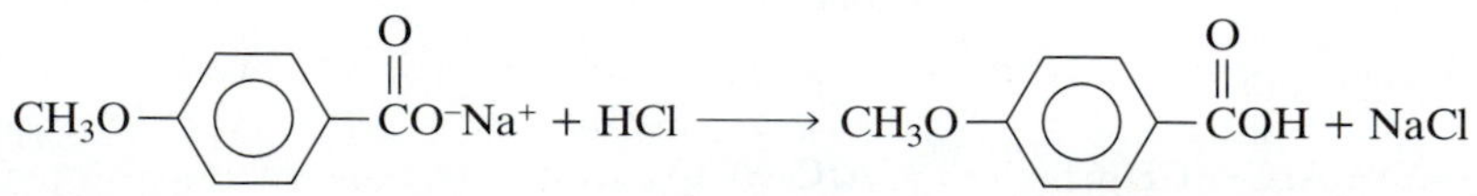

$$CH_3O\!-\!\langle\bigcirc\rangle\!-\!\overset{O}{\overset{\|}{C}}O^-Na^+ + HCl \longrightarrow CH_3O\!-\!\langle\bigcirc\rangle\!-\!\overset{O}{\overset{\|}{C}}OH + NaCl$$

4-methoxybenzoic acid

Table 42.1 Physical properties

	M.W.	m.p.	b.p.	d
4'-methoxyacetophenone	150.2	38	154^{26}	
4-methoxybenzoic acid	152.2	185	280	
chloroform	119.4	−63.5	62	1.484

Note: m.p. and b.p. are in °C, density is in g/mL

Directions

The reaction should be performed under a hood to keep chloroform and chlorine vapors out of the lab. If this is not possible, it should be carried out in a reflux apparatus fitted with a sodium hydroxide gas trap [OP-22].

> **4-Methoxybenzoic acid is a mild sensitizer that may cause contact dermatitis. Minimize contact with the product.**
> **Aqueous sodium hypochlorite can irritate the skin, eyes, and respiratory tract. Avoid contact and do not breathe vapors.**
> **The reaction mixture may evolve some chlorine gas, which irritates the eyes and respiratory tract, and chloroform, which is harmful if inhaled and is a suspected human carcinogen. Do not breathe the vapors from the reaction mixture.**

Safety Notes

chloroform

Reaction. Accurately weigh 10.0 mmol of 4'-methoxyacetophenone into a 125-mL Erlenmeyer flask. *Under the hood,* add 45 mL (~33 mmol) of 5.25% aqueous sodium hypochlorite (household bleach) and 6 drops of tricaprylmethylammonium chloride, then drop in a magnetic stirbar if you have one. Stir (or shake and swirl) the reaction mixture *vigorously* [OP-9] for 30 minutes or more in a 60° water bath [OP-6]. Test the reaction mixture with starch-iodide paper after 10 minutes and more often thereafter. If the test is negative (no immediate blue-black color) add more 5.25% NaOCl solution in small portions until the test is positive. When the reaction is complete, any oily liquid should have dissolved.

Take Care! Avoid contact with the bleach, do not breathe its vapors.

Observe and Note: What evidence do you see for a reaction?

Separation. Test the reaction mixture with starch-iodide paper and add acetone a few drops at a time with swirling until the test is negative (no blue-black color). When the reaction mixture has cooled to room temperature, transfer it to a separatory funnel and wash it with two 10-mL portions of solvent grade ethyl ether [OP-19]. (**Waste Disposal:** Place the ether in the appropriate solvent recovery container.) Be sure to save the aqueous layer, which contains the product. Add enough 3 *M* hydrochloric acid (about 10 mL or more) to the aqueous layer with stirring to bring the pH of the solution down to 2. Cool the reaction mixture [OP-7] in an ice/water bath until crystallization is complete. Collect the product by vacuum filtration [OP-12], washing it on the filter with several portions of cold water.

Stop and Think: What does the acetone do?

Stop and Think: What is the ether removing from the reaction mixture?

Purification and Analysis. Purify the product by recrystallization from 60% aqueous ethanol [OP-23]. Dry it to constant mass and measure its mass and melting point [OP-28]. Record an infrared spectrum of the product [OP-34] or obtain a spectrum from your instructor.

Waste Disposal: Pour the filtrates down the drain.

Report. Your report should include a statement of the problem and an account of how you applied scientific methodology to solve it. Calculate the percent yield of the product and interpret its IR spectrum as completely as you can.

Exercises

1 Write a balanced equation for the reaction undergone by acetone during the separation step.

2 Ketones can also be chlorinated under acidic conditions. Could the haloform reaction be carried out successfully in the presence of an acidic catalyst such as HCl? Why or why not?

3 Describe and explain the possible effect on your results of the following experimental errors or variations. (a) The laundry bleach you used was from an old bottle and contained 3.5% NaOCl. (b) You didn't wash the reaction mixture with ethyl ether; (c) You used benzophenone in place of 4'-methoxyacetophenone.

4 Following the format in Appendix V, construct a flow diagram for the synthesis of 4-methoxybenzoic acid.

5 Which of the following compounds should give a positive iodoform test?

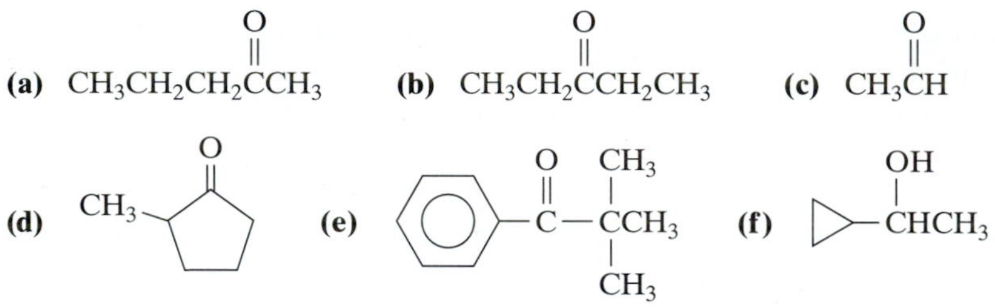

(a) $CH_3CH_2CH_2\overset{\displaystyle O}{\overset{\|}{C}}CH_3$ (b) $CH_3CH_2\overset{\displaystyle O}{\overset{\|}{C}}CH_2CH_3$ (c) $CH_3\overset{\displaystyle O}{\overset{\|}{C}}H$

(d)   (e)   (f)

6 During the haloform reaction, each successive hydrogen atom of the methyl group is replaced by chlorine more rapidly than the previous one. Explain why.

7 (a) The time-averaged permitted emission limit (PEL) for chloroform in air set by the Occupational Safety and Health Administration (OSHA) is 2 ppm by volume. What is the minimum mass of chloroform that would have to vaporize at 25°C and 1 atm in a laboratory having a volume of 300 m^3 in order to attain a concentration in air of 2.0 ppm? (b) If each student in this laboratory treats 10 mmol of 4'-methoxyacetophenone with laundry bleach and allows all of the chloroform produced to vaporize into the lab, what is the minimum number of students it would take to exceed the PEL?

Other Things You Can Do

(Starred projects require your instructor's permission.)

*1 Observe a spontaneous reaction of benzaldehyde and identify the product in Minilab 35.

*2 Prepare iodoform by the following procedure. Dissolve 0.75 g of potassium iodide in 5 mL of water in a small Erlenmeyer flask and add 0.2 mL of acetone. Then add, in small portion with stirring or shaking, 15 mL of bleach solution (5.25% NaOCl). Let the mixture stand for 5 minutes with stirring or shaking, cool it in ice, and collect the product by vacuum filtration. Dry the iodoform and obtain its melting point. (It can be crystallized from methanol if desired.)

 3 Starting with sources listed in the Bibliography, write a research paper about general anesthetics. Describe the applications and physiological effects of some modern general anesthetics, giving their chemical structures and their advantages and disadvantages.

Electronic Effect
of a *para*-Iodo Substituent

Reactions of Aromatic Amines. Preparation of Aryl Halides. Aromatic Nucleophilic Substitution. Carboxylic Acids. Diazonium Salts. Linear Free-Energy Relationships.

Operations:

OP-6   Heating
OP-7   Cooling
OP-9   Mixing
OP-12 Vacuum Filtration
OP-21 Drying Solids
OP-23 Recrystallization
OP-28 Melting Point

Before You Begin

1 Read the experiment, read or review the operations as necessary, and write an experimental plan.
2 Calculate the mass of 10.0 mmol of *p*-aminobenzoic acid, of 10.0 mmol of sodium nitrite, and of 15 mmol of potassium iodide. Calculate the theoretical yield of *p*-iodobenzoic acid.
3 For each acid listed in Table 43.1, calculate the mass needed to prepare 8.0 mL of a 0.10 *M* solution.

Scenario

Si Starr, an energetic professor of physical chemistry at Miskatonic University, is using molecular orbital theory to predict the electronic effects of various substituents on reactions such as the ionization of aromatic carboxylic acids. Such substituent effects can be expressed quantitatively by empirical parameters called sigma values. To test the validity of his results, Professor Starr wants to compare his calculated sigma values with experimentally measured sigma values. Sigma values have been determined for most of the common substituents, but Starr has been unable to locate sigma values for iodo substituents (—I) on a benzene ring. Your assignment is to prepare *p*-iodobenzoic acid, measure its pK_a value, and use that to determine the sigma value for a *para*-iodo substituent.

Applying Scientific Methodology

After reading the experiment you can develop a tentative hypothesis about the electronic effect of a *p*-iodo substituent. The sigma value you obtain will show whether or not your hypothesis is correct.

Linear Free-Energy Relationships

In general chemistry, you learned that electrons in a covalent bond tend to migrate toward the more electronegative atom, building up its electron density at the expense of the less electronegative atom. The situation is not always so simple in conjugated organic compounds, where the electrons of unshared pairs or in pi bonds may migrate *away* from a more electronegative atom into the "electron sink" of a delocalized pi-electron system. For example, lone-pair electrons from an oxygen-containing substituent such as methoxyl ($-OCH_3$) tend to migrate toward and overlap with the pi-electron cloud of a benzene ring, as illustrated for anisole (methoxybenzene). The net electronic effect of a substituent is indicated by an empirical (experimentally measured) parameter called its *sigma value*. Electron-withdrawing groups have positive sigma values and electron-donating groups have negative ones. The measured sigma value of $-OCH_3$ is -0.27, indicating that it has a net electron-donating effect at the *para* position of a benzene ring, even though oxygen is considerably more electronegative than carbon.

Consider a general substituent Z at the *para* position of a substituted benzoic acid. If Z is more electronegative than carbon, it will tend to withdraw electrons through the sigma bond framework of the benzene ring by an *inductive effect*, and this should increase the strength of the acid by stabilizing its conjugate base. Conversely, a substituent that is more electropositive than carbon can donate electrons inductively and weaken an acid. But inductive effects decrease rapidly with distance, and because a *para* substituent is remote from the reaction site (the COOH group), its inductive effect may be quite small.

Substituents that have unpaired electrons and are adjacent to a conjugated system tend to donate electrons by a *resonance effect*. Electron donation by resonance can cause a buildup of electron density at locations within the ring and on conjugated substituents, as illustrated by the following resonance structures for the conjugate base of a substituted benzoic acid:

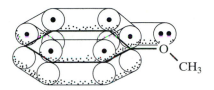

Overlap of lone-pair electrons with pi system in anisole

Z—⟨◯⟩—COOH

⇅

Z⇌⟨◯⟩—COO⁻ + H⁺

Conjugate base is stabilized by electron withdrawal transmitted through sigma bonds

Z—⟨◯⟩—COOH ⇌ H⁺ +

Conjugate base is destabilized by electron donation transmitted through the pi-electron system

In this example, the conjugate base is destabilized by a concentration of negative charge near the reaction site. Destabilization of the conjugate base shifts the ionization equilibrium to the left and thus weakens the acid. Resonance effects are transmitted freely throughout a conjugated system and do not decrease significantly with distance. Substituents such as the methoxyl group can donate electrons by resonance but withdraw them inductively, and thus may either weaken or strengthen an acid depending on the relative importance of the two effects. Thus a *meta*-methoxyl substituent increases the acidity of a substituted benzoic acid by an inductive effect, but a *para*-methoxyl substituent decreases its acidity by resonance.

The pK_a of an acid is defined as the negative logarithm of its ionization constant:

$$pK_a = -\log K_a$$

For example, the pK_a of phosphorous acid, which has a K_a value of 1.0×10^{-2}, is 2.00. At a given temperature, the pK value for any equilibrium reaction is directly proportional to its standard free-energy change, $\Delta G°$. This can be shown by using the above definition to rewrite the thermodynamic equation that relates $\Delta G°$ to the equilibrium constant:

$$\Delta G° = -(2.303 \cdot RT) \log K = (2.303 \cdot RT)\, pK$$

The value of $\Delta G°$ for a reaction is a measure of its tendency to go to completion, so pK must measure the same tendency. Negative pK_a and $\Delta G°$ values are observed for strong acids that dissociate completely, or nearly so. Most organic acids have positive pK_a values because they are only partly dissociated in ionizing solvents; a high positive pK_a value indicates a weak acid and a low one a stronger acid.

Suppose we compare the pK_a values for a pair of acids that differ in only one respect—that one acid has a substituent at a site where the other has only a hydrogen. Since each acid's pK_a is a measure of its acid strength, the difference between their pK_a values must measure the substituent's effect on acid strength. For example, pK_a for the ionization of benzoic acid in water at 25°C is 4.19, whereas pK_a for *p*-nitrobenzoic acid is 3.41. The difference between these pK_a values, or $4.19 - 3.41 = 0.78$, is a quantitative measure of the *para*-nitro substituent's acid-strengthening effect. Similarly, *p*-toluic acid (p-CH$_3$—C$_6$H$_4$COOH) has a pK_a value of 4.36; subtracting this value from the pK_a for benzoic acid gives a difference of -0.17, indicating that the methyl substituent has a small acid-weakening effect. The difference between the pK_a value for the unsubstituted acid and that for the substituted acid, which we shall call ΔpK, is a measure of the substituent's ability to donate or withdraw electrons.

$$\Delta pK = pK_a \text{ of unsubstituted acid} - pK_a \text{ of substituted acid} \quad \textbf{(1)}$$

Because substituents that give negative ΔpK values weaken acids, they must be electron donors, whereas those that give positive ΔpK values are electron acceptors. The numerical value of ΔpK in either case measures the magnitude of the substituent's effect.

The ionization of benzoic acid in water at 25°C has been selected as the reference reaction for measuring the effects of substituents. The ΔpK values for the benzoic acid system are called *substituent constants* and are symbolized by the Greek letter sigma (σ).

$$\sigma = \Delta pK(\text{benzoic acids}) = pK_a(\text{bz}) - pK_a(\text{Zbz}) \quad \textbf{(2)}$$
$$(\text{bz} = \text{benzoic acid}; \text{Zbz} = \text{substituted benzoic acid})$$

Table 43.1 (on page 343) lists the reported sigma values for some common *para* substituents.

Don't confuse a negative ΔpK value, which is associated with an acid-weakening substituent, with a negative pK_a value, which indicates a strong acid.

The effect of a substituent will usually change if the substrate or the reaction type is changed, or even if the reaction conditions are varied. For example, the pK_a values for phenylacetic acid and *p*-nitrophenylacetic acid in water are 4.28 and 3.85, respectively; thus ΔpK for the *para*-nitro group on this substrate is 0.43, which is considerably less than its ΔpK of 0.78 in the benzoic acid system. We can use the ratio of these ΔpK values to estimate the relative effect of a substituent on the ionization of phenylacetic acid and its derivatives, keeping in mind that σ is ΔpK (benzoic acids):

$$\frac{\Delta pK \text{ (phenylacetic acids)}}{\sigma} = \frac{0.43}{0.78} = 0.55$$

This shows that the nitro group has only about half the acid-strengthening effect on phenylacetic acid that it does on benzoic acid, because it is farther from the reaction site in phenylacetic acid. Comparing the effects of a large number of substituents on the same reaction yields an "averaged" $\Delta pK/\sigma$ value of 0.49 for the ionization of phenylacetic acid. We call this value the *reaction constant*, which is symbolized by the Greek letter rho (ρ). Rearranging the equality $\rho = \Delta pK/\sigma$ leads to the *Hammett equation*,

$$\Delta pK = \rho\sigma \qquad\qquad \textbf{(3)}$$

which can also be written as

$$pK_a(\text{unsubstituted acid}) - pK_a(\text{substituted acid})$$

$$= \log \frac{K_a(\text{substituted acid})}{K_a(\text{unsubstituted acid})} = \rho\sigma$$

According to the Hammett equation, the effect that any substituent (Z) will have on the pK of a particular reaction can be estimated by multiplying the sigma value for Z (the parameter that measures the electronic effect of Z on a reference reaction) by the rho value for the reaction in question (the parameter that measures the sensitivity of the reaction to the electronic effects of substituents). Because of the proportionality between pK and $\Delta G°$ mentioned previously, the Hammett equation is called a *linear free-energy relationship*.

By definition, the reaction constant (rho value) for the ionization of benzoic acids in water is 1; reactions that are more sensitive to substituent effects than the reference reaction have rho values greater than 1, and reactions that are less sensitive have rho values between 0 and 1. Negative rho values are observed for reactions in which electron donors increase the extent of reaction and electron acceptors decrease it, rather than the other way around. The rho value for a particular reaction may change if the solvent or other reaction conditions are changed. Just as a car's power is needed more on a steep hill than on the level, substituents have the greatest effect when a reaction is most difficult and the least effect when it is easiest.

Understanding the Experiment

In this experiment, you will first prepare *p*-iodobenzoic acid from *p*-aminobenzoic acid, which is also known as PABA. By measuring the pK_a value of *p*-iodobenzoic acid, you will then determine the sigma value for the *para*-iodo substituent. An aryl iodide can be prepared by treating the corresponding aromatic amine with nitrous acid in the presence of a mineral acid

p-nitrobenzoic acid

p-nitrophenylacetic acid

such as HCl, then heating the resulting diazonium salt with aqueous potassium iodide.

$$ArNH_2 + HONO + HCl \rightarrow ArN_2^+ + Cl^- + 2H_2O$$

$$ArN_2^+ + Cl^- + KI \rightarrow ArI + N_2 + KCl$$

Although some displacement reactions of diazonium salts may involve free radical intermediates, others appear to be S_N1 reactions in which the diazonium salt loses nitrogen (as N_2) to form an intermediate aryl carbocation (Ar^+). A nucleophilic species then combines with the carbocation to form the product, or another intermediate that yields the product after a proton exchange.

The diazonium salt of a primary aromatic amine can be prepared by dissolving or suspending the amine in an aqueous acid such as dilute HCl, cooling the solution to 5°C or below, and adding aqueous sodium nitrite. Sodium nitrite reacts with some of the acid to generate nitrous acid according to the equation

$$NaNO_2 + HCl \rightarrow HONO + NaCl$$

An excess of nitrous acid can bring about undesirable side reactions, so it is important to weigh the PABA and sodium nitrite accurately. The diazonium salt solution must be kept cold because diazonium salts can react with the solvent at elevated temperatures. When this solution is heated with a solution of potassium iodide, *p*-iodobenzoic acid separates as a fine precipitate that may take some time to filter. Using a large Buchner funnel and an efficient aspirator will help speed up the process.

To determine the pK_a of *p*-iodobenzoic acid in 95% ethanol, you will first dissolve some of the acid in 95% ethanol and obtain two 2-mL portions of the resulting solution. You will add just enough dilute KOH to one of these portions to convert all of the acid it contains to the conjugate base, according to the following reaction equation.

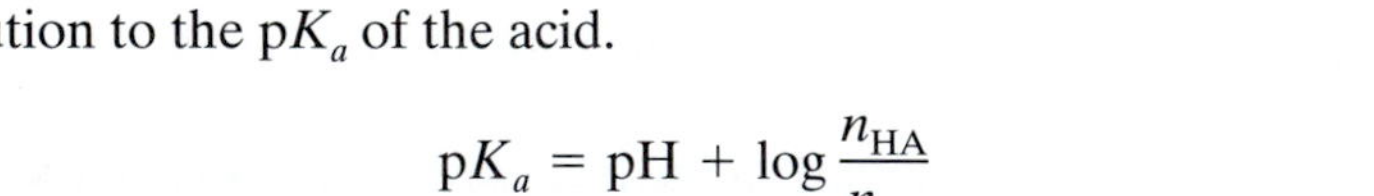

$$I-\langle C_6H_4 \rangle-COOH + OH^- \longrightarrow I-\langle C_6H_4 \rangle-COO^- + H_2O$$

You will then combine this neutralized portion with the second 2-mL portion containing the unneutralized acid, giving a solution that contains equimolar amounts of the acid and its conjugate base. Equation 4, where n_{HA} = moles of acid and n_{A^-} = moles of conjugate base, relates the pH of the solution to the pK_a of the acid.

$$pK_a = pH + \log \frac{n_{HA}}{n_{A^-}} \tag{4}$$

When n_{HA} and n_{A^-} are equal, the logarithmic term equals zero, so $pK_a = pH$.

By the same procedure, you and your coworkers will measure pK_a values for most or all of the acids listed in Table 43.1, from which you can calculate their ΔpK values. Graphing the ΔpK values for the acids against the sigma values for the substituents yields a *Hammett plot* of equation 3 from which you should be able to determine the sigma value for the *para*-iodo substituent and the rho value for the ionization of benzoic acids in 95% ethanol.

Reactions and Properties

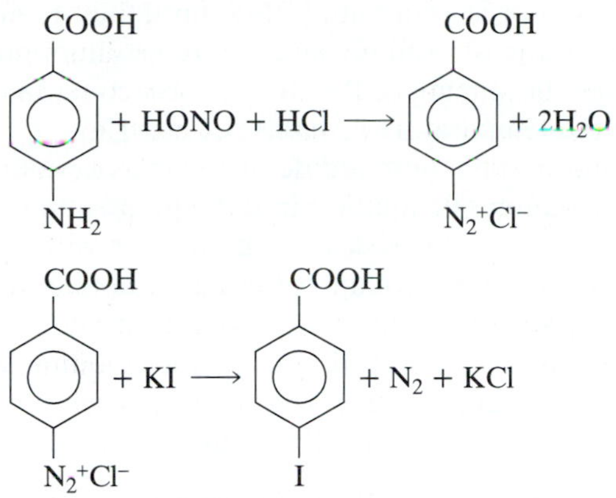

COOH, + HONO + HCl → COOH + 2H₂O

(structures as drawn: *p*-aminobenzoic acid reacting to form the diazonium salt)

COOH + KI → COOH + N₂ + KCl

(diazonium salt reacting with KI to form *p*-iodobenzoic acid)

Table 43.1 Molecular weights and sigma values for *para*-substituted benzoic acids

Acid	M.W.	Substituent (Z)	σ
p-aminobenzoic acid	137.1	—NH₂	−0.66
p-hydroxybenzoic acid	138.1	—OH	−0.37
p-anisic acid	152.2	—OCH₃	−0.27
p-toluic acid	136.2	—CH₃	−0.17
benzoic acid	122.1	none	0.00
p-chlorobenzoic acid	156.6	—Cl	0.23
terephthalic acid	166.1	—COOH	0.45
p-nitrobenzoic acid	167.1	—NO₂	0.78

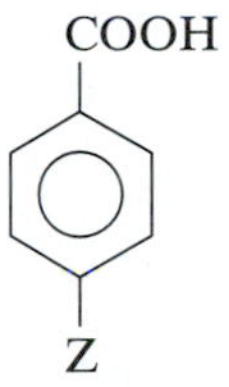

(structure shown: benzoic acid ring with COOH at top and Z at bottom)

Table 43.2 Physical properties

	M.W.	m.p.
p-aminobenzoic acid	137.1	189
p-iodobenzoic acid	248.0	270
potassium iodide	166.0	681
sodium nitrite	69.0	271

Directions

A. *Preparation of* p-*Iodobenzoic Acid*

Sodium nitrite may cause a fire if mixed with combustible materials. Keep it away from such materials.
Benzoic acid and the substituted benzoic acids in Table 43.1 may be harmful if inhaled, ingested, or absorbed through the skin. Minimize contact with the acids and do not breathe their dust.

benzoic acid

Safety Notes

Stop and Think: Why is it important to keep the temperature low?

Observe and Note: Describe what happens as the reaction proceeds.

Take Care! A vigorous reaction with considerable foaming will occur.

Stop and Think: What gas is responsible for the foaming?

Waste Disposal: The filtrates can be poured down the drain.

Reaction. Weigh 10.0 mmol of *p*-aminobenzoic acid (PABA) into a 50-mL Erlenmeyer flask, add 10 ml of 3 *M* hydrochloric acid, and warm the solution gently [OP-6] until the PABA dissolves. Dissolve 10.0 mmol of sodium nitrite in 10 mL of distilled water in another small flask or beaker. Cool both solutions in an ice/water or ice/salt bath [OP-7] until their temperatures are 5°C or below. To form the diazonium salt, add the sodium nitrite solution with swirling or magnetic stirring [OP-9] to the flask containing the PABA, slowly enough that the temperature remains below 10°C.

Dissolve 15 mmol of potassium iodide in 100 mL of water in a 400-mL beaker. Pour the diazonium salt solution into the potassium iodide solution with stirring (rinse the reaction flask into the beaker with a little water). Heat the reaction mixture on a steam bath or in a boiling water bath for 10–15 minutes [OP-6], stirring occasionally. Cool the reaction mixture in an ice/water bath. If the mixture has a red-violet color of iodine, add saturated aqueous sodium bisulfite drop by drop with stirring until the reddish color has disappeared (the suspension will still be a tan or brown color).

Separation and Purification. Collect the crude *p*-iodobenzoic acid by vacuum filtration [OP-12] using a Buchner funnel, washing it on the filter with cold water. Purify the product by recrystallization from 95% ethanol [OP-23], using decolorizing carbon (preferably pelletized Norit) to remove most of the color. Dry the product to constant mass [OP-21] and measure its mass. At your instructor's request, measure its melting point [OP-28], but do not use an ordinary mineral oil bath because you will have to heat the mineral oil well above its flash point.

B. *Measurement of* pK_a *Values*

This part can be performed by teams of 2–4 students, if desired. Each student should carry out pH measurements using his or her *p*-iodobenzoic acid and two or more additional acids from Table 43.1. Prepare an approximately 0.10 *M* solution of *p*-iodobenzoic acid in 95% ethanol by dissolving 0.20 g of your purified product in 8.0 mL of 95% ethanol. Use a volumetric pipet to transfer two 2.00 mL portions of this solution to two test tubes. Add a drop of phenolphthalein indicator solution to *one* of the test tubes, then neutralize *that* solution to a light pink end point using ~0.1 *M* KOH in 95% ethanol added drop by drop from a Pasteur pipet (about 2 mL will be required). Keep the solution well mixed with a stirring rod during the titration. If you go past the end point, discard the solution and titrate another 2.00 mL portion of your 0.10 *M p*-iodobenzoic acid solution. Combine the contents of both test tubes (the neutralized and unneutralized solutions) in a labeled test tube or small beaker that will accomodate the pH meter probe, and mix them thoroughly with a stirring rod. Obtain (or prepare) 0.10 *M* solutions (in 8.0 mL of 95% ethanol) of the other acids whose pK_a values are to be measured, and use the same procedure to prepare half-neutralized solutions of those acids as well (be sure to label the test tubes to correspond to the acids). Then measure the pH values of all of the solutions you have prepared. Report the pH values you have measured to your coworkers (or instructor), and obtain the pH values for the remaining acids from them. Turn in the remaining *p*-iodobenzoic acid to your instructor.

Stop and Think: How is the pK_a of each acid related to the pH of its solution?

Waste Disposal: Dispose of the solutions as directed by your instructor.

Report. Your report should include a statement of the problem and an account of how you applied scientific methodology to solve it. Tabulate the

pK_a and ΔpK values for all the acids and construct a Hammett plot. From your plot, determine the sigma value for the *para*-iodo substituent and the rho value for the ionization of benzoic acids in 95% ethanol. Describe the electronic effect of the *para*-iodo substituent, and tell whether an inductive or a resonance effect can best account for your results. Explain how you arrived at your conclusions in each case.

Exercises

1 Is the ionization of benzoic acids in 95% ethanol more or less sensitive to substituent effects than its ionization in water? Explain why.

2 Derive Equation 4 (p. 342) from the equilibrium constant expression for the ionization of an acid, HA.

3 Describe and explain the possible effect on your results of the following experimental errors or variations. (a) You misread the label on a bottle of sodium nitrate and used it in place of sodium nitrite. (b) When titrating the *p*-iodobenzoic acid solution, you added enough dilute NaOH to turn it dark red. (c) You misread the directions and neutralized both portions of the *p*-iodobenzoic acid solution before combining them for the pH measurement.

4 Following the format in Appendix V, construct a flow diagram for the synthesis of *p*-iodobenzoic acid.

5 (a) A student, Bea Wilder, heated the diazonium salt solution before adding potassium iodide and isolated a white solid with a melting point of 215°C. What product did she obtain instead of *p*-iodobenzoic acid? (b) Propose a mechanism for the reaction that formed it.

6 Using resonance structures, explain why *m*-methoxybenzoic acid is more acidic than benzoic acid, whereas *p*-methoxybenzoic acid is less acidic than benzoic acid.

7 Predict the acid ionization constant (K_a) of *p*-trifluoromethylbenzoic acid in water and in 95% ethanol, given that a *para*-trifluoromethyl substituent ($-CF_3$) has a sigma value of 0.54.

8 (a) Would you expect the rho value for the ionization of substituted phenols ($Z-C_6H_4OH$) in water to be greater than or less than 1? Explain. (b) Give examples of reactions that you would expect to have negative rho values.

Other Things You Can Do

(Starred projects require your instructor's permission.)

*1 As an alternative to the method described in part **B**, you can measure the pK_a of *p*-iodobenzoic acid (or the other acids) by carrying out a pH titration. Dissolve approximately 1 mmol of the acid in 20 mL of 95% ethanol. Titrate this solution with ~0.05 *M* KOH in 95% ethanol, adding about 1 mL at a time and measuring the pH after each addition. When the pH begins to rise rapidly, add the KOH in 0.5-mL increments and continue to titrate until the pH begins to level off. Plot pH versus volume of base, mark the inflection point of the curve, and determine the volume of base at that point. (The inflection point of a curve is the point where it changes direction; in this case, it is the point where the slope of the curve reaches its maximum and begins to decrease.) The pH reading at exactly half that volume (the half-neutralization point) equals the pK_a of the acid.

*2 Determine the relative strengths of some common organic acids and bases by carrying out Minilab 36.

 3 Write a research paper about the natural sources, physiological functions, and commercial uses of p-aminobenzoic acid (PABA), starting with sources listed in the Bibliography.

Synthesis and Identification of an Unknown Carboxylic Acid

Reactions of Acid Anhydrides. Reactions of Carbon-Carbon Double Bonds. Preparation of Carboxylic Acids. Qualitative Analysis.

Operations:

OP-6   Heating
OP-9   Mixing
OP-11 Gravity Filtration
OP-12 Vacuum Filtration
OP-21 Drying Solids
OP-28 Melting-Point Determination

Before You Begin

1 Read the experiment, read or review the operations as necessary, and write an experimental plan.

2 Calculate the mass of 25.0 mmol of maleic anhydride and of 27.5 mmol of zinc.

Scenario

Tetrahydrofuran (THF) is manufactured by the catalytic hydrogenation of maleic anhydride.

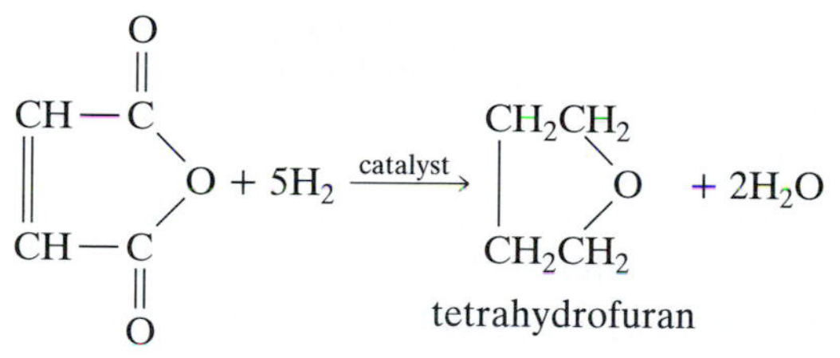

Reduction of maleic anhydride to tetrahydrofuran

Willy Hackett, an inexperienced graduate student at Miskatonic University, was attempting to develop an improved way of preparing this useful solvent as part of his thesis research project. He reasoned that if zinc and hydrochloric acid were combined in the presence of maleic anhydride, they would react to form hydrogen, which would then reduce the maleic anhydride to tetrahydrofuran.

$$Zn + 2HCl \rightarrow H_2 + ZnCl_2$$

He thought this procedure would avoid most of the hazards associated with the use of hydrogen gas, as well as the need for expensive apparatus and catalysts. So he dissolved maleic acid in boiling water, stirred in some zinc followed by concentrated HCl, and was pleased when the reaction mixture started fizzing vigorously, as expected. But when the reaction mixture cooled down, a white solid crystallized from solution. Since tetrahydrofuran is a low-boiling liquid, Willy realized that something had gone wrong.

Not wishing to admit to his research professor that the experiment had failed, Willy decided to study the white solid as part of his research project. But first he must find out what it is. Your supervisor learned of his predicament and agreed to have the solid analyzed. Preliminary tests suggest that it is a carboxylic acid. Your assignment is to identify the unknown carboxylic acid.

Applying Scientific Methodology

After reading the experiment you can formulate a hypothesis about the structure of the unknown carboxylic acid by considering the structure of the starting material and the composition of the reaction mixture. You will test your hypothesis by determining the melting point and equivalent weight of the acid, which you will use to identify it.

Maleic Anhydride and Reid's Reaction

Maleic anhydride is manufactured by air oxidation of benzene in the presence of a vanadium oxide catalyst:

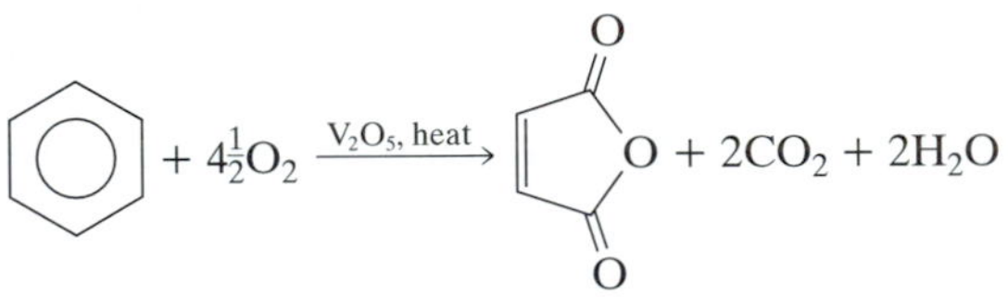

maleic
anyhdride

Because both air and benzene are cheap and abundant, maleic anhydride is an attractive starting material for many other marketable chemicals. Besides being used in the production of tetrahydrofuran, maleic anhydride is used to manufacture agricultural chemicals, polyester resins, surface coatings, dye intermediates, pharmaceuticals, lubricant additives, and other commercially significant products. As noted in Experiment 32, it is also an excellent dienophile that forms Diels–Alder adducts with most conjugated dienes.

The chemical transformation described in the Scenario is an example of a little-known reaction that was first reported by Dr. E. Emmett Reid. When Reid added maleic anhydride to a mixture of granular zinc and boiling water, he noticed an immediate reaction and the separation of a white precipitate. After acidifying the reaction mixture with hydrochloric acid, he recovered the product and identified it as a carboxylic acid. This result was not published until 1972, when it was mentioned—almost as an after-thought—in a "chemical autobiography" describing Reid's 76 years of research. Reid's reaction works only with maleic anhydride and certain other unsaturated carboxylic acids and anhydrides, including maleic acid, fumaric acid, cinnamic acid, and butynedioic acid. The reaction with cinnamic acid requires a zinc-mercury amalgam rather than pure zinc.

Understanding the Experiment

You will synthesize the unknown acid by adding granular zinc to a hot aqueous solution of maleic anhydride, whose infrared spectrum is shown in Figure 32.1 (page 256). Because powdered maleic anhydride reacts quite rapidly with atmospheric moisture, it is usually manufactured in the form of briquettes, which must be pulverized before use. If maleic anhydride is provided in powdered form, you should open the reagent jar only momentarily and replace the cap immediately after you have removed the amount needed. After the initial reaction is complete, concentrated hydrochloric acid is added to liberate the final product and dissolve any excess zinc. Because the product is somewhat soluble even in cold water, the solution should be concentrated to increase the yield. The product should require no further purification after it crystallizes from the cold solution.

The *equivalent weight* of an acid, sometimes called its neutralization equivalent, can be defined as the mass of the acid that provides a mole of protons (H^+) in a neutralization reaction. For example, 1 mole of acetic acid (CH_3COOH, M.W. = 60) and 1/2 mole of oxalic acid ($HOOCCOOH$, M.W. = 90) both contain 1 mole of acidic hydrogens that can be removed as H^+ ions by a strong base. Thus the equivalent weight of acetic acid is 60, but the equivalent weight of oxalic acid is 45—half of its molecular weight. If the structure of an acid is known, its equivalent weight can be calculated by dividing its molecular weight by the number of acidic hydrogens per molecule.

The equivalent weight of an unknown acid is usually determined by titrating an accurately weighed sample of the acid with a standardized solution of sodium hydroxide. The volume of the base required for the titration and its molar concentration are used to calculate the number of moles of base neutralized by the acid, which is equal to the number of moles of protons that the acid provided. For example, suppose 0.222 g of an unknown carboxylic acid with a melting point of 210°C was titrated with 23.2 mL of 0.115 *M* NaOH. Multiplying the molarity of the base by the volume used (in liters) gives the number of moles of base and protons. Dividing the mass of the acid by the number of moles of protons then yields the equivalent weight of the acid, which is about 83.

maleic acid

fumaric acid

cinnamic acid

$$HOOCC \equiv CCOOH$$

butynedioic acid

Note that equivalent weight, like molecular weight, is dimensionless.

$$\text{moles of } H^+ = \text{moles of NaOH} = \frac{0.115 \text{ mol}}{1 \text{ L}} \times 0.0232 \text{ L}$$

$$= 2.67 \times 10^{-3} \text{ mol}$$

$$\text{equivalent weight} = \frac{\text{mass of acid}}{\text{moles of } H^+} = \frac{0.222}{2.67 \times 10^{-3}} = 83.2$$

The melting point of this unknown compound is too high for a carboxylic acid with a molecular weight near 83, so the unknown acid must have two or more carboxyl groups and a molecular weight that is some multiple of 83. A

O
‖
COH

COH
‖
O

phthalic acid

glance at Table 7 in Appendix VI indicates that the unknown is probably phthalic acid, whose molecular weight is 166.

The unknown acid you will synthesize in this experiment is also listed in Table 7 in Appendix VI, so you should be able to identify it after you have measured its equivalent weight and melting point. With your instructor's permission, you can obtain the IR or ^{1}H NMR spectrum of the acid to help confirm its identity.

Properties

Table 44.1 Physical properties

	M.W.	m.p.	b.p.
maleic anhydride	98.1	53	202
zinc	65.4	419	907 d

Note: m.p. and b.p. are in °C, density is in g/mL

Directions

Safety Notes

maleic anhydride

hydrochloric acid

Maleic anhydride is corrosive and toxic and can cause severe damage to the eyes, skin, and upper respiratory tract. Wear gloves, avoid contact, and do not breathe the dust. If you must powder maleic anhydride briquettes, do it under the hood and wear safety goggles and protective clothing. Hydrochloric acid is poisonous and corrosive. Contact or inhalation can cause severe damage to the eyes, skin, and respiratory tract. Wear gloves and dispense under a hood; avoid contact and do not breathe the vapors. The reaction of zinc with hydrochloric acid produces hydrogen gas, which is highly flammable. Keep the reaction mixture away from flames. The product may irritate the eyes, skin, or respiratory tract. Avoid contact with it and do not breathe its dust.

Take Care! Avoid contact with maleic anhydride, do not breathe its dust.

Observe and Note: What evidence is there that a reaction is occurring?

Take Care! Wear gloves, avoid contact with HCl, do not breathe its vapors.

Reaction. Dissolve 25.0 mmol of maleic anhydride in 15 mL of distilled water by heating the water to boiling [OP-6] in an Erlenmeyer flask. Remove the flask from the heat source and immediately add 27.5 mmol of 40-mesh zinc in three or four portions, stirring or swirling [OP-9] after each addition. Let the flask stand for 15 minutes with magnetic stirring or occasional swirling. *Under the hood,* slowly add 5 mL of concentrated HCl with stirring or swirling.

Separation. When the zinc (or most of it) has dissolved, heat the mixture to boiling under the hood. Any white solid that formed during the reaction should dissolve. Filter the hot solution by gravity through fluted filter paper [OP-11]. Boil the filtrate under the hood until its volume has been reduced to about 5 mL (stop boiling if it becomes cloudy). Cover the beaker with a watch glass and set it aside to cool to room temperature. Then cool the beaker in an ice/water bath until crystallization is complete. Collect the product by vacuum filtration [OP-12], washing it on the filter with a little cold acetone. Air dry the product on the filter for a few minutes, then dry it [OP-21] to constant mass.

Analysis. Measure the mass and melting point [OP-28] of the thoroughly dried product. Then weigh about 0.20 g of product to the nearest milligram and dissolve it in 25 mL of water in a 125-mL Erlenmeyer flask. Add 2 drops of phenolphthalein solution and titrate with a standardized ~0.2 M NaOH solution.

Report. Your report should include a statement of the problem and an account of how you applied scientific methodology to solve it. Calculate the equivalent weight of your unknown acid and identify it by referring to Table 7 in Appendix VI. Write the structure of the product and a balanced equation for its formation, and give its systematic and common names. Calculate the theoretical yield and the percentage yield of the product.

Waste Disposal: Pour filtrate down the drain.

Observe and Note: Record the exact concentration of the NaOH solution in your lab notebook.

Exercises

1 Estimate the amount of product that was lost in the filtrate, assuming that the volume of the reaction mixture was 5 mL when you filtered it. Compare this with the amount of product you would have lost if you had not boiled off most of the water. The solubility of the product is 6.8 g/100 mL at 20°C.

2 What was wrong with the graduate student's idea (see the Scenario) that the reaction of maleic acid with zinc and aqueous HCl would yield tetrahydrofuran?

3 Describe and explain the possible effect on your results of the following experimental errors or variations. (a) You used phthalic anhydride (1,2-benzenedioic anhydride) instead of maleic anhydride. (b) You forgot to boil down the initial filtrate in the separation step. (c) You forgot to record the concentration of the NaOH solution (which was 0.255 M) so you assumed a concentration of exactly 0.2 M.

4 Following the format in Appendix V, construct a flow diagram for the synthesis you carried out in this experiment.

5 Propose structures for the products you would obtain by treating fumaric acid, cinnamic acid, and butynedioic acid with excess zinc and HCl. Write balanced equations for the reactions.

6 Suppose that 0.196 g of an unknown carboxylic acid is titrated using 19.3 mL of 0.196 M aqueous NaOH. What is the probable identity of the unknown if its melting point is 132±5°C? (Use Table 7 in Appendix VI.)

7 Describe the ^{1}H NMR spectrum you would expect from the product of this experiment, giving peak areas, multiplicities, and approximate chemical shifts for all signals.

8 Write balanced equations for the reactions of maleic anhydride with water, cyclohexanol, aniline, and isoprene (2-methyl-1,3-butadiene).

Other Things You Can Do

(Starred projects require your instructor's permission.)

*1 Record the infrared spectrum of the product and compare it with the spectrum of the starting material in Figure 32.1 (page 256). Interpret both spectra as completely as you can.

*2 After you predict the ^{1}H NMR spectrum you would expect from the product (see Exercise 7), record its spectrum in DMSO-d_6 or another suitable solvent and see how accurate your prediction was.

*3 Observe the relative hydrolysis rates of some esters in Minilab 37.

 4 Write a research paper about the manufacture and uses of maleic anhydride, maleic acid, and fumaric acid, starting with sources listed in the Bibliography.

Preparation of the Insect Repellent
N,N-Diethyl-*m*-Toluamide
EXPERIMENT 45

Reactions of Carboxylic Acids. Preparation of Amides. Nucleophilic Acyl Substitution. Acid Chlorides. Infrared Spectrometry.

Operations:

OP-6   Heating
OP-7   Cooling
OP-9   Mixing
OP-10 Addition
OP-13 Extraction
OP-14 Evaporation
OP-16 Column Chromatography
OP-19 Washing Liquids
OP-20 Drying Liquids
OP-22 Drying and Trapping Gases
OP-32 Gas Chromatography
OP-34 Infrared Spectrometry
OP-35 (optional) Nuclear Magnetic Resonance Spectrometry

Before You Begin

1 Read the experiment, read or review the operations as necessary, and write an experimental plan.
2 Calculate the mass of 30.0 mmol of *m*-toluic acid, the mass of 25.0 mmol of diethylamine hydrochloride, and the theoretical yield of *N,N*-diethyl-*m*-toluamide.

Scenario

Skeeters 'N Such manufactures insect repellents, including *Bug Off*, whose main active ingredient is *N,N*-diethyl-*m*-toluamide (deet). Its current process for the manufacture of deet requires the use of diethyl ether as a solvent and a 2:1 molar ratio of nucleophile (diethylamine) to substrate. To reduce production costs, Skeeters 'N Such wants your Institute to develop a synthetic method that uses a cheaper solvent, such as water, and a 1:1 or lower nucleophile:substrate ratio. Your supervisor thinks a procedural variation called the Schotten–Baumann reaction may accomplish both of these objectives, but is not sure whether the product will be pure enough to meet specifications. Your assignment is to see whether or not deet can synthesized in at least 95% purity using the Schotten–Baumann procedure.

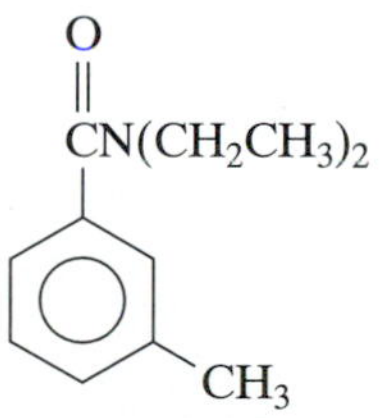

deet
(*N,N*-diethyl-*m*-toluamide)

Applying Scientific Methodology

You should develop a working hypothesis based on the Scenario, which will be tested when you analyze the product by gas chromatography and infrared spectrometry.

Chemical Mosquito Evasion

a hungry mosquito

Is there anyone on earth who hasn't cringed upon hearing the high-pitched whine of a hungry mosquito? Besides their capacity to torment us, these bloodthirsty insects have a well-earned reputation for spreading diseases including malaria, yellow fever, dengue, and viral encephalitis. Numerous methods of controlling mosquitoes have been developed, but the hardy creatures can withstand a variety of adverse conditions. Mosquitoes have been known to breed in the hot alkaline volcanic pools of Uganda, and even in a tank of hydrochloric acid in India! Since we are not likely to eradicate mosquitoes from the earth anytime soon, we must resign ourselves to living with them. This fate is made more tolerable by the availability of effective insect repellents. Among the best of these is *N,N*-diethyl-*m*-toluamide, better known as deet, which is the major ingredient of most commercial mosquito repellents manufactured in North America.

Mosquito "repellents" don't really repel mosquitoes in the same way that a disagreeable odor might repel a human from its source. Instead, they appear to jam the insect's sensors so that it can't find its victim. Warm objects generate convection currents in the air around them; warm living objects also evolve carbon dioxide, which alerts a mosquito to the presence of a blood source and starts it on its flight. This flight is initially random, but when the insect encounters a warm, moist stream of air it moves toward the source, which is generally a living object. Unless the object takes rapid evasive action, the mosquito follows the convection current until it makes contact—unless it gets squashed first.

When you (the intended victim) are protected by an effective insect repellent, the mosquito still knows you're around but is unable to find you. This is because the repellent prevents the insect's moisture sensors from responding normally to the high humidity of your convection current. Ordinarily, when a mosquito passes from warm, moist air into drier air, its moisture sensors send fewer signals to its central nervous system, causing it to turn back into the air stream. By blocking these sensors, the repellent reduces the signal frequency and convinces the mosquito that it is heading into drier rather than moister air; so it turns away before landing. Individual variations in diet, skin chemistry, or other factors such as the color of one's clothing (dark colors attract mosquitos) may help explain why some unlucky individuals are eaten alive by mosquitoes and others escape unscathed. For example, thiamine (Vitamin B_1) taken orally is excreted through the skin, where it acts as an insect repellent.

The molecular features that make a compound a good insect repellent are not well understood at this time. Repellents occur in nearly every chemical family and exhibit a large variety of molecular shapes. A number of *N,N*-disubstituted amides similar to deet are represented (see Figure 45.1), including the diethylamide of thujic acid, a constituent of the western red cedar that may be partly responsible for that tree's resistance to insect attack. Other effective repellents include esters such as dimethyl phthalate

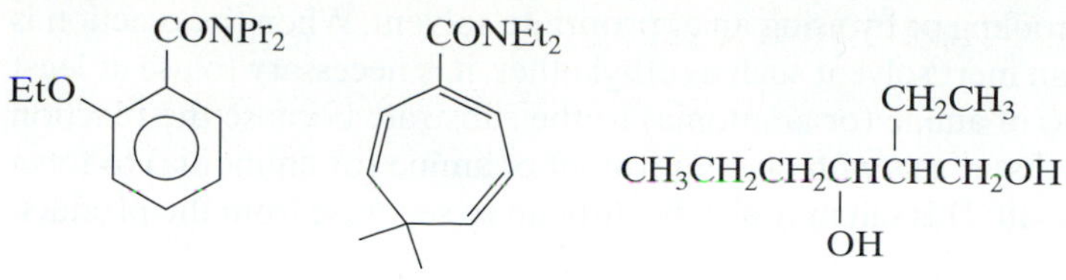

CONPr$_2$
EtO

N,N-dipropyl-2-
ethoxybenzamide

CONEt$_2$

thujic acid
diethylamide

CH$_2$CH$_3$
CH$_3$CH$_2$CH$_2$CHCHCH$_2$OH
OH

2-ethyl-1,3-hexanediol

CH$_2$OH
CH$_3$CH$_2$CH$_2$CH$_2$CCH$_2$CH$_3$
CH$_2$OH

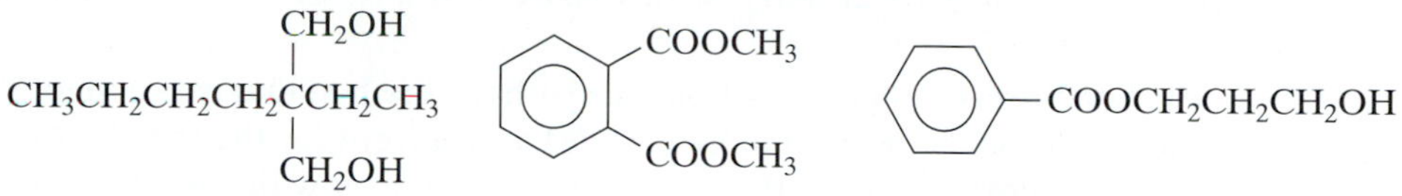

2-butyl-2-ethyl-1,3-
propanediol

COOCH$_3$
COOCH$_3$

dimethyl
phthalate

—COOCH$_2$CH$_2$CH$_2$OH

1,3-propanediol
monobenzoate

Et = CH$_3$CH$_2$—   Pr = CH$_3$CH$_2$CH$_2$—

Figure 45.1 Some typical insect repellents

and diols such as 2-ethyl-1,3-hexanediol. When more is learned about the structural features that make a substance act as a repellent, it should be possible to develop even more effective and convenient repellents for long-term protection against insect bites.

Understanding the Experiment

Amides are usually prepared by treating a carboxylic acid derivative with ammonia or with a primary or secondary amine. Acid anhydrides or esters are sometimes used as the acid derivative, but acyl chlorides are the most useful for preparing the widest variety of amides.

General reactions for preparing amides

$$\underset{\parallel}{R-\overset{O}{C}}-Z + NH_3 \longrightarrow R\overset{O}{\underset{\parallel}{C}}NH_2 + HZ$$

$$\underset{\parallel}{R-\overset{O}{C}}-Z + R'NH_2 \longrightarrow R\overset{O}{\underset{\parallel}{C}}NHR' + HZ$$

$$\underset{\parallel}{R-\overset{O}{C}}-Z + R'-\underset{\underset{R''}{|}}{N}H \longrightarrow R\overset{O}{\underset{\parallel}{C}}N\underset{\underset{R''}{|}}{R'} + HZ$$

Z = Cl, OR, OCOR, etc.

Because of the high reactivity of acyl chlorides, the reactions are usually rapid and exothermic—so much so that, in many cases, the rate must be controlled by cooling or by using an appropriate solvent. When the reaction is carried out in an inert solvent such as ethyl ether, it is necessary to use at least a 2:1 mole ratio of amine (or ammonia) to the substrate because the reaction produces HCl that reacts with one equivalent of amine (or ammonia) to form an ammonium salt. This salt may also be difficult to separate from the product.

Example of reaction in inert solvent

$$\underset{\text{RCCl}}{\overset{\text{O}}{\parallel}} + 2R'NH_2 \longrightarrow \underset{\text{RCNHR}'}{\overset{\text{O}}{\parallel}} + R'NH_2^+Cl^-$$

Schotten-Baumann method for preparing amides

$$\underset{\text{RCCl}}{\overset{\text{O}}{\parallel}} + R'NH_2 + NaOH \longrightarrow$$

$$\underset{\text{RCNHR}'}{\overset{\text{O}}{\parallel}} + NaCl + H_2O$$

Reaction of an amine hydrochloride with a base

$$RNH_3^+Cl^- + NaOH \longrightarrow$$

$$RNH_2 + NaCl + H_2O$$

The Schotten–Baumann reaction is a synthetic method that uses aqueous sodium hydroxide (or potassium hydroxide) as a solvent for the reactions of certain acyl chlorides. The sodium hydroxide combines with the HCl produced during the reaction, making it unnecessary to add excess amine for that purpose. Some acyl chlorides, particularly low-molecular-weight aliphatic ones, hydrolyze rapidly in water to form carboxylic acids and thus are unsuitable candidates for the Schotten–Baumann reaction. However, aromatic and long-chain aliphatic acyl chlorides are nearly insoluble in water and therefore hydrolyze much more slowly. The small amount of acyl chloride lost by hydrolysis can be compensated for by using a small excess of this reactant. The presence of aqueous alkali in the reaction mixture also makes it possible to use an amine hydrochloride in place of the free amine, since the amine hydrochloride reacts with base to liberate the amine *in situ*. Since many amines are volatile, corrosive, and quite unpleasant to handle, the use of the comparatively well-behaved amine salt is a definite advantage.

The acyl chlorides used for preparing amides are generally prepared from the corresponding carboxylic acids. Several reagents can be used for this transformation, each with its advantages and disadvantages. Thionyl chloride ($SOCl_2$) has the great advantage that all of the inorganic reaction products (HCl and SO_2) are gases that can be easily removed from the acyl chloride, which can thus be used for further reactions without purification.

Methods for preparing acyl chlorides

$$\underset{\text{RCOH}}{\overset{\text{O}}{\parallel}} + PCl_5 \longrightarrow \underset{\text{RCCl}}{\overset{\text{O}}{\parallel}} + POCl_3 + HCl$$

$$3\underset{\text{RCOH}}{\overset{\text{O}}{\parallel}} + 2PCl_3 \longrightarrow 3\underset{\text{RCCl}}{\overset{\text{O}}{\parallel}} + 3HCl + P_2O_3$$

$$\underset{\text{RCOH}}{\overset{\text{O}}{\parallel}} + SOCl_2 \longrightarrow \underset{\text{RCCl}}{\overset{\text{O}}{\parallel}} + SO_2 + HCl$$

In this experiment you will prepare *m*-toluoyl chloride by heating *m*-toluic acid with excess thionyl chloride and then treating the acid chloride with diethylamine (from the hydrochloride) in aqueous NaOH to obtain *N,N*-diethyl-*m*-toluamide. The excess thionyl chloride from the first step is used up by reacting with sodium hydroxide in the second. Because corrosive gases are released, you will need to carry out the reaction under a fume hood or use a gas trap to keep them out of the atmosphere. Since the acid chloride is relatively insoluble in the aqueous reaction mixture, efficient mixing is necessary to provide adequate contact between the phases. The detergent sodium lauryl sulfate helps disperse the acyl chloride into smaller particles, increasing the area of contact between phases and thus increasing the reaction rate.

After the product is isolated by extraction with ethyl ether, it will be purified by column chromatography, then analyzed by gas chromatography to assess its purity. By comparing the infrared spectrum of your product with those of the starting materials in Figure 45.2, you should see evidence indicating whether or not the desired reaction has taken place.

Reactions and Properties

A.

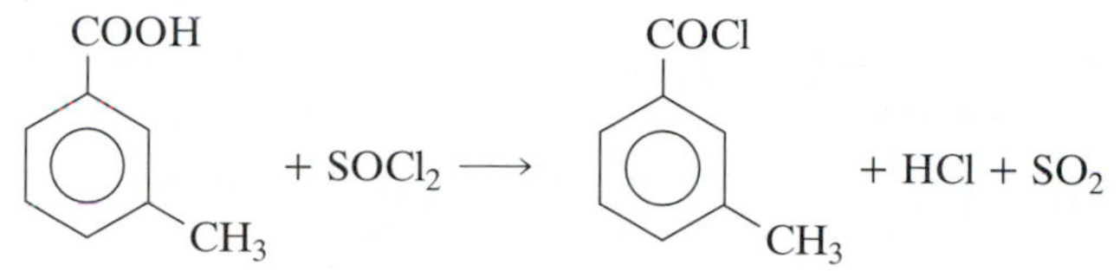

m-toluic acid + $SOCl_2$ ⟶ m-toluoyl chloride + HCl + SO_2

B. $(CH_3CH_2)_2NH_2{}^+Cl^- + NaOH$
    diethylamine
    hydrochloride

⟶ $(CH_3CH_2)_2NH + NaCl + H_2O$
    diethylamine

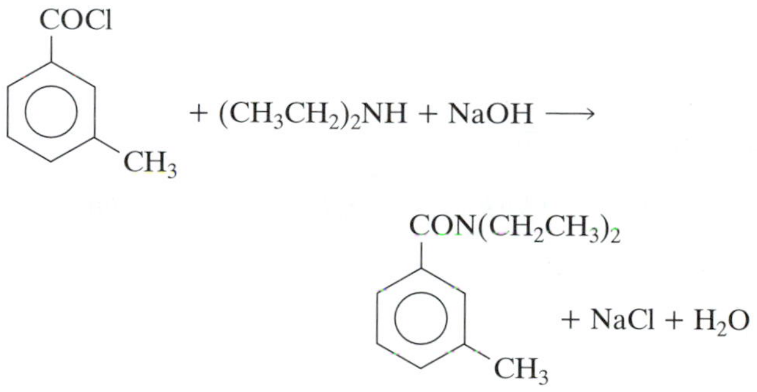

m-toluoyl chloride (COCl) + $(CH_3CH_2)_2NH + NaOH$ ⟶

$CON(CH_2CH_3)_2$ ... deet + $NaCl + H_2O$

Table 45.1 Physical properties

	M.W.	m.p.	b.p.	d
m-toluic acid	136.2	113		
thionyl chloride	119.0		79^{746}	1.65
m-toluoyl chloride	154.6		86^{5}	1.173
diethylamine hydrochloride	109.6	227–30		
N,N-diethyl-*m*-toluamide	191.3		160^{19}	0.996

Note: m.p. and b.p. are in °C, density is in g/mL

m-Toluic acid

3057.4	1590.5	1217.0
2678.2	1417.6	933.7
1690.2	1312.3	746.1

Diethylamine $CH_3CH_2NHCH_2CH_3$

2964.2	1377.6	1046.3
2742.8	1326.6	776.2
1451.5	1138.2	727.5

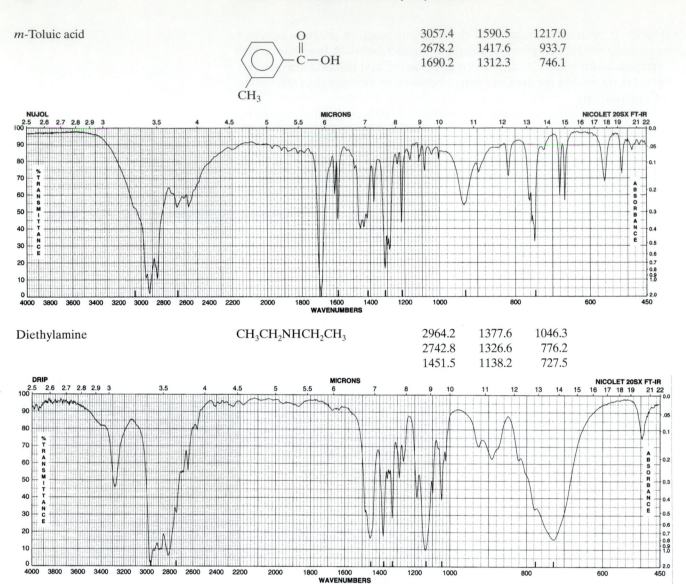

Figure 45.2 IR spectra of the starting materials (Reproduced from *The Aldrich Library of FT-IR Spectra, Edition II,* with the permission of the Aldrich Chemical Company.)

m-Toluic acid

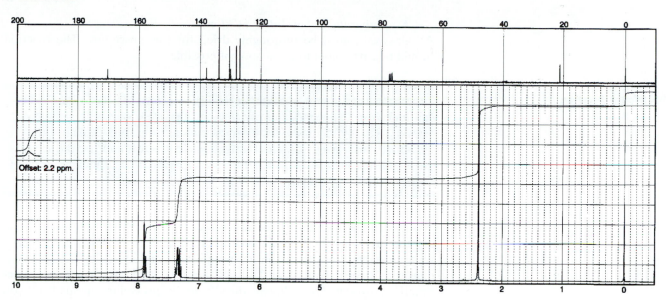

Diethylamine

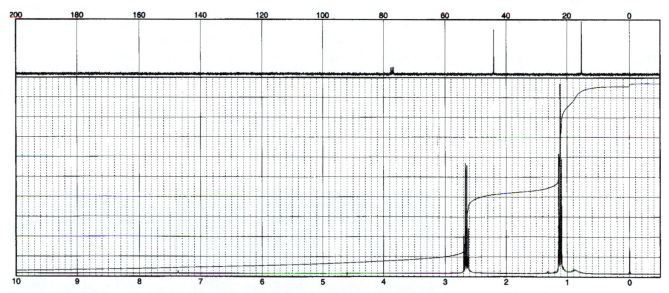

Figure 45.3 NMR spectra of the starting materials. (Reproduced from *The Aldrich Library of ^{13}C and ^{1}H FT-NMR Spectra* by C. J. Pouchert and J. Behnke, with the permission of the Aldrich Chemical Company.)

Directions

A. *Preparation of* m-*Toluoyl Chloride*

All glassware must be thoroughly dried for this procedure. The reaction should be carried out under a hood, if possible.

> **Thionyl chloride and *m*-toluoyl chloride are both corrosive and lachrymatory (tear inducing) and can damage the eyes, skin, and respiratory system. Thionyl chloride decomposes violently on contact with water to produce corrosive gases. Wear gloves, work under a hood, avoid contact with thionyl chloride and the reaction mixture, keep them away from water, and do not breathe their vapors.**

Stop and Think: What reactions would take place in a gas trap?

Assemble an apparatus for addition under reflux [OP-10] by fitting a clean, *dry* 100-mL round-bottomed flask with a Claisen connecting tube, placing a separatory/addition funnel on its straight arm, and attaching a reflux condenser to its bent arm. (*Note:* Use only a reflux condenser in a 25-mL reaction flask if you will be following the alternative procedure in part **B**.) If you cannot perform the experiment under a fume hood, equip the reflux condenser with a gas trap containing dilute NaOH [OP-22]. Be sure the apparatus is securely clamped so that you can remove and re-attach the reaction flask safely. Detach the reaction flask and measure 30.0 mmol of *m*-toluic acid into it, then drop in a stirbar (if you have one) or boiling chips. *Under the hood* add 2.6 mL (~36 mmol) of thionyl chloride, then re-attach the flask to the reaction apparatus. If you must take the reaction flask out of the hood, stopper it first. Heat the reaction mixture gently under reflux [OP-6] for at least 20 minutes. Evolution of gases should stop when the reaction is complete. Cool the reaction mixture in an ice/water bath to 10°C or below while you prepare the diethylamine solution in part **B**.

Take Care! Wear gloves, avoid contact with thionyl chloride, do not breathe its vapors.

B. *Preparation of* N,N-*Diethyl-*m-*toluamide*

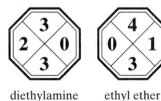

diethylamine   ethyl ether

> **Diethylamine hydrochloride irritates the skin, eyes, and respiratory tract, so avoid contact and inhalation. In the reaction mixture it forms diethylamine, which is corrosive and has toxic vapors.**
> **Ethyl ether is extremely flammable and may be harmful if inhaled. Avoid breathing vapors, keep away from flames and hot surfaces.**
> **Hexane is very flammable and may irritate the eyes and respiratory tract. Avoid breathing vapors and keep away from flames and hot surfaces.**

Take Care! Wear gloves, avoid contact with diethylamine hydrochloride, do not breathe its vapors.

Stop and Think: What is the purpose of the sodium lauryl sulfate?

Reaction. (*Note:* See the alternative procedure if no magnetic stirrers are available.) Measure 35 mL of 3.0 *M* aqueous sodium hydroxide into an Erlenmeyer flask and cool the flask in an ice/water bath [OP-7] for 5 minutes or more. *Under the hood*, add 25.0 mmol of diethylamine hydrochloride, in small portions with stirring or swirling. This amine salt will be converted to diethylamine as it is added. Then mix in 0.1 g of sodium lauryl sulfate. Transfer this solution to the separatory/addition funnel, using a few milliliters of water in the transfer. Keep the reaction flask containing *m*-toluoyl chloride in the ice/water bath and start the stirrer [OP-9]. Add the diethyl-amine solution to the reaction flask over a period of about 5 minutes (about 6–8 mL per minute). Then heat and stir the reaction mixture using a steam bath or boiling water bath for 15 minutes or more. At this point the acid chloride odor should be gone and the reaction mixture should be basic to litmus paper. (Go on to the *Separation* procedure.)

Alternative Reaction Procedure. If magnetic stirrers are not available, pre-pare the diethylamine solution (containing sodium lauryl sulfate) as described, then transfer the acid chloride solution to a *dry* separatory/addition funnel *under the hood* and support the funnel over the flask containing the cooled diethylamine solution. Add the acid chloride in ~1 ml portions, swirling the Erlenmeyer flask in an ice bath after each addition until the reaction sub-sides. Stopper the reaction flask and shake/swirl it vigorously at room tem-perature for several minutes. Then loosen the stopper (don't remove it) and heat the mixture using a steam bath or boiling water bath for 15 minutes with continued shaking. At this point the acid chloride odor should be gone and the reaction mixture should be basic to litmus paper.

Take Care! Be sure the stopcock is closed

Take Care! Wear gloves, avoid con-tact with the reaction mixture.

Separation. Allow the reaction mixture to cool to room temperature, transfer it to a separatory funnel, and extract it [OP-13] with three 20-mL portions of solvent grade ethyl ether. Wash the combined ether extracts [OP-19] with 30 mL of 1 *M* HCl followed by 30 mL of saturated aqueous sodium chloride. Dry the ether solution [OP-20] over anhydrous magne-sium sulfate. Evaporate the ether under vacuum [OP-14] using a cold trap, with gentle heating if necessary.

Take Care! Keep ether away from ignition sources.

Waste Disposal: Pour the aqueous layers down the drain.

Waste Disposal: Place recovered ether in the appropriate solvent recovery container.

Purification. (*Note:* The product can also be purified by vacuum distillation [OP-26] using small-scale apparatus.) Pack a chromatography column [OP-16] with about 25 g of activated alumina, adding a 1-cm layer of clean sand at the top. Make sure the surfaces of the alumina and sand are as level as pos-sible. Obtain about 50 mL of hexanes, a *dry*, tared 50-mL Erlenmeyer flask, and a 50-mL beaker. Set the beaker under the column. Dissolve the crude deet (*N,N*-Diethyl-*m*-toluamide) in a minimum volume of hexanes (5 mL or less). Add about 10 mL of hexanes to the column and let it drain until the eluent surface just enters the top of the sand layer, then immediately add the deet solution to the column. Use a little hexanes to rinse the flask that contained the deet and add the rinse liquid to the column. When the liquid surface again begins to enter the sand layer, fill the column nearly to the top with hexanes. Continue to add eluent during elution to keep its level rea-sonably constant. As the deet passes onto the column it should form a yel-low band; when the bottom of that band nears the bottom of the alumina layer replace the beaker by the tared flask and collect all of the eluate until the top of the yellow band has passed out of the column. Evaporate the hexanes from the deet solution under vacuum with gentle heating [OP-14].

Take Care! Keep hexanes away from ignition sources.

Analysis. Measure the mass of the *N,N*-diethyl-*meta*-toluamide and ana-lyze it by gas chromatography [OP-32]. Record the infrared spectrum of the neat liquid [OP-34] or obtain a spectrum from your instructor. At your instructor's option, record its ^{1}H NMR spectrum as well [OP-35].

Waste Disposal: Put any hexanes that remains in the column and other recovered hexanes in the appropriate solvent recovery con-tainer.

Report. Your report should include a statement of the problem and an account of how you applied scientific methodology to solve it. Estimate the percent purity of the product from its gas chromatogram. Compare your spectrum with the IR spectra of the reactants in Figure 45.2, and describe the evidence indicating that the expected reaction has taken place.

Exercises

1 If you recorded the product's ¹H NMR spectrum, compare it with the spectra of the reactants in Figure 45.3 and try to account for any similarities or differences. Interpret the spectrum as completely as you can.

2 (a) Write balanced equations for the gas trap reactions between aqueous sodium hydroxide and both HCl and SO_2. (b) Write balanced equations for the reactions of excess *m*-toluoyl chloride and thionyl chloride with aqueous sodium hydroxide.

3 Describe and explain the possible effect on your results of the following experimental errors or variations. (a) The reaction flask you used for the preparation of *m*-toluoyl chloride was wet. (b) You used 1.0 *M* NaOH rather than 3.0 *M* NaOH as the reaction solvent for part **B**. (c) You didn't stir or shake the reaction mixture in part **B**.

4 Following the format in Appendix V, construct a flow diagram for the synthesis of deet.

5 Write reasonable mechanisms for (a) the reaction of *m*-toluic acid with thionyl chloride and (b) the reaction of *m*-toluoyl chloride with diethylamine.

6 The insect repellent *N*-butylacetanilide is used to treat clothing for fleas and ticks. Outline a synthesis of this compound starting with aniline.

7 (a) Calculate the total volume of 3 *M* NaOH required for part **B**, including the amount used up in combining with excess reactants. (b) Calculate the percentage excess of NaOH that was actually used.

8 A student misread the label on a reagent bottle and used a 50% NaOH solution in this experiment instead of the 3 *M* (11%) solution. Very little deet was obtained from the ether layer after extraction, but acidification of the aqueous layer yielded a white solid that melted at 112°C. Identify the solid and write balanced equations showing how it formed.

9 Hair is made up of the protein keratin, which is a polyamide containing many α-amino acid residues. Hair is often responsible for clogged drains, which can be unclogged by drain cleaners that contain lye (sodium hydroxide). Explain how such drain cleaners get rid of hair.

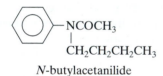

N-butylacetanilide

Other Things You Can Do

(Starred projects require your instructor's permission.)

*1 With your instructor's permission, test the effectiveness of your product as a mosquito repellent. Deet is a mild irritant and your product might contain harmful impurities, so do not apply it directly to your skin. Instead, prepare a 15% solution of the deet in isopropyl alcohol and use it to saturate a piece of absorbent cheesecloth. When the treated cloth is dry, take it to a mosquito-infested area, and drape it over one arm. Place an untreated piece of cheesecloth on the other arm to serve as a control, and see which arm is targeted by more mosquitoes.

*2   Prepare another amide, benzamide, by the procedure in Minilab 38.

3   Deet repels insects, but there are other chemicals that attract them. Starting with sources listed in the Bibliography, write a research paper about some of the pheromones that function as insect attractants and tell how they can be used for insect control.

EXPERIMENT 46

Synthesis of 2-Acetylcyclohexanone by the Stork Enamine Reaction

Reactions of Carbonyl Compounds. Preparation of Dicarbonyl Compounds. Nucleophilic Acyl Substitution. Enamines. Infrared Spectrometry.

Operations:

OP-6   Heating
OP-10 Addition
OP-13 Extraction
OP-14 Evaporation
OP-20 Drying Liquids
OP-22 Drying and Trapping Gases
OP-25 Simple Distillation
OP-26 Vacuum Distillation
OP-34 Infrared Spectrometry

Before You Begin

1 Read the experiment, read *Water Separation* in OP-25, read or review the other operations as necessary, and write an experimental plan.
2 Calculate the mass and volume of 100 mmol of cyclohexanone, 130 mmol of morpholine, 120 mmol of triethylamine, and 110 mmol of acetic anhydride. Calculate the theoretical yield of 2-acetylcyclohexanone.

Scenario

beehive

In a beehive there is only one queen bee, the fertile female that produces all of the eggs from which succeeding generations of bees are propagated. Scientists have discovered that queen pheromone, a "chemical messenger" produced by the queen honeybee, is responsible for her preeminence. Queen pheromone not only keeps the worker bees from rearing new queens, it also inhibits the development of their ovaries and attracts male bees to the queen for mating. Midge Miller, a Miskatonic University entomologist, believes she has isolated the major active ingredient of queen pheromone. To support Dr. Miller's claim, the substance must be prepared in the laboratory and tested to see whether it produces the same effects on honeybees as natural queen pheromone. That requires some synthetic chemists, so M.U. has contacted your Institute for assistance. Your supervisor believes it should be possible to synthesize this substance, (2*E*)-9-oxo-2-decenoic acid, from 7-oxooctanoic acid, which can be prepared by the alkaline hydrolysis of 2-acetylcyclohexanone. First you will need to make some 2-acetylcyclohexanone. It has been synthesized by direct acetylation of cyclohexanone in the presence of boron trifluoride, but the yield is low, so your

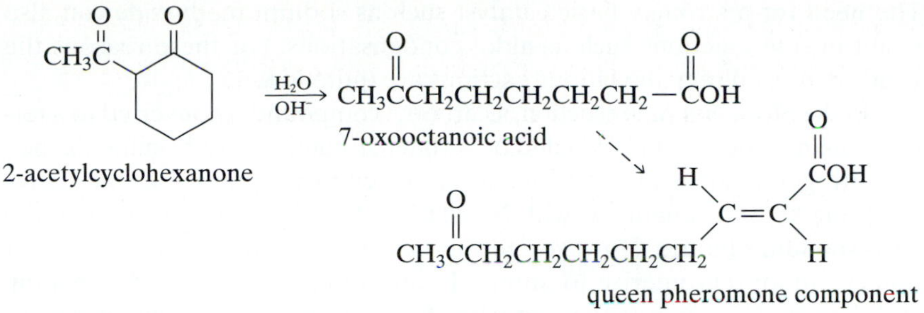

Proposed synthetic route to queen pheromone from 2-acetylcyclohexanone

supervisor developed an alternative procedure based on the Stork enamine reaction. Your assignment is to see whether or not 2-acetylcyclohexanone can be prepared using the Stork reaction.

Applying Scientific Methodology

The scientific problem described in the Scenario involves a straightforward preparation, but another problem discussed in the experiment deals with the structure of 2-acetylcyclohexanone. After reading the experiment you should develop a hypothesis regarding the structure of this compound, then test your hypotheses by obtaining its infrared spectrum.

Enamines and the Stork Reaction

In 1954, Gilbert Stork and his coworkers sent a letter to the editor of the *Journal of the American Chemical Society* that began with the sentence, "We have discovered a new method for the alkylation and acylation of ketones." Their brief communication caused a stir in the world of organic chemistry because it announced a synthetic tool of exceptional utility, which came to be known as the Stork enamine reaction.

Before the development of the Stork reaction, carbonyl compounds were alkylated directly by a reaction in which an alpha proton is removed by strong base to form an enolate ion, which acts as a carbon nucleophile toward alkyl halides. But multiple alkylation can occur, with a second alkyl group usually ending up on the same alpha carbon atom as the first, as illustrated for the direct alkylation of cyclohexanone by methyl bromide.

Direct alkylation of cyclohexanone

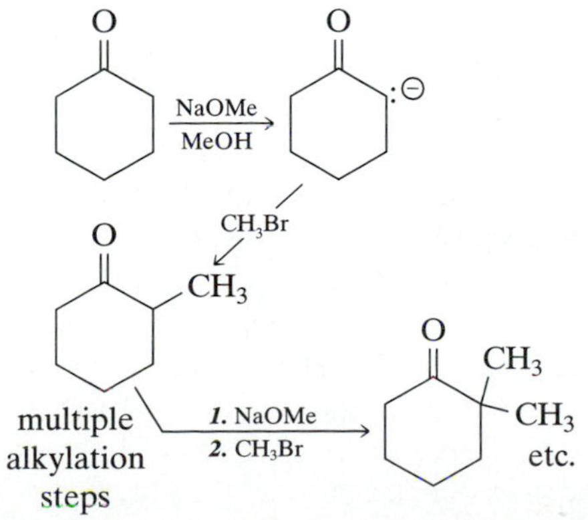

Tautomeric equilibria of enols and enamines

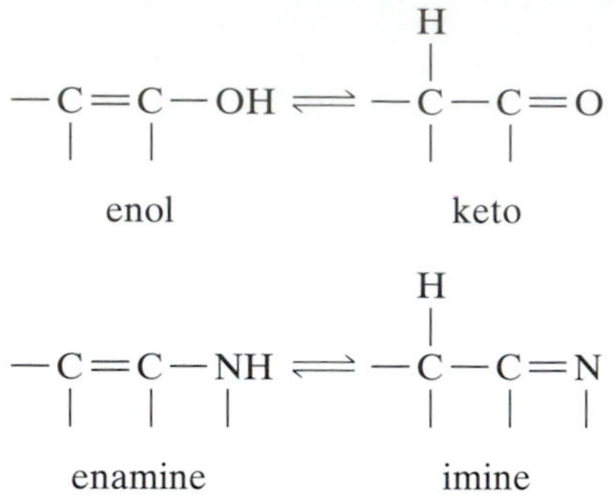

enol   keto

enamine   imine

Formation of an enamine

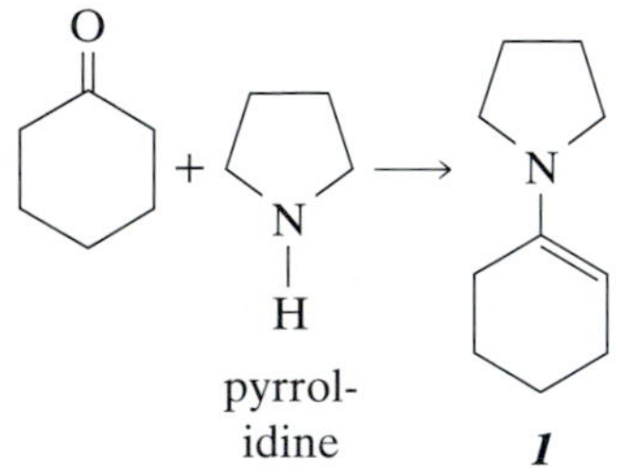

pyrrol-
idine   **1**

Alkylation and hydrosis of an enamine

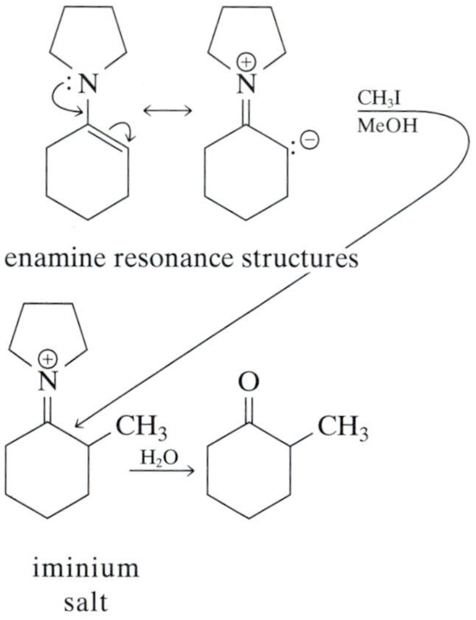

enamine resonance structures

iminium
salt

The need for a strongly basic catalyst such as sodium methoxide can also result in side reactions such as aldol condensations. For these reasons the yields of most direct alkylation reactions are quite low.

In the Stork enamine reaction, a carbonyl compound is converted to a tertiary enamine before it is alkylated or acylated. Enamines are 1-aminoalkenes, the nitrogen analogs of enols. Just as enols tend to isomerize to the corresponding ketones, enamines with N—H bonds are in equilibrium with the corresponding imines. However, tertiary enamines, which have no N—H bonds, cannot tautomerize to imines. In an example of the Stork reaction, cyclohexanone is first condensed with the secondary amine pyrrolidine to yield the tertiary enamine 1-pyrrolidinocyclohexene (**1**). This enamine is then alkylated under neutral reaction conditions to yield a monoalkylated iminium salt, which is hydrolyzed to the corresponding 2-alkyl ketone. Since one of the enamine's two resonance structures has a negative charge on the alpha carbon atom, the alkylation step can be viewed as a nucleophilic substitution reaction with a carbanion acting as the nucleophile. Enamines thus provide an indirect method for preparing alkyl (and acyl) derivatives of carbonyl compounds, in which the enamine functional group both activates the molecule and prevents unwanted side reactions.

Enamines can also be acylated by acid chlorides and anhydrides. This provides a unique method for lengthening a carbon chain by six atoms. For example acetyl chloride, which is prepared from acetic acid, reacts readily with 1-morpholinocyclohexene (2) to yield, after hydrolysis of the acylated enamine, 2-acetylcyclohexanone. Treatment of this diketone with strong base results in ring cleavage to 7-oxooctanoic acid, which can be reduced by a Wolff–Kischner reaction to octanoic acid.

Enamine synthesis of octanoic acid

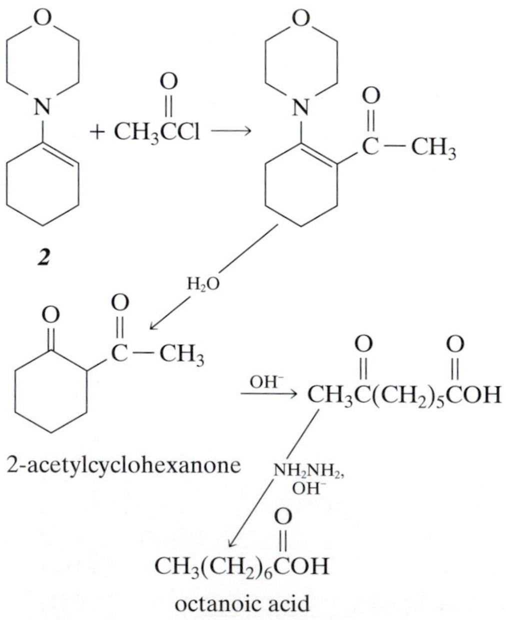

2

2-acetylcyclohexanone

$$CH_3C(CH_2)_5COH$$

$$CH_3(CH_2)_6COH$$

octanoic acid

In this way a two carbon acid, acetic acid, is (in effect) converted to an eight carbon acid. Other carboxylic acids can undergo the same chain-lengthening process via their acyl chlorides or anhydrides.

Understanding the Experiment

Because this experiment involves three reflux periods and an overnight reaction period, it is important that you plan your lab time carefully. You should be ready to start the overnight reaction by the end of the first lab period, if at all possible.

The formation of a ketone enamine is usually carried out in the presence of a catalyst such as *p*-toluenesulfonic acid. Although pyrrolidine reacts faster with ketones than morpholine, morpholine enamines give higher yields in acylation reactions, so morpholine will be used in this experiment. The reaction is carried out under reflux in toluene. To prevent premature hydrolysis of the enamine, water that forms during the reaction is distilled as a toluene-water azeotrope into a water separator. Some morpholine may distill along with the water, so a small excess is used in this experiment.

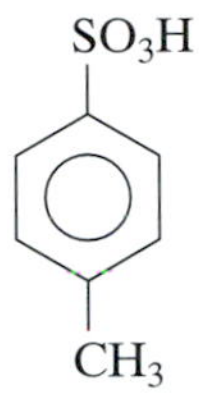

The acylation of 1-morpholinocyclohexene will be carried out, using acetic anhydride as the acylating agent, by simply heating the reactants in dichloromethane and allowing the solution to stand. Triethylamine is added to neutralize the acetic acid evolved during the reaction. The acylated enamine is then hydrolyzed by heating it with water under reflux, and the product is purified by vacuum distillation.

You can analyze the product by recording its infrared spectrum. The infrared spectra of cyclohexanone and acetic acid are shown in Figure 46.1 for comparison. Most β-diketones exist partly or predominantly in an enolic form, as illustrated for 2,4-pentanedione.

$$SO_3H$$

p-toluenesulfonic acid

Keto and enol forms for 2,4-pentanedione

$$CH_3CCH_2CCH_3 \rightleftharpoons CH_3C=CHCCH_3$$

keto form   enol form

The infrared spectrum of 2-acetylcyclohexanone should tell you whether this compound exists mainly in the diketo form or the enolic form. The carbonyl band of the diketo form appears in the usual frequency range for a ketone, but as a doublet with two closely spaced peaks. The IR spectrum of the enol tautomer of a diketone contains the expected O—H and C=O stretching bands, but hydrogen bonding between the enolic OH group and the carbonyl oxygen alters their appearance and locations considerably. The O—H band broadens and moves into the 3200–2500 cm^{-1} region; sometimes this band is so broad and low that it is hard to identify. At the same time, the carbonyl band broadens and moves from its normal position (~1715 cm^{-1}) to the 1650–1580 cm^{-1} region.

Reactions and Properties

A. cyclohexanone + morpholine $\xrightarrow{\text{TsOH}}$ 1-morpholinocyclohexene + H_2O

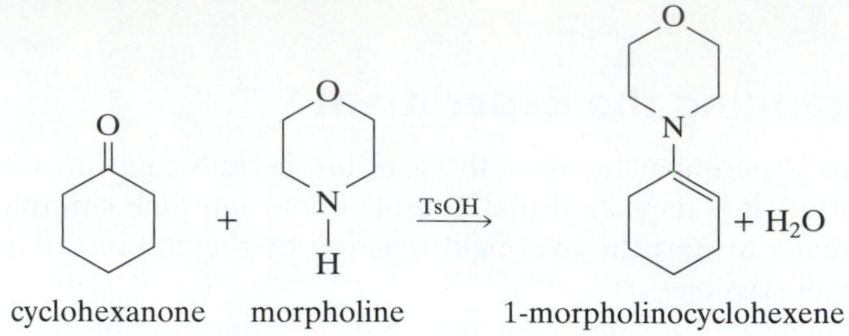

B.

2-acetylcyclohexanone

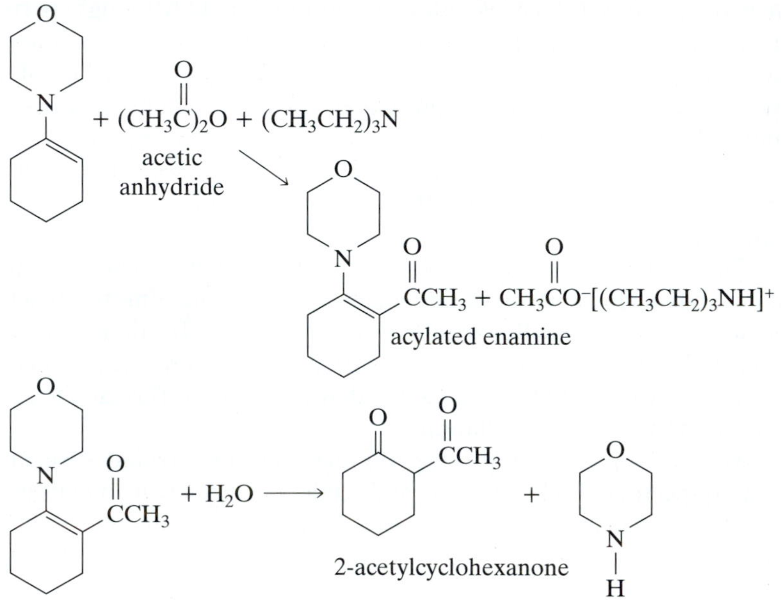

Table 46.1 Physical properties

	M. W.	m.p.	b.p.	d
cyclohexanone	98.2	−16	156	0.948
morpholine	87.1	−5	128	1.000
toluene	92.15	−95	111	0.867
p-toluenesulfonic acid	172.2	105	140[20]	
1-morpholinocyclohexene	167.3		119[10]	
acetic anhydride	102.1	−73	140	1.082
triethylamine	101.2	−115	89	0.728
2-acetylcyclohexanone	140.2		112[18]	1.078

Note: m.p. and b.p. are in °C, density is in g/mL

Cyclohexanone

2937.8	1311.2	908.4
1714.0	1221.7	749.8
1449.5	1118.8	489.7

Acetic acid

$$CH_3COH$$

3275.1	1714.5	1294.1
3043.0	1412.9	628.5
2631.9	1360.8	480.8

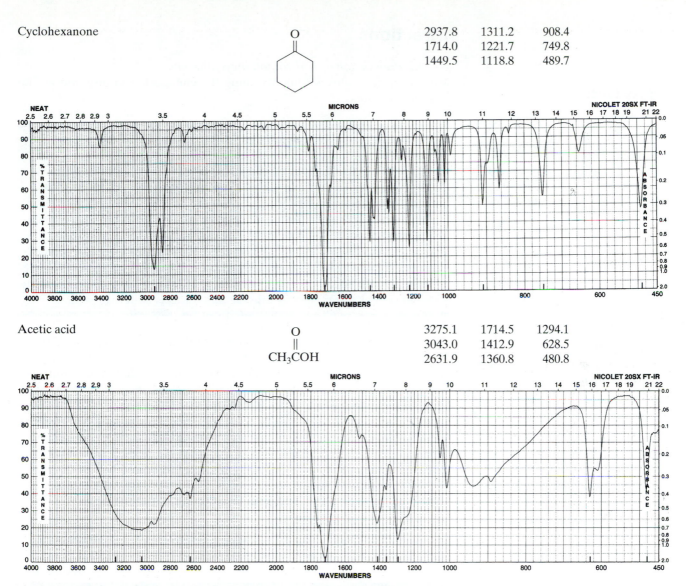

Figure 46.1 IR spectra of cyclohexanone and acetic acid (Reproduced from *The Aldrich Library of FT-IR Spectra, Edition II,* with the permission of the Aldrich Chemical Company.)

Safety Notes

toluene   morpholine

p-toluenesulfonic   cyclohexanone
acid

Take Care! Wear gloves, avoid contact with toluene and morpholine and do not breathe their vapors.

Take Care! Avoid contact with *p*-toluenesulfonic acid, do not breathe its dust.

Stop and Think: Where did the water in the separator come from?

Waste Disposal: Place the distillate and toluene from the water separator in an appropriate solvent recovery container.

Safety Notes

acetic
anhydride

triethylamine   dichloromethane

Directions

A. *Preparation of 1-Morpholinocyclohexene*

Wear protective gloves while handling the chemicals and dispense all liquids under a hood.

> Toluene is flammable, and inhalation, ingestion, or skin absorption may be harmful. Avoid contact and do not breathe vapors.
> Morpholine is flammable and can cause severe damage to the skin, eyes, and respiratory tract. Wear gloves, dispense from a hood, avoid contact, and do not breathe vapors.
> *p*-Toluenesulfonic acid can cause serious damage to the skin, eyes, and respiratory tract. Avoid contact and do not breathe dust.
> A vacuum distillation apparatus may implode if any of its components are cracked or otherwise damaged. Inspect the parts for damage and have your instructor check your apparatus. If possible, it is best to work behind a hood sash or safety shield while the apparatus is under vacuum.

Assemble a reflux apparatus equipped with a water separator [OP-25, Figure E 10], and fill the water separator to the side arm with toluene. In the reaction flask, combine 100 mmol of cyclohexanone, 130 mmol of morpholine, 0.10 g of *p*-toluenesulfonic acid, and 20 mL of toluene. Heat the reaction mixture under gentle reflux [OP-6] for at least an hour until the volume of water in the separator remains constant for at least 10 minutes. Remove the toluene and excess morpholine by simple distillation [OP-25] or by vacuum distillation (**Take Care!** See Safety Notes.) from a hot water bath [OP-26]. Toluene boils at about 20°C and morpholine at 35°C at 20 torr. If the 1-morpholinocyclohexane cannot be used the same day, keep it in a tightly stoppered flask (sealed with Parafilm) in the refrigerator.

B. *Preparation of 2-Acetylcyclohexanone*

If possible, the reaction should be carried out under a hood; otherwise use a gas trap containing dilute NaOH [OP-22].

> Acetic anhydride can cause severe damage to skin and eyes, its vapors are very harmful if inhaled, and it reacts violently with water. Use gloves and a hood, avoid contact with the liquid, do not breathe its vapors, and keep it away from water.
> Triethylamine is toxic, flammable, corrosive, and lachrymatory (tear-producing) and can cause severe damage to the eyes, skin, and respiratory tract. Use gloves and a hood, avoid contact with the liquid, do not breathe its vapors, and keep it away from flames or hot surfaces.
> Dichloromethane may be harmful if ingested, inhaled, or absorbed through the skin. There is a possibility that prolonged inhalation of dichloromethane may cause cancer. Minimize contact with the liquid and do not breathe its vapors.

Reaction. Assemble an apparatus for addition and reflux [OP-10] with the 1-morpholinocyclohexene in the reaction flask, attaching a calcium chloride drying tube [OP-22] to the reflux condenser. *Under the hood*, prepare a solution containing 120 mmol of triethylamine in 100 mL of dry dichloromethane. In a separate container prepare another solution containing 110 mmol of acetic anhydride in 50 mL of dry dichloromethane. Add the triethylamine solution to the reaction flask and place the acetic anhydride solution in the addition funnel. Heat the mixture under reflux [OP-6] using a 50°C water bath while adding the acetic anhydride solution drop by drop over a period of 30 minutes or more. Carefully transfer the reaction mixture to a clean, dry Erlenmeyer flask, stopper the flask tightly to keep out moisture, and let it stand until the next lab period (or for at least 24 hours).

Evaporate the dichloromethane on a hot water bath [OP-14], using a cold trap to recover the solvent. Cool the residue to room temperature or below, add 25 mL of water, and heat the mixture under reflux [OP-6] for 30 minutes to hydrolyze the acylated enamine.

Separation. Extract the reaction mixture [OP-13] with three 20-mL portions of dichloromethane. Dry the combined organic layers over anhydrous magnesium sulfate [OP-20], and evaporate the solvent [OP-14] using a cold trap.

Purification and Analysis. Purify the product by vacuum distillation [OP-26]. (**Take Care!** See Safety Notes.) 2-Acetylcyclohexanone should distill above 100°C under aspirator vacuum; place any forerun that distills below that temperature in a designated waste container. Weigh the product, record its infrared spectrum [OP-34], and turn it in to your instructor.

Report. Your report should include a statement of the problems and an account of how you applied scientific methodology to solve them. Calculate the percent yield of 2-acetylcyclohexanone. Interpret the infrared spectrum of the product as completely as you can. Decide whether 2-acetylcyclohexanone exists mostly in the diketo form or an enolic form, and if the latter predict which enolic structure (of two possibilities) it should have.

Take Care! Wear gloves, avoid contact with triethylamine, dichloromethane, and acetic anhydride; do not breathe their vapors.

Stop and Think: What is the purpose of the triethylamine?

Waste Disposal: Put all recovered dichloromethane in an appropriate solvent recovery container.

Waste Disposal: Pour the aqueous layer down the drain.

Exercises

1 Why is it necessary to separate water from the reaction mixture in part **A**? Write an equation for the reaction that would occur if it weren't removed.

2 Describe and explain the possible effect on your results of the following experimental errors or variations. (a) In part **A**, you forgot to fill the water separator with toluene. (b) You used equimolar amounts of cyclohexanone and morpholine in part **A**. (c) You omitted the triethylamine in part **B**.

3 Following the format in Appendix V, construct a flow diagram for the synthesis of 2-acetylcyclohexanone.

4 Acylation of enamines differs from their alkylation reaction in that an acylated iminium salt loses a proton to form an enamine before hydrolysis. Propose detailed mechanisms for the acylation of 1-morpholinocyclohexene by acetic anhydride and the hydrolysis of the resulting enamine.

5 Propose a mechanism for the cleavage of 2-acetylcyclohexanone to the salt of 7-oxononanoic acid in aqueous KOH.

6 Propose a synthesis for 9-oxo-2-decenoic acid from 7-oxooctanoic acid. Do you think your synthesis would yield mainly the (2*E*) isomer? Explain.

7 Outline enamine syntheses for the following compounds, starting with cyclohexanone and using any other needed reagents.

(a) 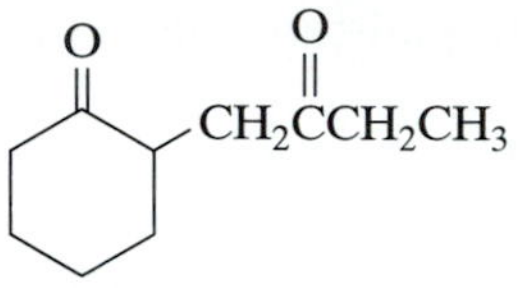 [structure: cyclohexanone with $CH_2CCH_2CH_3$ side chain bearing a ketone O]   (b) 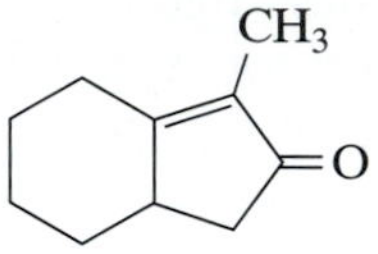 [bicyclic structure with CH_3 and $=O$]   (c) $HOC(CH_2)_5C(CH_2)_5COH$ (with three O groups)

Other Things You Can Do

(Starred projects require your instructor's permission.)

*1 Record the ^1H NMR spectrum of 2-acetylcyclohexanone in deuterochloroform and interpret it as completely as you can. Use the sweep offset control to locate any enolic proton signal, which should occur around 15δ. This signal may be quite broad as a result of proton exchange.

*2 Assess the purity of your 2-acetylcyclohexanone by gas chromatography.

3 Write a research paper about the pheromones that control behavior among social insects such as ants and honeybees, starting with sources listed in the Bibliography.

Synthesis of Dimedone and Measurement of Its Tautomeric Equilibrium Constant

EXPERIMENT 47

Reactions of α,β-Unsaturated Carbonyl Compounds. Preparation of Dicarbonyl Compounds. Nucleophilic Addition to Carbon-Carbon Double Bonds. Condensation Reactions. Decarboxylation. Enolate Ions. Reaction Equilibria. NMR Spectrometry.

Operations:

OP-6   Heating
OP-12 Vacuum Filtration
OP-14 Evaporation
OP-21 Drying Solids
OP-22 Drying and Trapping Gases
OP-23 Recrystallization
OP-28 Melting Point
OP-35 Nuclear Magnetic Resonance Spectrometry

Before You Begin

1 Read the experiment, read or review the operations as necessary, and write an experimental plan.
2 Calculate the mass and volume of 25.0 mmol of dimethyl malonate and the theoretical yield of dimedone.

Scenario

Harry Lingo, a punctilious chemistry editor for the Fulcourt Press, is faced by the dilemma of how to deal with compounds that exist in two tautomeric forms. In a soon-to-be published encyclopedia of organic compounds, he wants to list such a compound under the name corresponding to the major species present. For example, acetylacetone, which goes by the IUPAC name 2,4-pentanedione, contains four times more enol than keto tautomer, so he plans to list this compound under the name 4-hydroxy-3-penten-2-one.

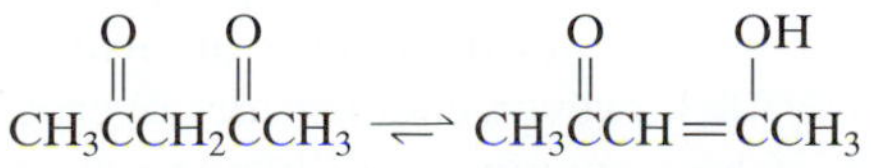

$$CH_3CCH_2CCH_3 \rightleftharpoons CH_3CCH=CCH_3$$

20% keto   80% enol

acetylacetone

Keto-enol equilibrium for dimedone

β-diketone　　　　enolic
　　　　　　　　ketone

The organic compound known by the common name dimedone is represented as a β-diketone in most textbooks of organic chemistry. But many β-diketones contain a substantial amount of enolic tautomer, so Mr. Lingo has asked your Institute for help in determining the most appropriate structure and name for dimedone. Your assignment is to prepare dimedone, measure its tautomeric equilibrium constant to see whether the predominant species present is the β-diketone or the enolic ketone, and give this species an appropriate IUPAC name.

Applying Scientific Methodology

Your working hypothesis should include an "educated guess" about the structure of dimedone. You will test your hypothesis by analyzing its ^{1}H NMR spectrum.

~100% keto　　0.00025% enol

acetone

Vitamin C
(ascorbic acid)

Tautomers and Life

Enols have long been known as unstable intermediates in reactions involving carbonyl compounds, such as the bromination of acetone. Although acetone exists overwhelmingly as the keto tautomer, it is the minute amount of the enolic form present that actually reacts with bromine under acidic conditions. Other enols are considerably more stable than that of acetone. For example, ethyl 3-oxobutanoate (ethyl acetoacetate) contains about 7.5% enol at equilibrium, and 2,4-pentanedione (acetylacetone) contains about 80% enol. Certain natural compounds such as Vitamin C, an enediol, exist almost entirely in the enolic form. The Kekulé structures of phenols are enol-like, suggesting that some phenols may exist in keto forms. For example, phlorglucinol (1,3,5-trihydroxybenzene) forms carbonyl-type derivatives with hydroxylamine and similar derivatizing agents, which indicates that its phenolic form may exist in equilibrium with a triketone.

phlorglucinol　　　　　　　　　　　　　　　phlorglucinol
　　　　　　　　　　　　　　　　　　　　　　oxime

A molecule's preference for one of several possible tautomeric structures might appear to be a matter of interest only to the molecule itself, and perhaps a few highly specialized scientists, but in fact it plays a crucial role in living systems. Ordinarily the keto form of a phenolic compound is the less stable tautomer by far, since its formation involves a loss of resonance energy. In the early 1950s, however, chemists discovered that certain biological amines (bases) such as guanine and thymine exist mainly in the keto forms at the pH of physiological systems. This discovery provided Francis Crick and James D. Watson with the key to the structure of DNA, and thus to the genetic code. The four bases guanine, thymine, adenine, and cytosine,

Watson's book The Double Helix *[Bibliography L67] provides a fascinating account of the discovery of the DNA structure.*

keto   enol   keto   enol

guanine   thymine

which are attached to the polyester backbone of a nucleic acid molecule, are responsible for both the transmission of genetic traits (by DNA) and the synthesis of proteins (by RNA) in living beings. Only the keto forms of guanine and thymine allow for the formation of adenine-thymine and guanine-cytosine base pairs, and such base pairing is essential for nucleic acid molecules to function. If these two bases existed in enolic forms rather than keto forms, life as we know it might be impossible!

Understanding the Experiment

In this experiment you will synthesize dimedone starting with the α,β-unsaturated ketone 4-methyl-3-penten-2-one, commonly known as mesityl oxide. This synthesis involves four distinct reactions that take place in sequence (see page 376). The first reaction is a *Michael addition* of dimethyl malonate to mesityl oxide. Combining sodium methoxide with dimethyl malonate generates an enolate ion that undergoes nucleophilic addition to the carbon-carbon double bond of mesityl oxide, yielding a keto diester (**1**). In the presence of sodium methoxide, the methyl group alpha to the keto carbonyl group loses a proton, forming another enolate ion that attacks the carbonyl carbon of one of the ester functions and displaces its methoxyl group. The result of this Claisen-type *cyclization* reaction is a diketo ester (**2**) with a six-membered ring. Both of these reactions occur spontaneously in the same reaction mixture. *Hydrolysis* of the ester yields a carboxylic acid, which when heated undergoes *decarboxylation* to form dimedone.

A literature procedure in *Organic Syntheses* [Bibliography B20] for the synthesis of dimedone requires the *in situ* preparation of sodium ethoxide by adding metallic sodium to absolute ethanol. Since this operation can be hazardous, commercial sodium methoxide in methanol will be used instead. The enolate ion of dimethyl malonate, $[CH(COOCH_3)_2]^-$ forms immediately when sodium methoxide is added to dimethyl malonate. This species reacts rapidly and exothermically with mesityl oxide, which should therefore be added in small portions. The Michael addition and the subsequent cyclization reaction should go virtually to completion during a one hour reflux period. The cyclic ester is then hydrolyzed with boiling sodium hydroxide, and the solution is acidified to form the carboxylic acid, which loses carbon dioxide on heating. The resulting dimedone is isolated by vacuum filtration and recrystallized from aqueous acetone.

One of the best ways to detect and analyze enol content is by ^{1}H NMR spectrometry. The enolic OH protons of α,β-diketones absorb far downfield, with chemical shift values in the $10-16\ \delta$ range. Enols also show vinylic proton (H—C=C) absorption in the usual range for these signals, about 4.5–6.0 δ. The keto form of an α,β-diketone can usually be recognized by the signal of the protons alpha to both carbonyl groups [H—C(C=O)$_2$]; these protons absorb near 3.5 δ. Since there are two such protons for every enolic vinyl proton in the enol form of dimedone, the equilibrium constant for dimedone's keto-enol equilibrium can be determined from the integrated areas of these two signals using the following formula:

$$K = \frac{[\text{enol}]}{[\text{keto}]} = \frac{2 \times \text{area of the H—C=C signal}}{\text{area of the H—C(C=O)}_2 \text{ signal}}$$

Enols are stabilized by hydrogen-bonding interactions with other polar species, so the equilibrium constant might vary depending on the solvent used. Infrared spectrometry can also be used to detect the existence of enols, as described in Experiment 46. The "Other Things You Can Do" section describes some additional studies you can carry out with your product.

Reactions and Properties

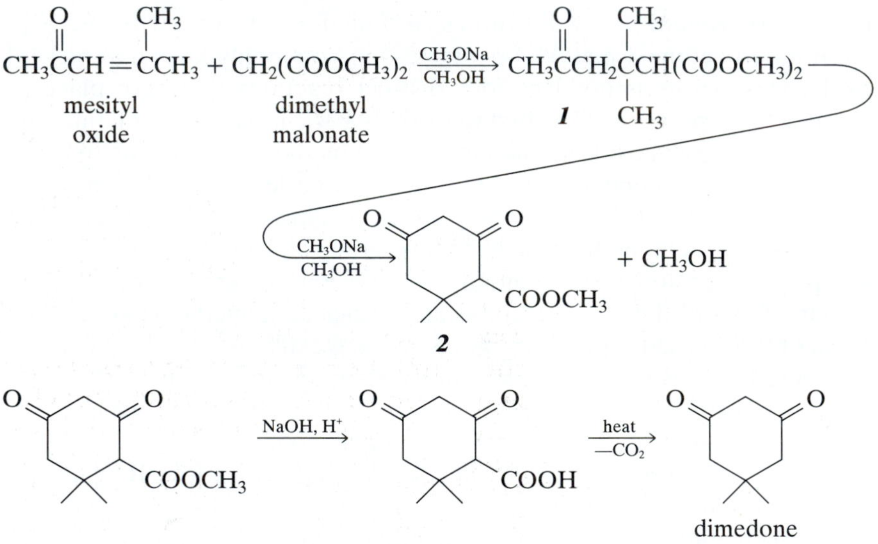

Note: All of the cyclohexanedione derivatives shown can also be represented by enolic forms.

Table 47.1 Physical properties

	M.W.	m.p.	b.p.	d
mesityl oxide	98.2	−52	129	0.858
dimethyl malonate	132.1	−62	181	1.154
sodium methoxide	54.0			
methanol	32.0	−94	65	0.791
dimedone	140.2	151		

Note: m.p. and b.p. are in °C, density is in g/mL. A 25% solution of sodium methoxide in methanol has a density of 0.945 g/mL.

Mesityl oxide (4-methyl-3-penten-2-one)

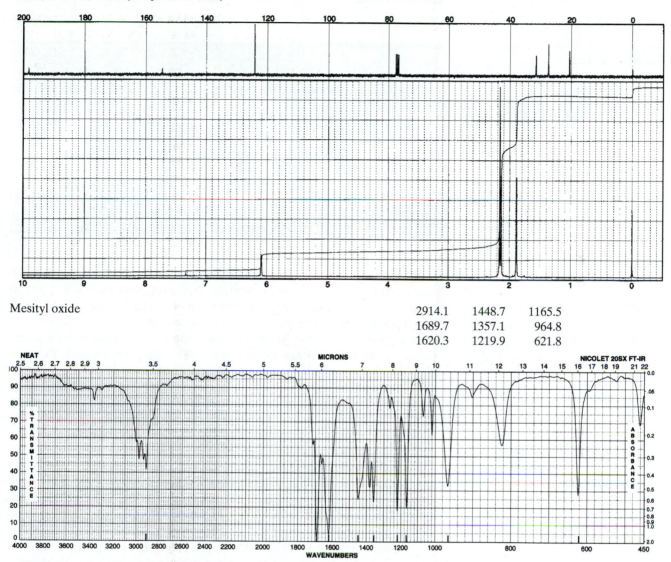

Mesityl oxide

2914.1	1448.7	1165.5
1689.7	1357.1	964.8
1620.3	1219.9	621.8

Figure 47.1 NMR and IR spectra of mesityl oxide (NMR spectra from *The Aldrich Library of ^{13}C and ^{1}H FT-NMR Spectra* by C. J. Pouchert and J. Behnke; IR spectrum from *The Aldrich Library of FT-IR Spectra, Edition II,* both with the permission of the Aldrich Chemical Company.)

Directions

Safety Notes

mesityl oxide

methanol

> The sodium methoxide/methanol solution is corrosive, toxic, and flammable. Contact or inhalation can cause severe damage to the skin, eyes, and respiratory tract. Wear gloves, avoid contact, do not inhale vapors, and keep away from flames.
>
> Mesityl oxide is a lachrymator and a strong irritant; inhalation and skin absorption are harmful. Avoid contact, do not breathe vapors. Mesityl oxide forms explosive peroxides on standing, so it should never be distilled to dryness.
>
> Deuterochloroform is harmful if inhaled or absorbed through the skin, and it is a suspected human carcinogen. Avoid contact with the liquid and do not breathe its vapors.

Take Care! Wear gloves, avoid contact with the solution, and do not breathe its vapors.

Stop and Think: What is the solid?

Take Care! Avoid contact with mesityl oxide, do not breathe its vapors.

Waste Disposal: Place recovered methanol in the appropriate solvent recovery container.

Stop and Think: Why did you have to remove the methanol before this step?

Take Care! Foaming may occur.

Stop and Think: What is the gas?

Waste Disposal: Pour the filtrate down the sink.

Waste Disposal: Place the filtrate in an appropriate solvent recovery container.

Take Care! Avoid contact with CDCl$_3$, do not breathe its vapors.

Reaction. Equip a dry reflux apparatus [OP-6] with a drying tube [OP-22]. Place 25.0 mmol of dimethyl malonate in the reaction flask, then mix in 6.0 mL (~26 mmol) of a 25% solution of sodium methoxide in methanol. Any solid that forms when sodium methoxide is added should dissolve on heating. Drop in a stirbar or boiling chips, attach the reflux condenser, and heat the solution just to boiling on a hot water bath or steam bath. Remove the heat source and add 3.0 mL (~26 mmol) of recently distilled mesityl oxide through the condenser in several small portions over a period of 2–3 minutes, swirling or stirring after each addition. Then heat the reaction mixture under gentle reflux [OP-6] for one hour. Remove the reflux condenser, connect the reaction flask to a trap and aspirator, and evaporate the methanol [OP-14] under vacuum using a steam bath or hot water bath. There may be some bumping during the evaporation, so use fresh boiling chips and heat gently. Stop the evaporation when the solid residue is still moist but all standing liquid has been removed.

Add 20 mL of 3.0 *M* aqueous sodium hydroxide to the reaction flask and heat the mixture under reflux [OP-6] for another hour to hydrolyze the intermediate ester. Pour the warm reaction mixture, with stirring, into a 150-mL beaker containing 15 mL of 6.0 *M* hydrochloric acid. *Under the hood*, carefully boil the mixture for 10 minutes or more until no more gas is evolved. If an oil forms during this step, it should crystallize on cooling. Cool the beaker in an ice bath until precipitation is complete.

Separation. Collect the product by vacuum filtration [OP-12] and air dry it.

Purification and Analysis. Purify the partly dried dimedone by recrystallization from 50% aqueous acetone [OP-23]. Dry [OP-21] and weigh the dimedone, and measure its melting point [OP-28]. Record the ^{1}H NMR spectrum [OP-35] of dimedone in deuterochloroform, or obtain a spectrum from your instructor. Turn in the product to your instructor.

Report. Your report should include a statement of the problem and an account of how you applied scientific methodology to solve it. On your NMR spectrum, identify the enol H—C=C and keto H—C(C=O)$_2$ signals. Calculate the keto-enol equilibrium constant, decide which species predominates in the solution, and give it an IUPAC name.

Exercises

1 (a) Compare the ^{1}H NMR spectrum of your product to that of mesityl oxide shown in Figure 47.1, and interpret both spectra as completely as you can. (b) Interpret the IR spectrum of mesityl oxide as completely as you can. Account for the wavenumber of the carbonyl band, comparing it with the usual value of ~1715 cm^{-1} for an aliphatic ketone.

2 (a) Write a mechanism for the Michael addition of dimethyl malonate to mesityl oxide in the presence of sodium methoxide/methanol. (b) Write a mechanism for the cyclization step in the synthesis of dimedone. (c) Write a mechanism for the decarboxylation step of this synthesis, showing the structure of the transition state.

3 Describe and explain the possible effect on your results of the following experimental errors or variations. (a) You neglected to evaporate methonol from the reaction mixture; (b) The mesityl oxide bottle was contaminated with water; (c) The bottle labeled "dimethyl malonate" actually contained dimethyl maleate.

4 Following the format in Appendix V, construct a flow diagram for the synthesis of dimedone.

5 What inexpensive starting material do you think is used to prepare mesityl oxide commercially? Outline the synthesis of mesityl oxide from this starting material.

6 A rather confused student, Ina Fogg, hydrolyzed the intermediate ester with 6 M HCl rather than 3 M NaOH and then boiled the reaction mixture with 50% NaOH instead of 6 M HCl. Filtration of the cooled solution yielded only a little dimedone. Finally realizing her mistake, she acidified the filtrate and a white solid precipitated, but its melting point was different from that of dimedone. What was this solid and how did it form?

7 Draw structures of some by-products that might be present in your product if you neglected to use a drying tube in the first step of the synthesis of dimedone. Write reaction pathways for their formation.

8 When heated with sodium ethoxide in ethanol, the diethyl esters of butanedioic acid and heptanedioic acid both yield products having six-membered rings. Write structures for both products and mechanisms for their formation.

9 Write a mechanism for the following reaction of dimedone with 3-buten-2-one.

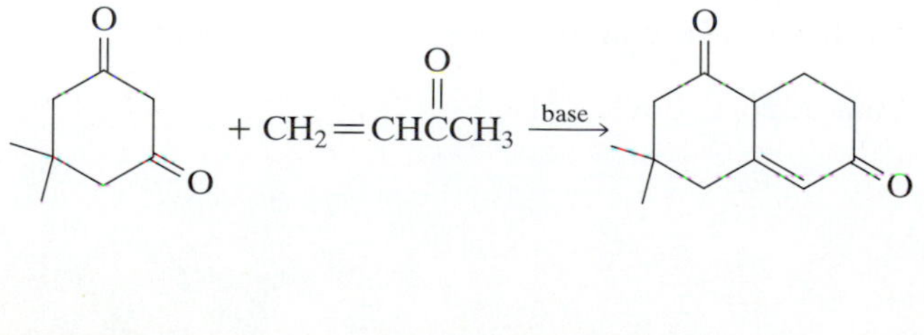

Other Things You Can Do

(Starred projects require your instructor's permission.)

*1 Record ^{1}H NMR spectra of dimedone in solvents other than deutero-chloroform, such as DMSO-d_6, and compare the keto-enol equilibrium constants for the different solvents. Try to explain any differences.

*2 Record the infrared spectrum of dimedone in a KBr disc or Nujol mull and then in a chloroform-d solution [OP-34]. Look for absorption bands arising from the enolic form and the diketo form in both spectra, and compare the relative amounts of enol (see Experiment 46). Try to explain any differences.

*3 Use your product to prepare dimedone derivatives of benzaldehyde in Minilab 39.

4 Like the synthesis of dimedone from mesityl oxide, the Robinson annulation reaction involves a Michael addition followed by a ring-forming condensation reaction. Starting with sources listed in the Bibliography, write a research paper about the Robinson annulation, describing some of its applications in the synthesis of terpenoids and steroids.

Preparation of Para Red
and Related Azo Dyes

Reactions of Amines. Reactions of Diazonium Salts. Preparation of Azo Compounds. Electrophilic Aromatic Substitution.

Operations:

OP-7   Cooling
OP-9   Mixing
OP-12 Vacuum Filtration
OP-21 Drying Solids

Before You Begin

1   Read the experiment, read or review the operations as necessary, and write an experimental plan.
2   Calculate the mass of 10.0 mmol of *p*-nitroaniline, the mass of 10.0 mmol of 2-naphthol, and the theoretical yield of Para Red. Be prepared to calculate the mass of 10.0 mmol of any other coupling component or diazo component and the theoretical yield of any other azo dye.

Scenario

The despotic King of Erewhon (see Experiment 22) has just been overthrown in a coup led by Sergeant Obmar, a soldier of fortune from California. In his honor the provisional government has authorized a new flag, consisting of a navel orange on a field of blue. But the first flags they ordered were sent back because the orange was the wrong color—it looked more like a pink grapefruit. Acting President Gib Retsim has asked your Institute to come up with a dye that will produce just the right shade of orange. Your supervisor knows how to make "American Flag Red," an orange-red dye the color of the stripes in the United States flag, and thinks that by tinkering with its molecular structure your project team should be able to come up with a suitable orange dye for the new flag.

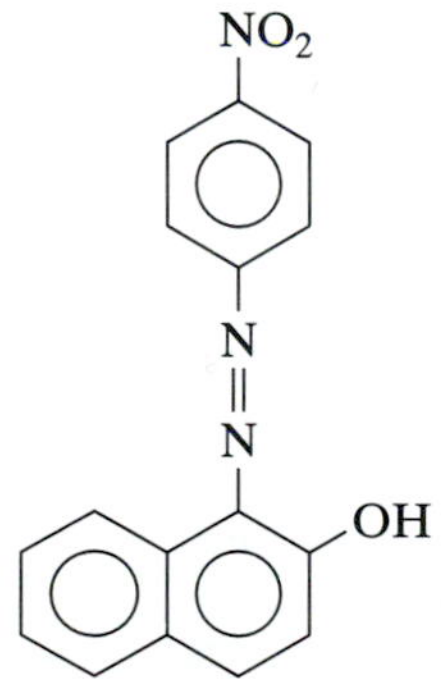

Applying Scientific Methodology

Based on the information in the experiment, you should develop a hypothesis regarding a pair of reactants from Table 48.1 that you think might produce a dye the color of a navel orange (perhaps one provided by your instructor). Alternatively, members of your project group can get together during a preceding lab period and select a different reactant pair for each member to work on.

"American Flag Red"
(Para Red)

Dyes and Serendipity

In the Persian fairy tale *The Three Princes of Serendip*, the title characters were forever discovering things they were not looking for at the time—thus, *serendipity* is the aptitude for making happy discoveries by accident. The preparation of the first commercially important synthetic dye by William Henry Perkin in 1854 is a good example of a serendipitous discovery in science. During the nineteenth century quinine was the only drug known to be effective against malaria, and it could be obtained only from the bark of the cinchona tree, which grew in South America. French chemists had isolated pure quinine from cinchona bark in 1820, but the inaccessibility of the tree made natural quinine very expensive. Perkin, then an 18-year-old graduate student working for the eminent German chemist August Wilhelm von Hofmann, realized that anyone who could make synthetic quinine might well become rich and famous. Perkin knew nothing about the molecular structure of quinine—structural organic chemistry was in its infancy in the mid-1800s—but he knew its molecular formula, $C_{20}H_{24}N_2O_2$. So he prepared some allyltoluidine ($C_{10}H_{13}N$), apparently thinking that two molecules of allyltoluidine plus three oxygen atoms minus a molecule of water would magically yield $C_{20}H_{24}N_2O_2$—quinine!

Perkin's idea

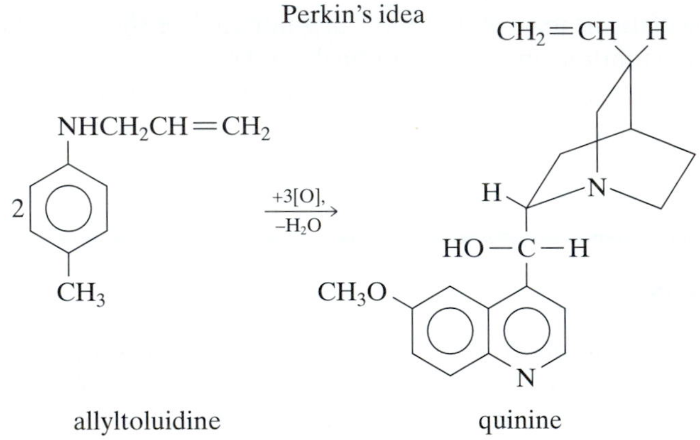

allyltoluidine   quinine

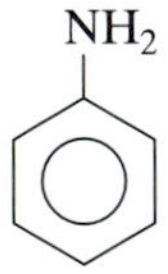

aniline

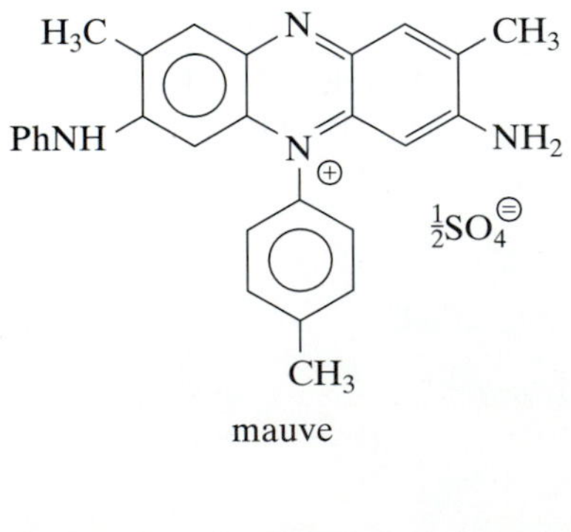

mauve

Of course Perkin had attempted the impossible—the molecular structure of allyltoluidine bears no resemblance to that of quinine, which was not synthesized until 1940. But he hopefully oxidized allyltoluidine with potassium dichromate and came up with a reddish-brown precipitate that he quickly realized was not quinine. Most chemists would have thrown out the stuff and started over, but it had properties that interested Perkin, so he decided to try the same reaction with a simpler base, aniline. This time he obtained a black precipitate that, when extracted by ethanol, formed a beautiful purple solution that greatly impressed some of the local dyers. Perkin knew a good thing when he saw it, so he promptly gave up his study of chemistry and went into the business of manufacturing "aniline purple," or mauve as the dye soon came to be known. Ironically, Perkin became rich and famous by *failing* to synthesize quinine. His dyestuffs plant was so successful that he was able to retire at the age of 36 and devote the rest of his life to pure research.

While Perkin was getting the synthetic dye industry under way, other chemists were experimenting with aniline and the vast array of other compounds that could be extracted from coal tar. One of these was a brewery chemist named Peter Griess, who took time off from the brewing of Alsopps' Pale Ale to discover the azo dyes. Undiscouraged by the fact that many of the diazo compounds he prepared had a tendency to explode, Griess did some fundamental research into the diazotization of aromatic amines and went on to discover the coupling reaction by which virtually all azo dyes are now synthesized. Aniline Yellow and Bismark Brown were synthesized in the 1860s, and the production and use of azo dyes grew rapidly thereafter.

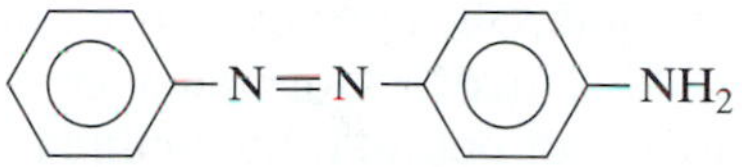

Aniline Yellow

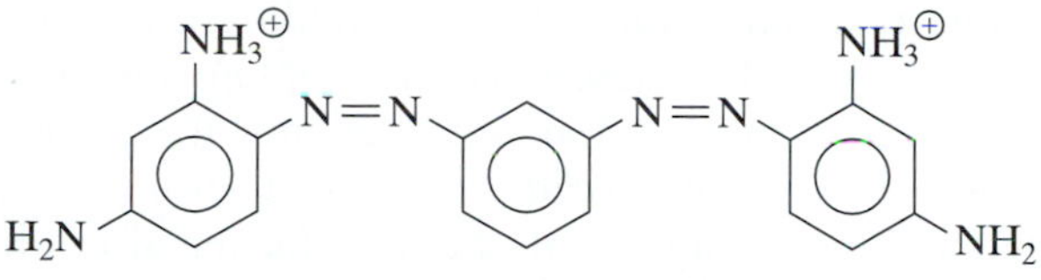

Bismark Brown

The advancement of organic chemistry as a science is due, in large part, to the discovery that chemists could prepare synthetic colors that were in many ways superior to the natural ones.

Understanding the Experiment

In this experiment you will prepare the azo dye Para Red and at least one other azo dye, and use your products to dye cloth. Para Red, which is made from *para*-nitroaniline and 2-naphthol, was once called "American Flag Red" because it was used to dye the cloth used for the stripes in the American flag.

Most azo dyes can be prepared by combining a diazonium salt (the *diazo component*) with an activated aromatic compound (the *coupling component*). The preparation of the dye involves two stages, known as *diazotization* and *coupling*. In the diazotization stage, a primary aromatic amine reacts with nitrous acid (HONO), to form the diazonium salt. The nitrous acid is generated *in situ* from sodium nitrite and a mineral acid. The reaction is carried out at a low temperature because diazonium salts react with water to form phenols and other by-products at higher temperatures. In the coupling stage, the diazonium salt is added to a solution of the coupling component, which is usually a phenol or another aromatic amine. Phenols couple most readily in mildly alkaline solutions, whereas amines react best in acidic solutions. However, too low a pH will prevent an amine from reacting by causing protonation of the amino group, whereas too high

Diazotization

$$ArNH_2 \xrightarrow{\text{HONO}} Ar - N_2^+$$
$$\text{diazonium salt}$$

Coupling

$$Ar - N_2^+ + H - Ar' \xrightarrow{-H^+}$$
$$\text{coupling component}$$

$$Ar - N = N - Ar'$$
azo compound

(Ar' must contain an activating group such as $-OH$ or $-NR_2$.)

a pH will cause the diazonium salt to change to a diazotate ion, which is incapable of coupling.

Formation of unreactive species at low and high pH

$$\text{Low pH: ArNR}_2 \xrightleftharpoons{\text{H}^+} \text{ArNHR}_2^+$$

$$\text{High pH: ArN}_2^+ \xrightleftharpoons{\text{OH}^-} \text{ArN}=\text{N}-\text{O}^-$$
$$\text{diazotate ion}$$

The coupling reaction is an electrophilic aromatic substitution reaction with the diazonium salt acting as the electrophile. Because the diazo group, N_2^+, is only weakly electrophilic, the coupling component must contain strongly activating groups such as OH or NR_2 in order for the reaction to occur.

To prepare the diazonium salt, a primary aromatic amine is dissolved in about 2.5 equivalents of dilute hydrochloric acid (or other suitable acid) and the solution is cooled to 5°C or below. If the amine is insoluble, the diazotization is carried out in suspension, with stirring. Aqueous sodium nitrite is then added and the solution is tested for excess nitrous acid using starch-iodide paper.

The coupling reaction is carried out by adding the diazo compound, with cooling and stirring, to a solution of a coupling component in dilute acid or base. If the coupling component is a phenol, it is dissolved in about 2 equivalents of 1 M sodium hydroxide and cooled before adding the diazonium salt solution. If necessary, the pH can be adjusted after the addition to obtain a better yield of azo dye. If the coupling component is an amine, it should be dissolved in 1 equivalent of 1 M HCl. After the diazo component is added and coupling is complete, the solution is neutralized to litmus paper by adding 3 M aqueous sodium carbonate (slowly to minimize foaming). The dye is then cooled in an ice bath before filtering. Some azo dyes decompose on heating, so they should be dried at room temperature.

Cloth can be dyed by several different processes. In the *direct process*, the dye is dissolved in water, the solution is heated, and the cloth is immersed in the hot solution. The dye molecules attach themselves to the cloth fibers by direct chemical interactions. In the *disperse process*, a water-insoluble dye is suspended in water and a small amount of a carrier substance is added. The carrier dissolves the dye and carries it into the fibers. In the *ingrain process* a dye is synthesized right inside the fiber, usually by the combination of a diazonium salt with a coupling component. The cloth is immersed in a solution of one of the components, allowed to dry, then immersed in a solution of the other component to develop the color. The comparatively small molecules of the separate components diffuse into the spaces between the fibers, but once they have combined within the fiber to form the dye, the larger dye molecules are trapped there. You will be using the ingrain method of dyeing in this experiment, but you can experiment with different dyeing methods as described in "Other Things You Can Do."

The color of a dye depends on the wavelengths of light it absorbs. If a dye absorbs light at a certain wavelength, the color perceived by the human eye consists of the remaining visible wavelengths that are reflected. The color of each dye you prepare in this experiment should be some shade of yellow, orange, or red; a yellow color is usually associated with absorption of light of shorter wavelengths, and a deep red color with absorption of light of longer wavelengths. The light-absorbing portion of a dye molecule, called a *chromophore*, is a conjugated system of delocalized pi electrons. The chromophoric system of Para Red, for example, includes the benzene and naphthalene rings, the two doubly bonded nitrogen atoms that connect them, and the unsaturated nitro group. In general, the more extended the chromophore, the longer the wavelength of the light it absorbs. Thus Para Red, with 21 atoms in its chromophore, absorbs light of considerably longer wavelengths than does Aniline Yellow, with only 14 atoms in its chromophore.

Certain saturated substituents called *auxochromes* can, in effect, extend a conjugated system by resonance. Auxochromes such as the OH, OCH_3, and NR_2 groups have one or more pairs of nonbonded electrons that they can share with a chromophore, thereby increasing the wave length of the light it absorbs. Such groups exert the greatest effect if they are *ortho* or *para* to the —N=N—group so that they can participate in resonance with the rest of the chromophore.

Reactions and Properties

Equations are given for the preparation of Para Red only.

Diazotization

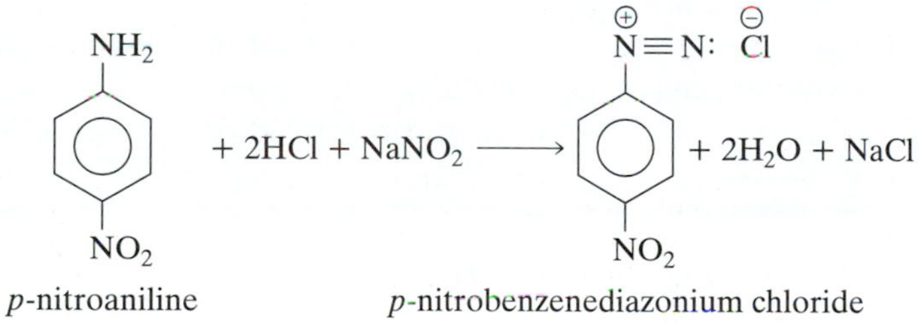

p-nitroaniline   *p*-nitrobenzenediazonium chloride

Coupling

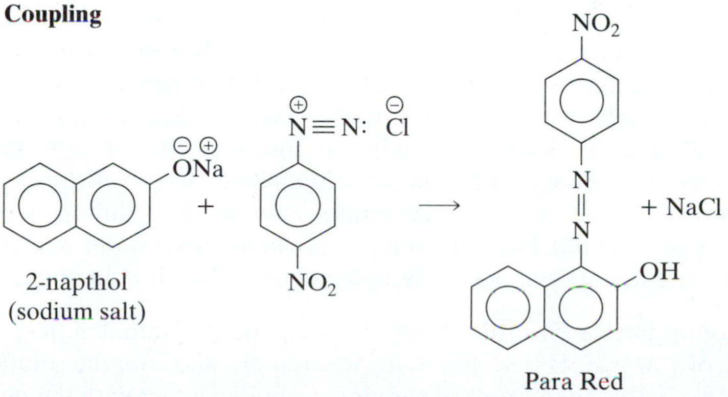

2-napthol
(sodium salt) NO_2

Para Red

Table 48.1 Suggested diazo and coupling components

Diazo component	M.W.	Coupling component	M.W.
aniline	93.1	aniline	93.1
m-anisidine	123.2	*N*-methylaniline	107.2
m-nitroaniline	138.1	*N,N*-dimethylaniline	121.2
m-toluidine	107.2	*m*-phenylenediamine	108.1
p-anisidine	123.2	phenol	94.1
p-nitroaniline	138.1	1-naphthol	144.2
p-toluidine	107.2	2-naphthol	144.2
		resorcinol	110.1

Note: Select one diazo component and one coupling component for each dye you prepare.

Directions

Each student should prepare Para Red and at least one other dye. With the instructor's permission, students can work in small groups with each person preparing a different dye. Any diazo component from Table 48.1 can be used in combination with any coupling component from that table.

> **Aromatic amines are very harmful if inhaled, ingested, or absorbed through the skin. Some aromatic amines are suspected carcinogens. Wear gloves, dispense under a hood, avoid contact, and do not inhale vapors. Most phenols are harmful if inhaled, ingested, or absorbed through the skin, and some are very corrosive, causing severe irritation or damage to skin and eyes. Some phenols are suspected carcinogens. Wear gloves, avoid contact, and do not inhale dust or vapors.**
> **Some azo dyes may be carcinogenic; wear gloves and avoid contact with the products.**

Safety Notes

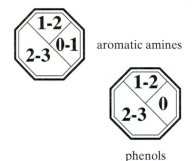

aromatic amines

phenols

Take Care! Wear gloves, avoid contact with the amine, do not breathe its vapors.

Observe and Note: What evidence is there that a reaction is occurring?

Stop and Think: Why is the sodium salt of the phenol used for coupling rather than the phenol itself?

A. *Preparation of an Azo Dye*

1. Diazotization of an Aromatic Amine. Mix 10.0 mmol of the diazo component with 8.0 mL of 3 *M* HCl. If the diazo component doesn't dissolve completely, heat the solution gently, adding up to 10 mL of water to get most or all of it in solution. Cool this solution to 5°C in an ice/water or ice/salt bath [OP-7] with manual or magnetic stirring [OP-9]. The amine salt may precipitate as you cool the solution, but it will diazotize satisfactorily if the reaction mixture is well stirred. Continue to stir as you add 10 mL of freshly prepared 1 *M* sodium nitrite at a rate slow enough that the temperature remains below 10°C during the addition. Test the solution with starch-iodide paper; if necessary, add enough sodium nitrite drop by drop to give a positive test (blue-black color). Accurately divide the solution into two equal parts (**d1** and **d2**), keeping both parts cold in an ice water bath. Go to step *2a* if the coupling component is a phenol or to *2b* if it is an amine.

2a. Coupling with a Phenol. Dissolve or suspend 10.0 mmol of the phenol in 20 mL of 1 *M* NaOH (use 40 mL for resorcinol), and cool the solution in an ice/water bath. (**Take Care!** Wear gloves, avoid contact with the phenol, do not breathe its dust or vapors.) Accurately divide the solution into two equal parts (**c1** and **c2**). Slowly add diazonium salt solution **d1** to coupling component solution **c1**, with stirring, and leave the mixture in the ice bath for 15 minutes or more until crystallization is complete. (Keep solution **d2** cold and save it and **c2** for part **B**.) If little or no colored solid appears, adjust the pH with dilute HCl or NaOH to induce coupling. Collect the azo dye by vacuum filtration [OP-12], washing it on the filter with water. Dry your azo dye at room temperature [OP-21], weigh it, and turn it in to your instructor.

2b. Coupling with an Amine. Dissolve or suspend 10.0 mmol of the aromatic amine in 10 mL of 1 *M* HCl (use 20 mL for *m*-phenylenediamine), and cool the solution in an ice/water bath. Accurately divide the solution into two equal parts (**c1** and **c2**). Slowly add diazonium salt solution **d1** to coupling component solution **c1**, with stirring, and leave the mixture in the cold bath for 15 minutes or more. (Keep solution **d2** cold and save it and **c2** for part **B**.) Neutralize the solution to litmus with 3 *M* aqueous sodium carbonate, then allow it to stand in the cold bath until crystallization is complete. Collect the azo dye by vacuum filtration [OP-12], washing it on the filter with water. Dry your azo dye at room temperature [OP-21], weigh it, and turn it in to your instructor.

B.   *Dyeing a Cloth by the Ingrain Process*
Dilute coupling component solution **c2** to 50 mL with water, and soak a piece of clean white cloth in it for 2–3 minutes. Remove the cloth with forceps or a pair of stirring rods, blot it between towels to remove most of the water, and hang it up to dry. Dilute diazonium salt solution **d2** to 50 mL with ice water, insert the dry cloth, and agitate the solution with a stirring rod for a few minutes to dye the cloth uniformly. If your coupling component was an aromatic amine, dip the cloth briefly into a little 3 *M* sodium carbonate solution. Remove the cloth and dry it as before. If there is a navel orange handy, compare its color to that of your dyed cloth. Turn in the dyed cloth with your report.

Report. Your report should include a statement of the problem and an account of how you applied scientific methodology to solve it. Write balanced equations for all the syntheses of azo dyes you carried out. Calculate the percent yield of each dye prepared in part **A**, and describe the color of both the dye itself and the dyed cloth. Prepare a table listing the colors of the cloths dyed by the azo dyes prepared by your group or lab section.

Observe and Note: What evidence is there that a reaction is occurring?

Stop and Think: Why may adjusting the pH help induce coupling?

Take Care! Wear gloves, avoid contact with the azo dye.

Take Care! Wear gloves, avoid contact with the amine, do not breathe its vapors.

Observe and Note: What evidence is there that a reaction is occurring?

Stop and Think: What is the purpose of the neutralization step? How should it affect the color of the dye?

Take Care! Wear gloves, avoid contact with the azo dye.

Stop and Think: What is happening on the cloth to account for your observations?

Exercises

1   Discuss the effects of structural features (such as substituents and chromophore size) on the color of the dyes prepared by your group.
2   Write a mechanism for the coupling reaction of *p*-nitrobenzenediazonium chloride with 2-naphthol.
3   Describe and explain the possible effect on your results of the following experimental errors or variations. (a) You forgot to cool the solution of your diazo component before adding aqueous sodium nitrite. (b) Your coupling component was a phenol, but you followed

the procedure in *2b* to couple it. (c) You tried to dye a cloth by dipping it into a solution of aniline in 3 *M* HCl, drying it, then dipping it into a solution of 2-naphthol in 1 *M* NaOH.

4 Following the format in Appendix V, construct a flow diagram for your synthesis of Para Red in part **A**.

5 Mel A. Droyt was trying to prepare *p*-dimethylaminoazobenzene (Butter Yellow) by coupling 10 mmol of *N,N*-dimethylaniline with an equimolar amount of aniline. He added 20 mL of 1 *M* sodium nitrite to the diazo component. Mixing this solution with the coupling component yielded some Butter Yellow along with a pale yellow oil. (a) What did he do wrong, and what was the yellow oil? (b) Write a balanced equation for its formation.

6 Kay Serra was attempting to prepare chrysoidine by coupling 10 mmol of *m*-phenylenediamine with an equimolar amount of aniline. To her surprise, she had to add 20 mL of 1 *M* sodium nitrite in the diazotization step before the solution turned starch-iodide paper blue. After pouring the diazonium salt solution into the solution of the other component, she recovered a dark-colored precipitate that was not chrysoidine. The following day the filtrate contained another precipitate that she identified as resorcinol. (a) What did she do wrong, and what was the structure of the azo dye she synthesized? (b) Write balanced equations for its synthesis and for the reaction that formed resorcinol.

7 Why is it important to keep the temperature low during diazotization and coupling? Give the structure of the product that might form if the solution is heated during the diazotization of *p*-nitroaniline, and write an equation for its formation.

8 Why does coupling of *p*-nitrobenzenediazonium chloride occur mainly *para* to the $-N(CH_3)_2$ group of *N,N*-dimethylaniline but *ortho* to the $-OH$ group of 2-naphthol?

9 Outline syntheses of Aniline Yellow and Bismark Brown (p. 383) starting with benzene.

chrysoidine

Other Things You Can Do

(Starred projects require your instructor's permission.)

Take Care! Wear gloves, avoid contact with H_2SO_4.

*1 Use your dye for direct dyeing by suspending 0.5 g of the dye in 100 mL of hot water and acidifying the mixture with a few drops of concentrated sulfuric acid. Immerse pieces of wool or cotton cloth in the mixture for 5 minutes or more. Remove the cloth, rinse it with water, and let it dry. Adjusting the pH of the dyeing mixture with dilute HCl or NaOH may give better results in some cases.

*2 Use your dye for disperse dyeing by suspending 0.5 g of the dye in 100 mL of hot water and stirring in 0.1 g of biphenyl (the carrier) and 2–3 drops of liquid detergent. Then immerse a piece of cloth made of Dacron or another polyester in the mixture, and heat the solution on a steam bath for 15–20 minutes. Remove the cloth and let it dry.

*3 Dissolve 3–5 mg of an azo dye in 10 mL of 95% ethanol. If any solid remains undissolved, filter the solution. Record the ultraviolet-visible spectrum of the dye [OP-36] over the tungsten lamp range (~800–350 nm), diluting the solution with more ethanol if necessary. Compare the λ_{max} values of different dyes and try to explain some of the differences you observe. Note that such comparisons are meaningful only if the bands you are comparing arise from the same kind of electronic transition; such bands should be similar in appearance and intensity.

*4 In Minilab 40, see what happens when you diazotize anthranilic acid and heat the resulting solution.

5 Starting with sources listed in the Bibliography, write a research paper about the chemistry and uses of food colorings. Outline syntheses for some azo dyes that have been used as food colorings and discuss the controversy surrounding such dyes as FD&C Red No. 2.

EXPERIMENT 49

Reaction of Phthalimide with Sodium Hypochlorite

Reactions of Amides and Imides. Nucleophilic Acyl Substitution. Functional Derivatives of Carboxylic Acids. Molecular Rearrangements.

Operations:

OP-6   Heating
OP-7   Cooling
OP-8   Temperature Monitoring
OP-9   Mixing
OP-12 Vacuum Filtration
OP-21 Drying Solids
OP-23 Recrystallization
OP-28 Melting Point

Before You Begin

1 Read the experiment, read or review the operations as necessary, and write an experimental plan.
2 Calculate the mass of 34.0 mmol of phthalic anhydride, of 17.0 mmol of urea, and of 20.0 mmol of phthalimide. Calculate the theoretical yield of $C_7H_7NO_2$.

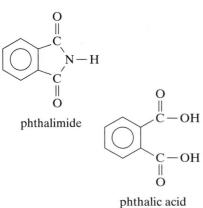

phthalimide

phthalic acid

Scenario

Otto Fökus, a nearsighted chemistry professor from Miskatonic University, was attempting to hydrolyze some phthalimide to phthalic acid when he mistook a bottle of chlorine bleach for a similar bottle containing aqueous sodium hydroxide. By the time he took a close look at the label and realized his error, he had already added some bleach to the reaction mixture. Thinking he might still salvage the experiment, he then added the designated amount of sodium hydroxide and heated the reaction mixture, but the crystalline solid he obtained had the wrong melting point for phthalic acid. Curious about this unexpected result, Professor Fökus sent a sample of the product to a commercial laboratory for analysis. The analytical report showed that the product contained nitrogen but no chlorine.

Analysis of Sample	
Carbon:	61.30%
Hydrogen:	5.15%
Nitrogen:	10.23%
Chlorine:	0.00%

Based on the product's elemental composition and an estimate of its molecular weight, the professor determined that its molecular formula is $C_7H_7NO_2$. Unfortunately, 18 compounds with that formula are listed in the *CRC Handbook of Chemistry and Physics* alone. Because of M.U.'s inadequate science equipment budget, Professor Fökus doesn't have access to the instruments needed to determine the structure of this compound, so he has asked your Institute for help. Your supervisor suspects that the reaction of phthalamide with sodium hypochlorite may involve some kind of molecular rearrangement. Your assignment is to carry out the reaction of phthalimide with sodium hypochlorite and sodium hydroxide and identify the product, using its melting point and your chemical intuition. You may first have to prepare phthalimide from a less expensive starting material, phthalic anhydride.

Applying Scientific Methodology

The following sections on the Curtius rearrangement and the reactions of imides contain some clues that should lead you to a reasonable hypothesis regarding the structure of $C_7H_7NO_2$, which is one of the compounds listed in Table 49.1. You will test your hypothesis by measuring the melting point of the product.

The Curtius Rearrangement

Acyl azides are compounds containing the azide ($-N_3$) functional group on a carbonyl carbon. They can be prepared by treating acid chlorides with sodium azide (NaN_3). When an acyl azide is heated, it evolves gaseous nitrogen to yield an isocyanate by a reaction called the Curtius rearrangement. At one time it was thought that the rearrangement took place via an electron-deficient intermediate called a *nitrene*, as illustrated for benzoyl azide. This reaction resembles a carbocation rearrangement except that the migrating group moves to an electron-deficient nitrogen atom instead of to a carbon atom. In this example, migration of a phenyl group to the nitrogen atom would restore its missing pair of electrons and yield a stable product.

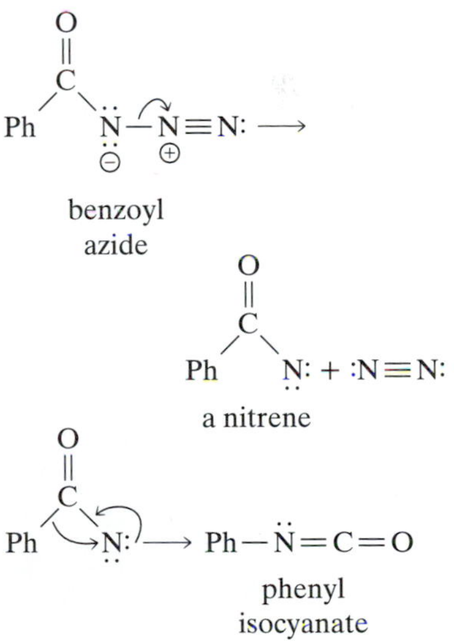

   More recent studies of such rearrangements suggest that free nitrenes are probably not involved. The rearrangement step is thought to proceed by a concerted mechanism such as the following, where Z is a good leaving group (such as $-N_3$) and R an alkyl or aryl group:

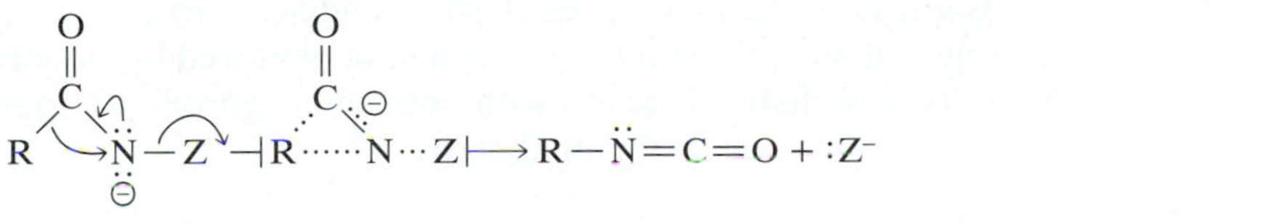

Nitrene mechanism for the Curtius rearrangement

benzoyl azide

a nitrene

phenyl isocyanate

Thus the migration to a *potentially* electron-deficient nitrogen occurs *as* the leaving group is being ejected, not after it has already left to form a

Reactions of isocyanates with
hydroxylic solvents

$$RN{=}C{=}O + R'OH \longrightarrow RN\overset{\displaystyle O}{\overset{\|}{H}}COR'$$

a urethane

$$RN{=}C{=}O + HOH \longrightarrow RN\overset{\displaystyle O}{\overset{\|}{H}}COH$$

a carbamic
acid

$$\longrightarrow RNH_2 + CO_2$$

*The carbon dioxide is converted to car-
bonate ion under the alkaline reaction
conditions used in some rearrangements
of this type.*

nitrene. Depending on the reaction conditions, the resulting isocyanate may
be isolated as such, or it may react further with the solvent. If the Curtius
rearrangement is run in an alcoholic solution, for example, the alcohol adds
to the C=N bond of the isocyanate to form a urethane. In water, a carbamic acid is formed at first, but it loses carbon dioxide spontaneously to
yield an amine as the final product.

Imides and their Reactions

Imides are nitrogen analogs of carboxylic anhydrides, compounds containing a —CONRCO— functional group where R is H, alkyl, or aryl. The
chemistry of imides resembles that of amides and other acyl compounds.
For example, imides can undergo nucleophilic acyl substitution reactions
with good nucleophiles.

Nucleophilic acyl substitution with an imide (H⁺ may
be obtained from water or another solvent)

$$RC\overset{\displaystyle O}{\overset{\|}{}}NHC\overset{\displaystyle O}{\overset{\|}{}}R' \xrightarrow{Nu:} \xrightarrow{H^+} RC\overset{\displaystyle O}{\overset{\|}{}}Nu + R'C\overset{\displaystyle O}{\overset{\|}{}}NH_2$$

The amide formed in such a reaction may react further under appropriate
conditions. Thus phthalimide is hydrolyzed by aqueous sodium hydroxide to
phthalamic acid (as the sodium salt); if the reaction mixture is heated for a
long time with concentrated NaOH, phthalic acid is obtained:

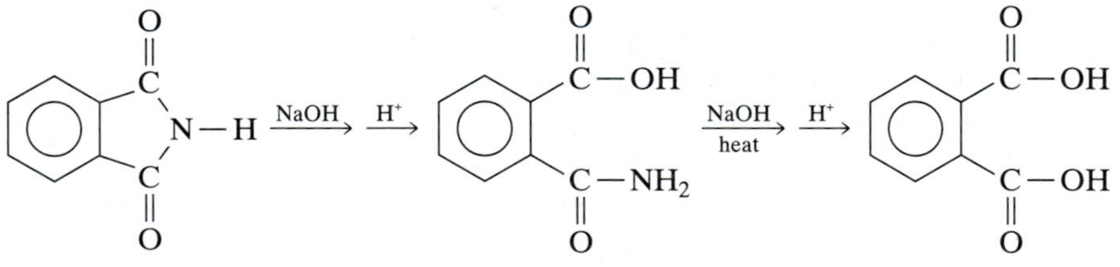

phthalimide   phthalamic acid   phthalic acid

Because of the two carbonyl groups adjacent to nitrogen, the N—H
hydrogen of an imide is acidic enough to be removed by moderately strong
bases. Thus phthalimide reacts with potassium hydroxide or potassium carbonate to form potassium phthalimide.

Reaction of phthalimide with potassium carbonate

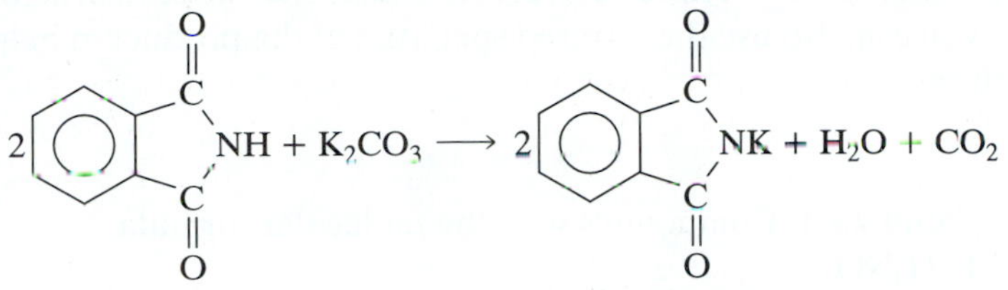

potassium
phthalimide

Imides and most other compounds with N—H bonds can be chlorinated by sodium hypochlorite. For example, sodium hypochlorite converts indole to *N*-chloroindole by the following reaction:

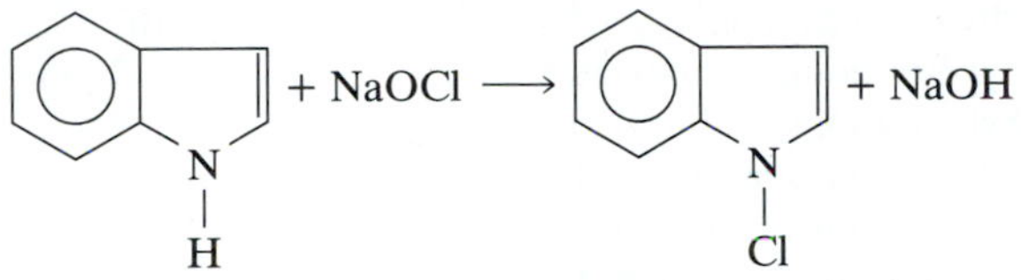

N-chloroindole

Understanding the Experiment

In this experiment you will first prepare phthalimide by heating phthalic anhydride with urea, unless the phthalimide is provided. Then you will carry out the reaction of phthalimide with sodium hypochlorite and sodium hydroxide and attempt to identify the product.

During the preparation of phthalimide, the reaction mixture expands to about three times its initial volume near the end of the reaction period, due to the rapid evolution of carbon dioxide. For this reason you should use a reaction flask large enough to allow for the volume increase.

You will carry out the reaction of phthalimide by heating it with a chlorine laundry bleach and aqueous sodium hydroxide. Most chlorine bleaches of this type contain about 5.25% NaOCl by mass, in water. Thus a commercial product such as Clorox or Javex is a convenient, inexpensive source of sodium hypochlorite in solution. When the reaction is complete, the resulting basic solution should contain the sodium salt of the product, which is precipitated with acetic acid after most of the sodium hydroxide has been neutralized with hydrochloric acid. Considerable foaming may occur during the acidification step, but it can be reduced by keeping the reactants cool and adding the acid slowly. The solid product is separated from the reaction mixture by vacuum filtration and purified by recrystallization.

The product will be one of the compounds listed in Table 49.1, some of whose structures are shown in Figure 49.1. After reading the experiment, you should be able to propose one or more reasonable structures for the product and a mechanism for its formation. The melting point of the product should then lead you to the correct structure. With your instructor's permission, you can also use the infrared spectrum of the product to help verify its structure.

Table 49.1 Compounds with the molecular formula $C_7H_7NO_2$

Compound	m.p.
2-hydroxybenzaldoxime*	63
3-hydroxybenzaldoxime	90
benzohydroxamic acid*	131–132
2-aminobenzoic acid*	146–147
3-aminobenzoic acid	174
4-aminobenzoic acid	188–189
2-hydroxybenzamide*	142
3-hydroxybenzamide	170.5
4-hydroxybenzamide	162
methyl 3-pyridinecarboxylate*	42–43
methyl 4-pyridinecarboxylate	8.5
1-methyl-3-pyridinecarboxylic acid*	218
2-hydroxy-5-nitrosotoluene	134–135
5-hydroxy-2-nitrosotoluene*	165
α-nitrotoluene	(b.p. 225–227)
2-nitrotoluene*	−10
3-nitrotoluene	16
4-nitrotoluene	54.5

Note: The structures of the compounds designated by asterisks are shown in Figure 49.1.

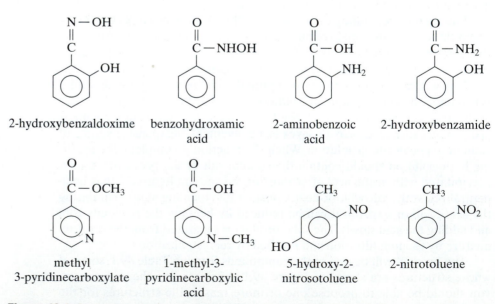

2-hydroxybenzaldoxime   benzohydroxamic acid   2-aminobenzoic acid   2-hydroxybenzamide

methyl 3-pyridinecarboxylate   1-methyl-3-pyridinecarboxylic acid   5-hydroxy-2-nitrosotoluene   2-nitrotoluene

Figure 49.1 Structures of some representative compounds from Table 49.1

Reactions and Properties

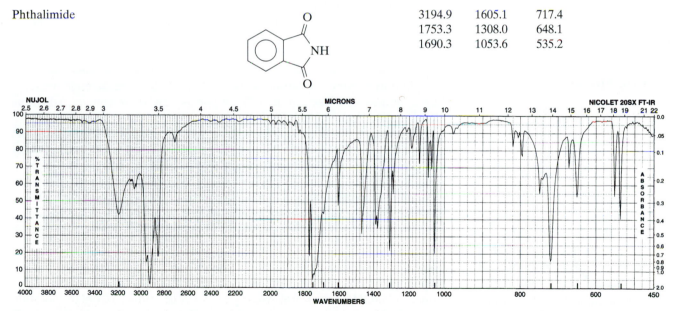

Table 49.2 Physical properties

	M.W.	m.p.
phthalic anhydride	148.1	132
urea	60.1	135
phthalimide	147.1	238
sodium hydroxide	40.0	322
sodium hypochlorite	74.4	

Note: m.p. and b.p. are in °C, density is in g/mL. 5.25% sodium hypochlorite has a concentration of about 0.74 M.

Phthalimide

3194.9	1605.1	717.4
1753.3	1308.0	648.1
1690.3	1053.6	535.2

Figure 49.2 IR spectrum of phthalimide (Reproduced from *The Aldrich Library of FT-IR Spectra, Edition II*, with the permission of the Aldrich Chemical Company.)

Directions

A. *Preparation of Phthalimide*
This part can be omitted if commercial phthalimide is provided.

Safety Notes

phthalic anyhdride

Take Care! Avoid contact with phthalic anhydride, do not inhale its dust.

Take Care! Do not heat oil above its flash point; keep flames away.

Stop and Think: What causes the frothing?

Intimately mix 34.0 mmol of pure phthalic anhydride with 17.0 mmol of urea and put the mixture into a 100-mL boiling flask. Insert a wide-bore condenser (a distilling column works fine) but do *not* connect it to a water line. Heat the reaction flask in a 130–135°C oil bath [OP-6]. Within about 20 minutes, the mixture should suddenly froth up and become nearly solid. At this point, stop heating the oil bath but leave the reaction flask in the oil while it cools down. Add about 5 mL of cold water and break up the solid. Collect the product by vacuum filtration [OP-12], washing it with a little cold water. Dry the phthalimide [OP-21] and measure its mass and melting point [OP-28].

B. *Reaction of Phthalimide with Sodium Hypochlorite*
If you obtain less than 20.0 mmol of phthalimide from part **A**, scale down the quantities of other reactants and solvents proportionately.

Safety Notes

sodium hydroxide

hydrochloric acid

acetic acid

Reaction. Measure 30 mL (~22 mmol) of 5.25% aqueous sodium hypochlorite into a 250-mL Erlenmeyer flask, then add 10 mL of 8 *M* (25%) aqueous NaOH. (**Take Care!** Wear gloves, avoid contact.) Cool the solution below 5°C in an ice/water or ice/salt bath [OP-7]. Add 20.0 mmol of finely powdered phthalimide and stir or swirl vigorously to mix the reactants [OP-9]. Add 6 mL more of the 8 *M* NaOH solution and again stir or swirl to mix. Monitor the temperature of the reaction mixture [OP-8], which should rise slowly as the solid dissolves, then more rapidly. When the temperature stops rising, heat the reaction mixture on a hot plate or steam bath to 80°C, with stirring or occasional swirling. Keep it at that temperature for 5 minutes, then let it stand for about 10 minutes at room temperature.

Under the hood, cool the reaction mixture in an ice/water bath, then carefully stir in 10 mL of concentrated hydrochloric acid. Test the solution with pH paper, and slowly add *just* enough additional concentrated HCl to bring its pH down to 10. Do not add too much HCl or the product may dissolve. Still under the hood, add 3.3 mL of glacial acetic acid *slowly* with continuous stirring or swirling. (Don't let the reactants foam out the top of the flask.) Let the reaction mixture stand in the ice/water bath until precipitation is complete.

Separation. Collect the precipitate by vacuum filtration [OP-12], and wash it on the filter with ice water until the odor of acetic acid is gone.

Purification and Analysis. Purify the product by recrystallization from boiling water [OP-23], using decolorizing carbon (preferably pelletized Norit) as necessary to remove colored impurities. Dry the product [OP-21] at room temperature, and measure its melting point [OP-28].

Report. Calculate the percent yield of phthalimide and of the final product. Give the name and structure of the product, and tell how you arrived at your conclusion. Propose a detailed mechanism that explains how phthalimide is converted to the product, showing the transition state for any rearrangement step.

Take Care! Wear gloves, avoid contact with HCl, do not breathe its vapors.

Stop and Think: What should you do if this happens?

Take Care! Wear gloves, avoid contact with acetic acid, do not breathe its vapors

Stop and Think: What do your observations tell you about the nature of the reaction?

Waste Disposal: Pour the filtrates down the drain.

Exercises

1   What gas was responsible for the foaming when you acidified the reaction mixture in part **B**? Write a balanced equation for the reaction that liberated the gas.

2   (a) A student misread the directions and acidified the reaction mixture to a pH of 2 with HCl, and was surprised when there was no precipitate to filter. Explain what went wrong and write an equation for the reaction that caused the problem. (b) How could the student have recovered the product and salvaged the experiment?

3   Describe and explain the possible effect (if any) on your results of the following experimental errors or variations. (a) The phthalimide from part **A** wasn't dried completely and it was allowed to sit for a week before it was used in part **B**. (b) The laundry bleach came from an old bottle and its NaOCl concentration was about 4 percent. (c) You used 3.3 mL of 3.0 *M* acetic acid rather than glacial acetic acid, and there was no foaming after you added all of it.

4   Following the format in Appendix V, construct a flow diagram for the synthesis you carried out in part **B**.

5   (a) When 35 mg of the product from this experiment was mixed with 0.46 g of camphor, the melting point of the mixture was found to be 157°C. Calculate the approximate molecular weight of the product if the melting point of pure camphor is 179°C and its freezing-point depression constant (K_f) is 40 K·kg·mol^{-1}. (b) Show how the molecular formula $C_7H_7NO_2$ can be derived using this result and the analytical data given in the Scenario.

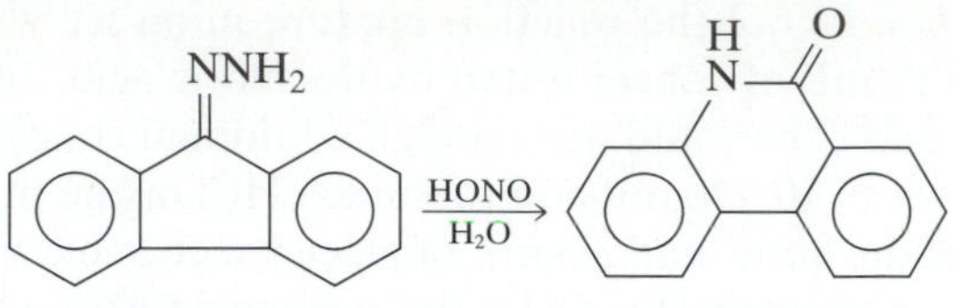

9-fluorenone   phenanthridone
hydrazone

6 When 9-fluorenone hydrazone is treated with sodium nitrite in aqueous sulfuric acid, it rearranges to form phenanthridone. Propose a detailed mechanism that explains this reaction, showing the transition state for the rearrangement step. (*Hint:* What happens to amino groups in an acidified solution of sodium nitrite?)

7 (a) When a famous German chemist warmed a solution of *N*-bromoacetamide in base, he recognized an unmistakable pungent odor that slowly faded and was replaced by the strong ammonia-like odor of an escaping gas. What were the two substances that his nose told him were there? Write a mechanism explaining the formation of both. (b) Who was the chemist, and what reaction had he discovered?

Other Things You Can Do

(Starred projects require your instructor's permission.)

*1 Record the infrared spectrum [OP-34] of the product and use it to assist you in identifying the product. Compare its spectrum with that of phthalimide in Figure 49.2, and interpret it as completely as you can.

*2 Carry out a molecular rearrangement of benzophenone oxime in Minilab 41.

3 Starting with sources listed in the Bibliography, write a research paper about the use of potassium phthalimide in the synthesis of amines and amino acids. Include a discussion of the phthalimido-malonic ester method and the Gabriel synthesis, and illustrate synthetic routes to specific products using each method.

Identification of an Unknown Amine   EXPERIMENT 50

Reactions of Amines. Infrared Spectrometry. Qualitative Analysis.

Operations:

OP-21 Drying Solids
OP-23 Recrystallization
OP-25 Simple Distillation
OP-28 Melting Point
OP-29 Boiling Point
OP-34 Infrared Spectrometry

Before You Begin

Read the experiment, read or review the operations as necessary, and write an experimental plan.

Scenario

The city health department of Arkham, Mass., has been deluged with reports of bad-tasting city water, which may have been responsible for several reported cases of illness. Arkham's municipal water supply comes from the Miskatonic River that winds through the town. A routine water analysis indicates the presence of a basic, nitrogen-containing compound, apparently an amine. The three most likely sources of the pollution, all upstream of the town, are (1) a dye factory that uses aniline and other aromatic amines as raw materials; (2) a chemical specialties company that synthesizes aliphatic amines for the manufacture of surfactants, corrosion inhibitors, and antioxidants; (3) an abandoned cemetery that was recently flooded, possibly causing contamination by the aliphatic diamines putrescine and cadaverine. With some help from the chemistry faculty at nearby Miskatonic University, health department personnel have obtained a sample of the unknown amine, and the Arkham town council has now asked your Institute for help in identifying the source of this water contaminant. Your assignment is to identify the unknown amine and thereby determine its source.

Applying Scientific Methodology

As you carry out the experiment you should develop provisional hypotheses about the nature and identity of your unknown, which you will test—and perhaps reject or revise—as you gather additional experimental evidence. Your conclusion should, if possible, be consistent with all of the experimental evidence you obtain. If any evidence is not consistent with your conclusion, you should be able to explain why.

Biological Amines

Certain families of organic compounds, such as aldehydes and esters, tend to be associated with the pleasant aromas of fruits and perfumes. Amines, on the other hand, are more often associated with the unpleasant smells of body wastes, not-so-fresh fish, and decaying flesh. The amine family includes deadly poisons such as coniine, a component of the poison hemlock that killed Socrates, and dangerous drugs such as LSD, heroin, and methamphetamine. But the same family also includes some highly beneficial members that we couldn't get along without.

Some amines are produced by the enzyme-catalyzed breakdown of proteins and their component amino acids in decaying plant or animal material. For example, bacteria containing the enzymes called amino acid decarboxylases bring about the degradation of the amino acids ornithine and lysine to putrescine (1,4-butanediamine) and cadaverine (1,5-pentanediamine), respectively.

$$^+H_3NCH_2CH_2CH_2\underset{\underset{NH_2}{|}}{\overset{\overset{O}{\|}}{C}}HCO^- \xrightarrow[-CO_2]{enzyme} H_2NCH_2CH_2CH_2CH_2NH_2$$

ornithine   putrescine

$$^+H_3NCH_2CH_2CH_2CH_2\underset{\underset{NH_2}{|}}{\overset{\overset{O}{\|}}{C}}HCO^- \xrightarrow[-CO_2]{enzyme} H_2NCH_2CH_2CH_2CH_2CH_2NH_2$$

lysine   cadaverine

As their common names suggest, these amines are responsible for much of the objectionable odor of decaying flesh. The next homolog in this family of diamines, 1,6-hexanediamine, is more often associated with hosiery, camping gear, and outdoor clothing. Along with hexanedioic acid (adipic acid), it is a monomer used in the preparation of nylon 6,6, a commercially important polyamide.

$$n\,H_2NCH_2CH_2CH_2CH_2CH_2CH_2NH_2 + n\,HO\overset{\overset{O}{\|}}{C}CH_2CH_2CH_2CH_2\overset{\overset{O}{\|}}{C}OH \xrightarrow[-H_2O]{\Delta}$$

1,6-hexanediamine   hexanedioic acid

$$-[HNCH_2CH_2CH_2CH_2CH_2CH_2NH-\overset{\overset{O}{\|}}{C}CH_2CH_2CH_2CH_2\overset{\overset{O}{\|}}{C}]_n-$$

nylon 6,6

The family of amines called *phenethylamines*, whose parent compound is 2-phenylethylamine ($C_6H_5CH_2CH_2NH_2$), has many biologically active members including body regulators such as epinephrine (adrenaline), useful drugs such as pseudoephedrine, and dangerous street drugs such as methamphetamine. Epinephrine is synthesized in the body as the end product of an important biosynthetic pathway that starts with the amino acid L-phenylalanine (see Figure 50.1).

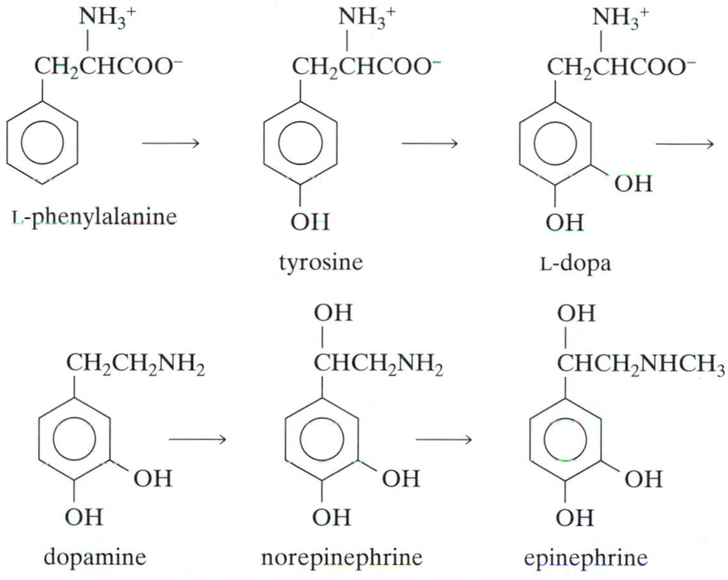

Figure 50.1 Biosynthetic pathway to epinephrine

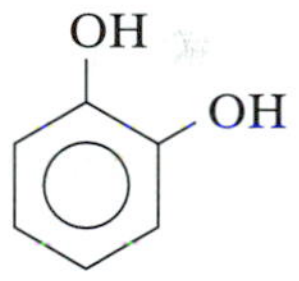

Along this pathway are two other amino acids, tyrosine and L-dihydroxyphenylalanine (L-dopa), and the three *catecholamines* dopamine, norepinephrine, and epinephrine. Catecholamines are so named because they can be regarded as derivatives of the phenol 1,2-dihydroxybenzene, whose common name is catechol. L-Dopa, which is converted to dopamine in brain tissue, is used to help reduce the characteristic tremors and other abnormal movements characteristic of Parkinson's disease. The disease is apparently associated with a deficiency of dopamine in the brain, causing an imbalance of the chemicals responsible for transmitting nerve impulses.

Norepinephrine is one of the body's major neurotransmitters, and is responsible for the transmission of nerve impulses along the sympathetic (adrenergic) nervous system. Epinephrine, a hormone produced in the adrenal gland, increases the heart rate and blood pressure by constricting blood vessels, and also dilates bronchial passageways to permit free breathing. The "adrenaline rush" sometimes experienced by athletes and people in dangerous situations is produced by the release of epinephrine in response to stress, including intense emotion or fear. Both norepinephrine and epinephrine have been used therapeutically, the former to maintain blood pressure for persons in shock and the latter to treat allergies and stimulate the heart during heart attacks.

Another phenethylamine, ephedrine, occurs naturally in several species of leafless green-stemmed shrubs of the genus *Ephedra*. Twigs from the

catechol

Chinese shrub ma huang (*Ephedra sinica*) have been used for more than 5000 years to treat asthma and a variety of other ailments. Like epinephrine, ephedrine constricts the walls of blood vessels and can also dilate bronchial tubes. Such properties have led to the widespread use of ephedrine and its diastereomer pseudoephedrine as decongestants for people suffering from asthma, sinus congestion, allergies, and even the common cold.

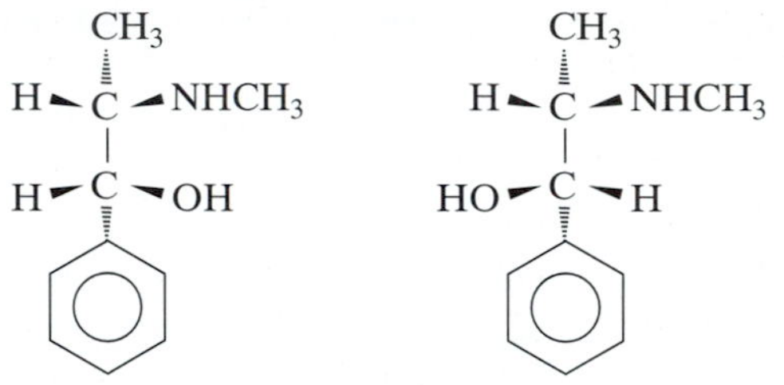

(1*R*,2*S*)-ephedrine   (1*S*,2*S*)-pseudoephedrine

Ephedrine has also been used widely in dietary supplements that are claimed to help people lose weight, feel more energetic, and develop their muscles. However, overdoses of ephedrine can cause heart attacks, strokes, seizures, and sometimes death, so the U.S. Food and Drug Administration recently limited the amount of ephedrine that can be included in any dietary supplement and banned its use in weight loss and body building products.

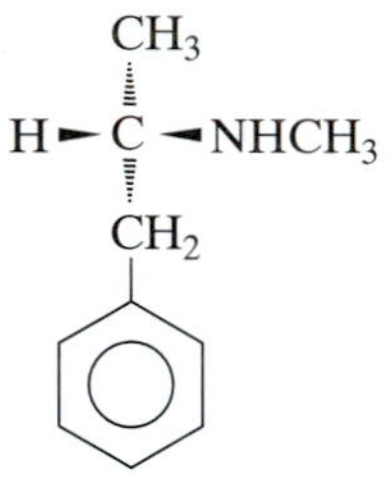

The illicit drug methamphetamine has a structure very similar to that of ephedrine but is considerably more dangerous. Before its addictive potential was recognized, methamphetamine—also known as "speed"— was prescribed as an appetite suppressant and to treat depression. It is a nervous system stimulant said to generate a feeling of confidence and mental alertness, but its side effects include hallucinations, paranoia, and death. Epinephrine and methamphetamine both appear to stimulate the release of norephineprine in the body, thereby increasing the transmission of nerve impulses.

methamphetamine

Understanding the Experiment

This section provides a general discussion of most of the procedures you will follow to identify your unknown. For more detailed information about the interpretation of test results, see the appropriate sections in Part IV. For information about the interpretation of infrared spectra, see Operation 34. Locations of classification tests and derivative preparation procedures are given in Table 50.1.

Since your unknown amine may be impure, you should purify it by distillation before you perform any chemical tests or record its spectra. Solids can be purified by recrystallization, but see your instructor first to find out if an unknown solid requires purification. Measure the boiling point or melting point of your unknown as accurately as you can since your list of possibilities will be based on the value you obtain. A preliminary examination of your unknown may provide some clues that will help you identify it, as will its solubility behavior in water and 5% HCl.

Since tertiary amines are listed on a different table and have different derivatives than primary or secondary amines, it is important to classify your amine correctly. Many amines can be classified with reasonable accuracy using a simple color test, the quinhydrone test. Before you can interpret this test, you must known whether your amine is aromatic or aliphatic; aliphatic amines are basic enough to dissolve in a buffer solution having a pH of 5.5, while most aromatic amines are not. To confirm your tentative classification you should carry out the Hinsberg test, which is based on the properties of the products formed when an amine reacts with an arenesulfonyl chloride in aqueous NaOH. The arenesulfonamide formed by a primary amine is soluble in the reagent and precipitates when HCl is added, while the arenesulfonamide formed by a secondary amine is insoluble in the reagent and does not dissolve when HCl is added. Most tertiary amines leave an insoluble residue that dissolves when HCl is added.

You can also classify an amine from its infrared spectrum and characterize it further. A typical primary amine has a medium-intensity two-pronged N—H stretching band near 3350 cm^{-1}, a medium to strong N—H bending band near 1615 cm^{-1}, and a strong, broad N—H bending band in the vicinity of 800 cm^{-1}. A typical secondary amine has a single weak N—H stretching band near 3300 cm^{-1} and a strong, broad N—H bending band around 715 cm^{-1}. A tertiary amine has no N—H bonds so its spectrum shows none of these bands, but it may have a recognizable C—N stretching band in the 1340–1020 cm^{-1} region. The position of the C—N band, which is present for primary and secondary amines as well, depends on the structure of the amine; for aromatic amines it is around 1340–1250 cm^{-1}, and for aliphatic amines around 1250–1020 cm^{-1}. The spectrum of an aromatic amine should also exhibit characteristic Ar—H stretching and bending bands, as described in OP-34.

Your unknown amine will be one of the amines listed in Tables 5 and 6 in Appendix VI. After performing the classification tests and interpreting your IR spectrum as completely as possible, you should be able to prepare a short list of possibilities from the appropriate tables. You should then be able to identify your unknown by preparing a derivative whose melting point will distinguish it from all other amines with similar boiling or melting points. If possible, avoid preparing derivatives whose melting points for the compounds on your short list are too low, too close together, or not listed for some of the possibilities. Keep in mind that incomplete purification and insufficient drying of the derivative can lower its melting point, so follow the directions for preparing the derivative carefully and be sure the product is completely dry before you measure its melting point. If you have difficulty preparing a certain derivative, or if the melting point of the derivative you prepare does not eliminate all the possibilities but one, you should prepare a second derivative.

When you think you have gathered enough evidence to identify your unknown with some certainty, you are free to write down your conclusion. But keep in mind that your evidence should be sufficient to convince your instructor—and yourself—that your conclusion is correct.

Reactions and Properties

General equations for derivative preparations for amines are given on pages 560–562. General equations for classification tests are given on the pages listed in Table 50.1.

Table 50.1 Classification tests for amines

No.	Test	Page
C-4	basicity test	541
C-15	Hinsberg's test	546
C-20	quinhydrone test	549

Directions

Safety Notes

> **Aromatic amines are very harmful if inhaled, ingested, or absorbed through the skin. Many aliphatic amines are flammable and can cause serious burns on skin and eyes, and are harmful if inhaled, ingested, or absorbed through the skin. Some amines are suspected carcinogens. Avoid contact with your unknown, do not inhale its vapors, and keep it away from flames.**
>
> **Safety information for chemicals used in classification tests and derivative preparations is included with the corresponding procedures in Part IV.**

Take Care! Wear gloves, avoid contact with the amine, do not breathe its vapors

Preliminary Work. Obtain an unknown amine from your instructor and record its code number in your laboratory notebook. If the unknown is a liquid, purify it by simple distillation and record its distillation boiling range and median boiling temperature [OP-25], then carry out a micro boiling point determination on the purified liquid [OP-29]. If it is a solid, measure its melting point [OP-28]. Describe the physical state, general appearance, and any other notable characteristics of the compound in your lab notebook. Carry out an ignition test as described on p. 532.

Stop and Think: Why might its odor disappear?

Solubility Tests. Test the solubility of the unknown in water as described on p. 535. If it is water soluble, test its aqueous solution with red litmus paper, then add 5% HCl dropwise to see whether its odor (if any) disappears. If it is water insoluble, test its solubility in 5% HCl as described on p. 536.

Classification of the Amine. Classify the unknown amine as aliphatic or aromatic using the basicity test (classification test C-4). Then classify it as primary, secondary, or tertiary using the quinhydrone test (C-20) and/or Hinsberg's test (C-15). Obtain the infrared spectrum [OP-34] of your unknown and use it to confirm your classifications. In your lab notebook, list all compounds from the appropriate table (Table 5 or 6) in Appendix VI that have melting or boiling points within ±10°C of your observed value, and record their melting or boiling points and the melting points of the derivatives listed. At your instructor's option, show him or her your list; the instructor may approve the list if it includes your unknown or suggest additional work if it doesn't.

Detection of Structural Features. Write the structure of every compound on your list and consider whether additional classification tests, such as Beilstein's test (C-5), would help you select the most likely possibilities. Inspect your infrared spectrum to find out what you can about any structural features or secondary functional groups such as aromatic rings, alkoxyl groups, or nitro groups. At this point you should be able to prepare a "short list" of compounds by eliminating the least likely possibilities. Keep in mind, however, that classification tests and infrared wave numbers are not infallible indicators of molecular structure, so it may be necessary later to reconsider some of the compounds you eliminated from your short list.

Preparation of a Derivative. Select the derivative that should best differentiate the compounds on your short list. If your amine is primary or secondary, you can prepare one or more of the following derivatives for which reagents are available: benzamide (D-8), *p*-toluenesulfonamide (D-9), phenylthiourea (D-10) or picrate (D-12). If your amine is tertiary, you can prepare a methiodide (D-11) or picrate (D-12). Purify the derivative by recrystallization as described in the appropriate procedure [OP-23], dry it thoroughly [OP-21], and obtain its melting point [OP-28]. Deduce the identity of your unknown from the derivative melting point and all other relevant evidence, and justify your conclusion based on the evidence.

Report. Your report should include a statement of the problem and an account of how you applied scientific methodology to solve it. Tabulate your data and results, including your observations for all tests. Justify your conclusion based on the evidence you accumulated.

Exercises

1 Interpret any infrared spectrum you obtained as completely as you can.
2 (a) Write balanced equations for the reactions involved in all of the classification tests in which you obtained a positive result. (b) Write balanced equations for the reaction(s) involved in your derivative preparation(s).
3 Describe and explain the possible effect on your results of the following experimental errors or variations. (a) When you carried out the Hinsberg test, you accidentally used 3 *M* HCl in place of 3 *M* NaOH. (b) You mistakenly classified a tertiary amine as secondary and tried to prepare a *p*-toluenesulfonate derivative. (c) While performing the basicity test on a water-insoluble unknown, you mistook a sodium acetate solution for the acetate-acetic acid buffer.
4 Following the format in Appendix V, construct a flow diagram showing the process you followed to identify your unknown.

5   The basicity test differentiates aromatic and aliphatic amines based on their solubility in a pH 5.5 buffer. Given that K_b for aniline is 4.2×10^{-10} and K_b for cyclohexylamine is 5.0×10^{-4}, calculate the ratio of amine salt to dissolved amine for both compounds in such a buffer, and explain the difference in their solubility behavior.

6   An unknown liquid boiling around $185\pm5°C$ dissolves in 5% HCl but is insoluble in water and in a pH 5.5 buffer. Shaking the unknown with *p*-toluenesulfonyl chloride in aqueous NaOH produces a clear solution, which when acidified yields a white precipitate. Assuming that the unknown is listed in Appendix VI, give its name and draw its structure.

7   An unknown liquid boiling around $185\pm5°C$ dissolves in 5% HCl and in a pH 5.5 buffer but is insoluble in water. Shaking the unknown with *p*-toluenesulfonyl chloride in aqueous NaOH produces a white precipitate that doesn't dissolve in dilute HCl. Assuming that the unknown is listed in Appendix VI, give its name and draw its structure.

8   Write mechanisms for the following reactions, which are used in chemical tests and derivative preparations. (a) The reaction of aniline with benzoyl chloride in aqueous NaOH; (b) The formation of the methiodide of triethylamine. (c) The reaction of diethylamine with benzenesulfonyl chloride in aqueous NaOH.

Other Things You Can Do

(Starred projects require your instructor's permission.)

*1   Obtain and interpret the ^1H or ^{13}C NMR spectrum [OP-35] of your unknown.

2   Starting with sources listed in the Bibliography, write a research paper describing the structures, biological functions, and therapeutic uses (if any) of some phenethylamines.

Preparation and Mass Spectrum of 2-Phenylindole

Reactions of Carbonyl Compounds. Preparation of Heterocyclic Amines. Fischer Indole Synthesis. Phenylhydrazones. Mass Spectrometry

Operations:

OP-37 Mass Spectrometry
OP-6   Heating
OP-8   Temperature Monitoring
OP-12 Vacuum Filtration
OP-21 Drying Solids
OP-23 Recrystallization
OP-24 Sublimation
OP-28 Melting Point

Before You Begin

1 Read the experiment and OP-37, read or review the operations as necessary, and write an experimental plan.
2 Calculate the mass and volume of 10.0 mmol of acetophenone and the theoretical yield of 2-phenylindole.

Scenario

The indole family of heterocyclic amines has a shady reputation because many of its members, such as psilocybin and LSD, have mind-altering properties that have made them popular recreational drugs. But other indoles such as tryptophan (an assential amino acid) and serotonin (a brain-regulating chemical) are necessary for normal body functioning. The Third Millenium Pharmaceutical Company is interested in developing new synthetic indole derivatives for use as legitimate drugs. To better focus their search for promising drugs, they are conducting a preliminary study of the effect of ring substitution on physiological activity. Virtually all physiologically active natural indole derivatives have a substitutent at the #3 position and none at the #2 position of the indole ring system, so the physiological effects of 2-substituted indoles have not been thoroughly explored. This may be because they have no significant physiological effects, but before coming to that conclusion the drug development division at Third Millennium wants to obtain and test a representative sample of 2-substituted indoles whose structures are known with certainty. Your assignment is to prepare a sample of 2-phenylindole and authenticate its structure by obtaining its mass spectrum.

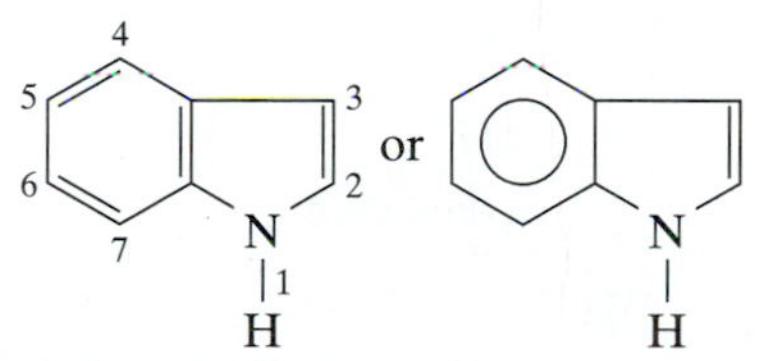

indole

Applying Scientific Methodology

After reading the experiment you should be able to predict some features of the mass spectrum of 2-phenylindole, including the relative intensities of the $M + 1$ and $M + 2$ peaks. Such predictions can be incorporated into a hypothesis, which will be tested when you obtain its mass spectrum.

Of Toads and Toadstools

The indoles constitute a large family of natural and synthetic compounds with extraordinary properties and functions. Indoles as a group display an ambivalent nature, as illustrated by the parent compound; indole has an odor of fine jasmine in dilute solutions but smells like feces when undiluted. 3-Methylindole is a major product of the digestive putrefaction of proteins and is mainly responsible for the repulsive odor of feces, giving rise to its common name, skatole—from scat, meaning animal droppings. However, skatole is also found in cabbage sprouts, tea, and even lilies!

In Mexico during the sixteenth century, Spanish conquistadors observed the Aztecs using little brown mushrooms called *teonanacatl* ("flesh of the gods") in their religious ceremonies. According to the Spanish friar Bernardino de Sahagun, "They ate these little mushrooms with honey, and when they began to be excited by them, they began to dance, some singing, others weeping.... Some saw themselves dying in a vision and wept; others saw themselves being eaten by a wild beast; others imagined that they were capturing prisoners in battle, that they were rich, that they possessed many slaves, that they had committed adultery and were to have their heads crushed for the offense." Naturally the Spanish friars disapproved of these ceremonies and banned them. This only led to the formation of cults that consumed the mushrooms in secret, until ethnomycologist R. Gordon Wasson revealed the existence of such practices in the 1930s. The little brown mushrooms are from several different species of the genus *Psilocybe*. Their hallucinogenic constituents include psilocin and psilocybin, both of which are derivatives of tryptamine, the parent compound of many physiologically active indoles. A similar tryptamine derivative, bufotenin, is the active principle of cohoba snuff, which is inhaled by certain Indians of South America and the Caribbean area because it produces hallucinations and intoxication. Bufotenin has also been isolated from toads of the genus *Bufo* and "toadstools" such as the poisonous mushroom *Amanita porphyria*.

skatole

magic mushrooms

tryptamine

psilocybin

psilocin

bufotenin

Other indole derivatives are more beneficial, such as the essential amino acid tryptophan, the plant growth hormone 3-indoleacetic acid (also called heteroauxin), the beautiful dye indigo, and the bioregulator serotonin. Tryptophan is important as a source of serotonin and the B-vitamin nicotinamide, both of which are formed during its metabolism. Because the human body cannot biosynthesize aromatic compounds, tryptophan and the other aromatic amino acids (phenylalanine and tyrosine) must be obtained from food. 3-Indoleacetic acid promotes the enlargement of plant cells and is the principal natural growth regulator for many plants. Indigo was one of the first natural dyes to be prepared synthetically; its structure was determined in 1883 by German organic chemist Adolph von Baeyer, after 18 years of research. Although the function of serotonin in the human body is not fully understood, it appears to play an important role in mental processes. Some researchers see it as a mind stabilizer that helps to preserve sanity. The resemblance between serotonin and such mind-altering drugs as bufotenin is striking; some of the psychological activity of these drugs may arise from their interference with the action of serotonin.

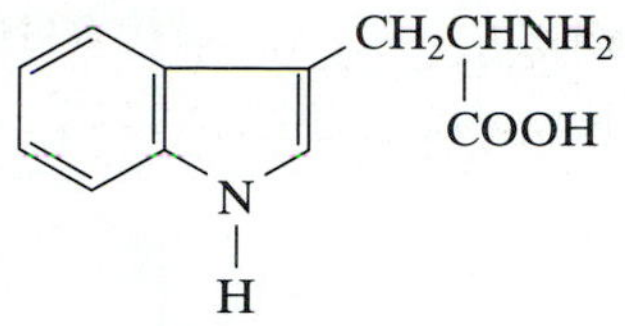

Understanding the Experiment

The preparation of 2-phenylindole by the Fischer indole synthesis involves the acid-catalyzed cyclization of a phenylhydrazone with the loss of a molecule of ammonia.

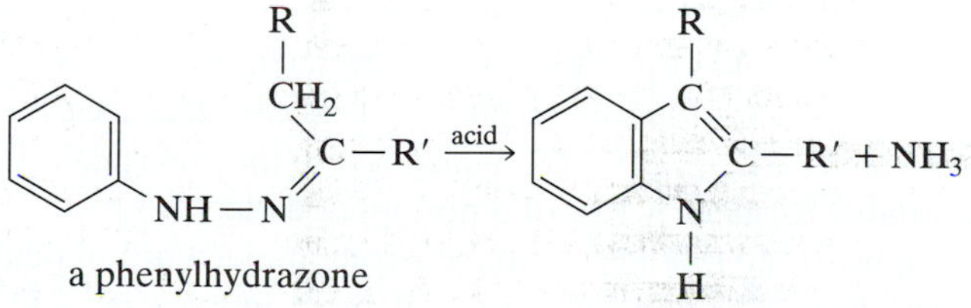

a phenylhydrazone

The phenylhydrazone can be prepared by combining phenylhydrazine (or a substituted phenylhydrazine) with an aldehyde or ketone of the structure RCH_2COR', where the R groups can be alkyl, aryl, or hydrogen. The generally accepted mechanism for this indolization reaction, proposed by Robinson and Robinson in 1918, includes these steps, as shown on the next page.

1 A tautomeric shift of a proton from carbon to the β-nitrogen
2 Protonation of that nitrogen
3 A concerted electron shift that forms a C—C bond to the ring and breaks the N—N bond
4 Another tautomeric shift of a proton from carbon to nitrogen
5 Nucleophilic attack by nitrogen on a doubly bonded carbon atom, followed by the loss of a proton
6 Loss of ammonia to yield an aromatic pyrrole ring

tryptophan

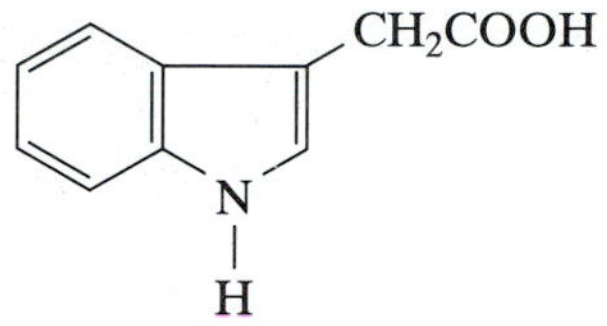

3-indoleacetic acid

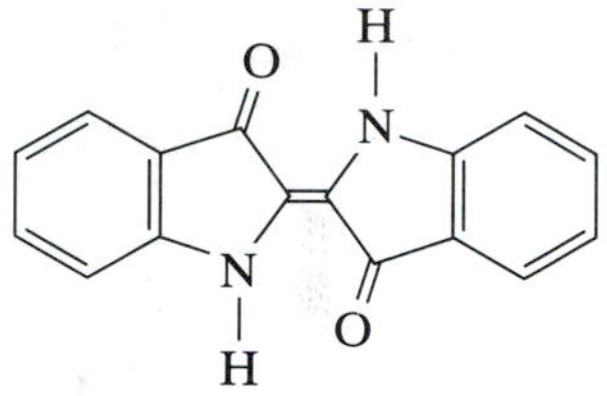

indigo

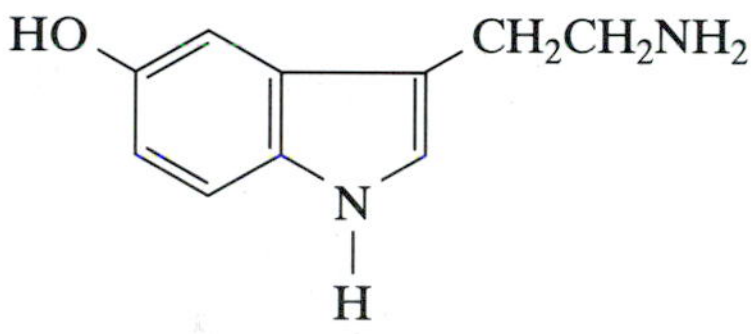

serotonin

Robinson mechanism

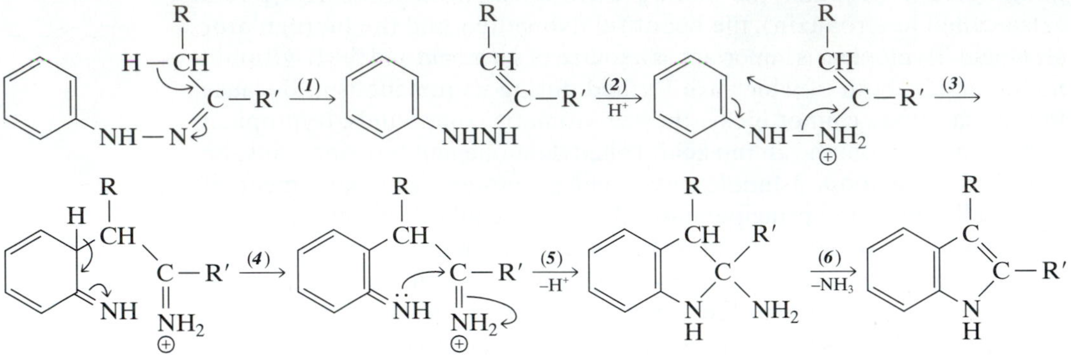

In this experiment you will synthesize 2-phenylindole by first converting acetophenone to acetophenone phenylhydrazone, then heating acetophenone phenylhydrazone with the acid catalyst polyphosphoric acid. Because acetophenone phenylhydrazone is heat and light sensitive, you should dry it at room temperature and store it in the dark. Its reaction in polyphosphoric acid is exothermic and requires only a few minutes to go to completion, after which the reaction mixture is poured into water to dissolve the acid catalyst and precipitate crude 2-phenylindole. You can purify 2-phenylindole by recrystallization from a mixed solvent—either ethanol/water or toluene/hexane—followed by vacuum sublimation. If you use toluene/hexane, the product should be thoroughly dried before recrystallization. Recrystallization from ethanol/water is complicated by the fact that the crystals dissolve quite slowly in ethanol; it is advisable to use a reflux condenser to keep the solvent from boiling away. About 25 mL of absolute ethanol should be sufficient to dissolve the product.

Read OP-37 if you are not familiar with the principles and terminology of mass spectrometry.

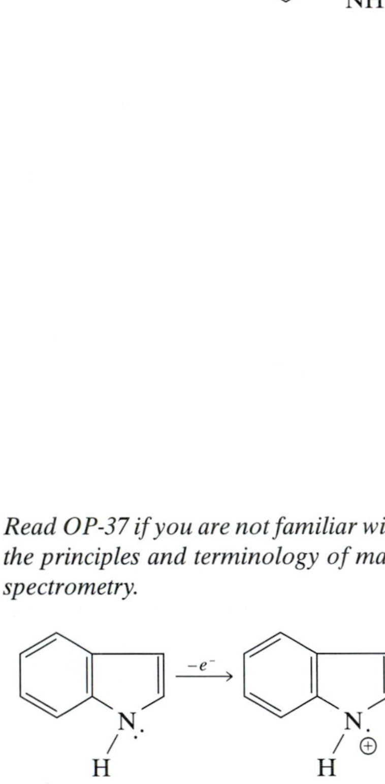

formation of the molecular ion

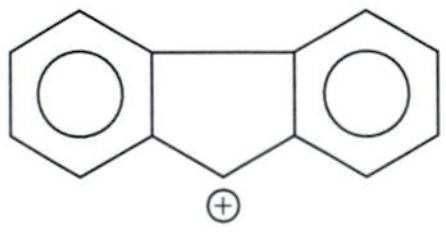

fluorenyl cation

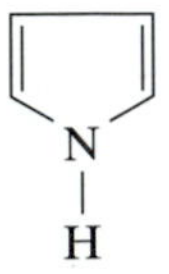

pyrrole

The mass spectrum of indole (Figure 51.1) is characterized by an intense molecular ion peak (M^+, $m/e = 117$), which is also the base peak. The molecular ion is probably formed by loss of a nonbonded electron from the nitrogen atom. Most substituted indoles also have strong molecular ion peaks. Neutral fragments lost from the molecular ions of most indoles include HCN and CH_2N, forming ions with m/e values of M - 27 and M - 28, respectively. The loss of CH_2N from 1-, 2-, and 3-phenylindoles leaves a $C_{13}H_9^+$ ion, which could be a fluorenyl cation. If so, the molecular ion must undergo considerable "scrambling" (intramolecular rearrangement) before it ejects CH_2N. Phenylindoles may lose the phenyl-substituted analogs of HCN (C_6H_5CN) and CH_2N (C_6H_5CHN) to form additional daughter ions. Because the molecular ion peak of 2-phenylindole is quite strong, its M + 1 and M + 2 peaks are also prominent, making it easy to compare their intensities with the expected values. The mass spectra of some indoles also show "half-mass" peaks that result from the formation of ions that have a 2+ charge. For example, the doubly charged molecular ion of indole itself gives an M^{++} peak with an m/e value of 58.5.

The infrared spectra of heteroaromatic compounds (see "Other Things You Can Do") are similar to those of analogous aromatic compounds, with some additional bands arising from the heteroatoms. Thus pyrroles and indoles, like other aromatic compounds, show ring-stretchig bands in the 1650–1300 cm^{-1} region and C—H out-of-plane bending bands in the 910– 665 cm^{-1} region. Most pyrrole rings are distinguished by a characteristically strong, broad C—H bending band near 740 cm^{-1}. Heteroaromatic amines with N—H bonds absorb in the 3500–3220 cm^{-1} region. The N—H stretching band of indoles occurs near 3450 cm^{-1}.

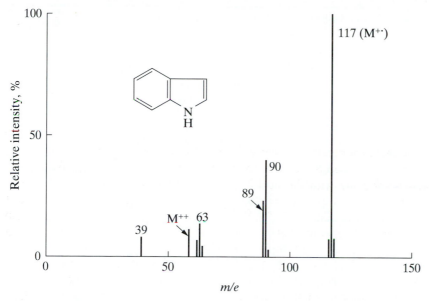

Figure 51.1 Mass spectrum of indole (Reproduced from *Mass Spectrometry of Heterocyclic Compounds*, by Q. N. Porter and J. Baldas. Copyright © 1971 by John Wiley & Sons, Inc. Reprinted by permission of John Wiley & Sons, Inc.)

Reactions and Properties

A.

phenylhydrazine   acetophenone

acetophenone
phenylhydrazone

B.

2-phenylindole

Table 51.1 Physical properties

	M.W.	m.p.	b.p.	d
phenylhydrazine	108.15	20	243 (115^{10})	1.099
acetophenone	120.2	20.5	202	1.028
acetophenone phenylhydrazone	210.35	106		
2-phenylindole	193.25	189	250^{10}	

Note: m.p., b.p. are in °C, density is in g/mL

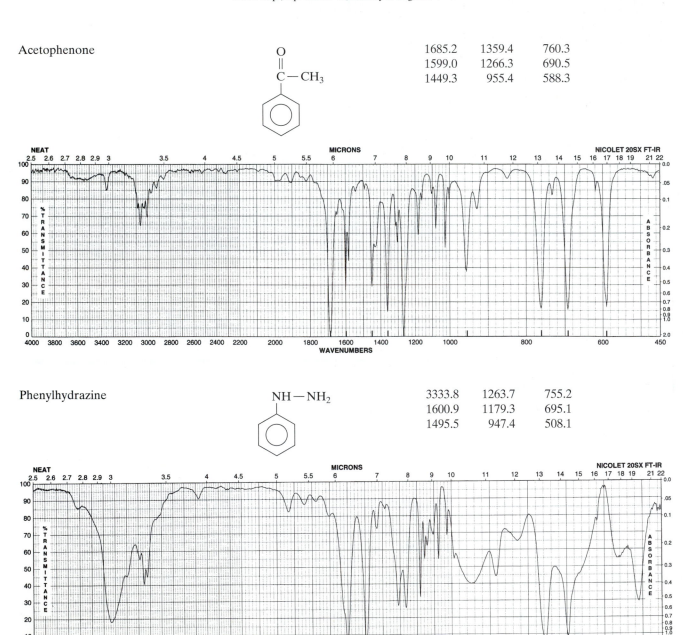

Acetophenone

1685.2	1359.4	760.3
1599.0	1266.3	690.5
1449.3	955.4	588.3

Phenylhydrazine

NH—NH$_2$

3333.8	1263.7	755.2
1600.9	1179.3	695.1
1495.5	947.4	508.1

Figure 51.2 IR spectra of the starting materials (Reproduced from *The Aldrich Library of FT-IR Spectra, Edition II*, with the permission of the Aldrich Chemical Company.)

Directions

Acetophenone phenylhydrazone should be used soon after it is prepared, or else stored in a cool, dark place.

A. *Preparation of Acetophenone Phenylhydrazone*

> **Phenylhydrazine is corrosive and very toxic if inhaled, ingested, or absorbed through the skin; contact may cause painful skin eruptions in sensitive individuals. It is also a suspected carcinogen. Use gloves, dispense under a hood, avoid contact, and do not breathe vapors.**
>
> **Acetic acid causes chemical burns that can seriously damage skin and eyes; its vapors are highly irritating to the eyes and respiratory tract. Dispense under a hood, avoid contact, and do not breathe vapors.**

In a test tube or small Erlenmeyer flask, dissolve 10.0 mmol of acetophenone in 5 mL of 95% ethanol. Add 1.0 mL (~10 mmol) of freshly distilled phenylhydrazine followed by 2 drops of glacial acetic acid, then swirl to mix. Add a boiling chip and heat the reactants gently over a steam bath or in a boiling water bath for 15 minutes, taking care not to boil away the ethanol (add more if necessary). Cool the mixture in ice and collect the acetophenone phenylhydrazone by vacuum filtration [OP-12] when crystallization is complete. If necessary, scratch the sides of the test tube to promote crystallization. Wash the product on the filter with 1 *M* hydrochloric acid followed by 2 mL of ice-cold 95% ethanol, and let it air dry. Wearing protective gloves, blot the product as dry as possible between large filter papers.

B. *Preparation of 2-Phenylindole*

> **Polyphosphoric acid can cause serious burns, particularly to the eyes, so avoid contact with skin and eyes.**
>
> **Some indoles are suspected of causing cancer; minimize your contact with the product.**

Reaction. Measure about 10 mL of polyphosphoric acid into a dry 20-cm test tube. (Polyphosphoric acid can be warmed gently over a steam bath to make it easier to pour.) Clamp the tube so that the portion containing the liquid extends inside a steam bath [OP-6], removing enough rings to let the test tube just slip through. Clamp a thermometer in the liquid [OP-8], near one wall of the test tube but not touching it, to allow room for manual stirring (don't use the thermometer as a stirring rod!). Warm the polyphosphoric acid to 50°C, then mix in the acetophenone phenylhydrazone with a stirring rod and heat the reactants, with stirring. When the temperature of the reaction mixture rises to 100°C, turn off the steam. Have a cold water bath handy to keep the temperature below 125°C. When the temperature begins to drop, turn on the steam and heat the mixture for 10–15 minutes with occasional stirring.

Separation. Cautiously pour the warm reaction mixture into 30 mL of ice water, using more water for the transfer, and stir until all the polyphosphoric

Safety Notes

phenylhydrazine   acetophenonone

Take Care! Wear gloves, avoid contact with phenylhydrazine and acetic acid, do not breathe their vapors

Impure phenylhydrazine is dark reddish-brown and will give an inferior product.

Stop and Think: What is the purpose of the 1 *M* HCl?

Waste Disposal: Pour the filtrate down the drain.

Safety Notes

Take Care! Avoid contact with the acid.

Stop and Think: What evidence is there that a chemical reaction is occurring?

Waste Disposal: Pour the filtrate down the drain.

Stop and Think: Should the product be more soluble in toluene or hexane?

Waste Disposal: If you used toluene/hexane, put the filtrate in the designated solvent recovery container. Otherwise pour it down the drain.

acid has dissolved. Collect the crude 2-phenylindole by vacuum filtration [OP-12], washing it on the filter with cold methanol, and let it air dry.

Purification and Analysis. Recrystallize the product from either ethanol/water or toluene/hexane, following the procedure for recrystallization from mixed solvents [OP-23], and using about 0.1 g of decolorizing carbon (preferably pelletized Norit). Dry the 2-phenylindole at room temperature [OP-21] and weigh it. Purify the product further by vacuum sublimation [OP-24]. Your instructor may suggest that you sublime only a small amount of the product for analysis. Measure the mass (if requested) and melting point [OP-28] of the sublimed 2-phenylindole. Record a mass spectrum [OP-37] of the product or obtain one from your instructor. Turn in the product to your instructor.

Report. Your report should include a statement of the problem and an account of how you applied scientific methodology to solve it. Calculate the percent yield of 2-phenylindole after recrystallization. On the mass spectrum, locate the molecular ion peak, the M + 1 and M + 2 peaks, and all other peaks whose intensities are 10% or more of the base peak intensity. Derive the molecular formula of 2-phenylindole from its structure, use the formula to calculate the expected intensities of the M + 1 and M + 2 peaks, and compare the experimental and theoretical values. Characterize as many of the other peaks as you can, in each instance giving the formula of the species lost from the molecular ion and the formula of the resulting daughter ion.

Exercises

1 (a) Rewrite the general mechanism given in the "Methodology" section so that it applies to the synthesis of 2-phenylindole, and sketch the activated complexes for steps **3** and **5**. (b) Write a mechanism for the reaction that forms acetophenone phenylhydrazone, showing the role of acetic acid.

2 Outline syntheses of 1-phenylindole and 3-phenylindole from appropriate starting materials.

3 Describe and explain the possible effect on your results of the following experimental errors or variations. (a) You rinsed the test tube used for part **A** with acetone and failed to remove all the acetone. (b) You used benzophenone rather than acetophenone in part **A**. (c) You used phenylhydrazine hydrochloride ($C_6H_4NH_2NH_3^+Cl^-$) rather than phenylhydrazine in part **A**.

4 Following the format in Appendix V, construct a flow diagram for the synthesis of 2-phenylindole.

5 Draw structures for the indoles that would be obtained from the phenylhydrazones of the following carbonyl compounds: (a) acetone, (b) phenylacetaldehyde (2-phenylethanal), (c) cyclopentanone, (d) pyruvic acid ($CH_3COCOOH$), and (e) camphor.

6 When the phenylhydrazone of isobutyrophenone (2-methyl-1-phenyl-1-propanone) is heated to 150°C with polyphosphoric acid, it

yields a mixture of the two compounds shown. Propose a mechanism that explains the formation of each product.

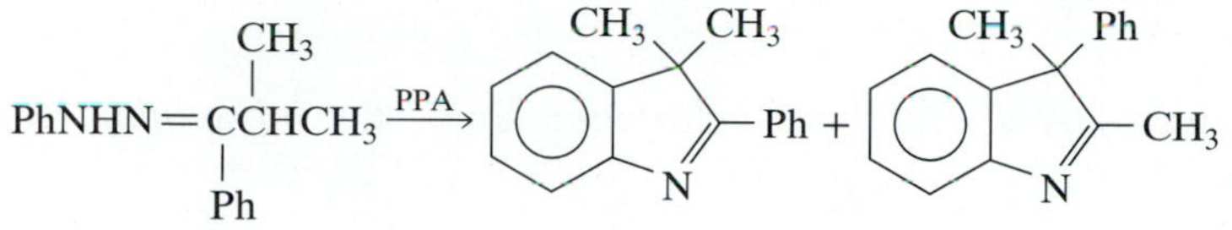

7 The total synthesis of strychnine, accomplished by Robert B. Woodward in 1954, began with the preparation of 2-(3,4-dimethoxyphenyl) indole. Show how this compound can be prepared starting with catechol (1,2-dihydroxybenzene).

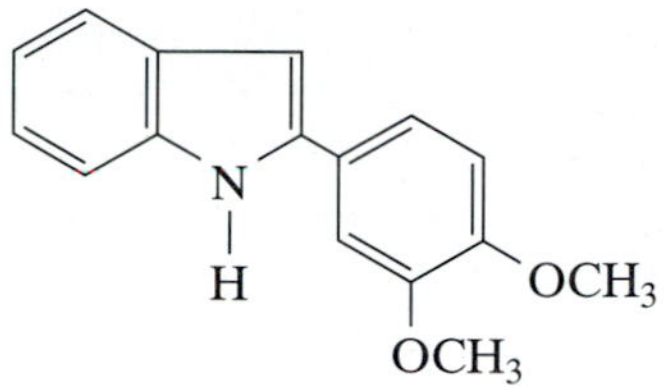

2-(3,4-dimethoxyphenyl)
indole

Other Things You Can Do

(Starred projects require your instructor's permission.)

*1 Record the infrared spectrum of 2-phenylindole. Interpret this spectrum as completely as you can, compare it with the spectra of the starting materials shown in Figure 51.2, and describe the evidence suggesting that the expected reaction has taken place.

*2 Prepare some indigo and use it to dye cloth in Minilab 42.

3 Starting with sources listed in the Bibliography, write a research paper about natural and synthetic plant growth hormones (auxins), describing their sources and chemical structures and giving examples of their applications.

EXPERIMENT 52

Nucleophilic Strength and Reactivity in S$_N$Ar Reactions

Reactions of Aryl Halides. Nucleophilic Aromatic Substitution. Reaction Kinetics.

Operations:

> OP-9   Mixing
> OP-36 Ultraviolet-Visible Spectrometry

Before You Begin

Read the experiment, read part b of OP-36 on colorimetry, read or review the other operations as necessary, and write an experimental plan.

Scenario

Organic chemists rely on handbooks and other sources of compiled chemical information to help them design their experiments. For example, a nucleophile's *nucleophilic constant*, which measures the nucleophile's ability to displace a leaving group from a specific kind of substrate, could help a chemist estimate how long it would take to carry out a given nucleophilic substitution reaction. Minnie Colfax, a chemical information specialist for the Fulcourt Press (see Experiment 47), is trying to compile a list of accurate nucleophilic constants, but some of these constants have not been reported in the chemical literature, while others were measured many years ago by questionable methods and need to be re-evaluated. Your project group's mission is to evaluate the nucleophilic constants for two heterocyclic amines, morpholine and piperidine, in an S$_N$Ar reaction, and to determine which is the better nucleophile.

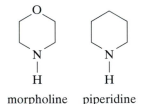

morpholine   piperidine

Applying Scientific Methodology

After reading the experiment, you should formulate a hypothesis predicting which amine you expect to be the better nucleophile. Your hypothesis will be tested when you compare the relative values of their nucleophilic constants.

Nucleophilicity and the S$_N$Ar Reaction

Nucleophilic aromatic substitution can occur (1) through an S$_N$1 mechanism, as in some substitution reactions of aryldiazonium salts; (2) by an elimination-addition mechanism involving an aryne intermediate; or (3) by a bimolecular addition-elimination route called the S$_N$Ar reaction. In an S$_N$Ar reaction, the nucleophile attacks an activated aromatic ring to form an intermediate complex, which then loses a leaving group to yield the product.

The nature of the intermediate complex in the S$_N$Ar reaction has long been the subject of speculation. The first evidence of its structure was obtained in 1898 when Jackson and Boos mixed picryl chloride (2,4,6-tri-nitrochlorobenzene) with sodium methoxide in methanol and isolated a red salt, which was converted by ethanol to another red salt. J. Meisenheimer prepared the second salt by two different methods, treating either 2,4,6-trinitroanisole with potassium ethoxide or 2,4,6-trinitrophenetole with potassium methoxide. He was then able to assign the salt the structure labeled **1** in the following reaction scheme.

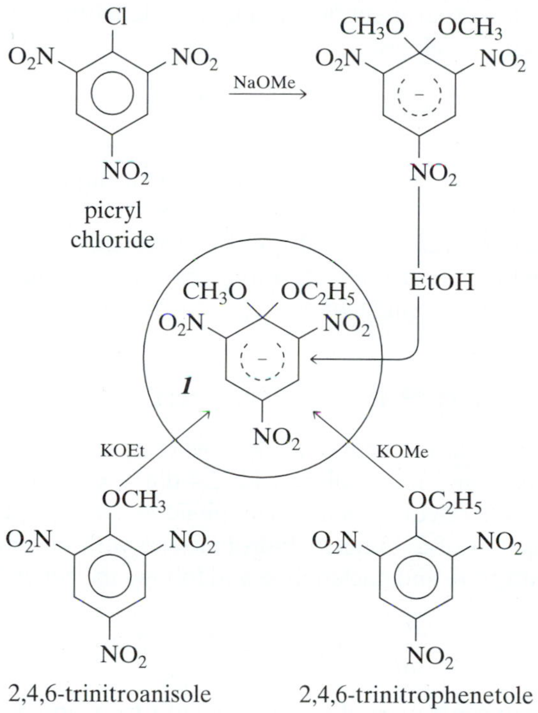

2,4,6-trinitroanisole   2,4,6-trinitrophenetole

Negatively charged aromatic species analogous to **1** are called Meisenheimer complexes, and many such complexes have been prepared and characterized. A large body of experimental evidence indicates that the intermediates in S$_N$Ar reactions are Meisenheimer complexes, explaining why only aromatic compounds that are activated by electron-withdrawing substituents, such as nitro groups, undergo S$_N$Ar reactions easily. Such substituents remove excess electron density from the aromatic ring, stabilizing the intermediate complex and thus facilitating its formation.

The mechanism of most S$_N$Ar reactions appears to be a simple two-step process involving the initial addition of the nucleophile to form a Meisenheimer complex, followed by the departure of the leaving group. If the first step of the reaction is rate limiting, the nucleophilic strength of the

General S$_N$Ar mechanism

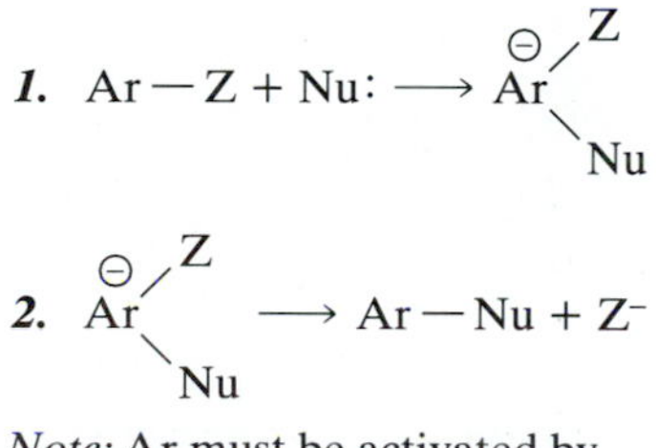

1. $Ar-Z + Nu: \longrightarrow Ar\overset{\ominus}{\underset{Nu}{\diagup Z}}$

2. $Ar\overset{\ominus}{\underset{Nu}{\diagup Z}} \longrightarrow Ar-Nu + Z^-$

Note: Ar must be activated by electron-withdrawing groups; Z is the leaving group and Nu: the nucleophile.

Table 52.1 Nucleophilic constants for various nucleophiles

Nucleophile	n
CH_3OH	0.00
F^-	2.7
Cl^-	4.37
pyridine	5.23
NH_3	5.50
aniline	5.70
Br^-	5.79
CH_3O^-	6.29
$(CH_3CH_2)_3N$	6.66
$(CH_3CH_2)_2NH$	7.0
pyrrolidine	7.23
piperidine	7.30
I^-	7.42

Note: n values are measured relative to methanol, with methyl iodide as the substrate.

reactant—its ability to donate its electron pair to the substrate and form a sigma bond—should affect the reaction rate. The *Swain–Scott equation,*

$$\log \frac{k}{k_0} = ns \tag{1}$$

relates reaction rates to the strength of the nucleophile (n) and the sensitivity of the substrate to nucleophilic substitution (s). It has been applied widely to nucleophilic substitution reactions with aliphatic substrates, and some values of the nucleophilic constant n for reactions of various nucleophiles with methyl iodide are given in Table 52.1. Attempts to use such correlations for S_NAr reactions have met with less success because changing the substrate or solvent often changes the order of nucleophilic strength. However, an equation of this kind can be useful in comparing the nucleophilic strengths of various reactants with reference to the same class of substrates. For this experiment, we define a nucleophilic constant (n_{Ar}) for the S_NAr reaction as follows:

$$n_{Ar} = \frac{1}{s} \log \frac{k}{k_0} \tag{2}$$

The reaction of 2,4-dinitrochlorobenzene with ammonia in absolute ethanol, for which the rate constant (k_0) is 4.0×10^{-6} L mol^{-1} s^{-1} at 25°C, is used as the reference reaction for Equation **2**. For this substrate and nucleophile, s is 1 and n_{Ar} is zero by definition. The rate constant k is for a reaction involving the nucleophile whose n_{Ar} value is being determined.

Understanding the Experiment

In this experiment, you and your coworkers will measure the rates of the reactions of piperidine and morpholine with 2,4-dinitrochlorobenzene and use the rate constants to calculate their nucleophilic constants, as defined in equation **2**. The reaction of 2,4-dinitrochlorobenzene with an amine is second-order in both substrate and nucleophile and follows the general rate equation

$$\frac{dx}{dt} = k(S_0 - x)(N_0 - 2x) \tag{3}$$

where x is the concentration of the product at time t, and S_0 and N_0 are the initial concentrations of substrate and nucleophile, respectively. The computations can be simplified considerably if the experiment is carried out with the initial concentration of nucleophile being just twice that of 2,4-dinitrochlorobenzene. Equation **3** then becomes $dx/dt = 2k(S_0 - x)^2$, and the integrated rate equation is

$$\frac{1}{(S_0 - x)} = 2kt + \frac{1}{S_0} \tag{4}$$

By measuring the concentration of the product, x, at regular intervals during the reaction, you can calculate the term on the left side of Equation **4**, which, when plotted versus time, should yield a straight line with slope $2k$.

The products of the S_NAr reactions are yellow-orange and absorb strongly in the visible region around 380 nm, so the concentration of each product will be determined indirectly by measuring the absorbance of 380-nm light by aliquots that are removed from the reaction mixture at various times. Since the absorbances are determined at a single wavelength, you can use a nonrecording spectrophotometer. Each aliquot must first be quenched by adding dilute acid to stop the reaction; this is done so that the product concentration will remain constant until you are ready to take the absorbance readings. The concentration term in Equation **4**, $S_0 - x$, can be shown to be proportional to $A_\infty - A$, where A_∞ is the absorbance when the reaction is 100% complete, and A is the absorbance at time t. Therefore

$$\frac{1}{(S_0 - x)} = \frac{q}{(A_\infty - A)} \qquad (5)$$

where q is a proportionality constant. The "infinity" value of the absorbance, A_∞, can be obtained by warming the reaction mixture to complete the reaction and measuring the absorbance of the resulting solution. The value of q can then be calculated from the relationship $q = A_\infty/S_0$. Note that S_0 is the initial substrate concentration in the *reaction mixture*, not in the stock solution you will use to prepare the reaction mixtures.

Because of their different reaction rates, the amines will be used in different initial concentrations so that both reactions will be about half complete after 30 minutes. The absorbance values of the products at 380 nm are high enough that you will have to dilute your solutions to obtain readings in a convenient range.

Reaction and Properties

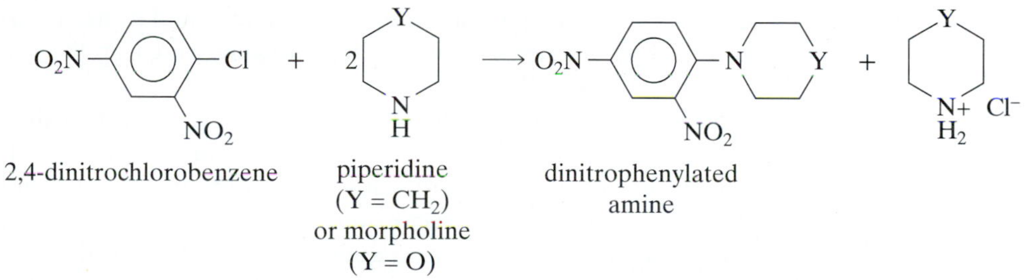

2,4-dinitrochlorobenzene   piperidine   dinitrophenylated
                                 (Y = CH$_2$)   amine
                              or morpholine
                                 (Y = O)

The extra mole of amine combines with HCl liberated during the reaction.

Table 52.2 Physical properties

	M.W.	m.p.	b.p.	d
2,4-dinitrochlorobenzene	202.6	53	315	
morpholine	87.1	− 5	128	1.000
piperidine	85.2	− 9	106	0.861

Directions

All glassware must be clean and dry. If equipment is limited, you may be asked to work in pairs or larger groups. Each student or group should have a timer or a watch that measures in seconds. Stock solutions should be dispensed from burets or bottles with dispensing pumps.

> **2,4-Dinitrochlorobenzene is poisonous and skin contact may cause unpleasant and persistent dermatitis. Wear gloves and avoid contact with the 2,4-dinitrochlorobenzene solution.**
>
> **Morpholine and piperidine are toxic and corrosive, capable of causing severe damage to the skin, eyes, and respiratory system. Avoid contact with their solutions.**
>
> **Ethanol is very flammable, so keep ethanolic solutions away from flames or hot surfaces.**

Safety Notes

2,4-dinitro-chlorobenzene

morpholine   piperidine

Take Care! Wear gloves and avoid contact with the stock solutions.

A. *Kinetic Run A— Reaction of 2,4-Dinitrochlorobenzene with Piperidine* Label seven clean, dry 4-dram screwcap vials (or 15-cm test tubes) from A1 to A7. Using a 10-mL volumetric pipet, accurately measure 10.0 mL of the quenching solution (0.5 M sulfuric acid in 50% ethanol) into each vial and cap the vials. (**Take Care!** Do not pipet by mouth!) Measure about 20 mL of absolute ethanol into a 25-mL volumetric flask, then accurately measure 2.00 mL of the 0.20 M stock solution of 2,4-dinitrochlorobenzene in ethanol and add it to the flask. Have your stopwatch or timer ready, then measure 2.00 mL of the 0.40 M piperidine/ethanol stock solution into the volumetric flask, and start the timer (or record the starting time) when about half the solution has drained from the pipet. Without delay, fill the flask to the mark with absolute ethanol, stopper and shake it, and pour the contents into a 50-mL Erlenmeyer flask (the reaction flask). Use a clean, dry 1-mL volumetric pipet to withdraw a 1.00-mL aliquot of the reaction mixture, and transfer it to vial A1 as you record the quenching time to the nearest second. Cap the vial and seal the reaction flask with Parafilm.

Stop and Think: Why does this stop the reaction?

Stop and Think: What evidence is there that a reaction is occurring?

Stir or occasionally swirl [OP-7] the contents of the reaction flask during the reaction period. Approximately every 5 minutes, rinse the 1-mL volumetric pipet with the reaction mixture, withdraw a 1.00-mL aliquot, and transfer it to the next quenching vial, recording the quenching time. Repeat this process until you have withdrawn and quenched a total of six aliquots. Then warm the reaction flask (still sealed with Parafilm) in a 50°C water bath for at least 2 hours, or let it stand for at least 48 hours at room temperature, to bring the reaction to completion. Pipet a final 1.00-mL aliquot into vial A7 for the A_∞ solution.

Waste Disposal: Dispose of the remaining reaction mixture as directed by your instructor.

Take Care! Do not pipet by mouth!

Take Care! Wear gloves and avoid contact with the stock solutions.

B. *Kinetic Run B— Reaction of 2,4-Dinitrochlorobenzene with Morpholine* Label seven clean, dry 4-dram screwcap vials (or 15-cm test tubes) from B1 to B7. Using a 10-mL volumetric pipet, accurately measure 10.0 mL of the quenching solution (0.5 M sulfuric acid in 50% ethanol) into each vial and cap the vials. Measure about 17 mL of absolute ethanol into a 25-mL volumetric flask and accurately measure 5.00 mL of the 0.20 M stock solution of 2,4-dinitrochlorobenzene in ethanol and add it to the flask. Have your stopwatch or timer ready, then measure 2.00 mL of the 1.0 *M* morpholine/ethanol stock solution into the volumetric flask, and start the timer (or record the starting time) when about half the solution has drained from the pipet. Without delay, fill the flask to the mark with absolute ethanol, stopper and shake it, and pour the contents into a 50-mL Erlenmeyer flask (the reaction flask). Use a clean, dry 1-mL volumetric pipet to withdraw a 1.00-mL aliquot of the reaction mixture and transfer it to vial B1 as you record the quenching time to the nearest second. Cap the vial and seal the reaction flask with Parafilm.

Stir or occasionally swirl [OP-7] the contents of the reaction flask during the reaction period. Approximately every 5 minutes, rinse the 1-mL volumetric pipet with the reaction mixture, withdraw a 1.00-mL aliquot, and transfer it to the next quenching vial, recording the quenching time. Repeat this process until you have withdrawn and quenched a total of six aliquots. Then warm the reaction flask (still sealed with Parafilm) in a 50°C water bath for at least 2 hours, or let it stand for at least 48 hours at room temperature, to bring the reaction to completion. Pipet a final 1.00-mL aliquot into vial B7 for the A_∞ solution.

Stop and Think: What evidence is there that a reaction is occurring?

Waste Disposal: Dispose of the remaining reaction mixture as directed by your instructor.

C. *Absorbance Measurements*

Obtain as many clean, dry 4-dram screwcap vials (or 15-cm test tubes) as you have solutions to analyze and number them to correspond to the quenched solutions. Using a volumetric pipet, accurately measure 10.0 mL of 95% ethanol into each vial. Then accurately measure 0.40 mL of each quenched solution into the corresponding vial, swirl to mix the contents, and cap the vial.

Stop and Think: Why is this step necessary?

Zero the spectrophotometer [OP-36] at 380 nm and set the 100% transmittance control using a cuvette filled with 95% ethanol. Make sure the sample cuvette is positioned correctly, then record the transmittance of each solution from each vial. Use the same instrument to analyze all of your solutions, including the A_∞ solution.

Waste Disposal: Dispose of all solutions as directed by your instructor.

Report. Your report should include a statement of the problem and an account of how you applied scientific methodology to solve it. Calculate the absorbance of each solution, and use your data to compute $1/(S_0 - x)$ for each aliquot. For each amine, plot these values versus time and determine the slope of the line, or use a calculator (or computer) with a linear regression program to calculate the slope. Determine the second-order rate constant for each reaction and use it to calculate the n_{Ar} value for each amine. Arrange morpholine, piperidine, and ammonia ($n_{Ar} = 0$) in order according to their n_{Ar} values and try to explain their relative nucleophilicities.

Exercises

1   (a) Write a detailed mechanism for the reaction of piperidine with 2,4-dinitrochlorobenzene. (b) The rate constants for the reactions of piperidine with 2,4-dinitrochlorobenzene and with 2,4-dinitrobromobenzene are virtually identical. Identify the rate-determining step in your mechanism from (a) and explain your reasoning.

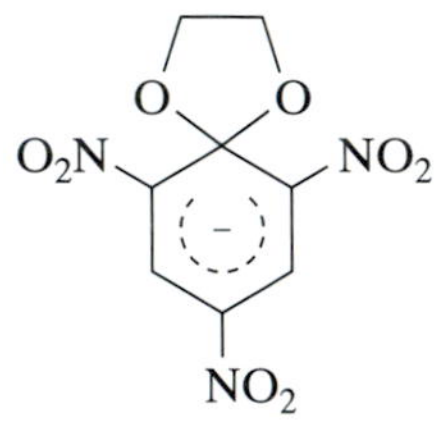

a spiro Meisenheimer complex

2 Explain why the reaction of 2,4-dinitrochlorobenzene with piperidine or morpholine is second order even though there are three reactant species in the overall equation for the reaction.

3 Describe and explain the possible effect on your results of the following experimental errors or variations. (a) The reaction flask you used in the kinetic runs was rinsed with water and not dried. (b) The stockroom was out of 2,4-dinitrochlorobenzene, so the lab assistant substituted chlorobenzene. (c) The student assistant who prepared the stock solutions couldn't find any piperdine so she used piperdine hydrochloride instead.

4 (a) Derive equation 5 using Beer's law and evaluate the constant q in terms of Beer's law parameters. (b) Derive Equation 4 from Equation 3.

5 A *spiro* species contains rings that have one carbon atom in common. Outline a synthesis of the spiro Meisenheimer complex shown, using 1-chloro-2,4,6-trinitrobenzene and any appropriate inorganic and organic reactants.

6 (a) Calculate the concentrations of the dinitrophenylamines in the solutions used for the spectrophotometry infinity reading. (b) If you know the path length of the cuvette you used, calculate the molar absorptivities of these products at 380 nm.

7 Explain why quenching the reaction mixture with H_2SO_4 stops the S_NAr reaction, giving equations for any reactions involved.

Other Things You Can Do

(Starred projects require your instructor's permission.)

*1 Measure the reaction rates for one of the amines at different temperatures (0°C, 20°C, and 40°C, for example). Then plot ln k versus $1/T$ to determine the activation energy of the reaction, based on the following form of the Arrhenius equation:

$$\ln k = -\frac{E_{act}}{RT} + \ln A \qquad (R = 8.31 \text{ J mol}^{-1}\text{K}^{-1})$$

*2 Carry out one of the reactions in other solvents, such as 2-propanol and aqueous ethanol, to measure the effect of solvent polarity on the rate constants. If you make up your own solutions, it is essential to wear protective gloves and avoid contact with 2,4-dinitrochlorobenzene and the amines.

3 Read the paper by Bunnett, Garbisch, and Pruitt, *J. Am. Chem. Soc.* **1957**, *79*, 385. Then tell what is meant by the "element effect" in reactions of 1-substituted 2,4-dinitrobenzenes, and explain how it was used to elucidate the mechanism of the S_NAr reaction.

Structure of an Unknown D-Hexose

Reactions of Monosaccharides. Preparation of Alditols. Reduction Reactions. Structure Determination.

Operations:

OP-6   Heating
OP-7   Cooling
OP-9   Mixing
OP-12 Vacuum Filtration
OP-21 Drying Solids
OP-23 Recrystallization
OP-28 Melting Point
OP-31 Optical Rotation

Before You Begin

1 Read the experiment, read or review the operations as necessary, and write an experimental plan.
2 Calculate the mass of 20.0 mmol of $C_6H_{12}O_6$ and the mass of 7.5 mmol of sodium borohydride ($NaBH_4$).

Scenario

S. A. Tucker, a food chemist working for the Global Food Research Foundation, is studying the chemical composition of foods from around the world. In Hunza, an isolated state in a high valley of the Hindu Kush mountains, Tucker obtained a yogurt made from yak's milk that is said to contribute to the exceptional longevity of the inhabitants. The yogurt contains a simple sugar that has the same molecular formula as glucose, $C_6H_{12}O_6$, but is only half as sweet. Your assignment is to determine the structure of this unknown D-hexose by using chemical methods.

Applying Scientific Methodology

After reading the experiment you may be able to venture an "educated guess" about the identity of the unknown D-hexose and express as a hypothesis. You should then be able to predict the results of each experimental procedure you will carry out, and see if your actual results agree with your predictions.

Sweet Molecules

One of the most apparent properties of the sugars is their sweetness. Fructose is the sweetest known sugar—1.8 times sweeter (in its crystalline form) than sucrose—but other substances are sweeter than any of the sugars.

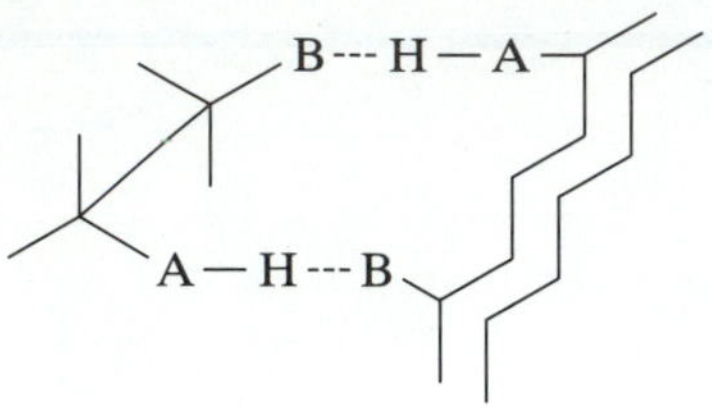

Interaction of a sweet molecule with the receptor site, according to the AH,B theory

Cyclamates such as sodium cyclohexylsulfamate are approximately 30 times sweeter than sucrose, saccharin is nearly 500 times sweeter, and 1-*n*-propoxy-2-amino-4-nitrobenzene (P-4000) is estimated to be 4000 times sweeter (see Figure 53.1).

A number of theories and models have been devised to explain the sweet taste sensation. According to the so-called AH,B theory, all sweet molecules contain an AH,B couple that is necessary to bind the compound to the taste bud receptor site. A and B are electronegative atoms such as oxygen or nitrogen (but sometimes halogen or even carbon) that must be 0.25–0.40 nm apart in order to interact (by hydrogen bonding) with the receptor site, which is assumed to possess a similar AH,B couple. In sugars, AH and B are assumed to be an OH group and the oxygen atom of an adjacent OH group, respectively. For effective interaction, the OH groups should be in a *gauche* conformation because in an *anti* conformation they would be too far apart to interact with the receptor site, and in a *syn* conformation they would tend to hydrogen-bond intramolecularly rather than with the receptor site.

Gauche conformation in a sugar molecule

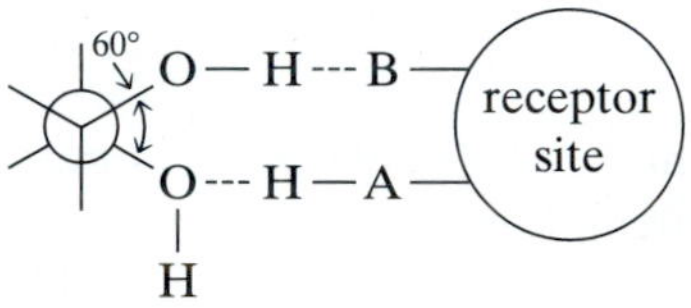

Proposed AH and B sites for various sweet molecules are illustrated in Figure 53.1.

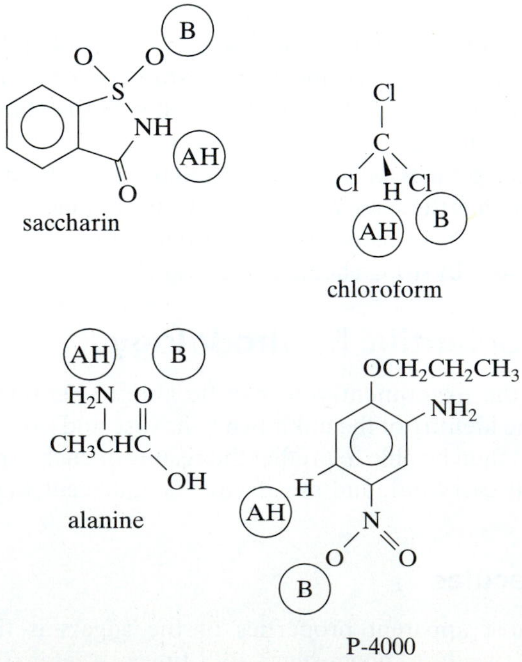

Figure 53.1 AH and B sites in sweet molecules

Another possible requirement for the sweet taste sensation is a hydrophobic or "greasy" site, X, on the sweet molecule. In the Keir tripartite model, this site is estimated to be about 0.35 nm from A and 0.55 nm from B in the triangular grouping illustrated.

X⋯⋯⋯⋯⋯⋯⋯⋯⋯⋯ 0.35 nm ⋯⋯⋯⋯ A

"greasy"
site

0.55 nm ⋯⋯⋯⋯⋯⋯ B

Keir tripartite model

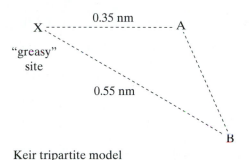

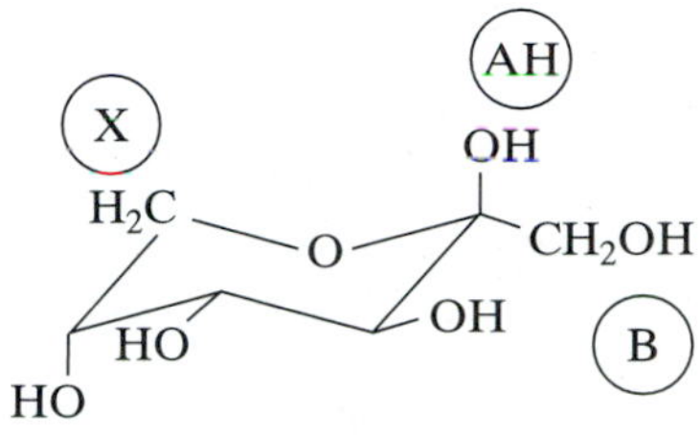

β-D-fructopyranose

In the crystalline form of fructose, β-D-fructopyranose, the AH, B, and X sites are considered to be the C-1 OH, the CH$_2$OH oxygen atom, and the ring methylene group, respectively. An even more complex model, the Tinti–Nofre model, postulates no fewer than eight sites, including the three of the Keir tripartite model. Not all of the sites are utilized by every sweet molecule, but it is assumed that the more sites a molecule occupies, the greater its potential for sweetness.

Unfortunately, models proposed to explain the sweet taste sensation tend to be so general that they predict sweetness for many compounds that are actually tasteless or bitter, or so specific that only a few related molecules can satisfy their requirements for sweetness. Clearly, scientists do not yet have a complete understanding of the origin of the sweet taste.

Understanding the Experiment

In this experiment you will determine the structure of an unknown D-hexose by (1) performing a simple chemical test that distinguishes aldohexoses from ketohexoses; (2) reducing the unknown with sodium borohydride and measuring the optical rotation of the product; and (3) converting the unknown to a phenylosazone whose melting point you will measure.

Both aldohexoses and ketohexoses are dehydrated in dilute acid to 5-(hydroxymethyl)-2-furaldehyde, which reacts with resorcinol to form a red condensation product.

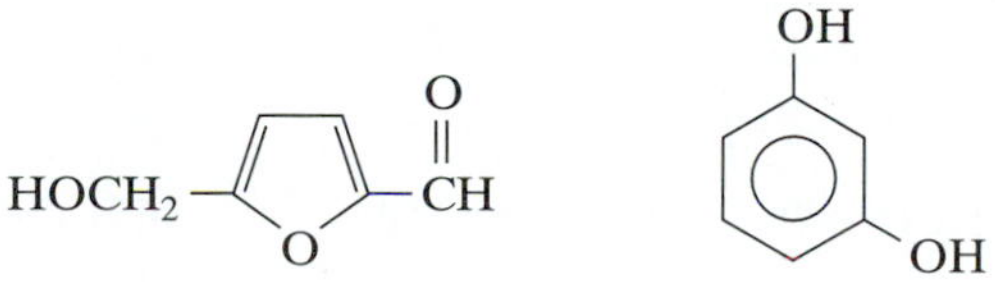

HOCH$_2$ —⟨ furan ⟩— CH=O   resorcinol (OH, OH on benzene ring)

5-(hydroxymethyl)-2-furaldehyde   resorcinol

Ketohexoses dehydrate more rapidly than aldohexoses, making it possible to differentiate them with Seliwanoff's reagent, a solution of resorcinol in dilute HCl.

Reduction of a monosaccharide with sodium borohydride converts it to an alditol by reducing its carbonyl group. D-Glucose is converted to the alditol D-glucitol. This alditol, having no plane of symmetry, is optically active.

D-glucose D-glucitol

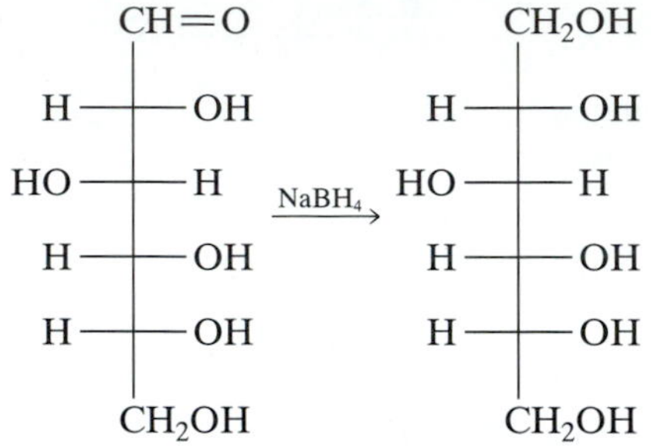

D-Allose is converted to D-allitol, which has a symmetry plane and is therefore optically inactive.

D-allose D-allitol

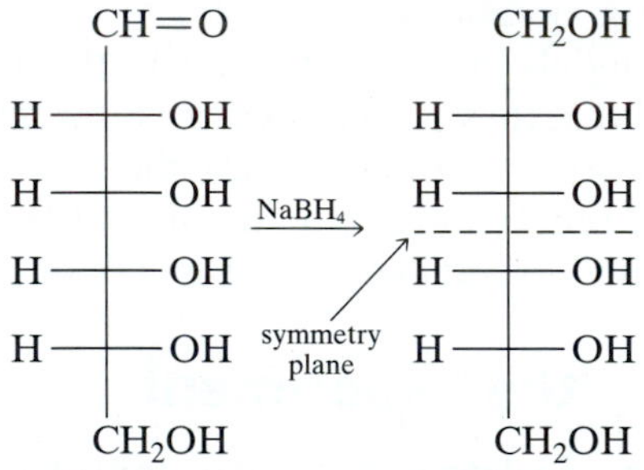

Thus the optical activity or inactivity of the alditol formed from reduction of a monosaccharide can yield structural information about the monosaccharide itself, thereby narrowing down the number of possible structures.

Converting monosaccharides to their phenylosazones in effect destroys any differences at the #1 and #2 carbon atoms of their molecules by changing both —CHOH—CHO and —CO—CH_2OH to the same structural unit:

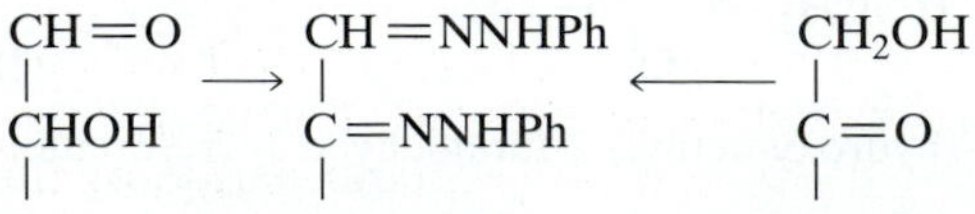

For example, D-glucose, its epimer D-mannose, and D-fructose are all converted to the same phenylosazone.

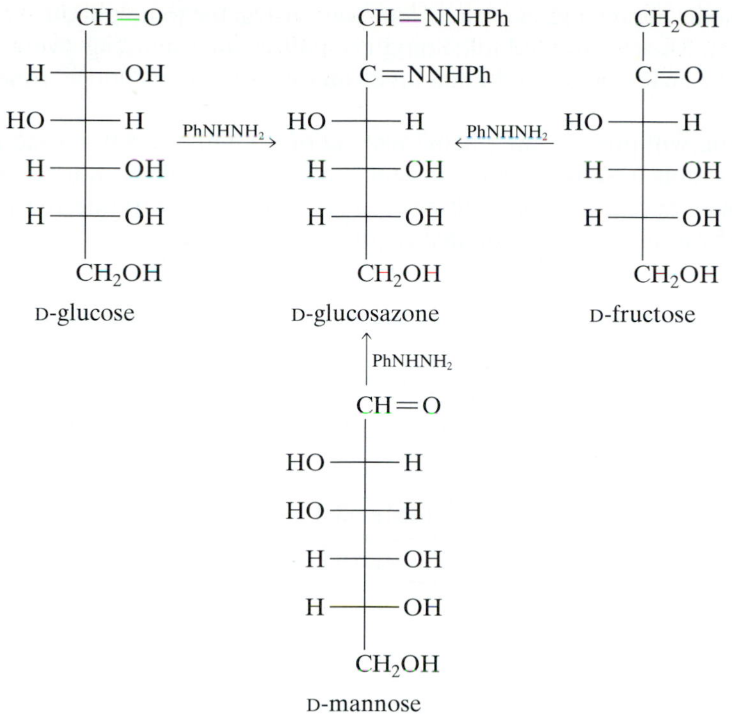

CH=O   CH=NNHPh   CH₂OH

H——OH   C=NNHPh   C=O

HO——H   PhNHNH₂→   HO——H   ←PhNHNH₂   HO——H

H——OH   H——OH   H——OH

H——OH   H——OH   H——OH

CH₂OH   CH₂OH   CH₂OH

D-glucose   D-glucosazone   D-fructose

↑PhNHNH₂

CH=O

HO——H

HO——H

H——OH

H——OH

CH₂OH

D-mannose

Only compounds that differ in structure from carbon #3 on down the chain will yield different phenylosazones. Among the D-hexoses there are four C-3 to C-6 structural units that yield different phenylosazones; these are illustrated in Figure 53.2 with Rosanoff symbols.

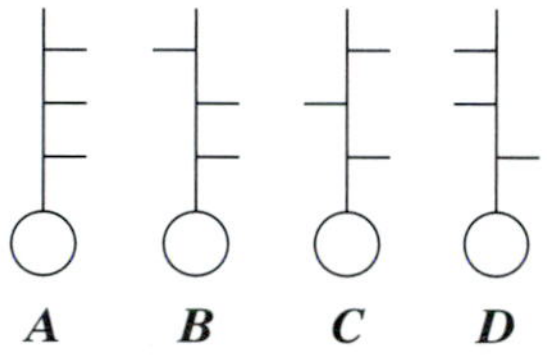

Figure 53.2 C-3 to C-6 structural units of D-hexoses.

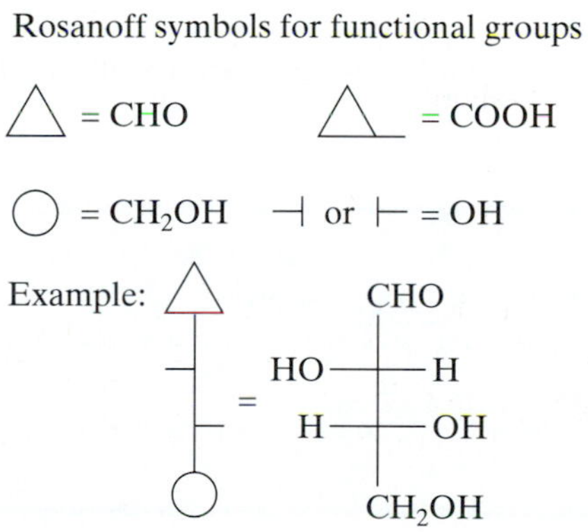

Rosanoff symbols for functional groups

△ = CHO   △ = COOH

○ = CH₂OH   ⊣ or ⊢ = OH

Example: △ = CHO

HO——H

H——OH

CH₂OH

Thus D-glucose, D-mannose, and D-fructose yield the same phenylosazone because they all contain the same C-3 to C-6 structural unit, **B**.

You will reduce the unknown monosaccharide by mixing an aqueous solution of the unknown with a 50% excess of sodium borohydride in aqueous NaOH. Like many sugars, the alditol tends to form supersaturated solutions and crystallizes slowly from solution. If your alditol refuses to form crystals after extended cooling of the reaction mixture in ice water, you may be able to obtain a seed crystal from your instructor to help induce crystallization. Most of the alditol should crystallize after standing overnight or until the next lab period. At that time you can measure its optical rotation in water.

You will prepare the phenylosazone of the unknown monosaccharide by a simple test tube reaction and measure its melting point. Because osazones decompose on melting, you will have to determine the melting point by a special technique that requires rapid heating.

Reactions and Properties

See "Understanding the Experiment" for reactions of representative hexoses.

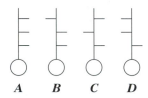

Table 53.1 Phenylosazone melting points

Structural unit	Phenylosazone m.p.
A	178
B	205
C	173
D	201

Directions

Take Care! The reagent is acidic, so avoid contact.

Stop and Think: Is the D-hexose an aldohexose or a ketohexose?

Waste Disposal: Dispose of the solution as directed by your instructor.

Safety Notes

A. *Seliwanoff's Test*
Dissolve 10 mg of the D-hexose in 1.0 mL of water in a small test tube. Then add 2.0 mL of Seliwanoff's reagent, swirl to mix, and heat the test tube in a boiling water bath for 2 minutes (no longer). Development of a deep red color within 2 minutes is a positive test for a ketohexose. Aldohexoses give a positive test with additional heating or after standing for some time.

B. *Preparation of the Alditol*

> **Sodium borohydride is corrosive and can react violently with concentrated acids, oxidizing agents, and other chemicals. Aqueous sodium borohydride solutions with pH values below 10.5 have been known to decompose violently, so be sure your reaction mixture is sufficiently alkaline. Avoid contact, do not breathe the dust, and keep NaBH$_4$ away from other chemicals.**

Reaction. Mix 20.0 mmol of the unknown D-hexose with 10 mL of water in a small beaker (it will not dissolve completely). Dissolve 7.5 mmol of sodium borohydride in 4.0 mL of 1.0 M NaOH in a 25-mL Erlenmeyer flask and drop in a stirbar, if you have one. Cool both mixtures in an ice/water bath [OP-7] for about 5 minutes. Start the stirrer [OP-9] and use a calibrated dropper to add a small portion (~1 mL) of the D-hexose suspension to the sodium borohydride solution every minute or two (if you don't have a stirrer, swirl after each addition). Stir the suspension before each addition, and wait until the reaction mixture has become clear before the next addition. If its temperature rises above 25°C, cool it in the ice bath for a minute or so before continuing. When the addition is complete, let the reaction mixture stand at room temperature, with stirring or occasional swirling, for 20 minutes. *Under the hood*, add 6 *M* HCl drop by drop until foaming stops and the reaction mixture turns blue litmus paper red.

Separation and Purification. Cool the reaction mixture in ice water and induce crystallization of the product, if necessary. Seal the flask with Parafilm and let the reaction mixture stand in a cool place overnight or longer to allow complete crystallization. Decant the water as completely as possible and replace it by 8 mL of cold absolute ethanol. Break up the mass of crystals and collect the product by vacuum filtration [OP-12]. Dry the product [OP-21] and measure its mass.

Analysis. Prepare an aqueous solution containing about 0.5 g of the alditol per 25 mL of solution. You may have to heat the mixture gently to dissolve the alditol. Measure the optical rotation of this solution [OP-31], using pure water as the blank.

C. *Preparation of the Phenylosazone*

> **Phenylhydrazine hydrochloride is very toxic by ingestion, inhalation, and skin absorption, and it is a suspected carcinogen. Wear gloves, avoid contact, and do not breathe the dust.**

Dissolve 0.20 g of the unknown D-hexose in 4 mL of water in a 15-cm test tube, and stir in 0.40 g of phenylhydrazine hydrochloride, 0.60 g of sodium acetate trihydrate (or 0.36 g of anhydrous sodium acetate), and 0.40 mL of saturated aqueous sodium bisulfite. (**Take Care!** Wear gloves, avoid contact with phenylhydrazine hydrochloride.) Seal the test tube with Parafilm and leave it in a boiling water bath for 30 minutes [OP-6], with occasional shaking. Add 5 mL of water and cool the reaction mixture in an ice bath. Collect the phenylosazone by vacuum filtration [OP-12], washing it on the filter with a little ice-cold methanol. Recrystallize the product from an ethanol-water mixture [OP-23], dissolving it in the ethanol first. When you collect it by vacuum filtration, wash it on the filter with ice-cold methanol. Dry the phenylosazone thoroughly [OP-21] at room temperature. Pulverize enough of the dry solid for two melting-point tubes [OP-28], and measure the temperature, T_1, at which the first sample melts with rapid heating (a rise of 10–20°C per minute). After the melting-point apparatus has cooled below T_1, adjust it for a temperature rise of 3–6°C per minute and insert the second capillary just as the temperature reaches T_1. Record the temperature T_2 at which this sample becomes completely liquid as the melting point of the phenylosazone.

Report. Your report should include a statement of the problem and an account of how you applied scientific methodology to solve it. If the optical rotation of the alditol is essentially equal to that of the blank (±0.2°), assume that it is optically inactive; otherwise, calculate its specific rotation. Deduce the structure of the D-hexose and explain your reasoning. Write equations for the reactions you carried out, showing the open-chain structures of the products. Calculate the percentage yields. Draw the Haworth

Take Care! Hydrogen is evolved, so keep flames away.

Stop and Think: What is the source of the hydrogen?

Waste Disposal: Unless directed otherwise, pour the decanted water and the ethanol down the drain.

Stop and Think: Is the D-hexose optically active or optically inactive?

Safety Notes

phenylhydrazine

Waste Disposal: Place the filtrate in a designated waste container.

Waste Disposal: Unless directed otherwise, pour the filtrate down the drain.

Stop and Think: Which two C-3 to C-6 structural units might the D-hexose contain? Which one can you eliminate based on your results in part **B**?

(flat-ring) structure of your D-hexose, and use the formula index of the *Merck Index* or another reference work to find its common name.

Exercises

1 (a) Draw structures of all the monosaccharides that would yield the same alditol as your D-hexose. (b) Draw structures of all the monosaccharides that would yield the same phenylosazone as your D-hexose. (c) Draw structures of all the alditols that would be obtained by reduction of all possible D-hexoses, and indicate whether each one is optically active or optically inactive.

2 It is believed that the C-4 OH group and the C-3 oxygen atom of an aldopyranose, or the C-1 OH group and the CH_2OH oxygen atom of a ketopyranose, can interact as an AH,B couple with the sweet taste receptor site. Build molecular models of β-D-glucopyranose and the most stable pyranose ring form of the D-hexose you identified, and use them to explain why your D-hexose is not a sweet as glucose.

3 Describe and explain the possible effect on your results of the following experimental errors or variations. (a) You had to leave the lab while the Seliwanoff test solution was in a boiling water bath, and when you returned the test solution was red. (b) In part **B**, you dissolved the $NaBH_4$ in 1.0 M HCl rather than 1.0 M NaOH. (a) In part **C**, you rinsed the test tube with acetone and didn't dry it out completely.

4 Following the format in Appendix V, construct a flow diagram for the synthesis of the alditol in part **B**.

5 α-D-Mannose has a specific rotation of $+29.3°$ and β-D-mannose has a specific rotation of $-17.0°$. If either anomer is dissolved in water and allowed to stand until equilibrium is reached, the specific rotation of the equilibrium mixture is $+14.2°$. Calculate the percentage of each anomer in the equilibrium mixture.

6 (a) Draw the structures of the other D-hexoses that will yield the same phenylosazone as D-allose. (b) Draw the structures of the L-hexoses that will yield the same alditol as D-glucose.

Other Things You Can Do

(Starred projects require your instructor's permission.)

1 Measure the optical rotation of a solution containing about 1 g (accurately weighed) of the unknown D-hexose in 25 mL of aqueous solution. *Under the hood*, mix 2 drops of concentrated aqueous ammonia into the solution, and again measure its optical rotation. If the optical rotation changes with time, wait until it equilibrates before taking a final reading, then calculate the specific rotation of the equilibrium mixture. Using the *Merck Index* or another reference book, find the specific rotations in water of the α and β anomers of the D-hexose and calculate the percentage of each in the equilibrium

Take Care! Avoid contact with NH_3, do not breathe its vapors.

mixture. Draw chair-form pyranose rings for both anomers, predict which one should be more stable, and tell how the calculated composition of the equilibrium mixture verifies your prediction (if it does).

2 Carry out reactions of some monosaccharides with phenols in Minilab 43.

3 After referring to articles in *J. Chem. Educ.* **1995**, *72*, pages 671–683, write a research paper about sweet compounds and the theory of the sweet taste sensation.

EXPERIMENT 54

Fatty Acid Content of Commercial Cooking Oils

Reactions and Preparation of Carboxylic Esters. Transesterification. Fats and Oils.

Operations:

OP-6   Heating
OP-13 Extraction
OP-14 Evaporation
OP-20 Drying Liquids
OP-32 Gas Chromatography

Before You Begin

Read the experiment, read or review the operations as necessary, and write an experimental plan.

Scenario

The Consumer's Advocate publishes *Caveat Emptor,* a periodical that rates consumer products and exposes inaccurate or misleading advertising (see Experiment 8). This organization is currently conducting a study of commercial vegetable oils to see whether their composition is consistent with advertising claims and to assess their relative dietary quality in terms of their effect on blood cholesterol. TCA's technical director, Patsy Haven, has asked your Institute for help in evaluating some of the vegetable oils sold in your geographic area. Your project group's assignment is to determine the fatty acid composition of each vegetable oil provided and to calculate the ratio of unsaturated to saturated fatty acids as a measure of their potential effect on cardiovascular health.

Applying Scientific Methodology

If you are given the names of the vegetable oils in advance of the lab period, try to predict their relative desirability with respect to cardiovascular health. Express your predictions as a hypothesis, to be be tested when you obtain the results of your project group's analyses.

Fatty Acids and Health

Cooking fats and oils are glyceryl esters of long-chain carboxylic acids called fatty acids. (An "oil" in this sense is essentially a fat that is liquid at room temperature.) Fats and oils have been put to many uses throughout human history. Excavation of an Egyptian tomb more than 5000 years old yielded several earthenware vessels containing substances identified as palmitic acid and stearic acid, most likely formed by the breakdown of palm oil and beef or mutton tallow, which were placed there as provisions for the deceased.

$$CH_3(CH_2)_{14}COOH$$
palmitic acid

$$CH_3(CH_2)_{16}COOH$$
stearic acid

The Egyptians used olive oil as a lubricant when moving huge stones for their building projects, and a mixture of fat and lime as axle grease for their war chariots. They may also have originated the art of painting with oils or waxes, using pigments mixed with natural waxes to paint portraits on their mummy cases. The Roman scholar Pliny the Elder described a process for making soap by boiling goat fat with wood ashes, then treating the pasty mass with sea water to harden it. But the art of making soap from animal fats must have originated much earlier; the Phoenicians of 600 B.C. were already trading soap to the Gauls, who later introduced it to Rome. Soap factories have been excavated at Pompeii, the Roman city that was buried under a blanket of volcanic ash. The same eruption killed Pliny, whose scientific curiosity impelled him to study the erupting Vesuvius up close.

Modern research has revealed the importance of fats and oils in nutrition and is just beginning to clarify their role in human disease. Linoleic acid, for example, is an essential component of the human diet because it is not synthesized in the body. Polyunsaturated fatty acids (PFAs) such as linoleic acid are believed to be involved in the biosynthesis of the prostaglandins, which help control blood pressure and muscle contraction.

$$\underset{\text{linoleic acid}}{CH_3(CH_2)_4\overset{\overset{\displaystyle H}{|}}{C}=\overset{\overset{\displaystyle H}{|}}{C}CH_2\overset{\overset{\displaystyle H}{|}}{C}=\overset{\overset{\displaystyle H}{|}}{C}(CH_2)_7\overset{\overset{\displaystyle O}{||}}{C}OH}$$

They play a vital role in the functioning of biological membranes and have been implicated in the occurrence or prevention of such illnesses as atherosclerosis, cancer, and multiple sclerosis. Recent research suggests that PFAs of the $\omega 3$ and $\omega 6$ families are especially effective in preventing the buildup of cholesterol in arteries, which is a major cause of heart disease. The ω designation here refers to the distance of the last double bond from the methyl end of the fatty acid chain, as illustrated. Certain $\omega 3$ PFAs found in fish oils, called eicosapentaenoic acid and docosahexaenoic acid, appear to be even more beneficial than the PFAs in vegetable oils. Besides lowering serum cholesterol levels, they also reduce the tendency of blood to clot.

With respect to cardiovascular health, the desirability of a cooking oil is related to the kinds and proportions of fatty acid residues that are incorporated into its triglyceryl ester molecules. Animal fats and coconut oil contain a relatively high proportion of saturated fatty acids (SFA), which raise blood cholesterol levels in humans and are believed to increase the risk of heart disease. Vegetable oils and fish oils that are high in polyunsaturated fatty acids apparently have the opposite effect, lowering the amount of cholesterol in the blood by hastening its excretion. Monounsaturated

$$CH_3CH_2CH=CH\ldots$$
$\overset{\omega 1}{\bigcirc}\overset{\omega 2}{\bigcirc}\overset{\omega 3}{\bigcirc}$

fatty acids (MFA), which occur in some vegetable oils, also help lower total cholesterol and are probably at least as beneficial as the PFAs. Some studies suggest that, unlike PFAs, they lower total cholesterol without lowering the amount of HDL (high density lipoprotein), also known as "good cholesterol."

Fatty acids are named and symbolized by several different systems. Most of the common fatty acids have trivial names that are still widely used, in part because of their simplicity compared to the systematic names. Linoleic acid, for example, is named *cis*-9-*cis*-12-octadecadienoic acid under the IUPAC system. Trivial names do not reveal the structures of the corresponding acids, however, so several shorthand notations have been adopted. Oleic acid can be represented by *c*-9-18: 1, where *c* means *cis*, 9 refers to the position of the double bond (numbering from the COOH group), 18 is the total number of carbon atoms in the chain, and 1 is the number of double bonds. Linoleic acid is represented by *c,c*-9, 12-18:2 using this system. Since all of the acids to be studied in this experiment have *cis* double bonds, the *c* prefix will be omitted here.

$$\underset{\text{oleic acid (9-18:1)}}{CH_3(CH_2)_7 - \overset{H}{\underset{|}{C}} = \overset{H}{\underset{|}{C}} - (CH_2)_7COOH}$$

$$CH_3(CH_2)_{16}COOCH_2 \quad CH_2OH$$
$$CH_3(CH_2)_{16}COOCH \quad\ CHOH$$
$$CH_3(CH_2)_{16}COOCH_2 \quad CH_2OH$$

a triglyceride,   glycerol
tristearin

$$CH_3(CH_2)_{16}COO^-Na^+$$

a soap, sodium stearate

$$CH_3(CH_2)_{16}COOCH_3$$

methyl ester of stearic acid
(methyl stearate)

Understanding the Experiment

The cooking oils to be analyzed in this experiment are triglycerides; that is, esters of the trihydroxy alcohol glycerol containing three fatty acid residues. A mixture of fatty acid salts can be obtained by hydrolyzing a triglyceride in the presence of a base. This is the process used in soap formation, with the acid salts comprising the soap. Alternatively, a triglyceride can be transesterified with an alcohol such as methanol to yield a mixture of fatty acid methyl esters (FAME). Since the esters are much more suitable for GLC analysis than the acid salts or the free acids themselves, the second method will be used in this experiment. The fatty acid composition of each cooking oil can then be derived from the percentages of the corresponding esters.

When methyl esters of saturated fatty acids are analyzed on a suitable gas chromatography column, their retention times (T_r) increase with the length of the carbon chain according to the relationship log $T_r \propto$ number of carbons. The retention times of unsaturated fatty acid esters do not coincide with those of saturated esters. On a nonpolar liquid phase their retention times are shorter than those of the saturated esters of the same chain length, whereas on a polar liquid phase they are longer. The actual value of T_r depends on the number and positions of the double bonds. By comparing the retention times of a large number of such esters, a series of equivalent-chain-length (ECL) values has been worked out for various esters on different liquid phases. The ECL for a *saturated* ester is the same as its actual chain length—for instance, 14.0 for myristic acid (as the methyl ester), 18.0 for stearic acid, etc. The ECL for an *unsaturated* ester on a polar liquid phase is greater than its actual chain length—for instance, 18.43 for methyl oleate (9-18: 1) and 19.22 for methyl linoleate (9,12-18: 2), both analyzed on a diethylene glycol succinate (DEGS) liquid phase. If the retention times for two or more saturated esters are known, a calibration curve can be constructed by plotting the logarithms of their retention times against their chain lengths, as shown in Figure 54.1. To identify an unknown fatty acid ester, one can calculate the logarithm of its retention time, read its ECL value off the graph, and compare that value with the ECL values of known methyl esters (see Table 54.1).

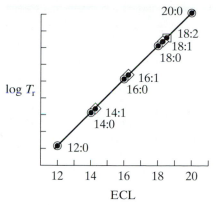

Figure 54.1 Plot of log T_r versus equivalent chain length for fatty acid methyl esters on a DEGS liquid phase

Table 54.1 Shorthand notation and ECL values (on DEGS liquid phase) for representative fatty acids

Fatty acid	Shorthand notation	ECL	Type
lauric	12:0	12.00	SFA
myristic	14:0	14.00	SFA
myristoleic	9-14:1	14.71	MFA
palmitic	16:0	16.00	SFA
palmitoleic	9-16:1	16.55	MFA
stearic	18:0	18.00	SFA
oleic	9-18:1	18.43	MFA
linoleic	9,12-18:2	19.22	PFA
linolenic	9,12,15-18:3	20.12	PFA
behenic	22:0	22.00	SFA

Note: ECL values are for the corresponding methyl esters

In order for ECL values to provide an accurate means of identifying the esters, the same type of column and carrier gas should be used in all cases. Even then, the experimental values will not usually correspond exactly to literature values because of differences in such factors as the kind of column support, the amount of liquid phase, the age of the column, and the experimental conditions. In addition, some peaks may overlap in complex mixtures so that two or more columns must be used for their separation, and deviations from linearity in the log T_r plot may occur when short-chain fatty acids are analyzed. The relatively simple mixtures of fatty acids that you will encounter in this experiment should not present any serious experimental difficulties, however.

The transesterification of the oil to its methyl esters will be carried out by hydrolyzing the triglycerides in methanolic sodium hydroxide and esterifying the resulting fatty acids in methanol with a boron trifluoride catalyst. After the methyl esters have been isolated from the reaction mixture, they will be mixed with an internal standard, methyl heptadecanoate, and analyzed by gas chromatography. Heptadecanoic acid (17:0) does not occur naturally, so the 17:0 peak from its ester will not interfere with other component peaks. To minimize the effect of retention time drift due to changes in operating parameters, relative retention values should be calculated by dividing the retention time for each peak by that of the 17:0 peak. Once the 17:0 peak has been located, those of the 16:0 and the 18:0 esters can be identified by the fact that the log T_r interval between their peaks and 17:0 is the same. (The 16:0 ester often gives the first strong peak on the chromatogram, making it easy to recognize.) Then a log T_r versus ECL plot can be prepared, from which the other peaks can be identified. Since the peak area for each component should be proportional to its mass, the percentage composition of the mixture can be determined from peak areas with good accuracy.

Assuming that saturated fatty acids have a negative effect on cardiovascular health, and that both monounsaturated and polyunsaturated fatty acids have a positive effect, you can use the ratio (MFA + PFA)/ SFA as a rough indicator of the "healthfulness" of your cooking oil compared to the other oils analyzed in your laboratory section.

Reactions and Properties

$$\begin{matrix} R'COOCH_2 & & & R'COOCH_3 \\ | & & & + \\ R''COOCH + 3CH_3OH & \xrightarrow[BF_3]{NaOH} & R''COOCH_3 + CH_2CHCH_2 \\ | & & & + \\ R'''COOCH_2 & & & R'''COOCH_3 \end{matrix}$$

(R groups can be alike or different.)

Table 54.2 Physical properties

	M.W.	m.p.	b.p.	d
methanol	32.0	−94	65	0.791
boron trifluoride-methanol complex	131.9		59^4	
boron trifluoride	67.8	−127	−100	

Note: m.p. and b.p. are in °C, density is in g/mL

The complex has the composition $BF_3 \cdot 2CH_3OH$.

Figure 54.2 IR spectra of saturated and polyunsaturated 18-carbon methyl esters (Reproduced from *The Aldrich Library of FT-IR Spectra, Edition II*, with the permission of the Aldrich Chemical Company.)

Methyl stearate

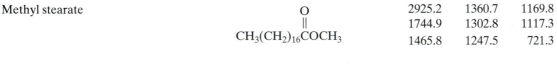

2925.2	1360.7	1169.8
1744.9	1302.8	1117.3
1465.8	1247.5	721.3

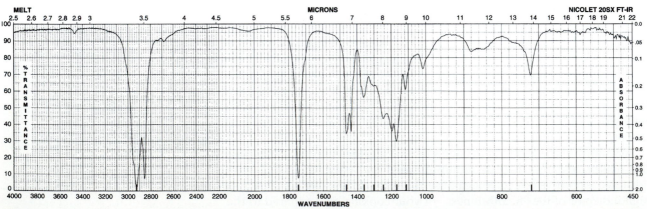

Methyl linolenate

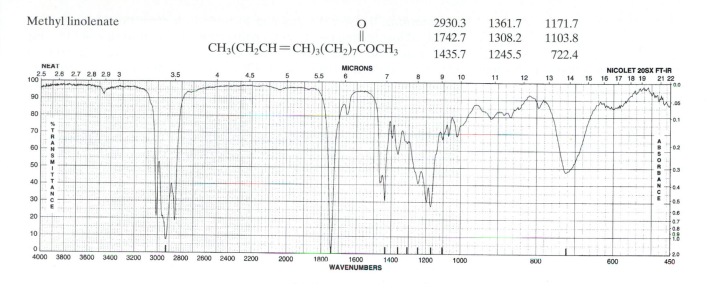

$$CH_3(CH_2CH{=}CH)_3(CH_2)_7\overset{\overset{\displaystyle O}{\displaystyle \|}}{C}OCH_3$$

2930.3	1361.7	1171.7
1742.7	1308.2	1103.8
1435.7	1245.5	722.4

Directions

(Adapted from a procedure in the *Journal of Chemical Education* **1974**, *51*, 406 with permission.)

A selection of commercial cooking oils may be provided, or students may be asked to bring their own cooking oils. Different kinds of oils should be analyzed such as canola, corn, olive, peanut, safflower, soybean, and sunflower oil. All glassware and reagents should be dry for this experiment.

Safety Notes

> The methanolic sodium hydroxide solution is flammable and toxic. Avoid contact, do not breathe vapors, and keep it away from flames.
> The boron trifluoride/methanol solution is flammable, corrosive, and toxic and can cause severe damage to the eyes, skin, and respiratory tract. Use gloves and a hood, avoid contact, and do not breathe vapors.
> Petroleum ether is extremely flammable. Keep it away from flames and hot surfaces.

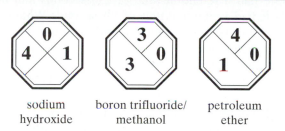

sodium hydroxide   boron trifluoride/methanol   petroleum ether

Reaction. In a clean, *dry* 15-cm test tube, combine 0.15 g of a commercial cooking oil with 5.0 mL of 0.5 *M* methanolic sodium hydroxide solution. Heat the reaction tube over a steam bath or in a boiling water bath [OP-6] until the oil has completely dissolved (about 3–5 minutes). *Under the hood*, add 6.0 mL of a 12.5% (weight/volume) solution of boron trifluoride in methanol, seal the test tube with Parafilm, and boil the mixture gently over the steam bath or in a hot water bath for 2 minutes.

Take Care! Avoid contact with the NaOH solution and do not breathe its vapors.

Take Care! Wear gloves, avoid contact with the BF₃ solution, do not breathe its vapors.

Stop and Think: What is left behind in the aqueous layer?
Waste Disposal: Pour the aqueous layer down the drain.
Waste Disposal: Place the recovered petroleum ether in the designated solvent recovery container.

Separation. Transfer the reaction mixture to a separatory funnel using 30 mL of low-boiling (~35–60°C) petroleum ether and rinsing out the test tube with some of the solvent. Add 20 mL of saturated aqueous sodium chloride and shake to extract [OP-13] the methyl esters into the petroleum ether layer. Dry the petroleum ether layer with anhydrous magnesium sulfate [OP-20], filter it by gravity [OP-11], and evaporate the solvent completely [OP-14].

Analysis. Obtain a gas chromatogram [OP-32] of a 1–2 μL sample of the fatty acid methyl ester mixture on a DEGS/Chromosorb W or similar column. Add a drop of a 50% solution of methyl heptadecanoate in dichloromethane to the remaining mixture, and obtain its gas chromatogram. Measure the areas of all peaks on the first gas chromatogram. Identify the 17:0 peak and measure the retention times of all peaks on the second gas chromatogram. Calculate relative retention times and log T_r values for all the components, and use these data to tentatively identify several of the methyl esters. For these esters, plot log T_r versus their ECL values and try to fit the rest of your data to the plot, redrawing it as necessary to get the best straight line. (If your data points don't fit the plot, you probably misidentified the esters. Make another guess and try again.) Identify the esters and calculate the percentage of each fatty acid in your cooking oil.

Report. Your report should include a statement of the problem and an account of how you applied scientific methodology to solve it. Calculate the total percentages of saturated, monounsaturated, and polyunsaturated fatty acids and the ratio of unsaturated to saturated fatty acids. Tabulate the class results and rank the different oils according to their expected effect on cardiovascular health.

Exercises

1 Write the structures of the fatty acids whose triglycerides were in the oil you analyzed and give their IUPAC names.

2 On the gas chromatogram of a mixture of methyl esters, methyl palmitate had a retention time of 150 seconds and methyl heptadecanoate had a retention time of 198 seconds. What is the probable identity of a methyl ester with a retention time of 363 seconds on the same chromatogram?

3 Describe and explain the possible effect on your results of the following experimental errors or variations. (a) The reaction test tube contained water. (b) You heated the oil with methanolic sodium hydroxide but forgot to add the boron trifluoride/methanol solution. (c) You didn't evaporate all of the petroleum ether from your reaction mixture.

4 (a) Following the format in Appendix V, construct a flow diagram for the synthesis. (b) Explain why there was no glycerol peak on the gas chromatogram of the methyl ester mixture, even though glycerol was a product of the hydrolysis.

5 Why was the Lewis acid boron trifluoride used to catalyze the transesterification reaction rather than hydrochloric acid?

6   Calculate the (MFA + PFA)/SFA ratio for palm oil, whose fatty-acid content is approximately 9% linoleic, 2% myristic, 40% oleic, 45% palmitic, and 4% stearic acid.

7   (a) Neat's-foot oil consists almost entirely of the triglycerides of oleic and palmitic acids. How many different triglycerides of these two acids can it contain? (b) Draw structures for the triglycerides that contain two oleic acid units and one palmitic acid unit.

8   Outline a synthesis of the detergent sodium lauryl sulfate from glyceryl trilaurate (trilaurin).

$$CH_3(CH_2)_{10}CH_2OSO_2O^-Na^+$$
sodium lauryl sulfate

Other Things You Can Do

(Starred projects require your instructor's permission.)

*1   Record an infrared spectrum of your methyl ester mixture and compare it with the spectra of the 18-carbon methyl esters in Figure 54.2. Try to account for significant similarities and differences in the spectra and point out any bands that show evidence of unsaturation.

*2   Isolate the fat trimyristin from nutmeg in Minilab 44.

*3   Make some soap using a phase-transfer catalyst in Minilab 45.

4   Starting with sources listed in the Bibliography, write a research paper about the health implications of various types of fats and oils. Include a description of the HDL and LDL forms of cholesterol and a discussion of the role they play in heart disease.

EXPERIMENT 55   # Structure of an Unknown Dipeptide

Reactions of Peptides. Nucleophilic Aromatic Substitution. Amino Acids. Structure Determination.

Operations:

> **OP-18** Paper Chromatography
> OP-3   Using Glass Rod and Tubing
> OP-13 Extraction
> OP-14 Evaporation
> OP-17 Thin-Layer Chromatography
> OP-19 Washing liquids

Before You Begin

Read the experiment and operation OP-18, read or review the other operations as necessary, and write an experimental plan.

Scenario

Snake venoms have been used in medicine as anticoagulants, which help dissolve blood clots or prevent their formation, and for other purposes. Natural Nostrums, a pharmaceutical company that manufactures drugs based on natural substances, anticipates some medical uses for the venom of the black boomslang, a deadly African tree snake. An unfortunate accident has befallen their native snake hunter, so their supply of natural boomslang venom has dried up. Before they can prepare a synthetic version of the venom, a polypeptide, they need to determine the sequence of amino acids in its polypeptide chain. Ferdie Lance, the company's snake venom expert, has used an enzyme called cathepsin C to break down the boomslang venom into dipeptide units. Now each dipeptide unit has to be identified so that Dr. Lance can reconstruct the amino acid sequence in the polypeptide. Your project group's assignment is to determine the structures of the unknown dipeptides.

Applying Scientific Methodology

After you analyze the hydrolysis mixture from your assigned dipeptide using paper chromatography, you can formulate a tentative hypothesis about the structure of the dipeptide. Your hypothesis will be tested by TLC analysis of a hydrolysis mixture from the dinitrophenylated dipeptide.

Mushrooms, Black Mambas, and Memory Molecules

The naturally occurring polypeptides and proteins range in size from tripeptides such as glutathione to complex proteins having molecular weights on the order of 1 million; their variations in structure and function cover just as broad a range.

$$\overset{O}{\underset{\overset{|}{COO^-}}{\underset{|}{^+H_3NCHCH_2CH_2\overset{\|}{C}NHCHCNHCH_2CO^-}}}$$

glutathione

Some of them, such as the keratins of hair, horns, nails, and claws, act as biological building materials and have little or no biological activity. Others function as hormones, enzymes, antibiotics, or toxins, exerting a profound effect on biochemical reactions.

Bradykinin, sometimes known as the "pain molecule," is released by enzymatic cleavage of plasma glycoprotein whenever tissues are damaged.

Arg — Pro — Pro — Gly — Phe — Ser — Pro — Phe — Arg
bradykinin

It causes the sensation of pain by bonding to certain receptors on nerve endings. Peptides similar to bradykinin are present in wasp venom; other kinins act as hormones, stimulating a variety of physiological responses such as the contraction or relaxation of smooth muscles and the dilation of blood vessel walls.

Oxytocin and vasopressin are both secreted by the posterior pituitary gland. Their structures are identical except for two amino acid residues, but their functions are entirely different. Vasopressin increases retention of water in the kidneys and is used as an antidiuretic in treating a form of diabetes that is characterized by excessive urine flow. Oxytocin intensifies uterine contractions during childbirth and is used clinically to induce labor. Both are cyclic compounds (cyclopeptides) that have a tripeptide side chain.

Poisonous mushrooms of the genera *Amanita* and *Galerina* contain cyclopeptides known as amatoxins and phallotoxins. These toxins are extremely potent; as little as one "death cap" mushroom, *Amanita phalloides*, can kill an adult. The cinema version of mushroom poisoning—in which the unlucky victim eats the mushrooms, turns pale, rises from the table clutching his throat, and drops dead—is fallacious. The victim is usually not aware that he or she has been poisoned for 8–24 hours after eating the mushrooms. Following a day or so of violent cramps, nausea, vomiting, and other unpleasant symptoms, the patient appears to recover and may even be sent home from the hospital. Death from kidney or liver failure often follows in several days. α-Amanitin (p. 442) and the other amatoxins are believed to attack the nuclei of liver and kidney cells, depleting their nuclear RNA and inhibiting the synthesis of more RNA so that protein synthesis stops and the cells die.

Larger polypeptides, containing from 60 to 74 amino acid residues, are found in snake venoms such as those of the African black mamba and the Indian cobra (Figure 55.1). Most of these venom toxins have long chains that are cross-linked by four or more cystine disulfide bridges. The venom toxin of the black mamba is probably the most potent—it can kill a mouse in less than 5 minutes.

One of the most exciting areas of polypeptide biochemistry is the study of "memory molecules." Researchers have discovered that an animal's learned behavior can be forgotten when a substance that interferes with

See Table 55.1 on page 443 for the names and abbreviations of some amino acids. The textbook for your lecture course should contain abbreviations for amino acids not listed in Table 55.1.

$$\begin{array}{l}
Cys—Tyr—Ile \\
\ |\qquad\qquad\quad| \\
\ S\qquad\qquad\quad| \\
\ |\qquad\qquad\quad| \\
\ S\qquad\qquad\quad| \\
\ |\qquad\qquad\quad| \\
Cys—Asn—Gln \\
\ | \\
Pro—Leu—GlyNH_2
\end{array}$$

oxytocin

$$\begin{array}{l}
Cys—Tyr—Phe \\
\ |\qquad\qquad\quad| \\
\ S\qquad\qquad\quad| \\
\ |\qquad\qquad\quad| \\
\ S\qquad\qquad\quad| \\
\ |\qquad\qquad\quad| \\
Cys—Asn—Gln \\
\ | \\
Pro—Arg—GlyNH_2
\end{array}$$

vasopressin

The Roman emperor Claudius is said to have died soon after eating mushrooms, but they were probably edible Amanita caesarea *mushrooms that had been laced with poison by his wife, Agrippina.*

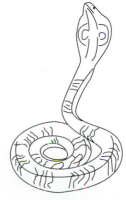

Indian Cobra

$$\alpha\text{-amanitin}$$

peptide synthesis is injected into the animal's brain at a certain time. This suggests that peptide synthesis is involved in the consolidation of long-term memory. For example, a peptide called scotophobin was isolated from the brains of rats that had been conditioned to fear the dark.

Ser—Asp—Asn—Asn—Gln—Gln—Gly—Lys—Ser—Ala—Gln—Gln—Gly—Gly—TyrNH$_2$
scotophobin
(from the Greek *scotos*, dark; and *phobos*, fear)

When scotophobin was injected into unconditioned rats, they also became afraid of the dark! This raises the intriguing possibility that breaking the chemical memory code could enable chemists to synthesize the peptides corresponding to any kind of learning experience. Perhaps someday it will be possible to get an injection of Biology 101 or Philosophy 416 rather than absorbing the subject matter in the classroom!

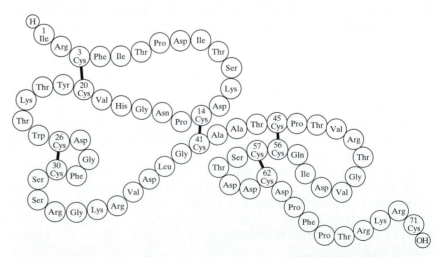

Figure 55.1 Venom toxin of the Indian cobra (*Naja naja*) (Reprinted from *Biochemical and Biophysical Research Communications*, **1973**, *55*, 435, by permission.)

Understanding the Experiment

The structures of long-chain polypeptides, such as those found in the venom of poisonous snakes, can be determined by breaking them down into shorter chains of amino acids and analyzing the fragments. For example, an enzyme called cathepsin C hydrolyzes a polypeptide into dipeptide units starting at the *N*-terminal amino acid residue (the one at the end of the chain terminated by an amino group), and the enzyme trypsin breaks a polypeptide chain after each lysine or arginine residue. By analyzing each fragment from these and other reactions, a biochemist can locate overlapping sequences of amino acids, from which the entire primary structure of the polypeptide can be deduced.

In this experiment you will determine the structure of an unknown dipeptide that contains one or more of the amino acids in Table 55.1. To do so, you will first identify the amino acids your dipeptide contains by hydrolyzing it and analyzing the resulting amino acid mixture by paper chromatography. The dipeptides and amino acid derivatives you will be using are quite expensive; therefore, the experiment will be carried out on a micro scale, using only a milligram of the dipeptide for each reaction. You will use a Pasteur pipet for volume measurements and solvent extractions and sealed capillary tubes for the hydrolysis reactions. It takes some practice to fill and seal a capillary tube properly, so it is recommended that you practice the technique with water before using the dipeptide and DNP-dipeptide solutions. The capillary method is very reliable when performed correctly, but there is always the chance that an improperly sealed tube will break or leak during the overnight hydrolysis period. To prevent the considerable loss of time that would result from a ruined sample, you should prepare at least three tubes for each hydrolysis reaction.

The paper chromatography of amino acids can be carried out in a multitude of developing solvents; no one solvent system is the best under all circumstances. The 2-propanol : formic acid: water (16 : 1 : 4) solvent system used in this procedure is suitable for most amino acids, but certain combinations of amino acids may be separated more readily with a different solvent system such as 1-propanol : ammonia (7 : 3) or 2-butanone : propionic acid : water (15 : 5 : 6). The amino acids are colorless, so their spots will be made visible by using a solution of ninhydrin, which reacts with most amino acids (proline is an exception) to form a blue-violet product.

To work out the complete structure of your dipeptide you must determine the sequence in which the amino acids appear in the dipeptide. This can be done using a dinitrophenylation procedure developed by Frederick Sanger, who won the Nobel Prize for chemistry in 1956 for determining the structure of insulin and again in 1980 for his work on nucleic acid structures. 2,4-Dinitrofluorobenzene (DNFB) is an unusually reactive aryl halide that undergoes nucleophilic substitution reactions with unprotonated amino groups. Because most peptides exist as zwitterions in neutral solutions, sodium bicarbonate is used to raise the pH of the peptide solution enough to deprotonate the amino group, allowing it to react with DNFB. This reaction yields a dipeptide with a dinitrophenyl (DNP) substituent attached to the *N*-terminal amino acid residue, as shown in the "Reactions and Properties" section. In effect, the DNP group functions as a "label" for the *N*-terminal residue. In the alkaline solution, the DNP-substituted dipeptide exists in the anionic form with a polar —COO⁻ group. Extracting this solution with

Table 55.1 Names and abbreviations of selected amino acids

Name	Abbreviation
histidine	His
lysine	Lys
serine	Ser
aspartic acid	Asp
glycine	Gly
threonine	Thr
alanine	Ala
proline	Pro
tyrosine	Tyr
valine	Val
phenylalanine	Phe
leucine	Leu

The solvent ratios are volume ratios. Thus a 1-propanol:ammonia (7 : 3) solvent system can be prepared by mixing 70 mL of 1-propanol with 30 mL of concentrated aqueous ammonia.

ninhydrin

2,4-dinitrophenyl group
(DNP)

ether removes the less polar DNFB but leaves the DNP-dipeptide in the aqueous layer. Lowering the pH with hydrochloric acid then protonates the carboxyl group, allowing the DNP-dipeptide to be extracted from the aqueous layer by ethyl ether and isolated by evaporating the ether. It is important to remove the last traces of ether before hydrolysis so the residue is dissolved in acetone, which is also evaporated. The DNP-dipeptide is then hydrolyzed in aqueous hydrochloric acid, yielding the free *C*-terminal amino acid (the one at the end of the chain terminated by a carboxyl group) and a dinitrophenyl derivative of the *N*-terminal amino acid.

DNP-amino acids are more easily separated on silica gel than on paper, so your product will be analyzed by thin-layer chromatography using known DNP-amino acids as standards for comparison. Because the DNP derivatives are light sensitive, their chromatograms should be developed in a location away from strong light. The yellow spots formed by the DNP-amino acids are easy to locate but they fade with time, so you should circle them soon after the TLC plate is developed. Dinitrophenol is a possible by-product of the dinitrophenylation procedure. However, its yellow spot is usually evident only when an alkaline developing solvent is used. If TLC analysis of your unknown produces two or more distinct spots, exposing the chromatogram to vapors of hydrochloric acid should bleach out the dinitrophenol spot.

Reactions and Properties

A. Dinitrophenylation of dipeptide

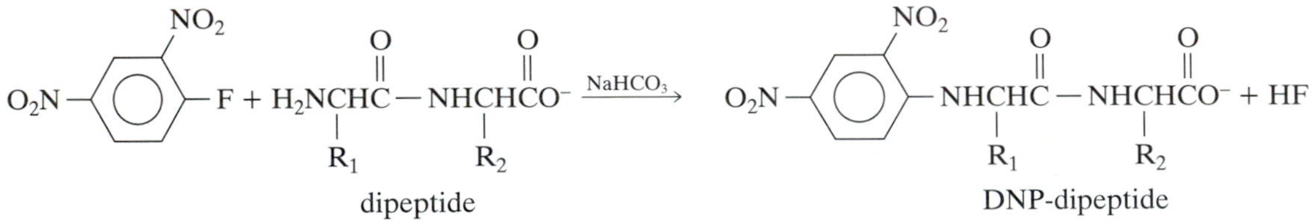

dipeptide   DNP-dipeptide

R_1, R_2 = amino acid side chains

B. Hydrolysis of dipeptide and DNP-dipeptide

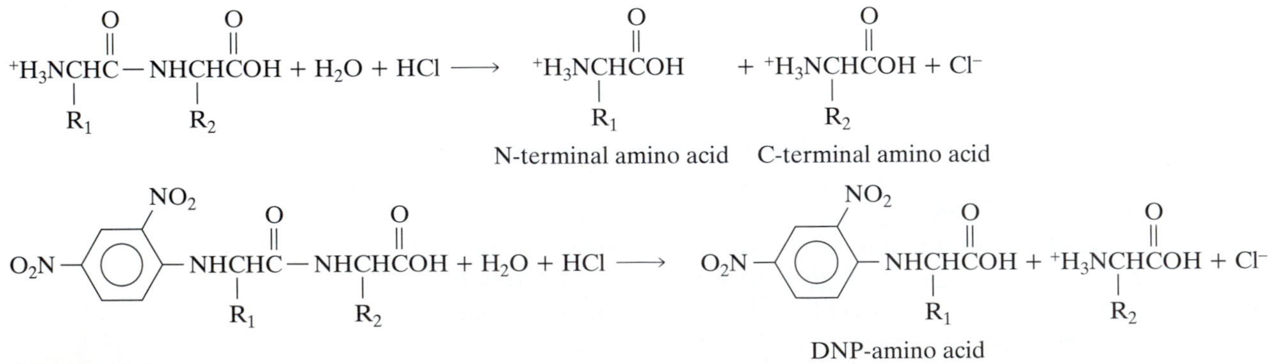

N-terminal amino acid   C-terminal amino acid

DNP-amino acid

Table 55.2 Physical properties

	M.W.	m.p.	b.p.
dinitrofluorobenzene	186.1	27.5–30	178[15]

Note: m.p. and b.p. are in °C

Directions

Use a Pasteur pipet for all volume measurements; the pipet should deliver about 35–45 drops per milliliter. Organize your time efficiently so that you can have your hydrolysis tubes prepared and in the oven by the end of the first laboratory period.

Safety Notes

A. *Dinitrophenylation of the Dipeptide*

In a small conical centrifuge tube or a 7.5-cm test tube, combine about 1 mg of your unknown dipeptide with 4 drops of distilled or deionized water, 1 drop of 4% aqueous sodium bicarbonate, and 8 drops of a 5% solution of 2,4-dinitrofluorobenzene in ethanol. Cover the tube with Parafilm and shake it occasionally over a 1 hour period. Every 10 minutes, check the pH by dipping the closed end of a melting-point capillary into the solution and touching it to a strip of narrow-range pH paper. As necessary, add a drop or two of 4% sodium bicarbonate solution to keep the pH between 8 and 9.

Take Care! Wear gloves, avoid contact with the DNFB solution.

Add 10 drops of water and 10 drops of 4% aqueous sodium bicarbonate to the reaction mixture. Wash the reaction mixture [OP-19] two or three times with ethyl ether, each time using a volume of ether approximately equal to the total volume of solution. Carry out each washing operation by (1) adding the ether to the centrifuge tube; (2) drawing both layers into a Pasteur pipet and ejecting the mixture forcefully into the centrifuge tube to mix the layers, repeating this operation six times or more; (3) cooling the mixture in ice water and waiting for any emulsion to settle; (4) removing the ether (top) layer with the Pasteur pipet. The last ether layer should be colorless; wash the reaction mixture with more ether if it is not. Add enough 6 M hydrochloric acid (2 or 3 drops are usually sufficient) to the aqueous layer to bring its pH down to 1 or 2. Then extract [OP-13] the aqueous solution with two portions of ethyl ether by the same technique you used to wash it, and combine and save the extracts. Be careful not to include any of the aqueous layer when you withdraw the ether layers.

Stop and Think: Why is important to keep the pH in this range?
Take Care! Keep ether well away from flames.
Stop and Think: What is being removed by the ether?

Take Care! Avoid contact with the ether layers.
Waste Disposal: Place the combined ether layers in a designated solvent recovery container.
Stop and Think: Why is it necessary to acidify the solution?
Waste Disposal: Pour the aqueous layer down the drain.

Use a dry Pasteur pipet to distribute the ether solution equally among three wells on a porcelain spot plate, and let the ether evaporate *completely* under the hood. Dissolve each residue in 2 drops of acetone and let the acetone evaporate under the hood. Then dissolve each residue in 4 drops of 6 M hydrochloric acid (you can use the closed end of a capillary melting point tube for stirring) and prepare at least three hydrolysis tubes by method 1 if you are using capillary tubes that are sealed at one end, or by method 2 if you are using open-ended capillary tubes.

Method 1: Warm a capillary melting point tube by holding it near its open end and moving it in and out of the "cool" part of a Bunsen burner flame for several seconds, then immediately insert the open end into the liquid in one

Take Care! Don't burn yourself!

of the wells. As the tube cools (blowing on it may help), it should begin to fill with liquid. When 2–3 cm of liquid is inside the tube, take the open end out of the liquid in the well, and, holding the open end just above horizontal, cool the closed end under a cold water tap. Holding the tube upright near the open end, carefully strike it at a point below the liquid level with your fingernail until nearly all the liquid is in the bottom of the tube. Capillary tubes are fragile, so don't tap the tube too hard or it will break. Holding the closed end so that your fingers just cover the liquid, rotate the tube as you move it rapidly in and out of the cool part of the burner flame for several seconds. (The tube should be inserted lengthwise into the flame so that most of the empty part is heated.) Immediately seal the open end [OP-3] by rotating it in the inner cone of the flame, near its apex. Because of the vacuum created by preheating the tube, its end should collapse and bend over as it seals.

Method 2: Insert one end of an open capillary tube into one of the wells and draw in (by capillary action) about 2–3 cm of liquid. Hold the tube horizontally and tap it so that the liquid is in the middle, then seal one end by rotating it in the outermost edge of the flame. Hold the open end up and strike the lower part of the tube with your fingernail until nearly all the liquid is in the bottom, then seal the other end as described for method 1.

Put the capillary tubes in a small test tube labeled with your name and "DNP-amino acid."

B. *Hydrolysis of the Dipeptide and the DNP-Dipeptide*
Dissolve about 1 mg of your unknown dipeptide in 10 drops of 6 *M* hydrochloric acid. Use this solution to prepare three or more capillary tubes by one of the methods described in part **A**. Put the capillary tubes in a small test tube labeled with your name and "amino acids." Leave this test tube and the one from part **A** in a 100–110°C oven overnight or longer.

C. *Separation and Analysis of the Amino Acids and the DNP-Amino Acid*

Safety Notes

> **Both chromatographic developing solvents contain flammable or toxic liquids with harmful vapors, and chloroform is a suspected carcinogen. Use a hood, avoid contact, do not breathe vapors, and keep flames away.**

2-propanol

formic acid

chloroform

acetic acid

Take Care! Avoid contact, do not breathe vapors, keep flames away.

Remove the hydrolysis tubes from the oven and let them cool. *Under a hood*, prepare a large beaker (or another suitable developing chamber) for paper chromatography [OP-18] by adding enough of the 2-propanol : formic acid : water (16 : 1 : 4) solvent system to form a 1-cm layer on the bottom. Cover the beaker with plastic wrap and let the system equilibrate for at least an hour before using it. Prepare a similar developing chamber for the TLC analysis [OP-17] using the chloroform : acetic acid : *t*-pentyl alcohol (23 : 1 : 10) solvent system. Cover the beaker with plastic wrap and let the system equilibrate for at least a 30 minutes before using it.

Take Care! Avoid contact and do not breathe vapors.

Use a sharp triangular file or other glass cutter to open one of the cooled amino acid capillary tubes just above the liquid level, and empty it into one well of a spot plate. Carefully evaporate the solvent [OP-14] in a dry air stream under the hood and dissolve the residue in 1 drop of water. Obtain a sheet of Whatman #1 chromatography paper that is at least 12 cm long (in the direction of development) and wide enough to accommodate the standard amino acid solutions as well as your unknown solution. Spot the paper in two (or more) places with the amino acid solution, then spot it with the amino acid standards. Develop the chromatogram under a hood in the 2-propanol : formic acid : water solvent system and let it dry under the hood, then spray the paper lightly with the ninhydrin spray reagent. Develop the color by heating the paper in a 100–110°C oven for about 10 minutes. Measure the R_f values of all the spots.

Waste Disposal: Place the chromatography solvent in a designated solvent container.

Open one of the DNP-amino acid capillary tubes and empty the contents into a small test tube or centrifuge tube, washing it down with 10–15 drops of water. Extract [OP-13] the aqueous solution twice with ethyl ether as described in part **A**, using a volume of ether equal to the volume of the solution for each extraction. Combine the extracts. Transfer the ether solution to a well on a spot plate and let the ether evaporate under the hood. Dissolve the residue in 3 drops of acetone, and immediately use this solution to spot a silica gel TLC plate in two places. Spot the plate with the standard solutions of DNP-amino acids and develop it under a hood with the chloroform : acetic acid : *t*-pentyl alcohol solvent system in a location away from strong light. Let the developed TLC plate dry under the hood. If your unknown solution forms more than one yellow spot, momentarily hold the chromatogram over an open bottle of concentrated HCl *under a hood*. Circle the yellow spots as soon as the developed plate is dry; you can use an ultraviolet lamp to make them more clearly visible. Measure the R_f values of all the spots.

Waste Disposal: Unless you plan to confirm the identity of the *C*-terminal amino acid (see "Other Things You Can Do"), pour the aqueous layer down the drain.

Waste Disposal: Place the chromatography solvent in a designated solvent container.

Take Care! Do not look directly at the lamp.

Waste Disposal: Place any unused capillary tubes in a sharps collector.

Report. Your report should include a statement of the problem and an account of how you applied scientific methodology to solve it. Tabulate the R_f values from your chromatograms and identify the two amino acids and the DNP-amino acid. Draw the structure of the unknown dipeptide, name it, and tell how you arrived at your conclusion.

Exercises

1   Write equations for the reactions undergone by your dipeptide during the dinitrophenylation and hydrolysis steps.
2   Tell whether the dipeptide you identified could be obtained by using cathepsin C to break down the polypeptide from cobra venom illustrated in Figure 55.1, and explain your answer. The *N*-terminal amino acid is the one labeled "1."

3 Describe and explain the possible effect on your results of the following experimental errors or variations. (a) You failed to check the pH of the dinitrophenylation reaction mixture and it fell below 6. (b) You didn't add 6 *M* HCl to the dinitrophenylation reaction mixture before extracting it with ether. (c) You dissolved the DNP-dipeptide in 6 *M* NaOH rather than 6 *M* HCl before transferring it to the hydrolysis tubes.

4 Following the format in Appendix V, construct a flow diagram for the synthesis and hydrolysis of the DNP-dipeptide.

5 When you extracted the DNP-dipeptide hydrolysate with ether, why wasn't your *C*-terminal amino acid extracted along with your DNP-amino acid? Write equations explaining your answer.

6 (a) Propose a detailed mechanism for the reaction of 2,4-dinitrofluorobenzene with the dipeptide you identified. (b) With reference to the mechanism, explain why the reaction would not take place at low pH.

7 A student, Will Bobble, accidentally added aqueous sodium bisulfate instead of sodium bicarbonate to the dinitrophenylation reaction mixture just before the first ether extraction. His DNP-dipeptide hydrolysate produced no spots when analyzed by TLC. What happened to his product, and why?

8 Another student, Bea Wilder, decided to save time by adding a large excess of sodium bicarbonate at the beginning of the dinitrophenylation reaction instead of adding smaller amounts during the reaction. When she developed her TLC plate with a solvent system containing butanol and ammonia, a large yellow spot showed up that didn't match any of the DNP-amino acid standards. What compound formed the yellow spot? Write an equation for its formation.

9 (a) Some amino acids, such as lysine and tyrosine, yield dinitrophenylated derivatives even when they are not at the end of a peptide chain. Explain, giving structures for the DNP derivatives. (b) Such derivatives do not usually interfere with the identification of the *N*-terminal DNP-amino acids since they are not extracted from the aqueous hydrolysis mixture at pH 1 by ethyl ether. Explain.

$$^{+}H_3N - CH - COO^{-}$$
$$|$$
$$CH_2CH_2CH_2CH_2NH_2$$

lysine

$$^{+}H_3N - CH - COO^{-}$$
$$|$$
$$CH_2$$

OH

tyrosine

Other Things You Can Do

(Starred projects require your instructor's permission.)

*1 To confirm the identity of the *C*-terminal amino acid, evaporate the reserved aqueous layer from the DNP-dipeptide hydrolysate, add a drop of water to the residue, and identify the *C*-terminal amino acid by paper chromatography as described.

*2 Isolate the protein casein from milk as described in Minilab 46.

3 Starting with sources listed in the Bibliography, write a research paper about primary structure determination of polypeptides and proteins, describing the applications of such reagents and enzymes as phenyl isothiocyanate, dansyl chloride, trypsin, chymotrypsin, cathepsin C, and cyanogen bromide.

Multistep Synthesis of Benzilic Acid from Benzaldehyde

Reactions of Carbonyl Compounds. Preparation of Carboxylic Acids.
Nucleophilic Addition. Oxidation. Molecular Rearrangements.

Operations:

OP-6   Heating
OP-9   Mixin
OP-11   Gravity Filtration
OP-12   Vacuum Filtration
OP-21   Drying Solids
OP-23   Recrystallization
OP-28   Melting Point
OP-34   Infrared Spectrometry

Before You Begin

1 Read the experiment, read or review the operations as necessary, and write an experimental plan.
2 Calculate the mass and volume of 150 mmol of benzaldehyde and the theoretical yields of benzoin, benzil, and benzilic acid expected from that much benzaldehyde.

Scenario

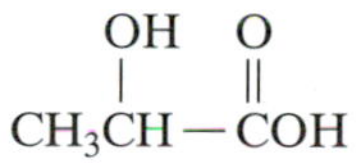

The alpha-hydroxy acids (AHAs) include lactic acid, which forms in milk as it sours and is also produced in muscles and blood after vigorous physical activity. Certain AHAs are used in anti-aging "wrinkle creams" that soften the skin and smooth out fine wrinkles and roughness. Golden Age Sundries is an organization that develops and markets various consumer goods for older people, and the success of AHAs in combating some effects of aging has led them to support research exploring the possible anti-aging benefits of other α-hydroxy acids. Your supervisor has just received a grant from them to investigate one of the AHAs, benzilic acid.

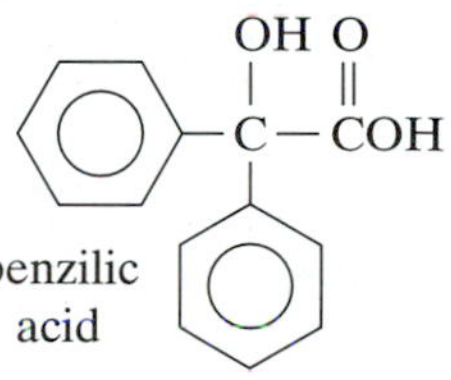

Benzilic acid can be prepared from an inexpensive starting material, benzaldehyde, by a three-step synthesis that involves a reaction called the benzoin condensation, an oxidation step, and a molecular rearrangement (see the "Reactions and Properties" section for the equations). The benzoin condensation has traditionally been carried out using cyanide ion as a catalyst, but because of the toxicity of cyanide your supervisor has decided to try Vitamin B_1 instead. Your assignment is to find out whether or not the vitamin is a suitable catalyst for the benzoin condensation and, if it is, to convert the benzoin you obtain to benzilic acid.

Applying Scientific Methodology

The scientific problem is outlined in the Scenario, and you can solve it by the successful completion of the multistep synthesis described.

Justus von Liebig and the Bitter Almond Tree

Chemical warfare is generally thought of as being a recent and uniquely human invention, but, as with many such inventions, nature beat us to it. An otherwise unexceptional insect, the millipede *Apheloria corrugata*, discourages predators with a dose of poison gas powerful enough to kill a mouse. Many trees in the rose family (such as the cherry, apple, peach, plum, and apricot) insure their continued survival by protecting their seeds and foliage with cyanide-containing substances. There are cases on record of human fatalities from eating the seeds of these species; one man who considered apple seeds a delicacy died after eating a cup of them at one sitting. The chemical weapon in each case is a natural cyanohydrin, mandelonitrile, which can be decomposed by enzymes or stomach acids into benzaldehyde and lethal hydrogen cyanide.

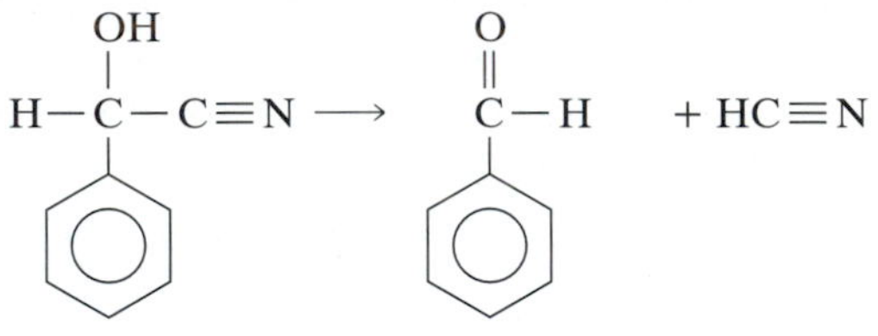

mandelonitrile   benzaldehyde   hydrogen cyanide

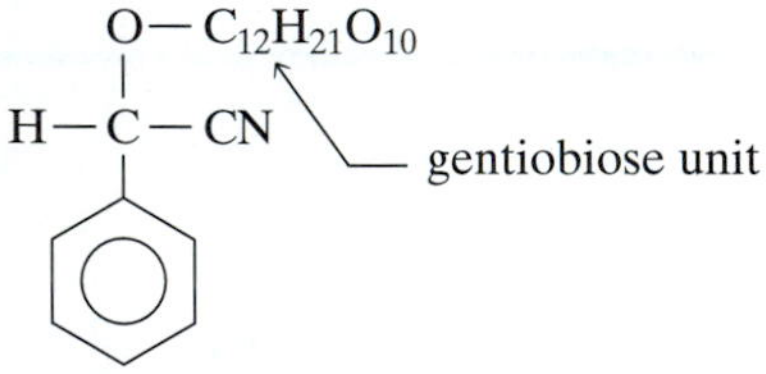

amygdalin

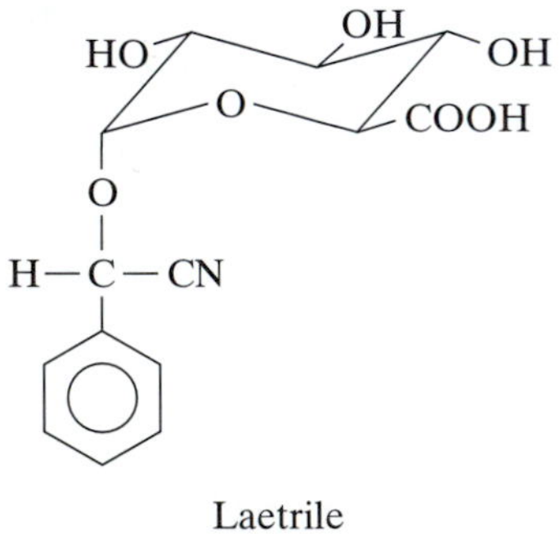

Laetrile

In most cyanogenetic (cyanide-forming) plants, mandelonitrile is present as a carbohydrate derivative called a glycoside. The most common of these glycosides is amygdalin, which is an acetal of mandelonitrile and the carbohydrate gentiobiose. Although amygdalin itself is not toxic, it can be broken down enzymatically under certain conditions to mandelonitrile and its decomposition product, hydrogen cyanide. Closely related to amygdalin is the controversial cancer drug Laetrile, which allegedly kills malignant cells by releasing hydrogen cyanide (or possibly mandelonitrile) at the site of the malignancy.

One of the most prolific sources of amygdalin is the bitter almond, which—unlike the sweet varieties used for human consumption—is grown for the oil that can be pressed from its seed kernels. The familiar "maraschino cherry" odor of almond oil comes from benzaldehyde formed by the breakdown of amygdalin. Both synthetic benzaldehyde and natural almond oil are used to flavor such food and beverage products as amaretto, cappucino, and marzipan. In the early 1800s, almond oil was purified by washing it with aqueous base to extract acids. This process yielded a small amount of a white solid that was later identified as benzoin. Friedrich Wöhler and Justus von Liebig studied the reaction in 1832 and discovered that benzoin formation is due to the catalytic action of sodium cyanide (formed when HCN from amygdalin reacted with the base) on benzaldehyde. Surprisingly, cyanide ion is almost the only substance that catalyzes this reaction, so the early discovery of the benzoin condensation depended on the coincidental production of both cyanide and benzaldehyde when bitter almond oil was washed with base.

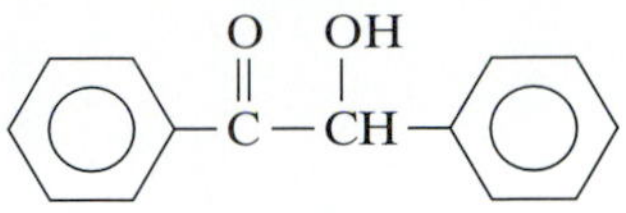

Not long after his work on the benzoin condensation, von Liebig discovered yet another unusual reaction. When benzil is treated with potassium hydroxide it rearranges to yield (upon acidification) benzilic acid. This reaction, called the benzilic acid rearrangement, is the oldest known molecular rearrangement; the prototype of a general class of rearrangements to electrophilic carbon atoms. Its starting material can made from benzoin, the serendipitous by-product of almond oil production.

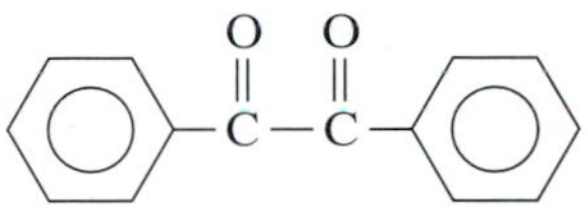

benzoin

benzil

Understanding the Experiment

In this experiment you will convert benzaldehyde to benzoin by the benzoin condensation, oxidize benzoin to benzil, and then convert benzil to benzilic acid. Because this is a multistep synthesis, you should try to keep material losses at a minimum in each step; otherwise, your overall yield will be quite low. It is important that the benzaldehyde be pure because impure benzaldehyde generally contains benzoic acid, which will inhibit the condensation reaction. If there is any doubt about the purity of the benzaldehyde, it should be distilled before use.

For many years it was believed that only cyanide ion could catalyze the benzoin condensation, but recently thiamine hydrochloride (Vitamin B_1) has proven to be an effective catalyst as well.

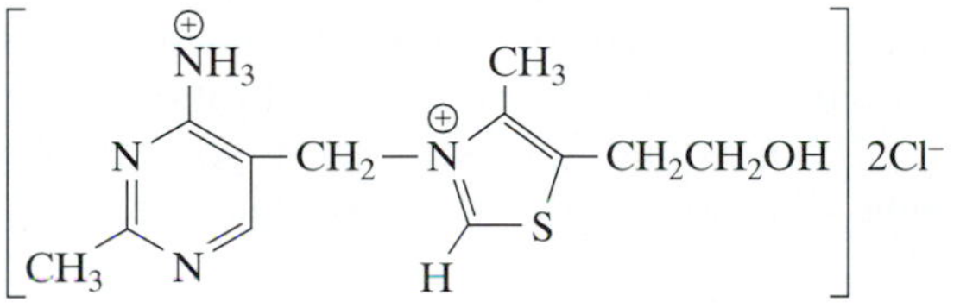

thiamine hydrochloride

It has the considerable advantage of being relatively hazard-free, unlike the very poisonous cyanides. In the presence of sodium hydroxide, thiamine hydrochloride loses two protons to form a nucleophilic species that attacks the carbonyl carbon of benzaldehyde (p. 452). The thiamine residue is sufficiently electron-withdrawing to increase the acidity of the adjacent hydrogen atom, allowing its removal by the base to yield a carbanion, which attacks another molecule of benzaldehyde. Loss of the thiamine residue from the product then yields benzoin. The benzoin condensation is carried out by simply heating a solution of benzaldehyde, sodium hydroxide, and

Major steps in the mechanism of the thiamine-catalyzed benzoin condensation

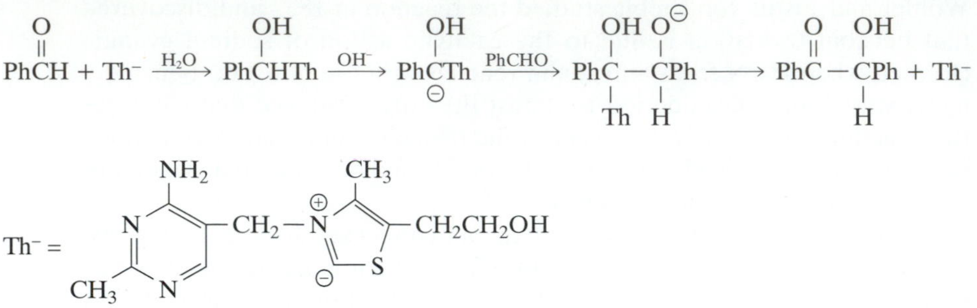

$$
\underset{\text{PhCH}}{\overset{\text{O}}{\parallel}} + \text{Th}^- \xrightarrow{\text{H}_2\text{O}} \underset{\text{PhCHTh}}{\overset{\text{OH}}{\mid}} \xrightarrow{\text{OH}^-} \underset{\underset{\ominus}{\text{PhCTh}}}{\overset{\text{OH}}{\mid}} \xrightarrow{\text{PhCHO}} \underset{\underset{\text{Th} \quad \text{H}}{\text{PhC} - \text{CPh}}}{\overset{\text{OH} \quad \text{O}^{\ominus}}{\mid \quad \mid}} \longrightarrow \underset{\underset{\text{H}}{\text{PhC} - \text{CPh}}}{\overset{\text{O} \quad \text{OH}}{\parallel \quad \mid}} + \text{Th}^-
$$

Th⁻ =

(structure: 4-amino-2-methylpyrimidine ring with NH₂, N, CH₃, N, connected by —CH₂—N⁺ to thiazole ring bearing CH₃, S⁻, —CH₂CH₂OH)

thiamine hydrochloride in an Erlenmeyer flask. Benzoin crystallizes from solution when the reaction mixture is cooled, and is collected by vacuum filtration. The 1-hour reaction time specified in the procedure is a bare minimum. If time permits, you should be able to improve your yield by increasing the reaction time and allowing plenty of time for crystallization.

Benzoin can be converted to benzil by mild oxidizing agents such as ammonium nitrate in the presence of a small amount of copper(II) acetate. The benzoin is apparently oxidized by direction action of copper(II) ions, which are reduced to copper(I) in the process. The ammonium nitrate then oxidizes the resulting copper(I) back to copper(II) ions.

When treated with alcoholic potassium hydroxide, benzil undergoes the molecular rearrangement that was discovered by von Liebig. This rearrangement involves nucleophilic attack by hydroxide ion on one carbonyl carbon, followed by migration of the adjacent phenyl group to the other carbonyl carbon (see Exercise 2). The reaction initially yields a suspension of potassium benzilate, which should dissolve when the reaction mixture is heated. Benzilic acid is then precipitated from the filtered solution by hydrochloric acid and purified by recrystallization from water.

You can gain additional experience in spectral interpretation by analyzing and comparing the infrared spectra of benzaldehyde and the products. These spectra should show bands from several different kinds of O—H, C—O, and C=O bonds, and in most of them you can observe aromatic C—H stretching vibrations with no interference from aliphatic C—H stretching bands.

Reactions and Properties

A. Benzoin condensation

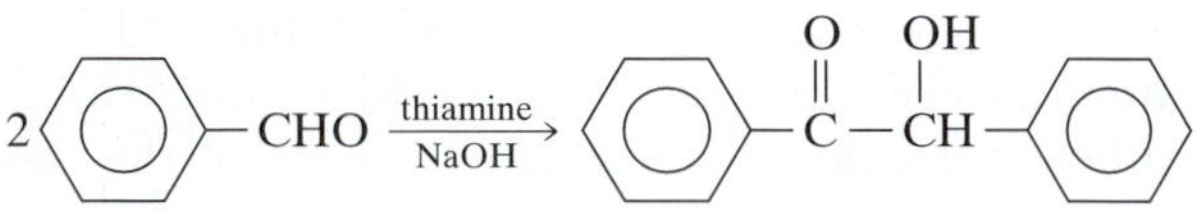

$$
2 \; \text{C}_6\text{H}_5 - \text{CHO} \xrightarrow[\text{NaOH}]{\text{thiamine}} \text{C}_6\text{H}_5 - \underset{\text{benzoin}}{\overset{\text{O} \quad \text{OH}}{\underset{}{\overset{\parallel \quad \mid}{\text{C} - \text{CH}}}} - \text{C}_6\text{H}_5}
$$

benzaldehyde   benzoin

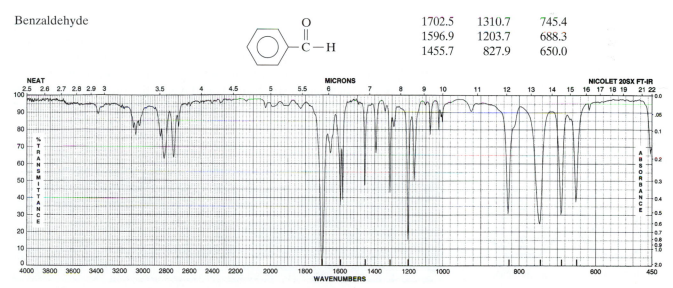

B. benzil

C. Benzilic acid rearrangement

potassium benzilate

benzilic acid

Benzaldehyde

1702.5	1310.7	745.4
1596.9	1203.7	688.3
1455.7	827.9	650.0

Figure 56.1 IR spectrum of benzaldehyde (Reproduced from *The Aldrich Library of FT-IR Spectra, Edition II*, with the permission of the Aldrich Chemical Company.)

Table 56.1 Physical properties

	M.W.	m.p.	b.p.	*d*
benzaldehyde	106.1	−26	178	1.042
thiamine hydrochloride	337.3	248 d		
benzoin	212.2	137		
benzil	210.2	95–6		
benzilic acid	228.2	151		
ammonium nitrate	80.0	170		
potassium hydroxide	56.1			

Note: m.p. and b.p. are in °C, density is in g/mL. Potassium hydroxide pellets are about 85% KOH.

Directions

A. *Preparation of Benzoin from Benzaldehyde*

> **Methanol is flammable and harmful if ingested, inhaled, or absorbed through the skin. Avoid contact with the liquid and do not breathe its vapors.**

Reaction. Dissolve 2.6 g of thiamine hydrochloride in 8 mL of water in a 125-mL Erlenmeyer flask. Add 30 mL of 95% ethanol and cool the solution in an ice bath. Add 5 mL of 3 *M* aqueous sodium hydroxide solution drop by drop, with stirring, slowly so that the temperature does not exceed 20°C. Add 150 mmol of *pure* benzaldehyde to this solution, and heat the mixture at 60°C with a hot water bath or other appropriate heat source for 1 hour or more [OP-6]. (Alternatively, you can let the reaction mixture stand at room temperature for 24 hours or more.)

Separation and Analysis. Cool the reaction mixture in an ice/water bath and scratch the inside of the flask to induce crystallization if necessary. When crystallization is complete, collect the benzoin by vacuum filtration [OP-12] and wash it on the filter with cold water, then with two 10-mL portions of ice-cold methanol. Dry [OP-21] and weigh the product before proceeding to part **B**. Measure its melting point [OP-28] and record its infrared spectrum [OP-34], or save enough product to do both later.

B. *Preparation of Benzil from Benzoin*

> **Ammonium nitrate may cause a fire or explosion if it is heated strongly or allowed to contact combustible materials. Keep it away from other chemicals or combustibles.**
> **Acetic acid causes chemical burns that can seriously damage skin and eyes. Its vapors are highly irritating to the eyes and respiratory tract. Wear gloves, dispense under a hood, avoid contact, and do not breathe vapors.**

Reaction. Multiply the mass of dry benzoin by 0.5 and combine that mass of ammonium nitrate with 35 mL of 80% (volume/volume) aqueous acetic acid in a boiling flask. Mix in the benzoin and 0.15 g of copper(II) acetate.

Safety Notes

benzaldehyde

methanol

Stop and Think: What impurity might benzaldehyde contain?

Waste Disposal: Dispose of the filtrate as directed by your instructor.

Safety Notes

ammonium
nitrate

acetic acid

Take Care! Avoid contact with acetic acid, do not breathe its vapors.

Attach a reflux condenser and a few boiling chips or a stirbar. With stirring or occasional shaking, heat the flask gently to start the reaction, which should be accompanied by vigorous evolution of gas. When the gas evolution subsides, heat the solution slowly to the boiling point, then heat it under gentle reflux [OP-6] for an hour or more.

Stop and Think: What is the gas and where did it come from?

Separation. Cool the reaction mixture to 50°C and pour it, with stirring, onto 75–100 mL of crushed ice in a beaker. Collect the benzil by vacuum filtration [OP-12] and wash it twice with cold water.

Waste Disposal: Pour the filtrate down the drain.

Purification and Analysis. Recrystallize the benzil from 95% ethanol [OP-23], washing the yellow crystals with 50% aqueous ethanol. Dry it at 75°C or below [OP-21], and measure its mass. Measure its melting point [OP-28] and record its infrared spectrum [OP-34], or save enough product to do both later.

Waste Disposal: Pour the filtrate down the drain.

C. *Preparation of Benzilic Acid from Benzil*

> **The solution of potassium hydroxide in ethanol is flammable, toxic, and corrosive; it can cause severe damage to the eyes, skin, and respiratory tract. Wear gloves, avoid contact, and do not breathe vapors.**
> **Hydrochloric acid is poisonous and corrosive. Contact or inhalation can cause severe damage to the eyes, skin, and respiratory tract. Wear gloves and dispense under a hood, avoid contact, and do not breathe the vapors.**

Safety Notes

potassium
hydroxide

hydrochloric
acid

Reaction. For every gram of dry benzil, measure out approximately 2.5 mL of aqueous 6 *M* potassium hydroxide and 3.0 mL of 95% ethanol. Combine the benzil with the liquids in a round-bottomed flask, add boiling chips or a stirbar, and heat the mixture under gentle reflux [OP-6] for 15 minutes.

Separation. Pour the *hot* reaction mixture, with stirring, into 100 mL of water and let the mixture stand for a few minutes. Then warm it to 50°C, with stirring [OP-9], to dissolve the potassium benzilate. There may be a colloidal suspension of unreacted benzil or by-products at this point, but most of the solid should dissolve. If it does not, add more warm water. Add about 0.5 g of decolorizing carbon (preferably pelletized Norit) and 0.5 g of filtering aid (Celite), stir or swirl the mixture at 50°C for 2 minutes, and filter the hot solution by gravity through fluted filter paper [OP-11]. *Under the hood*, carefully add 15 mL of concentrated hydrochloric acid to 100 mL of crushed ice in a beaker. Add 10–15 mL of the potassium benzilate solution with stirring and scratch the sides of the beaker until crystals begin to form. Then add the rest of the solution slowly with continuous stirring. When the addition is complete, test the solution with pH paper. If the pH is higher than 2, add enough HCl to bring it down to 2. Cool the solution in ice and collect the benzilic acid by vacuum filtration [OP-12], washing it on the filter with cold water.

Take Care! Wear gloves, avoid contact with HCl, do not breathe its vapors.

Waste Disposal: Pour the filtrate down the drain.

Purification and Analysis. Recrystallize the benzilic acid [OP-23] from boiling water; then dry and weigh it. Measure the melting point [OP-28] of the benzilic acid and record its infrared spectrum [OP-34], or obtain the spectrum from your instructor.

Waste Disposal: Pour the filtrate down the drain.

Report. Your report should include a statement of the problem and an account of how you applied scientific methodology to solve it. Calculate the percent yield for each step of the synthesis and the overall percent yield of benzilic acid. Try to account for significant losses.

Exercises

1 Interpret the infrared spectra of the products as completely as you can and discuss the evidence suggesting that each reaction step has taken place as expected.
2 Propose a mechanism for the rearrangement of benzil to potassium benzilate in the presence of potassium hydroxide.
3 Describe and explain the possible effect on your results of the following experimental errors or variations. (a) The bottle of benzaldehyde for part **A** contained some white solid. (b) The label on the copper acetate bottle for part **B** read "cuprous acetate." (c) The pH of the solution in part **C** was 6 when you vacuum filtered it.
4 Following the format in Appendix V, construct a flow diagram for the synthesis of benzilic acid from benzaldehyde.
5 Write a detailed mechanism for the cyanide-catalyzed condensation of benzaldehyde, and explain the role played by the cyanide.
6 Write a synthetic pathway illustrating the preparation of compound **1** from a suitable five-carbon starting material, using the reactions described in this experiment.
7 Benzil reacts with urea in the presence of sodium hydroxide to form the sodium salt of 5,5-diphenylhydantoin (Dilantin Sodium), a powerful anticonvulsant used to treat epilepsy. By analogy with the benzilic acid rearrangement, propose a mechanism for this reaction.
8 In the presence of a little sulfuric acid, benzilic acid reacts with acetone to form a compound with the molecular formula $C_{17}H_{16}O_3$. Propose a structure for this product and write a balanced equation for the reaction.

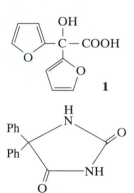

1

5,5-diphenylhydantoin

Other Things You Can Do

(Starred projects require your instructor's permission.)

*1 Test both your crude and purified benzil for the presence of unreacted benzoin by dissolving a few crystals of each in 1 mL of ethanol and adding a drop of dilute sodium hydroxide solution. Test a little benzoin in the same way and compare the results. Shake each solution with air, let them stand, and record your observations.
*2 Carry out another reaction of benzaldehyde in Minilab 35.
3 Starting with sources listed in the Bibliography, write a research paper about the use of alpha-hydroxy acids, Retin-A, and other skin care products to treat skin conditions, giving some of the advantages and disadvantages of each.

Using the Chemical Literature in an Organic Synthesis

Multistep Synthesis. Searching the Chemical Literature.

Before You Begin

1 Read the experiment and Appendix VII, *The Chemical Literature*, and familiarize yourself with the layout of the Bibliography.

2 In your laboratory notebook, outline a synthetic pathway leading to the target compound from readily available starting materials or the starting material(s) recommended by your instructor. For the chemicals you will be using and the target compound, list information concerning their physical properties, purification, safety and health hazards, and disposal. If the synthesis requires laboratory techniques with which you are not familiar, find information about these techniques in the literature, describe them, and sketch any special apparatus required for them. Locate any published spectra or other data that will help you characterize the target compound and list their sources.

3 In your laboratory notebook, write a complete description of your plan for the synthesis, including a detailed procedure for each synthetic step (with safety precautions) and a description of the methods you propose to use to analyze the purity and verify the structure of the product. The procedure should be designed to yield about 2–5 g of the final product unless your instructor indicates otherwise. Your writeup should also include a concise description of the experimental methodology. Submit your laboratory notebook (or duplicate pages) to your instructor for approval well in advance of the scheduled lab period. You should also discuss your plans with the instructor to make sure that the necessary chemicals and equipment are available, that the procedure can be carried out safely under the existing conditions, and that you have a good understanding of the experiment and any potential problems that might arise.

Scenario

Your supervisor has a backlog of chemicals that need to be prepared for various clients, but doesn't have time to write up the procedures for you. You are on your own!

Applying Scientific Methodology

In the previous experiments, the scientific problem to be solved was specified and a course of action was laid out for you. For this experiment, you will have to operate like a research chemist—find a suitable scientific problem to work on, plan a course of action, work out detailed experimental procedures, carry out your experimental plan in the laboratory, and evaluate your results. Throughout, you will be expected to apply what you have already learned about organic synthesis in this course and what you can learn from the chemical literature.

Planning Your Synthesis

Your instructor will either suggest a target compound for you to synthesize or allow you to choose your own, subject to approval. Your instructor will also inform you whether you will work individually or in teams. If you are to select your own target compound, you should consult your instructor before coming to a final decision since your plans may be either too ambitious or not challenging enough. You can search for ideas in chemical periodicals such as the *Journal of Chemical Education*, in other laboratory manuals, or in many of the sources described herein (see Appendix VII for more detailed information on these and other sources). Preferably, the target should be a compound that you have an interest in and really *want* to synthesize, such as an artificial flavor ingredient, a cosmetic ingredient, a perfume ingredient, an insect repellent, a plant growth hormone, an analgesic drug, a pheromone, an artificial sweetener, a sunscreen, an antibiotic drug, a local anesthetic, a fluorescent dye, a chemoluminescent compound—the possibilities are limitless!

Sources from the Bibliography are referred to herein by category and number. For example, Organic Syntheses *[B20] is the 20th entry under Category B,* Organic Reactions and Syntheses.

Once you have selected a target compound, you need to outline a synthetic pathway for making it from some readily available starting material. To help you work out a synthetic pathway, you can consult such sources as *Guidebook to Organic Synthesis* [Bibliography, B14], *Organic Synthesis: The Disconnection Approach* [B28], and *Principles of Organic Synthesis* [B17], as well as your own lecture textbook.

After you develop one or more promising synthetic pathways, you should learn more about the reactions required for the synthesis. *Modern Synthetic Reactions* [B12] and advanced organic chemistry texts by March [H5] and Carey/Sundberg [H1] describe a number of synthetic reactions, and also give literature references to specific synthetic procedures. *Organic Reactions* [B19] is a comprehensive source of information about specific reactions, and also includes some representative synthetic procedures as well as numerous references to literature procedures. It describes the applications and limitations of many synthetic reactions and may help you tailor a synthetic procedure to fit the particular starting materials you will be working with. Many reactions and compound types are described in depth in review articles from such journals as *Chemical Reviews* and *Angewandte Chemie (International Edition in English)*.

Two excellent sources of detailed information about chemical reagents are Fieser's *Reagents for Organic Syntheses* [B9] and *Encyclopedia of Reagents for Organic Synthesis* [B21]. General reference books such as *The Merck Index* [A4], *Lange's Handbook* [A8], the *CRC Handbook of Chemistry and Physics* [A15], the *Dictionary of Organic Compounds* [A11], and the *Aldrich Catalog* [A1] can be consulted for information about the physical properties of the reagents, as well as other chemicals (including solvents) involved in your synthesis.

The *Aldrich Catalog* also gives information about specific hazards associated with chemicals. More detailed information about chemical hazards can be found in *Sax's Dangerous Properties of Industrial Materials* [C4], the *Sigma-Aldrich Library of Chemical Safety Data* [C3], and other sources from category C of the Bibliography. You should also obtain and read the Material Safety Data Sheets (MSDS) for any hazardous chemicals you will be working with. Information about safe laboratory practices may be found in *Prudent Practices in the Laboratory* [C6], which also provides information about dealing with laboratory wastes.

There are many useful collections of detailed synthetic procedures. One of the best (and the first you may want to consult) is *Organic Syntheses* [B20], which provides a representative selection of carefully tested procedures. Other useful if less comprehensive sources include *Vogel's Textbook of Practical Organic Chemistry* [B29], *Organic Functional Group Preparations* [B25], and other works from category B of the Bibliography, as well as a variety of basic and advanced organic chemistry lab textbooks. For more information about these and other useful sources on organic reactions and syntheses, as well as suggested literature searching strategies, read the three-part series "Information Sources for Organic Chemistry" [K5].

If you find a procedure that requires more (or less) of a reactant than you have, you will have to scale down (or scale up) the procedure for your use. For example, if a literature procedure calls for 25 g of acetophenone and you have only 6.0 grams, you should multiply the quantities of all reactants, reagents, catalysts, solvents, or other chemicals by the scaling factor 0.24, which is the ratio of 6.0 g to 25 g. You will also need to reduce the size of the apparatus accordingly, and in some cases you may be able to reduce the reaction time. Most scaled-down syntheses do not give as high a percent yield as the original procedure. For example, losing 0.1 g of product during a transfer has a greater effect on the percent yield of a small-scale experiment than it does when you have more product to work with. Other factors such as differences in the rate of heat transfer or surface-contact area may also cause differences in the outcome of a scaled-down experiment, but most procedures can be scaled down to a moderate extent without difficulty.

If you can't locate a synthetic procedure for a specific synthetic conversion you plan to carry out, you may have to adapt a procedure for the same reaction of a closely related compound or search for a suitable procedure in the literature. Adapting a procedure is often just a matter of recalculating the masses of the reactants. For example, suppose you want to adapt a literature procedure for the $NaBH_4$ reduction of cyclohexanone to the same reaction of 4-methylcyclohexanone. If the original procedure used 3.00 g (30.5 mmol) of cyclohexanone (M.W. = 98.2) you will need to use about 3.43 g (30.5 mmol) of 4-methylcyclohexanone (M.W. = 112.2) while keeping the quantities of other reactants the same. Such an adaptation may also require changes in such things as the size of the glassware you use, the amount or kind of recrystallization solvent needed, and the boiling range for a distillation, so it is important to identify changes in the properties of the reactants and products that necessitate procedural changes.

References to literature procedures for organic syntheses are given in many of the sources noted, such as *Organic Reactions* and the textbooks by March and Carey/Sundberg. *Theilheimer's Synthetic Methods of Organic Chemistry* [B29] is an extensive guide to synthetic procedures that is updated each year. Each entry gives a brief outline of the procedure and cites its source. Other useful if less complete sources include *Synthetic Organic Chemistry* [B32], which covers many of the older reactions,

Comprehensive Organic Transformations [B13], and *Compendium of Organic Synthetic Methods* [B11].

The most comprehensive guides to the literature of organic chemistry are *Chemical Abstracts (CA)* [J3] and *Beilstein* [A3]. With some practice, you can use them to find information about every aspect of your synthesis. For example, to find out more about your target compound, you should first find its *CA* index name and registry number in the *Index Guide*, which appears with the collective indexes that are published periodically by *Chemical Abstracts*. Then you can conduct a search of *Chemical Abstracts*—from an online database, if one is available—to locate articles that refer to your compound. See Appendix VII for more information about the use of *CA*, *Beilstein*, and online searches.

If you locate a specific procedure for the synthesis of your target compound in the primary literature, you should not assume that it will be suitable for your purposes as written. It may call for unavailable or excessively hazardous chemicals or suggest techniques and apparatus that are not appropriate to your laboratory situation. Primary literature procedures tend to be rather condensed, so you will probably have to "fill in the blanks" from your own experience or by consulting other sources.

If you are unfamiliar with some of the laboratory techniques required for a synthesis, consult *Guide for the Perplexed Organic Experimentalist* [D3], the appropriate volume in Weissberger's *Technique of Organic Chemistry* [D6], or other references from Categories D and E of the Bibliography. If you need to purify a solvent or another chemical, see *Purification of Laboratory Chemicals* [D5] or another source from Category D.

Once you have synthesized your target compound (or what you think is your target compound) you will have to confirm its identity by measuring one or more physical properties, such as its melting point or boiling point, and obtaining and interpreting at least one kind of spectrum. Refer to the appropriate sources in Category F if you need information about spectral analysis and the interpretation of spectra. *Spectrometric Identification of Organic Compounds* [F20] is a good place to start. There are many sources of published spectra such as the *Aldrich Library of FT-IR Spectra* [F18] and the *Aldrich Library of ^{13}C and ^{1}H FT-NMR Spectra* [F16]. Additional information about spectrometric and "wet" chemical methods for identifying organic compounds can be found in sources from Category G. You may also need to assess the purity of your product by gas chromatography, TLC, HPLC, or another method. Refer to Category E of the bibliography for appropriate sources. For additional help in carrying out a literature search, consult Appendix VII and appropriate sources listed in Category K of the Bibliography.

Directions

Independent synthesis projects are intended for advanced or honors students who have mastered the major operations described in Part V. These directions only furnish general guidelines regarding your synthesis. You must provide the procedural details and have them approved by your instructor. Keep detailed notes of your work and observations as described in Appendix II.

Except when you have reliable information to the contrary, assume that all chemicals you will use are flammable and are hazardous by ingestion, inhalation, and skin absorption. Wear gloves and use a hood whenever possible, avoid contact with the chemicals, do not breathe their dust or vapors, and keep them away from ignition sources and other chemicals that might react with them.

If necessary, purify any starting materials, reagents, or solvents that appear to be insufficiently pure. Carry out the synthesis in the laboratory according to the procedures you have developed. Purify the product, weigh the thoroughly dried product, and calculate the percentage yield. Measure appropriate physical constants of the product and assess its purity by an appropriate chromatographic method, if requested. Confirm the identity of the product by one or more spectrometric methods and/or by preparing one or more derivatives.

Report. Write up your report as if it were a scientific paper being submitted to a professional publication such as the *Journal of Organic Chemistry*. Such papers are traditionally written in an impersonal, third-person style (for instance, "the solution was stirred" rather than "I stirred the solution"). Your report should include the following items, unless your instructor directs otherwise.

1 A brief but descriptive title
2 Your name(s) and affiliation
3 A brief abstract that summarizes the principal results of the work
4 An introductory section (usually untitled) that provides a concise statement of the purpose and possible applications of the work, supported by descriptions of related work from the literature. If your synthesis differs significantly from those reported in the literature, tell how and why.
5 A section titled "Experimental Methods" that gives enough detail about your materials and methods so that another experienced worker could repeat your work. Give current *Chemical Abstract* index names and registry numbers (see Appendix VII) for important starting materials and the product, and provide information about their purity if possible. Note any significant hazards and safety precautions in a separate paragraph labeled "Caution."
6 A section titled "Results and Discussion" (or two separate "Results" and "Discussion" sections) that summarizes the important experimental results and points out any special features, limitations, or implications of your work. You should include any data (including spectral parameters) that will help justify your conclusion. You may also wish to suggest different approaches to the problem or areas that require further study.
7 A section titled "Conclusions" that states any conclusions you can draw from your work, based on the evidence presented.
8 A section titled "References" that gives complete citations for all literature sources referred to in the report.

Your instructor may suggest additions to or modifications of this list. *The ACS Style Guide* [L11] provides a more detailed discussion of these components, as well as helpful suggestions about writing style.

Exercises

1 Herbert C. Brown was awarded the Nobel Prize for chemistry in 1979 for his work with reagents for organic syntheses, principally organoboranes. Give the structure of Brown's reagent 9-BBN (9-borabicyclo[3.3.1]nonane) and describe some of its applications and characteristics, giving equations where appropriate. Cite several recent papers reporting the use of 9-BBN.

2 The synthesis of dodecahedrane has been described as the "Mount Everest of alicyclic chemistry." (a) Give its *Chemical Abstracts* index name and registry number and draw its structure. (b) Locate the paper in which the synthesis of dodecahedrane was first reported. Give its title, the authors and their affiliation, and a standard literature citation showing where and when the paper appeared. (c) Summarize the salient points of the synthesis in your own words, specifying the starting material, the number of synthetic steps required, the yield of dodecahedrane, and the purification method. Tell how dodecahedrane was characterized, and report any spectral parameters. (d) Find a systematic name for dodecahedrane that is different from the *CA* index name. (e) Quote and explain the reference to an ancient Greek philosopher that appears in one of the previously listed papers.

3 (a) Cite the paper in which the use of pyridinium chlorochromate to oxidize alcohols to carbonyl compounds was first reported. (b) Tell how pyridinium chlorochromate is prepared, describe a typical experimental procedure for oxidation of a primary alcohol to an aldehyde, and discuss the stoichiometry of the reaction. (c) Report on any hazards associated with the use of pyridinium chlorochromate, and describe safe disposal procedures for the reagent.

4 (a) Give a concise definition of the Knoevenagel condensation. (b) Describe typical experimental conditions for conducting a Knoevenagel condensation between an aldehyde and diethyl malonate. (c) Find and summarize a detailed procedure for the synthesis of ethyl coumarin-3-carboxylate using this reaction. (d) Describe the Doebner modification of the Knoevenagel condensation, giving at least one example.

5 (a) Describe any hazards associated with the use of thionyl chloride, and describe proper handling precautions and disposal procedures for this reagent. (b) Tell how thionyl chloride can be purified for use as a chemical reagent. (c) Cite a paper in which thionyl chloride was used to convert an amino acid to an ester in one step, and briefly describe the experimental conditions.

6 (a) Give the current *Chemical Abstracts* index name and registry number for (+)-camphor. (b) Draw a structure for (+)-camphor that shows its absolute configuration. (c) What is the melting point of the oxime of (+)-camphor? (d) Find an infrared spectrum for camphor and give the wave numbers of the major absorption bands. (e) Tell where you can find information about camphor in *Beilstein*. (f) Tell how most synthetic camphor is currently produced, giving equations for the reactions.

7 (a) Give the *CA* index name of the compound whose *CA* registry number is [5543-57-7]. (b) Give the trivial name of this compound and describe its major application. (c) Tell where information about this compound can be found in *Beilstein*.

8 For each of the following abbreviated names, provide the full name of the journal, any names under which it was previously published, the year in which it was first published (as volume 1, under a current or previous name), the language or languages in which it is published, and a nearby library that carries it. (a) *Acc. Chem. Res.* (b) *Helv. Chim. Acta*, (c) *Dokl. Akad. Nauk SSSR*, (d) *Chem. Ber.*

9 Find detailed synthetic procedures for the following compounds, briefly describe the experimental conditions, and write equations for the relevant reactions: (a) hexaphenylbenzene, (b) vanillic acid, (c) 1,2-cyclononadiene, (d) octadecanedioic acid.

10 For each of the following compounds, find and reproduce as many different kinds of published spectra (or spectral parameters) as you can: (a) mandelic acid, (b) resorcinol, (c) exaltone, (d) bourbonal, (e) testosterone.

A Research Project in Organic Chemistry

Research in Chemistry.

Before You Begin

1 Read the experiment and review the "Scientific Methodology" section of the Introduction (pp. 2-5). Review Appendix VII as necessary.
2 Write a research proposal based on one of the projects discussed in this experiment or on a project that you have developed yourself.
3 In your laboratory notebook, write a complete description of your research plans, including a detailed procedure for each experiment you intend to perform, with safety precautions. Submit your research proposal and lab notebook (or duplicate pages) to your research mentor.

Scenario

Your job with the Consulting Chemists Institute is nearly finished. Your supervisor, having a high regard for your abilities and potential, has encouraged you to consider making chemistry your career. To obtain employment as a professional chemist, you will probably have to earn a master's degree or a Ph.D. in chemistry. One of the requirements for your degree is the completion of an independent research project. You have decided to get a head start by carrying out an undergraduate research project under the direction of your supervisor or another faculty mentor. Now you must select a project to work on.

Applying Scientific Methodology

For this experiment, you will have to operate like a professional chemist—find a suitable scientific problem to work on, develop a hypothesis, plan a course of action, work out detailed experimental procedures, follow your experimental plan in the laboratory, evaluate your results, and arrive at a conclusion. Throughout, you will be expected to apply what you have already learned about organic chemistry in this course and what you can learn from the chemical literature.

Conducting Scientific Research

Conducting worthwhile research in chemistry requires a curious and logical mind, efficient work habits, superior laboratory skills, a good understanding of chemical concepts, and a great deal of perserverence. There is no single method to be followed in scientific research; different scientists often employ vastly different approaches. But the outline of scientific methodology in the Introduction to this book lists the important steps in a scientific research project.

1 Define the problem
2 Plan a course of action
3 Gather evidence
4 Evaluate the evidence
5 Develop a hypothesis
6 Test the hypothesis
7 Reach a conclusion
8 Report the results

First you need to decide what problem to investigate. Some examples of possible research problems are discussed in the next section, but you may wish to select your own problem to work on, after consultation with your mentor. It should, above all, be some aspect of organic chemistry that interests you intensely; without such motivation it will be difficult for you to expend the time and effort required to complete a research project successfully. If possible, it should be in an area in which your mentor, or other chemists at your institution, have considerable interest and expertise. That will help you benefit from the guidance and enthusiasm they can provide. One way to come up with ideas for a research project in organic chemistry is to think about things that puzzled you during the lab or lecture course, such as unexplained lab observations, theories or mechanisms that don't account for certain experimental results, incomplete or inconsistent explanations of phenomena, and gaps in textbook descriptions of some aspect of organic chemistry. After consulting the chemical literature you may find that there are published explanations for such phenomena—no textbook or course can cover the entire field of organic chemistry—but in the process of looking for explanations you may find other problems worth exploring. You can also get ideas by looking through chemistry periodicals such as the *Journal of Chemical Education*, which occasionally contains articles describing research-type experiments, and review journals or research reports such as *Angewandte Chemie (International Edition in English)*, *Chemical Reviews*, or *Annual Reports on the Progress of Chemistry, Section B* [J2]. When you do find an article dealing with a subject that interests you, be sure to check the references in that article to locate more articles on the same subject. The important thing is to read and learn as much as you can about the areas that interest you; the more you learn, the easier it is to select a worthwhile problem.

Sources from the Bibliography are referred to herein by category and number. For example, Organic Syntheses *[B20] is the 20th entry under Category B,* Organic Reactions and Syntheses.

Before you start working on any kind of scientific problem, you need to find out what research has already been carried out that is related to the problem. Therefore you should do a thorough search of *Chemical Abstracts* [J3], and perhaps of other sources such as *Beilstein* [A3] as well. See Appendix VII for information about both of these references, and about online searches of *Chemical Abstracts*. If, after a literature search, you learn that a problem you were planning to work on has already been "solved" satisfactorily, there is little point in working on the same problem. If only certain aspects of a problem have been investigated, you may decide to modify your objectives and concentrate on the areas where original research can still be performed. And learning what has already been done on problems releated to your own will help you fine-tune your objectives and plan your course of action.

The course of action you develop will depend on the kind of problem you select. If the problem requires a multistep synthesis, for example, you should read Experiment 57, *Using the Chemical Literature in an Organic Synthesis*. In any case, you will need to consult the chemical literature to help you design your experimental approach and work out detailed procedures for all experimental work. Start by reading the appropriate sections of Appendix VII and consulting the Bibliography to find out about some of the sources that are available to you. You may also want to consult one or more of the sources in category K of the Bibliography for more complete and detailed information about the chemical literature.

If you will need special chemicals, supplies or equipment, first check with your mentor or the stockroom manager to find out if they are available. If chemicals must be purchased, check *Chem Sources*, which lists the chemicals manufactured by hundreds of chemical companies, or consult the *Aldrich Catalog* [A1] or another chemical catalog. If supplies need to be purchased, check *The VWR Scientific Products Catalog*, the *Fischer Catalog*, or another chemical supplies catalog. Your plans and procedures should be written up in detail in your laboratory notebook, summarized on a form like the one in Figure 58.1, and submitted to your instructor for approval.

While you are carrying out your research project you should faithfully record your ideas, experimental work, significant observations, results, and other aspects of your research in your laboratory notebook. The importance of taking good notes and suggestions for maintaining and using a lab notebook effectively are covered in *Writing the Laboratory Notebook* [L35].

Unless your instructor indicates otherwise, you should report your results as if they were to be published in a scientific journal—and they could be, as the results of undergraduate research projects are sometimes accepted for publication. See the Directions for a description of the format to be followed.

Suggested Research Projects

Many of these projects are open-ended and can lead to further investigations not suggested in the outlines. Although some of them may not lead directly to new or unique contributions to the sum of chemical knowledge, others could yield hitherto unknown facts and new discoveries. Each project is outlined only briefly so that you will have considerable latitude in developing and pursuing a course of action. Each project names one or more key references that will often have citations to other articles on the same subject.

1   Isolation and Testing of a Natural Growth Inhibitor
Key References: *J. Chem. Educ.* **1977,** *54*, 156. K. Paech and M. V. Tracey, *Modern Methods of Plant Analysis*, Vol. 3 (Berlin: Springer-Verlag, 1955).

Name _____ Date submitted _____

Name of Project_____

Describe the scientific problem and indicate its significance.

State what you expect to accomplish and, in general terms, how you plan to accomplish it.

Describe, in approximate chronological order, specific experiments you expect to perform. In each case give a rough estimate of the time you expect each experiment to take.

Describe any special chemicals, supplies, and equipment that will be needed. If any must be purchased, estimate their cost.

Approved by (mentor) _____ on (date) _____

Figure 58.1 Research Project Summary Page

juglone

It has been speculated that juglone, a natural quinone that occurs in walnut trees, acts as a chemical defense agent for walnut seedlings, favoring their growth at the expense of competing species, particulary plants of the heath family (genus *Ericaceae*).

You could isolate juglone or another substance that affects plant growth from a natural source and test its effect on the growth of various species or varieties of plants. Such substances could also be modified chemically to see what effect, if any, the modifications have on their growth-regulating properties.

2   Analysis of an Essential Oil

Key References: *J. Chem. Educ.* **1994**, *71*, A146. *J. Chem. Educ.* **1969**, *46*, 846.

Essential oils can be obtained from many plants by steam distillation, expression, and other isolation techniques. They can also be purchased in some health food stores and other stores that sell herbs and new age supplies. You could obtain an essential oil and analyze it by gas chromatography to see if it contains a few major components that might be easily separated and identified. Its components could be separated by preparative gas chromatography, preparative HPLC, or another method. You may then be able to identify one or more components by IR, NMR, GC-MS, or some other spectrometric method.

3   Medicinal Components of Indigenous Wild Plants

Key References: S. Foster and J. A. Duke, *A Field Guide to Medicinal Plants: Eastern and Central North America.* (Boston: Houghton Mifflin, 1990). Ikan, R. *Natural Products: A Laboratory Guide,* 2nd ed. (San Diego: Academic Press, 1991).

Many indigenous wild plants were used by Native Americans for medicine and some still have medicinal uses. You could select a plant that appears to show significant pharmacalogical activity and attempt to isolate and identify one or more of its components. Keep in mind that the plant you select must be abundant enough to provide the large amounts of plant material that may be needed. In some cases it may be possible to isolate major components by extraction or steam distillation, analyze them by gas chromatography, separate one or more components using basic laboratory operations, and identify them by IR or NMR spectrometry. But since most plants contain many diverse and complex constituents, it may be necessary to use advanced instrumental methods such as high-performance liquid chromatography (HPLC) or gas chromatography-mass spectrometry (GC-MS) to separate and identify the components of a species you have selected. Unless you have access to such instruments, you may have to analyze the essential oils or extracts of several different species using gas chromatography or HPLC until you find one that is simple enough to work with.

4   The Effect of Molecular Modification on Odor

Key References: *J. Chem. Educ.* **1992**, *69*, A43. R. W. Moncrieff, *The Chemical Senses, 3rd ed.* (Cleveland, Ohio: CRC Press, 1967).

Relatively minor modifications in molecular structure can have drastic and often unpredictable effects on odor. Select an organic compound that has a discernible odor and two or more functional groups, such as *trans*-cinnamaldehyde or 4-hydroxyacetophenone, and modify its molecules in various ways to find out how such modifications affect its odor.

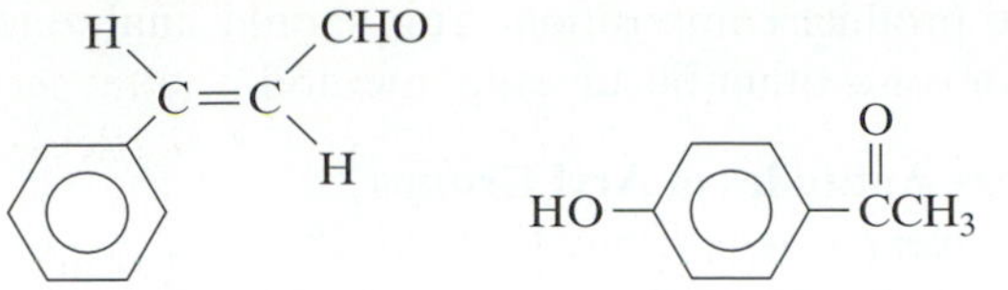

trans-cinnamaldehyde   4-hydroxyacetophenone

Note that the presence of several functional groups may cause complications in certain reactions unless one of them is protected. See *Protective Groups in Organic Synthesis* [B10] or a related source for information about the use of protective groups.

5   Gas Chromatographic Analysis of Gasoline
Key References: *J. Chem. Educ.* **1996**, *73*, 1056. *J. Chem. Educ.* **1976**, *53*, 51.

The compositions of different grades of gasoline and gasoline from different suppliers can vary considerably. Most gas stations label their gasoline accurately, but a few have been known to sell unleaded regular gasolines as premium gasolines, at premium prices. Some suppliers add certain oxygenates (oxygen-containing components) to their gasoline, such as methyl *t*-butyl ether (MTBE), which is a subject of environmental concern. You can analyze samples of gasoline from different sources by gas chromatography to look for oxygenates or for differences in their gas chromatographic hydrocarbon profile.

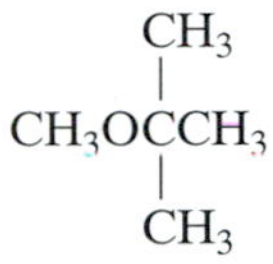

$$CH_3OCCH_3$$

MTBE

6   Stereochemisty of Addition Reactions
Key Reference: *J. Chem. Educ.* **1990**, *67*, 554.

Most bromine addition reactions appear to proceed through a bromonium ion intermediate that yields *anti* addition exclusively, but certain unsaturated substrates, such as *trans*-anethole, yield substantial amounts of *syn*-addition products. You can investigate the effect of substrate structure on stereochemistry by selecting one or more unsaturated compounds, carrying out a bromine addition reaction, and analyzing the product. You can explore the effect of reaction conditions on stereochemistry by using different brominating reagents, varying the solvent, or varying the reaction temperature. You might also extend your study to stereoselective reactions other than bromine addition.

7   Alkylation of a Bidentate Nucleophile
Key Reference: *J. Chem. Educ.* **1990**, *67*, 611.

Under different reaction conditions, the alkylation of sodium saccharin with ethyl iodide can yield varying proportions of *N*-ethylsaccharin and *O*-ethylsaccharin (see Experiment 20).

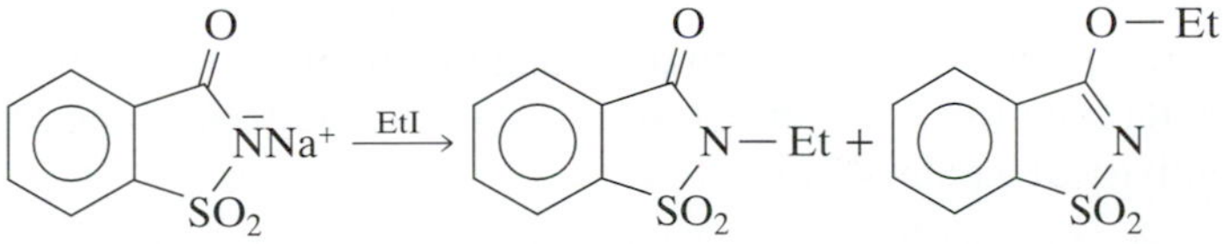

sodium saccharin   *N*-ethylsaccharin   *O*-ethylsaccharin

You can explore the effects of different factors such as the metal cation, alkylating agent, substrate, reaction temperature, and use of a phase-transfer catalyst on the product composition. You should analyze the product by HPLC, NMR, or some other intrumental method.

8   Migratory Aptitudes of Aryl Groups

Key Reference: *J. Chem. Educ.* **1971**, *48*, 257.

The pinacol rearrangement involves the migration of an alkyl or aryl group, presumably by means of a bridged intermediate.

You can investigate the factors that contribute to migratory aptitude by preparing pinacols in which either of two different groups (R and R') can migrate, and analyzing the products of their pinacol rearrangement reactions. By carrying out a reasonable number of reactions using carefully selected substrates, you should be able to arrange different groups in order of their migratory aptitude. The pinacols can be prepared from appropriate ketones and the products analyzed by NMR, gas chromatography, or another instrumental method.

9   Stability of Endocyclic and Exocyclic Double Bonds

Key Reference: *J. Chem. Educ.* **1973**, *50, 372.*

The stability of an alkene depends on the number of alkyl substitutents attached to the double-bonded carbon atoms. In general, the more alkyl substituents there are, the more stable the alkene. The stability of a cyclic alkene can also vary with the location of the double bond; in general, double bonds that are within the ring (endocyclic) are more stable than those external to the ring but involving a ring carbon (exocyclic). You can investigate the stability of double bonds in one or more ring systems by carrying out elimination reactions (such as dehydration or dehydrohalogenation) on *tertiary* substrates (**1**) that can yield both endocyclic and exocyclic double bonds without rearrangement.

By varying certain features of the substrate, such as ring size and the kinds of substituents on the starred carbon atom, you can explore the factors that affect double-bond stability. Alcohol substrates can be made by Grignard reactions with cyclic ketones; alkyl halides and other substrates can be made from the alcohols. The products can be analyzed by gas chromatography, HPLC, NMR, or another appropriate method.

10   Synthesis and Activity of an Insect Pheromone

Key References: *J. Chem. Educ.* **1991**, *68*, 71. *J. Chem. Educ.* **1986**, *63*, 1014. *J. Chem. Educ.* **1984**, *61*, 927.

Insect pheromones are "chemical messengers" that attract other insects for mating, inform them of danger, help them find their way, and perform a variety of other functions. A number of pheromone syntheses are reported in the chemical literature. You can select a pheromone that interests you, synthesize it, and test its effect on the target insect (if you can obtain specimens). You may also be able to determine whether one stereoisomer is more effective than another or than a racemic mixture of the pheromone, and you can investigate the effect of impurities or molecular modifications on the activity of a pheromone.

11   Effect of Phase-Transfer Catalysts on a Reaction

Key References: E. V. Dehmlow and S. S. Dehmlow, *Phase Transfer Catalysis,* 3rd ed. (New York: VCH, 1993). W. P. Weber and G. W. Gokel, *Phase Transfer Catalysis in Organic Synthesis* (New York: Springer-Verlag, 1977).

Bridged intermediate in the pinacol rearrangement

Note: R, R' can be alkyl, aryl, or hydrogen; Z is leaving group such as OH or Br

Phase-transfer catalysts accelerate many two-phase reactions by helping reactants or intermediates cross the phase boundary and come into contact with one another. You should select an organic reaction that is carried out in two phases, preferably one for which phase-transfer catalysis has not been investigated extensively. Perform one or more examples of the reaction with and without a quaternary ammonium or phosphonium salt to see whether or not the onium salt catalyzes the reaction. If it does, you can use different onium salts and observe their effects on the reaction rate to determine their relative catalytic effectiveness. You can also vary substrates, solvents, and other factors to observe the effect of such changes.

12   Synthesis of a New Organic Compound
Key Reference: *J. Amer. Chem. Soc.* **1974**, *94*, 4024.

To prove the structure of α-pinene, Adolf von Baeyer started by oxidizing it to pinonic acid with potassium permanganate.

pinonic acid

This reaction has been shown to proceed in high yield using "purple benzene," a solution of potassium permanganate in benzene containing a crown ether. Pinonic acid is a relatively uncommon compound containing two reactive functional groups. It should therefore be possible, using it as the starting material, to prepare a completely new organic compound—one that has never been reported in the chemical literature (you will need to conduct a thorough literature search to establish this). Benzene is hazardous and the crown ether needed for making purple benzene is quite expensive, so you should develop an alternative procedure for preparing pinonic acid. Your new compound should not be a simple functional group derivative (such as an ester or a phenylhydrazone) of pinonic acid; its preparation should require at least two synthetic steps. Note that some reagents may react with both functional groups of pinonic acid unless one of them is protected. See *Protective Groups in Organic Synthesis* [B10] or a related source for information about the use of protective groups. Once you have synthesized and purified the new compound, you will need to prove its structure by spectrometric and/or chemical methods.

Directions

Research projects are intended for advanced or honors students who have mastered the major operations described in Part V. These directions only furnish general guidelines to help you complete your project. You must provide the procedural details and have them approved by your instructor. Keep detailed notes of your work and observations as described in Appendix II.

Safety Notes

> Except when you have reliable information to the contrary, assume that all chemicals you will use are flammable and are hazardous by ingestion, inhalation, and skin absorption. Wear gloves and use a hood whenever possible, avoid contact with the chemicals, do not breathe their dust or vapors, and keep them away from ignition sources and other chemicals that might react with them.

Begin your research project by following the plan and procedures you have developed. Keep in mind that you are venturing into unknown territory, so your results may not turn out as you planned. If so, try to benefit from unexpected outcomes by considering what you might learn from them. You may then decide to revise your research objectives in order to investigate your findings. In any case, be flexible enough to follow new leads or even switch to a different research project if your original research plan seems unproductive. Meet with your mentor regularly to discuss your findings and receive guidance. Provide him or her with duplicate copies of your lab notebook pages. Continue to read everything you can find that relates to your research; this may help you solve a problem that has hampered your research or suggest new avenues for you to explore.

Report. Write up your report as if it were a scientific paper being submitted to a professional publication such as the *Journal of Organic Chemistry*. Your report should include the following items, unless your instructor directs otherwise.

1 A brief but descriptive title.

2 Your name(s) and affiliation.

3 A brief abstract that summarizes the principal results of the work.

4 An introductory section (usually untitled) that provides a concise statement of the purpose and possible applications of the work, supported by descriptions of related work from the literature.

5 A section titled "Experimental Methods" that gives enough detail about your materials and methods so that another experienced worker could repeat your work. Give current *Chemical Abstract* index names and registry numbers (see Appendix VII) for important starting materials and products, and provide information about their purity if possible. Note any significant hazards and safety precautions in a separate paragraph labeled "Caution."

6 A section titled "Results and Discussion" (or two separate "Results" and "Discussion" sections) that summarizes the important experimental results and points out any special features, limitations, or implications of your work. You should include any data (including spectral parameters) that will help justify your conclusion. You may also wish to suggest different approaches to the problem or areas that require further study.

7 A section titled "Conclusions" that states any conclusions you can draw from your work, based on the evidence presented.

8 A section titled "References" that gives complete citations for all literature sources referred to in the report.

Your instructor may suggest additions to or modifications of this list. *The ACS Style Guide* [L11] provides a more detailed discussion of these components, as well as helpful suggestions about writing style.

PART III

Minilabs

The minilabs are short, self-contained experiments that ordinarily take no more than an hour or two to complete. They can be used to supplement the experiments in Parts I and II, and to fill in gaps that can occur when there is a long reaction time or an experiment that is completed well before the end of the lab period.

Making Useful Laboratory Items

Before You Begin: Read OP-3, "Using Glass Rod and Tubing."

   A flat-bottomed stirring rod comes in handy when you are trying to dissolve lumps of a solid, pulverize crystals for a melting-point measurement, wash a solid on a Buchner funnel, or stir a solution manually. You may need a boiling point tube to carry out a micro boiling point determination, and in preparing one you will learn how to seal and fire polish glass tubing. You can get some practice bending glass tubing by preparing a solvent trap, which is used to prevent water from backing up into a filter flask from an aspirator, and to recover volatile solvents that are being evaporated.

Directions

Safety Notes

> **Be careful not to burn or cut yourself while working with glass rod and tubing. See OP-3 for safe glassworking procedures.**

Your instructor should demonstrate the correct glassworking techniques. Prepare as many of the following items as requested, and have your instructor inspect and approve the items when you are finished.

Flat-Bottomed Stirring Rod. Cut a 20-25 cm length of 5-mm or 6-mm soft glass rod (see OP-3). Flatten one end so that it flares out to a diameter of 10 mm or more by holding that end in a burner flame until it is soft and incandescent and quickly pressing the hot end (hold it vertically) onto the metal base of a ring stand. When it has cooled, round off the other end in a burner flame.

Boiling-Point Tube. Obtain a piece of 5-mm O.D. (outside diameter) soft glass tubing and carefully seal it at one end. When it has cooled, cut it to a length of 8-10 cm and fire polish the open end. Test the tube as described in OP-3 to make sure it is sealed. Save it for use in Experiment 8 and later experiments.

Solvent Trap. Using a 250-mL (or larger) thick-walled bottle, 8-mm soft glass tubing, and a rubber stopper to fit, construct a solvent trap like the one shown in Figure C4 of OP-12. One length of glass tubing should be about 15 cm long and the other should be about half that long. Using a flame spreader, bend both pieces of tubing smoothly so that the shorter arm on each is ~3 cm long, and fire polish the ends. Bore two holes of appropriate diameter in the rubber stopper and insert the longer arm of each tube, using glycerol as a lubricant. When this assembly is inserted in the bottle, the lower end of the long tube should be 5 cm or so from the bottom; the short tube has only to extend through the stopper. Wrap the bottle with transparent tape to prevent possible injury from implosion; a heavy duty plastic tape is recommended.

Take Care! Hold the glass close to the stopper and protect your hands.

Extraction of Iodine by Dichloromethane

Before You Begin: Read OP13b, "Small-Scale Extraction."

Many extractions involve colorless solutions, making it impossible to observe the transfer of a solute from one layer to the other. In this minilab the solute (iodine) is colored so you will be able to see and describe what is going on as the solute is extracted from an aqueous solution by an organic solvent. Iodine is nearly insoluble in water, so potassium iodide is added to the aqueous solution to increase its solubility. Because iodide ions combine chemically with I_2 in water, the iodine color in the aqueous and organic layers will be different (brown in one and violet in the other). You should interpret any color in either layer as indicating the presence of iodine in that layer.

Directions

Record your observations carefully in your laboratory notebook, as you will have to explain them in your report.

Safety Notes

> **Dichloromethane may be harmful if ingested, inhaled, or absorbed through the skin. There is a possibility that prolonged inhalation of dichloromethane may cause cancer. Minimize contact with the liquid and do not breathe its vapors.**

Support a conical centrifuge tube in a test tube rack and add 2 mL of 0.50 M iodine/potassium iodide solution. Hold the centrifuge tube at an angle and use a Pasteur pipet to add 2 mL of dichloromethane down the side of the tube. The dichloromethane layer should slide underneath the colored aqueous layer with a minimum of mixing. Record the color of each layer and the intensity of the color (dark, very light, etc.). Cap the tube and shake it vigorously, with occasional venting, until the color of the aqueous layer does not change with further shaking [OP-13]. Record the color of each layer and the intensity of the color. Let the centrifuge tube stand until there is a sharp interface (dividing line) between the liquid layers. Use a Pasteur pipet to remove all of the dichloromethane layer without getting any of the aqueous layer into the pipet, and transfer it to a small Erlenmeyer flask. Now add a fresh 2 mL portion of dichloromethane to the centrifuge tube, shake the capped tube as before, and record your observations.

Stop and Think: What happened that caused the color changes you observed?

Stop and Think: Why is the color of each layer lighter than after the first extraction?

Waste Disposal: Place the combined dichloromethane layers in the chlorinated solvents waste container and pour the aqueous layer down the drain.

Report. Describe the colors of the two liquid phases before mixing and before and after each extraction. Account for the color changes and for any differences in the color intensity of the layers after the first and second extractions.

Purification of an Unknown Compound by Recrystallization

MINILAB 3

Before You Begin: Read OP-23d, "Choosing a Recrystallization Solvent," and read or review the other operations as necessary.

When you are purifying a solid substance by recrystallization, the solid must be soluble in the boiling solvent but quite insoluble in the same solvent when it is cold. In this minilab you will select a suitable solvent for the recrystallization of an unknown solid by carrying out some preliminary solubility tests in different solvents.

Directions

Safety Notes

Inhalation, ingestion, or absorption through the skin of the solvents and the unknown solid may be harmful. Avoid contact with them and do not breathe their vapors.

Follow the directions in "Choosing a Recrystallization Solvent" for testing the solvents provided. Obtain about 2 g of an unknown solid and select a suitable recrystallization solvent for it by testing its solubility in hot and cold water, ethanol, hexane, 2-butanone, and any other solvents suggested by your instructor. Accurately weigh about 1.0 g of the unknown solid [OP-4]. Recrystallize it from the solvent you selected [OP-23]. Dry the purified solid to constant mass [OP-21] and weigh it. Measure the melting points of the impure and purified solids [OP-28] and turn in the purified solid to your instructor.

Waste Disposal: Place the filtrate in an appropriate solvent recovery container (water and ethanol filtrates can be flushed down the drain).

Report. Calculate the percent recovery and report your results in tabular form.

MINILAB 4

Developing and Testing a Hypothesis

Before You Begin: Read (or reread) the "Scientific Methodology" section starting on page 2.

In this minilab you will observe and attempt to explain the effect of a ferric chloride solution on solutions of seven organic compounds. Record your observations carefully and think about their significance. After noting what happens to the first three solutions, you will develop a hypothesis to account for your observations. You will then test your hypothesis by adding $FeCl_3$ to two more solutions, modify your original hypothesis as necessary to account for your observations, and repeat this procedure with two more solutions until you arrive at a hypothesis that accounts for all of your observations.

Directions

> **Methyl salicylate irritates the skin and eyes; avoid contact.**

Use 1-mL graduated pipets or calibrated Pasteur pipets to add 0.5 mL of distilled water and 0.5 mL of 95% ethanol to each of eight numbered test tubes. (You can calibrate a Pasteur pipet by accurately measuring a specified volume of water, carefully drawing it into the pipet so there are no air gaps, and marking the liquid level with a glass-marking pen.) Add one drop of (1) methyl salicylate to the first test tube, about 10 mg each of (2) salicylic acid, (3) aspirin, (4) acetaminophen, (5) phenacetin, and (6) menthol to test tubes two through six, and one drop of (7) benzyl alcohol to test tube number seven. Add two drops of 1% ferric chloride solution to test tube number eight, which will serve as a control.

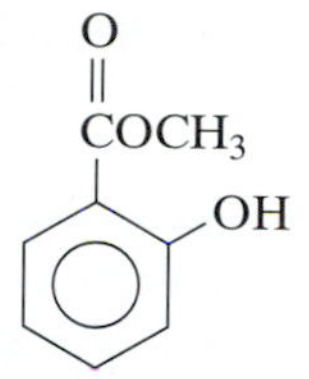

methyl salicylate 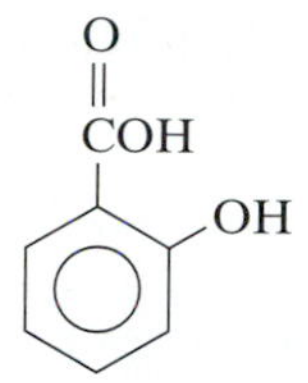   salicylic acid 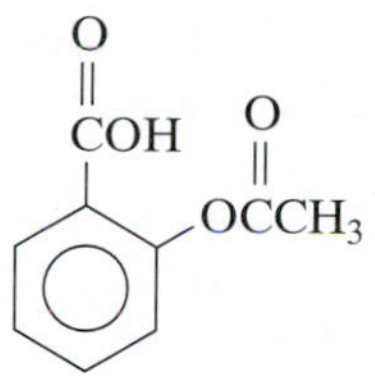   aspirin

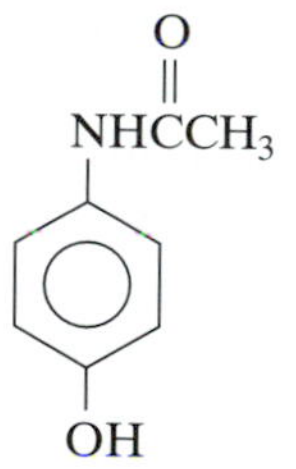

acetaminophen 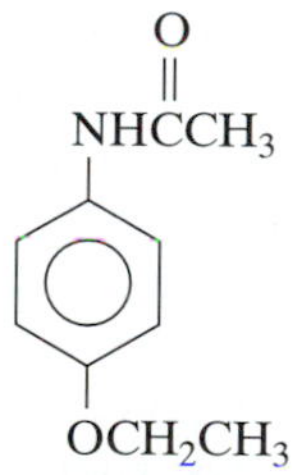   phenacetin 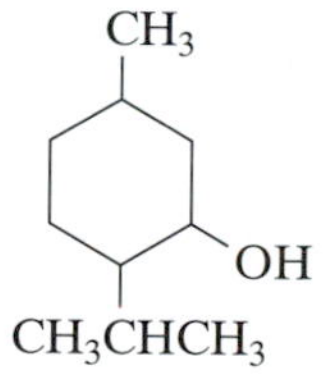   menthol 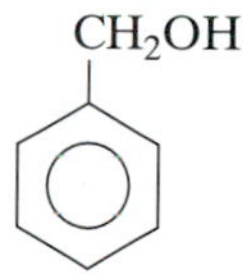   benzyl alcohol

Add two drops of 1% $FeCl_3$ to each of the first *three* test tubes, then **stop** and record your observations, comparing the contents of these test tubes to the control. Based on the structures of the compounds in these test tubes, formulate a hypothesis to account for your results.

Add two drops of 1% $FeCl_3$ to each of the next two test tubes, 4 and 5, then **stop** and record your observations. Do the results support your original hypothesis? If not, formulate a new hypothesis or revise your original one to explain them.

Add two drops of 1% $FeCl_3$ to each of the next two test tubes, 6 and 7. Do the results support your current hypothesis? If not, formulate a new hypothesis or revise your current one to explain them.

Report. Report your observations in a table. Write down your hypothesis from each stage of the procedure, telling why it changed from one step to the next (if it did), and state your final conclusion.

Observe and Note: Which of the test tubes show evidence of a reaction?

Waste Disposal: Dispose of the solutions as directed by your instructor.

MINILAB 5

Preparation of Acetate Esters

Before You Begin: Read or review the operations as necessary.

The pleasant aromas of many esters make them popular ingredients of flavorings and perfumes. In this minilab you will prepare some acetate esters by treating the corresponding alcohols with acetyl chloride, and then compare their odors.

$$\underset{\text{acetyl chloride}}{CH_3\overset{\displaystyle O}{\overset{\|}{C}}-Cl} + ROH \longrightarrow \underset{\text{acetate ester}}{CH_3\overset{\displaystyle O}{\overset{\|}{C}}-OR} + HCl$$

Directions

This minilab can be performed as a group project with each student responsible for preparing one ester.

Safety Notes

Acetyl chloride is very corrosive and its vapors are irritating and toxic. It reacts violently with water and some alcohols. Use gloves and a hood, avoid contact, do not breathe vapors, and keep away from water.

Take Care! Possible violent reaction. Wear gloves, avoid contact, and do not breathe vapors.

Waste Disposal: Pour the aqueous layers down the drain. Dispose of the esters as directed by your instructor

Obtain as many clean, *dry* test tubes as there are alcohols and number them consecutively. Add 10 drops or 0.3 mL of the following alcohols to the corresponding test tubes, plus any other alcohols specified by your instructor: ethanol, 1-propanol, 3-methyl-1-butanol, 1-octanol, benzyl alcohol. *Under the hood, slowly* add 10 drops or 0.3 mL of acetyl chloride to each of the five reaction tubes (use of an automatic pipet is recommended). Let the solutions stand for 3 minutes or more as you add 2 mL of water to each of five (or more) additional numbered test tubes. Then pour the contents of each reaction tube into the test tube having the same number. Extract the contents of each test tube with 2 mL of ethyl ether, using small-scale extraction techniques [OP-13]. *Under the hood,* evaporate the ether from each test tube under a stream of dry air or nitrogen [OP-14]. Do not apply heat or evaporate the contents to dryness. Transfer a drop of each ester to a spot plate and cautiously observe the odor of each ester. (Note whether the odor changes with time; if it does the ether may not have entirely evaporated.)

Report. Describe and compare the odors of the esters as best you can. You may be able to associate an odor with the odor of something familiar. Write the formula and give the name of each ester you prepared.

Gas Chromatographic Analysis of Commercial Xylene

Before You Begin: Read or review OP-32 on gas chromatography.

You will be analyzing a commercial xylene mixture containing *ortho-*, *meta-*, and *para*-xylenes.

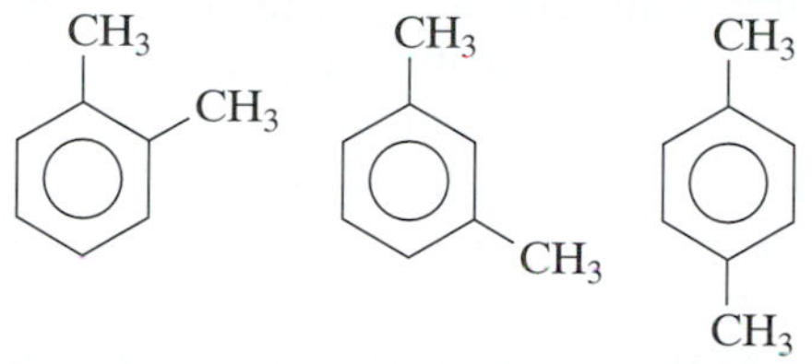

ortho-xylene   *meta*-xylene   *para*-xylene

Commercial xylene is used in microscopy, as a solvent, and in the manufacture of starting materials for the preparation of polyester fibers.

Directions

> **Xylenes are flammable, toxic by inhalation and ingestion, and irritating to skin and eyes. Minimize contact with the xylene mixtures and do not breathe their vapors.**

Safety Notes

A standard mixture containing *ortho-*, *meta-*, and *para*-xylene will be provided. Record the percentage of each component from the label on the bottle. Obtain gas chromatograms of (1) the standard mixture and (2) a commercial xylene mixture by injecting 1.0 μL of each mixture, using the column and temperature settings suggested by your instructor. From the relative retention times and peak areas, identify the component corresponding to each peak in the gas chromatogram of the standard. Record the retention time and calculate the detector response factor for each component. Identify the peak on the commercial xylene chromatogram corresponding to each component in the standard solution. Measure all peak areas and calculate the mass percent of each component in the commercial xylene.

Report. Report your results in a table showing the retention time, peak area, detector response factor, and mass percent of each component. Show your calculations and include your gas chromatograms with your report.

Isolation of Caffeine from No-Doz Tablets

MINILAB 7

CH₃—N, O, CH₃ structure

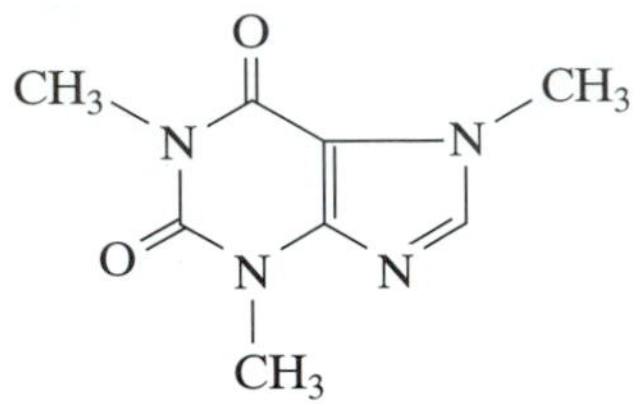

caffeine

Before You Begin: Read or review the operations as necessary.

No-Doz tablets are used by truck drivers, students studying for exams, and other people who need to stay awake at night. Their active ingredient is caffeine, which, as a component of coffee, tea, cocoa, and many soft drinks, is probably the most frequently consumed drug in the world. Caffeine stimulates the cerebral cortex and, in small doses, can improve concentration and coordination. It also acts as a heart stimulant and makes skeletal muscles less susceptible to fatigue. As anyone with a bad case of "coffee nerves" can attest, caffeine also has its down side. Too much caffeine may cause headaches, heart irregularity, muscular trembling, irritability, and insomnia. In this minilab you will use hot ethanol to extract caffeine from No-Doz tablets.

Directions

Place two 200 mg No-Doz (or a generic equivalent) tablets in a 50-mL Erlenmeyer flask and crush them with the flat end of a flat-bottomed stirring rod (see Minilab 1). Add 10 mL of 95% ethanol and boil the mixture gently for 3 minutes or more by swirling it over a steam bath or other flame-free heat source [OP-6]. Filter the *hot* solution by gravity [OP-11] using a fast fluted filter paper and a *heated* powder funnel. Wash the solid on the filter paper with 4 mL of hot 95% ethanol and combine the filtrates. Under the hood, concentrate the solution [OP-14] to a volume of 4 mL by heating it gently under a stream of dry air or nitrogen. If the solution is cloudy or contains a precipitate, heat it until it becomes clear. Let the solution cool at room temperature until a good crop of crystals is obtained, then cool it in ice water for five minutes. Collect the caffeine by vacuum filtration on a Hirsch funnel [OP-12], washing it twice with cold 50% ethanol. Dry it at room temperature [OP-21] and measure its mass and melting point [OP-28]. At your instructor's request you can purify the caffeine by small-scale sublimation [OP-24] and measure the melting point [OP-28] of the purified product.

Report. Calculate the percent recovery of caffeine from No-Doz tablets.

MINILAB 8

A Missing Label Puzzle

Before You Begin: After reading the minilab, devise a method for distinguishing the two dry-cleaning solvents.

Most dry-cleaning solvents are hydrocarbon mixtures or chlorinated hydrocarbons that, because of their low polarity, dissolve greasy stains in clothing. Many chlorinated solvents that were once used for dry cleaning, such as carbon tetrachloride, are no longer allowed because of their toxicity. Tetrachloroethene ($Cl_2C=CCl_2$), sold under such trade names as Perclene, is less toxic than most other chlorinated solvents. Mineral spirits, a mixture of petroleum hydrocarbons, is also used widely in dry-cleaning.

For this minilab you will assume that the labels have fallen off a bottle of tetrachloroethene and a bottle of mineral spirits, and you will devise a method of telling the solvents apart. Your method cannot be based on odor differences. No instruments such as balances or refractometers are allowed. You can use only the equipment in your locker and basic lab utilities, plus anything else your instructor is willing to provide (check in advance if you're not sure what will be available). The simplest method may be the best—a few minutes of thought can often save hours of experimental work.

Directions

Safety Notes

> **Tetrachloroethene is an eye and skin irritant and may be harmful by inhalation. It is also a suspected carcinogen. Avoid contact and do not breathe vapors.**
> **Mineral spirits is very flammable and inhalation may be harmful. Do not breathe vapors, keep flames away.**

In the laboratory you will find two containers of "dry-cleaning solvent," labeled **A** and **B**. One of the solvents is tetrachloroethene and the other is mineral spirits. According to the *Merck Index*, tetrachloroethene has a boiling point of 121°C, a density of 1.623 g/mL, and a refractive index of 1.5055. Assuming that the hydrocarbons in mineral spirits have properties similar to those of the alkanes in Experiment 8 (except for boiling point), devise an experiment for finding out which solvent is in which container. Then carry out your experiment and report your results to your instructor. Don't reveal your method or results to others—let them think for themselves, as you did.

Report. Report the identity of **A** and **B** and tell how you arrived at your conclusion.

Paper Chromatography of Dyes in Commercial Drink Mixes

MINILAB 9

Before You Begin: Read OP-18 on paper chromatography.

Many *f*oods, *d*rugs, and *c*osmetics are colored with F, D & C dyes to give them eye appeal or, in the case of foods, to give them the color we associate with the food in question. For example, the artificial flavors used to prepare a grape drink may be completely colorless, so dyes are added to give the drink a purple "grape" color. A number of former F, D & C dyes have been removed from the market because they were found to be toxic or suspected of causing cancer or birth defects. For example, F, D & C Red No. 2, which was once the most widely used food dye, was banned in 1976 because it is a suspected carcinogen. At present there are only eight F, D & C dyes allowed for unrestricted food use, the four most popular being Blue No. 1, Red No. 40, Yellow No. 5, and Yellow No. 6. In this minilab you will analyze some powdered drink mixes (such as Kool-Aid) to see which F, D & C dyes they contain. The four dyes mentioned can be identified by their colors and R_f values. Some foods also contain F, D & C Green No. 3 and Violet No. 2, which can be identified by the colors of their spots alone.

Directions

Safety Notes

> Ethanol and 1-butanol are flammable and 1-butanol may cause eye or skin irritation. Ammonia can cause severe skin and eye irritation. Avoid contact with the developing solvent, do not breathe its vapors, and keep flames away.

Take Care! Avoid contact with the developing solvent, do not breathe its vapors.

Stop and Think: Why not use a pen?

Waste Disposal: Place the developing solvent in a designated solvent recovery container.

You will be provided with a selection of drink mix flavors each containing one or more of the food dyes listed in Table M1. *Under the hood*, place enough developing solvent (a 1:1:1 mixture of 95% ethanol, 1-butanol, and *2 M* aqueous ammonia) in a suitable developing chamber (such as a 600-mL beaker) to cover its bottom to a level of about 5 mm. Cover it with a lid or plastic wrap, swirl it to agitate the liquid, and let it sit to equilibrate while you prepare the chromatogram [OP-18]. Label the wells of a spot plate to correspond to the drink mix flavors, using a one- or two-letter code. Place a *small* amount of each drink mix powder in a well of the spot plate (about enough to cover the rounded bottom of the well), then add room temperature water drop by drop with stirring to each tube until the powder dissolves. Rinse your stirring rod before going on to the next spot plate.

Obtain an approximately 11 × 22 cm rectangle of Whatman #1 chromatography paper and use a pencil to draw a starting line along one of its long sides, about 1.5 cm from that side. Using a different capillary micropipet (or toothpick) for each spot, spot the solutions 1.5–2.0 cm apart along the starting line. If space permits, apply two or three spots of varying concentration for each solution. Make each spot about 2 mm in diameter and write its code letter underneath it with a pencil. Develop the chromatogram in the developing chamber and let it air dry (don't forget to mark the solvent front). Identify as many dyes in your drink mix flavors as you can using information from Table M1.

Report. Prepare a table showing the color and your experimental R_f value for each spot in each drink mix. Report the composition of each drink mix and explain how you arrived at your conclusions. Turn in your chromatogram with your report.

Table M1 Colors and approximate R_f values for F, D & C dyes

F, D & C dye	Color	R_f
Yellow #5 (tartrazine)	bright yellow	0.31
Yellow #6 (sunset yellow)	orange	0.58
Red #40 (allura red)	bright red	0.62
Blue #1 (brilliant blue)	turquoise blue	0.67

Note: R_f values are for the solvent mixture 95% ethanol : 1-butanol : 2 M aqueous ammonia (1 : 1 : 1)

Gas Chromatographic Analysis of an Essential Oil from Orange Peel

MINILAB 10

This minilab is based on an article published in the *Journal of Chemical Education* **1994,** *71,* A146.

Before You Begin: Read OP-13d, "Liquid-Solid Extraction," and read or review the other operations as necessary.

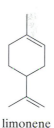

limonene

The orange tree originated in China and was first cultivated in the United Stated by Franciscan monks in the part of Spanish North America that is now California. Orange oil is used medicinally and also to flavor foods, drinks, and confections. It is colored by β-carotene and contains a number of minor components, but its major component is limonene. Limonene is a ubiquitous terpene that occurs in a multitude of other essential oils, including the oils of lemon, caraway, and dill. Limonene exists in two enantiomeric (mirror image) forms, (R)-$(+)$-limonene and (S)-$(-)$-limonene, which have different odors.

In this minilab you will isolate orange oil by extraction of grated orange peel with hexanes (a mixture of isomeric six-carbon alkanes) and analyze it by gas chromatography to determine the approximate percentage of limonene it contains. If your instructor has samples of (R)-$(+)$-limonene and (S)-$(-)$-limonene available, you can compare their odors to that of your extract to find out which enantiomer you have. Some people are unable to distinguish the enantiomers by odor; if you are one of them, ask a coworker for help. If a suitable preparative GC column is available, you can also isolate pure limonene and record its infrared spectrum as described in *J. Chem. Educ.* **1994,** *71,* A146.

Directions

Hexanes are flammable and their vapors may be harmful. Avoid inhalation, keep flames away.

Safety Notes

On a piece of wax paper or aluminum foil, grate the rind of an orange until you have about 1 gram. Measure the mass of the grated orange rind and transfer it to a small Erlenmeyer flask. To extract the orange oil [OP-13], use a flat-bottomed stirring rod (see Minilab 1) to crush and mix the peel with 4 mL of hexanes for 3 minutes or more. Then use a Pasteur pipet to transfer the extract to a drying pipet—another Pasteur pipet containing about 0.5 g of anhydrous sodium sulfate supported on a cotton plug [OP-20]. Collect the extract in a tared screwcap vial. Repeat the extraction using another 4 mL of hexanes, drying the extract as before and combining it with the first extract. Use about 0.5 mL of hexanes to rinse the transfer pipet and filter pipet into the vial. *Under the hood*, remove the solvent by evaporation in a dry stream of air or nitrogen [OP-14]. Weigh the orange oil; if its mass decreases as you measure it, resume the evaporation and weigh it again. Compare the odor of the orange oil to that of authentic samples of (R)-$(+)$-limonene and (S)-$(-)$-limonene, if available. Record a gas chromatogram of the orange oil [OP-32], estimate the percentage of limonene in it by assuming the same detector response factor for all components, and calculate the percent recovery of limonene from orange peel.

Stop and Think: What would cause the mass to decrease as you measure it?

Report. Report your data and results, showing your calculations. If you can, identify the limonene in orange peel as (R)-$(+)$-limonene or (S)-$(-)$-limonene and draw its stereochemical structural formula.

MINILAB 11

Identification of an Unknown Felt Tip Pen Ink by TLC

This minilab is based on an article published in the *Journal of Chemical Education* **1983,** *60,* 232.

Before You Begin: Read or review OP-17 on thin-layer chromatography.

Unlike food colorings, which are made from a very limited selection of approved dyes (see Minilab 9), pen inks are made from a wide variety of dyes and the dye mixtures in pen inks from different manufacturers can vary greatly. These dyes can be separated by thin-layer chromatography (TLC), producing chromatograms with characteristic patterns of colored spots. For this reason forensic scientists often use TLC to identify pen inks from documents associated with a criminal or civil case. For example, matching the ink from a pen in the suspect's possession with ink in the signature on a forged document provides evidence that can result in the suspect's conviction.

In this minilab you will attempt to identify the ink from an unknown felt tip pen by comparing its TLC pattern with those of known pen inks. The pens will have fine points so they can be used to spot the TLC plate directly. You will develop the chromatogram using a 1-butanol : water : ethanol : acetic acid (120 : 40 : 20 : 1) mixed solvent or another suitable developing solvent.

Directions

Safety Notes

> The developing solvent is flammable and harmful by inhalation, inges-
> tion, and skin absorption. Avoid contact, do not breathe vapors, keep
> flames away.

Under the hood, prepare a developing chamber using a 1-L beaker or other
suitable container as described in OP-17, using enough developing solvent
to cover its bottom to a level of about 5 mm. Cover it with a lid or plastic
food wrap, swirl it to agitate the liquid, and let it sit to equilibrate while you
prepare the TLC sheet [OP-17]. Obtain a 10 × 10 cm silica gel TLC sheet
and lightly pencil a line about 1.5 cm from the bottom, taking care not to
touch the surface of the adsorbent with your fingers. Using the felt tip pens
provided, carefully apply spots placed about 1.5 cm from each edge and
1 cm or more apart, leaving spaces open for the unknown. Each spot should
be no more than 2 mm in diameter, so practice your spotting technique on a
piece of paper before you spot the TLC sheet. Then apply two or more spots
from an unknown felt tip pen ink provided by your instructor. Applying the
unknown spots between known spots will make the inks easier to compare.
Be sure to label each spot (in pencil) with a 1–2-letter code identifying the
corresponding felt tip pen. Develop the TLC sheet in the pre-equilibrated
developing chamber and let it air dry (don't forget to mark the solvent
front). For each pen ink, including the unknown, describe the pattern of colored spots you observe. Then identify the unknown pen ink by matching its
pattern with that from one of the known pen inks.

Waste Disposal: Place the developing solvent in a designated solvent
recovery container.

Report. Describe your observations and explain how you arrived at your
conclusion. Turn in your chromatogram with your report.

Optical Rotation of Turpentine

MINILAB 12

Before You Begin: Read or review OP-31, "Optical Rotation."

North American turpentine contains (+)-α-pinene and (−)-β-pinene,
whose structures are shown in Experiment 12. In this minilab you will measure the optical rotation of a commercial turpentine (or a simulation
thereof) to estimate the percentages of (+)-α- and (−)-β-pinene it contains.

Directions

Safety Notes

> Turpentine and its components are flammable and irritate the skin and
> eyes; inhalation of their vapors may be harmful. Minimize contact with
> the turpentine and do not breathe its vapors.

Waste Disposal: Place the ethanol solutions in an appropriate solvent recovery container.

Stop and Think: What equation in OP-31 can you use to do this?

Using a 25-mL volumetric flask, prepare a solution containing about 2.5 g of turpentine (accurately weighed) in absolute ethanol. Measure the optical rotation of the solution and of an absolute ethanol blank using an accurate polarimeter [OP-31]. Prepare solutions of $(+)$-α-pinene and $(-)$-β-pinene in the same way and measure their optical rotations. Calculate the specific rotation of each solution. Assuming that it contains only $(+)$-α-pinene and $(-)$-β-pinene, calculate the percentage of each constituent in the turpentine.

Report. Show your calculations and prepare a table with your data and results.

MINILAB 13

The Structures of Organic Molecules

Before You Begin: Familiarize yourself with the different kinds of formulas used to represent organic compounds and know how to name organic compounds by the IUPAC system.

In this minilab you will study some molecular models to help you better understand the structure and geometry of organic molecules, and to learn how to write and interpret different kinds of formulas that are used to represent them.

Directions

When you arrive at the lab you will find numbered molecular models of representative organic compounds distributed around the laboratory. Start at one of the models and proceed around the laboratory until you have studied all of them. Unless your instructor indicates differently, assume that atoms are represented by balls of the following colors: carbon—black; hydrogen—white or yellow; oxygen—red; nitrogen—blue; chlorine—green; bromine—orange or brown; sulfur—yellow. The type of covalent bond connecting two atoms is indicated by the number of connectors; one for a single bond, two for a double bond, and three for a triple bond. For each model, write the following in your lab notebook, along with the number of the model:

1 IUPAC name (the name may be provided in some cases)
2 Empirical formula
3 Molecular formula
4 Condensed structural formula
5 Expanded structural formula (Kekulé structure)
6 Bond-line formula

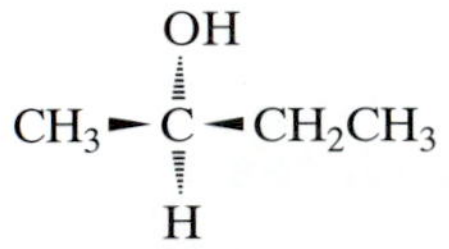

OH
|
CH₃ — C — CH₂CH₃
|
H

stereochemical
formula

Examples of these types of formulas are shown on the next page for 2-methylpropane. For certain molecular models you may be asked to write a stereochemical formula using either a flying wedge, ball-and-stick, or Fischer projection. A flying wedge projection is illustrated for (R)-2-butanol. Your instructor may request special formulas in some cases, such as chair-ring formulas for substituted cyclohexanes.

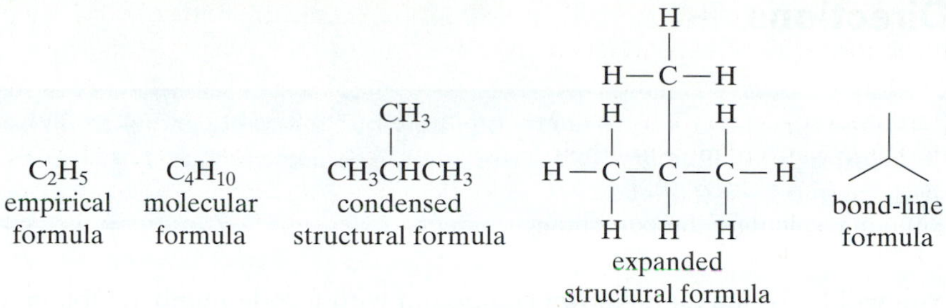

$$\begin{array}{ccccc}
& & CH_3 & & \\
& & | & & \\
C_2H_5 & C_4H_{10} & CH_3CHCH_3 & & \\
\text{empirical} & \text{molecular} & \text{condensed} & & \text{bond-line} \\
\text{formula} & \text{formula} & \text{structural formula} & & \text{formula}
\end{array}$$

expanded
structural formula

formulas for 2-methylpropane

Report. Write the formulas for each numbered model neatly in your lab report. Give the IUPAC name for the compound corresponding to each model; if a compound is chiral, designate it as (*R*) or (*S*) as well.

Who Else Has My Compound?

This minilab is based on an article published in the *Journal of Chemical Education* **1995,** *72,* 1120.

Before You Begin: Develop an experimental plan based on the equipment and chemicals available to you, and read or review the appropriate operations as necessary.

Few, if any, scientists perform their work in isolation. Successful science requires cooperation and collaboration among scientists who exchange information, ideas, and even materials. For example, suppose you isolated a natural product that was not previously reported in the chemical literature, but shortly afterward a Japanese chemist reported a natural product (isolated from a different source) whose spectra and properties appear to be the same as yours. You would then have to exchange data and samples with the Japanese chemist to establish whether your compounds were, in fact, the same.

In this minilab you will receive an unknown organic compound, knowing that at least one other student in your laboratory has the same compound. You will be expected to learn enough about your compound so that, by sharing your findings with other students, you can find another unknown that appears to match your own. You and the student having that unknown should then confirm that your compounds are identical by, for example, having student **A** obtain data for his or her unknown that student **B** has already obtained for the other unknown and vice versa. If more than two students share the same unknown, you may eventually identify yourself as part of a larger group of students. Since the outcome of the experiment depends on the efforts of all the participants, your individual contribution is important for the success of the group.

Your instructor may advise you about the techniques you should use, or you may be asked to plan your own experimental strategy, given the chemicals and equipment available.

Directions

Safety Notes

> **Assume that the unknown compound may be flammable and harmful by inhalation, ingestion, and skin absorption. Avoid contact, do not breathe vapors, and keep flames away.**

You will be issued an unknown compound with a code number. Obtain as much information about your compound as you think is necessary to characterize it, given the equipment and chemicals available. Then approach other students with your findings (they may approach you first) to locate another student who appears to have the same compound you do. When you find one or more such students, perform any additional work necessary to determine whether or not your compounds are, in fact, identical.

Report. Report the names and unknown numbers of other students whose unknowns are identical to your own and tell how you arrived at your conclusion. Report all of your findings in an appropriate format and turn in any spectra or chromatograms you recorded.

MINILAB 15

Isomers and Molecular Structure

Before You Begin: As necessary, review material from your lecture textbook about structural (constitutional) isomers and stereoisomers.

Just as zoology is the study of animals and botany the study of plants, organic chemistry can be regarded as the study of carbon-containing molecules. Certain plants and animals are familiar to everyone because we see them all around us, but molecules are not. We can't *see* a molecule—the best we can do is to imagine what one looks like, and molecular models can help us do that. All models have their limitations. A plastic model of a Sopwith Camel is a far cry from the real thing, but it can show us what that WWI biplane looked like in three dimensions. A real molecule is obviously not made of colored balls, but an accurate molecular model can depict the three-dimensional structure of a molecule far better than any drawing. In this minilab, you will use molecular models to help you see—in three dimensions—the structural and geometric differences between isomers.

Isomers are different compounds that share the same molecular formula. This means that the same set of atoms can be combined in different ways to form molecules with different structures. In this minilab you will apply a "hands-on" approach to the study of isomers and molecular structure. You will be issued a set of atoms from a molecular model kit, and will be expected to construct molecular models for as many isomers as you can in the time allowed. The most popular molecular models are the ball-and-stick type, in which balls of different colors represent different kinds of atoms and rigid or flexible connectors represent the bonds that hold the atoms together in molecules. Except for nitrogen, each ball is ordinarily drilled with a number of holes equal to the normal *covalence* of the corresponding atom—the number of covalent bonds formed by the neutral atom. Thus balls representing carbon atoms—colored black—are drilled with four holes arranged tetrahedrally, whereas those representing oxygen atoms have two holes, and those representing hydrogen atoms have one. The balls for nitrogen usually have four holes to allow the construction of ammonium compounds, although the normal covalence of nitrogen is three. Building models of all the isomers that have a given molecular formula is thus a matter of putting the appropriate colored balls together in all possible combinations, using just enough connectors to fill the holes. Often a large number of isomers will share the same molecular formula, and it may take a considerable amount of ingenuity to find them all!

Directions

This minilab can be done individually or in groups. Your instructor will issue you or your group a set of atoms and will provide connectors, including flexible connectors for constructing strained and multiple bonds. Unless informed otherwise, assume that atoms are represented by balls of the colors given in Table M2.

Table M2 Colors of model kit balls and normal covalences of atoms

Atom	Color of ball	Normal covalence
bromine	orange or brown	1
carbon	black	4
chlorine	green	1
hydrogen	white or yellow	1
nitrogen	blue	3
oxygen	red	2
sulfur	yellow	2

Construct molecular models for as many isomers as you can based on the normal covalences in Table M2. Write a structural formula representing each model, showing all multiple bonds. Make sure that each model does, in fact, represent a different compound. Some of your molecular models may exhibit stereoisomerism; in that case you can count each *cis-trans* isomer or enantiomer as a different compound. At your instructor's option, you can obtain additional sets of atoms and use them to prepare isomers as before.

Report. Draw neat structural formulas for all the different molecular models you constructed. Designate any geometric isomers as *cis* or *trans* and any enantiomers as (*R*) or (*S*), and show their stereochemistry clearly.

Reactivities of Alkyl Halides in Nucleophilic Substitution Reactions

Before You Begin: Draw the structures of all substrates you will be using in this reaction and classify them as 1°, 2°, 3°, aryl, or benzyl. Identify the nucleophile, substrate, and leaving group in the general equations for reactions **1** and **2**.

In this minilab you will compare the relative reactivities of different alkyl halides with two different reagents: sodium iodide in acetone and silver nitrate in ethanol. General equations for these reactions are

1 $RX + NaI \xrightarrow{\text{acetone}} RI + NaX \qquad (X = Cl \text{ or } Br)$

2 $RX + AgNO_3 + EtOH \longrightarrow ROEt + AgX + HNO_3$

(In the second reaction, additional products such as alkenes may be formed.) For each reagent the occurrence of a reaction is indicated by the formation of a precipitate; sodium iodide (which is insoluble in acetone) in reaction **1** and a silver halide in reaction **2**. Your initial objective is to discover the mechanism of each reaction. Based on your conclusions, you will predict the relative reactivities of other substrates and test your predictions. In order to accomplish these objectives, you will have to think about the significance of your observations as you make them so that your results from one set of reactivity measurements can lead to predictions about another set of measurements.

Directions

Safety Notes

> **The alkyl bromides are all toxic and have harmful vapors, and some of them are suspected carcinogens. Avoid contact and do not breathe their vapors.**

A. *Reactions in Sodium Iodide/Acetone*

Directions for conducting reaction **1** follow. Be sure to label the reaction tubes so you know which substrate each contains.

1. Carry out reaction **1** using the following substrates: 2-bromobutane, 2-bromo-2-methylpropane, 1-bromobutane. Based on your results, decide whether the reaction in NaI/acetone is S_N1 or S_N2.

2. Predict the reactivities of bromocyclohexane and benzyl bromide (or other substrates provided by your instructor) in NaI/acetone relative to the substrates in step 1. Then carry out reaction **1** using these substrates.

Stop and Think: Are your results as you predicted? If not, how can you explain them?

Reaction 1. Obtain as many clean, dry 10-cm test tubes as there are alkyl halides to test. Measure 1 mL of 15% sodium iodide in acetone into each test tube. Add 2 drops of a different alkyl halide to each test tube and stopper and shake the test tubes. Observe them closely and record the time needed for any precipitate to form. After 5 minutes, put any test tubes that do *not* contain a precipitate into a 50°C water bath and leave them there for 6 minutes. Then cool the test tubes to room temperature and note the formation of any precipitate.

Waste Disposal: Place the sodium iodide test solutions in a designated waste container.

B.   *Reactions in Silver Nitrate/Ethanol*

Directions for conducting reaction **2** follow. Be sure to label the reaction tubes so you know which substrate each contains.

1.   Carry out reaction **2** using the following substrates: 2-bromobutane, 2-bromo-2-methylpropane, 1-bromobutane. Based on your results, decide whether the reaction in AgNO$_3$/ethanol is S$_N$1 or S$_N$2.

2.   Predict the reactivities of benzyl bromide and bromobenzene (or other substrates provided by your instructor) in AgNO$_3$/ethanol relative to the substrates in step 1. Then carry out reaction **2** using these substrates.

Reaction 2. Obtain as many clean, dry 10-cm test tubes as there are alkyl halides to test. Measure 2 mL of a 0.1 *M* solution of silver nitrate in ethanol into each test tube. Add a drop of a different alkyl halide to each test tube and stopper and shake the test tubes. Observe them closely and record the time needed for any cloudiness or precipitate to form. After 5 minutes, heat any solutions that do *not* contain a precipitate for 2–3 minutes in a boiling water bath and note the formation of any precipitate.

Stop and Think: Are your results as you predicted? If not, how can you explain them?

Report. Write balanced equations for the reactions undergone by each halide and write mechanisms for both if the reactions undergone by 2-bromobutane. Arrange the alkyl bromides in order of reactivity in each reaction, classify each reaction as S$_N$1 or S$_N$2, and explain your reasoning. If either reaction is S$_N$1, use your results to predict the relative stabilities of the carbocations formed by each substrate. If any substrates appear to be equally reactive (or unreactive), show their carbocations as being of approximately equal stability.

Waste Disposal: Place the silver nitrate test solutions in a designated waste container.

An S$_N$1 Reaction of Bromotriphenylmethane

MINILAB 17

Before You Begin: Read or review the operations as necessary.

The traditional Williamson synthesis of ethers involves the reaction of an alkyl halide with the sodium salt of an alcohol or phenol.

$$RX + R'ONa \rightarrow ROR' + NaX$$

An alkoxide ion (R'O$^-$) is strongly nucleophilic and displaces halide ion by an S$_N$2 mechanism. Because alkoxide ion is also a strong base, tertiary alkyl halides tend to undergo E2 elimination under these conditions, producing alkenes rather than ethers. In the presence of weaker nucleophiles such as alcohols, most tertiary alkyl halides form alkenes by an E1 mechanism, along with some S$_N$1 substitution products. But both E2 and E1 elimination are impossible for bromotriphenylmethane (trityl bromide) and the triphenylmethyl (trityl) carbocation is very stable, so this *tertiary* halide reacts readily with alcohols to form ethers but no by-product alkenes. In this minilab you will treat bromotriphenylmethane with ethanol to form an ether, ethoxytriphenylmethane, by an S$_N$1 reaction.

Stop and Think: Why?

Reaction of bromotriphenylmethane with ethanol

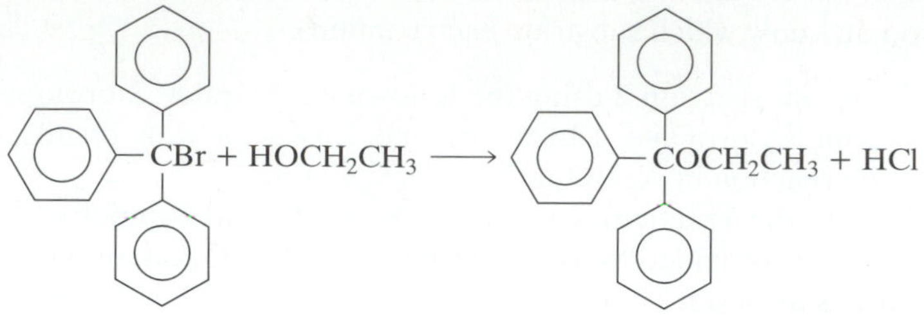

trityl bromide   ethoxytriphenylmethane
                                      (trityl ethyl ether)

Directions

Safety Notes

> **Bromotriphenylmethane is harmful if inhaled or absorbed through the skin. Avoid contact and do not breathe dust.**
> **Since the reaction generates some gaseous hydrogen bromide, carry it out under a hood. Avoid contact with the reaction mixture and do not breathe its vapors.**

You can use purified bromotriphenylmethane from Experiment 25 in this minilab, scaling down the quantities if necessary. *Under the hood*, mix 0.50 g of bromotriphenylmethane with 5.0 mL of absolute ethanol in a test tube. Add a boiling chip and boil the mixture gently over a steam bath or hot water bath until no more HBr is evolved (test by holding moist blue litmus paper over the test tube); replace any ethanol that evaporates. Remove the boiling chip, let the reaction mixture cool to room temperature, induce crystallization if necessary, and cool it further using an ice/water bath. Then collect the ethoxytriphenylmethane by vacuum filtration on a Hirsch funnel [OP-12]. Dry the product [OP-21] and measure its mass and melting point [OP-28].

Report. Report the percent yield of your preparation and write a detailed mechanism for the reaction. Explain why E1 elimination, which often competes with S_N1 substitution, does not occur during this reaction.

Preparation and Properties
of a Gaseous Alkene

MINILAB 18

Before You Begin: Predict the structure of the gaseous alkene that you will be preparing.

Alkenes can be prepared by elimination reactions in which a molecule of hydrogen halide (HX), halogen (X_2), or water is removed from a substrate. In this experiment you will be carrying out an acid-catalyzed elimination reaction of 2-methyl-2-propanol to form a gaseous alkene with the molecular formula C_4H_8. You will then test your alkene with bromine in dichloromethane (CH_2Cl_2 is the solvent) and with aqueous $KMnO_4$, looking for evidence of a reaction in each case. You will also test its flammability to see whether it reacts with atmospheric oxygen. You should then be able to explain your observations and write equations for the reactions involved (you may have to consult your lecture textbook for help in some cases).

2-methyl-2-propanol

Directions

Sulfuric acid causes chemical burns that can seriously damage skin and eyes. Wear gloves and avoid contact.
Keep the gas you collect away from flames except when you are testing its flammability.

Safety Notes

Fill four 15-cm test tubes with water, stopper them tightly, and invert them in a 1-L beaker or pneumatic trough that is about three-quarters filled with water. Remove the stoppers under water so that no water flows out of the test tubes, and leave them inverted in the beaker or trough. Measure 1.0 mL of 2-methyl-2-propanol (*t*-butyl alcohol) into a 15-cm Pyrex test tube, add 5 drops of concentrated sulfuric acid, swirl to mix, and drop in a boiling chip. (If necessary, warm the reagent bottle of 2-methyl-2-propanol over a steam bath until the alcohol flows freely.)

Take Care! Wear gloves, avoid contact with H_2SO_4.

Assemble the apparatus illustrated in Figure M1, with the reaction tube clamped to a ring stand and the free end of the gas delivery tube under water. (Alternatively, you can use clear plastic tubing in place of the bent glass tube.) Move one of the water-filled test tubes so that its mouth is over the outlet of the gas delivery tube. Heat the reaction tube gently on a steam bath or hot water bath [OP-6] until a steady stream of gas bubbles emerges from the tubing and fills the test tube. As each test tube fills with gas, quickly remove and stopper it and replace it with a water-filled test tube. Repeat this until all the test tubes are filled, then remove the heat source and cool the reaction tube in cold water to stop the generation of gas.

Take Care! Keep flames away from the collected gas.

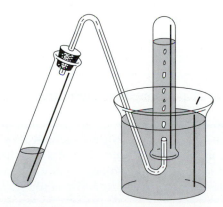

Figure M1 Apparatus for gas generation

Observe and Note: What happens?
Waste Disposal: Dispose of the contents of the test tubes as directed by your instructor.

Discard the gas in the first test tube, which contains air. Add 5 drops of 0.1 *M* aqueous potassium permanganate to the second tube, then stopper and shake it. Add 5 drops of 1 *M* bromine in dichloromethane to the third tube, then stopper and shake it. Light one end of a wood splint with a match or burner flame and carefully lower the burning end into the fourth tube. Record all observations in your laboratory notebook.

Report. Write the structure of the gas and a balanced equation for its synthesis from 2-methyl-2-propanol. At your instructor's request, write a detailed mechanism for this reaction. Describe and explain your observations during the tests, and write a balanced equation for the chemical reaction that occurred in each of them.

MINILAB 19

Addition of Iodine to α-Pinene

This minilab was adapted from a procedure in *J. Chem. Educ.* **1977**, *54, 228.*

Before You Begin: Review material in your lecture textbook about ring strain and carbocation rearrangements.

In *All Creatures Great and Small,* the author, James Herriot, describes how veterinarians once disinfected wounds in large animals. The veterinarian would pack the wound with iodine, splash some turpentine on it, and stand back. The result was a violent reaction that forced iodine into the wound amid a cloud of purple vapor, accompanied by an equally violent reaction from the animal. The major component of turpentine is α-pinene, an alkene with a carbon-carbon double bond in a six-membered ring and adjacent to the highly strained four-membered ring highlighted here. Although alkenes readily undergo addition reactions with chlorine and bromine, most of them are unreactive with iodine. If α-pinene is responsible (in part) for the reaction of turpentine with iodine, it must possess some special structural feature that makes it more reactive than most other alkenes. You will investigate the effect of treating α-pinene with iodine and compare it to the effect of treating a control compound, cyclohexene, with iodine. Then you will attempt to explain your observations by proposing a mechanism for the reaction and predicting its products.

α-pinene   cyclohexene

Directions

This experiment may be performed by your instructor, as a demonstration.

Safety Notes

α-**Pinene is flammable and irritates the skin and eyes; inhalation of its vapors may be harmful. Minimize contact and do not breathe vapors.
The reaction with iodine may be violent; conduct it under a fume hood.
Wear safety goggles and adequate protective clothing.**

Weigh 0.5 g of iodine into each of two 15-cm test tubes. *Under the hood*, add 1.0 mL of cyclohexene to the first test tube in one portion. Then add 1.0 mL of α-pinene to the second test tube in one portion and step back quickly.

Report. Deduce a mechanism that explains the exothermicity of the α-pinene reaction, assuming a carbocation as the initial intermediate. Write an equation for this reaction showing two likely products (neither is a vicinal diiodide). Tell how your mechanism explains your experimental observations.

Take Care! Wear goggles and protective clothing.

Observe and Note: What happened?

Waste Disposal: Dispose of the mixture in both test tubes as directed by your instructor.

Unsaturation in Commercial Products

Before You Begin: For as many of the commercial products as you can, try to predict whether or not the product should be unsaturated or contain unsaturated components.

A substance is *unsaturated* if it contains fewer hydrogen atoms than are possible based on its molecular structure. Compounds having multiple carbon-carbon bonds are unsaturated because they can add more hydrogen.

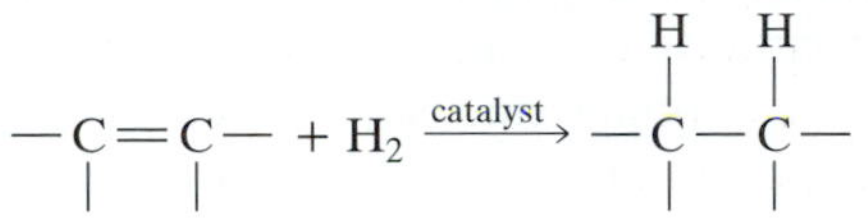

$$-C=C- \ + \ H_2 \ \xrightarrow{\text{catalyst}} \ \overset{\overset{\text{H}}{|}}{-C} - \overset{\overset{\text{H}}{|}}{C}-$$

Since bromine also adds to most carbon-carbon multiple bonds, a solution of bromine in carbon tetrachloride has often been used to test for unsaturation. If the compound tested is unsaturated, the red-brown color of the bromine will fade as the bromine is consumed.

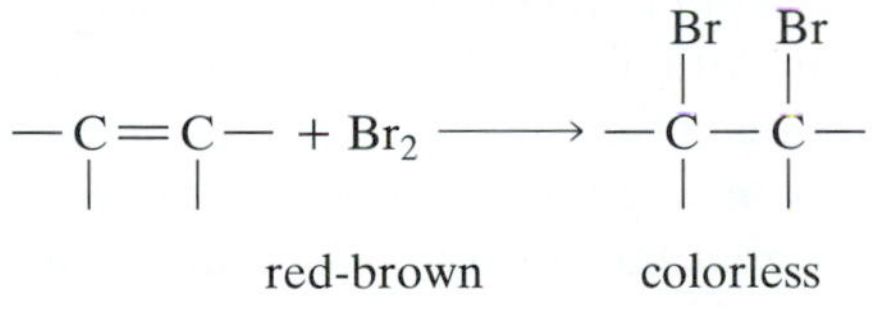

$$\underset{\text{red-brown}}{-C=C-} \ + \ Br_2 \ \longrightarrow \ \underset{\text{colorless}}{\overset{\overset{\text{Br}}{|}}{-C} - \overset{\overset{\text{Br}}{|}}{C}-}$$

Unfortunately carbon tetrachloride is toxic and carcinogenic, so you will use a saturated solution of bromine in water instead. When this solution is shaken with a solution of an unsaturated compound in dichloromethane, bromine migrates to the dichloromethane layer, where it can react. Thus decolorization of the lower dichloromethane layer upon shaking indicates unsaturation, and the amount of bromine water it takes to produce a color in the lower layer is a measure of the degree of unsaturation.

Directions

Test as many of the following commercial products as possible; your instructor may add or remove some.

butter	canola oil
corn oil	dry-cleaning solvent
linseed oil	margarine
mineral oil	olive oil
paint thinner (mineral spirits)	rubber cement
rubbing alcohol	safflower oil
turpentine	vegetable shortening (Crisco, etc.)

Have ready enough clean 13×100 mm test tubes (with stoppers) to test each commercial product provided, and label each test tube appropriately. Calibrate a Pasteur pipet by accurately measuring 0.5 mL of water, carefully drawing it into the pipet (no air bubbles in the tip), and marking the liquid level with a glass-marking pen. Dissolve one drop of each commercial product (or about 30 mg of a solid) in 0.5 mL of dichloromethane. *Under the hood* add 0.5 mL of saturated bromine water to each test tube using the calibrated Pasteur pipet. Stopper and shake each test tube vigorously for 10 seconds. Record the volume of bromine water added to each test tube during this and all future additions. If the dichloromethane layer in any test tube has a red-orange color, classify the commercial product it contains as saturated and set it aside. Add another 0.5 mL portion of bromine water to each of the remaining test tubes and shake as before. Repeat this process with any test tubes whose dichloromethane layer is clear after shaking, until the dichloromethane layer of every test tube is a red-orange color. Record the total volume of bromine water you added to each test tube.

Report. Classify each commercial product as saturated or unsaturated and compare your results with your predictions. Considering the volume of bromine water you added to each test tube as a rough measure of the unsaturation of its contents, arrange the commercial products in order of degree of unsaturation. Note that there may be several commercial products with about the same degree of unsaturation. Using sources such as those cited in the Bibliography, find out what kinds of unsaturated compounds may be present in each of the unsaturated substances. Draw structural formulas for representative components when you can.

Safety Notes

Take Care! Avoid contact with dichloromethane and bromine water and do not breathe their vapors.

Observe and Note: What happens?

Waste Disposal: Place the dichloromethane layers in a designated solvent recovery container.

Free-Radical Bromination
Of Hydrocarbons

MINILAB 21

Before You Begin: Try to predict the structures and relative stabilities of the free-radical intermediates that will form when each of the arenes reacts with bromine, and from that predict the relative reactivities of the arenes.

The side chain of an arene such as toluene can undergo free-radical halogenation in the presence of light, as illustrated for the bromination of toluene. The rate of such a reaction depends on the stability of the intermediate free radical. As a rule, the more stable a free radical is, the faster it will form. In this minilab you will determine the relative stabilities of the intermediates by measuring the relative reactivities of the corresponding substrates. You will then attempt to explain their relative stabilities.

Directions

Safety Notes

Dichloromethane may be harmful if inhaled or absorbed through the skin, and it is a suspected carcinogen. Minimize contact with the liquid and do not breathe its vapors.
Bromine is toxic and corrosive and its vapors are very harmful. Avoid contact with the bromine solution and do not breathe its vapors.
Toluene and the other aromatic hydrocarbons are flammable and have harmful vapors. Do not inhale their vapors.

In each of four clean, *dry*, labeled 13×100 mm test tubes, mix 2 mL of dichloromethane with 0.5 mL of one of the following aromatic hydrocarbons: toluene, ethylbenzene, isopropylbenzene, *t*-butylbenzene. Recording the time of addition to the nearest minute, add 1 mL of freshly prepared $0.5\ M$ bromine in dichloromethane to each test tube, then stopper and shake it. Set the test tubes in a well-lighted location such as a window sill, and record the approximate time it takes for each solution to become colorless. Observe the solutions closely for the first 5 minutes or so, and then at intervals during the lab period. If two or more solutions are not completely colorless by the end of the lab period, describe the relative intensity of their colors at that time.

Report. Arrange the four hydrocarbons in order of their reactivity toward bromine, most reactive first, and compare your results with your predictions. Write a balanced equation for each reaction (assuming monobromination), and give the structure of the free-radical intermediate. Arrange the free radicals in order of stability and explain their relative stabilities. Write a mechanism for the bromination of toluene.

Take Care! Avoid contact with dichloromethane, the arenes, and the bromine solution; do not breathe their vapors.

Waste Disposal: Place the bromine/dichloromethane solutions in a designated solvent recovery container.

The Nylon Rope Trick

Before You Begin: Review the reactions of acyl chlorides with amines in your lecture textbook.

In a chemical magic show the "magician" will sometimes combine two immiscible solutions in a beaker and draw out a seemingly endless rope of nylon. This "nylon rope trick" is based on the fact that reactants separated by an interface between two immiscible solvents will meet at that interface, where the product will appear. If the product is a polymer that—like nylon—forms strong fibers, it can be pulled from the interface like a rope. If the rope is drawn out slowly enough, the nylon that is removed is continually replenished at the interface until one of the reactants is used up.

In this minilab the reactants are 1,6-hexanediamine (hexamethylenediamine) and decanedioyl chloride (sebacoyl chloride). The product, because it is made from a 6-carbon diamine and a 10-carbon acyl dichloride, is called Nylon 6,10.

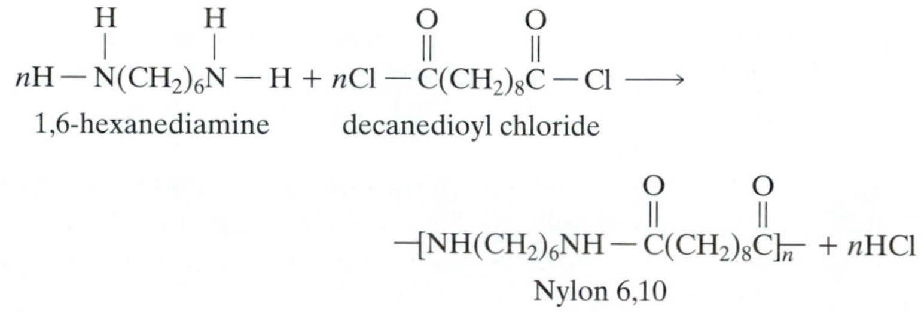

$$n\mathrm{H-N(CH_2)_6N-H} + n\mathrm{Cl-C(CH_2)_8C-Cl} \longrightarrow$$

1,6-hexanediamine   decanedioyl chloride

$$-[\mathrm{NH(CH_2)_6NH-C(CH_2)_8C}]_n- + n\mathrm{HCl}$$

Nylon 6,10

Directions

> Acid chlorides are corrosive and harmful by inhalation and skin absorption. Avoid contact with decanedioyl chloride and do not breathe its vapors.
>
> 1,6-Hexanediamine is corrosive and harmful by inhalation. Avoid contact and do not breathe its vapors.
>
> Dichloromethane may be harmful if inhaled or absorbed through the skin, and it is a suspected carcinogen. Minimize contact with the liquid and do not breathe its vapors.
>
> Wear protective gloves throughout this experiment. Avoid touching the wet polymer with your hands; if you do, wash them immediately with soap and warm water.

Take Care! Wear gloves, avoid contact with decanedioyl chloride, dichloromethane, and 1,6-hexanediamine; do not breathe their vapors.

Under the hood, dissolve 1.0 mL of pure decanedioyl chloride in 50 mL of dichloromethane in a 150-mL beaker. Use a tall-form beaker if they are available. Combine 10 mL of 1 *M* aqueous sodium hydroxide with 15 mL of water in a small beaker, then stir in 0.55 g of pure 1,6-hexanediamine until it dissolves. Tilt the beaker that contains the dichloromethane solution and slowly pour the 1,6-hexanediamine solution down the side, taking care not to mix the layers. Use a small spatula to free the polymer film from the side of the beaker if necessary, and use a piece of copper wire bent at one end like a fishhook to hook the film at its center. Pull the film up slowly and continuously to form a strand of Nylon 6,10, loop it around a cardboard tube (such as the core of a paper towel or toilet tissue roll), and rotate the tube to wind it out of the solution. When no more nylon rope can be drawn, stir the solution in the beaker vigorously to form additional polymer. Unwind the nylon strand into a beaker containing about 100 mL of 50% ethanol, add the recovered polymer, and stir gently to wash the Nylon 6,10. Then decant the wash solvent and lay the polymer on a paper towel to dry (blotting it between two towels will reduce the drying time). Weigh the dry Nylon 6,10 and turn it in to your instructor.

Report. Calculate the percent yield of Nylon 6,10 based on the amount of 1,6-hexanediamine you used.

Nucleophilic Substitution Rates of Alcohols

MINILAB 23

Before You Begin: Classify each alcohol as primary, secondary, or tertiary and try to predict their relative reactivities with the $HCl/ZnCl_2$ reagent.

In acidic solutions, different alcohols can react with various nucleophiles by either an S_N1 or S_N2 mechanism. The S_N1 mechanism involves loss of water from the protonated alcohol to form a carbocation, which then combines with the nucleophile. The S_N2 reaction involves nucleophilic attack on the protonated alcohol to yield the product directly.

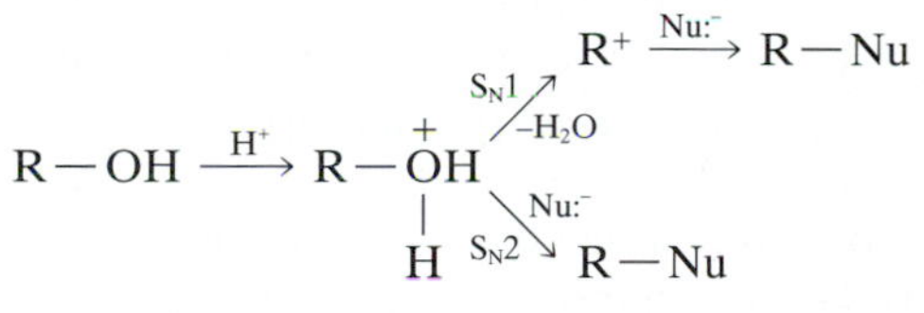

$$R-OH \xrightarrow{H^+} R-\overset{+}{\underset{\underset{H}{|}}{O}}H \begin{array}{c} \overset{S_N1}{\nearrow}_{-H_2O} R^+ \xrightarrow{Nu:^-} R-Nu \\ \\ \underset{S_N2}{\searrow}_{Nu:^-} R-Nu \end{array}$$

In this minilab you will compare the reactivities of different kinds of alcohols with hydrochloric acid in the presence of zinc chloride, a Lewis acid catalyst.

$$R-OH + HCl \xrightarrow{ZnCl_2} R-Cl + H_2O$$

The $ZnCl_2$ apparently coordinates with the oxygen atom of the alcohol, converting its —OH (a very poor leaving group) to a better leaving group,

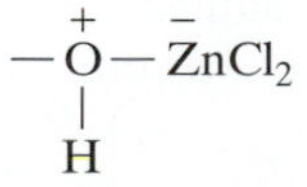

$$-\overset{+}{\underset{\underset{H}{|}}{O}}-\overset{-}{ZnCl_2}$$

You will use your results to decide whether the reaction appears to proceed by an S_N1 or S_N2 mechanism.

Directions

Safety Notes

> **Hydrochloric acid is very corrosive. Contact or inhalation can cause severe damage to the eyes, skin, and respiratory tract. Zinc chloride is toxic and corrosive, with harmful fumes. Avoid contact with the HCl/ZnCl$_2$ reagent and do not breathe its vapors. Wear gloves and work under a hood.**

Take Care! Wear gloves, avoid contact with the ZnCl$_2$/HCl reagent, do not breathe its vapors.

Stop and Think: What is happening to cause the changes you observed?

Waste Disposal: Put the contents of your test tubes in a designated waste container.

Label three clean, dry 10-cm test tubes, and measure 10 drops of (1) 1-butanol, (2) 2-butanol, and (3) 2-methyl-2-propanol into the respective test tubes. *Under the hood,* add 2.0 mL of ZnCl$_2$/HCl reagent to each test tube, stopper and shake each tube vigorously for 5 seconds, and let them stand at room temperature. Look for any evidence of a reaction, recording your observations during the first 15 minutes or so and at intervals throughout the lab period.

Report. Compare your results with your predictions. Explain your observations, decide whether the reaction occurs by an S_N1 or S_N2 mechanism (assume that all of the alcohols react by the same mechanism, if they react), and explain how you arrived at your conclusion. Write balanced equations and mechanisms for any reactions for which you saw evidence.

MINILAB 24

Photoreduction of Benzophenone to Benzopinacol

Before You Begin: Read or review the operations as necessary.

When a benzophenone molecule absorbs light of an appropriate wavelength, one of two paired ground state electrons on its oxygen atom jumps to an antibonding pi orbital without changing its spin state. This puts the molecule into a *singlet* excited state, in which both electrons have opposite spins. The singlet state molecule can undergo another transition to a *triplet* excited state, in which both electrons have the same spin.

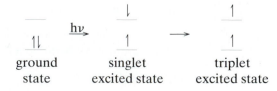

A triplet state benzophenone molecule is very reactive and behaves like a diradical. It can even strip a hydrogen atom from a molecule of 2-propanol (a very unlikely reducing agent), which then loses a second hydrogen atom to ground state benzophenone and is converted to a molecule of acetone.

Photochemical reduction of benzophenone to benzpinacol

1.
$$\text{PhC} \overset{\text{O*}}{\underset{\text{Ph}}{\Vert}} + \text{PhC} \overset{\text{O}}{\underset{\text{Ph}}{\Vert}} + \text{CH}_3\text{CHCH}_3 \overset{\text{OH}}{\longrightarrow} 2\text{PhC} \overset{\text{OH}}{\underset{\text{Ph}}{\vert}} + \text{CH}_3\text{CCH}_3 \overset{\text{O}}{\Vert}$$

benzophenone
(* = triplet state)

2.
$$2\text{PhC} \overset{\text{OH}}{\underset{\text{Ph}}{\vert}} \cdot \longrightarrow \text{PhC} \overset{\text{OH}}{\underset{\text{Ph}}{\vert}} - \text{CPh} \overset{\text{OH}}{\underset{\text{Ph}}{\vert}}$$

benzopinacol

This leaves two identical diphenylhydoxymethyl radicals that can combine to yield a molecule of benzopinacol. A possible by-product of this reaction is benzhydrol [PhCH(OH)Ph], which can form when a diphenylhydoxymethyl radical abstracts a hydrogen atom from 2-propanol. In this minilab you will find out whether the light available in your laboratory will bring about the conversion of benzophenone in 2-propanol to benzopinacol. Bases can catalyze cleavage of the product, so a drop of acetic acid is added to neutralize any base present in the reaction mixture.

Directions

> **Acetic acid causes chemical burns and its vapors are highly irritating to the eyes and respiratory tract. Dispense under a hood, avoid contact, and do not breathe vapors.**

Safety Notes

Combine 1.0 g of benzophenone with 5.0 mL of 2-propanol (isopropyl alcohol) in a test tube, and warm the mixture over a steam bath to dissolve the benzophenone. *Under the hood,* add a drop of glacial acetic acid, then stopper the test tube tightly and shake it. Place it on a windowsill where it will receive direct sunlight, if possible; otherwise place it close to a strong artificial light source. After a week or more, collect the precipitate (if there is any) by vacuum filtration on a Hirsch funnel [OP-12], washing it on the filter with cold ethanol. Dry the product [OP-21], and measure its melting point [OP-28]. Leave the filtrate in sunlight for another week or so to see if any additional product forms; if it does, collect and dry it also. Weigh the product and turn it in to your instructor.

Take Care! Avoid contact with acetic acid, do not breathe its vapors.

If the laboratory is very cold, some benzophenone may precipitate before it reacts.

Waste Disposal: Place the filtrate in a designated solvent recovery container.

Report. Provide evidence that your product is benzopinacol, and not benzhydrol or unreacted benzophenone, by looking up the properties of these compounds in *The Merck Index.* Write a balanced equation for the reaction and calculate the percent yield of the product.

Oxidation of Alcohols by Potassium Permanganate

MINILAB 25

$$-\overset{\displaystyle |}{\underset{\displaystyle |}{C}} \overset{\displaystyle O-H}{\underset{\displaystyle H}{}} \xrightarrow[-2H]{\text{oxidizing agent}} -\overset{\displaystyle |}{C}\overset{\displaystyle O}{}$$

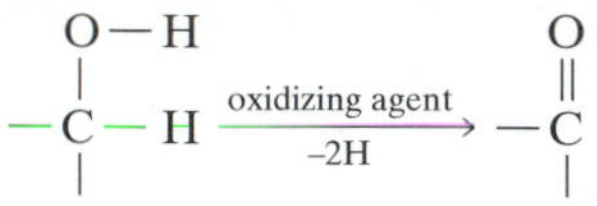

Before You Begin: Classify the alcohols as primary, secondary, and tertiary. Review the reactions of alcohols in your lecture textbook.

Primary and secondary alcohols can be converted to aldehydes and ketones by certain oxidizing agents. Tertiary alcohols are not converted directly to carbonyl compounds by oxidizing agents because such a reaction would require cleavage of a carbon-carbon bond. In this minilab you will treat several alcohols with neutral potassium permanganate, then with potassium permanganate acidified with HCl, and compare their reactivities. You will also observe the effect of adding some sulfuric acid to the reaction mixture.

Directions

Safety Notes

> **Sulfuric acid causes chemical burns that can seriously damage skin and eyes; avoid contact.**

Clean and number five small test tubes. Add 2 mL of 0.05% aqueous potassium permanganate to each test tube. Use test tube #1 as a control and add two drops of the following alcohols to the other test tubes: #2, methanol; #3, ethanol; #4, 2-propanol; #5, 2-methyl-2-propanol. Stopper and shake each test tube for 10 seconds and record your observations immediately after shaking, then after standing for 5 and 10 minutes, respectively. To any solution that shows little evidence of reaction (compared to the control) after 10 minutes, add 3 drops of 3 M HCl, shake, and observe at the same intervals as before. To any solution that shows little evidence of reaction 10 minutes after the HCl addition, add 2 drops of concentrated sulfuric acid, shake, and observe at the same intervals as before.

Stop and Think: What is the solid that forms?

Take Care! Avoid contact with H_2SO_4.

Waste Disposal: Place the permanganate test solutions in a designated solvent recovery container.

Report. Arrange the alcohols in order of reactivity toward $KMnO_4$ and describe the effect of substrate structure on reactivity. Based on alcohol reactions you have studied, explain any reaction of a tertiary alcohol for which you saw evidence. Draw the structure of the organic product of each reaction. Write a balanced equation for the reaction of methanol with $KMnO_4$ under acidic reaction conditions (you can balance it using the half-reaction method taught in general chemistry).

Preparation of a Fluorescent Dye   **MINILAB 26**

Before You Begin: Calculate the molar amounts of resorcinol and phthalic anhydride used and decide which one is the limiting reactant.

Certain dyes and indicators such as phenolphthalein and fluorescein resemble the triphenymethane dyes (see Experiment 31) in that their molecules contain three benzene rings attached to a central carbon atom. In this minilab you will prepare fluorescein by a Lewis-acid catalyzed reaction of resorcinol (1,3-dihydroxybenzene) with phthalic anhydride.

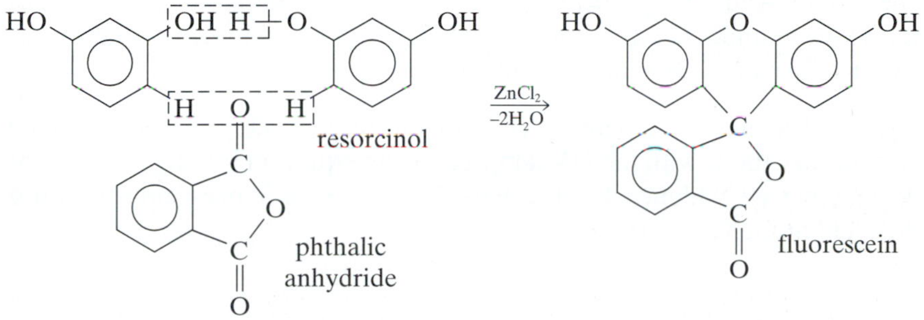

You will then convert the fluorescein to its basic form, disodium fluorescein, by dissolving it in dilute NaOH, and irradiate the resulting solution with ultraviolet light.

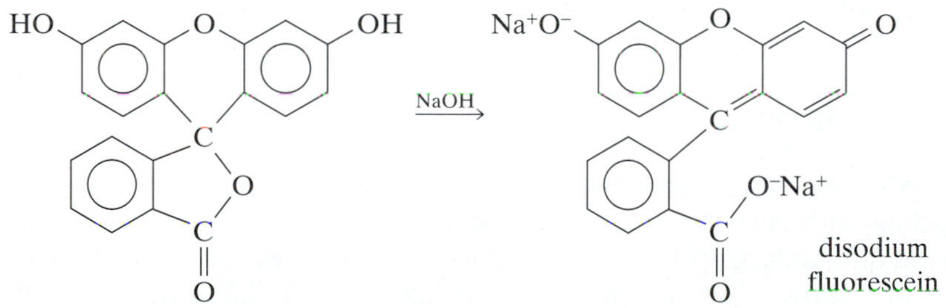

Directions

Zinc chloride is toxic and corrosive, with harmful fumes. Resorcinol and phthalic anhydride are severe skin and eye irritants. Wear gloves and use a hood during this experiment. Avoid contact with zinc chloride, resorcinol, and phthalic anhydride, and do not breathe their fumes or dust.

Safety Notes

Under the hood, measure 0.32 g of anhydrous zinc chloride into a 10-cm test tube, and heat it gently over a burner flame until no more bubbles of water vapor are evolved from the molten salt. Tip and rotate the tube while stirring

Take Care! Wear gloves, avoid contact, do not breathe fumes.

Take Care! Avoid contact with phthalic anhydride and resorcinol.

Take Care! Do not use your thermometer as a stirring rod!

Fluorescein stains are very hard to remove, so avoid spilling any and keep it off your skin and clothing.

Waste Disposal: Pour the filtrates down the drain.

Take Care! Do not look directly at the UV light source.

Waste Disposal: Place the solution in a designated waste container.

the melt so that the zinc chloride coats the sides of the test tube as it cools. Add 0.44 g of phthalic anhydride and 0.72 g of resorcinol, and heat the test tube cautiously until the temperature of the melt is 180°C. Keep it near that temperature, stirring constantly with a stirring rod, until no more bubbles are evolved. Let the mixture cool for a minute or so; then add 5 mL of 2 *M* hydrochloric acid and heat the liquid to boiling with constant stirring. Use your stirring rod to break up the solid mass and continue stirring for several minutes to dissolve zinc salts and unreacted starting materials. Collect the fluorescein by vacuum filtration in a Hirsch funnel [OP-12], grind it fine in a mortar, treat the solid with another 5-mL portion of 2 *M* HCl as before, and filter it. Dry the fluorescein [OP-21] and weigh it.

Dissolve about 10 mg (0.010 g) of fluorescein in 10 mL of 0.1 *M* aqueous sodium hydroxide. Irradiate the solution with an ultraviolet lamp in a darkened room. Add 2 *M* hydrochloric acid drop by drop until the appearance of the solution under the UV lamp changes markedly and record your observations. Turn in the remaining product to your instructor.

Report. Calculate the percent yield of fluorescein. Describe and explain your observations with the UV lamp, giving an equation for any reaction for which you saw evidence. Explain why the color of a fluorescein solution is different at high and low pH.

Diels–Alder Reaction of Maleic Anhydride and Furan

MINILAB 27

Before You Begin: Read or review the operations as necessary. Calculate the theoretical yield of the adduct, looking up any needed properties in a chemistry reference book. Referring to Experiment 32, propose a hypothesis regarding the stereochemistry of the adduct.

Aromatic compounds sometimes behave like conjugated dienes when reacting with certain dienophiles. Furan is an aromatic heterocyclic compound that undergoes Diels–Alder reactions as if it had the diene resonance structure shown. Furan forms a bicyclic adduct with maleic anhydride that can have either an *endo* or an *exo* stereochemistry.

furan   resonance structure
              for furan

furan   maleic   *endo*-adduct   *exo*-adduct
         anyhdride

The transition state leading to an *endo* adduct is stabilized by pi electron overlap, causing the *endo* adduct to form faster. When a Diels–Alder reaction is under *kinetic control*, meaning that the product that forms faster is favored, the *endo* adduct is the major or sole product. However *endo* adducts are generally less stable than *exo* adducts, so when a Diels–Alder reaction is under *thermodynamic control*, meaning that the more stable product is favored, the *exo* adduct should be the major or sole product.

In this minilab you will prepare the Diels–Alder adduct of maleic anhydride and furan, measure its melting point, and decide whether the reaction is under kinetic or thermodynamic control. Since the reaction takes several days, you should prepare the reaction mixure during the lab period before the one in which you plan to isolate and analyze the product. The melting point of the *endo* adduct is 81°C and the melting point of the *exo* adduct is 114°C. The adduct tends to decompose near its melting point, so measure an approximate melting point (T_1) at a rapid rate, then obtain a more accurate melting point by raising the temperature slowly and inserting the melting point capillary only when the temperature reaches T_1.

Stop and Think: What products do you think the adduct forms when it decomposes?

Directions

Maleic anhydride is corrosive and toxic and can cause severe damage to the eyes, skin, and upper respiratory tract. Wear gloves, avoid contact, and do not breathe the dust. If you must powder maleic anhydride briquettes, do it under the hood and wear safety goggles and protective clothing.
Ethyl ether and hexanes are extremely flammable and may be harmful if inhaled. Do not breathe their vapors and keep them away from flames and hot surfaces.
Furan is flammable and harmful by inhalation, ingestion, and skin contact. Avoid contact, do not breathe its vapors, and keep flames away.

Safety Notes

Dissolve 1.20 g of finely divided maleic anhydride in 10 mL of anhydrous ethyl ether in a conical centrifuge tube (or small test tube) by warming the mixture gently on a steam bath or hot water bath. Replace any ether that evaporates. When the solution is cool add 1.00 mL of furan (accurately measured). Stopper the reaction tube and cap it, wrapping the cap with Parafilm to prevent evaporation of ether, and leave the reaction mixture in a designated location for several days or until the next lab period. Collect the crystallized adduct by vacuum filtration using a Hirsch funnel [OP-12], and air dry it on the funnel. Recrystallize it from hexane-ethyl acetate [OP-23] by heating it *just* to boiling in 5 mL of hexanes, then adding enough warm ethyl acetate (~1–3 mL) to the mixture, with stirring, to dissolve the solid. When collecting the product by vacuum filtration, wash it on the filter with cold hexanes. Dry [OP-21] the adduct and measure its mass and melting point [OP-28].

Report. Calculate the percent yield of the adduct. Write an equation for the decomposition reaction undergone by the adduct at its melting point. Tell whether the adduct is *endo* or *exo* and propose an explanation for your result, writing equations for any relevant reactions.

Take Care! Wear gloves, avoid contact with maleic anhydride and furan, do not breathe their dust or vapor.

Take Care! Keep flames away from ether.

Waste Disposal: Place the filtrate in the appropriate solvent recovery container.

Take Care! Keep flames away from the solvents.

Waste Disposal: Place the filtrate in a designated solvent recovery container.

Stop and Think: Was your hypothesis correct?

Identification of an Unknown Arene by NMR Spectrometry

MINILAB 28

Before You Begin: Read or review OP-35 on NMR spectrometry.

NMR spectrometry is one of the most valuable tools at the disposal of the organic chemist. From the ^{1}H and ^{13}C NMR specta of a complex organic compound, a chemist skilled in spectral interpretation can often derive its complete molecular structure. In this minilab you will record the ^{1}H NMR spectrum of a relatively simple aromatic hydrocarbon (arene) or another type of compound selected by your instructor. At your instructor's option, you may also record its ^{13}C NMR spectrum. You will then use your spectrum (or spectra) to derive the molecular structure of the unknown compound. Before you begin, make sure you understand how to interpret NMR spectra by reading the appropriate sections in OP-35 and, as necessary, relevant sections from your lecture textbook.

Directions

Your instructor may elect to assign you a compound with a different molecular formula than the one specified here.

Safety Notes

> The aromatic hydrocarbons are flammable and have harmful vapors. Do not inhale their vapors.
> Deuterochloroform is harmful if inhaled or absorbed through the skin, and it is a suspected human carcinogen. Avoid contact with the liquid and do not breathe its vapors.

Take Care! Avoid contact with $CDCl_3$, do not breathe its vapors.

Waste Disposal: Place the deuterochloroform solution in the designated solvent recovery container. Place any remaining unknown in the designated waste container.

You will be issued a small amount of an unknown arene with the molecular formula $C_{10}H_{14}$. As directed by your instructor, record an integrated ^{1}H NMR spectrum [OP-35] of the compound in deuterochloroform using TMS as a reference compound. If requested, record its ^{13}C NMR spectrum as well, or obtain the spectrum from your instructor. Letter the signals (a, b, c, etc.) from right to left and construct a table listing the chemical shift (δ), signal area, and multiplicity for each signal. For each signal determine the number of protons giving rise to the signal, the number of neighboring protons (except for Ar**H** signals), and the proton type (aromatic, benzylic, etc.), and include this information on your table. Use your table to deduce the structure of your unknown compound.

Report. Give the structure of your unknown arene and explain, in detail, how you arrived at it. Include your table and turn in your NMR spectrum (or spectra) with your report.

Interpretation of a Mass Spectrum

Before You Begin: Read OP-37 on mass spectrometry. If you are to analyze an acetate ester of an unknown alcohol, also read the spectral interpretation rules provided by your instructor.

When an organic compound is injected into a mass spectrometer, its molecules are bombarded by a stream of high-energy electrons. When a high-energy electron encounters a molecule, it can dislodge one of the molecule's electrons, producing a charged *molecular ion*, $M^{\cdot+}$. Each molecular ion can then fragment in a variety of ways, giving rise to an array of *daughter ions*. A mass spectrum of a compound is a record of all the molecular and daughter ions arising from the fragmentation of its molecules. Each peak on a mass spectrum corresponds (usually) to one kind of ion, whose position indicates the mass of the ion (actually its mass/charge ratio) and whose intensity indicates the relative abundance of the ion.

In this minilab you will obtain the mass spectrum of the major component of an unknown liquid using a gas chromatograph-mass spectrometer (GC-MS), or the mass spectrum of a pure liquid using a mass spectrometer. Following your instructor's directions and the procedure described here, you will then interpret your mass spectrum as completely as you can.

Directions

This minilab may be performed by teams of three or more students, each working with a different unknown, if desired.

Safety Notes

> **Assume that the unknown liquid is flammable and harmful by inhalation, ingestion, and skin absorption. Avoid contact, do not breathe vapors, and keep flames away.**

You will receive a liquid in a capped vial for analysis. Alternatively, you can use an acetate ester of an unknown alcohol prepared as described in Minilab 5. Your instructor will then show you how to operate the mass spectrometer or GC/MS instrument [OP-37]. Unless directed otherwise, inject a 2-3 μL head-space sample into the injection port and obtain a printout of the mass spectrum corresponding to the liquid's major GC peak. Identify the base peak and molecular ion peak and determine the molecular weight of your compound. Analyze as many peaks as you can by assigning structures for the species that may be responsible for them, including (for example) species arising from α-cleavage and McLafferty rearrangements. At your instructor's request, determine the molecular formula of the compound and attempt to derive its structure, working with the other members of your team.

Report. Summarize your spectral data and interpretations in a table, and report any other results and your conclusions. Explain how you arrived at your conclusions. Write mechanisms for the formation of as many species as you can.

Nitration of Naphthalene

Before You Begin: Read or review the operations as necessary. Review the mechanisms of electrophilic aromatic substitution reactions, as needed.

Polynuclear aromatic hydrocarbons such as naphthalene can be nitrated by the same methods as benzene derivatives, including the well-known "mixed acid" method that utilizes a mixture of nitric acid and sulfuric acid. The electrophile is the nitronium ion, NO_2^+, which is formed at low concentration by a reaction of the two acids. The mononitration of naphthalene could lead to either one of two products, 1-nitronaphthalene or 2-nitronaphthalene.

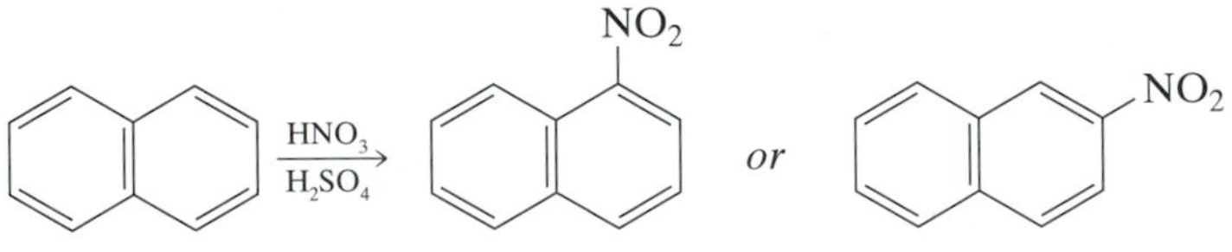

1-nitronaphthalene
m.p. 61° C

2-nitronaphthalene
m.p. 79° C

The major product of the reaction you will carry out is the kinetic product, the one that forms from the more stable carbocation intermediate (also called an arenium ion). By drawing all possible resonance structures for the carbocation intermediate leading to each of these products, you should be able to predict the major product. You will then carry out the nitration of naphthalene, identify the product from its melting point, and find out whether your prediction was correct. Because it is difficult to purify the product completely, your melting point may be somewhat lower than the listed value.

Directions

Safety Notes

> **Sulfuric acid and nitric acid can cause very serious burns and they react violently with water and other chemicals. Nitric acid produces toxic nitrogen dioxide fumes during the reaction. Use gloves and a hood, avoid contact, and do not breathe vapors.**
> **The nitronaphthalene is a suspected carcinogen; avoid contact.**
> **The mixture of hexanes is very flammable; keep flames away.**

If the mixed-acid solution has not been prepared, make it by measuring 1.0 mL of concentrated nitric acid into a conical centrifuge tube or test tube, cooling it in ice, and cautiously mixing in 1.0 mL of concentrated sulfuric acid.

Measure 2.0 mL of the mixed acid nitrating solution into a conical centrifuge tube or test tube. (**Take Care!** Wear gloves, avoid contact, do not breathe vapors.) In small portions add 1.0 g of finely divided napthalene to the mixed acid, stirring or shaking after each addition and cooling as necessary to keep the temperature around 45–50°C. Then stir or shake the reaction mixture in a 60°C water bath for 20 minutes. After the reaction mixture cools to room temperature, stir it into 50 mL of ice-cold water. Wait for the yellow product to solidify, then decant (pour out) the liquid carefully, leaving the solid behind. *Under the hood*, boil the solid with 15 mL of fresh water for 10 minutes or more, then cool the mixture in ice and collect the product by vacuum filtration [OP-12]. Dry the crude product [OP-21] and measure its mass. Recrystallize about 0.2 g of the product [OP-23] by heating it under reflux [OP-6] for 5 minutes or more with 10 mL of hexanes and filtering the hot solution, leaving any undissolved oil behind. If there is an appreciable amount of oil in the boiling flask, heat it under reflux with another 5 mL or so of hexanes and filter the hot solution as before, combining the filtrates. When crystallization is complete, collect the product by vacuum filtration. Measure the melting point [OP-28] of the purified nitronaphthalene and decide which isomer it is.

Report. Calculate the percent yield of the crude nitronaphthalene and give the structure of the product. Write a mechanism for the reaction, showing all resonance structures for the intermediate, and explain its orientation.

Take Care! Do not use the thermometer as a stirring rod.

Waste Disposal: Dispose of the liquid as directed by your instructor.

Waste Disposal: Pour the filtrate down the drain.

Take Care! Keep flames away from hexanes.

Waste Disposal: Place the filtrate and any remaining oil in a designated solvent recovery container.

Stop and Think: Was your prediction correct? If not, explain why not in your report.

Preparation of Trityl Cations by the Friedel–Crafts Reaction

MINILAB 31

Before You Begin: Review the mechanism of Friedel–Crafts alkylation in your lecture textbook.

A simple chemical test for aromatic hyrocarbons is performed by pipeting a solution of the unknown compound in trichloromethane (chloroform) over some freshly sublimed aluminum chloride. If the unknown is aromatic, a characteristic color (other than light yellow) will appear. The test is based on the Friedel–Crafts reaction of the aromatic hydrocarbon with trichloromethane, which, because it has three chlorine atoms, reacts with three molecules of the aromatic hydrocarbon to form a triarylmethane. The triarylmethane then loses a hydride ion to one of several different carbocation intermediates, yielding a colored triarylmethyl (trityl) carbocation that is structurally related to the triphenylmethane dyes discussed in Experiment 31.

$$3ArH + CHCl_3 \xrightarrow{AlCl_3} Ar_3CH + 3HCl$$
$$Ar_3CH + R^+ \longrightarrow Ar_3C^+ + RH \quad (R^+ = Ar_2CH^+, etc.)$$

In this experiment you will carry out the Friedel–Crafts reaction on a test tube scale with several aromatic hydrocarbons and observe the colors of the products. You will use only a few drops of trichloromethane for each reaction, but because it is a suspected carcinogen you must handle it with gloves, under a hood.

Directions

At your instructor's option, this minilab can be carried out by teams of several students. Your instructor may add or remove some test compounds.

> **Aluminum chloride reacts violently with water, skin and eye contact can cause painful burns, and inhaling the dust or vapors is harmful. Wear gloves, avoid contact, do not inhale dust or vapors, and keep it away from water.**
> **Trichloromethane is harmful if inhaled, swallowed, or absorbed through the skin, and it is a suspected human carcinogen. Wear gloves and work under a hood. Avoid contact with the liquid and do not breathe its vapors. The aromatic hydrocarbons have harmful vapors; avoid inhalation.**

Take Care! Wear gloves, avoid contact with trichloromethane and aluminum chloride, do not breathe their vapors

Take Care! Be sure there are no flammable solvents in the vicinity.

Observe and Note: What happens? Describe your observations.

Stop and Think: Are all of the test compounds aromatic?

Waste Disposal: Empty the test tubes that contain the trichloromethane solutions into a designated waste container.

Label five or more clean, *dry*, small test tubes and add a drop of one of the following liquids, or about 20 mg of a solid, to each test tube: toluene, limonene, naphthalene, anthracene, biphenyl. *Under the hood,* add 8 drops of trichloromethane to each test tube. *Under the hood,* carefully measure about 0.10 g of anhydrous aluminum chloride into a clean, *dry* 10-cm test tube. *Under the hood,* heat the test tube gently over a burner flame until the aluminum chloride forms a thin layer of sublimed solid up the sides of the tube. When it is cool, pipet the toluene solution down one side that is coated with the white solid. Repeat the test using the remaining solutions. You should be able to use the same $AlCl_3$ layer to test several compounds; prepare a fresh layer on another clean, dry test tube as necessary.

Report. Tabulate and explain your observations, telling which compounds are aromatic. Write equations and a detailed mechanism for the reaction of toluene with trichloromethane, assuming *para* substitution.

MINILAB 32

Air Oxidation of Fluorene to 9-Fluorenone

Before You Begin: Read or review the operations as necessary.

The side chains of arenes can be oxidized by powerful oxidizing agents such as potassium permanganate and chromium(III) oxide/acetic anhydride, as illustrated for the following reactions of toluene.

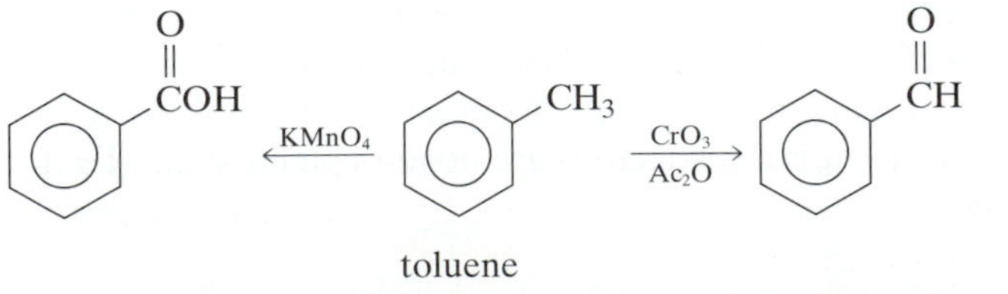

toluene

The methylene side chain of fluorene, however, can be oxidized to a carbonyl group in basic solution by a much milder oxidizing agent, air.

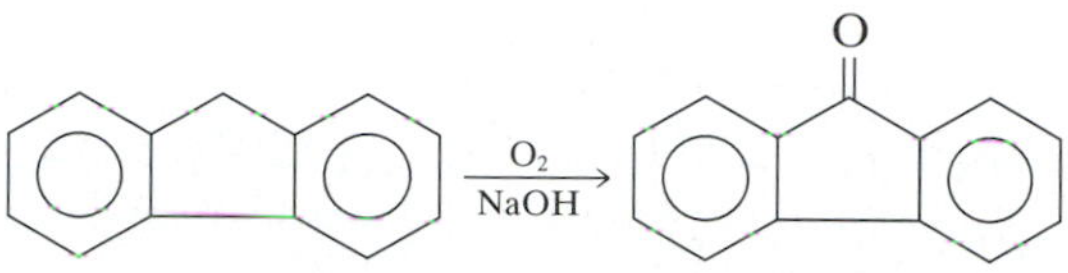

fluorene   9-fluorenone

This reaction, which yields the aromatic ketone 9-fluorenone, is ordinarily quite slow, but with vigorous stirring and a phase-transfer catalyst it can be carried out in about an hour.

Directions

> **Sodium hydroxide is toxic and corrosive, causing severe damage to skin, eyes, and mucous membranes. Wear gloves and avoid contact with the NaOH solution.**

Safety Notes

Mix 0.80 g of fluorene with 8 mL of heptane in a 125-mL Erlenmeyer flask. Add 4 mL of 10 M (~30%) aqueous sodium hydroxide. Add 10 drops of tricaprylmethylammonium chloride (Aliquat 336) and stir the mixture vigorously for an hour or more, using a 1-inch or larger magnetic stirbar. The stirring rate should be high enough to produce a froth on the surface of the reaction mixture. Transfer the reaction mixture to a 15-mL conical centrifuge tube (or a 15-cm test tube), cool it in ice, and carefully remove the aqueous (lower) layer with a Pasteur pipet. Collect the crude fluorenone by vacuum filtration on a Hirsch funnel [OP-12], washing it on the filter with a little 1 M HCl, then with water. Dry it [OP-21] at room temperature or in an oven at 60°C. Recrystallize [OP-23] the 9-fluorenone from cyclohexane and measure its mass and melting point [OP-28].

Take Care! Keep flames away from heptane.

Take Care! Wear gloves and avoid contact with NaOH.

Report. Calculate the percent yield of 9-fluorenone. Suggest a reason why fluorene is oxidized by O_2 much more rapidly than toluene, given that oxidation by O_2 is ordinarily a free-radical reaction.

Take Care! Keep flames away from cyclohexane.

Waste Disposal: Place the filtrates containing heptane and cyclohexane in a designated solvent recovery container.

A Nucleophilic Addition-Elimination Reaction of Benzil

MINILAB 33

Before You Begin: Read or review the operations as necessary. Review the reactions of carbonyl compounds with ammonia derivatives in your lecture textbook.

Like the Wittig reaction (see Experiment 40), the reaction of an ammonia derivative with a carbonyl compound involves a nucleophilic addition step, followed by an elimination step that yields a double bond.

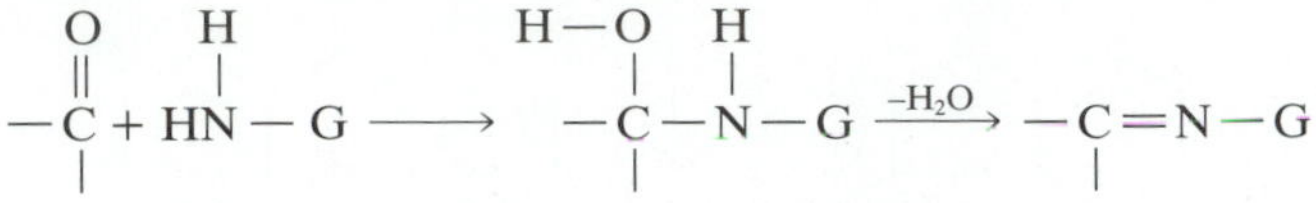

nucleophilic addition elimination

G = R, Ar, NHAr, OH, etc.

Benzil has two carbonyl groups so it can react with certain ammonia derivatives that have two NH_2 groups to form cyclic compounds. In this minilab you will carry out the reaction of benzil with 1,2-benzenediamine (*o*-phenylenediamine) to prepare the heterocyclic amine 2,3-diphenylquinoxaline.

benzil 1,2-benzenediamine 2,3-diphenylquinoxaline

Directions

Safety Notes

1,2-Benzenediamine can cause dermatitis and serious eye damage. Wear gloves and avoid contact.

Take Care! Wear gloves, avoid contact with 1, 2-benzenediamine.

Waste Disposal: Pour the filtrates down the drain.

In a conical centrifuge tube or small test tube dissolve 0.50 g of benzil in 2.0 mL of warm 95% ethanol. In a small beaker or test tube dissolve 0.25 g of 1,2-benzenediamine in 2.0 mL of 95% ethanol and add this solution to the benzil solution. Heat the reaction mixture in a 50°C water bath for 20–30 minutes. Add enough water dropwise to just saturate the warm solution (watch for a slight cloudiness that persists), then cool the reaction tube in an ice/water bath until crystallization is complete. Recover the product by vacuum filtration [OP-12]. Recrystallize the 2,3-diphenylquinoxaline from ethanol-water mixed solvent [OP-23]. Dry the purified product [OP-21] and measure its mass and melting point [OP-28]. The pure compound should melt at 125–126°C.

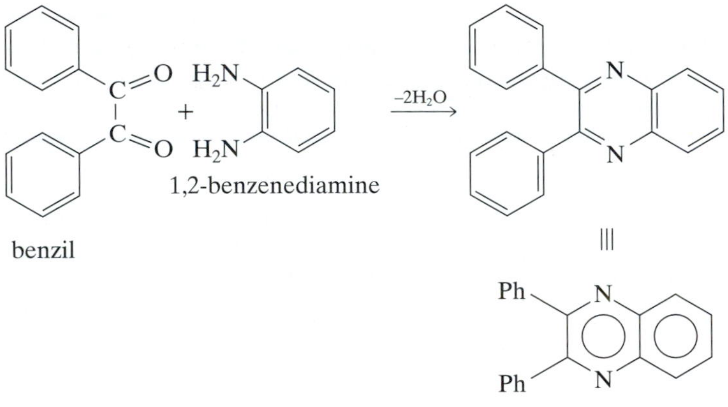

Report. Write a detailed mechanism for the reaction of benzil with 1,2-benzenediamine.

Preparation of Aldol Condensation Products

This minilab is based on an article published in the *Journal of Chemical Education* **1987**, *64*, 367.

Before You Begin: Read or review the operations as necessary. Review the mechanisms of aldol condensation reactions. Be prepared to calculate the mass and volume of 2.00 mmol of any aldehyde and 8.0 mmol of any ketone in Table M3.

As described in Experiment 41, aromatic aldehydes react with aliphatic ketones in the presence of base to form condensation products. In the presence of excess aldehyde, each mole of ketone can condense with two moles of the aldehyde, as illustrated for the following reaction of benzaldehyde and acetone.

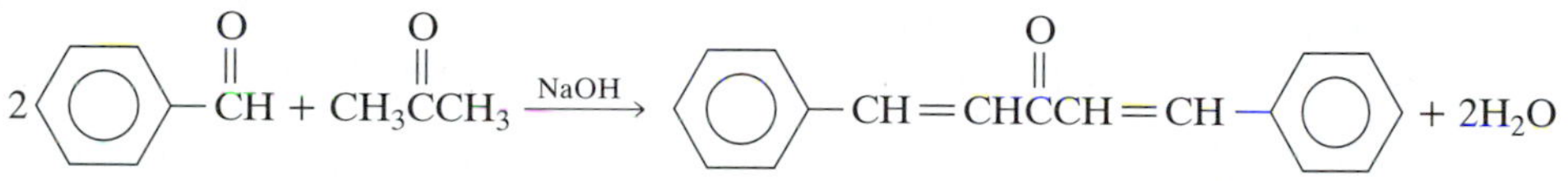

$$2\ C_6H_5-\overset{O}{\overset{\|}{CH}} + CH_3\overset{O}{\overset{\|}{C}}CH_3 \xrightarrow{NaOH} C_6H_5-CH{=}CHC{CH}{=}CH-C_6H_5 + 2H_2O$$

dibenzalacetone

In this minilab you will be able to select your reactants from a list containing four aldehydes and four ketones, making it possible to synthesize up to 16 aldol condensation products. If you work in a group, the group members can select combinations of reactants that will enable them, by sharing information, to learn something about the effect of structure on reactivity in aldol condensation reactions

Table M3 Aldehydes and ketones for the aldol condensation reactions

	Compound	Formula weight	Density
aldehydes	benzaldehyde	106.1	1.045
	4-methylbenzaldehyde	120.2	1.019
	4-methoxybenzaldehyde	136.2	1.119
	trans-cinnamaldehyde	132.2	1.050
ketones	acetone	58.1	0.791
	cyclopentanone	84.1	0.951
	cyclohexanone	98.2	0.948
	4-methylcyclohexanone	112.2	0.916

Directions

With your instructor's permission you can work in groups of four to six or more students, with each student in a group using a different set of reactants.

> **Assume that the aldehydes and ketones are flammable and harmful by inhalation, ingestion, and skin absorption. Minimize contact and avoid inhaling their vapors.**

Safety Notes

Select one aldehyde and one ketone from Table M3, or from a list provided by your instructor. If you are working in a group, select a set of reactants that is not being used by any other member of the group. Accurately measure 2.00 mmol of your ketone into a test tube and dissolve it in 4.0 mL of 95% ethanol. Add 8.0 mmol of your aldehyde (it can be measured by volume) and 3.0 mL of 1 *M* NaOH. Let the reaction mixture stand at room temperature for 10 minutes with occasional shaking. If little or no precipitate has formed after 10 minutes, heat the reaction mixture in a boiling water bath until precipitation appears complete. Be careful that it does not boil over. Let the reaction tube cool to room temperature, then cool it further in an ice/water bath and collect the aldol condensation product by vacuum filtration [OP-12], washing it on the filter with 2 mL of cold 4% acetic acid/95% ethanol (v/v) followed by 2 mL of cold 95% ethanol. Dry the crude product and measure its mass and melting point [OP-28]. At your instructor's request, recrystallize [OP-23] some of the product from a suitable solvent before you measure its melting point. Either 95% ethanol, toluene, or 2-butanone may be a suitable recrystallization solvent.

Observe and Note: Record the time it takes for a precipitate to appear.

Stop and Think: What is the purpose of the acetic acid?

Waste Disposal: Pour the filtrate down the drain.

Waste Disposal: Place the filtrate in the appropriate solvent recovery container.

Take Care! Avoid contact with toluene and 2-butanone, do not breathe their vapors.

Report. Calculate the percent yield of your preparation and report your melting point. Give the structure of your product and an equation and detailed mechanism for its formation. If you worked in a group, prepare a table giving the structure of each product and its mass, melting point, and percent yield. Based on the yields and other information gathered from the members of your group, propose a hypothesis to account for the effect of aldehyde structure on reactivity in these reactions.

MINILAB 35

A Spontaneous Reaction of Benzaldehyde

Before You Begin: After reading the minilab, develop a hypothesis about the possible identity of the product. Then gather information and prepare a plan of action that will enable you to confirm its identity, if your hypothesis is correct.

In this minilab you will observe a spontaneous reaction of benzaldehyde, isolate the product, and identify it by any means at your disposal. If you are not sure about the availability of certain equipment, test solutions, etc., ask your instructor. As in many other experiments of this type, a little thought and preparation can save you hours of experimental work, so plan ahead.

Directions

Benzaldehyde and the product are mild irritants; avoid contact with them.

Put about 15 mL of clean dry sand into a small beaker. Stir in just enough benzaldehyde to to moisten the sand—there should be no standing liquid on top. Bend a pipe cleaner into a U shape and push both ends into the sand so that is stands upright. Label the beaker with your name and set it aside until the next lab period. Scrape off the substance that has formed and recrystallize it [OP-23] from boiling water. Dry the product [OP-21] and identify it by any means at your disposal.

Report. Give the structure of the product and write a balanced equation for its formation. Describe your experimental work and explain how you arrived at your conclusion.

Safety Notes

Waste Disposal: Pour the filtrate down the drain.

Stop and Think: What is there for the benzaldehyde to react with?

Acid/Base Strengths of Organic Compounds

MINILAB 36

Before You Begin: Review sections of your organic or general chemistry lecture texts about the effect of structure on acid/base strength and about writing resonance structures, and try to predict the relative acid/base strengths of the compounds listed.

In your general chemistry course you learned some rules for predicting acid/base strengths. For example, the strength of an acid depends to some extent on the periodic table location of the atom to which a proton is bonded; thus compounds with H—O bonds tend to be more acidic than those with H—N or H—C bonds, and less acidic than those with H—S bonds. Conversely, oxygen bases are less basic than comparable nitrogen bases and more basic than comparable sulfur bases. But most organic acids have O—H bonds and most organic bases are nitrogen bases, so such rules are not of much help to an organic chemist. To predict the strength of an organic acid, you must consider the stability of its conjugate base. Increasing the stability of an acid's conjugate base shifts its acid/base equilibrium to the right, making the acid more acidic.

Stop and Think: What are the general rules that lead to such predictions? Consult your general chemistry textbook if you don't remember.

$$H\text{—}A + H_2O \rightleftharpoons H_3O^+ + A^-$$
$$\text{acid} \qquad\qquad\qquad \text{conjugate base}$$

The conjugate base of an acid, if it has a negative charge (as most do), is stabilized by anything that tends to disperse (spread out) the charge. Thus electron-withdrawing groups and pi electron systems that allow delocalization of a negative charge by resonance can both stabilize a conjugate base, thereby increasing the acidity of the corresponding acid.

Uncharged bases such as ammonia and its derivatives may also be stabilized by resonance and electron withdrawal, but stabilizing the base and not its conjugate acid shifts the acid/base equilibrium to the left, *decreasing* the strength of the base.

$$B: + H_2O \rightleftharpoons B-H^+ + OH^-$$
base   conjugate acid

In this minilab you will be assessing the acid/base strengths of some organic compounds by measuring the pH values of their aqueous solutions. Before the lab period you should try to predict the relative acid/base strengths of the compounds listed by, for example, guessing which compounds may be acidic and writing all possible resonance structures (if there are any) leading to dispersal of the negative charge of their conjugate bases.

Directions

Your instructor may add or remove some test compounds.

Safety Notes

> **Some of the organic compounds are quite toxic, corrosive, or have other hazardous properties. Although the solutions are very dilute, minimize contact with them.**

You will be provided with 0.1 *M* solutions in 50% ethanol of such organic compounds as benzenesulfonic acid, benzoic acid, benzyl alcohol, benzylamine, benzyltrimethylammonium hydroxide, *p*-cresol, and *p*-toluidine.

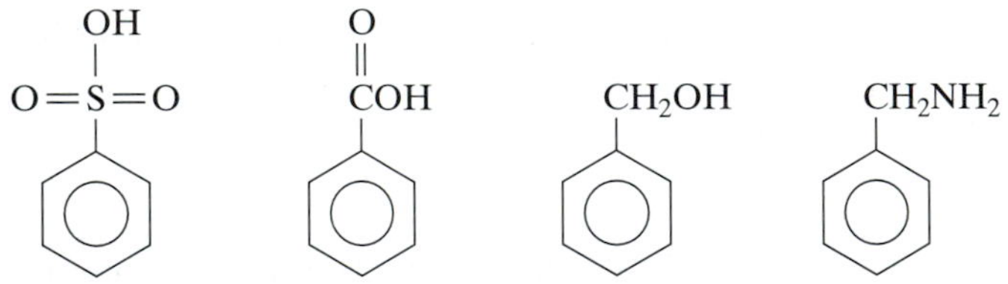

benzenesulfonic   benzoic acid   benzyl alcohol   benzylamine
acid

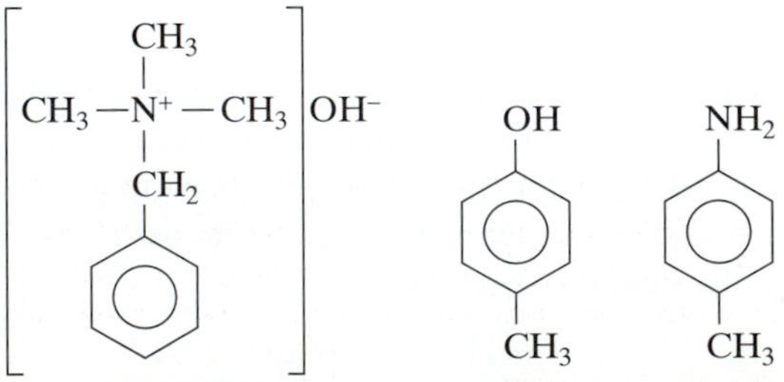

benzyltrimethylammonium   *p*-cresol   *p*-toluidine
hydroxide

Transfer 5 drops of each solution to a well of clean spot plate (label all wells) and determine its pH by using pH paper having a range from pH 1 to 12 or better. Alternatively, add a drop of universal indicator to each well, or measure enough of each solution into a microwell plate to use a pH meter with a mini probe. Estimate the pH of each solution. Assuming that benzyl alcohol is neutral, decide whether each compound other than the alcohol is acidic or basic (in 50% ethanol a neutral solution does *not* have a pH of 7).

Report. Report your results in a table. Compare your results with your predictions and try to account for any discrepancies. Write equations for the reactions of each of the acids and bases in water. Arrange the acidic compounds in order of acidity (most acidic first) and explain their relative acidities. Arrange the basic compounds in order of basicity (most basic first) and explain their relative basicities.

Hydrolysis Rates of Esters

MINILAB 37

Before You Begin: Review the mechanisms of acyl substitution reactions and try to predict the relative hydrolysis rates of the esters listed.

Under appropriate reaction conditions, an ester can be hydrolyzed to the corresponding carboxylic acid and alcohol.

$$\mathrm{R\overset{\overset{\displaystyle O}{\|}}{C}{-}OR' + H_2O \rightleftharpoons R\overset{\overset{\displaystyle O}{\|}}{C}{-}OH + R'OH}$$

The pH of the reaction mixture decreases as the acidic component is formed, providing a means of monitoring the reaction rate. In this minilab you will compare the hydrolysis rates of several esters by monitoring the pH values of their aqueous solutions as a function of time. Each reaction mixture will contain a small amount of NaOH, so the pH should be quite high at first and should decrease as carboxylic acid forms (initially as the acid salt). Based on your knowledge of the mechanisms of nucleophilic acyl substitution reactions, try to predict the relative hydrolysis rates of the esters. You can then compare your experimental results with your predictions.

Directions

Your instructor may add or remove some test compounds.

Safety Notes

The esters are flammable and inhalation and skin contact may be harmful. Avoid contact, do not breathe vapors, and keep flames away.

Place 2 mL of 50% aqueous ethanol, 2 drops of a universal indicator, and 2 drops of 1 *M* aqueous sodium hydroxide in each of four numbered test

Table M4 Colors and approximate pH values for Gramercy universal indicator

pH	Color
4	red
5	red-orange
5.5	orange
6	yellow-orange
6.5	yellow
7	yellow-green
7.5	green
8	dark green
8.5	blue-green
9	blue
9.5	violet
10	red-violet

tubes. Stopper and shake the test tubes. Add 5 drops of one of the following esters to each of the four test tubes: ethyl acetate, ethyl benzoate, ethyl formate, ethyl butyrate. Shake each test tube until the ester has dissolved and record the color of each solution. Heat the test tubes in a 50°C water bath for at least 30 minutes, recording the colors every 10 minutes. Use Table M4 or a color chart provided to estimate the pH of each solution at each time.

Report. Report your results (colors and pH values) in a table. Based on your results, arrange the esters in order of reactivity. Compare your results with your predictions and try to account for any discrepancies. Propose a hypothesis that accounts for the reactivity order you observed.

MINILAB 38

Preparation of Benzamide

Before You Begin: Read or review the operations as necessary.

The reaction of an acid chloride with ammonia is a traditional way of preparing primary amides. This is a nucleophilic acyl substitution reaction in which ammonia is the nucleophile. The reaction also produces HCl, which converts ammonia to ammonium ion.

$$RC\overset{\displaystyle O}{\overset{\|}{}}{-}Cl + NH_3 \longrightarrow RC\overset{\displaystyle O}{\overset{\|}{}}{-}NH_2 + HCl$$

$$HCl + NH_3 \longrightarrow NH_4^+Cl^-$$

Since ammonium ion is not nucleophilic, it will not react with the acid chloride, so the HCl must be neutralized in order for the reaction to go to completion. In Experiment 45 this was done using aqueous NaOH in a procedure called the Schotten–Baumann reaction. In this minilab it is accomplished by using excess ammonia to neutralize the HCl.

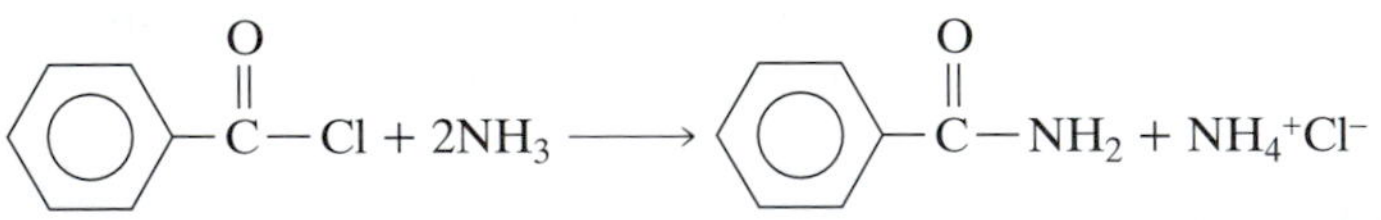

Directions

Safety Notes

> **Both concentrated ammonia and benzoyl chloride can cause serious burns to skin and eyes, and their fumes irritate the eyes and respiratory system. Wear gloves, use them only under the hood, and do not breathe their vapors.**

Under the hood, measure 2.0 mL of concentrated aqueous ammonia into a 15-cm test tube, and cool it well in an ice bath. *Wearing protective gloves, slowly* add 1.0 mL of benzoyl chloride to the cold ammonia with stirring [OP-9].(**Take Care!** Possible violent reaction.) Let the solution stand in the ice bath for 5 minutes, then collect the benzamide by vacuum filtration [OP-12], washing it on the filter with cold water. Recrystallize [OP-23] the benzamide from boiling water. Dry [OP-21] and weigh it, and measure its melting point [OP-28]. At your instructor's request, record the IR spectrum of the benzamide [OP-34].

Report. Write a mechanism for the reaction. If you recorded the IR spectrum of benzamide, compare it with the IR spectrum of *N,N*-diethyl-*m*-toluamide from Experiment 45 and discuss any significant similarities or differences.

Take Care! Wear gloves, avoid contact with ammonia and benzoyl chloride, do not breathe their vapors.

Waste Disposal: Pour the filtrates down the drain.

Synthesis of Dimedone Derivatives of Benzaldehyde

MINILAB 39

Before You Begin: Read or review the operations as necessary. Review the mechanisms of condensation and addition reactions of enolate ions.

Experiment 47 describes the preparation of dimedone (5,5-dimethyl-1,3-cyclohexanedione) by a synthesis that involves a Michael addition and a condensation reaction. In the presence of a base, dimedone itself reacts with aldehydes via a condensation reaction and a Michael addition, yielding a

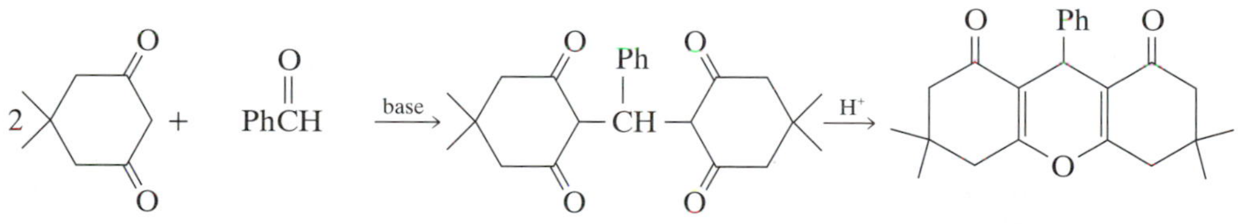

| dimedone | benzaldehyde | *bis*-dimedone derivative | octahydroxanthenedione derivative |

bis-dimedone derivative of the aldehyde. In the presence of an acid, the *bis*-dimedone derivative cyclizes to form an octahydroxanthenedione.
Both kinds of derivatives have been used to identify unknown aldehydes. In this minilab you will prepare the dimedone derivatives of benzaldehyde and test your mechanism-writing skills.

Directions

Piperidine is very toxic and corrosive. Avoid contact and do not breathe its vapors.

Safety Notes

Take Care! Avoid contact with piperidine, do not breathe its vapors.

Waste Disposal: Pour both filtrates down the drain.

Dissolve 0.10 g of benzaldehyde in 4.0 mL of 50% ethanol in a test tube. Stir in 0.30 g of dimedone and 1 drop of piperidine and boil the mixture gently for 5 minutes. If the hot solution is clear, add water drop by drop until it becomes cloudy. Cool the solution in an ice bath until crystallization is complete. Collect the product by vacuum filtration on a Hirsch funnel [OP-12], wash it with a little cold 50% ethanol, then dry [OP-21] and weigh it.

Dissolve 0.10 g of the *bis*-dimedone derivative in 4.0 mL of absolute ethanol with gentle heating. Stir in 1 mL of water and 2 drops of 6 *M* hydrochloric acid, and boil the mixture gently for 5 minutes. Add water drop by drop to the hot solution until it becomes cloudy, then cool it until crystallization is complete. Collect the octahydroxanthene derivative by vacuum filtration on a Hirsch funnel [OP-12], wash it with cold 50% ethanol, then dry [OP-21] and weigh it. Measure the melting points [OP-28] of both derivatives.

Report. Write balanced equations for the reactions and propose detailed mechanisms for the formation of both products. Remember that 1,3-diones such as dimedone can exist in both enolic and diketo forms.

A Diazonium Salt Reaction of 2-Aminobenzoic Acid

MINILAB 40

anthranilic acid
(2-aminobenzoic acid)

Before You Begin: Read or review the operations as necessary.

Diazonium salts of aromatic amines are usually prepared at low temperatures to prevent unwanted side reactions. In this minilab you will see what happens when a diazonium salt prepared from 2-aminobenzoic acid (anthranilic acid) is heated in solution. You should be able to deduce the structure of the product and propose a mechanism for its formation.

Directions

Safety Notes

> The product is a skin and eye irritant; avoid contact.

In a 25-mL Erlenmeyer flask, add 0.55 g of 2-aminobenzoic acid (anthranilic acid) to 3.0 mL of 3 *M* HCl with swirling or magnetic stirring [OP-9]. If necessary, heat the solution gently and add up to 4 mL of water to get the amine salt to dissolve. Cool the solution to 5°C or below in an ice/water bath, then slowly add 4.0 mL of freshly prepared 1 *M* sodium nitrite solution with stirring or swirling. Test the solution with starch-iodide paper; if the test is negative add just enough 1 *M* sodium nitrite to give a positive test. Drop in a boiling chip and boil the solution gently on a hot plate until no more gas is evolved. Cool it in an ice/water bath until crystallization is complete. Collect the product by vacuum filtration on a Hirsch funnel [OP-12] and

Stop and Think: What might the gas be?

purify it by recrystallization from boiling water [OP-23], using a little decolorizing carbon. Dry the product and measure its mass and melting point [OP-28]. Test the product with ferric chloride as described in Test C-13 of Part IV (p. 545).

Report. Describe and interpret the results of the FeCl$_3$ test. Deduce the structure of the product and verify your conclusion by looking up its melting point in an appropriate reference book. Write balanced equations for the reactions involved, and identify the gas that was evolved. Calculate the percent yield of the preparation.

Waste Disposal: Pour the filtrates down the drain.

Beckmann Rearrangement of Benzophenone Oxime

MINILAB 41

Before You Begin: Read or review the operations as necessary. Review the mechanisms of molecular rearrangements involving nitrogen atoms.

The Beckmann rearrangement of an oxime involves a migration of an alkyl or aryl group to an electron-deficient nitrogen atom. The transition state for the rearrangement step resembles that for a Curtius-type rearrangement, described in Experiment 49.

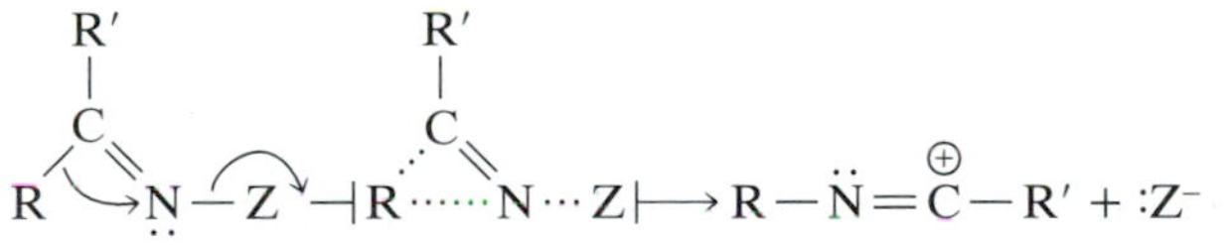

In this minilab you will prepare the oxime of benzophenone, carry out its Beckmann rearrangement to benzanilide, and work out a mechanism for the reaction.

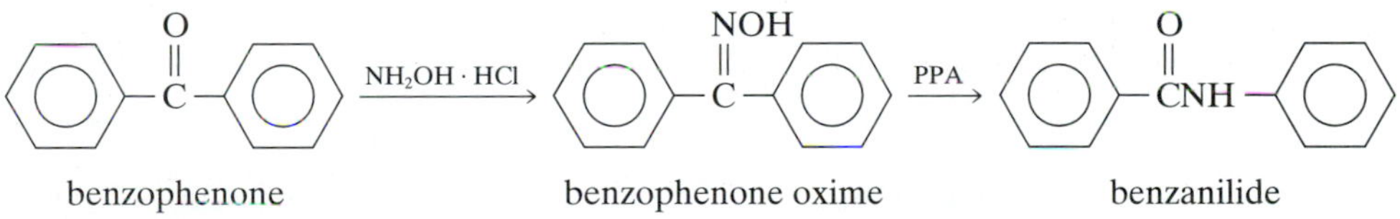

Directions

Polyphosphoric acid is corrosive; avoid contact.
Hydroxylamine hydrochloride is toxic and corrosive; wear gloves and avoid contact.

Safety Notes

Take Care! Avoid contact with the acid.

Take Care! Wear gloves, avoid contact with hydroxylamine hydrochloride.

Measure 2.5 mL of polyphosphoric acid into a 15-cm test tube. You may need to warm the viscous acid to 30°C or so before you can pour and measure it. Add 0.50 g of benzophenone and 0.60 g of hydroxylamine hydrochloride and clamp the reaction tube in a water bath. Heat the water to boiling and stir the reactants with a stirring rod until they are well mixed. Continue heating the reaction tube in the boiling water bath for 30–45 minutes (more if necessary) until frothing has stopped. Pour the warm reaction mixture into about 15 g of crushed ice in a beaker, using a little ice water to rinse out the reaction tube, and stir until the ice has melted. Collect the product by vacuum filtration on a Hirsch funnel [OP-12], washing it on the filter with ice water. Purify the benzanilide by recrystallization [OP-23] from 95% ethanol. Dry it [OP-21] and measure its mass and melting point [OP-28].

Report. Propose a detailed mechanism for the reaction, showing the transition state of the rearrangement step. Calculate the percent yield of benzanilide.

MINILAB 42   Dyeing with Indigo

Before You Begin: Read or review the operations as necessary.

The indigo plant, *Indigofera tinctoria*, is a shrub of the legume family that has been used to dye cloth since ancient times. Egyptian mummies were sometimes wrapped in cloth dyed blue with indigo. The leaves of the indigo plant contain a colorless natural product called indican, which is converted to indigo by a process that involves fermentation and oxidation. Indigo is an indole derivative that, in the presence of a reducing agent, changes to the colorless base-soluble compound leucoindigo.

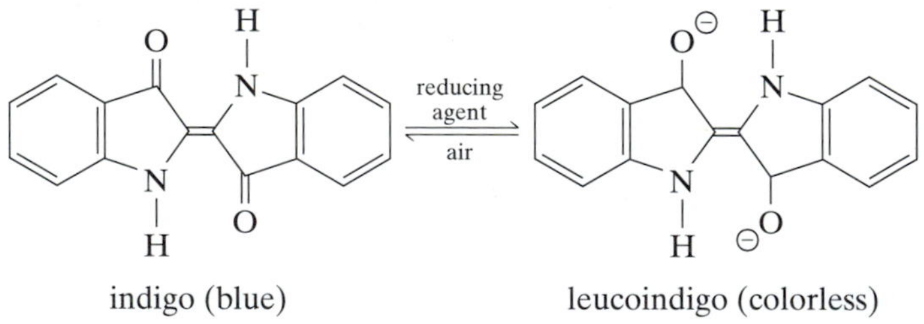

indigo (blue)   leucoindigo (colorless)

When a piece of cloth is dipped into a solution of leucoindigo and exposed to air, indigo is regenerated, dyeing the cloth the color of blue jeans. Indigo fades with time, but the popularity of faded blue jeans has turned this apparent deficiency into an advantage.

In this minilab you will prepare synthetic indigo by a condensation reaction of 2-nitrobenzaldehyde with acetone in an alkaline solution, then use it to dye a piece of cloth.

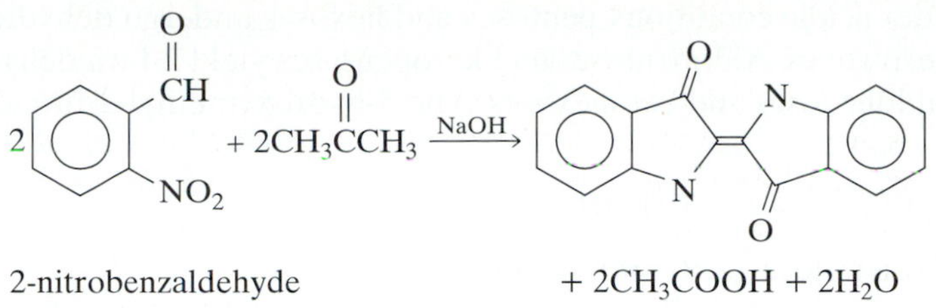

2-nitrobenzaldehyde

2 [structure: benzaldehyde with CH=O and NO_2 groups] $+ 2CH_3CCH_3 \xrightarrow{\text{NaOH}}$ [structure: indigo] $+ 2CH_3COOH + 2H_2O$

Directions

Indigo is a mild irritant and will turn the skin blue; wear gloves while working with indigo and its solutions.
2-Nitrobenzaldehyde is a mutagen (a substance that may cause genetic mutations), so avoid contact.

Safety Notes

Dissolve 0.50 g of 2-nitrobenzaldehyde in 2.0 mL of acetone in a conical centrifuge tube or test tube. Add water drop by drop until the solution becomes cloudy, then add a drop or so of acetone to clear it up. Add 20 drops of 1 *M* sodium hydroxide slowly, with stirring. The solution should warm up and turn dark brown. Let the mixture stand for 15 minutes or more. Then cool it in ice and collect the indigo by vacuum filtration on a Hirsch funnel [OP-12]. Wash the dye with a little ethanol, then with ethyl ether, and let it air dry.

Take Care! Avoid contact with 2-nitrobenzaldehyde.

Transfer the indigo to a 15-cm test tube, add 0.15 g of sodium dithionite, and grind the solids together with the end of a stirring rod until they are well mixed. Add 10 mL of 1 *M* aqueous sodium hydroxide and heat the mixture on a steam bath with stirring. When the indigo has dissolved, stopper and shake the tube until the indigo-blue color disappears (if the color persists, add a little more sodium dithionite). Immerse a small piece of cotton cloth in the solution with a stirring rod, stopper the tube immediately, and shake it for about 30 seconds. Remove the cloth, blot it dry between paper towels, and hang it up to dry. Note what happens when the test tube is left open for a time and then stoppered and shaken.

Take Care! Wear gloves.

Report. Calculate the percent yield of indigo. Describe and explain your observations during and after the dyeing step.

Reactions of Monosaccharides with Phenols

MINILAB 43

Before You Begin: Look up structures of the monosaccharides you will be testing in this experiment and classify each as an aldopentose, aldohexose, or ketohexose.

Under acidic conditions, pentoses and hexoses undergo dehydration to furan derivatives. Aldopentoses and ketopentoses yield 2-furaldehyde (furfural); aldohexoses and ketohexoses yield 5-hydroxymethyl-2-furaldehyde.

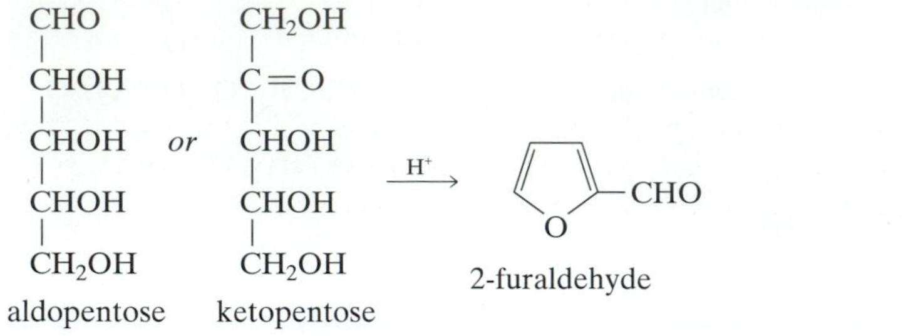

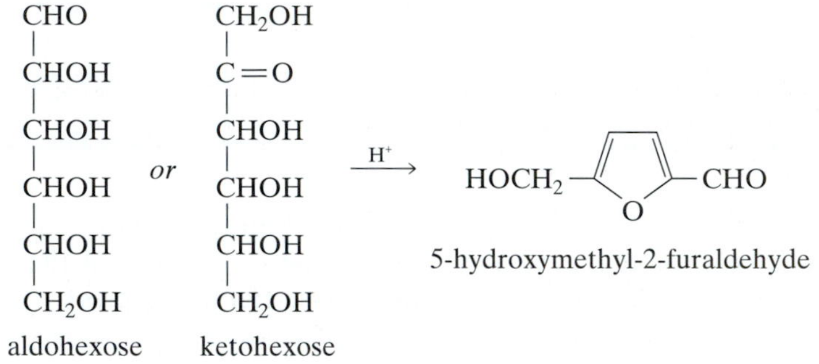

Both of these aromatic aldehydes react with phenols under acidic conditions to form colored products. For example, 2-furaldehyde condenses with 1-naphthol to yield compound **1**, which oxidizes to yield an intensely colored product, **2**.

Stop and Think: Why is compound **2** colored although compound **1** is not?

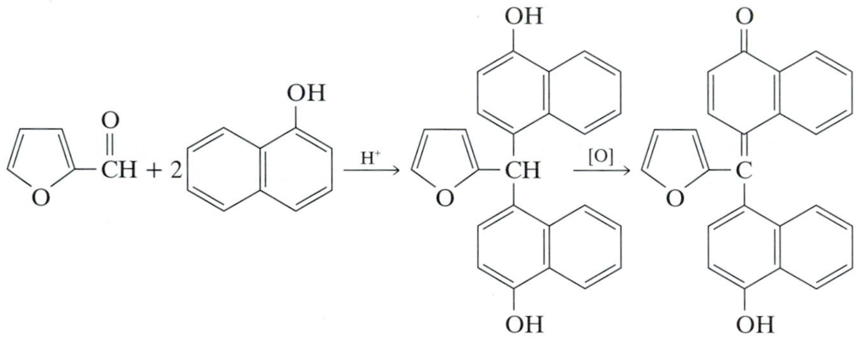

This reaction is the basis of the Molisch color test for pentoses and hexoses. In the presence of concentrated sulfuric acid and 1-naphthol, all pentoses yield compound **2** and all hexoses yield a similar derivative of 5-hydroxymethyl-2-furaldehyde. The phenols 1,3-dihydroxybenzene (resorcinol) and 3,5-dihydroxytoluene (orcinol) also give colored products analogous to **2** with furaldehydes.

In this minilab you will carry out the Molisch test on several monosaccharides and an unknown carbohydrate. You will also test these compounds

OH   CH₃

1,3-dihydroxybenzene   3,5-dihydroxytoluene

with acidic solutions of the other two phenols to find out what results they give with different classes of monosaccharides. Based on your observations, you should then be able to classify your unknown.

Directions

Your instructor may add more compounds to be tested, or substitute some.

Sulfuric acid causes chemical burns that can seriously damage skin and eyes. Wear gloves and avoid contact.
The 3,5-dihydroxytoluene solution contains concentrated hydrochloric acid, which is poisonous and corrosive. Contact or inhalation can cause severe damage to the eyes, skin, and respiratory tract. Wear gloves and dispense under a hood, avoid contact, and do not breathe the vapors.
The phenols are toxic and corrosive, and some are suspected carcinogens or teratogens. Although their solutions are very dilute, avoid contact.

Safety Notes

Have ready five clean, labeled 13 × 100 mm test tubes. Measure 1 mL of 1% aqueous solutions of D-fructose, D-glucose, D-xylose, and the unknown into four separate test tubes, and measure 1 mL of water into the fifth. To each test tube, add 2 drops of the Molisch reagent and swirl to mix. For each solution, tilt the test tube at an angle and *slowly* pour 1 mL of concentrated sulfuric acid down the side to form a separate layer below the aqueous layer. Observe the interface between the layers and record your observations in your lab notebook.

Take Care! Wear gloves, avoid contact with H_2SO_4 and the Molisch reagent.

Have ready five clean, labeled 13 × 100 mm test tubes. Measure 1 mL of 1% aqueous solutions of D-fructose, D-glucose, D-xylose, and the unknown into four separate test tubes, and measure 1 mL of water into the fifth. Add 2 mL of the 1,3-dihydroxybenzene solution to each test tube. Place all five test tubes into a beaker of boiling water and record your observations, including the time required for a color (if any) to develop.

Take Care! Avoid contact with the solution.

Have ready five clean, labeled 13 × 100 mm test tubes. Measure 1 mL of 1% aqueous solutions of D-fructose, D-glucose, D-xylose, and the unknown into four separate test tubes, and measure 1 mL of water into the fifth. Add 1 mL of the 3,5-dihydroxytoluene solution to each test tube. Place all five test tubes into a beaker of boiling water and record your observations, including the time required for a color (if any) to develop.

Take Care! The solution contains concentrated HCl; wear gloves, avoid contact.

Report. Report your results in a table and describe how the tests (except the Molisch test) can be used to distinguish different classes of monosaccharides. Then classify your unknown as an aldopentose, aldohexose, ketohexose, or none of these, and explain how you arrived at your conclusion.

Waste Disposal: Dispose of all test tube contents as directed by your instructor.

Write an equation for the reaction of 2-furaldehyde with 1,3-dihydroxy-benzene to form a product analogous to **1**, and propose a mechanism for this reaction.

MINILAB 44

Extraction of Trimyristin from Nutmeg

Before You Begin: Read or review the operations as necessary.

$$CH_3(CH_2)_{12}\overset{\displaystyle O}{\overset{\displaystyle \|}{C}}OH$$

myristic acid

Although most triglycerides found in plants are mixtures containing several different fatty acid residues, the triglyceride derived from nutmeg contains only one, myristic acid (tetradecanoic acid). In this minilab you will isolate trimyristin from ground nutmeg by hot solvent extraction with ethyl ether.

Directions

Safety Notes

> **Ethyl ether is extremely flammable and may be harmful if inhaled. Do not breathe the vapors and keep it away from flames and hot surfaces.**

Take Care! Keep flames away from ether, do not breathe its vapors.

Waste Disposal: Place any recovered ethyl ether in a designated solvent recovery container. Pour the ethanol filtrate down the drain.

Accurately weigh 2.0 g of finely ground nutmeg. Combine it with 10 mL of ethyl ether in a 25-mL boiling flask and heat the mixture gently under reflux [OP-6] for 45 minutes or more over a hot water bath or steam bath. Filter the mixture by gravity [OP-11], washing the nutmeg residue on the filter with a little ether and saving the filtrate. Evaporate [OP-14] the ether from the filtrate. Recrystallize [OP-23] the product from 95% ethanol. Dry [OP-21] the trimyristin, weigh it, and measure its melting point [OP-28].

Report. Draw the structure of trimyristin and calculate the percent recovery of trimyristin from nutmeg.

MINILAB 45

Preparation of a Soap Using a Phase-Transfer Catalyst

Before You Begin: Read or review the operations as necessary.

Soap has long been made by heating fats or oils with lye (sodium hydroxide) and other alkaline substances, including potash (potassium carbonate) from wood ashes. This alkaline hydrolysis process is called *saponification*, from the Latin *sapon*, meaning soap. Alcohol is sometimes used to speed up a soapmaking reaction by bringing the base and fat together in solution. However, the soap must then be salted out or the alcohol evaporated. In this minilab you will use a phase-transfer catalyst to bring the base and fat together, making the alcohol unnecessary. The fat will be a vegetable shortening such as Crisco, containing mainly palmitic, stearic, oleic, and linoleic acid residues.

Directions

Sodium hydroxide is toxic and corrosive, causing severe damage to skin, eyes, and mucous membranes. Wear gloves and avoid contact with the NaOH solution.

Safety Notes

Measure 10 g of vegetable shortening into a small beaker and heat it over a steam bath or boiling water bath until it melts. Stir in 4.5 mL of 8 *M* (~25%) aqueous sodium hydroxide and 2 drops of tricaprylmethylammonium chloride (Aliquat 336). Heat the mixture on the steam bath or boiling water bath, with stirring, for 20 minutes or more, until all the oily globules have disappeared. Cool the reaction mixture, then add 3 mL of water and use a flat-bottomed stirring rod to break up and wash the soap particles. Separate the soap by vacuum filtration [OP-12], washing it on the filter with several small portions of ice water. Dry [OP-21] the soap in a desiccator and weigh it.

Take Care! Wear gloves, avoid contact with the NaOH solution.

Report. Draw the structures of the major components of the soap you prepared. Write an equation for the reaction of glyceryl tristearate with NaOH.

Isolation of a Protein from Milk   MINILAB 46

Before You Begin: Read or review the operations as necessary.

Casein is a milk protein that contains all of the common amino acids and is particularly rich in the essential ones. Cheesemaking is based on the precipitation of casein from milk to form solid *curds*. The liquid *whey* is drained from the casein curds, which are concentrated by cooking, pressing, and salting, then aged to develop the characteristic flavor of the cheese. Casein, which exists in milk as a soluble calcium salt, precipitates at pH values below 4.6, so milk can be curdled by acids such as the lactic acid that forms during the natural souring of milk. In this minilab you will isolate casein from skim milk by curdling the milk with acetic acid. You will also carry out the *xanthoproteic test*, which is positive for proteins that contain aromatic amino acid residues such as tyrosine and tryptophan.

Directions

Safety Notes

Nitric acid is corrosive to eyes, skin, and mucous membranes and it can cause severe upper respiratory irritation. Wear gloves, use a hood, avoid contact, and do not breathe vapors.

Waste Disposal: Pour the liquid whey down the drain.

Take Care! Wear gloves, do not breathe vapors of HNO_3.

Stop and Think: What kind of reaction is occurring?

Mix 2.0 g (accurately weighed) of nonfat dry milk with 20 mL of water in a beaker. Heat the milk to 40°C with magnetic or manual stirring [OP-9], then add 10% (~1.7 *M*) aqueous acetic acid drop by drop with stirring until no more precipitate forms. Stir until the casein coagulates into an amorphous mass. Decant (pour out) the liquid and transfer the wet solid to a vacuum filtration apparatus [OP-12]. With the aspirator running, press the casein with a clean cork to squeeze out as much liquid as possible; then dry it further between filter papers. Transfer the casein to a small beaker, cover it with 95% ethanol, and use a flat-bottomed stirring rod to stir and crush it until it is finely divided. Collect the casein by vacuum filtration, wash it with two small portions of acetone, and let it dry [OP-21] at room temperature. Weigh the dry casein. *Under the hood,* add a drop of concentrated nitric acid to a small piece of casein in a test tube, and record your observations.

Report. Calculate the mass percent of casein in the dry milk. Interpret the result of the xanthoproteic test using nitric acid, and write an equation for the reaction of a tyrosine residue with nitric acid.

PART IV

Qualitative Organic Analysis

Part IV describes how to classify and identify unknown organic compounds by using chemical and spectrometric methods.

Qualitative Organic Analysis

Whenever an organic chemist synthesizes a new compound or isolates a previously unknown compound from a natural source, he or she must find out what it is. If the chemist has prepared the compound by a reaction whose outcome is reasonably predictable, its identification may require no more than a melting or boiling point and some spectra. But if the compound has been obtained by a process whose outcome is uncertain, or from a natural product or other source whose composition is unknown, the chemist will ordinarily follow a systematic course of action to identify it.

The systematic identification of organic compounds is called *qualitative organic analysis,* which chemists often shorten to "qual organic." As discussed in Experiment 39, you can identify an organic compound in somewhat the same way that a detective identifies the perpetrator of a crime. You should always be on the lookout for clues to the compound's identity, and for any evidence that will help you narrow down the roster of "suspects," the list of known compounds that you believe contains your unknown.

There is no single best way to identify an organic compound and no one path to follow. Although a useful approach is outlined in the qualitative analysis scheme in the margin, it is not necessary to follow it inflexibly. You should use your own judgment and initiative in choosing which tests to perform, which physical properties to measure, which derivatives to prepare, and which spectra to record. By keeping your eyes and mind open at all times, you may find clues to the structure or identity of your compound that suggest a more direct route to its identification.

A qual organic problem can provide an exciting and challenging experience for those who apply all of their skill and ingenuity to its solution. Besides testing your mastery of operations learned previously in the laboratory, it furnishes many practical applications of concepts learned during the lecture course in organic chemistry. Throughout the analysis of an unknown compound, you may apply your knowledge of functional group chemistry, nomenclature, acid-base equilibria, structure-property relationships, spectral analysis, organic synthesis, and many other areas of organic chemistry.

An Overview of Qual Organic

If you are doing Experiment 39, your unknown compound will be an alcohol, aldehyde, or ketone. If you are doing Experiment 50, your unknown will be an amine. If you are identifying a general unknown, the unknown will be from one of the following chemical families: *alcohols, aldehydes, ketones, amides, amines, carboxylic acids, esters, halides, aromatic hydrocarbons,* and *phenols.* Your instructor may also assign an unknown from a limited number of these families. An unknown may contain a subsidiary functional group, such as the carbon-carbon double bond in cinnamic acid or the nitro group in *p*-nitrophenol, but it will be classified as a member of one of the families listed.

The qualitative analysis scheme is divided into four parts, as outlined in the margin. The *preliminary work* generally involves purification of the

Qualitative Analysis Scheme

1. *Preliminary work*
   a. *Purification*
   b. *Measurement of physical constants*
   c. *Physical examination*
   d. *Ignition test*
2. *Classification*
   a. *Solubility tests*
   b. *Classification tests*
3. *Spectral analysis*
   a. *Infrared spectra*
   b. *NMR spectra*
4. *Identification*
   a. *List of possibilities*
   b. *Additional tests and data*
   c. *Preparation of derivatives*

Note: If the unknowns provided are all in a single family or a small group of families, part 2 can be omitted or modified.

unknown, measurement of its boiling point or melting point, a gross physical examination, and an ignition test. The *classification* phase involves solubility tests, chemical tests to identify functional groups, and additional chemical tests to classify alcohols and amines as 1°, 2°, or 3° and to detect such structural features as the $—COCH_3$ unit of a methyl ketone. *Spectral analysis* can be used to detect or confirm functional groups and to provide information about such structural features as conjugation and the position of substituents on benzene rings. The *identification* phase involves the preparation of a list of known compounds believed to contain the unknown, accumulation of additional evidence as needed to shorten the list, and synthesis of a derivative to identify the unknown as a compound on the short list. Your list of compounds can be based on the tables in Appendix VI, or on tables in another source recommended by your instructor.

If you have access to a computer simulation program such as SQUALOR (a simulation of the qualitative organic analysis laboratory experience), you can learn more about qualitative organic analysis and develop effective problem-solving strategies by identifying some "virtual unknowns." SQUALOR, which is structured like a computer game, offers practice in identifying nearly 100 unknowns from 11 different families.

Preliminary Work

Purification

Impurities in your unknown may cause inaccurate measurements of its physical constants, misleading results on classification tests, low derivative melting points, and other problems that will make it difficult to correctly identify the unknown. Light or dull colors (light yellow, tan, brown, black, etc.) may suggest that the unknown is impure, whereas intense yellows, oranges, reds, etc., are usually intrinsic colors. Compounds that are easily oxidized, such as aldehydes, aromatic amines, and phenols, are particularly likely to be impure. If your unknown is a liquid, you should purify it by simple distillation and record its boiling range. A liquid with a distillation boiling-point range greater than a 2°C may need to be redistilled and the high-boiling or low-boiling fractions discarded. If your unknown is a solid, you should first obtain its melting point to see whether or not it requires purification. Solids with melting point ranges of 2–3°C or more should ordinarily be purified by recrystallization from a suitable solvent. Purity can also be assessed by gas chromatography for a liquid and by TLC or HPLC for a solid.

Measurement of Physical Constants

The boiling point or melting point of your unknown is an important factor in its identification because it determines what compounds will be on your list of possibilities. If your value is inaccurate, your list may not even contain the name of your unknown. The median distillation boiling point of an unknown liquid may be sufficient for its identification, but obtaining a micro boiling point for confirmation is always a good idea. The melting point of a pure unknown solid should be measured as accurately as possible. Other physical constants, such as the density and refractive index of a liquid, may also be helpful, but you will have to obtain their literature values from an appropriate reference book.

Physical Examination

A simple physical examination of an unknown may provide some useful information about it. For instance, the fact that an unknown is a solid eliminates all organic compounds that are liquids at room temperature, and an intrinsic color suggests that chromophoric groups that have conjugated double bonds or rings are present. Odors can also provide clues to the identity of a compound, but since many organic compounds are quite toxic you should never smell them indiscriminately or in a way that would cause you to inhale appreciable amounts of vapor. Before you smell your unknown, it is a good idea to ask your instructor about its toxicity. The safest procedure is to hold the vial or other container at least six inches from your nose with the open end pointed away from you, then use your free hand to gently wave the vapors toward your nose, sniff briefly, and exhale.

Ignition Test

A compound's behavior on ignition can also provide some clues to its identity. As a rule, the higher the oxygen content of a compound, the bluer its flame; as hydrogen content increases, the flame becomes more yellow. For example, methanol burns with a bluer flame than 1-hexanol because it has a higher oxygen/hydrogen ratio. Most aromatic compounds burn with a sooty yellow flame. Organic compounds that do not burn in the ignition test may contain a high ratio of halogen to hydrogen or have a high molecular weight.

Directions

Record all of your observations and data in your laboratory notebook.

Safety Notes

> **Assume that all unknowns are flammable and harmful by inhalation, ingestion, and skin absorption. Do not inhale their vapors and avoid contact with eyes, skin, and clothing.**

A thermometer correction should be applied for temperatures over 200°C [see OP-28].

If your unknown is a liquid, purify it by simple distillation using small-scale apparatus [OP-25]. Collect the main fraction over a range of about 2°C and record the boiling range and median boiling point. If desired, carry out a micro boiling-point determination [OP-29] on the purified liquid.

If your unknown is a solid, measure its melting point [OP-28] accurately. If the melting point range is 2–3°C or higher, or if the solid shows evidence of impurities ("dirty" color, etc.) purify if by recrystallization [OP-23] from a suitable solvent or solvent mixture. Then dry the purified solid [OP-21] and measure its melting point.

Observe and describe the physical state and color of the unknown. With your instructor's permission, observe and describe its odor as well.

Under the hood, carry out an ignition test by placing a drop of an unknown liquid or about 25 mg of an unknown solid in a small evaporating dish and igniting it with a burning wood splint. If it burns, observe the color of the flame and whether it is clean or sooty.

Take Care! Make sure there are no flammable liquids nearby.

Classification

The traditional procedure for classifying unknowns involves the use of preliminary screening tests that categorize the unknowns into broad groups according to solubility properties, followed by classification tests to detect functional groups and other structural features. Infrared spectrometry can also be used to detect functional groups and structural features, so you may wish to read the section on Spectral Analysis (p. 552) before you carry out the classification tests. When you think you have determined the main functional group in your unknown, you can use its infrared (IR) spectrum to confirm your initial conclusion. If you prefer (and your instructor approves), you can use an IR spectrum to tentatively identify the main functional group and then perform one or more classification tests to confirm its identity.

Some of the classification tests in this section are used mainly to detect secondary functional groups and various structural features. You will find them most useful for shortening the list of possibilities that you construct in the Identification phase of your analysis (see p. 553). Since it is hard to know which of these tests will be useful until you know what compounds are on your list, you should construct the list as soon as you have confirmed your unknown's main functional group (unless it is an amine) and then perform the additional classification tests. If your unknown is an amine, you will have to classify it as 1°, 2°, or 3° before you can prepare your list.

Solubility Tests

The solubility behavior of a compound in appropriate solvents can be used to place it into one of the eight solubility classes listed in Figure Q1, page 534. For example, a compound that is insoluble in water and 5% HCl but soluble in 5% NaOH and 5% $NaHCO_3$ is placed in solubility class A_1, classifying it as a carboxylic acid. Most of the classes are broader than this. For example, class S_n includes a number of families of neutral compounds with relatively low numbers of carbon atoms.

In these tests, compounds that dissolve to the extent of about 30 mg per mL of the solvent are considered soluble. Solubility in water suggests the presence of at least one oxygen or nitrogen atom and a relatively low molecular weight. In the case of monofunctional oxygen or nitrogen compounds, the borderline for water solubility is usually around five carbon atoms. For example, 1-butanol (with four carbon atoms) is soluble, whereas 1-pentanol (with five carbon atoms) is not. Cyclic and branched compounds are usually more soluble than straight-chain compounds of the same carbon number; a phenyl group has about the same effect on solubility as a butyl group. Of the 10 families of organic compounds considered here, water solubility can be expected for low-molecular-weight alcohols, aldehydes, ketones, amides, amines, carboxylic acids, esters, and phenols. Most water-soluble carboxylic acids and aliphatic amines can be distinguished from the rest by testing their aqueous solution with blue and red litmus paper; the acids are placed in class **S**$_a$ and the amines in class **S**$_b$. Most phenols and aromatic amines are too weakly acidic or basic to give a positive litmus test, although there are some exceptions. Aldehydes that contain carboxylic acid impurities may give a false positive test with blue litmus. All water-soluble compounds that do not test positive to red or blue litmus are placed in class **S**$_n$.

Water-insoluble compounds that are soluble in 5% hydrochloric acid contain basic functional groups and are placed in class **B**. Of the families considered here, aliphatic and aromatic amines fall into this category. Some amines may form insoluble hydrochloride salts as they dissolve, so solubility behavior should be observed carefully to detect any change in the appearance of the unknown when it is shaken with the solvent.

Compounds that are insoluble in water and 5% hydrochloric acid but soluble in 5% sodium hydroxide contain acidic functional groups that give them pK_a values of ~12 or less. Acids with pK_a values of ~6 or less will also dissolve in 5% sodium bicarbonate, placing them in class **A**$_1$, which includes carboxylic acids. Carboxylic acids with 12 carbons or more may form relatively

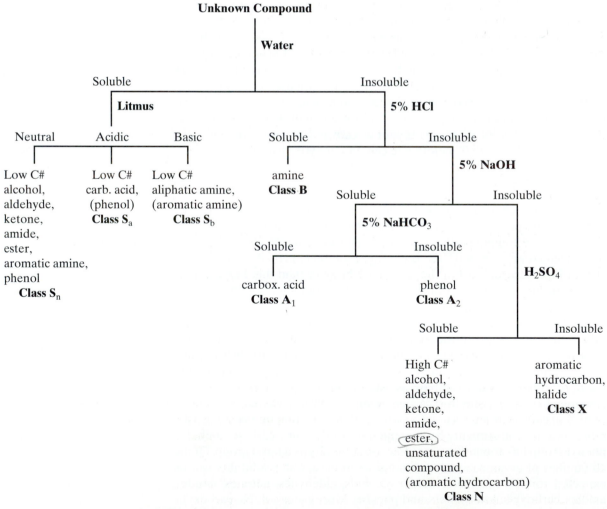

Figure Q1 Flow diagram for solubility tests.
Note: Compounds in parentheses are found only rarely in the solubility class indicated.

insoluble salts that yield a soapy foam when shaken with 5% NaOH. Compounds that are soluble in 5% NaOH but insoluble in 5% NaHCO$_3$ are placed in class **A$_2$**, which includes most phenols. Some other types of compounds, such as reactive esters, may react with 5% NaOH over time to form soluble reaction products, making it important to avoid delays in carrying out the test.

Compounds that are insoluble in all of the previous solvents but dissolve in or react with concentrated sulfuric acid are placed in solubility class **N**. This class includes high-molecular-weight alcohols, aldehydes, ketones, amides, and esters. Unsaturated compounds and some aromatic hydrocarbons—those containing several alkyl groups on the benzene ring—are also soluble in this reagent. Solution in sulfuric acid is accompanied by protonation of a basic atom (usually nitrogen or oxygen) or by some other reaction such as sulfonation, dehydration, addition to a multiple bond, or polymerization.

Compounds that are insoluble in all of these solvents (class **X**) include most aromatic hydrocarbons and the halogen derivatives of aliphatic and aromatic hydrocarbons.

An outline of the solubility scheme is given in Figure Q1. It is not necessary to test an unknown with every solvent, since if it dissolves in water, for example, it will also dissolve in aqueous solutions of HCl, NaOH, and NaHCO$_3$. In most cases, an unknown is not tested further once it is found to dissolve in a given solvent, since its solubility classification is established at that point. Compounds that are soluble in 5% NaOH, however, are also tested with 5% NaHCO$_3$ to determine whether they are carboxylic acids or phenols. It should be emphasized that solubility classifications can be misleading since there are exceptions and borderline cases. Therefore the solubility test results should be supported by other chemical or spectral evidence before a definite conclusion is drawn.

Directions

> **Sulfuric acid causes chemical burns that can seriously damage skin and eyes. Wear gloves and avoid contact.**

If the unknown is a liquid, calibrate a Pasteur pipet by measuring 0.10 mL of water into a test tube (use a graduated pipet or automatic pipet), carefully drawing the water into the Pasteur pipet (no air bubbles in the tip), and marking the meniscus with a glass-marking pen. Dry the pipet before you use it. (Alternatively, you can estimate the number of drops in 0.10 mL and measure the unknown by drops.) Measure 0.10 mL of the liquid into a 10-cm test tube and add 3.0 mL of water. Stopper and shake vigorously for 10 seconds or so, then observe the mixture carefully. If the mixture is homogeneous with no separate layer or suspended droplets of liquid, classify the unknown as soluble. Do not mistake air bubbles, which may form on shaking but should soon escape, for suspended liquid droplets. If the mixture looks cloudy or shows suspended droplets when shaken, or if a separate layer is visible above or below the water layer after standing, shake it again for a short while, then observe it. If the cloudiness, droplets, or separate layer disappear on shaking, classify it as soluble; otherwise classify it as insoluble.

Some reactions in cold concentrated sulfuric acid

$$RCHO \xrightarrow{H_2SO_4} R\overset{\textcircled{\scriptsize 1}}{C}HOH$$

$$RCOOH \xrightarrow{H_2SO_4} R\overset{\oplus}{C}(OH)_2$$

$$RCH_2OH \xrightarrow{H_2SO_4} RCH_2OSO_3H$$

$$RCH_2\underset{\underset{R''}{|}}{\overset{\overset{OH}{|}}{C}}-R' \xrightarrow{H_2SO_4} RCH=\underset{\underset{R''}{|}}{C}-R'$$

$$RCH=CHR' \xrightarrow{H_2SO_4} RCH_2\underset{\underset{OSO_3H}{|}}{C}HR'$$

Safety Notes

If the quantity of unknown is limited, you can scale down these amounts.

Take Care! Don't use your thumb for a stopper.

Take Care! Don't use your thumb for a stopper.

If the unknown is a solid, weigh out 90 mg of the solid and grind it to a fine powder on a clean watch glass using a clean flat-bottomed stirring rod. Mix the solid with 3.0 mL of water in a 10-cm test tube, and shake the test tube vigorously for 30 seconds or until the solid dissolves. If some of the solid remains undissolved, grind it against the bottom or sides of the test tube with your stirring rod and continue to shake and stir at intervals until the solid dissolves or until it is apparent that it will not completely dissolve. (It may take several minutes to dissolve some solids.)

If the unknown is *soluble* in water, test its aqueous solution with red and blue litmus paper. If there is no reaction to litmus but you suspect that the unknown may be an aromatic amine or a phenol, dissolve a little of the unknown in 5% HCl (for an amine) or 5% NaOH (for a phenol) and see if its odor disappears. Salts of amines and phenols are odorless. Decide whether the unknown should be in class S_n, S_a, or S_b and go on to the classification tests.

If the unknown is *insoluble* in water, test its solubility (using 0.10 mL of a liquid or 90 mg of a solid) in 3.0 mL of 5% HCl by the same procedure as described for water. If it does not dissolve completely in 5% HCl, separate it from the solvent by filtering it, removing it with a capillary pipet, or decanting the solvent. Then carefully neutralize the *solvent* to red litmus with dilute NaOH solution. If a precipitate, a separate liquid phase, or a cloudy solution appears upon neutralization, consider the unknown soluble. If it is *soluble* in 5% HCl, place it in class **B** and go on to the classification tests.

If the unknown is *insoluble* in 5% HCl, test its solubility (using 0.10 mL of a liquid or 90 mg of a solid) in 3.0 mL of 5% NaOH and in 3.0 mL of 5% $NaHCO_3$ by the same procedure as before. When testing with NaOH, notice whether the mixture foams on shaking, which could indicate a long-chain carboxylic acid. If the unknown does not dissolve completely, separate the residue as described for the HCl test and neutralize the solvent to blue litmus with dilute HCl. If a precipitate, a separate liquid phase, or a cloudy solution occurs upon neutralization, consider, the unknown soluble. If the unknown is *soluble* in 5% NaOH or in both solvents, place it in the appropriate class (A_1 or A_2) and go on to the classification tests.

Take Care! Wear gloves and avoid contact with H_2SO_4.

If the unknown is *insoluble* in all of the foregoing solutions, carefully test its solubility in 3.0 mL of concentrated sulfuric acid. Shake this mixture vigorously and look for any evidence of reaction such as the generation of heat, a distinct change in color, the formation of a precipitate, or the evolution of a gas. Solubility or a definite reaction is considered a positive test. The nature of any reaction that takes place may provide some clues to the identity of the unknown. Based on the results of this test, place the unknown into class **N** or **X**.

Classification Tests

The results of the solubility tests should reduce the number of families to which your compound may belong, limiting the number of classification tests that are needed to classify it into one of the remaining families. You should choose the tests carefully so that they will provide the information you need in a minimum number of steps. The tests used to detect functional groups are listed in Table Q1. Within families, functional class tests are listed in approximate order of simplicity and utility. It may be necessary to perform two (or

Table Q1 Functional class tests

Family	No.	Test	Comments
Alcohols	C-9	Chromic acid	Negative for 3° alcohols
	C-17	Lucas's test	Negative for 1° and high M.W. alcohols
	C-1	Acetyl chloride	Useful if other tests are inconclusive
Aldehydes	C-9	Chromic acid	Reacts more slowly than with alcohols
	C-23	Tollens's test	
Aldehydes and ketones	C-11	2,4-Dinitrophenyl-hydrazine	Ketones react with C-11, not with C-9 or C-23
Amides	C-2	Alkaline hydrolysis	Best for amides of ammonia or low M.W. amines
Amines		Solubility tests	Soluble in 5% HCl
	C-12	Elemental analysis	Detects N; use with solubility test results
Aromatic hydrocarbons	C-3	Aluminum chloride-chloroform	Should be confirmed by tests indicating absence of functional groups (such as halogens)
Carboxylic acids		Solubility tests	Soluble in 5% NaOH and 5% NaHCO$_3$
Esters	C-14	Ferric hydroxamate	
	C-2	Alkaline hydrolysis	Can be used to prepare derivatives
Halogenated hydrocarbons	C-5	Beilstein's test	Simple, not always reliable
	C-21	Silver nitrate/ethanol	Negative for vinyl and aryl halides
	C-10	Density test	Negative for monochloroalkanes
	C-12	Elemental analysis	Can distinguish Cl, Br, and I
Phenols	C-13	Ferric chloride	Positive for most (but not all) phenols
	C-8	Bromine water	Aromatic amines also react

Table Q2 Chemical tests that provide structural information

Family or Structural Feature	No.	Test	Application
Alcohols	C-17	Lucas's test	To classify alcohols as 1°, 2°, 3°, etc.
	C-16	Iodoform test	To detect -CH(OH)CH$_3$ groupings
Aldehydes	C-6	Benedict's test	To distinguish aliphatic from aromatic aldehydes
Ketones	C-16	Iodoform test	To detect -COCH$_3$ groupings
Amines	C-15	Hinsberg's test	To classify amines as 1°, 2°, or 3°
	C-4	Basicity test	To distinguish alkylamines from arylamines
	C-20	Quinhydrone	Complements C-15
Carboxylic acids	C-18	Neutralization equivalent	To determine the equivalent weight of an acid
Halogenated hydrocarbons	C-10	Density test	To distinguish aliphatic and aromatic Cl, Br, and I compounds
	C-22	Sodium iodide/acetone	To classify halides as 1°, 2°, 3°, etc.
	C-21	Silver nitrate/ethanol	Complements C-22
Unsaturation	C-7	Bromine test	To detect C$=$C and C$\equiv$C bonds
	C-19	Potassium permanganate	Complements C-7

more) tests for the same functional group in case the first test is inconclusive, or to use an infrared spectrum for confirmation.

You can use other classification tests, listed in Table Q2, to obtain additional information about your unknown. Some tests are used to further classify alcohols, amines, and alkyl halides. Others can distinguish aromatic compounds from aliphatic compounds in the same family, detect unsaturation, and detect certain groups of atoms such as —$COCH_3$ in methyl ketones. Tertiary amines are listed separately from primary and secondary amines in Appendix VI, so if you have an amine, you should classify it before you prepare your list of possibilities. If you have any other kind of compound, you can prepare your list first, then decide which of the tests in Table Q2 will help you shorten it.

Directions

Directions for performing all classification tests are given in this section by test numbers that are preceded by **C**. Unless otherwise indicated, solutions used for the classification tests are aqueous.

When each test is performed for the first time, it is advisable to run a *control* and a *blank* at the same time. Only by doing so will you know what to look for in deciding whether the test with the unknown is positive or negative. A control is a known compound that is expected to give a predictable result with the test reagent. A blank is run by combining all reagents as in the actual test but omitting the unknown. Unless otherwise indicated, all compounds suggested as controls in the following procedures should give a positive test.

The volumes of most liquid reactants are given by drops in the procedures; these should be delivered using medicine droppers or dropper bottles, not Pasteur pipets, which deliver smaller drops. Always use different droppers for the reagent and the unknown, and clean your droppers thoroughly after using them. Never insert your own droppers into reagent bottles, as you may contaminate the reagents.

Select the tests you want to perform, then turn to the appropriate procedures. Always read any Safety Notes before starting a test, and dispose of wastes as directed by your instructor.

C-1 *Acetyl Chloride.*

Safety Notes

> Acetyl chloride is very corrosive and its vapors are irritating and toxic. It reacts violently with water and some alcohols. Use gloves and a hood, avoid contact, do not breathe vapors, and keep away from water. *N,N*-Dimethylaniline and ammonia (procedure B) are harmful if inhaled or allowed to contact the skin. Use gloves and a hood, avoid contact, and do not breathe vapors.

Reaction: $CH_3COCl + ROH \rightarrow CH_3COOR + HCl$

Control: 1-butanol

A. *Under the hood,* cautiously add 10 drops of acetyl chloride to 10 drops of the unknown (0.4 g of a solid) in a test tube. Observe any evolution of heat; carefully exhale over the mouth of the test tube to see if a cloud of HCl gas

Take Care! Wear gloves, avoid contact with and inhalation of acetyl chloride.
Take Care! Possible violent reaction.

is revealed by the moisture in your breath. After a minute or two, pour the mixture into about 2 mL of water, shake it, and note any phase separation. Carefully smell the mixture for evidence of an ester aroma, which is usually pleasant and fruity.

Interpretation: Evidence of reaction (heat, HCl gas), especially if accompanied by phase separation and an ester-like odor, indicates a primary or secondary alcohol. Amines and phenols also react, but amines do not yield pleasant odors. Tertiary alcohols do not form esters by this procedure, but they should in the presence of a base such as *N,N*-dimethylaniline. If you suspect a tertiary alcohol, try variation **B**.

B. *Under the hood,* mix 5 drops of acetyl chloride with 10 drops of *N,N*-dimethylaniline. Cautiously add 5 drops of the unknown (0.2 g of a solid), warm the mixture on a 50° water bath for 15 minutes, then cool the mixture to room temperature. *Under the hood,* add 1 g of ice and 1 mL of concentrated ammonia, mix, and let stand. If a layer separates, remove it with a Pasteur pipet and test it for ester using the ferric hydroxamate test (**C-14**).

Take Care! Wear gloves, avoid contact with and inhalation of the reactants.

C-2 *Alkaline Hydrolysis.*

Use procedure **A** if your unknown may be an amide or procedure **B** if it may be an ester. Esters with boiling points higher than 200°C may be unreactive in aqueous NaOH.

Safety Notes

> **Sodium hydroxide is toxic and corrosive, causing severe damage to skin, eyes, and mucous membranes. Wear gloves and avoid contact with the NaOH solution.**

A. *Amides.*
Reactions: $RCONR'_2 + NaOH \rightarrow RCOONa + R'_2NH$
$RCOONa + H^+ \rightarrow RCOOH + Na^+$
$R' = H$, alkyl, or aryl

Control: benzamide

Place 0.1 g of the unknown (3 drops of a liquid) in a test tube containing 4 mL of 6 *M* sodium hydroxide. Secure a small piece of filter paper over the top of the tube and moisten it with 2 drops of 10% copper(II) sulfate. Boil the mixture for a minute or two and note any color change on the filter paper. Remove the paper and cautiously note the odor of the vapors while the solution is boiling. Acidify the solution with 6 *M* HCl; if a carboxylic acid precipitates, save it for use as a derivative.

Take Care! Wear gloves and avoid contact.

Interpretation: A blue color on the filter paper, accompanied by an ammonia or amine-like odor, indicates an amide. Amides of higher amines that do not turn the paper blue may nevertheless give an amine-like odor. Some amides will yield a precipitate or a separate liquid phase (the carboxylic acid) when the hydrolysis mixture is acidified. The characteristic odor of a carboxylic acid may also be observed. If the test is inconclusive, try increasing the reaction time or repeating the reaction at 200°C using 20% KOH in glycerine.

B. *Esters.*

Reactions: $RCOOR' + NaOH \rightarrow RCOONa + R'OH$

                $RCOONa + H^+ \rightarrow RCOOH + Na^+$

Control: butyl acetate

Take Care! Wear gloves, avoid contact with the NaOH solution.

Mix 1 mL of the unknown (1 g of a solid) with 10 mL of 6 *M* sodium hydroxide in a boiling flask, and heat the mixture under reflux [OP-6] for 30 minutes, or until the solution is homogeneous. Note whether the odor of the unknown is gone and (if the unknown was water insoluble) whether the organic layer has disappeared. If a separate organic layer or residue remains, heat the mixture longer until it disappears or until it is apparent that no reaction is taking place. Most esters boiling under 110°C will hydrolyze in 30 minutes; higher-boiling esters may take several hours. Cool the reaction mixture, remove any organic layer if there is one, and acidify the aqueous solution with 6 *M* sulfuric acid. If a carboxylic acid precipitates on acidification, save it for use as a derivative.

Interpretation: Evidence for an ester is indicated by disappearance of the organic layer (if any) and of the odor of the unknown during the reflux period, and the appearance of a precipitate or the odor of a carboxylic acid upon acidification. If acidification does not appear to yield a carboxylic acid, make the solution basic with 6 *M* sodium hydroxide and saturate it with potassium carbonate to see if an organic layer (the alcohol) separates. Note the odor of this layer, which should be different from that of the original ester.

C-3   *Aluminum Chloride and Chloroform.*

Safety Notes

Trichloromethane is harmful if inhaled or absorbed through the skin, and it is a suspected carcinogen in humans. Use gloves and a hood; avoid contact and do not breathe vapors.

Aluminum chloride reacts violently with water; skin and eye contact can cause painful burns, and inhaling the dust or vapors is harmful. Wear gloves, avoid contact, do not breathe vapors and keep water away.

Reaction: $ArH \xrightarrow{\text{CHCl}_3, \text{ AlCl}_3} Ar_3C^+$

Controls: toluene, biphenyl

Take Care! Wear gloves, avoid contact with and inhalation of trichloromethane and AlCl$_3$.

Under the hood, prepare a solution containing 3 drops of a solubility class **X** unknown (or 0.1 g of a solid) in 2 mL of dry trichloromethane (chloroform). Place about 0.2 g of anhydrous aluminum chloride in a dry test tube and heat it over a flame, angling the test tube so that the AlCl$_3$ sublimes onto the inner wall of the tube a few centimeters above the bottom. Allow the tube to cool until it can be held comfortably in the hand, then pipet a few drops of the solution down the side of the tube so that it contacts the aluminum chloride. Note any color change at the point of contact.

Interpretation: An intense color such as yellow-orange, red, blue, green, or purple indicates an aromatic compound. A light yellow color is inconclusive or negative.

C-4 *Basicity Test.*

Reaction:

$$RNH_2 + H^+ \xrightarrow{\text{pH 5.5}} RNH_3^+ \text{ (and similar reactions for 2° and 3° amines)}$$

Controls: *p*-toluidine, dibutylamine

If the unknown is water soluble, dissolve 4 drops (0.10 g of a solid) in 3 mL of water and measure the pH of the solution using pH paper or a universal indicator. If the unknown is insoluble in water, shake 4 drops (0.10 g of a solid) with 3 mL of a pH 5.5 acetate-acetic acid buffer in a stoppered test tube.

Interpretation: Most water-soluble aliphatic amines give pH values above 11, and water-insoluble aliphatic amines should dissolve in the buffer. Most water-soluble aromatic amines give pH values below 10, and water-insoluble aromatic amines do not dissolve in the buffer. Test C-8 can also be used to test for aromatic amines.

C-5 *Beilstein's Test.*

Control: chlorobenzene

Make a small loop in the end of a length of copper wire (10 cm or longer), and heat the loop to redness in a flame. Place a small amount of unknown on the loop and heat it in the nonluminous (blue) flame of a burner, near the lower edge.

Interpretation: A distinct green or blue-green flame indicates a halogen compound.

C-6 *Benedict's Test.*

Reaction: $RCHO + 2Cu^{2+} + 4OH^- \longrightarrow RCOOH + Cu_2O + 2H_2O$

Control: butanal (butyraldehyde)

Add 2 drops of the unknown (80 mg of a solid) to 2 mL of water and mix in 2 mL of Benedict's reagent. Heat the mixture to boiling. Observe whether a precipitate forms, and if one forms, what color it is.

Benedict's reagent contains copper(II) sulfate, sodium citrate, and sodium carbonate.

Interpretation: Aliphatic aldehydes generally produce a yellow to orange suspension or precipitate of copper(I) oxide; it may appear greenish in the blue solution. Most ketones and aromatic aldehydes do not react.

C-7 *Bromine Test.*

Safety Notes

> Dichloromethane may be harmful if inhaled or absorbed through the skin, and it is a suspected carcinogen. Avoid contact and do not breathe vapors. Bromine is toxic and corrosive and its vapors are very harmful. Avoid contact with the bromine solution and do not breathe its vapors.

Reaction:

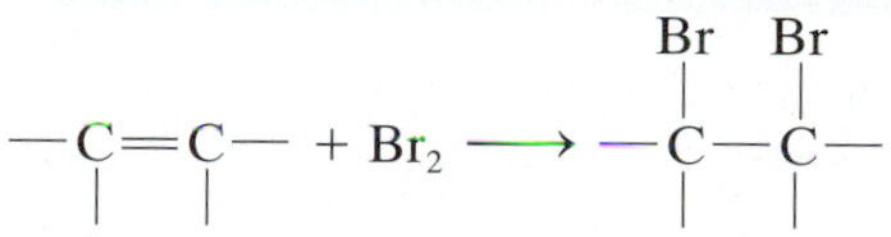

$$\overset{|}{-}C=C\overset{|}{-} + Br_2 \longrightarrow \underset{|}{\overset{\overset{\displaystyle Br}{|}}{-}C}-\underset{|}{\overset{\overset{\displaystyle Br}{|}}{C}}-$$

Control: cyclohexene

Under the hood, dissolve 1 drop of the unknown (40 mg of a solid) in 0.5 mL of dichloromethane. Add freshly prepared 1.0 *M* bromine in dichloromethane drop by drop, with shaking. Continue adding until the red-orange color of bromine persists *or* until about 20 drops have been added. Immediately after the addition, carefully exhale over the mouth of the test tube and observe whether a cloud of HBr gas appears.

Take Care! Avoid contact with and inhalation of dichloromethane and the bromine solution.

Interpretation: Decolorization of more than 1 drop of the bromine solution, without the evolution of HBr, indicates unsaturation (C=C or C≡C bonds). Aldehydes, ketones, amines, phenols, and other compounds react by substitution to evolve HBr.

C-8 *Bromine Water.*

Safety Notes

Bromine is toxic and corrosive and its vapors are very harmful. Avoid contact with the bromine water and do not breathe its vapors.

Reaction: Reaction is for phenol (aromatic amines and substituted phenols undergo similar reactions)

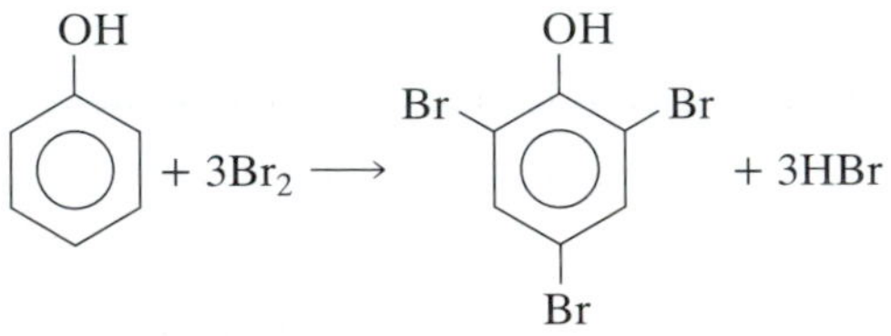

Control: phenol

Dissolve 3 drops of the unknown (0.1 g of a solid) in 10 mL of water. If it is insoluble in water, try adding just enough ethanol to bring it into solution. Measure the pH of the solution with pH paper. *Under the hood,* add saturated bromine water drop by drop until the bromine color persists. Watch for evidence of a precipitate.

Take Care! Avoid contact with and inhalation of Br_2.

Interpretation: Decolorization of the bromine, accompanied by simultaneous formation of a white (or nearly white) precipitate, indicates a phenol or aromatic amine. If the unknown is a phenol, the pH of the initial solution should be less than 7.

C-9 *Chromic Acid.*

Safety Notes

The chromic acid reagent is very corrosive and may be carcinogenic. Avoid contact.

Reactions:
1° Alcohol:

$$3RCH_2OH + 4CrO_3 + 6H_2SO_4 \rightarrow 3RCOOH + 2Cr_2(SO_4)_3 + 9H_2O$$

2° Alcohol:
$$3R_2CHOH + 2CrO_3 + 3H_2SO_4 \rightarrow 3R_2CO + Cr_2(SO_4)_3 + 6H_2O$$
Aldehyde:
$$3RCHO + 2CrO_3 + 3H_2SO_4 \rightarrow 3RCOOH + Cr_2(SO_4)_3 + 3H_2O$$

Controls: 1-butanol, butanal

Dissolve 1 drop of the unknown (40 mg of a solid) in 1 mL of reagent-grade acetone. (If there is any doubt about the purity of the acetone, test it with a drop of the reagent beforehand.) Add 1 drop of the chromic acid reagent and swirl, noting the time required for a positive test.

Interpretation: Formation of an opaque blue-green suspension within 2–3 seconds, accompanied by disappearance of the orange color of the reagent, indicates a primary or secondary alcohol. Aldehydes give the same result but react more slowly. With aliphatic aldehydes, the solution turns cloudy in about 5 seconds and the blue-green suspension forms within 30 seconds, while aromatic aldehydes require 30–90 seconds or longer to form the suspension. The generation of some other dark color, particularly with the color of the liquid remaining orange, should be considered a negative test.

Take Care! Avoid contact with the reagent.

The chromic acid reagent contains chromium (VI) oxide and concentrated sulfuric acid.

C-10 *Density Test.*

Controls: 1-chlorobutane, 1-bromobutane

Add a few drops of a solubility class **X** unknown to 1 mL of deionized water, stir gently, and note whether the unknown floats or sinks. If it sinks, measure the approximate density of the unknown by accurately pipetting 0.50 mL of the liquid into a tared vial and weighing the liquid on an accurate balance.

Interpretation: Of the compounds that are insoluble in cold, concentrated sulfuric acid, most aromatic hydrocarbons and monochloroalkanes will float, whereas aryl chlorides, polychloroalkanes, and all bromides and iodides will sink. Density ranges for some classes of organic halides are listed here.

Approximate density ranges for halogenated hydrocarbons:

Halide type	Density range
alkyl chloride (mono)	0.85–1.0
alkyl bromide (mono)	1.1–1.5
alkyl iodide (mono)	> 1.4
alkyl chloride (poly)	1.1–1.7
alkyl bromide (poly)	1.5–3.0
aryl chloride	1.1–1.3
aryl bromide	1.3–2.0
aryl iodide	> 1.8

C-11 *2,4-Dinitrophenylhydrazine.*

2,4-Dinitrophenylhydrazine (DNPH) is harmful if absorbed through the skin and it will dye your hands yellow. Wear gloves, avoid contact with the DNPH reagent, and wash your hands after using it.

Safety Notes

Reaction:

R, R′ = alkyl, aryl, or H

Controls: cyclohexanone, benzaldehyde

Dissolve 1 drop of the unknown (40 mg of a solid) in 1 mL of 95% ethanol; use more ethanol if necessary. Add this solution to 2 mL of the DNPH reagent. Shake and let the mixture stand for 15 minutes or until a precipitate

Take Care! Wear gloves, avoid contact with the reagent.

forms. If no precipitate has formed after 15 minutes, scratch the inside of the test tube.

Interpretation: Formation of a crystalline yellow or orange-red precipitate indicates an aldehyde or ketone. Some carbonyl compounds initially form oils that may or may not become crystalline; a few may require *gentle* heating, but overheating can cause oxidation of allylic or other reactive alcohols, resulting in a false positive test. Some aromatic compounds (hydrocarbons, halides, phenols, and phenyl esters) may form slightly soluble complexes with the reagent, and some alcohols may be contaminated with small amounts of the corresponding aldehyde or ketone. In such cases, the amount of precipitate should be quite small, and comparison with a control should show the difference between a positive test and a doubtful one. The color of the precipitate may give a clue to the structure of the carbonyl compound; unconjugated aliphatic aldehydes and ketones usually yield yellow precipitates, while aromatic and α,β-unsaturated aldehydes and ketones yield orange-red precipitates.

C-12 *Elemental Analysis.*

Rinsing the reaction tube with acetone and not drying it thoroughly may also result in a false positive test.

Safety Notes

> Sodium can cause serious burns and the sodium-lead alloy may react violently with some substances. Wear gloves, avoid contact, and keep Na(Pb) away from other chemicals. *Do not use this procedure with sodium metal;* it will react violently when water is added.
> The PNB reagent is harmful if inhaled or allowed to contact the skin. Wear gloves, avoid contact, and do not breathe vapors.

Reactions:

Nitrogen: $[C, N] \xrightarrow{\text{Na(Pb)}} NaCN \xrightarrow{\text{PNB}}$ purple color

Halogens: RX, ArX $\xrightarrow{\text{Na(Pb)}} NaX \xrightarrow{\text{AgNO}_3}$ **AgX** (X = Cl, Br, I)

Chlorine: $AgCl + 2NH_3 \longrightarrow Ag(NH_3)_2Cl$

Bromine: $2HBr + Cl_2 \longrightarrow 2HCl + Br_2$ (red−orange)

Iodine: $2HI + Cl_2 \longrightarrow 2HCl + I_2$ (purple)

Controls: acetamide (N), bromobenzene (Br)

Take Care! Wear gloves, avoid contact with the alloy.

Under the hood, place 0.25 g of 10% sodium-lead alloy in a small, *dry* test tube held vertically by a clamp. Melt the alloy with a burner flame and continue heating until the sodium vapor rises about 1 cm up the tube. Using a Pasteur pipet, add 2 drops of the unknown (or 10 mg of a solid) directly onto the molten alloy so that it does not touch the sides of the tube. Heat gently to start the reaction, remove the flame until the reaction subsides, then heat the tube strongly for a minute or two, keeping the bottom a dull red color. Let the tube cool to room temperature. Cautiously add 1.5 mL of water and heat gently for a minute or so until the excess sodium has decomposed and gas evolution ceases. Filter the solution through a Pasteur pipet with a small plug of cotton at the constriction [OP-11], wash the cotton with 1 mL of water, and combine the wash water with the filtrate. Use a rubber bulb to expel any liquid that adheres to the cotton. The filtrate should be colorless or just slightly yellow. If it is darker, repeat the fusion with stronger heating or more of the alloy.

To test for *nitrogen,* put 5 drops of the sodium fusion solution into a small test tube and add enough solid sodium bicarbonate, with stirring, to saturate it (a little excess solid should be present). Add 1 drop of this solution to a test tube containing 10 drops of PNB reagent (*p*-nitrobenzaldehyde in dimethyl sulfoxide) and note any color change.

To test for the *halogens,* acidify 10 drops of the sodium fusion solution with dilute nitric acid, boil it gently *under the hood* for a few minutes, add a drop or two of 0.3 M aqueous silver nitrate, and note the color and volume of any precipitate that forms. If a voluminous precipitate forms, let the precipitate settle (or centrifuge the solution) and remove the solvent using a Pasteur pipet with a cotton plug in the tip. Add 2 mL of 3 M aqueous ammonia to the solid, shake vigorously, and note your observations. To test further for bromine and iodine, acidify 1 mL of the original sodium fusion solution with 1 M sulfuric acid, boil for a few minutes, and add 0.5 mL of dichloromethane and a drop of freshly prepared chlorine water. Shake and look for a color in the dichloromethane layer.

Take Care! Wear gloves, avoid contact with and inhalation of PNB reagent.

Interpretation: In the PNB test, a purple color indicates the presence of nitrogen (green indicates sulfur). In the halogen tests, formation of a voluminous precipitate on addition of silver nitrate indicates that a halogen is present, and the color of the precipitate (a silver halide) may suggest which halogen: white for chlorine, pale yellow for bromine, and yellow for iodine. If only a faint turbidity is produced, it may be caused by traces of impurities or by incomplete sodium fusion. If the precipitate is silver chloride, it will dissolve in aqueous ammonia; silver bromide is only slightly soluble and silver iodide is insoluble. In the chlorine water test, a red-brown color is produced by elemental bromine and a violet color by elemental iodine.

C-13 *Ferric Chloride.*

Control: phenol

Dissolve 1 drop of the unknown (40 mg of a solid) in 1 mL of water, or in a water-alcohol mixture if it does not dissolve in water. Add two drops of *neutral* 1% ferric chloride solution. The neutral solution can be prepared by adding dilute NaOH to 1% FeCl$_3$ until a slight precipitate forms, then filtering the precipitate.

Interpretation: Formation of an intense red, green, blue, or purple color suggests a phenol or an easily enolizable compound. Some phenols do not react under these conditions. Many aromatic carboxylic acids form tan precipitates; aliphatic hydroxy acids yield yellow solutions.

C-14 *Ferric Hydroxamate Test.*

Safety Notes

Hydroxylamine hydrochloride is toxic and mutagenic and it can cause a form of anemia. Avoid contact with its solution.
Sodium hydroxide is toxic and corrosive, causing severe damage to skin, eyes, and mucous membranes. Wear gloves and avoid contact with the NaOH solution.

Reactions:

$$RC\overset{O}{\overset{\|}{-}}OR' + H_2NOH \longrightarrow RC\overset{O}{\overset{\|}{-}}NHOH + R'OH$$

$$3RCONHOH + FeCl_3 \longrightarrow (RCONHO)_3Fe + 3HCl$$
$$\text{ferric hydroxamate}$$

Control: butyl acetate

Before you use the ferric hydroxamate test, perform the following preliminary test. Dissolve 1 drop of the unknown (40 mg of a solid) in 1 mL of 95% ethanol. Add 1 mL of 1 *M* hydrochloric acid; then add two drops of 1% ferric chloride. If a definite color other than yellow results, the ferric hydroxamate test cannot be used. This test eliminates those phenols and enols that give colors with ferric chloride in acidic solution and that would therefore give a false positive result in the ferric hydroxamate test.

Mix 1 mL of 0.5 *M* ethanolic hydroxylamine hydrochloride with 5 drops of 6 *M* sodium hydroxide. Add 1 drop of the unknown (40 mg of a solid) and heat the solution to boiling. Allow it to cool slightly and add 2 mL of 1 *M* hydrochloric acid. If the solution is cloudy at this point, add enough 95% ethanol to clarify it. Add 5 drops of 1% ferric chloride solution and observe any color produced. If the color does not persist, continue to add the ferric chloride solution until the color becomes permanent.

Interpretation: A burgundy or magenta color that is distinctly different from the color obtained in the preliminary test indicates an ester.

C-15 *Hinsberg's Test.*

It is important to record all observations carefully during this test.

Safety Notes

> *p*-**Toluenesulfonyl chloride is toxic and corrosive. Use gloves and a hood, avoid contact, and do not breathe vapors.**

Reactions:

1° amine: $RNH_2 + ArSO_2Cl + 2NaOH \rightarrow ArSO_2NR^-Na^+ + NaCl + 2H_2O$
$ArSO_2NR^-Na^+ + HCl \rightarrow \textbf{ArSO}_2\textbf{NHR} + NaCl$

2° amine: $R_2NH + ArSO_2Cl + NaOH \rightarrow \textbf{ArSO}_2\textbf{NR}_2 + NaCl + H_2O$

3° amine: $R_3N + ArSO_2Cl \rightarrow$ no reaction
$R_3N + HCl \rightarrow R_3NH^+Cl^-$

Controls: butylamine, dibutylamine, tributylamine

Under the hood, mix 3 drops of the unknown (0.1 g of a solid) with 5 mL of 3 *M* sodium hydroxide in a test tube and add 0.2 g of *p*-toluenesulfonyl chloride. Stopper the tube and shake it intermittently for 3–5 minutes. Then remove the stopper and warm the solution over a steam bath, with shaking, for 1 minute. The solution should be basic at this point (if not, add more NaOH). If there is a solid or liquid residue in the test tube, separate it from the solution by vacuum filtration on a Hirsch funnel [OP-12] (if it is a solid)

Take Care! Avoid contact with hydroxylamine hydrochloride and the NaOH solution.

Take Care! Wear gloves, avoid contact with and inhalation of *p*-toluenesulfonyl chloride.

Benzenesulfonyl chloride (6 drops) can be used in place of p-*toluenesulfonyl chloride, but it is more hazardous and it tends to form oils.*

or with a Pasteur pipet (if it is a liquid) and test its solubility in water and (if it's insoluble) in 5% hydrochloric acid. Acidify the original solution with 6 M hydrochloric acid and, if no precipitate forms immediately, scratch the sides of the test tube and cool.

Interpretation: Formation of a white precipitate (a *p*-toluenesulfon-amide) when the reaction mixture is acidified indicates a *primary amine*. Most primary amines yield a clear solution after the initial reaction, but some form sodium salts or disulfonyl derivatives that precipitate during the reaction. The salts should dissolve in water, and amines that form disulfonyl derivatives should yield additional precipitate when the reaction mixture is acidified.

Most *secondary amines* yield a white solid that does not dissolve in water or 5% HCl. A liquid residue that is more dense than water and insoluble in 5% HCl may be a secondary amine's arenesulfonamide that has failed to crystallize.

Tertiary amines should not react; any residue will be the original liquid or solid amine, which should dissolve in dilute HCl. Water-soluble tertiary amines yield a clear solution that does not form a separate phase on acidification.

C-16 *Iodoform Test.*

Reactions: *1.* Methyl carbinols

$$\underset{\text{OH}}{\text{RCH}}-\text{CH}_3 + 4\text{I}_2 + 5\text{NaOH} \longrightarrow \underset{\text{O}}{\text{RC}}-\text{CI}_3 + 5\text{NaI} + 5\text{H}_2\text{O}$$

$$\underset{\text{O}}{\text{RC}}-\text{CI}_3 + \text{NaOH} \longrightarrow \underset{\text{O}}{\text{RC}}-\text{ONa} + \quad \text{CHI}_3$$
$$\text{(iodoform)}$$

2. Methyl ketones and acetaldehyde

$$\underset{\text{O}}{\text{RC}}-\text{CH}_3 + 3\text{I}_2 + 3\text{NaOH} \longrightarrow \underset{\text{O}}{\text{RC}}-\text{CI}_3 + 3\text{NaI} + 3\text{H}_2\text{O}$$

$$\underset{\text{O}}{\text{RC}}-\text{CI}_3 + \text{NaOH} \longrightarrow \underset{\text{O}}{\text{RC}}-\text{ONa} + \text{CHI}_3$$

Control: 2-butanone

Dissolve 3 drops of the unknown (or 0.1 g of a solid) in 2 mL of water in a large test tube. If the unknown is insoluble in water, dissolve it in 2 mL or more of methanol instead. Add 1 mL of 3 M sodium hydroxide solution, then add 0.5 M iodine-potassium iodide reagent drop by drop until the brown iodine color persists after shaking. If no yellow precipitate appears, place the test tube in a 60°C water bath and add more iodine-potassium iodide solution as necessary until the brown color remains after at least 2 minutes of heating. Then add 3 M NaOH drop by drop until the color just disappears (a light yellow color may remain). Remove the test tube from the water bath, add 10 mL of cold water, and let it stand for 15 minutes.

Interpretation: Formation of a yellow precipitate with the characteristic medicinal odor of iodoform is a positive test. If there is any doubt about the identity of the precipitate, its melting point ($\sim$121°C) can be measured. The test is positive for methyl ketones and acetaldehyde, and for alcohols that contain a —CH(OH)CH$_3$ grouping. Other compounds that may yield iodoform in this test include certain conjugated aldehydes such as acrolein and furfural, and some 1,3-dicarbonyl or dihydroxy compounds.

C-17   *Lucas's Test.*

This test is not applicable to most alcohols with boiling points higher than 140–150°C, or to solid alcohols.

Safety Notes

> **The Lucas reagent (ZnCl$_2$ in concentrated HCl) can cause serious burns, and inhaling the vapors is harmful. Use gloves and a hood, avoid contact, and do not breathe vapors.**

Take Care! Wear gloves, avoid contact with and inhalation of the Lucas reagent.

Take Care! Wear gloves, avoid contact with and inhalation of HCl.

Reaction: $\text{ROH} + \text{HCl} \xrightarrow{\text{ZnCl}_2} \text{RCl} + \text{H}_2\text{O}$

Controls: 1-butanol, 2-butanol, 2-methyl-2-propanol

Under the hood, place 2 mL of the Lucas reagent in a small test tube. Add 4 drops of the liquid unknown, stopper the tube immediately, and shake vigorously, taking note of any cloudiness or layer separation. Allow the mixture to stand for 15 minutes or more, observing it periodically for evidence of reaction. If the alcohol appears to be secondary or tertiary, repeat the test using concentrated HCl in place of the Lucas reagent.

Interpretation: *Tertiary alcohols* that are soluble in the Lucas reagent should turn the reagent cloudy almost immediately and soon form a separate layer of alkyl chloride. *Secondary alcohols* usually turn the clear solution cloudy in 3–5 minutes and form a distinct layer within 15 minutes. *Primary alcohols* do not react under these conditions. Most allylic and benzylic alcohols give the same result as tertiary alcohols, except that the chloride formed from allyl alcohol is itself soluble in the reagent and separates out only upon addition of ice water. High-boiling alcohols that are insoluble in the Lucas reagent cannot be tested because they will form a separate layer immediately. When tested with concentrated HCl, a tertiary alcohol should react within minutes, and a secondary alcohol should not react.

C-18   *Neutralization Equivalent.*

Reaction: $\text{RCOOH} + \text{NaOH} \rightarrow \text{RCOONa} + \text{H}_2\text{O}$

Control: hexanedioic acid (adipic acid)

Accurately weigh (to 3 decimal places) about 0.2 g of an unknown carboxylic acid and dissolve it in 50–100 mL of water, ethanol, or a mixture of the two, depending on its solubility. Titrate this solution with a standardized solution of $\sim$0.1 M sodium hydroxide using phenolphthalein as the indicator (or bromothymol blue if the solvent is ethanol). Calculate the neutralization equivalent (N.E.) of the acid using the formula shown.

$$\text{N.E.} = \frac{\text{mass of sample in mg}}{\text{mL of NaOH} \times \text{molarity of NaOH}}$$

Interpretation: The neutralization equivalent (equivalent weight) of a carboxylic acid is equal to its molecular weight divided by the number of carboxyl groups it has. For instance, the N.E. of adipic acid [$HOOC(CH_2)_4COOH$; M.W. = 146] is 73. A solid carboxylic acid that has an unusually low neutralization equivalent for its melting point probably contains more than one carboxyl group.

C-19 *Potassium Permanganate.*

Reaction: $3-C=C- + 2KMnO_4 + 4H_2O \longrightarrow$

$$3-\underset{|}{\overset{HO}{C}}-\underset{|}{\overset{OH}{C}}- + 2MnO_2 + 2KOH$$

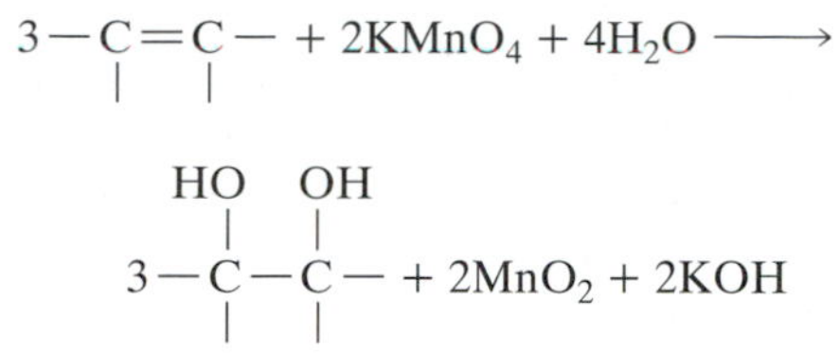

Control: cyclohexene

Dissolve 1 drop of the unknown (40 mg of a solid) in 2 mL of water or 95% ethanol (if ethanol is the solvent, run a blank). Add 0.1 *M* aqueous potassium permanganate drop by drop until the purple color of the permanganate persists, or until about 20 drops have been added. If a reaction does not take place immediately, shake the mixture and let it stand for up to 5 minutes. Disregard any decolorization that takes place after 5 minutes have elapsed.

Interpretation: Decolorization of more than one drop of the purple permanganate solution, accompanied by the formation of a brown precipitate (or reddish-brown suspension) of manganese dioxide, suggests unsaturation. The test is positive for most compounds containing double and triple bonds, except for conjugated alkadienes. Easily oxidizable compounds such as aldehydes, aromatic amines, phenols, formic acid, and formate esters also give positive tests. Most pure alcohols will not react in less than 5 minutes, but alcohols that contain oxidizable impurities may react slightly, so decolorization of only the first drop of potassium permanganate should not be considered a positive test.

C-20 *Quinhydrone.*

This test is not applicable to diaminobenzenes or nitro-substituted aromatic amines.

Safety Notes

The components of quinhydrone (hydroquinone and benzoquinone) are toxic and irritating. Avoid contact with the reagent.

Controls: butylamine, dibutylamine, tributylamine, aniline, *N*-methylaniline, *N,N*-dimethylaniline

This test should be run in conjunction with Test **C-4** or with an infrared spectrum that shows whether the amine is aliphatic or aromatic. Controls should be run for comparison since the colors are difficult to describe accurately. Shake 1 drop of an unknown alkylamine (30 mg of a solid) or 6 drops of an unknown arylamine (0.2 g of a solid) with 6 mL of water in a 15-cm test tube. If the amine dissolves, add 6 mL more of water; if not, add 6 mL of

Take Care! Avoid contact with the quinhydrone solution.

ethanol. Shake the mixture, add 1 drop of 2.5% quinhydrone in methanol and let it stand for 2 minutes or more. Compare the color of this solution with those of the controls.

Interpretation: Most amines from the following classes give the colors indicated. It is best to use this test in conjunction with the Hinsberg test or an IR spectrum for confirmation.

> 1° aliphatic: violet
> 2° aliphatic: rose
> 3° aliphatic: yellow
> 1° aromatic: rose
> 2° aromatic: amber
> 3° aromatic: yellow

C-21 *Silver Nitrate in Ethanol.*

Safety Notes

> **Silver nitrate is corrosive and toxic. Avoid contact with the reagent.**

Reaction: $RX + AgNO_3 + EtOH \rightarrow ROEt + \mathbf{AgX} + HNO_3$
        (Organic products other than ROEt are also formed.)

Controls: 1-chlorobutane, 2-chloro-2-methylpropane, 1-bromobutane, iodoethane

Take Care! Avoid contact with the AgNO₃ reagent.

Add 1 drop of the unknown (or 50 mg of a solid, dissolved in a little ethanol) to 2 mL of 0.1 *M* ethanolic silver nitrate. Shake the mixture and let it stand. If no precipitate has formed after 5 minutes, heat the solution to boiling and boil it for 30 seconds. If a precipitate forms, note its color and see if it dissolves when the mixture is shaken with 2 drops of 1 *M* nitric acid.

Interpretation: The reaction of a halide with silver nitrate occurs by an S_N1 mechanism, so the reactivity of a halide varies considerably with its structure. Alkyl iodides react faster than the corresponding bromides, which are more reactive than alkyl chlorides. For the same halogen, tertiary, allylic, and benzylic halides react fastest, followed by secondary and primary halides. Most aryl and vinyl halides are unreactive, except for aryl halides activated by two or more nitro groups. Most alkyl halides with more than one halogen atom on the same carbon (*gem* halides) are less reactive than the corresponding monohaloalkanes. The types of compounds that react at room temperature, react after boiling the solution, or do not react at all, are listed here. The color of the precipitate formed may indicate the type of halide responsible. Silver chloride is white, silver bromide is pale yellow or cream-colored, and silver iodide is yellow. These salts will not dissolve when the mixture is acidified. Some carboxylic acids and alkynes yield silver salts, but these should dissolve in the acidic solution.

Halides that react at room temperature:

> *Chlorides: 3°, allyl, benzyl.*
> *Bromides: 1°, 2°, and 3° alkyl (except geminal di- and tribromides); allyl, benzyl, CBr₄.*
> *Iodides: All aliphatic and alicyclic except vinyl.*

Halides that react upon heating:

> *Chlorides: 1° and 2° alkyl.*
> *Bromides: alkyl gem-di- and tribromides.*
> *Some activated aryl halides such as 2,4-dinitrohalobenzenes.*

Halides that do not react:

> *Chlorides: alkyl gem-di- and trichlorides, CCl₄.*
> *Most aryl and vinyl halides.*

C-22 *Sodium Iodide in Acetone.*

Reaction: $RCl + NaI \xrightarrow{\text{acetone}} RI + \mathbf{NaCl}$
        $RBr + NaI \xrightarrow{\text{acetone}} RI + \mathbf{NaBr}$

Controls: 1-chlorobutane, 2-chloro-2-methylpropane, 1-bromobutane

Place 1 mL of the sodium iodide/acetone reagent into a small test tube. Add 2 drops of a liquid halogen compound (or 0.1 g of a solid dissolved in the minimum volume of acetone). Shake the mixture and allow it to stand for 3 minutes, noting whether a precipitate or color forms. Disregard a precipitate that forms upon mixing but does not persist. If there is no precipitate at the end of this time, place the test tube in a 50°C water bath (replenish the acetone if some evaporates), and leave it there for an additional 6 minutes; then cool the mixture to room temperature and record your observations.

Interpretation: Certain alkyl chlorides and bromides react to precipitate sodium chloride or sodium bromide, which are insoluble in acetone. Bromides react faster than comparable chlorides, and the test is not applicable to iodides. Since the reaction involves an S_N2 displacement by iodide ion, halides of the same halogen react in the order methyl > primary > secondary > tertiary > aryl, vinyl. Cycloalkyl halides tend to react more slowly than the corresponding open-chain compounds and may give no precipitate even after heating. Vicinal dihalides (with two halogen atoms on adjacent carbon atoms) and some geminal halides (with two or more halogen atoms on the same carbon atom) undergo oxidation-reduction reactions to liberate iodine, which is red-brown in acetone, while precipitating the sodium halide. A summary of the results with various halides is given in the margin.

Halides that react at room temperature

> *1° alkyl bromides*
> *benzyl and allyl halides*
> *vic-dihalides*; CBr₄**
> *α-haloketones, -esters, and -amides*

Halides that react at 50°C:

> *1° and 2° alkyl chlorides*
> *2° and 3°alkyl bromides*
> *gem- di- and tribromides**

Halides that do not react:

> *3° alkyl chlorides*
> *aryl and vinyl halides*
> *cyclopropyl, cyclobutyl, and cyclo-*
> *hexyl halides*
> *gem-polychloro compounds (except*
> *benzyl and allyl)*

*Turns the solution red-brown

C-23 *Tollens's Test.*

> **Silver nitrate is corrosive and toxic. Avoid contact with its solutions. The Tollens reagent and the test solution must never be stored—explosive silver salts form on standing.**

Safety Notes

Reaction:

$$RCHO + 2Ag(NH_3)_2OH \rightarrow 2Ag + RCOONH_4 + H_2O + 3NH_3$$

Control: benzaldehyde

Prepare the reagent immediately before use as follows. Measure 2 mL of 0.3 M aqueous silver nitrate into a *thoroughly cleaned* test tube and add 1 drop of 3 M sodium hydroxide. Then add 2 M aqueous ammonia drop by drop, with shaking, until the precipitate of silver oxide just dissolves (avoid an excess of ammonia). Add 1 drop of the unknown (40 mg of a solid) to this solution, shake the mixture, and let it stand for 10 minutes. If no reaction has occurred by the end of this time, heat the mixture in a 35°C water bath for 5 minutes.

Take Care! Avoid contact with $AgNO_3$

Waste Disposal: Immediately after the test is completed, dissolve any residue in dilute nitric acid and place the solution in a designated waste container (*important!*).

Interpretation: Formation of a silver mirror on the inside of the test tube is a positive test for an aldehyde. If the tube is not sufficiently clean, a black precipitate or a suspension of metallic silver may form instead. Certain cyclic ketones (such as cyclopentanone), aromatic amines, phenols, and α-alkoxy- or α-dialkylaminoketones may also give positive tests.

$$\bar{\nu} = 1715 \text{ cm}^{-1}$$

$$\underset{\bar{\nu} = 1240 \text{ cm}^{-1}}{\text{R}-\text{C}-\text{O}-\text{H}}$$

$$\bar{\nu} = 3000 \text{ cm}^{-1}$$

Spectral Analysis

Infrared Spectra

To someone proficient in spectral interpretation, an infrared spectrum provides so much information that it may be equivalent to a number of classification tests. Because infrared absorption bands arise from the vibrational motions of specific chemical bonds, it is usually possible to identify or confirm the functional group(s) present in an organic compound from its infrared spectrum. For example, a compound whose infrared spectrum has a C=O stretching band at 1715 cm^{-1}, a C—O stretching band at 1240 cm^{-1}, and a very broad O—H stretching band centered at 3000 cm^{-1} is almost certainly a carboxylic acid. Characteristic infrared bands of the most common families of organic compounds are described in the Spectral Interpretation section of OP-34.

Other structural features can also be detected by using infrared spectrometry. If your unknown compound is an alcohol, amide, or amine, you should be able to classify it as 1°, 2°, or 3° by reading about the IR spectra of these families in OP-34. If your unknown may have a benzene ring or carbon-carbon double bond, you can confirm this and perhaps determine the kind of substitution on the ring or C=C bond by reading about the IR spectra of aromatic hydrocarbons or alkenes. You can also find out whether the C=O group of an aldehyde, ketone, ester, or carboxylic acid is conjugated with an aromatic ring or carbon-carbon double bond.

The best way to become proficient at infrared spectral interpretation is to read about and interpret actual spectra. Before attempting to interpret your IR spectrum you should read pp. 719–724 in OP-34 and examine the sample spectra on the pages that follow.

Directions

Obtain an infrared spectrum of the unknown [OP-34] as directed by your instructor. Interpret the spectrum as completely as you can. Use it to confirm any functional group suggested by the classification tests, and to detect any additional functional groups and other structural features.

NMR Spectra

Like infrared spectra, nuclear magnetic resonance (NMR) spectra can be used to detect certain functional groups, but they are more often used to provide detailed structural information about organic molecules. With practice, an organic chemist can deduce the structural units present in an organic compound, or even its entire molecular structure, from an NMR spectrum. See the section "Interpretation of ^{1}H NMR Spectra" in OP-35 for help.

Directions

Safety Notes

Assume that the NMR solvent is flammable and harmful by inhalation, ingestion, and skin absorption. Some NMR solvents, such as deuterochloroform, are also carcinogenic. Avoid contact and do not breathe their vapors.

With your instructor's permission and guidance, record an integrated ^{1}H NMR spectrum [OP-35] of your unknown in deuterochloroform or another suitable solvent, using a TMS reference standard. Measure and tabulate the NMR parameters (chemical shift, signal area, signal multiplicity, coupling constant), then use them to reconstruct as much of the molecular structure of your unknown as you can.

Identification

After you have determined the main functional group present in your unknown, you are ready to begin the identification phase of your analysis by preparing a list of possible compounds. This phase will be completed when you have obtained sufficient evidence to eliminate all of the possibilities but one.

List of Possibilities

At your instructor's option, you may either use the tables of selected organic compounds provided in Appendix VI of this book or consult a more complete listing such as the *CRC Handbook of Tables for Organic Compound Identification* [Bibliography, G4]. Most literature tabulations separate the compounds into solids and liquids, listing the solids in order of increasing melting point and the liquids in order of increasing boiling point. The dividing line between liquids and solids is generally near a melting point of 25°, but some compounds with melting points over 25° (Such as *t*-butyl alcohol) may be liquid at room temperature, particularly when impure. When in doubt, try freezing the compound in an ice/salt bath and estimating its melting point if it freezes.

Unless you are certain that your unknown is pure and that your boiling-point or melting-point determination is accurate, it is a good idea to list all compounds within about ±10° of your measured value, especially if your compound has a wide boiling-point or melting-point range, or if your boiling or melting point is 200° or higher. (Your instructor may suggest a different range.) Your list should be prepared as a table that indicates the boiling or melting points of the compounds and the melting points of all derivatives for the appropriate functional class for which reagents are available in your laboratory. Some derivative melting points may not be given in published tables, either because the compound does not form that derivative or because the derivative's melting point has not been reported in the literature. The structural formula for each compound should be drawn, using *Lange's Handbook of Chemistry*, *The Merck Index*, or another appropriate reference book if necessary. Additional data and information obtained from literature sources can be included in your table as well.

Directions

Referring to the appropriate table in Appendix VI (or another source recommended by your instructor), prepare a table listing each compound whose boiling point or melting point is within the applicable range. Give the boiling or melting point of each compound, its molecular structure, any other useful data or information you can find, and the melting points of possible derivatives. If requested, submit your list to your instructor, who may (or may not) elect to tell you if your unknown compound is on the list.

Additional Tests and Data

Once you have prepared your list of possibilities, you should be able to decide which compounds can be eliminated on the basis of spectral data and tests already performed, and what physical properties, chemical tests, or spectral bands will help you shorten the list further. If your list is already a short one, you may decide to prepare a derivative immediately. Otherwise, you may need to shorten it before you can decide on the most suitable derivative.

Directions

Carry out one or more of the following procedures to arrive at your "short list."

1 Refer to Table Q2 on page 537 and decide which classification test or tests will help you shorten your list. Then carry out the tests according to the directions provided.

2 Re-examine your infrared spectrum, looking for bands that may indicate the presence or absence of a subsidiary functional group or structural feature that is present in some of the compounds on your list but absent in others.

3 Record an ^{1}H NMR spectrum of your unknown, if you haven't already. Try to predict the kind of NMR spectrum that would be produced by the compounds on your list, and compare the predicted spectra with the actual one.

4 Measure the refractive index [OP-30] of your unknown at 20°C, or correct it to that temperature. Compare it with literature values for the refractive indexes of the compounds on your list. If the unknown is very pure, a thermostated refractometer that measures to the fourth decimal place can be used to eliminate most compounds whose refractive index values deviate by more than ±0.001 or so from the observed value.

5 Determine the density of your unknown by accurately weighing a precisely measured volume (1/2 mL or more) of your unknown [OP-5]. Compare it with literature values for the densities of the compounds on your list. Density cannot be measured as accurately as refractive index, but it is useful in distinguishing among compounds that have different structural features. For example, most aromatic compounds are more dense than aliphatic compounds of the same family, and density varies among halogenated hydrocarbons as described in Test **C-10**.

Preparation of Derivatives

In qualitative organic analysis, a *derivative* is a crystalline solid that can be prepared from nearly any compound of the same family by using a standardized procedure, and which upon purification gives a reproducible melting point that can be used to characterize the compound from which it was prepared. For example, an unknown carboxylic acid can be converted to an amide by treating it with thionyl chloride followed by ammonia.

$$\underset{\underset{\displaystyle RC-OH}{\|}}{O} \xrightarrow{SOCl_2} \underset{\underset{\displaystyle RC-Cl}{\|}}{O} \xrightarrow{NH_3} \underset{\underset{\displaystyle RC-NH_2}{\|}}{O}$$

If the unknown acid is hexanoic acid (b.p. = 205°C), its amide (hexanamide) should melt near 101°C, as shown in Table 7 of Appendix VI. The only other carboxylic acid listed that boils near 205°C is 2-bromopropanoic acid, but its amide melts at 123°C, making it easy to distinguish these compounds by means of their derivative melting points.

You will need to prepare at least one derivative to confirm the identity of your unknown. In some cases it may be necessary to prepare two or more derivatives to be certain of your identification. A suitable derivative should have a melting point between 50°C and 250°C, because solids melting below 50°C are not easily purified by recrystallization, and melting points higher than 250°C are hard to measure accurately. Its melting point should not be close to the melting point of the unknown itself, since it would then be hard to tell whether the product of a reaction is the derivative or the unreacted starting material. For example, preparing the amide of a compound believed to be 2-chlorobenzoic acid would not be a good idea because the amide has the same melting point (140°C) as the acid.

If possible, the derivative should be one whose melting point value will point to just one compound from your list of possibilities. For example, suppose your compound is a ketone whose boiling point is 168°, and you have prepared the list of possibilities shown in Table Q3. The oxime would not be a suitable derivative because melting points have not been reported for two of the possibilities, and two of the oximes melt below 50°C. Semicarbazones are reported for all the unknowns, but those for 2,6-dimethyl-4-heptanone and 2-octanone melt within a degree of each other. 2-Octanone might be distinguished by an iodoform test, but that would still leave 2-methyl- and 4-methylcyclohexanone, whose semicarbazones melt within 4° of each other—a little too close to distinguish them with certainty. (Your derivative might melt at 197°, for example.) The 2,4-dinitrophenylhydrazone is the preferred derivative since its melting points are well spread out.

The procedures whose locations are given in Table Q4 are suitable for preparing derivatives of most common organic compounds in the specified

Table Q3 Derivatives of selected ketones

Compound	b.p.	Oxime	Semicarbazone	2,4-Dinitro-phenylhydrazone
2-methylcyclohexanone	163	43	195	137
2,6-dimethyl-4-heptanone	168	210	121	92
3-methylcyclohexanone	169	...	180	155
4-methylcyclohexanone	169	37	199	130
2-octanone	172	...	122	58

Table Q4 Location of procedures for preparing derivatives

Family	Page
Alcohols	557
Aldehydes and Ketones	558
Amides	559
Primary and Secondary Amines	560
Tertiary Amines	562
Carboxylic Acids	563
Esters	565
Alkyl Halides	567
Aryl Halides	568
Aromatic Hydrocarbons	568
Phenols	569

class. In some cases, a variation of the procedure (the use of different reaction conditions or recrystallization solvents, etc.) may be required for satisfactory results. If you want to prepare a derivative that is not described in this section, see your instructor for permission and to find out whether the reagents are available. Procedures for additional derivatives, and alternative procedures for some of the derivatives used here, are given in qualitative analysis texts listed under category G of the Bibliography.

Sometimes the same procedure will be used with two different reagents to prepare two related derivatives, such as the phenylurethane and α-naphthylurethane of an alcohol. In such cases, select the reagent (phenyl isocyanate *or* α-naphthylisocyanate in this example) that will lead to the derivative that you wish to prepare. The relative quantities of unknown and reagent are usually not crucial, since any excess reagent is removed during the isolation and purification of the derivative. In cases where the quantities could affect the results, the quantity of the unknown is given in millimoles so that the required mass can be calculated using an estimate of its molecular weight.

In preparing a derivative you will be using small quantities of reactants and reagents, so you should use small-scale apparatus whenever feasible. Reactions are usually run in test tubes and only a few require heating under reflux [OP-6]. Vacuum filtration [OP-12], recrystallization [OP-23], and other operations are carried out on a small scale as described in the corresponding operation descriptions. When the recrystallization solvent is described as an ethanol-water mixed solvent or some other mixture, the procedure for recrystallization from mixed solvents should be used. The derivative should usually be dissolved in the less polar solvent and the solution saturated by adding the more polar solvent drop by drop. It should be understood that each derivative preparation requires a melting-point determination [OP-28] using the purified product.

When an ethanol-water mixed solvent is specified for recrystallization, use 95% ethanol.

As always, be careful to avoid contact with or inhalation of the reagents and unknown. Wearing gloves, protective clothing, and safety goggles is advisable. Many of the reagents are highly reactive and may react violently with water or other substances. All should be considered toxic, and many of them are corrosive, lachrymatory, or have other unpleasant properties.

Always read the Safety Notes before beginning any derivative preparation, and heed their warnings.

Directions

Derivatives of Alcohols

D-1 *3,5-Dinitrobenzoates,* p-*Nitrobenzoates.*

Safety Notes

The acid chlorides are corrosive and lachrymatory. Use a hood and avoid contact and vapors.

Reaction: $\text{ArCOCl} + \text{ROH} \rightarrow \text{ArCOOR} + \text{HCl}$

Under the hood, mix 0.20 g of pure 3,5-dinitrobenzoyl chloride *or* *p*-nitrobenzoyl chloride with 0.10 g of the alcohol. Heat the mixture over a *small* flame so that it is just maintained in the liquid state (do not overheat—decomposition will result). If the alcohol boils under 160°C, heat the mixture for 5 minutes; otherwise heat it for 10–15 minutes. Allow the melt to cool and solidify. Break it up with a stirring rod, and stir in 4 mL of 0.2 *M* sodium carbonate. Heat the mixture to 50–60°C with a steam bath or hot water bath, and stir it at that temperature for 30 seconds. Then cool it and collect the precipitate by small-scale vacuum filtration. Wash the precipitate several times with cold water, and recrystallize it from ethanol or an ethanol-water mixture. Derivatives of higher boiling (or melting) alcohols require a higher ethanol : water ratio in the recrystallization solvent.

Take Care! Avoid contact with the acid chloride and do not breathe its vapors.

D-2 *α-Naphthylurethanes, Phenylurethanes.*

The alcohol and glassware must be dry since moisture results in the formation of diphenylurea or di-α-naphthylurea, depending on the derivative you are preparing (these compounds melt at 241°C and 297°C, respectively). Note that tertiary alcohols do not form urethanes readily.

Safety Notes

The isocyanates are irritants and lachrymators. Use a hood, avoid contact, and do not breathe vapors.

Reaction:

$$\text{ArN}=\text{C}=\text{O} + \text{ROH} \longrightarrow \text{ArNH}\overset{\overset{\text{O}}{\|}}{\text{C}}-\text{OR}$$

Unless you are sure that your unknown alcohol is anhydrous, dry it with magnesium sulfate or sodium sulfate [OP-20], and dry a small test tube in an oven or over a flame. Stopper the test tube and allow it to cool. *Under the hood,* mix 5 drops of the alcohol (0.2 g of a solid) with 5 drops of phenyl isocyanate *or* α-naphthyl isocyanate. If no reaction takes place immediately, heat the

Take Care! Avoid contact with and inhalation of the isocyanates.

mixture in a 60–70°C water bath for 5–15 minutes. Cool the test tube in ice and scratch its sides, if necessary, to induce crystallization. Collect the precipitate by small-scale vacuum filtration. Recrystallize it from about 5 mL of high-boiling petroleum ether or heptane (filter the hot solution to remove high-melting impurities).

Derivatives of Aldehydes and Ketones

D-3 *2,4-Dinitrophenylhydrazones.*

Safety Notes

> **2,4-Dinitrophenylhydrazine is toxic and sulfuric acid is very corrosive. Avoid contact with the reagent.**

Reaction:

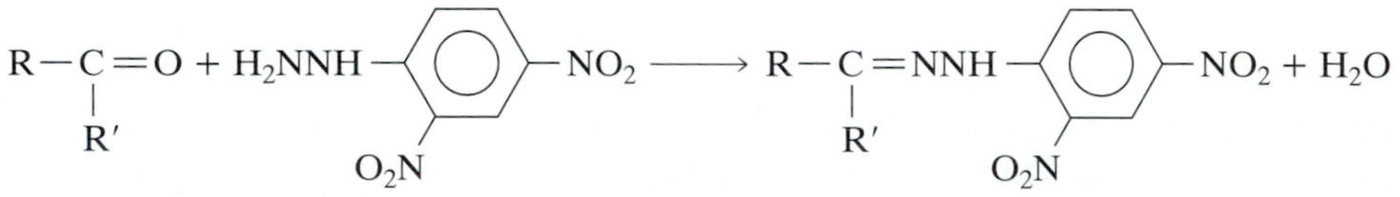

$$R, R' = \text{alkyl, aryl, or H}$$

Take Care! Avoid contact with the DNPH reagent.

Dissolve 0.10 g of the unknown in 4 mL of 95% ethanol. Add 3 mL of the 2,4-dinitrophenylhydrazine-sulfuric acid reagent and allow the solution to stand at room temperature until crystallization is complete. If necessary, warm the solution gently for a minute on a steam bath. If no precipitate appears after 15 minutes, or if precipitation does not seem complete, add water drop by drop to the warm solution until it is cloudy, heat to clarify, and cool. Collect the derivative by vacuum filtration, remove the filtrate, and wash the solid on the filter with 5 mL of cold 5% $NaHCO_3$ followed by cold water. Recrystallize it by dissolving it in 6 mL or less of boiling 95% ethanol and adding water (to a maximum of 2 mL) drop by drop. If the derivative does not dissolve in 6 mL of boiling ethanol, add ethyl acetate drop by drop until it goes into solution.

D-4 *Semicarbazones.*

Safety Notes

> **Semicarbazide hydrochloride is a suspected carcinogen. Avoid contact.**

Reaction:

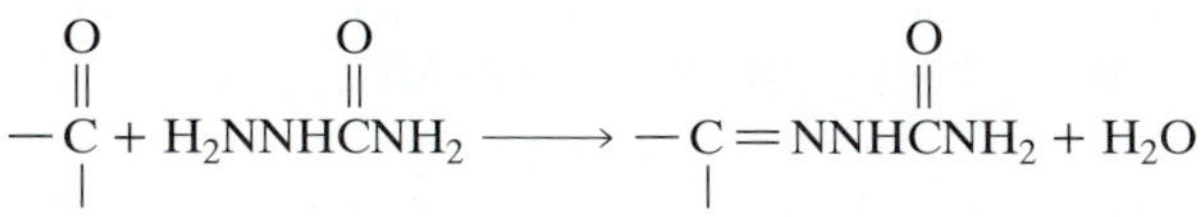

Take Care! Avoid contact with semicarbazide hydrochloride.

Mix together 0.20 g of semicarbazide hydrochloride, 0.30 g of sodium acetate, 2 mL of water, and 2 mL of 95% ethanol in a test tube (if the unknown is water soluble, omit the ethanol). Add 0.20 g of the aldehyde or ketone and stir to dissolve. If the mixture is cloudy, add more ethanol until it clears up.

Shake the mixture for a minute or two and let it stand, cooling it in ice if necessary to induce crystallization. If no crystals form, place the test tube in a boiling water bath for a few minutes, then allow it to cool. Collect the crystals by vacuum filtration, wash them with cold water, and recrystallize the product from alcohol or an alcohol-water mixture.

D-5 *Oximes.*

Oximes are suitable derivatives for most ketones and for some (though not all) aldehydes.

Safety Notes

Reaction:

$$\overset{O}{\underset{|}{-\overset{||}{C}}} + H_2N-OH \longrightarrow -C=N-OH + H_2O$$

Prepare the oxime following the procedure given for semicarbazones (**D-4**), using hydroxylamine hydrochloride in place of semicarbazide hydrochloride. It is usually necessary to heat the reactants with a steam bath or boiling water bath for 10 minutes or more. Adding a few milliliters of cold water to the reaction mixture may facilitate precipitation.

Take Care! Avoid contact with hydroxylamine hydrochloride.

Derivatives of Amides

The acid and amine portions of an amide can be obtained by hydrolysis, as described in **D-6**, and one or both of them characterized by a melting point or derivative preparation. Melting points of the carboxylic acids are given in the table for amides, Table 4 of Appendix VI. Melting points of amine and carboxylic acid derivatives are found in Tables 5–7. You can also prepare an amide derivative directly by Procedure **D-7**.

D-6 *Hydrolysis Products.*

If the amide is known to be primary, omit procedure **A**.

Safety Notes

Heat 0.60 g of the amide under reflux with 10 mL of 3 *M* sodium hydroxide for 15 minutes or more. If the alkaline hydrolysis classification test (**C-2**) suggests that the amide is difficult to hydrolyze, use 6 *M* NaOH or a longer reaction time, or both. Assemble an apparatus for small-scale simple distillation [OP-25]. Place 4 mL of 3 *M* hydrochloric acid in the receiver, a 25-mL round-bottomed flask. Since the resulting amine may be a gas, the receiving flask should be attached directly to the vacuum adapter outlet and the vacuum side arm should be connected to a gas trap containing dilute HCl. The gas trap can be omitted if an efficient fume hood is used. Distill the reaction mixture until about 6 mL of distillate has been collected.

Take Care! Avoid contact with arenesulfonyl chlorides and do not breathe their vapors.

A. *Characterization of the Amine. Under the hood,* add 10 mL of 3 *M* sodium hydroxide to the distillate; then add 0.60 g of *p*-toluenesulfonyl chloride *or* 0.40 mL of benzenesulfonyl chloride. Proceed according to the directions given in **D-9** for preparing arenesulfonamide derivatives of amines.

B. *Characterization of the Carboxylic Acid.* Carefully acidify the residue in the boiling flask with 6 *M* HCl. If a precipitate forms (the carboxylic acid), collect it by vacuum filtration, wash it, and recrystallize it from water, ethanol-water, or another suitable solvent. If no precipitate forms, you can prepare a *p*-nitrobenzyl derivative following the directions in **D-15**.

D-7 N-*Xanthylamides.*

This procedure is suitable only for primary amides.

Safety Notes

> Acetic acid can cause burns and its vapors are very irritating. Use gloves and a hood; avoid contact.

Reaction:

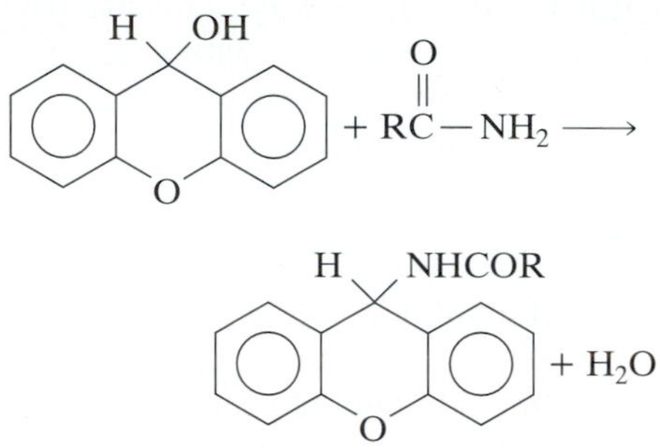

Take Care! Wear gloves, avoid contact with and inhalation of acetic acid.

Under the hood, dissolve 0.40 g of xanthydrol in 5.0 mL of glacial acetic acid and shake until most of the solid has dissolved. Decant the solution from any undissolved residue, add 0.20 g of the amide (or as close to 1.5 mmol as you can estimate), and warm the mixture in an 85°C water bath for 15–20 minutes. (If the amide is not soluble in acetic acid, dissolve it in a minimum volume of ethanol before adding it to the reaction mixture; then add 1 mL of water to the warm reaction mixture after the reaction is complete.) Let the solution cool until precipitation is complete. Collect the solid by vacuum filtration, and recrystallize it from an ethanol-water mixed solvent.

Derivatives of Primary and Secondary Amines

See **D-12** for directions for preparing picrate derivatives.

D-8 *Benzamides.*

Safety Notes

> Benzoyl chloride is corrosive, a lachrymator, and a possible carcinogen. Use gloves and a hood, avoid contact, and do not breathe vapors.

Reaction:

$$\underset{\underset{R'}{\overset{O}{\overset{\|}{PhC}}-Cl + R-\overset{}{\underset{\|}{N}}H} \longrightarrow \underset{\underset{R'}{\overset{O}{\overset{\|}{PhC}}-\overset{}{\underset{\|}{N}}-R + HCl}}{}}$$

(R' can be alkyl, aryl, or H.)

Under the hood, combine 0.20 g of the primary or secondary amine (or as close to 2.0 mmol as you can estimate) with 2.0 mL of 3 *M* sodium hydroxide. Then add 0.80 mL of benzoyl chloride drop by drop, with vigorous shaking. Continue to shake the stoppered test tube for about 5 minutes, then carefully neutralize the solution to a pH of 8 with 3 *M* HCl (use pH paper). Break up the solid mass with a stirring rod, if necessary, and collect the derivative by vacuum filtration. After washing the product thoroughly with cold water, recrystallize it from an ethanol-water mixed solvent.

Take Care! Wear gloves, avoid contact with and inhalation of benzoyl chloride.

D-9 *Benzenesulfonamides and* p-*Toluenesulfonamides.*

A precipitate that forms in the Hinsberg test (**C-15**) can be purified and used as a *p*-toluenesulfonamide derivative.

Arenesulfonyl chlorides are toxic and corrosive. Use a hood, avoid contact, and do not breathe vapors.

Safety Notes

Reactions:

$$RNH_2 + ArSO_2Cl \xrightarrow{\text{NaOH}} \xrightarrow{\text{HCl}} ArSO_2NHR$$

$$R_2NH + ArSO_2Cl \xrightarrow{\text{NaOH}} ArSO_2NR_2$$

Under the hood, combine in a test tube 0.20 g (or as close to 2.0 mmol as you can estimate) of the primary or secondary amine, 0.40 mL of benzenesulfonyl chloride *or* 0.60 g of *p*-toluenesulfonyl chloride, and 10 mL of 3 *M* sodium hydroxide. Stopper the test tube and shake it frequently over a period of 3–5 minutes. Remove the stopper and warm the tube gently over a steam bath or in a hot water bath for a minute or two. Let the solution cool.

Take Care! Avoid contact with arenesulfonyl chlorides, do not breathe their vapors.

A. If a precipitate forms, collect it by vacuum filtration, wash it with water, and recrystallize it from an ethanol-water mixed solvent.

B. If no precipitate forms on cooling, acidify the solution to pH 6 with 6 *M* HCl, cool the mixture to complete crystallization, collect the precipitate by vacuum filtration, wash it with water, and recrystallize it from an ethanol-water mixed solvent.

D-10 *α-Naphthylthioureas and Phenylthioureas.*

Protect the reactants and the reaction mixture from water.

The isothiocyanates are corrosive and toxic, and α-naphthylisothiocyanate is a lachrymator. Use gloves and a hood, avoid contact, and do not breathe vapors.

Safety Notes

Reaction:

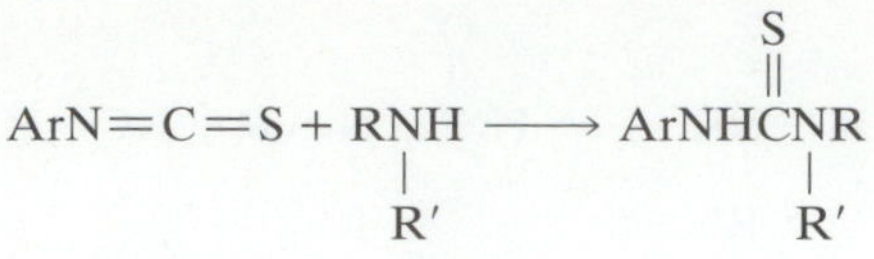

$$ArN=C=S + \underset{\underset{R'}{|}}{RNH} \longrightarrow ArNH\overset{\overset{S}{\|}}{C}\underset{\underset{R'}{|}}{NR}$$

(R' may be alkyl, aryl, or H.)

Take Care! Wear gloves and avoid contact with and inhalation of the isothiocyanate.

Under the hood, dissolve 0.20 g (or as close to 2.0 mmol as you can estimate) of the primary or secondary amine in 2.0 mL of absolute ethanol in a dry 15-cm test tube. Add 0.20 mL of phenylisothiocyanate *or* 0.30 g of α-naphthylisothiocyanate. Add a boiling chip and gently boil the mixture under the hood for 5 minutes, using a hot water bath or steam bath (insert a cold-finger condenser, if you have one). Replenish any ethanol that evaporate (setting a small funnel on the test tube will reduce evaporation). Let the reaction mixture cool and scratch the sides of the container to induce crystallization or cause any oil to crystallize. If no crystals form, boil the mixture gently (use a hood!) for another 20 minutes or so and again try to induce crystallization. Adding water drop by drop until the solution becomes cloudy may help.

Using a little 50% aqueous ethanol to facilitate transfer, collect the derivative by vacuum filtration and wash it once with 50% ethanol and once with 95% ethanol. If the solid appears impure, it may be necessary to extract impurities by boiling it with 4 mL of petroleum ether. Recrystallize the derivative from ethanol or an ethanol-water mixed solvent.

Derivatives of Tertiary Amines

D-11 *Methiodides.*

Safety Notes

> Iodomethane is toxic and irritant, and it is a suspected human carcinogen and mutagen. Use gloves and a hood, avoid contact, and do not breathe vapors.

Reaction: $RN_3 + CH_3I \longrightarrow R_3NCH_3^+I^-$
(The R groups may be alkyl, aryl, or a combination of the two.)

Take Care! Wear gloves, avoid contact with and inhalation of iodomethane.

Under the hood, mix 0.20 g of the tertiary amine with 0.20 mL of iodomethane in a test tube, and warm the mixture in a 50–60°C water bath for 5 minutes. Cool the mixture in ice, collect the product by vacuum filtration, and recrystallize it from alcohol, ethyl acetate, or a mixture of the two.

D-12 *Picrates.*

Picrates of aromatic hydrocarbons and other amines can also be prepared by this method. Picrates of many aromatic hydrocarbons dissociate when heated and therefore cannot be recrystallized.

Safety Notes

> In solid form, picric acid is unstable and can explode when subjected to heat or shock. It is also toxic and corrosive. Avoid contact with solutions containing picric acid.

Reaction:

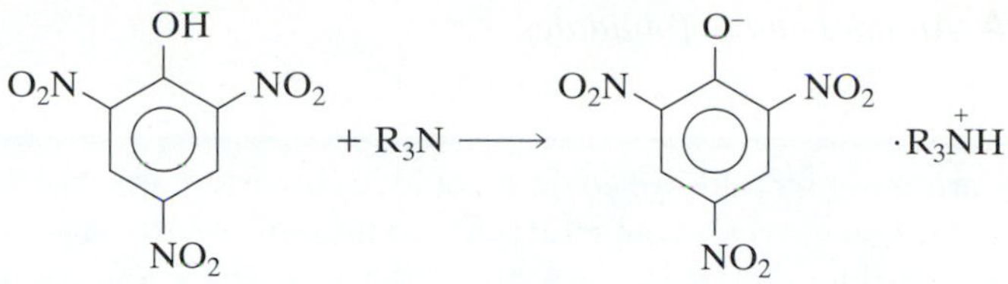

O$_2$N — C$_6$H$_2$(OH)(NO$_2$)$_2$NO$_2$ + R$_3$N $\longrightarrow$ O$_2$N — C$_6$H$_2$(O$^-$)(NO$_2$)$_2$NO$_2$ · R$_3$NH$^+$

Dissolve 0.20 g of the unknown in 5 mL of 95% ethanol and add 5.0 mL of a saturated solution of picric acid in ethanol. Heat the solution to boiling over a steam bath or in a boiling water bath, then let it cool to room temperature. Collect the crystals by vacuum filtration and wash them with cold ethanol or methanol. Recrystallize the derivative from ethanol, if necessary (some picrates are pure enough to use without recrystallization). Some picrates may explode when heated, so use caution when obtaining the melting point.

Take Care! Avoid contact with the picric acid solution.

Derivatives of Carboxylic Acids

D-13 *Amides.*

This derivative is suitable for aromatic acids and for most aliphatic acids that have six or more carbon atoms. Since Procedures **D-13** and **D-14** both involve preparation of the acid chloride, it is convenient to prepare one of the **D-14** derivatives with the same acid chloride solution.

Thionyl chloride and ammonia can both cause serious burns, and their vapors are highly irritating and toxic. Use gloves and a hood, avoid contact, and do not breathe vapors.
Acid chlorides are toxic and corrosive and their vapors are harmful. Use gloves and a hood, avoid contact, and do not breathe vapors.

Safety Notes

Reactions: RCOOH + SOCl$_2$ $\longrightarrow$ RCOCl + HCl + SO$_2$
RCOCl + NH$_3$ $\longrightarrow$ RCONH$_2$ + HCl

The reaction apparatus used for preparing the acid chloride must be dry. *Under the hood,* heat 0.80 g of the carboxylic acid under gentle reflux with 4.0 mL of thionyl chloride for 20–30 minutes, using a steam bath or hot-water bath. Cool the resulting acid chloride in an ice bath. Measure 2.0 mL of this solution and save the rest in a stoppered container to prepare an anilide or *p*-toluidide, if desired. (If you don't plan to prepare one of these derivatives, add all of the acid chloride to 10 mL of ammonia instead of 5 mL.) Still under the hood, add the 2 mL of acid chloride slowly, with constant stirring, to a beaker containing 5 mL of ice-cold concentrated aqueous ammonia. Let the reaction mixture stand for 5 minutes, separate the derivative by vacuum filtration, and recrystallize it from water or an ethanol-water mixture.

Take Care! Wear gloves, avoid contact with and inhalation of thionyl chloride and the acid chloride.

Take Care! Possible violent reaction! Wear gloves and avoid contact with and inhalation of NH$_3$.

D-14 *Anilides and* p-*Toluidides.*

Safety Notes

> The aromatic amines are toxic if inhaled, ingested, or absorbed through the skin. Use gloves, avoid contact, and do not breathe vapors.
> Acid chlorides are toxic and corrosive and their vapors are harmful. Avoid contact and do not breathe vapors.

Reactions: $RCOOH + SOCl_2 \longrightarrow RCOCl + HCl + SO_2$
$RCOCl + ArNH_2 \longrightarrow RCONHAr + HCl$

Take Care! Wear gloves, avoid contact with and inhalation of thionyl chloride and the acid chloride.

The reaction apparatus used for preparing the acid chloride must be dry. *Under the hood,* prepare the acid chloride as directed in Procedure **D-13**, using 0.40 g of the carboxylic acid and 2.0 mL of thionyl chloride, or use the reaction mixture saved from the amide preparation. Still under the hood, dilute the acid chloride with 2 mL of anhydrous ethyl ether. Then add in portions (shaking after each addition) a solution containing 0.80 g of *p*-toluidine *or* 0.70 mL of aniline dissolved in 10 mL of anhydrous ethyl ether. Continue the addition until the odor of the acid chloride has disappeared. If the odor persists after all the arylamine solution has been added, heat the mixture gently on a steam bath or in a boiling water bath for a minute or two. Wash the ether solution in a large test tube with 5 mL of 1.5 *M* HCl followed by 5 mL of water, removing the aqueous layers with a Pasteur pipet. Evaporate the solvent under vacuum. Recrystallize the derivative from water or an ethanol-water mixture.

Take Care! Wear gloves, avoid contact and inhalation.

D-15 p-*Nitrobenzyl Esters.*

Preparation of this derivative is recommended only when the previous derivatives are not suitable or when the carboxylic acid is obtained in aqueous solution. It is important that *p*-nitrobenzyl chloride not be present in excess, as it is difficult to remove from the product.

Safety Notes

> *p*-Nitrobenzyl chloride is corrosive, lachrymatory, and can cause blisters. Use gloves and a hood, avoid contact, and do not breathe vapors.
> The derivative may cause blistering. Use gloves and avoid contact.

Reaction:

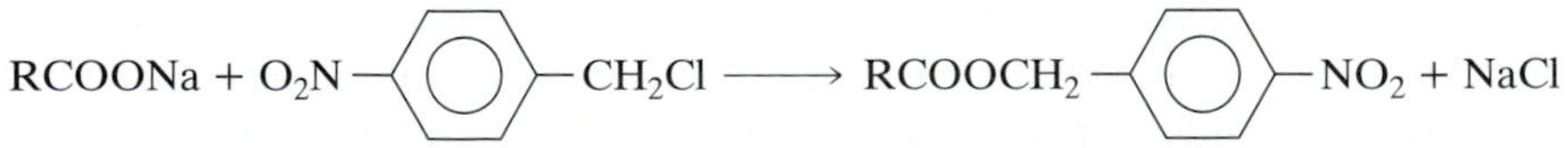

$$RCOONa + O_2N-\underset{}{\bigcirc}-CH_2Cl \longrightarrow RCOOCH_2-\underset{}{\bigcirc}-NO_2 + NaCl$$

If the carboxylic acid was obtained by hydrolysis of an unknown ester or amide, identify the alcohol or amine portion of the unknown first and estimate the molecular weight of the acid portion from your list of possibilities. Then estimate the amount of acid in the hydrolysis solution, and measure out enough of the solution to provide 2.0 mmol of the acid.

Mix an estimated 2.0 mmol of the carboxylic acid with 3 mL of water (or use the measured hydrolysis solution) and add a drop of phenolphthalein indicator solution. Add 3 M NaOH drop by drop until the solution turns pink. If necessary, heat to dissolve the acid salt, then add 2 drops or so of 1.5 M HCl to discharge the pink color. *Under the hood,* add 0.30 g (1.75 mmol) of *p*-nitrobenzyl chloride dissolved in 10 mL of 95% ethanol, and heat the mixture under gentle reflux for about 90 minutes (di- and triprotic acids should be heated for 2–3 hours). Allow the solution to cool to room temperature. If precipitation has not occurred by then, add 1 mL of water and scratch the sides of the reaction vessel. When crystallization is complete (it may take 20 minutes or more), collect the derivative by vacuum filtration and wash it twice with small portions of 5% sodium carbonate, then once with water. Recrystallize the derivative from ethanol or an ethanol-water mixed solvent. If the melting point of the product is close to 71°C (the melting point of *p*-nitrobenzyl chloride), the reaction may not have gone to completion.

Take Care! Wear gloves, avoid contact with and inhalation of *p*-nitrobenzyl chloride.

Take Care! Wear gloves, avoid contact with the derivative.

Derivatives of Esters

The acid and alcohol portions of an ester can be obtained by hydrolysis as described in **D-16** and one or both of them characterized by a melting point or derivative preparation. Melting points of solid carboxylic acids and alcohols are given in the table of esters, Table 8 of Appendix VI. Melting points of alcohol and carboxylic acid derivatives are given in Tables 1 and 7. You can also prepare a derivative from the ester itself by Procedure **D-17** or **D-18**. Procedure **D-17** forms a derivative of its carboxylic acid portion, and Procedure **D-18** forms a derivative of its alcohol portion.

D-16 *Hydrolysis Products.*

> **Hydrochloric acid is poisonous and corrosive. Use gloves and a hood, avoid contact, and do not breathe vapors.**

Safety Notes

Reactions: RCOOR′ + NaOH → RCOONa + R′OH
RCOONa + H⁺ → RCOOH + Na⁺

Carry out the hydrolysis of the ester according to the procedure in Test **C-2**, using 2–3 times the amounts given there. Acidify the reaction mixture *under the hood* using concentrated HCl. If a solid carboxylic acid precipitates, recrystallize it from water or an ethanol-water mixture and use it as a derivative, *or* convert it to one of the carboxylic acid derivatives. If no precipitate forms, make the solution basic to litmus with dilute sodium hydroxide and distill it until about 5 mL of distillate has been collected. Acidify the solution remaining in the boiling flask and either (a) use the solution to prepare a *p*-nitrobenzyl ester as described in **D-15** or (b) extract the solution with ether, then evaporate the ether to isolate the carboxylic acid for a derivative preparation.

The alcohol portion of the ester may separate if the distillate is saturated with potassium carbonate. It can be isolated by removing the aqueous layer with a Pasteur pipet or by extracting the distillate with ethyl ether and evaporating the ether. The alcohol can then be used to prepare an alcohol derivative.

Take Care! Wear gloves and avoid contact with and inhalation of HCl.

If the unknown may be an ester of a phenol, test a little of the precipitate for solubility in 5% NaHCO₃ (see Solubility Tests).

It may be advisable to purify the acid or alcohol before attempting to prepare a derivative.

D-17 N-*Benzylamides*.

This procedure works well only with methyl and ethyl esters. If your unknown may be an ester of a higher alcohol, heat 1.0 g of the ester under reflux with 5 mL of 5% sodium methoxide in methanol for 30 minutes and then evaporate the methanol. Use the resulting methyl ester in the following procedure.

Safety Notes

> **Benzylamine is corrosive and lachrymatory. Use gloves and a hood, avoid contact, and do not breathe vapors.**

Reactions:

$$RCOOR' + CH_3OH \xrightarrow{CH_3ONa} RCOOCH_3 + R'OH$$

$$RCOOCH_3 + PhCH_2NH_2 \xrightarrow{NH_4Cl} RCONHCH_2Ph + CH_3OH$$

$$RCOOCH_2CH_3 + PhCH_2NH_2 \xrightarrow{NH_4Cl} RCONHCH_2Ph + CH_3CH_2OH$$

Take Care! Wear gloves, avoid contact with and inhalation of benzylamine

Under the hood, combine 1.0 g of a methyl or ethyl ester with 3.0 mL of benzylamine and 0.10 g of powdered ammonium chloride. Heat the mixture under reflux for 1 hour. Cool the reaction mixture and wash it with a few milliliters of water, using a Pasteur pipet to remove the water. If no precipitate forms, add a drop or two of 3 *M* HCl and scratch the sides of the test tube. If a precipitate still does not form, transfer the mixture to an evaporating dish (use 1 mL of water in the transfer), and boil it for a few minutes to remove the excess ester. Collect the derivative by vacuum filtration, wash it with petroleum ether, and recrystallize it from an ethanol-water mixed solvent or from ethyl acetate.

D-18 *3,5-Dinitrobenzoates*.

This procedure is not satisfactory for esters of tertiary and some unsaturated alcohols.

Safety Notes

> **Sulfuric acid causes chemical burns that can seriously damage skin and eyes. Use gloves and avoid contact.**
> **3,5-Dinitrobenzoic acid is an irritant. Avoid contact.**

Reaction:

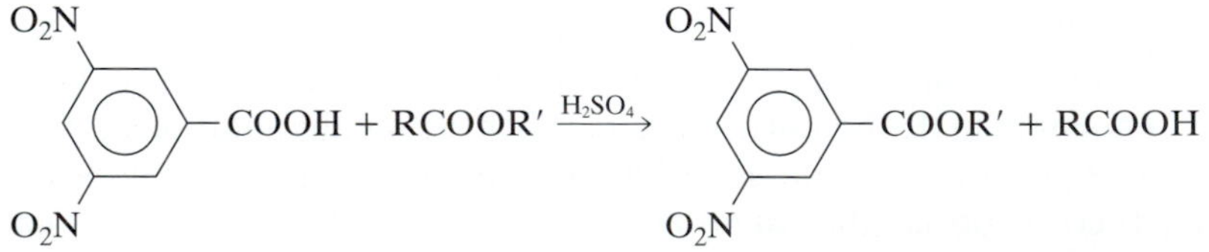

Take Care! Wear gloves and avoid contact with 3,5-dinitrobenzoic acid and H$_2$SO$_4$.

Combine 1.0 g of the ester with 0.80 g of 3,5-dinitrobenzoic acid and add a drop of concentrated sulfuric acid. Heat the mixture under reflux until the 3,5-dinitrobenzoic acid dissolves, then continue the reflux for an additional 30 minutes. Dissolve the cooled reaction mixture in 20 mL of ethyl ether. Wash the ether solution twice with 10-mL portions of 0.5 *M* sodium carbonate, then with water. Evaporate the ether and dissolve the residue (often an oil) in 2–3 mL of boiling ethanol. Filter the hot solution, add water until the mixture becomes cloudy, and cool the solution to crystallize the product. Recrystallize the derivative from an ethanol-water mixed solvent.

Take Care! Foaming will occur.

Derivatives of Alkyl Halides

In addition to the following derivative, density (Test **C-10**) and refractive index values are useful for characterizing alkyl and aryl halides.

D-19 S-*Alkylthiuronium Picrates.*

Tertiary alkyl halides will not form this derivative; chlorides and secondary bromides and iodides react slowly. If the unknown has been shown to be an alkyl chloride, use procedure **B**.

Safety Notes

In solid form, picric acid is unstable and can explode when subjected to heat or shock. It is also toxic and corrosive. Avoid contact with solutions containing picric acid.
Thiourea is toxic if absorbed through the skin. Avoid contact.

Reaction:

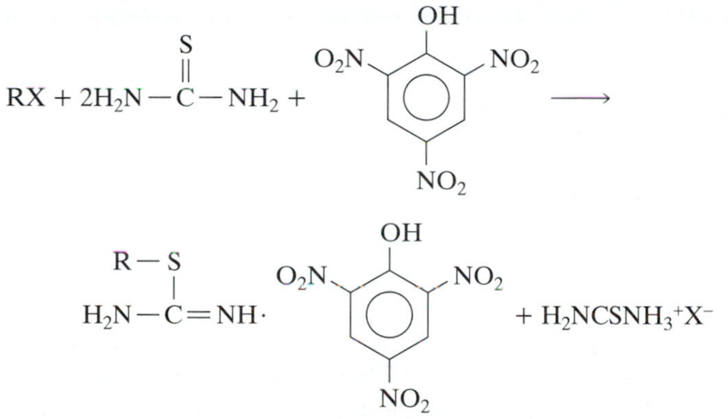

$$X = Cl, Br, I \quad R = alkyl$$

A. Mix together 0.50 g of the alkyl halide, 0.50 g of powdered thiourea and 5.0 mL of 95% ethanol. Heat the mixture under reflux for 30 minutes or more, add 5.0 mL of a saturated solution of picric acid in ethanol, and heat the mixture under reflux until a clear solution is obtained. Allow the solution to cool, collect the derivative by vacuum filtration, and recrystallize it from ethanol or an ethanol-water mixed solvent. If a derivative fails to form under these conditions, try procedure **B**.

Take Care! Avoid contact with thiourea and the picric acid solution.

Take Care! Do not allow the solution to evaporate to dryness as an explosion may result.

B. Mix together 0.20 g of the alkyl halide, 0.20 g of powdered thiourea, and 6 mL of ethylene glycol. Heat the solution with an oil bath (or other appropriate heat source) at 120°C for 30 minutes, add 2.0 mL of a saturated solution of picric acid in ethanol, and continue heating it at 120°C for 15 minutes. Add 6 mL of water and cool the mixture in an ice/water bath to assist crystallization. Collect the derivative

Take Care! Avoid contact with thiourea and the picric acid solution.

by vacuum filtration and recrystallize it from ethanol or an ethanol-water mixed solvent.

Derivatives of Aryl Halides

In addition to preparing the following derivative, you can characterize aryl halides that have alkyl side chains by oxidizing the side chain (Procedure **D-21**). Density (Test **C-10**) and refractive index values are also useful.

D-20   *Nitro Compounds.*

Procedure **A** yields mononitro derivatives of comparatively reactive aryl halides and aromatic hydrocarbons. Procedure **B** should yield dinitro or trinitro derivatives of reactive compounds and mononitro derivatives of unreactive ones. Most di- and trialkylbenzenes yield trinitro derivatives by procedure **B**, while monoalkylbenzenes yield dinitro derivatives. Since di- and trinitro derivatives are often easier to purify than mononitro derivatives, procedure **B** is usually preferred.

Safety Notes

> **Sulfuric acid, nitric acid, and (especially) fuming nitric acid are toxic, highly reactive, and very corrosive. Use gloves and a hood, avoid contact, and do not inhale vapors or any brown fumes that may be generated during the reaction.**

Reaction: $\text{ArH} + \text{HNO}_3 \xrightarrow{\text{H}_2\text{SO}_4} \text{ArNO}_2 + \text{H}_2\text{O}$
(and di- or trinitro derivatives in some cases)

Take Care! Wear gloves, avoid contact with and inhalation of the acids.

Take Care! Possible violent reaction.

Take Care! Wear gloves, avoid contact with and inhalation of the acid.

A. *Under the hood,* mix 0.50 g of the unknown with 2.0 mL of concentrated sulfuric acid. Cautiously add 2.0 mL of concentrated nitric acid drop by drop with shaking or stirring. Then heat the mixture on a 60°C water bath for 10 minutes, shaking frequently. Cautiously pour the mixture onto 15 mL of cracked ice, with stirring. After the ice has melted, collect the precipitate by vacuum filtration, wash it with water, and recrystallize it from an ethanol-water mixed solvent. Repeated recrystallization may be necessary if the product melts at a lower temperature than expected, or over a wide range.

B. Follow procedure **A**, but use 2.0 mL of fuming nitric acid in place of concentrated nitric acid, and heat the solution for 10 minutes on a steam bath rather than a 60° water bath. Add the acid slowly to minimize the generation of brown fumes of nitrogen dioxide. If the product separates as an oil, it is probably a mixture of compounds with different numbers of nitro groups.

Derivatives of Aromatic Hydrocarbons

In addition to preparing the following derivative, you can characterize aromatic hydrocarbons by the preparation of picrates (**D-12**) and nitro compounds (**D-20**). The picrates of many aromatic hydrocarbons dissociate when heated and cannot be recrystallized.

D-21 *Aromatic Carboxylic Acids.*

The following procedure is for an aromatic hydrocarbon or aryl halide with one alkyl side chain. If the unknown may have more than one side chain, increase the quantities of the reagents in proportion to the anticipated number of side chains.

> **Sodium hydroxide is toxic and corrosive, and causes severe damage to skin, eyes, and mucous membranes. Use gloves and avoid contact with the NaOH solution.**
> **Potassium permanganate can react violently with oxidizable materials. Keep it away from other chemicals and combustibles.**

Safety Notes

Reaction: $ArR \xrightarrow{KMnO_4} ArCOOH$

Dissolve 1.5 g of potassium permanganate in 25 mL of water, and add 0.50 g of the aromatic hydrocarbon (or aryl halide) followed by 0.50 mL of 6 *M* aqueous sodium hydroxide. Heat the mixture under reflux for 1 hour or until the purple permanganate color has disappeared. Cool the reaction mixture to room temperature, acidify it with 6 *M* sulfuric acid, and boil it for a few minutes. If necessary, stir a little solid sodium bisulfite into the hot solution to destroy any unreacted manganese dioxide, keeping the solution acidic with 6 *M* H_2SO_4 during the treatment. Cool the mixture, collect the product by vacuum filtration, and recrystallize the carboxylic acid from water or an ethanol-water mixed solvent.

Take Care! Keep $KMnO_4$ away from other chemicals

Take Care! Wear gloves and avoid contact with the NaOH solution.

Derivatives of Phenols

D-22 *Aryloxyacetic Acids.*

The equivalent weight of an aryloxyacetic acid (and thus of the unknown phenol) can be determined by obtaining its neutralization equivalent (Test **C-18**).

> **Chloroacetic acid is corrosive and toxic. Use gloves, avoid contact, and do not breathe vapors.**
> **Sodium hydroxide is toxic and corrosive, and may cause severe damage to skin, eyes, and mucous membranes. Use gloves and avoid contact with the NaOH solution.**
> **Some aryloxyacetic acids, such as 2,4-dichlorophenoxyacetic acid (2,4-D), are suspected human carcinogens. Avoid contact with the derivative.**

Safety Notes

Reaction: $ArOH + ClCH_2COOH \rightarrow ArOCH_2COOH + HCl$

Under the hood, combine 0.50 g of the unknown with 0.70 g of chloroacetic acid and 3.0 mL of 8 *M* sodium hydroxide. Heat the mixture on a steam bath for 1 hour, then cool it to room temperature and add 6 mL of water. Acidify the solution to pH 3 with 6 *M* hydrochloric acid, extract it with

Take Care! Wear gloves, avoid contact with NaOH and chloroacetic acid, do not inhale the acid.

Take Care! Foaming will occur.
Take Care! Avoid contact with the derivative.

20 mL of ethyl ether, and wash the ether extract with 5 mL of cold water. Extract the derivative from the ether using about 10 mL of 0.5 M sodium carbonate, then acidify the sodium carbonate solution with 6 M hydrochloric acid. Collect the derivative by vacuum filtration and recrystallize it from water.

D-23 *Bromo Derivatives.*

Safety Notes

> **Bromine is toxic and corrosive, and its vapors are harmful. Use gloves and a hood, avoid contact, and do not breathe vapors.**

Reaction for phenol:

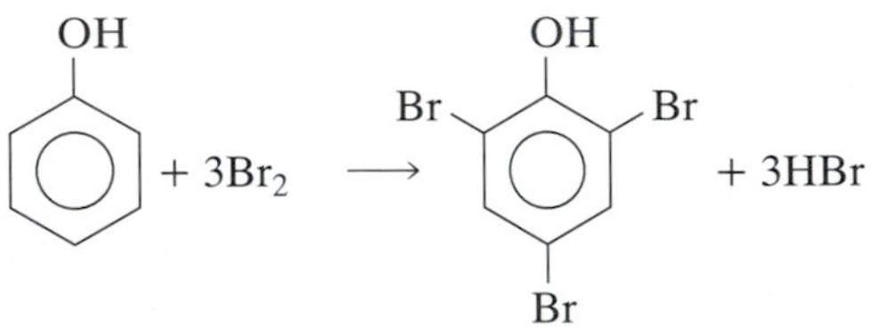

Take Care! Wear gloves and avoid contact with and inhalation of bromine.

Under the hood, prepare a brominating solution by dissolving 4.5 g of potassium bromide in 30 mL of water and *carefully* adding 1.0 mL (about 3 g) of pure bromine. Dissolve 0.30 g of the phenol in 6 mL of 50% ethanol; try 95% ethanol or acetone if it doesn't dissolve. Still under the hood, add the brominating solution drop by drop, with stirring, until the bromine color persists after shaking. Add 15 mL of water and shake the mixture, then collect the derivative by vacuum filtration. Wash the product with 1 M sodium bisulfite to remove excess bromine, and recrystallize it from ethanol or an ethanol-water mixture.

D-24 *α-Naphthylurethanes.*

Safety Notes

> **Pyridine is toxic, an irritant, and has an unpleasant odor. Use a hood, avoid contact, and do not breathe vapors.**
> **α-Naphthyl isocyanate is lachrymatory and a strong irritant. Use gloves and a hood, avoid contact, and do not breathe vapors.**

Reaction:

$$ArN{=}C{=}O + Ar'OH \longrightarrow ArNHC\overset{\displaystyle O}{\overset{\|}{-}}OAr'$$

Take Care! Avoid contact with and inhalation of pyridine.

Under the hood, carry out the procedure for preparing α-naphthylurethanes of alcohols (**D-2**), but add a drop of pyridine to catalyze the reaction. For particularly unreactive phenols, it may be helpful to add 1 mL of pyridine and a drop of 10% triethylamine in petroleum ether to the reaction mixture, and to heat the mixture at 70°C for about 30 minutes. If the urethane does not precipitate on cooling, add 1 mL of 0.5 M sulfuric acid.

Report

Your report should include a statement of the problem and an account of how you applied scientific methodology to solve it. It should also include the following information, or other information requested by your instructor.

Unknown number
Results of preliminary examination
Physical constants determined
Results of solubility tests
Results of classification tests
Spectra and spectrometric data
Interpretation of tests and spectra
Original list and short list of possibilities
Melting points of derivatives
Discussion and conclusion
Answers to assigned Exercises

Your report should reveal the thought processes that led you to your conclusion, as well as the physical and chemical information supporting it. For example, the "Interpretation" section should tell how your observations led you to make tentative assumptions about the nature of the functional group(s) or other structural features. You should also tell how you eliminated compounds from your original list of possibilities to arrive at your short list. When appropriate, you should tabulate your data and observations to present them clearly and concisely.

Exercises

1 Write balanced equations for the chemical reactions undergone by your unknown, including those involved in classification tests, derivative preparations, and all solubility tests except the one with water.

2 Indicate the solubility class to which each of the following is most likely to belong. (a) propanoic acid, (b) toluene, (c) *p*-anisidine, (d) *p*-cresol, (e) 3-methylpentanal, (f) *p*-toluic acid, (g) 2-chlorobutane, (h) butylamine, (i) ethylene glycol, (j) acetophenone

3 Deduce the functional class (chemical family) of each of the following unknowns and give any additional structural information suggested by the results. If sufficient information is provided, give one or more possible structures for the unknown, assuming that it is one of the compounds listed in Appendix VI. Explain the reasoning behind your conclusions.

   (a) Unknown A is insoluble in water but soluble in 5% NaOH. It decolorizes a dilute, neutral solution of potassium permanganate, yielding a brown precipitate. Addition of bromine water to a solution of the unknown yields a white precipitate.

   (b) Unknown B is in solubility class **B** and yields an infrared spectrum with no absorption band above 3100 cm^{-1}. The unknown is insoluble in a pH 5.5 buffer, and treatment with *p*-toluenesulfonyl chloride in aqueous sodium hydroxide leaves a liquid residue that is soluble in dilute HCl.

(c) Unknown C burns with a sooty, yellow flame and its infrared spectrum has strong bands at 1660, 3180, and 3370 cm^{-1}. When the unknown is heated with 6 M sodium hydroxide, a vapor is generated that turns red litmus paper blue. Acidification of the alkaline hydrolysis solution yields a white precipitate.

(d) Unknown D is insoluble in cold, concentrated sulfuric acid, and a drop of the unknown in water sinks to the bottom. A solution of the unknown in chloroform gives an orange-red color on sublimed aluminum chloride. It reacts immediately with both ethanolic silver nitrate and sodium iodide in acetone to form precipitates. Adding dichloromethane and chlorine water to its sodium fusion solution yields a red-orange color.

(e) Unknown E boils near 100°C, is water soluble, and reacts with acetyl chloride to yield a pleasant-smelling liquid. It dissolves in the Lucas reagent, giving a cloudy solution after 4 minutes. Treating the unknown with iodine in sodium hydroxide forms a yellow precipitate with a medicinal odor.

(f) Unknown F is water soluble and has an infrared spectrum with a strong band at 1691 cm^{-1}. It gives an orange-red precipitate with 2,4-dinitrophenylhydrazine reagent and a blue-green suspension with chromic acid reagent.

(g) Unknown G is insoluble in water and 5% NaHCO$_3$ and gives a negative test with 2,4-dinitrophenylhydrazine. When the unknown is heated with 6 M NaOH and the reaction mixture acidified, a vinegar-like odor is detected. When the reaction mixture is then made basic and distilled, extraction of the distillate yields a liquid that reacts with chromic acid.

(h) Unknown H is a water-insoluble solid that dissolves in 5% NaHCO$_3$. It decolorizes a solution of bromine in dichloromethane. An aqueous solution containing 0.195 g of the unknown is titrated to a phenolphthalein endpoint by 32.0 mL of 0.106 M NaOH.

(i) Unknown I is in solubility class **X**, burns with a clean yellow flame, and floats on water. It reacts with both silver nitrate/ethanol and sodium iodide/acetone after heating, but not at room temperature.

(j) Unknown J is a water-insoluble liquid and has strong infrared bands at 690, 746, and 1688 cm^{-1}. The unknown gives a precipitate with 2,4-dinitrophenylhydrazine reagent but no precipitate with either chromic acid reagent or iodine in sodium hydroxide.

4 Give reasonable mechanisms for the reactions involved in each of the following classification tests or derivative preparations: (a) the alkaline hydrolysis of N-ethylbenzamide; (b) the reaction of p-cresol with bromine water; (c) the preparation of the semicarbazone of acetophenone; (d) the reaction of ethyl acetate with hydroxylamine in alkaline solution; (e) the reaction of cyclohexylamine with p-toluenesulfonyl chloride in aqueous sodium hydroxide; (f) the reaction of 2-propanol in the Lucas test; (g) the reaction of 2-chloro-2-methylbutane with ethanolic silver nitrate; (h) the reaction of 1-bromobutane with sodium iodide in acetone; (i) the mononitration of toluene by nitric and sulfuric acids; (j) the preparation of the aryloxyacetic acid of p-cresol; (k) the formation of the 3,5-dinitrobenzoate of 2-butanol; (l) the iodoform reaction of acetophenone.

The Operations

Part V contains descriptions of all the laboratory operations employed in the experiments in this book.

A. Basic Operations

OPERATION 1

Cleaning and Drying Glassware

Clean glassware is essential for good results in the organic chemistry laboratory. Even small amounts of impurities can sometimes inhibit chemical reactions, catalyze undesirable side reactions, and invalidate the results of chemical tests or rate studies. Always clean your dirty glassware at the end of each laboratory period, or as soon as possible after the glassware is used. Otherwise, residues may harden and become more resistant to cleaning agents; they may also attack the glass itself, weakening it and making future cleaning more difficult. It is particularly important to wash out strong bases such as sodium hydroxide promptly because they can etch the glass permanently and cause glass joints to "freeze" tight. When glassware has been thoroughly cleaned, water applied to its inner surface should wet the whole surface and not form droplets or leave dry patches. Used glassware that has been scratched or etched may not wet evenly, however.

You can clean most glassware adequately by vigorous scrubbing with hot water and a laboratory detergent such as Alconox, using a brush of appropriate size and shape to reach otherwise inaccessible spots. A plastic trough or another suitable container can serve as a dishpan. A nylon mesh scrubbing cloth is useful for cleaning spatulas, stirring rods, beakers, and the outer surfaces of other glassware. Pipe cleaners or cotton swabs can be used to clean narrow funnel stems, eyedroppers, etc. Organic residues that cannot be removed by detergent and water will often dissolve in organic solvents such as acetone. Use organic solvents sparingly and recycle them, as they are much more costly than water (never use reagent-grade acetone for washing). After washing, always rinse the glassware thoroughly with water (a final distilled water rinse is a good idea) and check it to see if the water wets its surface evenly. If it doesn't pass this test, scrub it some more or use a cleaning solution such as Nochromix.

The easiest (and cheapest) way to dry glassware is to let it stand overnight in a position that allows easy drainage. You can dry the outer surfaces of glassware with a soft cloth, but don't dry any surfaces that will be in contact with chemicals this way because of the possibility of contamination. If a piece of glassware is needed shortly after washing, drain it briefly to remove excess water, then rinse it with one or two small portions of wash acetone and dry it in a gentle stream of clean, dry air. Compressed air from an air line may contain pump oil and moisture; it can be cleaned and dried as described in OP-22a. Keep in mind that thorough drying is not necessary if the glassware is to be used with aqueous solutions. In that case, it is usually sufficient to drain the glassware for a few minutes after rinsing.

Glassware that is to be used for a Grignard reaction or other highly moisture sensitive reaction must be dried thoroughly before use. If possible, clean the glassware during the previous lab period, let it dry overnight or longer, then dry it in an oven set at about 110°C for 15 minutes or longer. Assemble the apparatus and attach one or more drying tubes [see OP-22] as soon as possible after oven drying; otherwise, moisture will condense inside it as it cools. If the glassware must be cleaned the same day as it is used, rinse it with acetone after washing and flush it with clean, dry air to evaporate all the acetone before you put it in the oven. You can also dry glassware by playing a "cool" Bunsen burner flame over the surface of the assembled apparatus, but this practice should never be used in laboratories where volatile solvents such as ethyl ether are in use. It should be done only with the instructor's permission and according to his or her directions.

Take Care! Use tongs or heat-resistant gloves when handling hot glass.

Using Standard-Taper Glassware

Most ground-joint glassware used in organic chemistry is of the straight, standard-taper type with rigid joints. The size of a tapered joint is designated by two numbers, such as 19/22, in which the first number is the diameter at the top of the joint and the second is the length of the taper, measured in millimeters. The glassware in a commercial organic lab kit, or its equivalent purchased as separate parts, can be used to construct apparatus for many different laboratory operations, from heating under reflux to fractional distillation. Recommended setups for the various operations are usually illustrated in the instruction booklet that comes with each lab kit.

Lubricating Joints

For some operations such as vacuum distillations, glass joints other than the clear-seal type should be lubricated with a suitable stopcock/ground-glass joint grease. For most other operations, lubrication of glass joints is unnecessary except to keep a joint exposed to strong alkali from freezing. To lubricate a ground-glass joint, apply a thin layer of joint grease completely around the top half of the inner (male) joint. Do not lubricate the outer (female) joint. Be careful to keep grease away from the open end of the joint, where it may contact and contaminate your reaction mixture or product. When you assemble the components, press the outer and inner joints together firmly, with a slight twist, to form a seal around the entire joint, with no gaps. Grease should never extend beyond the joint inside the apparatus.

After disassembling the apparatus, remove the grease completely by using a suitable organic solvent. Petroleum based greases are easily removed with petroleum ether or hexanes; silicone greases can be removed by thorough cleaning with dichloromethane. An inner joint can be cleaned by wrapping a little cotton loosely around the end of an applicator stick, dipping it in the solvent, and wiping the joint with the moist cotton.

Assembling Glassware

Standard-taper joints are rigid, so a glassware apparatus must be assembled carefully to avoid strain that can result in breakage. First, place the necessary clamps and rings at appropriate locations on the ring stand (use two

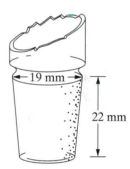

Figure A1 19/22 standard-taper joint

Take Care! Keep flames away from petroleum solvents. Avoid contact with dichloromethane and do not breathe its vapors.

ring stands for distillation setups). Then assemble the apparatus *from the bottom up, starting at the heat source.* Position the heat source on a ring or other support so that it can be removed easily when the heating period is over; otherwise it may continue to heat a reaction mixture or an empty distilling flask even after it is switched off, causing a danger of breakage, tar formation, or even an explosion. Clamp the reaction flask or boiling flask securely at the proper distance from the heat source.

As you add other components, clamp them to the ring stand(s), but do not tighten the clamp jaws completely until all of the components are in place and aligned properly. Use as many clamps as are necessary to provide adequate support for all parts of the apparatus. A vertical setup such as the one for addition under reflux [OP-10] requires at least two clamps for security because if the setup is bumped, the clamp holding the reaction flask may rotate and deposit your glassware on the lab bench—usually with very expensive consequences. Some vertical components, such as Claisen connecting tubes, need not be clamped if they are adequately supported by the component below. Nonvertical components such as distilling condensers should be clamped; otherwise they may be accidentally dislodged and fall. Distillation receivers, which may need to be replaced during a distillation, should be supported by a ring and wire gauze or another suitable support.

To clamp condensers and other components at an angle to a ring stand, an adjustable clamp with a wing nut on the shaft is required. This wing nut is tightened after the apparatus is aligned. A plastic or metal joint clip, or even a strong rubber band, can be used to hold some joints together. For example, a vacuum adapter can be secured to a condenser by stretching a rubber band around the tubulation on both, or by snapping a joint clip around the joint rim. Condensers and vacuum adapters should never be allowed to hang unsupported, even momentarily while you are assembling the apparatus.

When the clamps have been positioned so that all glass joints come together without applying excessive force, seat the joints (with a slight twist, if necessary) and tighten all the clamps. Examine each joint for possible gaps, then check to make sure that the apparatus is held securely by the clamp jaws and that the clamp holders are secured tightly to the ring stand(s).

Figure A2 summarizes the steps followed in assembling one kind of ground-glass apparatus. Most of the glassware setups you will be using are less complex than the one illustrated.

Disassembling Glassware

Disassemble (take apart) ground-joint glassware promptly after use, as joints that are left coupled for extended periods of time may freeze together and become difficult or impossible to separate without breakage. If a joint is frozen, you can sometimes loosen it by tapping it gently with the wooden end of a spatula or by applying steam to the joint while rotating the apparatus slowly, then pulling the components apart with a twisting motion. If this doesn't work, consult your instructor. Clean the glassware thoroughly [OP-1] and return each component to its proper location in the lab kit or to the stockroom.

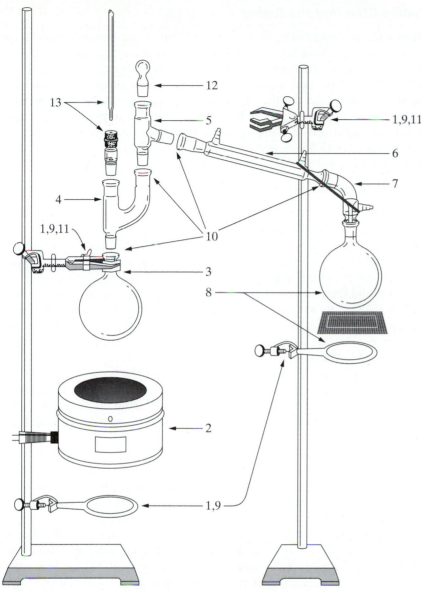

Steps

1 Position clamps, rings.
2 Position heat source.
3 Secure boiling flask (clamp tightly).
4, 5 Add Claisen, three-way connecting tubes.
6 Clamp condenser in place.
7 Attach vacuum adapter with rubber band or spring clamp.
8 Attach receiving flask, support with ring and wire gauze.
9 Readjust all clamps to align.
10 Press joints together.
11 Tighten clamps.
12 Add stopper.
13 Add thermometer adapter and position thermometer.

Figure A2 Steps in the assembly of ground-glass apparatus

Using Glass Rod and Tubing

OPERATION 3

Glass connecting tubes, stirring rods, and other simple glass items are required for certain operations in organic chemistry. Soft-glass rod and tubing can be worked easily with a Bunsen burner, but borosilicate glass (Pyrex, Kimax, etc.) may require the hotter flame provided by a Meker-type burner or an oxygen torch. To distinguish borosilicate from soft glass, dip the glass into turpentine or anhydrous glycerol; borosilicate glass will seem to disappear in the liquid.

Cutting Glass Rod and Tubing

Glass rods and tubes are cut by scoring them at the desired location and snapping them in two. Score the rod or tube by drawing a sharp triangular file (or other glass-scoring tool) across the surface at a right angle to the axis of the tubing. Often only a single stroke is needed to make a deep scratch in the surface; don't use the file like a saw. Moisten the scratch with water or saliva. Using a towel or gloves to protect your hands, place your thumbs about 1 cm apart on the side opposite the scratch and, while holding the glass firmly in both hands, press forward against the glass with your thumbs as you rotate your wrists outward (Figure A3).

Working Glass Rod and Tubing

The cut ends of a glass rod or tube should always be *fire polished* to remove sharp edges and prevent accidental cuts. To fire polish a glass rod or tube, hold it at a 45° angle to a burner flame (see OP-6 for directions on using a burner) and rotate its cut end slowly in the flame until the edge becomes rounded and smooth (Figure A4).

To round the end of a glass rod, rotate the rod in a burner flame, holding it at a 45° angle with its tip at the inner blue cone of the flame. To flatten one end of a glass rod, rotate it in the flame until it is incandescent and very soft, but not starting to bend, then press the rod straight down onto a hard surface such as the base of a ring stand.

To seal one end of a glass tube, rotate it in the inner blue cone of a flame (or its outer edge, for a capillary tube) until the soft edges come together and eventually merge. Remove the tube from the flame as soon as it is closed and immediately blow into the open end to obtain a sealed end of uniform thickness. Check for leaks by letting the tube cool to room temperature, connecting the open end to an aspirator or vacuum line with a length of rubber tubing, and placing the closed end in a test tube containing a small amount of dichloromethane. (**Take Care!** Avoid contact with dichloromethane and do not breathe its vapors.) If the tube is not properly sealed, the liquid will leak into it when you apply suction.

To bend glass tubing, place a flame spreader on the barrel of a Bunsen burner (or use a Meker-type burner). Hold the tubing over the burner flame parallel to the long axis of the flame spreader and rotate it constantly at a slow, even rate until it is nearly soft enough to bend under its own weight (Figure A5). (The flame will turn yellow as the glass begins to soften.) Remove the

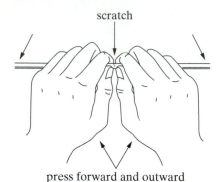

scratch

press forward and outward

Figure A3 Breaking glass tubing

Take Care! Don't burn yourself on the hot end of a glass rod or tube, or lay the glass onto combustible materials.

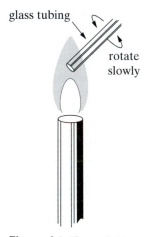

glass tubing

rotate slowly

Figure A4 Fire polishing

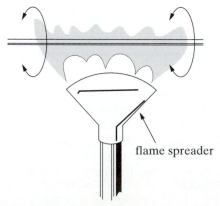

flame spreader

Figure A5 Bending tubing

hot tubing from the flame and immediately bend it to the desired shape with a firm, even motion and a minimum of force (if much force is required, the glass is not soft enough). Bend it in a vertical plane, with the ends up and the bend at the bottom; the bend should follow a smooth curve with no constrictions.

Inserting Glass Items into Stoppers

Improper insertion of glass tubing is one of the most frequent causes of laboratory accidents. The resulting cuts and puncture wounds can be very severe, requiring medical treatment and sometimes causing the victim to go into shock. Thermometers are particularly easy to break, especially at the scored immersion line.

Safety Notes

Cork borer

Sometimes you may have to insert a glass tube through a hole bored in a cork or rubber stopper. To bore a hole in a solid stopper, obtain a *sharp* cork borer that is slightly smaller in diameter than the objct to be inserted in the stopper. Lubricate its cutting edge with a little glycerine, then *twist* it through the stopper using a minimum of force (don't try to "punch" out the hole). Rotate the borer and stopper in opposite directions, checking the alignment frequently to make sure that the borer is going in straight. When the borer is about halfway through, twist it out and start boring from the opposite end of the stopper until the holes meet. You can remove the plug left inside the cork borer with a rod that comes with a set of cork borers.

To insert a length of glass tubing into a stopper, or a thermometer into a thermometer adapter, first lubricate the hole lightly with glycerol or another suitable lubricant; water may work if the hole is not too tight. You can use a cotton swab or applicator stick to apply the lubricant evenly. Protect your hands with gloves or a towel, then grasp the tube (or thermometer) close to the stopper (or thermometer adapter) and twist it through the hole with firm, steady pressure. Do not hold the tube too far from the stopper or the glass may break and lacerate your hand. Apply force directly along the axis of the tubing, as any sideways force may cause the tube to break. Using excessive force or forcing glass through a hole too small for it can also cause it to break. After the tube or thermometer is correctly positioned, rinse off the glycerol with water.

To remove a glass tube from a stopper or a thermometer from a thermometer adapter, wet the part of the glass that will pass through the stopper (or use glycerol if necessary), protect your hands with gloves or a towel, and twist the glass out with a firm, continuous motion. Hold the tube or thermometer close to the stopper, and avoid applying any sideways force that could cause it to break. If you can't remove a glass tube by this method, obtain a cork borer of a size that will just fit around the tube and twist it gently through the stopper until the glass can be removed.

Weighing

OPERATION 4

Most modern chemistry laboratories are equipped with electronic balances that display the mass directly, without any preliminary adjustments. If you will be using a different type of balance, your instructor will demonstrate its operation. Laboratory balances measure mass rather than weight, so when you "weigh" an object on a lab balance, you are actually measuring its mass. Most products obtained from a preparation are transferred to vials or other small containers, which should be *tared*—weighed empty—and then re-weighed after the product has been added. As a rule, the container should be weighed with its cap and label on and this *tare weight* recorded.

A balance is a precision instrument that can easily be damaged by contaminants, so avoid spilling chemicals on the balance pan or on the balance itself. If spillage does occur, clean it up immediately. If you spill a corrosive liquid or solid on any part of the balance, notify your instructor as well. Before you leave the balance area, replace the caps on all reagent bottles, return them to their proper locations if you obtained them elsewhere, and see that the area around the balance is clean and orderly.

Weighing Solids

Solids can be weighed in glass containers (such as vials or beakers), in aluminum or plastic weighing dishes, or on glazed weighing papers. Hygroscopic solids, those that absorb moisture from the atmosphere, should be weighed in closed containers. Filter paper and other absorbent papers should not be used for weighing, since a few particles will always remain in the fibers of the paper.

To weigh a sample of solid that is in a tared container, set the digital readout to zero by pressing the appropriate button, then place the container on the balance pan and read the mass of the container and its contents from the digital display. Record the mass in your laboratory notebook, including all digits after the decimal point. For example, if the balance reads 3.610 g, do not record the mass as 3.61 g, because zeroes following the decimal point are significant. Then subtract the mass of the container to obtain the mass of the solid.

To weigh a sample of solid that is not in a tared container, place a weighing container on the balance pan, press the tare button to zero the mass reading, transfer the solid to the weighing container, and read its mass from the digital display,

To measure out a specific quantity of a solid such as a solid reactant, place a weighing container on the balance pan and press the tare button to zero the digital display. Then use a spatula or scoopula to add the solid in small portions until the desired mass appears on the digital display.

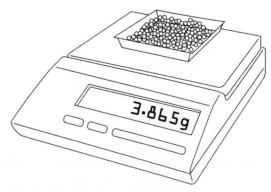

Figure A6 An electronic balance

Ordinarily you need not measure out the exact mass specified, but try not to deviate from the specified mass by more than 0.02 g or so. Since the theoretical yield of a preparation is based on the actual mass of a starting material, always use your measured mass, not the calculated or specified mass, for stoichiometric calculations.

Weighing Liquids

Unless they are very high-boiling, liquids should be weighed in screwcap vials or other closed containers to avoid losses by evaporation. The mass of a liquid sample in a tared or untared container is measured as described for a solid.

To weigh out a specific mass of a liquid from a reagent bottle, you can first measure the approximate quantity of the liquid by volume and then weigh that quantity accurately in an appropriate container. For example, if you need 3.71 g of 1-butanol ($d = 0.810$ g/mL) for an experiment, measure about 4.6 mL (3.71 g $\div$ 0.810 g/mL) of the liquid [OP-5] and transfer it to a tared container. If the measured mass is not close enough to 3.71 g, add or remove liquid with a clean Pasteur pipet or medicine dropper, then cap or stopper the container. If you only need a small amount of liquid, you can place the container on a balance pan, zero the balance, and add the liquid drop by drop until you reach the desired mass.

Measuring Volume

OPERATION 5

Several different kinds of volume-measuring devices can be used in the undergraduate organic chemistry laboratory. Relatively large volumes of liquids are measured using graduated cylinders whose capacity usually varies from 10 mL to 100 mL. Smaller volumes can be measured using pipets and syringes. Reagent bottles containing liquids may be provided with bottle-top dispensers that measure out a preset volume of the liquid. For some of the experiments in this book you will use a buret or a volumetric flask; their use is described in most general chemistry laboratory manuals.

Graduated Cylinders

Graduated cylinders are not highly accurate, but they are adequate for measuring specified quantities of solvents and wash liquids as well as liquid reactants that are present in excess. A small graduated cylinder can be used to premeasure a liquid that will then be weighed, as described in OP-4. The liquid volume should always be read from the bottom of the liquid meniscus as shown in Figure A7.

Bottle-Top Dispensers

The dispenser should have been pre-adjusted to deliver a designated volume of liquid, which can be read from the barrel of the dispenser. To use the dispenser, simply hold your container underneath the spout, raise the head

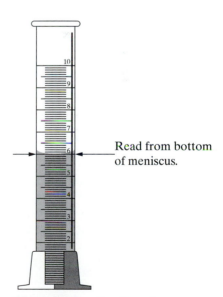

Read from bottom of meniscus.

Figure A7 Reading the volume contained in a graduated cylinder—in this case, 6.0 mL

of the dispenser as far as it will go, and release it so that it drops by gravity. Press down gently on the head to make sure that the dispenser has delivered the full volume of liquid, then touch the tip of the spout to an inside wall of your container to remove the last drop of liquid. If the liquid is the limiting reactant for a preparation, you should then weigh it accurately as described in OP-4.

Pipets

Suction is required to draw liquid into a pipet, but you should never pipet liquids by mouth because of the danger of ingesting toxic or corrosive liquids. A pipet pump is a simple and convenient suction device for filling measuring (Mohr) pipets. Other pipet fillers such as large rubber bulbs can be used for both measuring and volumetric pipets. To use a measuring pipet with a pipet pump of the type shown in Figure A8, first insert the wide (untapered) end of the pipet firmly into the opening at the bottom of the pump. Place the tip of the pipet in the liquid and rotate the thumb wheel back toward you with your thumb until the liquid meniscus has risen above the zero graduation mark. Slowly rotate the thumb wheel away from you until the meniscus drops to the zero mark. Measure the desired volume of liquid into a clean container by placing the pipet tip over the container and rotating the thumb wheel away from you until the meniscus drops to the graduation mark corresponding to the desired volume. If the pipet is one designated for use with a particular reagent bottle, the excess liquid can be returned to the bottle; otherwise it should be placed in another container or disposed of as directed by your instructor.

To use a volumetric pipet, obtain a bulb-type pipet filler that allows the liquid to drain by gravity (you can also use a pipet pump if it has a fast-release lever or if you remove it to drain the pipet). Use the bulb to fill the volumetric pipet to its calibration mark, then hold the pipet tip over a receiving container and let the liquid drain out until only a small amount of

*A convenient "homemade" pipetting bulb is described in J. Chem. Educ. **1974**, 51, 467.*

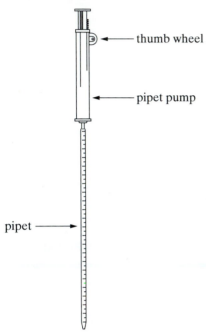

Figure A8 Pipet pump and measuring pipet

Figure A9 Reading the volume delivered from a measuring pipet—in this case, 0.30 mL

liquid is left in the tip. Do not force this liquid into the receiving container, since it is accounted for when the pipet is calibrated.

Automatic pipets (also called pipetters) provide a quick, convenient way of delivering the same volume of liquid with a high degree of reproducibility. A variable-volume automatic pipet can be set to a specified volume within a certain range of volumes, such as 100–1000 μL (0.100–1.000 mL). The volume is displayed on a digital display or analog scale, usually in microliters (μL). To prevent contamination, liquid is drawn into a disposable tip and never inside the pipet itself. Whenever the pipet is used for a different liquid, the volume is reset (if necessary) and a new pipet tip is installed. In your organic lab, an automatic pipet will ordinarily be set by the instructor or a lab assistant and designated for a specific liquid. Do *not* try to reset the volume or use it for a different liquid without explicit permission from your instructor.

To use an automatic pipet, see that the tip is attached securely, then insert the tip into the designated liquid to a depth of no more than 1 cm. Depress the plunger to the first *detent* (stop) position, when you will feel resistance to further movement. Slowly release the plunger to fill the pipet tip. Place the tip inside the receiving container and depress the plunger to the first detent position, pause a second or so, then push the plunger down to the second detent position (as far as it will go) and touch the tip to the inner wall of the receiving container to expel the last drop of liquid.

A calibrated Pasteur pipet with a latex rubber bulb can be used for very approximate measurements of small volumes. To calibrate the pipet, measure 0.50 mL of water into a small test tube or conical centrifuge tube using a graduated pipet or other device, carefully draw all of the liquid into the Pasteur pipet so that there are no air bubbles in the tip (this takes practice!), and mark the position of the meniscus with a glass-marking pen. Repeat this operation using 1.00 mL of water, or other volumes if desired. Additional Pasteur pipets of the same type can be calibrated by aligning them with the calibrated pipet and marking them at the same locations.

Figure A10 An automatic pipet

Syringes

Syringes are used to measure and deliver small volumes of liquid, often by inserting the needle through a rubber septum. To use a syringe, fill it by placing the needle tip in the liquid and slowly pulling out the plunger until the barrel contains a little more than the required volume of liquid. If there are air bubbles in the liquid, try to get rid of them by holding the syringe vertically with the needle up and tapping the barrel with your fingernail, or by expelling the liquid and filling the syringe again, more slowly. Then hold the syringe with the needle pointed upward and slowly push in the plunger to eject the excess liquid (collect it for disposal if requested) until the end of the plunger is at the appropriate graduation mark. Wipe off the tip of the needle with a tissue, place the needle tip into the receiving vessel or through the septum, and expel the liquid by gently pushing the plunger in as far as it will go. Clean the syringe immediately after use by repeatedly drawing acetone (or another volatile solvent) into it and expelling the solvent. Dry the syringe by pumping the plunger several times to expel excess solvent, then removing the plunger to let the barrel dry. If the syringe it to be used again shortly, you can dry it by drawing air through the barrel with an aspirator or vacuum line.

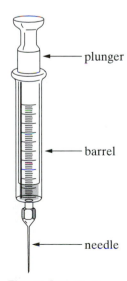

plunger

barrel

needle

Figure A11 Syringe

B. Operations for Conducting Chemical Reactions

Heating

a. Heat Sources

Many kinds of heating devices are available for such applications as heating reaction mixtures, evaporating volatile solvents, and carrying out distillations. The choice of a heat source for a particular application depends on such factors as the temperature required, the flammability of a liquid being heated, the need for simultaneous stirring, and the cost and convenience of the heating device.

Heating Mantles. A heating mantle is generally used to heat a round-bottomed flask during a reaction or distillation. A mantle can be used with a magnetic stirrer (see OP-9) and its heat output can be varied over a wide range, but its operating temperature cannot be monitored with a thermometer. Certain heating mantles, such as Thermowell ceramic flask heaters, are designed to heat round-bottomed flasks over a range of sizes, but with most mantles heating efficiency decreases and the chance of superheating increases when a small flask is heated in a large mantle. A heating mantle must be used in conjunction with a voltage regulating or time cycling ("on-off") heat control to vary the heat output.

You should never turn on an empty heating mantle or use it to heat an empty flask because that may burn out its heating element. If you spill any chemicals into the well of a heating mantle, particularly if it is hot, notify your instructor immediately.

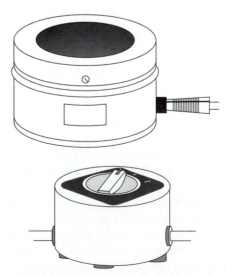

Figure B1 Heating mantle and heat control

To operate a heating mantle, first support it on a lab jack, a ring, a set of wood blocks, or another support so that it can be lowered and removed quickly if the heating rate becomes too rapid. Clamp the flask in place so that it is in direct contact with the well of the heating mantle. If you are heating a small flask in a larger mantle, filling the well with glass wool up to the flask's liquid level may help distribute the heat more evenly (this is said to be unnecessary with a Thermowell flask heater). Plug the mantle into the heat control unit—*never* directly into an electrical outlet—and adjust the heat control dial until the desired rate of heating is attained. Note that the dial controls only the rate of heating and cannot be set to a specific temperature. Because a heating mantle responds slowly to changes in the control setting, it is easy to overshoot the desired temperature by turning the control too high at the start. If this occurs, lower the mantle so that it is no longer in contact with the flask, reduce the dial setting, and allow sufficient time for the temperature to drop before raising the mantle again. Further adjustments may be needed to maintain heating at the desired rate. When you are done heating, lower the mantle, adjust the heat control dial to its lowest or "off" setting, and let the mantle cool down before you unplug it and return it to its original location.

Hot Plates. A hot plate can be used to heat most liquids in flat-bottomed containers such as Erlenmeyer flasks or beakers. It should *not* be used to heat low-boiling, flammable liquids that could splatter on the hot surface and ignite, or to heat round-bottomed flasks directly. Hot plates can also be used to heat oil baths, water baths, sand baths, and aluminum heating blocks while conducting chemical reactions and distillations. Hot plates with built-in magnetic stirrers are often used for reactions that require simultaneous stirring and heating.

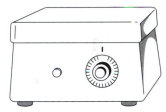

Figure B2 Hot plate

A hot plate is plugged into an electrical outlet and operated with a dial that controls the rate of heating. Heat-resistant gloves or beaker tongs should be kept handy so that you can quickly remove the container being heated when necessary. If you are using a hot plate with a heating bath or aluminum block, it is a good idea to prepare a calibration curve by measuring the equilibrium temperature at each dial setting (wait 10–15 minutes for the temperature to equilibrate at each setting) and plotting the temperature against the dial setting. Then you can adjust the hot plate for the desired operating temperature when you use the same bath or aluminum block again. Note that the volume of liquid or sand in a heating bath must be kept constant in order for the calibration to be valid.

Steam Baths. A steam bath is a safe, convenient heat source that is somewhat limited by the fact that it has only one operating temperature, 100°C. A steam bath is particularly useful for heating recrystallization mixtures, evaporating volatile solvents, and heating low-boiling liquids under reflux. It cannot be used to boil water or aqueous solutions. The condensation of steam in the vicinity of a steam bath may be a nuisance, but this can be reduced by using enough rings to bridge any gaps between a flask and the steam bath and by maintaining a slow rate of steam flow. Beyond a certain point there is no advantage to increasing the steam flow rate, since the steam temperature is constant. If the flask is placed correctly, heating is comparatively even and efficient, and the low operating temperature helps prevent the decomposition of heat sensitive substances.

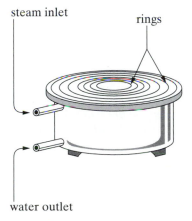

steam inlet

rings

water outlet

Figure B3 Steam bath

To operate a steam bath, first obtain two lengths of rubber tubing, attach one to the steam bath's *water outlet* tube and the other to the steam valve over the sink, and place the open ends of both rubber tubes in the sink. (If your steam bath has no water outlet tube, you will have to turn off the steam periodically to empty it of water.) Remove inner rings from the steam bath, leaving enough rings to safely support the container you wish to heat (unless it is supported by a clamp), but providing a large enough opening so that the steam will contact most of the container's base. If the container is a round-bottomed flask that is clamped to a ring stand, remove enough rings so that the flask can be lowered through the rings to about its midpoint, leaving the smallest possible gap between the inner ring and the flask.

Take Care! Avoid contact with the steam, which can cause serious thermal burns.

Directing the steam into the sink drain, open the steam valve fully until little or no water drips from the end of the rubber tube. Close the steam valve, connect it to the steam bath's *steam inlet* tube, then open it just enough to maintain the desired rate of heating with the container in place. You can adjust the heating rate by adding or removing rings, raising or lowering a clamped flask, and changing the steam flow rate. When you are done heating, turn off the steam valve completely and let the steam bath cool down. Then remove the rubber tubes, drain any water that remains in the steam bath, and put it and the rubber tubes back where you found them. (Don't leave rubber tubing in the sink!)

Hot Water Baths. Hot water baths are useful for evaporating volatile solvents (see OP-14) and in other applications that require gentle heating. Although special metal water baths that resemble steam baths are available, a beaker can be used for most applications. If accurate temperature control is not necessary, you can fill the beaker with preheated water from a hot water tap or another source. Adjust the temperature by adding hot or cold water; as the bath cools, siphon off some of the bath water and replace it by fresh hot water. For better temperature control, heat the water bath on a hot plate (or on a burner if there is no possibility of a fire) and monitor its temperature using a thermometer clamped with its bulb inside the beaker [OP-8]. The thermometer bulb should not touch the side or bottom of the beaker. For more uniform heating, the water bath can be stirred using a magnetic stirrer or hot plate-stirrer.

Oil Baths. An oil bath can supply uniform heating and precise temperature control, reducing the likelihood of decomposition and side reactions caused by local overheating, and its operating temperature can be measured easily with a thermometer. But oil baths are messy to work with and difficult to clean, and they can cause dangerous fires or severe burns. You should never heat an oil bath above the flash point of the heating oil, since above this temperature the vapors of the oil can be ignited by a spark or burner flame. Hot oil can cause severe injury if accidentally spilled on the skin—the oil, which is difficult to remove and slow to cool, remains in contact with the skin long enough to produce deep burns. Water should be kept away from hot oil baths since it causes dangerous splattering. Oil that contains water should not be used until the water is removed, and a bath liquid that is dark and contains gummy residues should be replaced.

Most oil fires can be extinguished by dry-chemical fire extinguishers or powdered sodium bicarbonate.

Mineral oil is probably the most commonly used oil bath liquid, but it presents a potential fire hazard and is hard to clean up. High molecular-weight polyethylene glycols such as Carbowax 600 are water soluble, which makes cleanup much easier, and can be used at comparatively high temperatures without appreciable decomposition. Some silicone oils can be used at even higher temperatures, but they are considerably more expensive. Flash points and other information about selected oil bath liquids are given here. Note that flash points may vary with composition; check the label or ask your instructor if you are not sure about the flash point of a specific oil bath liquid.

An oil bath can be heated by a coil of resistance wire, a power resistor, a Calrod heating element, or some other device that can be safely immersed in the bath liquid. External heat sources such as hot plates are also used, but they may cause a fire if the oil spills onto the hot surface. The output of a heating element is controlled by a variable transformer and the temperature of the bath is measured with a thermometer suspended in the liquid. A large porcelain casserole makes a convenient bath container, since it is less easily broken than a glass container and has a handle for convenient placement and removal.

To use an oil bath, first place it on a lab jack, a set of wood blocks, or some other support that will allow it to be lowered quickly when necessary. Do not set it on a ring support because of the danger of spilling hot oil when the ring is raised or lowered. See that the apparatus containing the reaction flask or boiling flask is clamped securely to a ring stand, then loosen (at the ring stand) the

Oil bath liquids:

Mineral oil
 flash pt. ~190°C, but varies with composition
 Potential fire hazard

Glycerol
 flash pt. 160°C
 Water soluble, viscous

Dibutyl phthalate
 flash pt. 171°C
 Viscous at low temperatures

Triethylene glycol
 flash pt. 165°C
 Water soluble

Polyethylene glycols (Carbowaxes)
 flash pt. varies with molecular weight
 Water soluble, some are solids at room temperature

Silicone oil, high temperature
 flash pt. 315°C, usable range −40°C–230°C
 Expensive. Decomposition products are very toxic.

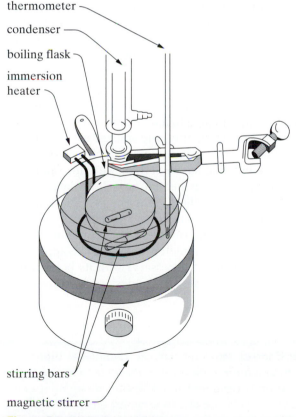

thermometer
condenser
boiling flask
immersion heater

stirring bars
magnetic stirrer

Figure B4 Oil bath assembly

Take Care! If the oil-bath liquid starts smoking, discontinue heating and use fresh oil or an oil-bath liquid with a higher flash point.

Take Care! Keep flames away from petroleum ether. Avoid contact with and inhalation of dichloromethane.

clamp holding the flask and lower it into the bath so that the liquid level inside it is 1–2 mm below the oil level. Clamp a thermometer so that its bulb is immersed in the oil but does not touch anything else in the oil bath. Drop in a stirbar, if desired, and use it to stir the bath gently; a smaller stirbar can be used to stir the flask contents (see Figure B4). Switch on the variable transformer (or the heat control dial of a hot plate) and adjust it until the desired temperature is obtained, then readjust it as needed to maintain that temperature. When you are done heating, turn off the heat and allow the oil bath to cool nearly to room temperature before you remove it. Transfer the oil to an appropriate container for reuse. Clean the bath container using a suitable solvent such as petroleum ether for mineral oil, dichloromethane for silicone oil, or water for glycerol and polyethylene glycol.

Heating Blocks

Aluminum heating blocks with holes or wells designed to accomodate small test tubes, flasks, reaction vials, and similar containers can be used for small-scale reactions and recrystallizations such as the ones used in organic qualitative analysis and in some of the Minilabs. The aluminum block is heated on a hot plate, usually with a thermometer to monitor the temperature. A mercury thermometer should be clamped to a ring stand with its bulb in the small hole drilled to accomodate it. A bimetallic thermometer with a dial face can also be used; it is inserted into a small hole provided in some heating blocks. The test tube, flask, or other apparatus containing the material to be heated is placed in a well of the appropriate size and the temperature is controlled by adjusting the hot plate dial. Unwieldy glassware setups should be clamped to a ring stand, but small test tubes and similar items can be supported adequately by wells of an appropriate size. A hot plate-stirrer can be used for simultaneous heating and stirring.

Sand Baths

Like an aluminum heating block, a sand bath is most useful for small-scale operations. Sand can be heated to a high temperature, but it heats and cools much more slowly than aluminum, and spillage can be a problem. The sand bath consists of a flat-bottomed container, such as a cylindrical crystallization dish or a Petri dish, that has been filled with fine sand to a depth of 1 cm or more. Such a sand bath is heated on a hot plate or hot plate-stirrer. Alternatively, the ceramic well of a Thermowell flask heater can be partly filled with sand and used as a sand bath. The temperature of a sand bath can be measured by clamping a thermometer to a ring stand and lowering it into the sand so that its bulb is completely covered. The apparatus being heated is supported (using clamps, if necessary) with its bottom immersed in the sand, and the sand bath is heated to the desired temperature.

Burners

Safety Notes

> **Always check to see that there are no flammable liquids in the vicinity before you light a burner. Never use a burner to heat a flammable liquid in an open container. Never leave a burner flame unattended; it may go out and cause an explosion due to escaping gas.**

Bunsen-type burners are simple and convenient to operate, but they present a serious fire risk in an organic chemistry lab where highly flammable solvents are often used. For that reason burners should be used mainly for operations that cannot be conducted with flameless heat sources, such as bending and fire polishing glass tubing.

To operate a typical burner with a needle valve at the base, first see that the gas is turned off at the main valve. Then close the needle valve by rotating the knurled wheel clockwise until you feel resistance (don't close it tightly). Open the needle valve a turn or two, then open the main valve and—without delay—ignite the burner with a burner lighter. If it doesn't light, rotate the barrel of the burner clockwise (or close the sleeve-type regulator) and try again. When the burner is lit, adjust the needle valve and rotate the barrel or sleeve to obtain a flame of the desired size and intensity. Rotating the barrel counterclockwise or opening the sleeve regulator to introduce more air produces a hotter, bluer flame. If you are using a burner to heat a nonflammable liquid in a beaker or other container, place the container on a ring support using a ceramic-centered wire gauze to spread out the flame and prevent superheating. The ring support should be positioned so that the bottom of the wire gauze is at the top of the inner blue cone of the flame, where it is hottest.

Other Heat Sources

Heating devices such as infrared heat lamps and electric forced-air heaters (heat guns) can be used in some heating applications. A heat lamp plugged into a variable transformer provides a safe and convenient way to heat comparatively low-boiling liquids. The boiling flask is usually fitted with an aluminum foil heat shield to concentrate the heat on the reaction mixture.

Figure B5 Heat lamp

b. Smooth Boiling Devices

When a liquid is heated at its boiling point, it may erupt violently as large bubbles of superheated vapor are discharged from the liquid; this phenomenon is called *bumping*. A porous object such as a boiling chip can prevent bumping by emitting a steady stream of small bubbles that breaks up the large vapor bubbles. Boiling chips (also called boiling stones) are made from pieces of alumina, marble, glass, teflon (PTFE), and other materials. Boiling chips made of teflon or another chemically resistant material should be used for heating strongly acidic or alkaline mixtures, since ordinary boiling chips may break down in such mixtures. Microporous carbon chips can be used when liquids are distilled under reduced pressure. Wooden applicator sticks can be broken in two and the broken ends used to promote smooth boiling in nonreactive solvents; they should not be used in reaction mixtures because of the possibility of contamination.

Unless you are instructed differently, you should always add boiling chips to an unstirred reaction mixture that will be heated under reflux (see section **c**), or to a liquid that will be distilled. Add the boiling chips before heating begins because the liquid may froth violently and boil over if you add them when it is hot. If you let a boiling liquid cool below its boiling point, add one or more fresh boiling chips if you intend to reheat it. Liquid is drawn into the pores of a boiling chip when boiling stops, reducing its effectiveness. Stirring causes turbulence that breaks up large bubbles, so boiling chips are not needed when a liquid is stirred constantly at the boiling point.

c. Heating under Reflux

Most organic reactions are carried out by heating the reaction mixture to increase the reaction rate. The temperature of a reaction mixture can be controlled in several ways, the simplest and most convenient being to use a reaction solvent that has a boiling point within the desired temperature range for the reaction. Sometimes a liquid reactant may itself be used as the solvent. The reaction is conducted at the boiling point of the solvent, using a *reflux condenser* to return the solvent vapors to the reaction vessel and prevent their escape. This process of boiling a reaction mixture and condensing the solvent vapors back into the reaction vessel is known as *heating under reflux* (or more informally as "refluxing"), where the word reflux refers to the "flowing back" of the solvent. Most reflux condensers have a water-cooled jacket to cool and condense the solvent vapors, but a jacketless *air condenser* may be used with a high-boiling solvent. Do not mistake a jacketed distilling column for a reflux condenser; the condenser has a smaller diameter.

General Directions for Heating under Reflux

Equipment and supplies:

- heat source
- reflux condenser
- two lengths of rubber tubing, each about 1 m long
- round-bottomed flask
- boiling chips

Safety Notes

> **Never heat the reaction flask before the condenser water is turned on; solvent vapors may escape and cause a fire or health hazard.**

Position the heat source at the proper location on a ring stand so that it can be lowered or removed quickly if the flask should break or the reaction become too vigorous. Select a round-bottomed flask of a size such that the reactants fill it about half full or less. For example, if the total volume of the reaction mixture will be 22 mL, use a 50 mL round-bottomed flask. Clamp this *reaction flask* securely to the ring stand at the proper location in relation to the heat source. Add the solvent and reactants to the reaction flask. Solids should be added through a powder funnel or a makeshift funnel constructed from a square of weighing paper, and liquids should be added through a stemmed funnel. Add a few boiling chips or a magnetic stirbar [see OP-9] and mix the reactants by swirling or stirring. Insert a reflux condenser into the flask, making sure the joint is tight. Put a clamp near the top of the condenser to keep the apparatus from toppling over if it is jarred, but don't tighten its jaws completely.

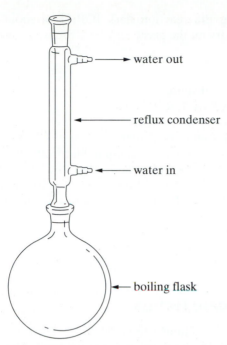

water out

reflux condenser

water in

boiling flask

Figure B6 Apparatus for heating under reflux

Connect the water inlet (the lower connector) on the condenser jacket to a cold-water tap with a length of rubber tubing, and run another length of tubing from the water outlet (the upper connector) to a sink, making sure that it is long enough to prevent splashing when the water is turned on. If the rubber tubing slips off when pulled with moderate force, replace it by tubing of smaller diameter, or secure it with wire or a tubing clamp. Turn on the water carefully so that the condenser jacket slowly fills with water from the bottom up, and adjust the water pressure so that a narrow stream flows from the outlet. The flow rate should be just great enough to (1) maintain a continuous flow of water in spite of pressure changes in the water line and (2) keep the condenser at the temperature of the tap water during the reaction. Excessively high water pressure may force the tubing off the condenser and spray water on you and your neighbors.

Turn on the heat source and adjust it to keep the solvent boiling gently; measure the reaction time from the time that boiling begins. A continuous stream of bubbles should emerge from the liquid, but bumping and excessive foaming should be avoided. Reflux has begun when liquid begins to drip into the flask from the condenser. The vapors passing into the condenser will then form a *reflux ring* of condensate that should be clearly visible. Below this point, solvent will be seen flowing back into the flask; above it, the condenser should be dry. If the reflux ring rises more than halfway up the condenser, reduce the heating rate or increase the water flow rate to prevent the escape of solvent vapors.

At the end of the reaction period, turn off the heat source and remove it from contact with the flask. Let the apparatus cool; then turn off the condenser water, and pour the reactants into a container suitable for the next operation. Clean the reaction flask [OP-1] as soon as possible so that residues do not dry on the glass.

Summary

1   Position heat source.
2   Clamp flask over heat source.
3   Add solvent and reactants.
4   Add boiling chips or stirbar.
5   Insert reflux condenser, clamp in place, attach tubing.
6   Turn on condenser water and adjust flow rate.
7   Start heating, adjust heat until reaction mixture boils gently.
8   Check position of reflux ring; readjust water flow or heating rate as necessary.
9   Turn off and remove heat source, let flask cool, transfer reaction mixture.
10   Disassemble and clean apparatus.

d. Small-Scale Reflux

Small amounts of reactants can be heated under reflux using a cold-finger condenser inserted into a test tube or small flask. As water passes through the condenser, it cools the surrounding area enough to condense the rising vapors. To prevent pressure buildup in the container, a stopper with a groove in one side (or a two-hole stopper) must be used. The reflux ring that appears on the sides of the container should be kept well below the top of the test tube or flask so that vapors do not escape. A small-scale reflux apparatus is used in essentially the same way as the one in Figure B6, except that cooling water goes into the upper connector of a cold-finger condenser and comes out the lower one.

Microcondensers are available that resemble an ordinary reflux condenser and are used with reaction vials or very small (~5–10 mL) reaction flasks.

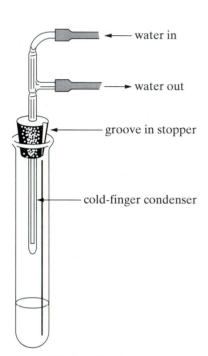

← water in

← water out

groove in stopper

cold-finger condenser

Figure B7 Small-scale apparatus for heating under reflux

OPERATION 7

Cooling

Some reactions proceed too violently to be conducted safely at room temperature, or involve reactants or products that decompose at room temperature. In such cases the reaction mixture is cooled with some kind of cold bath, which can be anything from a beaker filled with cold water to an electrically refrigerated cold bath. Cold baths are also used to increase the yield of crystals from a reaction mixture or recrystallization mixture, and for other purposes.

The cold bath container can be a beaker of suitable size, a crystallization dish, a large styrofoam cup, or a metal water bath. A well-insulated container such as a Dewar flask is preferred if the cold bath will be used for a long time or at a very low temperature. The useful life of a cold bath can be extended by wrapping the bath container with glass wool and placing it inside a larger container.

A number of cooling fluids and mixtures are used for cold baths. A mixture of ice and water can be used for cooling in the 0°–5°C range. The ice should be finely divided, and enough water should be present to just cover the ice, since ice alone is not an efficient heat-transfer medium. An ice-salt bath consisting of three parts of finely crushed ice or snow with one part sodium chloride can attain temperatures down to −20°C, and mixtures of $CaCl_2 \cdot H_2O$ containing up to 1.4 g of the calcium salt per 1 g of ice or snow can provide temperatures down to −55°C. In practice, these minimum values may be difficult to attain; the actual temperature of an ice-salt bath depends on the fineness of the ice and salt, the rate of stirring, and the insulating ability of the container. Temperatures down to −75°C can be attained by mixing small chunks of dry ice with acetone, ethanol, or another suitable solvent, using a Dewar flask. Liquid nitrogen, with a boiling point of −196°C, can be used when even lower temperatures are required.

Take Care! Never handle dry ice with bare hands.

Temperatures below −40°C cannot be measured using a mercury thermometer because mercury freezes at that temperature.

General Directions for Cooling

Equipment and supplies:

- cold bath container
- cooling fluid (or mixture)
- thermometer

Obtain a suitable cold bath container and fill it with the cooling fluid to a level depending on the size of the container to be immersed in the bath; when the container is immersed, the cooling fluid should fill the cold bath container about three-fourths full or more without overflowing. Clamp a thermometer so that the bulb is immersed in the cooling fluid but not touching the sides or bottom of the container [OP-8]. (For very low temperatures, a thermocouple is required.) Immerse the flask (or other container) to be cooled in the cooling bath so that the liquid level in the flask is below the cooling fluid level. Keep the contents of the cooling bath mixed by occasional stirring or swirling. Replace any ice that melts (or dry ice that sublimes) as needed to keep the temperature in the desired range. Additional salt may have to be added with the ice.

Temperature Monitoring

OPERATION 8

In the organic chemistry lab, thermometers are used to monitor the temperatures of heating baths, cooling baths, reaction mixtures, distillations, and for many other purposes. Such thermometers should have a range of at least −10°C to 260°C, and a wider range is desirable for some purposes. Most wide-range thermometers contain mercury, which is toxic and represents a safety hazard if a thermometer is broken, but wide-range thermometers containing nonhazardous, biodegradable liquids are also available.

You can monitor the temperature of a liquid or solution in a flask or other container using a thermometer clamped with its bulb entirely immersed in the liquid. If possible, the thermometer should be immersed down to an inscribed immersion mark, which is usually 76 mm from the bottom of the bulb. It should be held in place by a three-fingered clamp or a special thermometer clamp, or inserted into a stopper that is held by a utility clamp. The thermometer should not touch the side or bottom of the container or any apparatus that is placed inside the container

To monitor the temperature of a reaction mixture that will be stirred with a magnetic stirrer [OP-9], clamp a thermometer in the mixture so that it does not contact the stirbar (a large stirbar could break the thermometer bulb). If a magnetic stirrer is not available, you may have to hold the thermometer inside the reaction flask (usually an Erlenmeyer flask) as it is being swirled or shaken. Do this by holding the neck of the flask and nesting the thermometer stem in the "vee" between your thumb and index finger, so that the bulb of the thermometer is held securely inside the flask and continuously immersed in the liquid. With a little practice, you should be able to mix the contents of the flask quite vigorously without damage to the thermometer. If continuous mixing is not necessary, you can insert the thermometer each time you stop shaking or swirling the reaction flask, remove it as you record the temperature, and then resume mixing. Be sure the thermometer reading has come to rest before you record it. You should never use the thermometer itself for stirring because the bulb is fragile and breaks easily.

OPERATION 9

Mixing

Reaction mixtures are frequently stirred, shaken, or agitated in some other way to promote efficient heat transfer, prevent bumping, increase contact between the components of a heterogeneous mixture, or mix in a reactant as it is being added. If the reaction is being carried out in an Erlenmeyer flask, mixing can be accomplished by manual shaking and swirling or by using a stirring rod. A motion combining shaking with swirling is more effective than swirling alone. If the apparatus is not too unwieldy and the reaction time is comparatively short, ground-glass assemblies can sometimes be manually shaken for adequate mixing. This is most easily done by clamping the assembly *securely* to the ring stand and carefully sliding the base of the ring stand back and forth. But when more efficient and convenient mixing is required, particularly over a long period of time, it is necessary to use some kind of magnetic or mechanical stirring device.

Magnetic Stirring. A magnetic stirrer consists of an enclosed unit containing a motor attached to a bar magnet, which is underneath a metal or ceramic platform. As the bar magnet rotates, it in turn rotates a teflon coated stirring bar—a *stirbar*—inside a container placed on or above the platform. The rate of stirring is controlled by a dial on the stirrer. Since no moving parts extend outside of the stirrer, a reaction assembly that is to be stirred magnetically can be completely enclosed if necessary.

Magnetic stirrers can be used in conjunction with most heating mantles, oil baths, steam baths, and other heat sources that are constructed of nonferrous metals. A magnetic stirrer works well with a heating bath, since it can be used to stir the bath liquid and reaction mixture simultaneously (see Figure B4). When magnetic stirring is used during a reaction conducted under reflux, the heat source is set directly on the stirring unit and a stirbar is placed

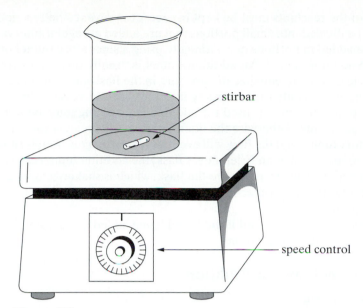

Figure B8 Magnetic stirrer

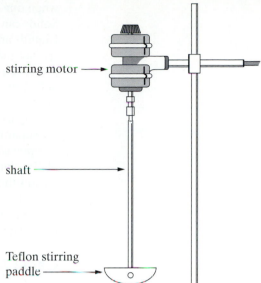

stirring motor

shaft

Teflon stirring
paddle

Figure B9 Mechanical stirrer

stirbar

speed control

in the reaction flask in place of boiling chips (the stirring action prevents bumping). The reaction flask must be positioned close enough to the bottom of the bath to allow sufficient transfer of magnetic torque from motor to stirring bar, and the stirbar in a heating bath (if one is being used) should be larger than one in the reaction mixture. The stirring motor should be started and cooling water for the reflux condenser turned on before heating is begun.

A hot plate-stirrer, which combines a hot plate with an integral magnetic stirrer, can also be used to simultaneously heat and stir a mixture. Flat-bottomed containers such as Erlenmeyer flasks can be heated and stirred directly on the hot plate, but round-bottomed flasks must be immersed in a heating bath that is heated by the hot plate.

Mechanical Stirring. A mechanical stirrer consists of a stirring motor connected to a paddle or agitator by means of a shaft that extends through the neck of the reaction vessel. A glass sleeve or bearing is used to align the shaft, which is ordinarily made of glass to reduce the likelihood of contamination. Mechanical stirrers exert more torque than magnetic stirrers and are preferred when viscous liquids or slurries must be stirred. A variety of stirring paddles made of Teflon, glass, and chemically resistant wire are available.

A slurry is a thick suspension of solid in a liquid.

Addition of Reactants

a. Standard-Scale Addition

In many organic preparations, the reactants are not all combined at the start of the reaction. Instead, one or more of them is added during the course of the reaction. This is necessary when the reaction is strongly exothermic or when one of the reactants must be kept in excess to prevent side reactions. Solids can be divided into small portions that are added at regular intervals. Liquids are added in portions or continually using a separatory funnel or a specialized addition funnel. An addition funnel is usually provided with a pressure-release tube to equalize the pressure in the flask and addition funnel, allowing its contents to flow freely into the reaction vessel. When an ordinary separatory funnel is used for addition, an opening must be left at the top for air to enter; otherwise the liquid outflow will create a vacuum in the separatory funnel and the flow will eventually stop. If a reaction is run in an open container such as an Erlenmeyer flask, the addition funnel can simply be clamped to a ring stand above the flask, which is shaken and swirled during the addition. For a reaction conducted under reflux, the following procedure should be used.

The stopper can be omitted if the liquid being added is quite involatile (for example, an aqueous solution).

General Directions for Addition

Equipment and supplies:

- boiling flask
- separatory/addition funnel
- stopper
- Claisen connecting tube
- reflux condenser

Assemble the apparatus shown in Figure B10, placing the addition funnel on the straight arm of the Claisen connecting tube so that it is directly over the reaction flask. Make sure the stopcock is closed, then place the liquid to be added in the addition funnel and the other reactants in the reaction flask. Unless the liquid in the addition funnel must be protected from atmospheric moisture, place a strip of filter paper between the stopper and the ground-glass joint of the addition funnel. If the liquid is moisture sensitive, insert a drying tube filled with drying agent [OP-22] in the addition funnel. Add the liquid either in portions or continuously, as specified by the directions in the experiment. For portionwise addition, add small portions of the liquid at regular intervals by opening the stopcock momentarily, with shaking or magnetic stirring to mix the reactants. For continuous addition, open the stopcock just far enough so that the liquid drips or drizzles slowly into the reaction flask, and adjust the stopcock to provide the desired rate of addition. Continuous addition is usually carried out dropwise (drop by drop), with magnetic stirring or periodic shaking to keep the reactants mixed.

The apparatus can be modified to provide for temperature monitoring or mechanical stirring if a three-necked flask is available. One neck of the flask is used for the addition funnel, one for the reflux condenser, and the third for another function.

b. Small-Scale Addition

Small amounts of reactants can be added to a reaction mixture using a Pasteur pipet, measuring pipet, or volumetric pipet. An addition pipet suitable for the addition of small quantities of liquid is illustrated in Figure B11. A measuring pipet can be used in place of the volumetric pipet shown. The

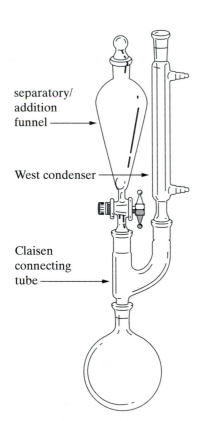

separatory/ addition funnel ——→

West condenser ——→

Claisen connecting tube ——→

Figure B10 Apparatus for addition and reflux

pinchcock valve is constructed by inserting a solid glass bead into a length of thin-walled rubber tubing. The assembly can be modified for addition under reflux by using a two-holed rubber stopper with a cold-finger condenser [OP-6d] in one hole. A gas trap or drying tube [OP-22] can be attached to the sidearm.

Additions on a very small scale can be carried out using a syringe, which is filled with a measured amount of the liquid to be added, then inserted through a rubber septum so that its needle is directly above the reaction mixture. During the reaction the liquid is added in small portions by depressing the plunger of the syringe.

General Directions for Small-Scale Addition

Assemble the addition pipet as shown in Figure B11. Insert the wide end of a medicine dropper tip into the pinchcock valve, squeeze a large rubber bulb to drive out the air, and place it over the narrow end of the medicine dropper tip. Insert the tip of the pipet into the liquid, then fill the pipet slowly to the calibration mark by "pinching" the valve at the glass bead. Remove the rubber bulb and place the addition pipet in a sidearm test tube or other suitable reaction vessel. (An ordinary test tube can be used if the stopper is notched for pressure release.) Add the liquid at the desired rate by pinching the valve while shaking or stirring the reactants.

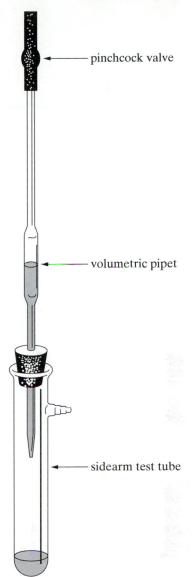

pinchcock valve

volumetric pipet

sidearm test tube

Figure B11 Addition pipet

C. Separation Operations

Gravity Filtration

Filtration is used for two main purposes in organic chemistry: (1) to remove solid impurities from a liquid or solution, and (2) to separate an organic solid from a reaction mixture or a crystallization solvent. In most instances, *gravity filtration* is preferred for the first operation and *vacuum filtration* for the second. In a gravity filtration, the liquid component of a liquid-solid mixture drains through a filter paper by gravity alone, leaving the solid on the paper. The filtered liquid, called the *filtrate*, is collected in an Erlenmeyer flask or other container. Gravity filtration is often used to remove drying agents from dried organic liquids or solutions and solid impurities from hot recrystallization solutions.

a. Standard-Scale Gravity Filtration

Gravity filtration of organic liquids can be carried out using a funnel with a short, wide stem (such as a powder funnel) and a relatively fast, fluted filter paper. Ordinary filter paper can be fluted (folded) as shown in Figure C1, but commercial fluted filter papers are available from chemical supply houses. Glass wool is sometimes used for very fast filtration of coarse solids. A thin layer of glass wool is placed over the outlet of a short-stemmed funnel, and the mixture to be filtered is poured directly onto the glass wool. Because fine particles will pass through, this method is most often used for prefiltration of mixtures that will be refiltered later.

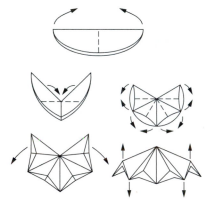

Figure C1 Making a fluted filter paper

General Directions for Gravity Filtration

Equipment and supplies:

- powder funnel
- bent wire or paper clip
- fluted filter paper
- Erlenmeyer flask or other collecting container

If you are filtering the solid-liquid mixture into a narrow-necked container such as an Erlenmeyer flask, support the funnel on the neck of the flask, with a bent wire or a paper clip between them to provide space for pressure equalization (see Figure C2). Otherwise, support the funnel in a ring or funnel support positioned directly over the collecting container. Open the fluted filter paper and insert it snugly into the funnel cone, trying to avoid flattening out any of its folds. Add the mixture fast enough to keep the filter paper two-thirds full or more throughout the filtration, but don't allow liquid to rise above the top of the filter paper. If the mixture contains a considerable amount of finely divided solid, let the solid settle and *decant* (pour out) the liquid carefully onto the filter paper so that most of the solid remains behind until the end of the filtration. Otherwise, the pores of the filter paper may become clogged and slow down the filtration. If you are filtering a solution,

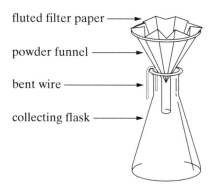

fluted filter paper —→

powder funnel —→

bent wire —→

collecting flask —→

Figure C2 Apparatus for gravity filtration

slurry (a thick suspension), which is poured into the filter under vacuum until a bed about 2–3 mm thick has been deposited. The solvent is then removed from the filter flask before continuing with the filtration. This technique cannot be used when the solid is to be saved, since it would then be contaminated with the filtering aid.

Unless otherwise instructed, you should always *wash* the solid on the filter paper with an appropriate solvent, usually the same as the one from which it was filtered. To reduce losses, the wash solvent should be cooled in ice water. For example, if the solid was filtered from an aqueous solution, use cold distilled water as the wash solvent. If it was filtered from a mixture of solvents, as in a mixed solvent recrystallization [OP-23c], you can use the component of that mixture in which the solid is least soluble. Sometimes a solid is washed a second time with a lower-boiling solvent to help it dry faster. The second wash solvent must be miscible with the first solvent, and the solid should not be appreciably soluble in it. Thus a solid filtered from an aqueous solution might be washed with water and then with methanol, and one filtered from toluene might be washed with toluene followed by petroleum ether.

General Directions for Vacuum Filtration

Equipment and supplies:

- Buchner funnel
- water trap (optional)
- thick-walled rubber tubing
- flat-bladed spatula
- filter flask
- filter paper
- flat-bottomed stirring rod
- wash solvent

Always use a filter trap when you are using a water aspirator if the filtrate is to be saved; otherwise the trap may be omitted. Clamp the filter flask and filter trap securely to a ring stand and connect them to an aspirator or vacuum line as shown in Figure C4. One bent glass tube on the trap should extend only a short way into it; connect this tube to the filter flask so that water cannot back up into the flask without filling the trap first. Use thick-walled rubber tubing for all connections. Insert a Buchner funnel with a snug-fitting rubber stopper into the filter flask, and place a circle of filter paper inside the funnel. The diameter of the filter paper should be slightly less than that of the perforated plate, so that the paper covers all the holes but does not fit too snugly; it should not extend up the sides of the funnel.

Moisten the filter paper with a few drops of the solvent that is present in the mixture being filtered, or one that is miscible with it. Open the aspirator tap or vacuum-line valve as far as it will go; operating an aspirator at a lower flow rate increases the chance of water backup. Direct the water stream from an aspirator into a large beaker or another container to prevent splashing. Add the filtration mixture in portions, keeping the funnel nearly full throughout. If the solid is finely divided, let it settle before you decant the liquid into the funnel, and transfer the bulk of the solid near the end of the filtration. Stir or swirl the filtration mixture near the end of the filtration to get most of the solid onto the filter. Transfer any remaining solid to the filter paper with a flat-bladed spatula, using a little of the filtrate or some cold wash solvent to facilitate the transfer. Leave the vacuum on until only an occasional drop of water emerges from the stem of the funnel. If you are using a trap with a pressure-release valve, open the valve to keep water from backing up into the system when the vacuum is turned off; otherwise disconnect the rubber tubing at the vacuum source.

Turn off the vacuum and add enough previously chilled wash solvent to cover the solid. Being careful not to disturb the filter paper, stir the mixture *gently* with a spatula or a flat-bottomed stirring rod until the solid is suspended in the liquid. Turn the vacuum on to drain the wash liquid; then turn it off and repeat the process with at least one more portion of chilled wash solvent. Try to work quickly to avoid dissolving an appreciable amount of solid in the wash solvent. After the last washing, leave the vacuum on for a few minutes to air dry the solid on the filter and make it easier to handle. Run the tip of a small flat-bladed spatula around the circumference of the filter paper to dislodge the *filter cake* (the compressed solid on the filter paper), then invert the funnel carefully over a square of glazed paper, a watch glass, a weighing dish, or another suitable container to remove the filter cake and filter paper. Use your spatula to scrape any particles remaining on the funnel and filter paper into the container. The filtrate should be disposed of as specified by the experimental directions or your instructor, and the solid should be dried by one of the methods described in OP-21.

Summary

1 Assemble apparatus for vacuum filtration.
2 Position and moisten filter paper, turn on vacuum.
3 Add filtration mixture in portions.
4 Transfer remaining solid to funnel.
5 Disconnect and turn off vacuum.
6 Wash solid on filter with cold solvent.
7 Air dry solid on filter paper.
8 Transfer solid to container, remove filtrate from filter flask.
9 Disassemble and clean apparatus.

b. Small-Scale Vacuum Filtration

Small quantities of solids can be filtered by the previously described procedure using a Hirsch funnel in a sidearm test tube or small filter flask. The perforated plate of a Hirsch funnel requires very small filter paper circles. These are available commercially, but they can also be cut from ordinary filter paper using a sharp cork borer on a flat cutting surface, such as the bottom of a large cork. The diameter of the cork borer should be slightly less than the diameter of the perforated plate.

Rule of Thumb: Use 1–2 mL of wash solvent per gram of solid.

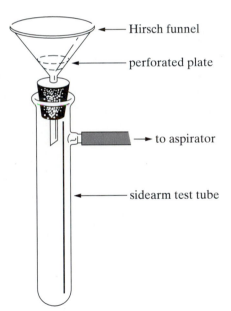

Hirsch funnel

perforated plate

to aspirator

sidearm test tube

Figure C5 Apparatus for small-scale vacuum filtration

OPERATION 13

Extraction

If you shake a bromine/water solution with a portion of dichloromethane, the red-brown color of the bromine fades from the water layer and appears in the dichloromethane layer as you shake. These color changes show that the bromine has been transferred from one solvent into another. The process of transferring a substance from a liquid or solid mixture to a solvent is called *extraction*, and the solvent is the *extraction solvent*. The extraction solvent is usually a low-boiling organic solvent that can then be evaporated [OP-14] to isolate the desired substance. In organic chemistry, extraction is often used (1) to separate an organic product from a reaction mixture or (2) to remove impurities from a solution of the product in an organic solvent. The second process is described in OP-19, Washing Liquids.

a. Liquid-Liquid Extraction

Principles and Applications

Liquid-liquid extraction is based on the principle that if a substance is soluble to some extent in two immiscible liquids, it can be transferred from one liquid to the other by shaking the two liquids together. For example, acetanilide is partly soluble in both water and ethyl ether. If a solution of acetanilide in water is shaken with a portion of ethyl ether, some of the acetanilide will be transferred to the ether layer. The ether layer, being less dense than water, separates above the water layer and can be removed and replaced with another portion of ether. When the fresh ether is shaken with the aqueous solution, more acetanilide passes into the new ether layer. This new layer can then be removed and combined with the first. By repeating this process enough times, virtually all of the acetanilide can be transferred from the water to the ether.

The ability of an extraction solvent, S_2, to remove a solute A from another solvent, S_1, depends on the partition coefficient (K) of solute A in the two solvents, as defined in Equation 1:

$$K = \frac{\text{concentration of A in } S_2}{\text{concentration of A in } S_1} \qquad \textbf{(1)}$$

In the example of acetanilide in water and ethyl ether, the partition coefficient is given by

$$K = \frac{[\text{acetanilide}]_{\text{ether}}}{[\text{acetanilide}]_{\text{water}}}$$

The larger the value of K, the more solute will be transferred to the ether layer with each extraction, and the fewer portions of ether will be required for essentially complete removal of the solute. A rough estimate of K can be obtained by using the ratio of the solubilities of the solute in the two solvents:

$$K \approx \frac{\text{solubility of A in } S_2}{\text{solubility of A in } S_1}$$

This approximate relationship can be helpful in choosing a suitable extraction solvent.

Extraction Solvents

Most extraction solvents are organic liquids that are used to extract nonpolar and moderately polar solutes from aqueous solutions, but water and aqueous solutions are sometimes used to extract polar solutes from organic solutions. For example, dilute aqueous NaOH can be used to extract carboxylic

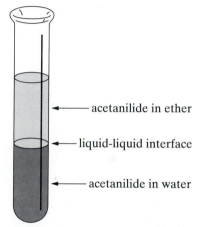

← acetanilide in ether

← liquid-liquid interface

← acetanilide in water

Figure C6 Distribution of a solute between two liquids

acids from organic solvents by first converting them to carboxylate salts, which are much more soluble in water and less soluble in organic solvents than the original carboxylic acids.

$$RCOOH + NaOH \rightarrow RCOO^-Na^+ + H_2O$$

If the carboxylic acid is sufficiently insoluble in water, it can be recovered by acidifying the aqueous extract to precipitate the acid, which is then recovered by vacuum filtration [OP-12]. Similarly, dilute aqueous HCl is used to extract basic solutes such as amines from organic solvents by first converting the amines to ammonium salts.

$$RNH_2 + HCl \rightarrow RNH_3^+ Cl^-$$

A good *organic* extraction solvent should be immiscible with water, dissolve a wide range of organic substances, and have a low boiling point so that it can be removed by evaporation after the extraction. The substance being extracted should be more soluble in the extraction solvent than in water, since otherwise too many steps will be required to extract it.

Ethyl ether is the most commonly used organic extraction solvent. It has a very low boiling point (34.5°C) and can dissolve both polar and nonpolar organic compounds. However, ethyl ether is extremely flammable and tends to form explosive peroxides on standing. Dichloromethane (methylene chloride) is more dense than water, which can simplify the extraction process, and it is not flammable. Dichloromethane has a tendency to form emulsions, which can make it difficult to separate cleanly, and it must be handled with caution because it is a suspected carcinogen. Some commonly used extraction solvents and their properties are listed in Table C1. Organic extraction solvents that have densities less than that of water (1.00 g/mL) will separate as the top layer during extraction of an aqueous solution; extraction solvents having densities greater than that of water will separate as the bottom layer.

Potential hazards should be considered in selecting and using an extraction solvent. Precautions must be taken with all organic solvents to minimize contact and inhalation of vapors. Organic solvents such as benzene,

Table C1 Properties of commonly used extraction solvents

Solvent	b.p., °C	d, g/ml	Comments
water	100	1.00	For extracting polar compounds, generally using a reactive solute such as NaOH or HCl
ethyl ether	34.5	0.71	Good general solvent; absorbs some water; very flammable
dichloromethane	40	1.34	Good general solvent; suspected carcinogen
toluene	111	0.87	For extracting aromatic and nonpolar compounds; difficult to remove
petroleum ether	~35–60	~0.64	For extracting nonpolar compounds; very flammable
hexane	69	0.66	For extracting nonpolar compounds; flammable

trichloromethane (chloroform), and tetrachloromethane (carbon tetrachloride), although they are good extraction solvents, should not be used in an undergraduate laboratory because of their toxicity and carcinogenic potential. Flames must not be allowed in the laboratory when highly flammable solvents, such as ethyl ether and petroleum ether, are in use.

Experimental Considerations

A standard-scale liquid-liquid extraction is ordinarily carried out in a *separatory funnel*, which has a stopcock at the bottom. This makes it possible to drain the lower liquid layer into a separate container, leaving the upper layer behind in the separatory funnel. Separatory funnels are very expensive and break easily. Never prop a separatory funnel on its base; support it on a ring, a funnel support, or some other stable support. If your separatory funnel has a glass stopcock, it should be lubricated by applying thin bands of stopcock grease on both sides, leaving the center (where the drain hole is located) free of grease to prevent contamination (see Figure C7). A teflon stopcock should *not* be treated with stopcock grease.

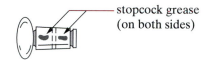

Figure C7 Lubricating a glass stockpock

The volume of extraction solvent and the number of extraction steps are sometimes specified in an experimental procedure. If they are not, it is usually sufficient to use a volume of extraction solvent about equal to the volume of liquid being extracted, divided into at least two portions. For example, 60 mL of an aqueous solution can be extracted with two 30-mL (or three 20-mL) portions of ethyl ether. Note that it is more efficient to use several small portions of extraction solvent than one large portion of the same total volume.

Rule of Thumb: Total volume of extraction solvent ≈ volume of liquid being extracted.

Under some conditions the liquid layers do not separate sharply, either because an *emulsion* forms at the interface between the two liquids or because droplets of one liquid remain in the other liquid layer. Emulsions can often be broken up by using a wooden applicator stick to stir the liquids gently at the interface or by mixing in some saturated aqueous sodium chloride solution (or enough solid NaCl to saturate the aqueous layer), then allowing the separatory funnel to stand open and undisturbed for a time. An applicator stick can be used to consolidate the liquid layers by rubbing or stirring any droplets or globs of liquid that form on the sides or bottom of the separatory funnel. An applicator stick can also be used to remove small amounts of insoluble "gunk" that sometimes form near the interface. Larger amounts of insoluble material can be removed by filtering the mixture through a loose pad of glass wool in a powder funnel. After each extraction, the extraction solvent should be transferred to a collecting flask, which should be kept stoppered to prevent evaporation and reduce any potential fire hazard.

An emulsion usually contains microscopic droplets of one liquid suspended in another.

General Directions for Extraction

Equipment and supplies:

- separatory funnel
- extraction solvent
- support for separatory funnel
- two flasks
- graduated cylinder
- ring stand

Support a separatory funnel on a ring of suitable diameter, equipped with some lengths of split rubber tubing to cushion the funnel and prevent damage.

(A special funnel support can also be used.) Close the stopcock by turning the handle to a horizontal position and add the liquid to be extracted through the top of the funnel. This liquid should be at room temperature (or below) to prevent vaporization of a volatile extraction solvent. Measure the required volume of extraction solvent using a graduated cylinder (the exact volume is not crucial), and pour it into the funnel. The total volume of both liquids should not exceed three-quarters of the funnel's capacity; if it does, obtain a larger separatory funnel or carry out the extraction in two or more steps, using a fraction of the liquid in each step.

Moisten the stopper with water and insert it firmly. Then pick up the funnel in both hands and partly invert it, with the right hand holding the stopcock (or the left hand if you are a southpaw) and the first two fingers of the left hand holding the stopper in place (see Figure C8). *Vent* the separatory funnel by slowly opening the stopcock to release any pressure buildup. Be sure that the stem of the funnel is pointed away from you and your neighbors when you are venting it. Close the stopcock, shake or swirl the funnel gently for a few seconds, then vent it again (be sure the funnel is inverted). Shake the funnel more vigorously (but not too vigorously if the solvent tends to form emulsions), with occasional venting, for 2–3 minutes. A combined shaking and swirling motion is more efficient than swirling alone. Venting should not be necessary after there is no longer an audible hiss of escaping vapors when the stopcock is opened.

Replace the funnel on its support, remove the stopper, and allow the funnel to stand until there is a sharp dividing line between the two layers.

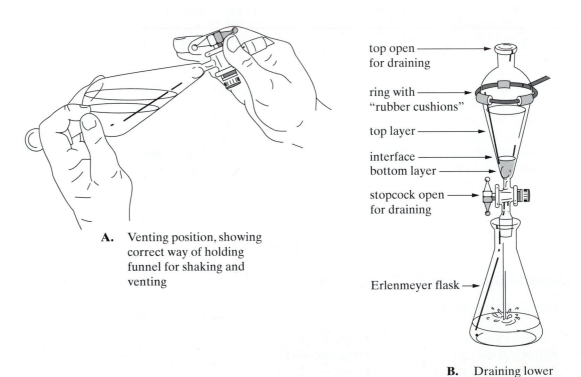

top open for draining

ring with "rubber cushions"

top layer

interface
bottom layer

stopcock open for draining

Erlenmeyer flask

A.   Venting position, showing correct way of holding funnel for shaking and venting

B.   Draining lower layer from separatory funnel

Figure C8 Extraction techniques

Drain the bottom layer into an Erlenmeyer flask (flask A) by opening the stopcock fully; partly close it to slow the drainage rate as the interface approaches the bottom of the funnel. When the interface just reaches the outlet, quickly close the stopcock to separate the layers cleanly. Follow procedure **1** if the extraction solvent is *more* dense than the liquid being extracted, and procedure **2** if it is *less* dense than the liquid being extracted.

   1 Extract the liquid that remains in the separatory funnel (the original top layer) with a fresh portion of the same extraction solvent, then drain the bottom layer as before and combine it with the first extract in flask A. Repeat the process, as necessary, with fresh extraction solvent. After the last extraction is finished and the bottom layer has been drained, pour the top layer out of the *top* of the separatory funnel into a separate container (flask B) and dispose of it as directed.

   2 Pour the liquid that remains in the separatory funnel (the original top layer) out the *top* of the separatory funnel into a separate container (flask B). Then return the liquid in flask A to the separatory funnel and extract it with a fresh portion of extraction solvent. Again drain the bottom layer into flask A and pour the top layer into flask B. Repeat the process, as necessary, with fresh extraction solvent, combining the extracts in flask B. After the last extraction, dispose of the contents of flask A as directed.

If there is any doubt about which layer should be discarded, save both until the correct one is identified. Mixing a drop or two of water with a little of each layer will establish which is the aqueous layer.

Summary

1. Add liquid to be extracted to separatory funnel.
2. Add extraction solvent, stopper funnel, invert, and vent.
3. Shake and swirl funnel, with venting, to extract solute into extraction solvent.
4. Remove stopper, drain lower layer into flask A.
   IF extraction solvent is more dense than water, GO TO 5.
   IF extraction solvent is less dense than water, GO TO 6.
5. Stopper flask A.
   IF another extraction step is needed, GO TO 2.
   IF extraction is complete, empty and clean separatory funnel; STOP.
6. Pour upper layer into flask B and stopper it.
   IF another extraction step is needed, return contents of flask A to separatory funnel, GO TO 2.
   IF extraction is complete, clean separatory funnel; STOP.

b. Small-Scale Extraction

A small-scale extraction can be performed when the volume of the solution to be extracted is less than 10 mL. To carry out a small-scale extraction, shake the aqueous solution with the extraction solvent in a capped conical centrifuge tube (or other suitable container). If necessary, spin the centrifuge tube in a centrifuge to facilitate layer separation. Insert a Pasteur pipet so that its tip is at the bottom of the centrifuge tube. Withdraw the *bottom* layer, taking care not to mix the layers, and transfer it to a test tube or other suitable container (A).

If the extraction solvent is *more* dense than the liquid being extracted, add fresh extraction solvent to the liquid in the centrifuge tube and extract it as before, transferring the extract (the bottom layer) to A. Repeat the process as necessary, combining all the extracts in A.

If the extraction solvent is *less* dense than the liquid being extracted, transfer the contents of the centrifuge tube to another container (B), return the contents of A to the centrifuge tube, and extract as before using fresh extraction solvent. After each extraction transfer the bottom layer to A, combine the top layer with the extracts in B, and return the contents of A to the centrifuge tube for the next extraction, as necessary.

c. Salting Out

Adding an inorganic salt (such as sodium chloride or potassium carbonate) to an aqueous solution containing an organic solute usually reduces the solubility of the organic compound in the water, and thus promotes its separation. This *salting out* technique is often used to help separate an organic liquid from its aqueous solution or to increase the amount of an organic solute transferred from the aqueous to the organic layer during an extraction. Usually enough of the salt is added to saturate the aqueous solution, which is stirred or shaken to dissolve the salt. The mixture is filtered by gravity if any undissolved salt remains, then transferred to a separatory funnel for separation or extraction.

d. Liquid-Solid Extraction

Some substances, particularly natural products, contain components that can be separated from a solid residue by extraction with a liquid solvent. A simple technique for liquid-solid extraction is to mix the solid intimately with an appropriate solvent using a flat-bottomed stirring rod in a beaker (or a mortar and pestle) and then remove the solid by gravity filtration [OP-11] or vacuum filtration [OP-12], repeating the process as many times as necessary. The mixing should be as thorough as possible; use the flat end of the stirring rod to press down or crush the solid in contact with the solvent, in order to extract more of the desired component. After each filtration, return the solid to the beaker for the next extraction. Combine the liquid extracts in a single collecting flask.

OPERATION 14

Evaporation

Evaporation is the conversion of a liquid to vapor at or below the boiling point of the liquid. Evaporation can be used to remove a volatile solvent, such as ethyl ether or dichloromethane, from a comparatively involatile liquid or solid. Complete solvent removal is used to isolate an organic solute after such operations as extraction [OP-13] or column chromatography [OP-16]. Partial solvent removal, or *concentration*, can be used to bring a recrystallization solution to its saturation point [see OP-23].

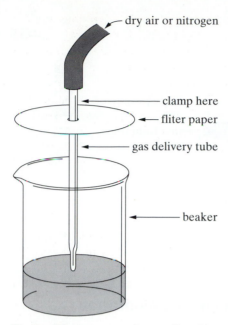

dry air or nitrogen

clamp here

fliter paper

gas delivery tube

beaker

Figure C9 Apparatus for small-scale evaporation

Experimental Considerations

Small quantities of solvent can be evaporated by leaving the solution in an evaporating dish under a fume hood, or by passing a slow stream of dry air or nitrogen over it. As shown in Figure C9, the solution can be protected from airborne particles by supporting a filter paper some distance above it. Larger quantities of solvents, including high-boiling solvents, can be removed by simple distillation [OP-25] or vacuum distillation [OP-26]. This procedure is most convenient when the residue is to be distilled immediately after the solvent is removed. Commercial *flash evaporators* are used to evaporate solvents rapidly under reduced pressure, but they are seldom available in undergraduate organic chemistry labs because of their high cost.

Small to moderate quantities of solvent can be evaporated using one of the setups pictured in Figure C10. The test tube or flask containing the liquid to be evaporated is usually heated gently with a water bath or steam bath [OP-6a], and an aspirator or vacuum line is used to reduce the pressure inside the apparatus, increasing the evaporation rate. Swirling the solution during evaporation speeds up the process and reduces foaming and bumping. A solvent trap similar to that pictured in Figure C4 [OP-12] should be connected between the evaporation vessel and the aspirator to collect the evaporated solvent, which should then be returned to a solvent recovery container. To recover a low-boiling solvent such as ethyl ether, you should immerse the solvent trap in an ice/water bath or other cold bath [OP-7].

Always check glassware for cracks, star fractures, and other imperfections that might cause them to implode under vacuum. Because of possible health and fire hazards, never evaporate an organic solvent by heating an open container outside a fume hood, and do so under a fume hood only with your instructor's permission. Even when using the method described here, you should know and allow for the hazards associated with each solvent.

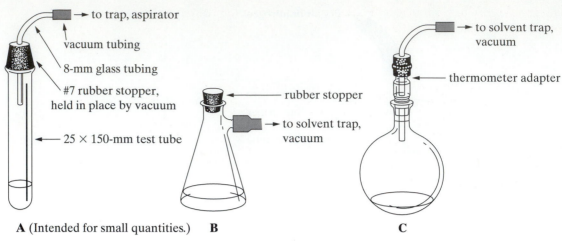

A (Intended for small quantities.)   **B**   **C**

Figure C10 Apparatus for solvent evaporation

General Direction for Evaporation

Equipment and supplies:

- evaporation vessel (Figure C10)
- solvent trap
- heat source
- rubber tubing
- aspirator

Assemble one of the three setups pictured in Figure C10 using heavy-walled rubber tubing that will not collapse under vacuum. Add the solution to be evaporated, stopper the evaporation vessel, and connect the apparatus to a solvent trap and vacuum source. Turn on the vacuum and heat the evaporation vessel over a steam bath or in a hot water bath [OP-6a], swirling it or stirring it magnetically [OP-9] throughout the evaporation to minimize foaming and bumping. Adjust the steam flow rate or water bath temperature to attain a satisfactory rate of evaporation; the liquid should boil gently but not foam up. Be ready to remove the evaporation vessel from the heat source immediately if it starts to foam up; otherwise your product may be carried over to the solvent trap.

If you are using a steam bath, wear gloves or use a towel to protect your hands.

Continue evaporating until all the solvent has been removed. At this point boiling will stop, the volume of the residue (which may be a solid or liquid) will not decrease with time, and the odor of the solvent will be gone. If you are not certain that all the solvent has evaporated, weigh the residue in an open container and reweigh it a few minutes later; if its mass decreases significantly between weighings, resume evaporation until its mass is constant between weighings.

If you only need to concentrate the solution, stop the evaporation when sufficient solvent has been removed.

When evaporation is complete, discontinue heating. Break the vacuum by detaching the vacuum hose, opening the pressure-release valve on the trap (if it has one), or sliding the stopper off the mouth of the test tube (for apparatus A); then turn off the vacuum source. Let the evaporation vessel cool down before you remove the residue. If the residue is a solid, remove it with a flat-bladed spatula. You can remove the last traces of residue by rinsing the container with a little volatile solvent (such as ethyl ether or dichloromethane) and allowing the solvent to evaporate under a hood or in a stream of dry air or nitrogen. Place the solvent from the trap in a designated solvent recovery container.

Summary

1 Assemble apparatus for solvent evaporation.
2 Add liquid, stopper evaporation vessel, connect to trap and vacuum source.
3 Turn on vacuum.
3 Apply heat with swirling or stirring until evaporation is complete.
4 Discontinue heating, turn off vacuum.
5 Recover residue and solvent.
6 Disassemble and clean apparatus.

Steam Distillation

Distillation of a mixture of two (or more) immiscible liquids is called *codistillation*. When one of the liquids is water, the process is usually called *steam distillation*. *Internal steam distillation* can be carried out by boiling a mixture of water and an organic material in a distillation apparatus (see Figure C11), causing vaporized water (steam) and organic liquid to distill into a receiver. *External steam distillation* is carried out by passing externally generated steam (as from a steam line) into a boiling flask containing the organic material (see Figure C12). The vaporized organic liquid is carried over into a receiver with the condensed steam.

Both kinds of steam distillation are used to separate organic liquids from reaction mixtures and natural products, leaving behind high-boiling residues such as tars, inorganic salts, and other relatively involatile components. Steam distillation is particularly useful for isolating the essential oils of plants from various parts of the plant, such as the clove oil steam distilled from clove buds in Experiment 10. Steam distillation is not very useful for the final purification of a liquid, however, because it cannot effectively separate components with similar boiling points.

Principles and Applications

When a *homogeneous mixture* of two liquids is distilled, the vapor pressure of each liquid is lowered by an amount proportional to the mole fraction of the other liquid present. This usually results in a solution boiling point that is somewhere between the boiling points of the separate components. For example, a solution containing equal masses of cyclohexane (b.p. = 81°C) and toluene (b.p. = 111°C) boils at 90°C.

When a *heterogeneous mixture* of two immiscible liquids A and B is distilled, each liquid exerts its vapor pressure more or less independently of the other. The total vapor pressure over the mixture (P) is thus approximately equal to the sum of the vapor pressures that would be exerted by the separate pure liquids (P_A° and P_B°) at the same temperature.

$$P \approx P_A^\circ + P_B^\circ$$

If you are not familiar with the principles of distillation, see OP-25.

This has several important consequences. First, the vapor pressure of a mixture of immiscible components will be *higher* than the vapor pressure of its most volatile component. Because raising the vapor pressure of a liquid or liquid mixture lowers its boiling point, the boiling point of the mixture will be *lower* than that of its most volatile (lowest boiling) component. Because the vapor pressure of a pure liquid is constant at a constant temperature, the vapor pressure of the mixture of liquids will be constant as well. Thus the boiling point of the mixture will remain constant throughout its distillation as long as each component is present in significant quantity.

For example, suppose you are distilling a mixture of the immiscible liquids toluene and water at standard atmospheric pressure (760 torr, 101.3 kPa). The mixture will start to boil when the sum of the vapor pressures of the two liquids is equal to the external pressure, 760 torr. This occurs at 85°C, where the vapor pressure of water is 434 torr and that of toluene is 326 torr. Because the vapor pressures of the two components are additive, the mixture distills considerably below the normal boiling point of either toluene (b.p. = 111°C) or water. According to Avogadro's law, the number of moles of a component in a mixture of ideal gases is proportional to its partial pressure in the mixture, so the mole fraction of toluene in the vapor should be about 0.43 (326/760) and that of water should be about 0.57 (434/760). (These calculations are approximate because the vapors are not ideal gases.) In other words, about 43 percent of the molecules in the vapor are toluene molecules. Because toluene molecules (M.W. = 92) are heavier than water molecules (M.W. = 18), they make a greater contribution to the total mass of the vapor. In 1.00 mol of vapor, there will be 0.43 mol of toluene and 0.57 mol of water, so the mass of toluene in the vapor will be about 40 g (0.43 mol × 92 g/mol), and the mass of water will be about 10 g (0.57 mol × 18 g/mol). The mass of one mole of the vapor is thus about 50 g, of which toluene makes up 40 g, or 80 percent. The liquid that collects in the receiver during a distillation—the *distillate*—is merely condensed vapor, so distilling a mixture of toluene and water will yield a distillate that contains about 80 percent toluene, by mass.

Because of its comparatively low molecular weight and its immiscibility with many organic compounds, water is nearly always one of the liquids used in a codistillation involving an organic liquid. The organic liquid must be insoluble enough in water to form a separate phase, and it cannot react with hot water or steam. As shown in Table C2, the higher the boiling point

Table C2 Boiling points and compositions of heterogeneous mixtures with water

Component A	b.p. of A	b.p. of mixture	mass % of A in distillate
toluene	111°	85°	80%
chlorobenzene	132°	90°	71%
bromobenzene	156°	95°	62%
iodobenzene	188°	98°	43%
quinoline	237°	99.6°	10%

Note: Component B is water

of the organic liquid, the lower will be its proportion in the distillate, and the closer the mixture boiling point will be to 100°C. Because the distillation boiling point is never higher than 100°C at 1 atm—well below the normal boiling points of most water-immiscible organic liquids—thermal decomposition of the organic component is minimized.

a. Internal Steam Distillation

An organic liquid can be separated by internal steam distillation using essentially the same procedure as for simple distillation [OP-25], except that additional water may need to be added during the distillation. If water is to be added, a setup like the one illustrated in Figure C11 should be used; otherwise, an ordinary simple distillation apparatus is adequate. If the organic distillate is quite volatile, a thermometer can be used to indicate when the end of the distillation is near. For example, with a toluene-water mixture, the temperature will rise rather rapidly from 85°C to about 100°C when the toluene is nearly gone. With less volatile materials, the temperature may be close to 100°C throughout the distillation, so a thermometer will be of little use.

General Directions for Internal Steam Distillation

Equipment and supplies:

- heat source
- condenser tubing
- condenser
- receiver
- three-way connecting tube
- thermometer and thermometer adapter (or stopper)
- ring stand, rings, clamps
- boiling flask
- vacuum adapter
- Claisen connecting tube (optional)
- separatory/addition funnel (optional)

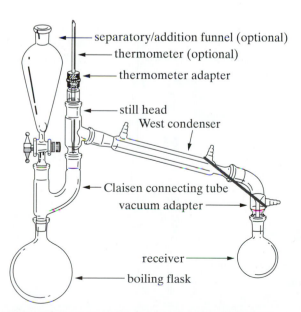

Figure C11 Apparatus for internal steam distillation

If it will be necessary to add more water during the codistillation, assemble the apparatus in Figure C11; otherwise, assemble the apparatus pictured in Figure E6 of OP-25. Add the mixture to be steam distilled, boiling chips or a stirbar, and enough water to fill the boiling flask about one-third to one-half full (unless enough water is already present). Turn on the stirrer (if you have one) and heat the flask with an appropriate heat source [OP-6a] to maintain a rapid rate of distillation. If necessary, add water to replace that lost during the distillation. Discontinue heating when the distillate contains no more of the organic component *and* the distillation temperature is about 100°C. The distillate should no longer be cloudy or contain droplets of organic liquid at this point.

b. External Steam Distillation

Externally generated steam is preferred for many codistillations, especially those involving solids or low-boiling liquids, because external steam produces a faster distillation rate and helps prevent bumping caused by solids and tars. The steam is usually obtained from a steam line; if another kind of steam generator is to be used, your instructor will show you how to use it. A *steam trap* is used to remove condensed water and foreign matter such as grease or rust from externally generated steam. A steam trap such as the one illustrated in Figure C12 is preferable, because it includes a valve for draining off excess water. With other kinds of traps, distillation may have to be interrupted periodically to drain the trap.

The boiling flask should be large enough so that the liquid will not fill it much more than half full throughout the distillation. Some steam will condense during the distillation, raising the water level in the boiling flask; during an extended distillation excessive water can be removed by external heating, if necessary. A Claisen connecting tube is used to help prevent mechanical transfer of liquids or particles from the boiling flask to the

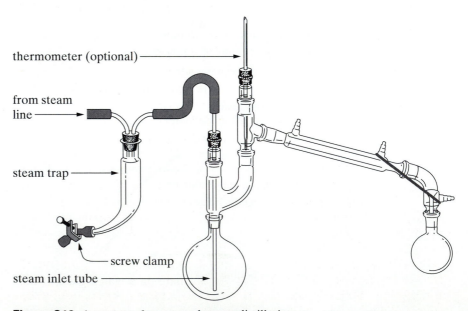

Figure C12 Apparatus for external steam distillation

receiver. The distillation should be carried out rapidly to reduce condensation in the boiling flask and to compensate for the large volume of water-laden distillate that may have to be collected to yield much of the organic component. Due to the rapid distillation rate and the high heat content of steam, efficient condensing is essential. The vacuum adapter should be cool to the touch throughout the distillation and no steam should escape from its outlet.

General Directions for External Steam Distillation

Equipment and supplies:

- heat source
- rubber tubing
- large boiling flask
- three-way connecting tube
- vacuum adapter
- thermometer and thermometer adapter (or stopper)
- steam trap (bent adapter, two-hole rubber stopper, bent glass tubes, rubber tubing, screw clamp)

- ring stand, rings, clamps
- steam inlet tube
- Claisen connecting tube
- condenser
- receiving flask

Assemble the apparatus pictured in Figure C12 using a large boiling flask and, as the steam inlet tube, a 6-mm O.D. (outer diameter) glass tube extending to within about 0.5 cm of the bottom of the flask. Position the boiling flask high enough so that external heat can be applied, if necessary. Connect the steam inlet tube to the steam trap, which should be clamped to a ring stand over a beaker, and see that the screw clamp on the steam trap is closed. Add the organic mixture and a small amount of water (unless the mixture already contains water) to the boiling flask, which should be no more than one-third full at the start. Turn on the condenser water so that it flows at a comparatively rapid rate (be sure the condenser hoses are tight!). Extend a rubber hose from the steam valve to the sink and turn on the steam until only a little water drips from the end of the hose. Turn off the steam and connect the hose to the steam trap.

Turn on the steam cautiously, and after distillation begins, adjust the steam flow to maintain a rapid rate of distillation. Check the vacuum adapter periodically; if it becomes warm, and especially if vapor begins to escape from its outlet, you should (a) turn down the steam; (b) increase the cooling water flow rate; and/or (c) cool the receiver in an ice/water bath. Check the connection between the condenser and three-way connecting tube frequently to make sure no vapor is escaping; this joint sometimes separates because of the violent action of the steam (you can use a joint clip to prevent this). Drain the trap periodically to remove condensed water. If the boiling flask begins to fill up excessively (it should not be much more than half full) heat it with a steam bath or other heat source to reduce condensation. If you must interrupt the distillation for any reason, open the steam trap valve (or raise the steam inlet tube out of the liquid) before turning off the steam.

When the distillate appears clear *and* the temperature is about 100°C, collect and examine a few drops of fresh distillate on a watch glass. If the fresh distillate is cloudy, contains oily droplets, or has a pronounced odor, continue distilling, and collect and examine the distillate at 5- or 10-minute

Take Care! Do not burn yourself with the live steam.

intervals. When the distillate is water-clear and you are certain the distillation is complete, open the steam-trap valve fully (or raise the steam inlet tube out of the liquid), then turn off the steam. This will keep the liquid in the boiling flask from backing up into the steam line.

The organic liquid can be separated from the distillate using a separatory funnel, or by extraction with ether or another suitable solvent [OP-13]. Extraction is necessary if the volume of the organic liquid is small compared to that of the water. If the aqueous layer is cloudy, you can saturate it with sodium chloride or other salt to "salt out" the organic liquid [OP-13c].

Summary

1 Assemble apparatus for external steam distillation.
2 Connect steam trap to steam-delivery tube.
3 Add water to boiling flask (if necessary), turn on condenser water.
4 Purge steam line, connect to steam trap, turn on steam.
5 Distill rapidly until distillate is clear; drain trap periodically.
6 Open steam-trap valve (or raise steam-inlet tube out of liquid), turn off steam.
7 Separate organic liquid from distillate.
8 Disassemble and clean apparatus.

OPERATION 16

Column Chromatography

If you touch the tip of a felt tip pen to a piece of absorbent paper, such as a coffee filter, and then slowly drip isopropyl rubbing alcohol onto the spot with a medicine dropper, the spot will spread and separate into rings of different color—the dyes of which the ink is composed. This is a simple example of *chromatography,* the separation of a mixture by distributing its components between two phases. The *stationary phase* (the coffee filter in this example) remains fixed in place while the *mobile phase* (the rubbing alcohol) flows through it, carrying components of the mixture along with it. The stationary phase acts as a "brake" on most components of a mixture, holding them back so that they move along more slowly than the mobile phase itself. Because of differences in such factors as the solubility of the components in the mobile phase and the strength of their interactions with the stationary phase, some components move faster than others and the components therefore become separated from one another.

Different types of chromatography can be classified according to the physical states of the mobile and stationary phases. In *liquid-solid* chromatography, which is used in most applications of column chromatography [OP-16] and thin-layer chromatography [OP-17], a liquid mobile phase filters down or creeps up through the solid phase, which may be cellulose, silica gel, alumina, or some other *adsorbent*. The adsorbent is a solid that holds onto molecules of different substances by surface attraction. In *liquid-liquid chromatography*, which is used in high performance liquid chromatography [OP-33], the mobile phase is usually an organic solvent and the stationary phase can be a high-boiling liquid that is adsorbed by or chemically bonded to a solid *support*. In *gas-liquid chromatography*, the most common type of gas chromatography [OP-32], the mobile phase is a gas such as helium that passes through a hollow or packed column containing a high-boiling liquid on a solid support. Gas chromatagraphy (GC) and high performance liquid chromatography (HPLC) are instrumental methods that are used primarily for analyzing mixtures, so they will be discussed in the Instrumental Analysis section of the Operations.

a. Liquid-Solid Column Chromatography

Principles and Applications

The usual stationary phase for liquid-solid column chromatography is a finely divided solid adsorbent, which is packed into a glass tube called the *column*. The mixture to be separated (the *sample*) is placed on top of the column and *eluted*—washed down the column—by the mobile phase, which is a liquid solvent or solvent mixture. Different components of the sample are attracted to the surface of the adsorbent more or less strongly depending on their polarity and other structural features. The more strongly a component is adsorbed, the more slowly it will move down the column, other factors being equal. So as the mobile phase (called the *eluent*) filters down through the adsorbent, the components of the sample spread out to form separate bands of solute, some passing down the column rapidly and others lagging behind.

For example, consider a separation of limonene and carvone on a silica gel adsorbent using hexane as the eluent. At any given time, a molecule of one component will either be adsorbed on the silica gel stationary phase or dissolved in the mobile phase. While it is adsorbed, the molecule will stay put; while dissolved, it will move down the column with the eluent. Molecules with polar functional groups are attracted to polar adsorbents such as silica gel and relatively insoluble in nonpolar solvents such as hexane. So a molecule of carvone, with its polar carbonyl group, tends to spend more time adsorbed on the silica than dissolved in the hexane. It will therefore pass down the column very slowly with this solvent. On the other hand, a nonpolar molecule of limonene is quite soluble in hexane and only weakly attracted to silica gel, so it will spend less time sitting still and more time moving than a carvone molecule. As a result, limonene molecules pass down the column rapidly and are soon separated from the slow-moving carvone molecules.

The separation attained by column chromatography depends on a number of factors including the nature of the components in the mixture, the quantity and kind of adsorbent used, and the polarity of the mobile phase. The lists on the next page show how strongly different functional groups are attracted to polar adsorbents and how strongly different adsorbents attract polar molecules. Table C4 (page 619) shows how different eluents compare in their ability to elute different components.

limonene   carvone

Approximate strength of adsorption of different functional groups on polar adsorbents

COOH	strongest
OH	
NH$_2$	
SH	
CHO	
C$=$O	
COOR	
OR	
C$=$C	
Cl, Br, I	weakest

Common chromatography adsorbents in approximate order of adsorbent strength

Alumina (Al$_2$O$_3$)	strongest
Activated carbon (polar) (C)	
Silica gel (SiO$_2$)	
Magnesia (MgO)	weakest

Note: Adsorbent strength varies with grade, particle size, and other factors

Experimental Considerations

Adsorbents. A number of different adsorbents are used for column chromatography, but alumina and silica gel are the most popular. Adsorbents are available in a wide variety of activity grades and particle size ranges; alumina can be obtained in acidic, basic, or neutral forms as well. The *activity* of an adsorbent is a measure of its attraction for solute molecules, the most active grade of a given adsorbent being one from which all water has been removed. The most active grade may not be the best for a given application, since too active an adsorbent may catalyze a reaction or cause bands to move down the column too slowly. Less active grades of alumina, for example, are prepared by adding different amounts of water to the most active grade (see Table C3). Since all polar adsorbents are deactivated by water, it is important to keep their containers tightly closed and to minimize their exposure to atmospheric moisture. Some samples should not be separated on certain kinds of adsorbents. For example, basic alumina would be a poor choice to separate a mixture containing aldehydes or ketones, which might undergo aldol condensation reactions on the column. Deactivated silica gel, although less active than alumina, is a good all-purpose adsorbent that can be used with most kinds of functional groups.

The amount of adsorbent required for a given application depends on the sample size and the difficulty of the separation. If the components of a mixture differ greatly in polarity, a long column of adsorbent should not be necessary, since the separation will be easy. The more difficult the separation, the more adsorbent will be needed. About 20–50 g of adsorbent per gram of sample is sufficient for most separations, but ratios of 200:1 or higher are occasionally required.

Table C3 Alumina activity grades

Grade	mass% water
I	0
II	3
III	6
IV	10
V	15

Eluents. In a column chromatography separation, the eluent acts primarily as a solvent to differentially remove molecules of solute from the surface of the adsorbent. In some cases, polar solvent molecules will also *displace* solute molecules from the adsorbent by becoming adsorbed themselves. If the solvent is too strongly adsorbed, the components of a mixture will spend most of their time in the mobile phase and will not separate efficiently. For this reason it is usually best to start with a solvent of low polarity and then (if necessary) increase the polarity gradually to elute the more strongly adsorbed components. Table C4 lists a series of common chromatographic solvents in order of increasing eluting power from alumina and silica gel. Such a listing is called an *eluotropic series*. The eluotropic series for a non-polar adsorbent is nearly the reverse of the one for alumina; less polar solvents are more effective eluents with such adsorbents.

Elution Techniques. Many chromatographic separations cannot be performed efficiently with a single solvent, so several solvents or solvent mixtures are used in sequence, starting with the weaker eluents (those near the top of the eluotropic series for the adsorbent being used). These eluents will wash down only the most weakly adsorbed components while strongly adsorbed solutes remain near the top of the column. By adding more powerful eluents, the remaining solute bands can then be washed off the column one by one.

In practice, it is best to change eluents gradually by using solvent mixtures of varying composition, rather than to change directly from one solvent to another. In *stepwise elution*, the strength of the eluting solvent is changed in stages by adding varying amounts of a stronger eluent to a weaker one. The proportion of the stronger eluent is increased more or less exponentially. For example, 5% dichloromethane in hexane may be followed by 15% and 50% mixtures of these solvents. According to one rule of thumb, the eluent composition should be changed after about three column volumes of the previous eluent have passed through; for example, if the packed volume of the adsorbent is 15 mL, then the eluent composition should be changed with every 45 mL or so of eluent.

Columns. There are many different kinds of chromatography columns, from a simple glass tube with a constriction at one end to an elaborate column with a porous plate to support the packing and a detachable base. A buret will suffice for most purposes, preferably one with a teflon stopcock, but the lack of a detachable base makes it difficult to remove the adsorbent afterward. If the column does not have a stopcock, the tip can be closed with a piece of flexible tubing equipped with a screw clamp. Unless the tubing is resistant to the eluents (polyethylene and teflon will not contaminate most solvents) it should be removed before elution begins.

In selecting a column for a particular chromatographic separation, first consider the amount of adsorbent needed for a given amount of sample and then choose a column that will completely contain the adsorbent with about 10-15 cm to spare. Ordinarily, the height of the column packing should be at least 10 times its diameter. If the column contains a porous plate to support the packing, no additional support is necessary; otherwise, the column packing should be supported on a layer of glass wool and clean

Table C4 Eluotropic series for alumina and silica gel

Alumina	Silica gel
pentane	cyclohexane
petroleum ether	petroleum ether
hexane	pentane
cyclohexane	trichloromethane
ethyl ether	ethyl ether
trichloromethane	ethyl acetate
dichloromethane	ethanol
ethyl acetate	water
2-propanol	acetone
ethanol	acetic acid
methanol	methanol
acetic acid	

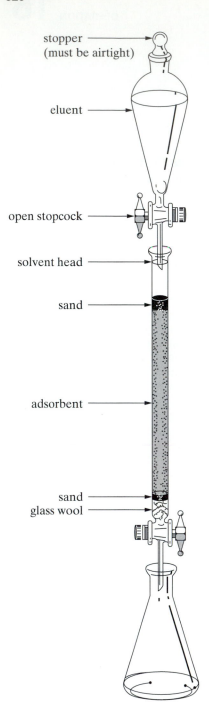

stopper
(must be airtight)

eluent

open stopcock

solvent head

sand

adsorbent

sand
glass wool

Figure C13 Packed column with continuous-feed reservoir

sand. Figure C13 shows a chromatography column (made from a buret) ready for operation.

Flow Rate. The rate of eluent flow through the column should be slow enough so that the solute can attain equilibrium, but not so slow that the solute bands will broaden appreciably by diffusion. For most purposes, a flow rate of between 5 and 50 drops per minute should be suitable—difficult separations require the slowest rates. The flow rate can be reduced by partly closing the stopcock on the column or by reducing the *solvent head*—the depth of the eluent layer above the adsorbent. The flow rate can be increased by opening the stopcock fully and maintaining a high solvent head.

Packing the Column. To achieve good separation with a chromatographic column, it must be packed properly. The packing must be uniform, without air bubbles or channels, and its surface must be even and horizontal. Columns using alumina are generally packed with the dry adsorbent; those using silica gel are packed with a slurry containing the adsorbent in a solvent. The following column-packing directions are for a column that lacks a porous plate; if there is one, omit the glass wool and sand support.

General Directions for Liquid-Solid Column Chromatography

Equipment and supplies:

- chromatography column
- column-packing solvent
- tapper
- adsorbent
- Pasteur pipet
- separatory/addition funnel (optional)

- powder funnel
- glass wool
- clean sand
- collectors (flasks, test tubes, vials, etc.)
- eluent(s)

Obtain an appropriate column and clamp it securely to a ring stand so that it is as nearly vertical as possible. Be sure the stopcock or screw clamp at the outlet of the column is closed. Construct a "tapper" by (for example) inserting one end of a pencil into a small one-hole rubber stopper. Measure out the amount of adsorbent you will need to prepare the column and keep it in a tightly closed container. Number and weigh the collectors you will be using to collect the eluent fractions. Then pack the column by one of the following methods.

Slurry-Packing the Column. Fill the column about half full with the least powerful eluting solvent to be used in the separation, or a solvent recommended in the experimental directions. Use a long glass rod to push a plug of glass wool to the bottom of the column and tamp the glass wool down gently to form a level surface and press out any air bubbles. Using a *dry* powder funnel, slowly pour in enough clean sand to form a 1-cm layer at the bottom of the column. As the sand filters down through the solvent, tap the column gently and continuously with the tapper so that the sand layer is uniform and level. The column should be tapped near the middle, where it is clamped, to avoid displacing it from the vertical.

Mix the measured amount of silica gel (or another suitable adsorbent) thoroughly with enough of the column-packing solvent to make a fairly thick, but pourable, slurry. Pour a little of this slurry into the column, with tapping, through the powder funnel, so that the adsorbent gently filters down to form a layer about 2 cm thick at the bottom. Then open the column outlet to let the solvent drain into a flask while you slowly add the rest of the slurry, tapping constantly to help settle and pack the adsorbent. If the slurry becomes too thick to pour, add more solvent to it. There should be enough solvent in the column so that the solvent level is always well above the adsorbent level; if necessary, add more. When all of the adsorbent has been added, close the outlet. The surface of the adsorbent should be as level as possible, so continue tapping until it has completely settled. Gently stirring the top of the solvent layer as the adsorbent is settling can also help form a level surface. Rinse down any adsorbent adhering to the sides of the column with the solvent, using a Pasteur pipet. If the column will not be used immediately, fill it with solvent and stopper it tightly, taking care not to disturb the adsorbent surface. No part of the column packing should ever be allowed to dry out—the entire column of adsorbent must be wet with solvent.

Packing the Column with Dry Adsorbent. Fill the column about two-thirds full with the column-packing solvent and add glass wool and a layer of sand as described previously. With the stopcock closed, pour enough alumina (or another suitable adsorbent) through a dry funnel, with tapping, to form a 2-cm layer at the bottom of the column. Then open the outlet and add the rest of the dry alumina as the solvent drains, tapping constantly so that the alumina settles uniformly. Finish preparing the column as described above for a slurry-packed column.

Separating the Sample. If the sample is a solid, dissolve it in a minimum amount of a suitable nonpolar solvent; use liquid samples without dilution. Open the column outlet until the solvent level comes down to the top of the adsorbent (no lower!), then close it. Use a Pasteur pipet to apply the sample around the circumference of the adsorbent, so that it spreads evenly over the surface (don't touch the adsorbent surface with the pipet tip). Then open the outlet until the liquid level again falls to the top of the adsorbent. Pipet a little of the initial eluent around the inside of the column to rinse down any adherent sample, and again open the outlet to bring the liquid level to the top of the adsorbent. Carefully add 3–5 mL of eluent, then sprinkle enough clean sand through the liquid to provide a uniform layer about 0.5 cm thick. (The sand is not essential, but it helps protect the adsorbent surface while eluents are being added.)

Clamp a separatory/addition funnel over the column and measure the initial eluent into it. Then add enough eluent to cover the adsorbent with 10–15 cm or more of liquid. If you have no suitable addition funnel to use as a reservoir, add the eluents through an ordinary funnel. Place a tared collector at the column outlet, open the outlet, and continue adding eluent to keep the liquid level nearly constant throughout the elution. (Alternatively, use a continuous-feed reservoir as shown in Figure C13, moistening the stopper with solvent to provide an airtight seal.) If you need to change eluents during the elution, allow the previous eluent to drain to the level of the sand before adding the next eluent.

If the components are colored or can be observed on the column by some visualization method (such as irradiation with ultraviolet light), change collectors each time a new band of solute begins to come off the column *and* when it has about disappeared from the column. If two or more bands overlap, collect the overlapping regions in separate collectors to avoid contaminating

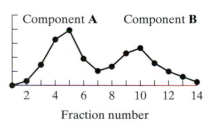

Figure C14 Elution curve

the purer fractions. When all of the desired components have been eluted from the column, evaporate the solvent [OP-14] from the purer fractions unless the experimental procedure indicates otherwise.

If the components are not visible and the procedure does not specify the fraction volumes, collect equal-volume fractions (usually 25–100 mL each) in tared collectors. Evaporate the solvent from each fraction, weigh the collectors and their contents, and plot the mass of each residue versus the fraction number to obtain an elution curve such as that illustrated in Figure C14. From the elution curve, you should be able to identify separate components and decide which fractions can be combined.

Summary

1 Pack column using appropriate adsorbent, solvent, and packing method.
2 Drain column to top of adsorbent, add sample, drain, rinse, drain again.
3 Add solvent and thin layer of sand, drain to top of sand.
4 Add eluent, put collector in place, open column outlet.
5 Elute sample, keeping eluent level nearly constant.
6 To change eluents, drain current eluent to top of sand, add next eluent.
   IF components are visible, GO TO 7.
   IF components are not visible, GO TO 8.
7 Change collectors when new band starts or ends and where bands overlap. GO TO 9.
8 Change collectors after a predetermined volume has been collected.
9 Evaporate and weigh pure fractions, combine like fractions.
10 Disassemble and clean apparatus, dispose of solvents as directed.

b. Small-Scale Column Chromatography

Very small amounts of sample can be separated or purified using a column consisting of a $5\frac{3}{4}$-inch Pasteur pipet filled with alumina or another adsorbant, as shown in Figure C15. As for standard column chromatography, the column of adsorbent must never be allowed to dry out once it is moistened with eluent.

To prepare a small-scale alumina column, first use a wooden applicator stick or stirring rod to push a small plug of cotton to the constriction in the pipet, then clamp the Pasteur pipet vertically to a ring stand. Add alumina slowly, tapping constantly with a pencil, until the adsorbent level is about 3 cm from the top of the pipet (or use the alternative method described next). Just before the column is to be used, put a collector beneath it and use another Pasteur pipet to add eluent to the top of this column until the entire column is moistened with eluent. Wait until the solvent level falls to the top of the adsorbent, add the sample, then add a little eluent from a Pasteur pipet when the sample level has fallen to the top of the adsorbent. Elute the sample by adding more eluent as necessary. Collect and evaporate the fractions as for standard column chromatography.

The column-packing method described is not suitable for silica gel. The column can also be packed by filling the Pasteur pipet about half full with eluent and slowly adding the adsorbent, with gentle tapping, as the solvent drains. If necessary, the solvent flow can be stopped temporarily by holding your fingertip on top of the pipet.

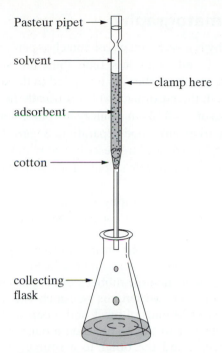

Pasteur pipet

solvent

clamp here

adsorbent

cotton

collecting
flask

Figure C15 Apparatus for small-scale column chromatography

c. Reversed-Phase Column Chromatography

Adsorbents are polar solids that attract polar compounds more strongly
than nonpolar ones, so nonpolar solutes are eluted from adsorbent-packed
columns more rapidly than polar ones. Another kind of stationary phase can
be prepared by coating particles of silica gel with a high-boiling nonpolar
liquid. With such a stationary phase, components of the sample are partitioned between the liquid mobile phase and the liquid layer of the stationary phase, where to *partition* a solute means to distribute it between two
phases. If the mobile phase is more polar than the stationary phase, the
usual order of elution will be reversed; that is, polar compounds will be
eluted before nonpolar ones. This general method is called *reversed-phase
chromatography*. The mobile phases for reversed-phase column chromatography are usually polar solvents or mixtures of such solvents as water,
methanol, and acetonitrile. Stationary phases can be prepared by coating a
specially treated (silanized) silica gel with a nonpolar liquid phase, such as a
hydrocarbon or silicone. Bonded liquid phases such as the ones described
for HPLC [OP-33], can also be used.

To prepare a column using coated silica gel, the liquid mobile and stationary phases are shaken together in a separatory funnel to saturate each
phase with the other, the stationary phase is stirred with the silica gel, and
the coated support is made into a slurry with the saturated mobile phase.
The column is slurry-packed, usually with stirring to make it more uniform
and to remove air bubbles. A separation is carried out by eluting it with the
saturated mobile phase, essentially as previously described for normal-
phase column chromatography.

d. Flash Chromatography

Flash chromatography is a variation of column chromatography that uses a single elution solvent and takes less time than the standard method. Pressurized nitrogen, air, or another gas is applied to the top of the column to force eluent through the adsorbent, which is usually finely divided silica gel with a particle size of ~40–63 μm. Since only one eluent is used, it must be selected carefully to ensure good separation. Separation is poor if the sample size is too large; the procedure described next works best with 0.25 g of sample or less, although it may be suitable for 1 g or so of an easily separated sample.

Commercial flash chromatography systems may require expensive columns, pumps, and flow controllers, so they are rarely available for use in undergraduate laboratories, but an inexpensive flash chromatography system can be constructed using the "homemade" flow controller shown in Figure C16. The flow controller is assembled by inserting the bottom of a small plastic T-tube into a one-hole rubber stopper that fits into the top of the column, then attaching two lengths of $\frac{3}{8}$-inch I.D. (inner diameter) Tygon tubing to the straight ends of the T and securing them with copper wires. One tube is attached to the gas inlet on a bunsen burner base (the barrel can be removed), and the other to a source of clean, dry, compressed air.

To carry out a separation using flash chromatography, pack a 50-mL buret (or other suitable column) to a depth of 15 cm or so with the appropriate adsorbent, using the eluent as the column-packing solvent (see the directions for packing a column). Drain the solvent to the top of the adsorbent, then introduce the sample onto the top of the adsorbent. Close the column outlet and add 25 mL or more of eluent to the column, taking care not to disturb the adsorbent surface. Insert the stopper of the flow controller into the top of the column with a firm twist to keep it from popping out (it should stay in place at the desired pressure). Open the needle valve on the burner base, then open the column outlet with a collector in place. Carefully turn on the air valve to pressurize the system, and adjust the needle valve so that the eluent level decreases at a rate of ~5 cm per minute. Collect and evaporate the fractions by the usual procedure.

This flash chromatography apparatus is described in J. Chem. Educ. **1992**, *69,* 939. *Pressure can also be applied with balloons, as described in J. Chem. Educ.* **1986**, *63,* 361.

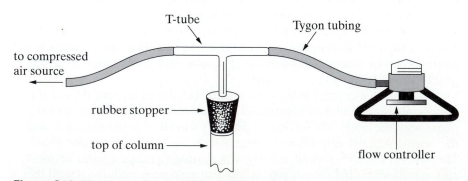

Figure C16 Flow controller for flash chromatography

Thin-Layer Chromatography

Principles and Applications

Like column chromatography, thin-layer chromatography (TLC) utilizes a solid adsorbent as the stationary phase and a liquid solvent as the mobile phase, but the mobile phase creeps *up* the adsorbent layer by capillary action rather than filtering down through it by gravity. A solid support, or *backing*, is coated with a thin layer of the adsorbent to make a *thin-layer plate* or *TLC plate*. The sample (or several samples) is dissolved in a suitable solvent and applied near the lower edge of the plate as several small spots. The plate is then *developed* by immersing its lower edge in a suitable mobile phase, the *developing solvent*. This solvent moves up the adsorbent layer by capillary action, carrying with it the components of each spot. In the process, the components are separated as described in OP-16 for column chromatography.

Although TLC is not useful for separating large amounts of a mixture, it is much faster than column chromatography and it can be carried out with very small sample volumes so that little is wasted. TLC provides better separation than the related technique of paper chromatography [OP-18], and it can be applied to a wider range of organic compounds. TLC is often used by organic chemists to identify unknown compounds, analyze reaction mixtures, determine the purity of products, and monitor various processes. For example, an experimenter can determine the optimum reaction time for a synthesis by obtaining thin-layer chromatograms of the reaction mixture at regular intervals and comparing the relative amounts of product, reactants and by-products on successive chromatograms. If a mixture is to be separated by column chromatography, TLC can be used to determine the best solvent for the separation, as described on page 627. The fractions eluted from a column can also be analyzed by TLC to determine which component is in each fraction; then the fractions containing the same component can be combined and evaporated. Finally, if a product is purified by recrystallization or another method, TLC analysis can quickly show whether the purified product still contains appreciable amounts of impurities.

Experimental Considerations

Adsorbents. The most commonly used adsorbents for TLC are silica gel, alumina, and cellulose. The adsorbent is more finely divided than that used in column chromatography, and it is provided with a *binder* such as polyacrylic acid to make it stick to the backing. It may also contain a *fluorescent indicator*, which makes most spots visible under ultraviolet light.

TLC Plates. Small "do-it-yourself" TLC plates can be prepared by dipping a glass microscope slide into a slurry of the adsorbent and binder in a suitable solvent, then allowing the solvent to evaporate. Such plates may give inconsistent results because of variations in the thickness of the adsorbent layer. More uniform TLC plates measuring 20×20 cm or larger are produced commercially with a wide variety of adsorbents, backings, and layer thicknesses. For example, a typical TLC plate suitable for use in undergraduate laboratories has a flexible plastic backing coated with a 200 μm-thick layer of silica gel mixed with a binder and a fluorescent indicator. TLC plates with metal and glass backings are also available. The adsorbent layer of a TLC plate is easily damaged, so it is important to avoid touching its coated surface and to protect the plate from foreign materials. A TLC adsorbent will pick up moisture when exposed to the atmosphere, making it less active; the adsorbent can be *activated* by heating TLC plates in a 110°C oven for an hour or so.

Spotting. A TLC plate is prepared for development by *spotting* it with solutions of the sample(s) to be analyzed, and often with standard solutions as well. The sample is dissolved in a suitable solvent to make an approximately 1% solution. The solvent should, if possible, be quite nonpolar and have a boiling point of 50–100°C. Column chromatography fractions and other solutions can often be used as is, if the solute is present at a concentration in the 0.2–2.0% range.

It is important to avoid touching the surface of the adsorbent while spotting because your fingerprints may hinder development, obscure developed spots, or be mistaken for spots upon visualization. You can avoid this by wearing thin disposable gloves while spotting the TLC plate. The spots should be positioned accurately, since incorrectly placed spots may run into one another or onto the edge of the adsorbent layer. This can be done with a transparent plastic ruler supported just above the surface of the plate so that it does not touch the adsorbent. The starting line is marked with a pencil on both edges, about 1.5 cm from the bottom of the plate (or 1.0 cm for a microscope slide plate). The spots should be located at least 1.5 cm from each edge of the plate and 1.0 cm from each other. Thus a 10 × 10 cm TLC plate can accomodate up to 8 spots, as shown in Figure C17.

Large, diffuse spots spread out too much for accurate results, so each spot should be as small and concentrated as possible. The spots are best applied with a microliter syringe or a capillary micropipet. Capillary micropipets can be obtained commercially, but suitable micropipets can be prepared by heating an open-ended melting-point capillary in the middle over a small flame and drawing it out to form a fine capillary about 4–5 cm long. The tube is allowed to cool, then scored and snapped apart in the middle to form two micropipets (see Figure C18). To spot a plate with such a micropipet, dip the narrow end into the solution, then gently touch the tip to the surface of the TLC plate, at the proper location, for only an instant. Do not dig a hole in the adsorbent surface; this will obstruct solvent flow and distort the chromatogram. Make several successive applications at each location, letting the solvent dry each time, delivering about 1–5 μL of solution to form a spot 1–3 mm in diameter. It may be worthwhile to try one, two, and three applications at three separate locations on the TLC plate to determine which quantity gives the best results. Too much solution can result in "tailing" (a zone of diffuse solute following the spot), "bearding" (a zone of diffuse solute preceding the spot), and overlapping of components. Too little solution will make it difficult to detect some of the components. Capillary micropipets can be reused a few times , but a different micropipet should be used for each solution.

When using a microliter syringe for spotting, deliver about 1 μL of solution with each application and make 2–3 applications for each spot.

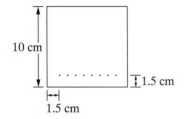

Figure C17 Spotted 10 × 10 cm TLC plate

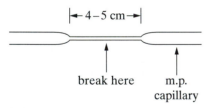

Figure C18 Drawing a capillary micropipet

You can practice your spotting technique on a used or damaged TLC plate.

Rinse the syringe with a suitable solvent and remove excess solvent by pumping the plunger gently a few times before filling it with a different solution.

Choosing a Developing Solvent. Solvents that are suitable eluents for column chromatography are equally suitable as TLC developing solvents; the eluotropic series of Table C4 may help you choose a solvent for a particular application. A quick way to find a suitable solvent is to spot a TLC plate with as many spots as you have solvents to test (you can use a grid pattern with the spots 1.5–2.0 cm apart), then apply enough solvent directly to each spot to form a circle of solvent 1–2 cm in diameter. Mark the circumference of each circle before the solvent dries. A solvent whose chromatogram (after visualization) shows well-separated rings, with the outermost ring about 50–75% of the distance from the center to the solvent front, should be satisfactory. It is preferable to use the least polar solvent that gives good separation. Hexane, toluene, dichloromethane, and methanol or ethanol (alone or in binary combinations) are suitable for most separations. If no single solvent is suitable, choose two mutually miscible solvents whose outermost rings bracket the 50–75% range (one with too little solute migration, the other with too much) and test them in varying proportions.

Development. TLC plates are developed by placing them in a *developing chamber* containing the developing solvent. A paper wick is often used to help saturate the air in the developing chamber with solvent vapors, which increases reproducibility and the rate of development. The developing chamber can be a jar with a screw-cap lid, a beaker covered with plastic film or aluminum foil, or a commercial developing tank with a lid. It should utilize the smallest available container that will accomodate the TLC plate, since a larger container takes longer to fill with solvent vapors. Development should be carried out in a place away from direct sunlight or drafts, to prevent temperature gradients. It may take 20 minutes or more to develop a 10×10 cm TLC plate; a microscope slide plate can often be developed in 5–10 minutes. The solvent should not be allowed to reach the top edge of the plate, since the spots will spread by diffusion once the solvent has stopped advancing. When the *solvent front* (the boundary between the wet and dry parts of the adsorbent) is within 5 mm or so of the top of the plate, the TLC plate should be removed from the developing chamber and the solvent front should be marked with a pencil before the plate has had time to dry. Any colored spots should be outlined with a pencil and the plate should be left to dry, preferably under a hood.

Visualization. If the spots are colored, they can be observed immediately; otherwise, they must be made visible (*visualized*) by some method. The simplest way to visualize many spots is to observe the TLC plate under ultraviolet light. Fluorescent compounds produce bright spots, and if the adsorbent contains a fluorescent indicator, compounds that quench fluorescence will show up as dark spots on a light background. The center of each spot should be marked immediately with a pencil, as the spots will disappear when the ultraviolet light is removed. If a spot is irregular, its "center of concentration," the midpoint of its most densely shaded region, is marked instead. It is also a good idea to outline each spot with your pencil.

See J. Chem. Educ. **1985**, 62, 156 for an alternative method of visualizing TLC plates with iodine.

Another general visualization procedure is to place the dry plate in a closed chamber such as a wide-mouthed jar with a screw-cap lid, add a few crystals of iodine, and gently heat the chamber on a steam bath so that the iodine vapors sublime onto the adsorbent. Most organic compounds, except saturated hydrocarbons and halides, form brown spots with iodine vapor. Unsaturated compounds may show up as light spots against the dark background. The iodine color fades in time, so the spots should be marked without delay.

See J. Chem. Educ. **1996**, 73, 358 for a description of the cotton ball procedure.

Spots can also be made visible by applying a visualizing reagent to the TLC plate. The visualizing reagent can be applied by spraying it onto the plate, dipping the plate into the reagent, or wiping the plate with a cotton ball that has been saturated with the reagent. A solution of phosphomolybdic acid in ethanol can be used to visualize most organic compounds; the spots appear after the plate is heated with a heat gun or in an oven. Other visualizing reagents are used for specific classes of compounds, such as ninhydrin reagent for amino acids and 2,4-dinitrophenylhydrazine reagent for aldehydes and ketones. All spraying should be done under a hood in a "spray box," which can be made from a large cardboard box with the top and one side removed. A thin spray is applied to the TLC plate from about 2 feet away, using an aerosol can or spraying bottle containing the visualizing reagent. Large plates are sprayed by crisscrossing them with horizontal and vertical passes.

Analysis. The ratio of the distance a component travels up a TLC plate to the distance the solvent travels is called its R_f *value.*

$$R_f = \frac{\text{distance traveled by spot}}{\text{distance traveled by solvent}}$$

The R_f ("ratio to front") value of a spot is determined by measuring the distance from the starting line to the center of the spot and dividing this by the distance from the starting line to the the solvent front, both distances being measured along a line extending from the starting point of the spot to its final location. If the spot is irregular, its "center of concentration," the midpoint of its most densely shaded region, is used instead. The R_f value of a substance with a given mobile and stationary phase depends on the polarity of its functional groups and other structural features, so it is a physical property of the substance that can be used in its identification. R_f values for some compounds have been reported in the literature, using specified mobile and stationary phases. Reported R_f values can seldom be used to establish the identity of a substance, however, since they depend on a number of factors that are difficult to standardize such as the sample size, the thickness and activity of the adsorbent, the purity of the solvent, and the temperature of the developing chamber. The only way to be reasonably sure that a TLC unknown is identical to a known compound is to spot a solution of the known compound on the same TLC plate as the unknown. Even then, the identity of the unknown may have to be confirmed by an independent method. An unknown mixture is often analyzed on the same plate with a series of standard solutions, each containing a substance that may be one of the components of the mixture.

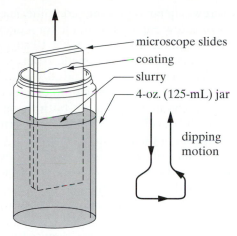

microscope slides
coating
slurry
4-oz. (125-mL) jar

dipping motion

Figure C19 Dipping a pair of stacked microscope slides

Directions for Preparing Microscope Slide TLC Plates

Several students should work together so they can use the same slurry. Clean as many microscope slides as you need TLC plates with detergent and water, then rinse them with distilled water and 50% aqueous methanol. After a slide has been cleaned, do not touch the surface where the slide will be coated; hold it by the edges or at the top. *Under the hood*, measure 100 mL of dichloromethane into a 4-oz. (125-mL) screw-cap jar, add 35 g of Silica Gel G with vigorous stirring or swirling, and shake the capped jar vigorously for about a minute to form a smooth slurry. Stack two clean microscope slides back-to-back, holding them together at the top. Without allowing the slurry to settle (shake it again, if necessary), dip the stacked slides into the slurry for about 2 seconds, using a smooth, unhurried, paddle-like motion (see Figure C19) to coat them uniformly with the adsorbent. Immerse the slides deeply enough so that only the top 1 cm or so remains uncoated. Touch the bottom of the stacked slides to the jar to drain off the excess slurry, let them air dry a minute or so to evaporate the solvent, then separate them and wipe the excess adsorbent off the edges with a tissue paper. Repeat with more slides, as needed; if the slurry becomes too thick, dilute it with dichloromethane. Activate the coated slides by heating them in a 110°C oven for 15 minutes. Slides with streaks, lumps, or thin spots in the coating should be wiped clean and redipped. If the coating is too fragile, try adding some methanol (up to one-third, by volume) to the slurry. Coated slides can be stored in a microscope slide box inside a dessicator.

Take Care! Avoid contact with dichloromethane, do not breathe its vapors.

General Directions for Thin-Layer Chromatography

Equipment and supplies:

- TLC plate(s)
- paper wick
- pencil and ruler
- capillary micropipets or syringe
- visualizer (sprayer, iodine chamber, UV lamp, etc.)
- developing chamber
- developing solvent
- standard solution(s)

Obtain a beaker, screw-cap jar, or another container large enough to hold the TLC plate. A 4-oz screw-cap jar is suitable for a microscope slide TLC plate; a 400-mL or 1-L beaker will hold a 6.7×10 cm or 10×10 cm TLC plate. Prepare and insert a paper wick made from filter paper or chromatography paper. The wick can be prepared by cutting a rectangular strip of paper 3–5 cm wide and long enough to extend in a "U" down one side of the developing chamber, along the bottom, and up the opposite side. Pour in enough developing solvent to form a liquid layer 5 mm deep on the bottom of the developing chamber. Cover the chamber with a screw cap, a square of plastic food wrap, or another suitable closure. Tip the beaker and slosh the solvent around to soak the liner with developing solvent. Let the developing chamber stand for 30 minutes or more to saturate the atmosphere inside the chamber with solvent vapors.

The liner may not be necessary if the developing chamber is allowed to stand for an hour or more after the developing solvent has been added.

Mark the starting line on a TLC plate and spot it with (1) the solution to be analyzed and (2) any standard solutions required, making spots 1–3 mm in diameter. Except on a microscope slide TLC plate, which can accomodate up to three evenly spaced spots, the spots should be at least 1.0 cm apart and the outermost spots should be 1.5 cm from the edges of the plate. When the spots are dry, put the developing chamber where it can sit undisturbed in a place away from direct sunlight or drafts—it shouldn't be moved after the TLC plate is inserted. Place the TLC plate in the chamber, with its spotted end down, so that it leans *across* the liner (perpendicular to it) with its top against the glass wall of the chamber. No part of a plastic- or aluminum-backed TLC plate should touch the liner because solvent can diffuse onto the adsorbent at that point. (A glass-backed plate may be allowed to touch the liner, however.) Cover the developing chamber without delay.

Observe the development frequently, and when the solvent front is within about 5 mm of the top of the plate, remove the plate from the developing chamber. Before the solvent evaporates, use a pencil to trace a line along the solvent front and mark the centers (or centers of concentration) of any visible spots. Let the plate dry thoroughly, and visualize the spots (if necessary) by one of the methods described, marking the center (or center of concentration) of each spot with a pencil. You should also outline each spot with your pencil. Measure the distance in millimeters from the starting line to the solvent front and to the center of each spot, and calculate the R_f value for each spot. If requested, make a permanent record of the chromatogram by photocopying or photographing it.

Summary

1   Put liner and developing solvent in developing chamber, cover and let stand.
2   Obtain or prepare TLC plate.
3   Spot TLC plate with solutions of the unknown and standards.
4   Place TLC plate in developing chamber, cover, and observe.
5   Remove TLC plate when development is complete.
6   Mark solvent front and visible spots, let plate dry.
7   Visualize and mark spots, as necessary.
8   Measure R_f values.
9   Clean up, dispose of solvent as directed.

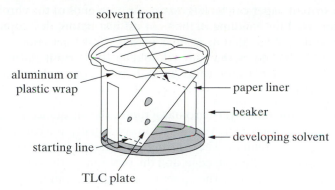

solvent front

aluminum or
plastic wrap

paper liner

beaker

developing solvent

starting line

TLC plate

Figure C20 Development of a TLC plate

Paper Chromatography

Principles and Applications

Paper chromatography is similar to thin-layer chromatography [OP-17] in practice but quite different in principle. Although paper consists mainly of cellulose, the stationary phase is not cellulose itself but instead the water that is adsorbed by it. Chromatography paper can adsorb up to 22% water, and the developing solvents usually contain enough water to keep it saturated. Development is carried out by passing a comparatively nonpolar mobile phase through the cellulose fibers, partitioning the solutes between the bound water and the mobile phase. Paper chromatography thus operates by a liquid-liquid partitioning process rather than by adsorption on the surface of a solid.

Since only polar compounds are appreciably soluble in water, paper chromatography is most frequently used to separate polar substances such as amino acids and carbohydrates. Manufactured chromatography paper is quite uniform and the activity of cellulose doesn't vary as much as the activity of most TLC adsorbents, so R_f values obtained by paper chromatography are generally more reproducible than those from thin-layer chromatography. However, the resolution of spots is often poorer and the development times are usually much longer. Nevertheless, paper chromatography is a useful analytical technique that can accomplish a variety of separations.

Experimental Considerations

Since many of the experimental aspects of paper chromatography are similar or identical to those for thin-layer chromatography, you should read the appropriate parts of OP-17 for additional information about experimental techniques.

Paper. Various grades of chromatography paper, such as Whatman #1 Chr, are manufactured in rectangular sheets, strips, and other convenient shapes. For good results, the chromatography paper must be kept clean. The bench top or other surface where the paper is handled should be covered with a sheet of butcher paper or some other liner. The chromatography paper should be held only by the edges or along the top; alternatively, an extra strip of paper can be left on one or both ends of the chromatography paper, used for handling it, then cut off just before development. The paper should be cut so that its grain is parallel to the direction of development. For most purposes, the paper should be 10–15 cm high (in the direction of development) and wide enough to accommodate the desired number of spots.

Spotting. As for thin-layer chromatography, the substance(s) to be analyzed should be dissolved in a suitable solvent, usually at a concentration of about 1% (mass/volume). The starting line should be marked in pencil about 2 cm above the bottom edge, and the positions of the spots can also be marked lightly with pencil. The spots tend to spread out more with paper chromatography than with TLC, so they should be applied farther apart— about 1.5–2.0 cm from each other and 2.0 cm from the edges of the paper, as shown in Figure C21. Spotting can be performed by the same techniques as for TLC using a capillary micropipet or syringe [see OP-17], but a round wooden toothpick is adequate for most purposes. During the spotting, nothing should touch the underside of the paper at the point of application. Each spot can be made with up to 10 μL of solution and should be 2–5 mm in diameter. Keep in mind that too little solution is usually better than too much. The solution can be applied in steps, drying the spot after each application. You should practice your spotting technique on a piece of filter paper before attempting to spot the chromatogram.

Developing Solvents. Nearly all paper chromatography developing solvents contain water, which is needed to maintain the composition of the aqueous stationary phase on the cellulose. As a rule, they also contain an organic solvent and (if necessary) one or more additional components to increase the solubility of the water in the organic solvent or provide an acidic or basic medium. A developing solvent for paper chromatography may be prepared by saturating the organic solvent(s) with water in a separatory funnel, separating the two phases, and using the organic phase for development. Alternatively, a monophase (single-phase) mixture with essentially the same composition as the organic phase of the saturated mixture can be used. Some typical monophase solvent mixtures are listed in Table C5. Most solvent mixtures should be made up fresh each time they are used and not kept more than a day or two.

Development. Narrow paper strips accommodating two or more spots can be developed in test tubes, bottles, cylinders, or Erlenmeyer flasks. A wider sheet can be rolled into a cylinder and developed in a beaker. A typical paper

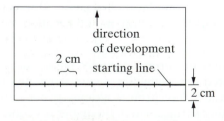

Figure C21 11 × 22 cm chromatography paper marked for spotting

Table C5 Some monophase solvent mixtures for paper chromatography

Solvents	Composition
2-propanol, ammonia, water	$9:1:2$
1-butanol, acetic acid, water	$12:3:5$
phenol, water	500 g phenol, 125 mL water
ethyl acetate, 1-propanol, water	$14:2:4$

Note: All solvent ratios are by volume

chromatogram is developed in much the same way as for TLC, but more time is required to saturate the developing chamber with solvent vapors.

Visualization. Visualization of spots by ultraviolet light is quite useful in paper chromatography since paper fluoresces dimly in a dark room and many organic compounds will quench its fluorescence. Paper chromatograms can also be visualized by spraying them or by dipping them into a solution of a suitable visualizing reagent.

Analysis. A substance responsible for a spot on a developed chromatogram is characterized by its R_f value, which is determined as illustrated in Figure C22. Spot migration distances are customarily measured from the starting line to the *front* of each spot, rather than to the center as for TLC. It is usually necessary to run a standard along with an unknown to identify the unknown. Whenever possible, the solvent and concentration should be the same for the standard as for the unknown.

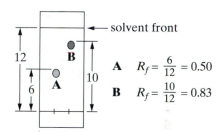

$$R_f = \frac{\text{Distance to leading edge of spot}}{\text{Distance to solvent front}}$$

Figure C22 Measuring R_f values on a paper chromatogram

General Directions for Paper Chromatography

Equipment and supplies:

- developing chamber
- chromatography paper
- capillary micropipets (or toothpicks, etc.)
- visualizing reagent and equipment
- developing solvent
- pencil and ruler
- solutions to be spotted

The following procedure can be used when a number of samples are to be spotted. For fewer samples, narrow strips of chromatography paper in jars, test tubes, or other appropriate containers can be used. Narrow strips can be folded in the middle or hung from a wire embedded in a cork for support.

Add enough developing solvent to a 600-mL beaker (or another developing chamber) to give a liquid depth of about 1 cm on the bottom. Cover the chamber tightly with plastic food wrap and slosh the solvent around in it for about 30 seconds. Allow sufficient time (usually an hour or more) for the solvent to saturate the developing chamber. While the developing chamber is equilibrating, obtain a sheet of chromatography paper and cut it to form an 11 × 22 cm rectangle (or another appropriate size). Without touching the surface of the paper, use a pencil to draw a starting line 2 cm from one long edge (the bottom edge) and lightly mark the positions for spots with a pencil, spacing them about 2 cm from each side and 1.5–2.0 cm apart. Spot the paper with the solutions and standards to be chromatographed; then roll

(cover omitted for clarity)

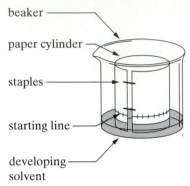

beaker

paper cylinder

staples

starting line

developing
solvent

Figure C23 Developing a paper
chromatogram

it into a cylinder with the starting line at the bottom and staple the ends
together, leaving a small gap between them (see Figure C23). Put the developing chamber in a place where it can sit undisturbed away from drafts and
direct sunlight. Uncover it, place the paper cylinder inside (spotted end
down), and cover it without delay. The paper must not touch the sides of the
chamber. Development may take an hour or more, depending on the developing solvent and the distance traveled.

When the solvent front is a centimeter or less from the top of the paper,
remove the cylinder and separate its edges, then accurately draw a line
along the entire solvent front with a pencil. If any spots are visible at this
time, outline them with a pencil (carefully, to avoid tearing the wet paper),
as they may fade in time. Again roll the paper into a cylinder, then stand it
on edge to air dry. When it is completely dry, visualize the spots by an appropriate method, if necessary. Measure the distance from the starting line to
the solvent front and to the leading edge of each spot, and calculate the R_f
value of each spot.

Summary

1 Add developing solvent to developing chamber, cover, and let stand.
2 Obtain chromatography paper and cut it to size.
3 Spot paper with solutions of the unknown and standards.
4 Place paper in developing chamber, cover, and observe.
5 Remove paper chromatogram when development is complete.
6 Mark solvent front and visible spots, let chromatogram dry.
7 Visualize and mark spots, as necessary.
8 Measure R_f values.
9 Clean up, dispose of solvent as directed.

D. Washing and Drying Operations

Washing Liquids

In practice, the process of washing liquids is identical to liquid-liquid extraction [OP-13], but its purpose is to remove impurities from an organic liquid, not to separate an organic substance from a reaction mixture or some other mixture. The organic liquid being washed may be a neat liquid (one without solvent) or a water-immiscible solvent containing a desired substance, such as the product of a reaction. In either case, the desired substance must not dissolve appreciably in the wash liquid or it will be extracted along with the impurities.

The *washing solvent* is usually water or an aqueous solution, although organic solvents such as ethyl ether can be used to remove low-polarity impurities from aqueous solutions containing polar solutes. Both water and saturated aqueous sodium chloride remove water-soluble impurities, such as salts and polar organic compounds, from organic liquids. Saturated sodium chloride is frequently used for the last washing before a liquid is dried [OP-20] because it removes excess water from the organic liquid by the salting-out effect [OP-13c]. It is preferred to water in some other cases because it helps prevent the formation of emulsions at the interface between the liquids.

Some aqueous washing solvents contain chemically reactive solutes that convert water-insoluble impurities to water-soluble salts, which then dissolve in the wash solvent. Aqueous solutions of bases such as sodium hydroxide, sodium carbonate, and sodium bicarbonate remove acidic impurities; aqueous solutions of acids such as hydrochloric acid and sodium bisulfate remove alkaline impurities; and aqueous sodium bisulfite removes certain aldehyde and ketone impurities by forming soluble bisulfite addition compounds.

Reactions in some chemically reactive washing solutions
$NaHCO_3$ + HA (acidic impurity) $\rightarrow Na^+A^-$ (soluble salt) + H_2O + CO_2
HCl + B (basic impurity) $\rightarrow BH^+Cl^-$ (soluble salt)
$NaHSO_3$ + RCHO (aldehyde impurity) $\rightarrow RCH(OH)SO_3^-Na^+$ (soluble salt)

When a chemically reactive washing solvent is used, it is usually advisable to perform a preliminary washing with water or aqueous sodium chloride to remove most of the water-soluble impurities. This may prevent a potentially violent reaction between the reactive wash liquid and the impurities.

The effectiveness of any extraction using a given total volume of solvent increases if the extraction is carried out in several steps. Unless otherwise indicated, a liquid should be washed in two or three stages using equal volumes of washing solvent for each stage, and the total volume of washing solvent should be roughly equal to the volume of the liquid being washed. For example, if you are washing 30 mL of an organic liquid, you can use two 15-mL or three 10-mL portions of the washing solvent. When several different washing solvents are used in succession, one or two washings with each

Rule of Thumb: Total volume of wash solvent ≈ volume of liquid being washed.

solvent may be sufficient, but the total volume of all the washing solvents should usually equal or exceed the volume of the liquid being washed.

General Directions for Washing Liquids

The procedure for washing liquids is essentially the same as that for liquid-liquid extraction [OP-13], except that the washing solvent is discarded and the liquid being washed is saved. Refer to the general directions for extraction for experimental details and a summary.

Combine the washing solvent and the liquid being washed in a separatory funnel. If the washing solvent contains sodium carbonate, sodium bicarbonate, or another reactive solute that generates a gas, stir it vigorously with the liquid being washed until gas evolution subsides. Otherwise, a pressure buildup might cause the stopper to pop out and your product to spray all over the lab. Then stopper the separatory funnel and shake it, very gently at first, with frequent venting. When you no longer hear a "whoosh" of escaping vapors upon venting, shake the separatory funnel more vigorously for 1–2 minutes, then set it on a support until the layers separate sharply.

If the liquid being washed is *less* dense than the washing solvent (for example, if it is an ether solution), drain the washing solvent after each washing and discard it, then add the next portion of washing solvent to the liquid being washed, which remains in the separatory funnel. If the liquid being washed is *more* dense than the washing solvent (for example, if it is a dichloromethane solution), drain the lower layer into a flask after each washing, then pour the washing solvent out the top of the separatory funnel and discard it. Return the liquid being washed to the funnel for the next washing, as necessary.

Drying Liquids

When an organic substance (or a solution containing it) is extracted from an aqueous reaction mixture, washed with an aqueous washing solvent, steam distilled from a natural product, or comes into contact with water in some other way, the substance or its solution will retain traces of water that must be removed before such operations as evaporation and distillation are carried out. Organic liquids and solutions can be dried by allowing them to stand in contact with a *drying agent*, which is then removed by decanting or filtration.

Drying Agents

Most drying agents are anhydrous (water-free) inorganic salts that form hydrates by combining chemically with water. For example, a mole of anhydrous magnesium sulfate can combine with up to seven moles of water to form hydrates of varying composition.

$$MgSO_4 + nH_2O \rightleftharpoons MgSO_4 \cdot nH_2O \quad (n = 1-7)$$

The effectiveness and general applicability of a drying agent depends on the following characteristics:

Table D1 Properties of commonly used drying agents

Drying Agent	Speed	Capacity	Intensity	Comments
magnesium sulfate	fast	medium	medium	good general drying agent, suitable for nearly all organic liquids
calcium sulfate (Drierite)	very fast	low	high	fast and efficient, but low capacity
sodium sulfate	slow	high	low	inefficient and slow, used for predrying; loses water above 32.4°C
calcium chloride	slow to fast	low to medium	medium	removes traces of water quickly, larger amounts slowly; reacts with many organic compounds
silica gel	medium	medium	high	good general drying agent, more expensive than most
potassium carbonate	fast	low	medium	cannot be used to dry acidic compounds
potassium hydroxide	fast	very high	high	used to dry amines, reacts with many other compounds; caustic

1 Speed: how fast drying takes place
2 Capacity: the amount of water absorbed per unit of mass
3 Intensity: the degree of dryness attained
4 Chemical inertness: unreactivity with substances being dried
5 Ease of removal

Ideally, a drying agent should be very fast and have a high capacity and intensity. It should not react with (or dissolve in) a substance being dried, and it should be easy to remove when drying is complete. Table D1 summarizes the properties of some common drying agents. As you can see in the table, there is no ideal drying agent, but magnesium sulfate is perhaps the best all-around drying agent. Because anhydrous magnesium sulfate may consist of a fine powder with a large surface area, it can adsorb much of your product, so it should be washed with fresh solvent to recover the adsorbed product. Other drying agents have advantages for specific applications. For example, sodium sulfate is often used to predry very wet solutions because it has a high capacity and is easy to remove. Final drying of a predried solution can be accomplished with a high intensity drying agent such as calcium sulfate. When very complete drying is required, a highly reactive drying agent such as sodium metal may be used, but sodium is too hazardous for routine use in an undergraduate laboratory.

a. Standard-Scale Drying of Liquids

Experimental Considerations

The choice of a drying agent depends, in part, on the properties of the liquid being dried and the degree of drying required. Solvents such as ethyl ether and ethyl acetate retain appreciable quantities of water, so they are often washed with saturated sodium chloride [see OP-19] to salt out some of the water before further drying. They should then be dried by a drying agent with a relatively high capacity, such as magnesium sulfate. Less polar solvents, such as hexane, petroleum ether, and dichloromethane, retain little water and can be

Rule of Thumb: Use about 1 g of drying agent per 25 mL of liquid.

Spent drying agent contains hydrates that reduce drying efficiency.

dried with calcium sulfate (Drierite) or calcium chloride. If a dichloromethane solution will later be evaporated, it may be sufficient to filter the solution through a cotton plug to remove water droplets (if there are any), since any remaining water will be removed as an azeotrope during evaporation. Calcium chloride reacts with many organic compounds that contain oxygen or nitrogen, including alcohols, aldehydes, ketones, carboxylic acids, phenols, amines, amides, and some esters, so it should not be used to dry such compounds or their solutions. Because of its convenience, sodium sulfate is often used when thorough drying is not necessary. Liquids can be dried by passing through a column of granular sodium sulfate supported on cotton or glass wool or they can be shaken with the sodium sulfate and separated by decanting.

The *quantity* of drying agent needed depends on the capacity and particle size of the drying agent and on the amount of water present. Usually about 1 g of drying agent should be used for each 25 mL of liquid; more should be added if the drying agent has a low capacity or a large particle size and if the liquid has a high water content. It is preferable to start with a small amount of drying agent and then add more if necessary, since using too much results in excessive losses by adsorption of liquid on the drying agent. The appearance of the drying agent when the drying time is up often suggests whether more drying agent is needed. As they take on water, magnesium sulfate clumps together in large crystals, calcium chloride displays a glassy surface appearance, and indicating (blue) Drierite changes color to pink. If much of a drying agent has changed as described after the initial drying period, more drying agent should be used or (preferably) the spent agent should be removed and replaced with fresh drying agent.

The *time* required for drying depends on the speed of the drying agent and the amount of water present. Most drying agents attain at least 80 percent of their ultimate drying capacity within 15 minutes, so longer drying times are seldom necessary. Five minutes is usually sufficient for magnesium sulfate or Drierite; 15 minutes is recommended for calcium chloride and sodium sulfate. When more complete drying is required, it is better to replace the spent drying agent or use a more efficient drying agent than to extend the drying time.

A drying agent should be removed as completely as possible when the drying period is over. A coarse-grained drying agent such as granular calcium chloride or Drierite can sometimes be removed by carefully decanting the liquid, but granular calcium chloride often contains fine powder that requires filtration. Most other drying agents can be removed by gravity filtration through a coarse fluted filter paper. When a solution (an ether extract, for example) is being dried, the drying agent should be washed with a little of the pure solvent to recover adsorbed solute that would otherwise be discarded with the spent drying agent. When a neat liquid is being dried, the drying agent can often be washed with dichloromethane or ethyl ether and the solvent evaporated to recover the adsorbed liquid.

General Directions for Drying Liquids

Equipment and supplies:

- Erlenmeyer flask
- stopper
- drying agent
- funnel
- fluted filter paper

If the liquid to be dried contains water droplets or a separate aqueous layer, remove the water using a Pasteur pipet or separatory funnel. If you are drying a small amount of neat liquid, it is a good idea to dissolve the liquid in a *carrier solvent* such as ethyl ether or dichloromethane and evaporate the solvent after drying; otherwise you may lose much of your product by adsorption on the drying agent.

Select an Erlenmeyer flask that will hold the liquid with plenty of room to spare and add the liquid. Weigh out the estimated quantity of a suitable drying agent and protect it from atmospheric moisture until you are ready to use it; be sure to cap its original container tightly. Add the drying agent to the liquid in the flask, stopper and swirl it, then set it aside for 5–15 minutes, swirling it frequently during the drying period. If a second (aqueous) phase forms during drying (this may happen when the drying agent is calcium chloride), remove it with a Pasteur pipet and add more drying agent.

When the drying period is up, examine the drying agent carefully. If most of it is spent, add more drying agent (or remove it and replace it with fresh drying agent) and continue drying for 5 minutes or more. When drying appears to be complete, remove the drying agent by decanting or gravity filtration [OP-11]. If you are drying a solution, wash the drying agent with a small volume of the pure solvent, and combine the wash liquid with the filtrate.

Summary

1 Select drying agent and weigh out estimated quantity needed.
2 Mix drying agent with liquid in flask, stopper and swirl, set aside.
3 Swirl occasionally until drying time is up.
   IF drying agent is spent or aqueous phase separates, GO TO 4.
   IF not, GO TO 5.
4 Remove aqueous phase or spent drying agent if necessary; add fresh drying agent. GO TO 3.
5 Filter or decant to remove drying agent.
   IF liquid being dried is a neat liquid, GO TO 6.
   IF liquid being dried is a solution, GO TO 7.
6 (Optional) Wash drying agent with an appropriate solvent, evaporate solvent, combine residue with dried liquid. GO TO 8.
7 Wash drying agent with fresh solvent, combine with dried solution.
8 Clean up, dispose of spent drying agent as directed.

b. Small-Scale Drying of Liquids

Small amounts of liquids can be dried (but not thoroughly) with a small *drying column* prepared by inserting a small plug of cotton or glass wool at the constriction of a $5\frac{3}{4}$-inch Pasteur pipet and adding enough granular anhydrous sodium sulfate to provide a 2–3 cm layer of the drying agent. Clamp the drying column vertically to a ring stand, place a suitable container under the outlet, and transfer the liquid to the drying column with another Pasteur pipet. If the liquid being dried is a solution, wash the sodium sulfate with a little fresh solvent and let the solvent drain into the container holding the dried liquid. Use a rubber bulb to gently expel any liquid remaining on the column into the container.

OPERATION 21

Drying Solids

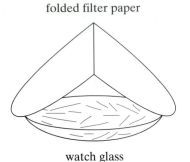

folded filter paper

watch glass

Figure D1 Covered watch glass for drying solids

Take Care! Don't blow the crystals away.

If you are not sure whether your product can be oven dried safely, consult your instructor.

Solids that have been separated from a reaction mixture or isolated from other sources usually retain traces of water or other solvents used in the separation. Solvents can be removed by a number of drying methods, depending on the nature of the solvent, the amount of material to be dried, and the melting point and thermal stability of the solid compound.

Experimental Considerations

Solids that have been collected by vacuum filtration [OP-12] are usually air dried on the filter by leaving the vacuum on for a few minutes after filtration is complete. Unless the solvent is very volatile, further drying is then required. Comparatively volatile solvents can be removed by simply spreading the solid out on a watch glass (covered to keep out airborne particles) and placing the glass in a location with good air circulation (such as a hood) for a sufficient period of time. Clamping an inverted funnel over the evaporating dish and passing a gentle stream of dry air or nitrogen over it will accelerate the drying rate.

Many wet solids can be spread out in a shallow ovenproof container and dried in a laboratory oven set at 110°C or another suitable temperature. The expected melting point of the solid should be at least 20°C above the oven temperature, and the solid should not be heat sensitive or sublime readily at the oven temperature. It is not unusual for a student to open an oven door and discover that the product he or she worked many hours to prepare has just turned into a charred or molten mass, or disappeared entirely. Aluminum weighing dishes and other commercially available containers made of heavy aluminum foil are usually suitable for oven drying because the aluminum conducts heat well, cools quickly, and is not likely to burn your fingers. Aluminum reacts with some acidic and basic compounds, which should be dried in porcelain or Pyrex containers such as watch glasses, evaporating dishes, and Petri dishes.

When time permits, the safest way to dry a solid is to leave it in a *desiccator* overnight or longer. A desiccator consists of a tightly sealed container partly filled with a *desiccant* (a drying agent) that absorbs water vapor, creating a moisture-free environment in which the solid should dry thoroughly. A *vacuum desiccator*, which can be connected to a vacuum line, combines the use of a desiccant with low air pressure for very fast, efficient drying. Desiccators such as the one in Figure D2 are available commercially. A simple "homemade" desiccator that can be used to dry small amounts of solid is illustrated in Figure D3. This desiccator consists of an 8-oz. (~250-mL) wide-mouthed jar with a screw cap, containing a 1-cm layer of solid desiccant at the bottom. The material to be dried is spread out in a suitable container such as a polystyrene or aluminum weighing dish or a drying tray made by folding a square of heavy aluminum foil. A small beaker or large vial can also be used, but the solid will dry more slowly in such a container. The container is then carefully set on top of the desiccant layer and the desiccator is capped until drying is complete. If there is much danger of the container tipping over, a wire screen can be cut to fit on top of the desiccant

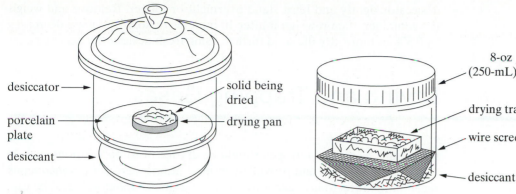

Figure D2 Commercial desiccator

Figure D3 "Homemade" desiccator

layer and provide a more stable surface. Except when a product is being added or removed, the desiccator must be kept tightly closed at all times.

Anhydrous calcium chloride is a good (if rather slow working) desiccant because it is inexpensive and has a high water capacity. Drierite (anhydrous calcium sulfate) is faster and more efficient than calcium chloride but has a much lower capacity. A combination of calcium chloride with a little indicating (blue) Drierite works better than either desiccant separately. The blue Drierite removes traces of moisture that calcium chloride cannot, and it turns pink when the desiccant is spent and needs to be replaced. Unless your instructor indicates otherwise, spent desiccant should be placed in a designated container for reactivation. Drierite can be reactivated by heating it in a 225°C oven overnight; calcium chloride should be heated at 250–350°C.

General Directions for Drying Solids

If the solid is quite wet, transfer it to a large circle of filter paper on a clean surface and blot it with another filter paper to remove excess water. Break up the solid with a large flat-bladed spatula and rub it against the filter paper with the blade of the spatula until it is finely divided and friable. You may have to blot it with fresh filter papers or pulverize it with the spatula blade to remove most of the water. Then use one of the following drying methods.

Drying in an Oven

Obtain a wide, shallow, ovenproof container such as an aluminum dish or watch glass, weigh it, and label it to prevent mixups. Spread the solid on the bottom of the container in a thin, uniform layer and place it in the oven, preferably where it is well separated from other containers. After 30 minutes or so, remove the container (use gloves for a glass container) and let it cool to room temperature. Weigh it, let it stand for 10 minutes or more, and weigh it again. If the mass has decreased by 0.5% or more, place it back in the oven until successive weighings made at 10-minute intervals differ by no more than 0.5%.

Drying in a Desiccator

Spread out the solid in a shallow, tared container such as a polystyrene weighing dish and place it in a desiccator containing fresh desiccant. Label the container if the desiccator will be used by other students. Close the desiccator tightly and let it stand overnight or longer. Remove and weigh the container, then reweigh it after 10 minutes or so. If the mass decreases by 0.5% or more, dry the solid further, using fresh desiccant if necessary.

Drying and Trapping Gases

Gases such as air and nitrogen should be dry for most applications such as evaporating solvents and providing an inert atmosphere for a nonaqueous reaction mixture. Many reactions, such as Grignard reactions, must be carried out in a dry atmosphere to prevent decomposition or side reactions brought about by contact with atmospheric moisture. Other reactions generate toxic gases that must be *trapped* to keep them out of the laboratory. Gases can be dried and moisture can be excluded from reaction mixtures using desiccants like the ones described in OP-21 for drying solids. Toxic gases can be trapped using reactive solutions that convert the gases to soluble salts.

a. Drying Gases

Many gases, such as nitrogen, are available in cylinders and can be purchased in a form that is dry enough for most applications. Some gases, especially compressed air obtained from a laboratory air line, should be dried before use. Air from an air line can be dried by passing it slowly through a large drying tube or a U-tube filled with a suitable desiccant. Plugs of cotton should be placed at both ends of the tube to support the desiccant and remove particles or other impurities from the air line. Indicating silica gel and granular alumina are very efficient desiccants; indicating Drierite and calcium chloride are satisfactory for many purposes.

b. Excluding Moisture from Reaction Mixtures

A *drying tube* containing a suitable desiccant (drying agent) is used to exclude moisture from an experimental setup during an operation such as refluxing or distillation. Calcium chloride, calcium sulfate (Drierite), granular alumina, and silica gel can be used for this purpose. Calcium chloride is the least efficient of these, but it is adequate in many cases. To assemble an apparatus for conducting a reaction in a dry atmosphere, use an applicator stick or glass rod to push a small plug of dry cotton or glass wool into the drying tube so that it covers the narrow opening at the bottom. Add the desiccant, tapping the drying tube with your finger to help it settle, and insert another plug of cotton or glass wool on top of the desiccant to prevent spills. Insert the drying tube in the top of a reflux condenser using a thermometer adapter, as shown in Figure D4. If the apparatus has other components that are open to the atmosphere, put a similar drying tube over each opening. A drying tube can be attached to the sidearm of a vacuum adapter with a short

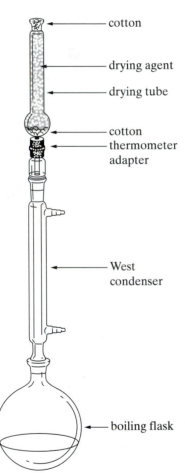

cotton

drying agent

drying tube

cotton
thermometer
adapter

West
condenser

boiling flask

Figure D4 Assembly for heating under reflux in a dry atmosphere

Take Care! Never stopper the drying tube, as this will result in a closed system that might explode or fly apart when heated.

length of rubber tubing. When a very dry atmosphere is required, the apparatus should be swept out with dry nitrogen or another dry gas introduced through a gas delivery tube.

c. Trapping Gases

The best way to keep toxic gases out of the laboratory air is to conduct all reactions under an efficient fume hood. If that is not possible or if the hood is not adequate, a gas trap containing a suitable gas-absorbing liquid should be used. Water alone will dissolve some gases effectively, but 1–2 *M* aqueous sodium hydroxide is generally used for acidic gases such as HBr or SO_2. This converts them to salts that dissolve in the water.

$$HBr + NaOH \rightarrow NaBr + H_2O$$
$$SO_2 + NaOH \rightarrow NaHSO_3$$

You can construct a simple gas trap by clamping an inverted narrow-stemmed funnel over a beaker containing a suitable gas-absorbing liquid, then lowering the funnel so that its rim just touches the surface of the liquid, as shown in Figure D5. Connect the gas trap to the reaction apparatus at any point that is open to the atmosphere (usually the top of a reflux condenser) using a short length of fire polished glass tubing inserted into a thermometer adapter or rubber stopper.

For small amounts of gases, a straight glass tube inserted into a large test tube containing the liquid may be adequate (see Figure D6). Clamp the outlet of the tube about a millimeter *above* the surface of the liquid to keep it from backing up into the reaction apparatus when gas evolution ceases or heating is discontinued.

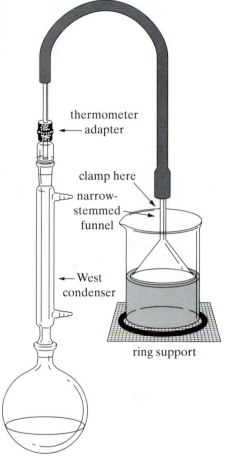

Figure D5 Apparatus for trapping gases during reflux

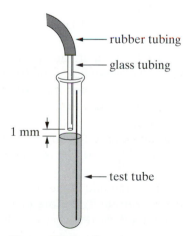

Figure D6 Small-scale gas trap

E. Purification Operations

Recrystallization

The simplest and most widely used operation for purifying organic solids is *recrystallization*. Recrystallization is so named because it involves dissolving a solid that (in most cases) had originally crystallized from a reaction mixture or other solution and then causing it to *again* crystallize from solution. In a typical recrystallization procedure the crude solid is dissolved by heating it in a suitable *recrystallization solvent*. The hot solution is then filtered by gravity and the filtrate is allowed to cool to room temperature or below, whereupon crystals appear in the saturated solution and are collected by vacuum filtration. The crystals are ordinarily much purer than the crude solid because most of the impurities either fail to dissolve in the hot solution, from which they are removed by gravity filtration, or remain dissolved in the cold solution, allowing them to be removed by vacuum filtration.

Recrystallization is based on the fact that the solubility of a solid in a given solvent increases with the temperature of the solvent. Consider the recrystallization from boiling water of a 5.00-g sample of salicylic acid contaminated by 0.25 g of acetanilide. The solubility of salicylic acid in water at 100°C is 7.5 g per 100 mL, so the amount of water required to just dissolve 5.00 g of salicylic acid at the boiling point of water is 67 mL.

$$5.00 \text{ g} \times \frac{100 \text{ mL}}{7.5 \text{ g}} = 67 \text{ mL of water}$$

All of the acetanilide impurity will also dissolve in the boiling water. If the solution is cooled to 20°C, where the solubility of salicylic acid is only 0.20 g per 100 mL, about 0.13 g of salicylic acid will remain dissolved.

$$67 \text{ mL} \times \frac{0.20 \text{ g}}{100 \text{ mL}} = 0.13 \text{ g of salicylic acid}$$

The dissolved salicylic acid will end up in the filtrate during the vacuum filtration; the remaining 4.87 g will crystallize from solution (if sufficient time is allowed) and be collected on the filter. The solubility of acetanilide in water is 0.50 g per 100 mL at 20°C, so up to 0.35 g of acetanilide can dissolve in 67 mL of water at 20°C.

$$67 \text{ mL} \times \frac{0.50 \text{ g}}{100 \text{ mL}} = 0.35 \text{ g of acetanilide}$$

This means that all 0.25 g of acetanilide in the crude product should remain in solution and end up in the filtrate. Therefore under ideal conditions, the recrystallization should yield a 97% recovery of salicylic acid uncontaminated by acetanilide.

$$\frac{4.87 \text{ g}}{5.00 \text{ g}} \times 100 = 97\% \text{ recovery}$$

This is a simplified description of a rather complex process; a number of factors may bring about results different from those calculated.

1 Crystals of the desired solid may adsorb impurities on their surfaces or trap them within their crystal lattice.
2 The solubility of one solute in a saturated solution of another solute is not usually the same as its solubility in the pure solvent.
3 Using only enough solvent to dissolve a solid may result in premature crystallization from the saturated solution, so extra recrystallization solvent is usually added to prevent this.

a. Standard-Scale Recrystallization

Experimental Considerations

For most experiments in this book, a suitable recrystallization solvent is specified; if the solvent is not specified, see section **d**, "Choosing a Recrystallization Solvent." In its simplest form, the recrystallization of a solid is carried out by dissolving the solid in the hot (usually boiling) recrystallization solvent and letting the resulting solution cool to room temperature or below to allow crystallization to occur. In practice, additional steps such as filtering or decolorizing the hot solution may be required. The size and purity of the crystals formed depends on the rate of cooling; rapid cooling yields small crystals and slow cooling yields large ones. The medium-sized crystals obtained from moderately slow cooling are usually the best because larger crystals tend to trap (*occlude*) more impurities, while smaller ones adsorb more impurities and take longer to filter and dry. Sometimes it is desirable to collect a second or third crop of crystals by concentrating [see OP-14] the *mother liquor* (the liquid from which the crystals are filtered) from the first crop. These crystals will contain more impurities than the first crop and may require recrystallization from fresh solvent. A melting-point determination [OP-28] or TLC analysis [OP-17] can be used to determine whether a recrystallized solid is sufficiently pure.

Filtering the Hot Solution. Some impurities in a substance being crystallized may be insoluble in the boiling solvent and should be removed by filtration after the desired substance has dissolved. Don't mistake such impurities for the substance being purified and add too much solvent in an attempt to dissolve them, because excess solvent will reduce the yield of crystals and may even prevent the substance from crystallizing at all. If, after most of the solid has dissolved, the addition of another portion of hot solvent does not appreciably reduce the amount of solid in the flask, that solid is probably an impurity—particularly if it is different in appearance from the remainder of the solid.

Undissolved impurities can be removed by filtering the hot solution through coarse fluted filter paper by the usual procedure for gravity filtration [OP-11], using a preheated funnel and collecting flask. A convenient way of preheating the apparatus is to set the funnel on the flask containing the hot recrystallization solvent, so that it is heated by boiling solvent vapors; this flask is then used as the collecting flask after unused solvent has been poured out. You can also heat the funnel in an oven or invert it inside a large beaker set on a steam bath. The filtration should be performed as quickly as possible to prevent premature crystallization. If a few crystals

Excess solvent can be removed by evaporation [OP-14].

form on the filter paper or in the funnel stem during filtration, dissolve them by pouring a small amount of hot recrystallization solvent through them. If a large quantity of solid crystallizes in the filter, scrape it into the filtrate and redissolve it by heating the mixture with a little additional recrystallization solvent; then filter the hot solution again. Redissolve any precipitate or cloudiness that forms in the collecting flask during the filtering operation by heating before you set the flask aside to cool.

Removing Colored Impurities. If a crude sample of a compound known to be white or colorless yields a recrystallization solution with a pronounced color, activated carbon (Norit) can often be used to remove the colored impurity. Pelletized Norit, which consists of small cylindrical pieces of activated carbon, is preferable to finely powdered Norit, which obscures the color and is difficult to filter. Let the hot solution cool down several degrees below the boiling point and stir in a *small* amount of pelletized Norit—about 0.1% of the mass of the crude solid. Stir or swirl the mixture for a few minutes, keeping it near the boiling point and adding more pelletized Norit if color remains. Then heat it to boiling and filter it by gravity through fluted filter paper.

Powdered Norit can be used in the same way as pelletized Norit, except that you will not be able to tell whether more is needed until the solution is filtered. If Norit particles pass through the filter paper during filtration, they can be removed (with some product loss) by vacuum filtration through a bed of a filtering aid such as Celite. To prepare such a bed, mix the filtering aid with enough low-boiling solvent (such as ethyl ether or dichloromethane) to form a thin slurry, then pour it onto the filter paper in a Buchner or Hirsch funnel with the vacuum turned on, until it forms a layer about 3 mm thick. When the Celite bed is dry, remove the solvent from the filter flask. Filter the hot solution through the Celite under vacuum, then wash the Celite with a little hot recrystallization solvent, saving the filtrate. If necessary, heat the filtrate and/or add more hot solvent to redissolve any solid that forms.

Inducing Crystallization. If no crystals form after a hot recrystallization solution is cooled to room temperature, the solution may be supersaturated. If so, crystallization can often be induced by one or more of the following methods. Rub the tip of a glass rod against the side of the flask with an up-and-down motion just above the liquid surface, so that it touches the liquid on the downstroke. If several minutes of scratching does not effect crystallization, drop a few *seed crystals* (crystals of the pure compound), if they are available, into the solution with cooling and stirring. If crystals still do not form, you may have used too much recrystallization solvent. In that case, concentrate the solution by evaporation [OP-14] until it appears cloudy or crystals appear, heat it until the cloudiness or crystals disappear (adding a little more solvent if necessary), and again let it cool. If this doesn't work, try heating the solution back to boiling and adding another solvent that is miscible with the first and in which the compound should be less soluble (for example, try adding water to an ethanol solution). Add just enough of the second solvent to induce cloudiness or crystals at the boiling point, then add enough of the original solvent to cause the cloudiness or crystals to disappear from the boiling solvent, and let it cool. As a last resort, you may

Take Care! Never add Norit to a solution at or near the boiling point—it may boil up violently.

have to remove all of the solvent by evaporation and try a different recrystallization solvent—but see your instructor for advice first.

Dealing with Oils and Colloidal Suspensions. When the solid being recrystallized is quite impure or has a low melting point, it may separate as an *oil* (a second liquid phase) on cooling. Oils are undesirable because even if they solidify on cooling the solid retains most of the original impurities. Some solids, such as acetanilide in water, separate as oils when a certain saturation level is exceeded. This can be prevented by using more recrystallization solvent. If you don't know the cause of the oiling, try the following procedures, in order. (1) Heat the solution until the oil dissolves completely, adding more solvent if necessary. Then cool it slowly with constant stirring, adding a seed crystal or two at the approximate temperature where oiling occurred previously. Seed crystals can sometimes be obtained by dissolving a little of the oil in an equal volume of a volatile solvent in a small, open test tube and letting the solvent evaporate slowly. If this is not successful, add about 25% more solvent and repeat the process. (2) Try to crystallize the oil by either (a) cooling the solution in an ice-salt bath, rubbing the oil with a stirring rod and adding seed crystals if necessary, or (b) removing all of the oil with a Pasteur pipet, dissolving it in an equal volume of a volatile solvent, and letting the solvent evaporate slowly in an open test tube. Then collect the solid by vacuum filtration and recrystallize it from the same solvent or a more suitable one, following method (1), if necessary, to prevent further oiling.

A *colloid* is a suspension of very small particles dispersed in a liquid or other phase. Colloids generally have a cloudy appearance and cannot be filtered through ordinary filter paper because the particles go through the paper. If a solid separates from a cooled solution as a colloidal suspension, the colloid can often be coagulated to form normal crystals by extended heating in a hot water bath or (if the solvent is polar) by adding an electrolyte such as sodium sulfate. Colloids can sometimes be prevented by treating a recrystallization solution with Norit as described, or by cooling the solution very slowly.

General Directions for Standard-Scale Recrystallization

Safety Notes

> **Unless you are informed otherwise, consider all recrystallization solvents (except water) to be flammable and harmful by ingestion, inhalation, and contact. Avoid contact with and inhalation of such solvents and keep them away from flames and hot surfaces.**

Equipment and supplies:

- two Erlenmeyer flasks
- graduated cylinder
- boiling sticks or boiling chips
- powder funnel
- small watch glass
- cold washing solvent
- filtering aid (optional)
- recrystallization solvent
- heat source
- flat-bottomed stirring rod
- fluted filter paper
- vacuum filtration apparatus [OP-12]
- Norit (optional)

If you know the approximate solubility of the solid in the recrystallization solvent, estimate the volume of solvent you will need to recrystallize it, adding about 10 percent extra to allow for error and evaporation. Otherwise, start with 5–10 mL per gram of solid and use more if needed. Measure the solvent into an Erlenmeyer flask—the *solvent flask*. Add a boiling stick or a few boiling chips, insert a powder funnel in the flask mouth, and heat the solvent to boiling with an appropriate heat source [OP-6] (see Figure E1). A steam bath is often preferred for volatile organic solvents; a hot plate can be used for water and high-boiling organic solvents. Place the solid to be purified in a second Erlenmeyer flask—*the boiling flask*—and add about half to three-quarters of the estimated volume of hot solvent, or 2–3 mL per gram of crude solid if the amount of solvent is not known. Heat the mixture *at the boiling point* with constant swirling or stirring, breaking up any large particles with a flat-bottomed stirring rod, until it appears that no more solid will go into solution with further boiling. If undissolved solid remains, add more hot solvent in small portions (5–10 percent of

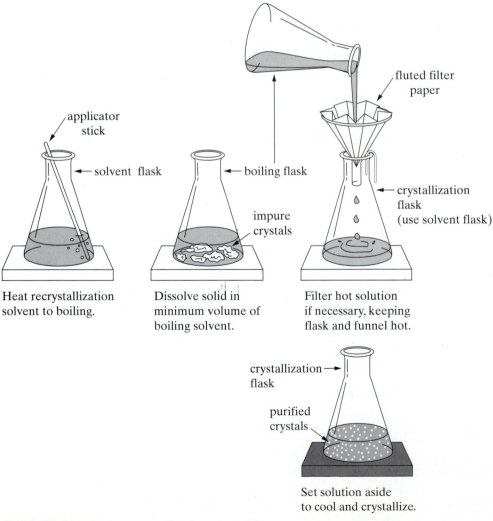

Heat recrystallization
solvent to boiling.

Dissolve solid in
minimum volume of
boiling solvent.

Filter hot solution
if necessary, keeping
flask and funnel hot.

Set solution aside
to cool and crystallize.

Figure E1 Steps in the recrystallization of a solid

the total solvent volume), heating the solution *at the boiling point* with swirling or stirring after each addition. Continue this process until (1) the solid is completely dissolved *or* (2) no more solid dissolves when a fresh portion of solvent is added and it appears that only solid impurities remain.

If the solution has a definite color but the pure product should not, see the section "Removing Colored Impurities." If the boiling solution contains no solid impurities, use the boiling flask as a crystallization flask and skip to the next paragraph. Pour any remaining solvent out of the solvent flask, which is now the *crystallization flask,* and return it to the heat source. Place a heated powder funnel on the neck of this flask (with a bent wire or paper clip between them), and insert a coarse fluted filter paper. Filter the hot solution while it is still near the boiling point, keeping any unfiltered solution hot throughout the filtration. If some crystals form in the filter paper or funnel stem, dissolve them into the crystallization flask with a little hot recrystallization solvent. If a considerable amount of solid has crystallized on the filter, redissolve it as described in the section "Filtering the Hot Solution."

Set the crystallization flask on the bench top or another nonconductive surface, cover it with a watch glass to keep out airborne particles, and let it stand undisturbed until crystallization is complete. If no crystals form by the time the solution reaches room temperature, see the section "Inducing Crystallization." If an oil separates, see "Dealing with Oils and Colloidal Suspensions." Once crystals have begun to form, allow at least 15 minutes at room temperature or below for complete crystallization (some compounds may require more time). You can increase the yield of crystals somewhat by cooling the mixture in an ice/water bath after a good crop of crystals is present, but their purity may decrease slightly as a result. Cooling the mixture before well-formed crystals are present may result in small, impure crystals that take longer to filter and dry.

Collect the crystals by vacuum filtration [OP-12] using a Buchner funnel and a clean, dry filter flask. Transfer any crystals remaining in the crystallization flask to the funnel with a little ice-cold recrystallization solvent (or another appropriate solvent) and use more of the cold solvent to wash the solid on the filter [see OP-12]. Air dry the crystals by leaving the vacuum on for a few minutes after the last washing, then dry them further [OP-21] as necessary. If an appreciable amount of solid (more than can be accounted for by solvent evaporation) forms in the filtrate during the vacuum filtration, crystallization was incomplete; cool the filtrate until crystallization is complete and refilter it.

Summary

1 Measure recrystallization solvent into solvent flask, heat to boiling.
2 Add some hot solvent to solid in boiling flask; boil with stirring.
3 Add more hot solvent in portions (as necessary) until solid dissolves.
   IF solution contains colored impurities, GO TO 4.
   IF solution contains undissolved impurities, GO TO 5.
   IF not, GO TO 6.
4 Cool below boiling point, stir in Norit, heat to boiling.
5 Filter hot solution by gravity.
6 Cover flask and set aside to cool until crystallization is complete.

7   Collect crystals by vacuum filtration; wash and air dry on filter.
8   Clean up, dispose of solvent as directed.

b. Small-Scale Recrystallization

Recrystallization of quantities ranging from a few tenths of a gram up to a gram or so should be carried out using small-scale apparatus to avoid excessive losses. Test tubes of the appropriate size can be used in place of Erlenmeyer flasks for dissolving and recrystallizing the solute; a hot water bath, sand bath, or heating block [OP-6a] can be used for heating the solvent and solution; and small-scale filtration equipment can be used to filter the hot solution and the crystals. The following procedure is suggested, but other kinds of equipment and techniques can also be used.

General Directions for Small-Scale Recrystallization

Prepare a *filter-tip pipet,* a short Pasteur pipet whose tip is plugged with a tiny piece of cotton, and heat it in an oven or elsewhere. A thin copper wire can be used to insert the cotton. Measure the recrystallization solvent into a test tube (the *solvent tube*), add a boiling chip, and heat it to boiling (see Figure E2). Place the solid in another test tube (the *boiling tube*) and use a short Pasteur pipet to transfer some hot solvent (about half of the estimated amount) from the solvent tube to the boiling tube. Stir the mixture in the boiling tube by twirling the round end of a small spatula in it while keeping it at the boiling point. Continue to add the hot solvent in small portions, boiling and stirring after each addition, until the solid has dissolved.

If the boiling tube does *not* contain insoluble impurities, use it as the crystallization tube and skip to the next paragraph. If it does, add a moderate excess of the recrystallization solvent (10–20%) to prevent premature crystallization, heat the mixture to boiling, and use the preheated filter-tip pipet to transfer the hot solution to a *crystallization tube* (use the emptied solvent tube) while leaving the impurities behind. Concentrate the solution to the saturation point by evaporating the excess solvent in a stream of dry air or nitrogen (use a hood). When the solution becomes cloudy or crystals start to form, add more recrystallization solvent drop by drop, with heating and stirring, until the solution clears up or the crystals dissolve.

Set the crystallization tube in a small Erlenmeyer flask and let the solution cool slowly to room temperature, then allow 10 minutes or more for complete crystallization. Cool it further in an ice bath to improve the yield, if desired. Collect the product by vacuum filtration on a Hirsch funnel [OP-12], washing it on the filter with a small quantity of cold recrystallization solvent. Air dry the crystals on the filter; then dry them further [OP-21], if necessary.

c. Recrystallization from Mixed Solvents

Solids that cannot be recrystallized readily from any single recrystallization solvent can usually be purified by recrystallization from a mixture of two compatible solvents such as those listed listed here. The solvents must be miscible in one another, and the compound should be quite soluble in one solvent and relatively insoluble in the other. If the composition of a suitable solvent mixture is known beforehand (such as 40% ethanol in water, for

Some compatible solvent pairs

ethanol–water
methanol–water
acetic acid–water
acetone–water
ethyl ether–methanol
ethanol–acetone
ethanol–petroleum ether
ethyl acetate–cyclohexane
toluene–petroleum ether

Table E1 Properties of common recrystallization solvents

Solvent	b.p.	f.p.	Comments
water	100	0	solvent of choice for polar compounds; crystals dry slowly
methanol	64	−94	good solvent for relatively polar compounds; easy to remove
95% ethanol	78	−116	excellent general solvent; often preferred over methanol because of higher boiling point
2-butanone	80	−86	good general solvent; acetone is similar but its boiling point is too low for most purposes
ethyl acetate	77	−84	good general solvent
toluene	111	−95	good solvent for aromatic compounds; high boiling point makes it difficult to remove from crystals
petroleum ether (high-boiling)	~60–90	low	a mixture of hydrocarbons; frequently used for nonpolar compounds;
hexane	69	−94	good solvent for nonpolar compounds; easy to remove
cyclohexane	81	6.5	good solvent for nonpolar compounds; freezes in some cold baths

Note: b.p. and f.p. (freezing point) are in °C. Solvents are listed in approximate order of decreasing polarity.

them for apparent yield and evidence of purity (absence of extraneous color, good crystal structure).

If no single solvent is satisfactory, choose one solvent in which the compound is quite soluble and another in which it is comparatively insoluble (the two solvents must be miscible). Dissolve 0.3 g (or less) of the solid in the first solvent with boiling and stirring, recording the amount required. Then add the second hot solvent drop by drop with boiling and stirring until saturation occurs. Cool the mixture and try to induce crystallization. If crystals form, examine them as before. If saturation cannot be attained or no crystals form, try another solvent pair. Once a suitable solvent or solvent pair has been identified, use the volume of solvent required to dissolve the measured mass of the solid to estimate the volume of solvent needed to dissolve all of the remaining solid.

Sublimation

Principles and Applications

Sublimation is a phase change in which a solid passes directly into the vapor phase without going through an intermediate liquid phase. Many solids that have appreciable vapor pressures below their melting points can be purified by (1) heating the solid to sublime it (convert it to a vapor), (2) condensing the vapor on a cold surface, and (3) scraping off the condensed solid. This method works best if impurities in the crude solid do not sublime appreciably. Sublimation is not as selective as recrystallization or chromatography, but it has some advantages in that no solvent is required and losses in transfer can be kept low.

Experimental Considerations

Sublimation is usually carried out by heating the *sublimand* (the solid before it has sublimed) with an oil bath, steam bath, or other uniform heat source and collecting the *sublimate* (the solid after it has sublimed and condensed) on a cool surface. For best results, the sublimand should be dry and finely divided, and the distance between the sublimand and the condenser should be minimized. A very simple sublimator consists of two nested beakers with a space of about 1–2 cm between them at the bottom. Good beaker combinations are 250 mL/400 mL and 400 mL/600 mL. The sublimand is spread out on the bottom of the outer beaker, and the condensing (inner) beaker is partially filled with ice water. More ice can be added periodically to maintain the temperature. As the outer beaker is heated, crystals of sublimate collect on the bottom of the condensing beaker. Figure E3 illustrates a sublimator operating on the same principle, except that an Erlenmeyer flask is used as a condenser and the temperature is controlled by flowing water.

Solids that do not sublime rapidly at atmospheric pressure may do so under vacuum. Figure E4A illustrates vaccum sublimator that can be assembled by fitting a 15×125 mm test tube snugly inside an 18×150 mm sidearm test tube using a thin rubber sleeve. The sleeve, which can be a rubber stopper bored out with a sharp cork borer that is just slightly smaller than the bottom of the stopper, must form a tight seal. Alternatively, a rubber O-ring cut from latex tubing can serve as the seal (see *J. Chem. Educ.* **1991,** *68,* A63). The inner condensing tube can be cooled by cold running water, as shown in the figure, or by ice water. Sublimation of a gram or more of solid in one of the makeshift sublimators described may take a long time unless the solid is quite volatile and the sublimation temperature is suffi-

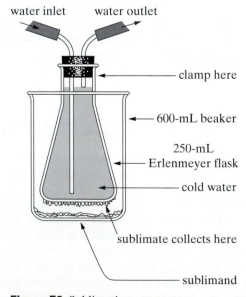

water inlet   water outlet

clamp here

600-mL beaker

250-mL
Erlenmeyer flask

cold water

sublimate collects here

sublimand

Figure E3 Sublimation apparatus

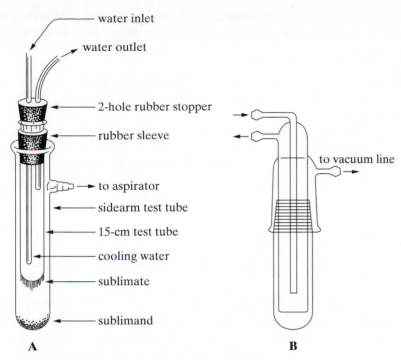

water inlet

water outlet

2-hole rubber stopper

rubber sleeve

to vacuum line

to aspirator

sidearm test tube

15-cm test tube

cooling water

sublimate

sublimand

A

B

Figure E4 Apparatus for vacuum sublimation

ciently high. Commercial vacuum sublimators such as the one in Figure E4B work more rapidly.

General Directions for Sublimation

Equipment and supplies:

- sublimation apparatus
- cooling fluid
- heat source
- flat-bladed spatula

Construct one of the sublimation setups described, or one suggested by your instructor. Powder the dry sublimand finely and spread it in a thin, even layer over the bottom of the beaker or other container. Assemble the apparatus, including a solvent trap supplied with a pressure release valve if you are using a vacuum sublimator. Turn on the cooling water (or partly fill a condensing beaker with ice water). If you are using a vacuum sublimator, turn on the vacuum. Heat the sublimand with an appropriate heat source until sublimate begins to collect on the condenser; then adjust the temperature to attain a suitable rate of sublimation without melting or charring the sublimand. If the condenser becomes overloaded with sublimate (especially if sublimate crystals begin to fall back into the sublimand) stop the sublimation and remove the sublimate as described in the next paragraph, then resume sublimation. If the sublimand hardens or becomes encrusted with impurities, stop the sublimation and grind it to a fine powder, then resume sublimation.

   When all of the compound has sublimed or only a nonvolatile residue remains, remove the apparatus from the heat source, open the pressure-release valve to break the vacuum if necessary, and let the apparatus cool.

Carefully remove the condenser (avoid dislodging any sublimate) and scrape the crystals into a suitable tared container using a flat-bladed spatula.

Summary

1 Construct sublimation apparatus, add sublimand.
2 Add cooling mixture or turn on cooling water; turn on vacuum if necessary.
3 Heat until sublimation begins, adjust heating to maintain good sublimation rate.
4 When sublimation is complete, stop heating, break vacuum if necessary, let cool.
5 Scrape sublimate into tared container.

<table>
<tr><td>OPERATION 25</td></tr>
</table>

Simple Distillation

Codistillation, the distillation of immiscible liquids, is discussed in OP-15.

A pure liquid in a container open to the atmosphere boils when its vapor pressure equals the external pressure, which is usually about 1 atmosphere (760 torr, 101.3 kPa). The vapor contains the same molecules as the liquid, so its composition is identical to that of the pure liquid. A mixture of two (or more) liquids with different vapor pressures will boil when the total vapor pressure over the mixture equals the external pressure, but the composition of the vapor will be different from that of the liquid itself, being richer in the more volatile component (the one with the higher vapor pressure). If this vapor is condensed into a separate receiving vessel, the condensed liquid will have the same composition as the vapor; that is, it will be enriched in the more volatile component.

The process of vaporizing a liquid mixture in one vessel and condensing the vapors into another is called *distillation*. The liquid mixture being distilled, the *distilland,* can be heated in a boiling flask—sometimes called the *pot*—using an apparatus such as the one in Figure E6 (p. 660). The vapors are condensed inside a water-jacketed condenser and the resulting liquid, the *distillate,* is collected in a receiver. The distilland is usually a mixture of two or more miscible liquids, but it may also be a liquid containing dissolved solids or a mixture of immiscible liquids. If the components of a mixture being distilled have sufficiently different vapor pressures, most of the more volatile component will end up in the receiver and most of the less volatile component(s) will remain behind in the pot. Thus distillation provides an effective means of purifying organic liquids.

The purity of the distillate increases with the number of vaporization-condensation cycles it experiences—the number of times it is vaporized and condensed—on its way to the receiver. *Simple distillation* involves only a single vaporization-condensation cycle. It is most useful for purifying a liquid that contains either involatile impurities or small amounts of higher- or lower-boiling impurities. *Fractional distillation* [OP-27] allows for several vaporization-condensation cycles in a single operation. It can be used to separate liquids with comparable volatilities and to purify liquids containing larger amounts of volatile impurities. *Vacuum distillation* [OP-26] is carried out under reduced pressure, which reduces the temperature of the distillation. It is used to purify high-boiling liquids and liquids that decompose when distilled at atmospheric pressure.

Principles and Applications

To understand how distillation works, consider a mixture of two ideal liquids with ideal vapors, which obey both Raoult's Law (Equation **1**) and Dalton's Law (Equation **2**).

$$\text{Raoult's law: } P_A = X_A \cdot P_A^o \qquad \textbf{(1)}$$

$$\text{Dalton's law: } P_A = Y_A \cdot P \qquad \textbf{(2)}$$

In these expressions P_A is the partial pressure of component A over the mixture, P_A^o is the equilibrium vapor pressure of pure A at the same temperature, X_A is the mole fraction of A in the liquid, and Y_A is the mole fraction of A in the vapor. Unfortunately, there are no *real* liquids that obey these laws perfectly, so we shall invent two imaginary hydrocarbons, *entane* (b.p. = 50°C) and *orctane* (b.p. = 100°C), named after two imaginary beings (Ents and Orcs) from J. R. R. Tolkien's *The Lord of the Rings* trilogy. If a mixture of entane and orctane is heated at normal atmospheric pressure, it will begin to boil at a temperature that is determined by the composition of the liquid mixture, producing vapor of a different composition. For example, an equimolar mixture of entane and orctane will start to boil at a temperature just above 66°C and the vapor will contain more than four moles of entane for every mole of orctane. The liquid and vapor composition of an entane-orctane mixture at any temperature can be calculated using Equations **3** and **4**, which are derived from Dalton's Law and Raoult's Law.

$$X_A = \frac{P - P_B^o}{P_A^o - P_B^o} \qquad \textbf{(3)}$$

$$Y_A = \frac{P_A^o}{P} X_A \qquad \textbf{(4)}$$

P is the total pressure over the mixture, assumed here to be 1 atm (760 torr).

For example, at 70°C the vapor pressure of orctane is 315 torr and that of entane is 1370 torr (see Table E2), so the mole fraction of entane in a distilland that boils at 70°C will be (from equation **3**),

$$X_{entane} = \frac{760 - 315}{1370 - 315} = 0.422$$

Its mole fraction in the vapor will be (from equation **4**),

$$Y_{entane} = \frac{1370}{760} \times 0.422 = 0.761$$

showing that the vapor (and thus the distillate) is considerably richer in entane than is the liquid. The vapor pressures and approximate liquid-vapor compositions for entane and orctane at this and other temperatures are given in the Table.

The key to an understanding of distillation is this: *The vapor over any mixture of volatile liquids contains more of the lower-boiling component than does the liquid mixture itself.* So at any time during a distillation, the liquid condensing into the receiver contains more of the lower-boiling component than does the liquid in the pot. As more of the lower-boiling component

Table E2 Equilibrium vapor pressures and mole fractions of entane and orctane at different temperatures

T, °C	Entane			Orctane		
	$P°$, torr	X	Y	$P°$, torr	X	Y
50	760	1.00	1.00	160	0.00	0.00
60	1030	0.67	0.90	227	0.33	0.10
70	1370	0.42	0.76	315	0.58	0.24
80	1790	0.24	0.57	430	0.76	0.43
90	2300	0.11	0.32	576	0.89	0.68
100	2930	0.00	0.00	760	1.00	1.00

Note: $P°$ = equilibrium vapor pressure of the pure liquid; X = mole fraction in liquid mixture; Y = mole fraction in vapor

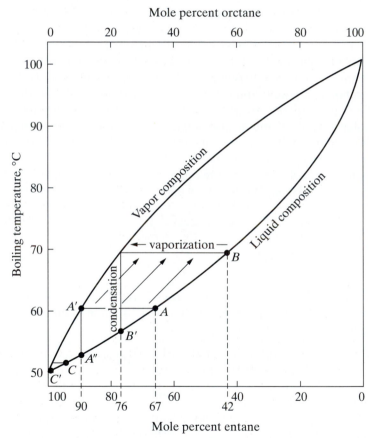

Figure E5 Temperature-composition diagram for entane-orctane mixtures

distills, the pot liquid becomes richer in the higher-boiling liquid, so by the end of the distillation, most of the lower-boiling liquid is in the receiver and most of the higher boiling liquid is in the pot.

The purification process is diagrammed in Figure E5, in which the liquid and vapor compositions are plotted against the boiling temperature of an entane-orctane mixture. Suppose we distill a mixture containing 2 moles of entane for every mole of orctane (67 mole percent entane). From the graph (and Table E2) you can see that such a mixture will boil at 60°C (point A) and that its vapor will contain 90 mole percent entane (point A'). Thus the distillate that is condensed from this vapor (point A") will be much richer in entane than was the original mixture in the pot. As the distillation continues, however, the more volatile component will boil away faster and the pot will contain progressively less entane. Thus the vapor will also contain less entane and the boiling temperature will rise. When the mole percent of entane has fallen to 42 (point B), the boiling temperature will have risen to 70°C and the distillate will contain only 76 mole percent entane (point B'). Only when nearly all of the entane has been distilled will the distillate be richer in the less volatile component; at 90°C, for example, more than two-thirds of the distillate will be orctane.

This example shows that the purification effected by simple distillation of a mixture of volatile liquids may be very imperfect. In the example, the distillate never contains more than 90 mole percent entane, and it may be considerably less pure than that, depending on the temperature range over which it is collected. If we start with a mixture containing only 5 mole percent orctane (point C), however, considerably better purification can be accomplished. The initial distillate will be 99 mole percent entane (point C') at 51°C, and if the distillation is continued until the temperature rises to 55°C, the final distillate will be 95 mole percent entane. The average composition of the distillate will lie somewhere between these values, so most of the orctane will remain in the boiling flask along with some undistilled entane, and the distillate will be relatively pure entane.

Thus simple distillation can be used to purify a liquid containing *small* amounts of volatile impurities if (1) the impurities have boiling points appreciably higher or lower than that of the liquid, and (2) the distillate is collected over a narrow range (usually 4–6°C), starting at a temperature that is within a few degrees of the liquid's normal boiling point.

a. Standard-Scale Simple Distillation

Experimental Considerations

Apparatus. Figure E6 illustrates a typical setup for standard-scale simple distillation. The size of the components should be consistent with the volume of the distilland—otherwise, excessive losses will occur. For example, suppose you recovered 20 mL of crude isopentyl acetate from Experiment 5 and decided to distill it from a 250-mL boiling flask. At the end of the distillation, the flask would be filled with 250 mL of undistilled vapor at the distillation temperature of 142°C (415 K); an ideal gas-law calculation shows that this is about 7.3 mmol of vapor.

$$n = \frac{PV}{RT} = \frac{(1.00 \text{ atm})(0.25 \text{ L})}{(0.0821 \text{ L atm mol}^{-1} \text{ K}^{-1})(415 \text{ K})} = 0.0073 \text{ mol}$$

The vapor would condense to about 1.1 mL of the liquid ester, which would not be recovered. By comparison, a 50-mL boiling flask will retain only one-fifth as much vapor and adsorb less than one-tenth as much liquid on its inner surface. Therefore using the smaller flask will cut your losses considerably.

Additional liquid losses occur in the still head, condenser, and any other parts of the apparatus that can trap vapors or adsorb liquid.

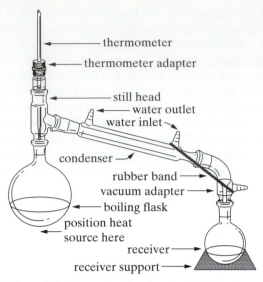

Figure E6 Apparatus for simple distillation

The boiling flask should be about one-third to one-half full of the distilland and should contain several boiling chips or a magnetic stirbar to prevent bumping [OP-6b]. The thermometer is inserted into the *still head* (a three-way connecting tube) through a stopper or thermometer adapter. It should be well centered in the still head (not closer to one wall than another) with the entire bulb below a line extending from the bottom of the sidearm. In other words, the *top* of the bulb should be aligned with the *bottom* of the sidearm, as illustrated in Figure E7. In this way, the entire bulb will become moistened by condensing vapors of the distillate. The placement of the thermometer is extremely important since otherwise the observed distillation temperature will be inaccurate (usually too low) and the distillate will be collected over the wrong temperature range.

The condenser should have a straight inner section of comparatively small diameter. A condenser used for heating under reflux can also be used for distillation, but do not use a jacketed distillation column for that purpose; its larger inner diameter results in less efficient condensation. The vacuum adapter should be secured to the condenser by a joint clip or a rubber band. If the distillate is quite volatile or hazardous, the receiver should be a ground joint flask that fits snugly on the vacuum adapter. If not, an open container can be used. If a ground-joint receiver is used, the vacuum adapter sidearm must not be plugged up—otherwise, heating the system will build up pressure that could result in an explosion and severe injury from flying glass.

Heating. Almost any of the heat sources described in OP-6a can be used for simple distillation. Heating mantles and oil baths are preferred because they provide reasonably constant, even heating over a wide temperature range. Steam baths can be used to distill some low-boiling liquids. The heat source should be positioned so that it can be easily removed at the end of the distillation. The heating rate should be adjusted to maintain gentle boiling in the pot and a suitable distilling rate. In most simple distillations, a dis-

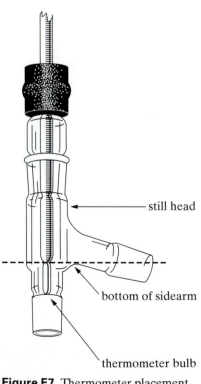

Figure E7 Thermometer placement

tilling rate of 1–3 drops per second (about 3–10 mL of distillate per minute) is recommended; higher rates may reduce the purity of the distillate. The temperature at the heat source should be kept relatively constant throughout a simple distillation. Temperature drops can cause the vapor level to fall, causing the observed boiling temperature to fluctuate and distillation to slow down or stop altogether. An excessive heating rate can reduce the efficiency of the distillation and cause decomposition or mechanical carryover of liquid to the receiver. If you are distilling a liquid over a wide boiling range, it may be necessary to increase the heating rate gradually to maintain the same distilling rate.

Boiling Range. An approximate boiling range for a distillation may be specified in the experimental procedure, as in Experiment 5, where it is designated as 136–143°C for the distillation of impure isopentyl acetate. Because the major impurity in that case is the more volatile isopentyl alcohol, the isopentyl acetate should distill below its normal boiling point of 142°C. (Using 143° as the end of the range allows for experimental error.) If the boiling range for a distillation is not given, you should collect the *main fraction* (the distilled liquid containing the desired component) over a relatively narrow boiling range (usually 4–6°C) that brackets the boiling point of the desired component. A liquid fraction that distills below the expected boiling range for the main fraction is called a *forerun*. It should be collected in a different container than the main fraction and disposed of as directed.

General Directions for Simple Distillation

Equipment and supplies:

- heat source
- clamps, ring stands
- thermometer adapter
- thermometer
- condenser tubing
- receiver(s)
- joint clip(s) or rubber band
- supports for heat source and receiver
- boiling flask
- three-way connecting tube (still head)
- condenser
- vacuum adapter
- boiling chips *or* stirbar and magnetic stirrer

Assemble the apparatus [OP-2] pictured in Figure E6 using a boiling flask of appropriate size and taking great care to position the thermometer correctly in the still head (Figure E7). If the liquid being distilled is quite volatile or hazardous, use a round-bottomed flask as the receiver; otherwise you can use an open container (ordinarily tared) such as an Erlenmeyer flask, graduated cylinder, or large vial. Remove the thermometer assembly and add the distilland to the boiling flask through a stemmed funnel, then drop in a few boiling chips or a stirbar. Replace the thermometer assembly and turn on the condenser water to provide a slow but steady stream of cooling water. Note that condenser water should flow *in* the lower end of the condenser and *out* the upper end.

Start the stirrer (if you are using one) and turn on the heat source, adjusting the heating rate so that the liquid boils gently and the reflux ring of condensing vapors rises slowly into the still head. Shortly after the reflux ring reaches the thermometer, the temperature reading should rise rapidly and vapors should begin passing through the sidearm into the condenser, coalescing into droplets that run into the receiving flask. As the first few droplets come over, the thermometer reading should rise to an equilibrium value and stabilize at that value. At this time, the entire thermometer bulb should be bathed in condensing liquid that drips off the end of the bulb into the pot. Record the temperature where the thermometer reading stabilizes; if it is lower than expected, recheck the thermometer placement. Distill the liquid at a rate of about 1–3 drops per second, monitoring the boiling temperature frequently throughout the distillation.

If the initial thermometer reading is below the expected boiling range, carry out the distillation until the lower end of the range is reached, collecting the forerun in the receiver, then quickly replace the receiver by another one. Try to make the switch quickly enough so that no distillate is lost. If the initial thermometer reading is within the expected boiling range, there is no need to change receivers. Continue distilling until the upper end of the expected boiling range is reached or until only a small volume of liquid remains in the boiling flask. Turn off and remove the heat source before the boiling flask is completely dry; heating a dry flask might cause tar formation or even an explosion. Disassemble and clean the apparatus as soon as possible after the distillation is completed.

Summary

1 Assemble distillation apparatus.
2 Add distilland and boiling chips or stirbar.
3 Turn on condenser water.
4 Commence heating; adjust heat so that vapors rise slowly into still head.
5 Record temperature after distillation begins and thermometer reading stabilizes.
   IF initial temperature is below expected boiling range, GO TO 6.
   IF initial temperature is within expected boiling range, GO TO 7.
6 Collect forerun, change receivers at low end of boiling range.
7 Distill until temperature reaches high end of boiling range or until pot is nearly dry.
8 Turn off and remove heat source.
9 Disassemble and clean apparatus, dispose of forerun (if any) and residue in pot.

b. Small-Scale Simple Distillation

If you distill a comparatively small quantity of a liquid ($\sim$1–10 mL) in a conventional distillation apparatus, losses due to trapping of vapors and adsorption of liquid in the apparatus can comprise a large fraction of your product mass. You can minimize such losses by distilling the liquid in a special small-scale distillation apparatus. If this kind of apparatus is unavailable, you can at least reduce your losses by using the smallest available boiling flask and replacing the condenser by a cooling bath surrounding the receiver, as shown in Figure E8. The vacuum adapter should be secured by a wire or joint clip because a rubber band exposed to the heat of the vapors may break. If a bulky heat source is used, it may be necessary to twist the vacuum adapter and receiver at an angle to the still head to allow room for

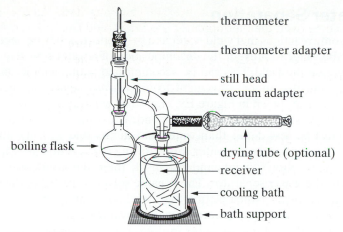

thermometer
thermometer adapter
still head
vacuum adapter
boiling flask
drying tube (optional)
receiver
cooling bath
bath support

Figure E8 Apparatus for small-scale distillation

Both flasks must be clamped or otherwise supported.

the heat source. In order for the cooling bath to work properly, the ground-glass joint between the receiver (which cannot be an open container) and vacuum adapter must be tight. The receiving flask should be clamped in place or secured by a joint clip.

If the distilland boils above 150°C and the receiver is well insulated from the heat source, it should be possible to omit the cooling bath. Cold tap water is a suitable coolant for liquids boiling near 100°C or above; ice water should be used for liquids distilling between 50° and 100°C. Below 50°C an ice-salt bath [see OP-7] should be used. When an ice or ice-salt bath is used, the vacuum adapter sidearm should be connected to a drying tube [OP-22b] so that moisture will not condense inside the receiver.

Distillation is carried out as described in the General Directions for simple distillation except that the distillation rate should be reduced to 1 drop per second or lower. If vapors begin to come out the vacuum adapter outlet, cooling is not sufficient; use a cooling bath or add ice (or salt) to the one you are already using. If you need to change receivers, stop the distillation and remove the heat source first.

c. Distillation of Solids

Low-melting solids can be distilled using the apparatus pictured in Figure E8 or with a special apparatus designed for that purpose. Unless the solid has a boiling point below 150°, a cooling bath should not be necessary. Have a heat gun or heat lamp handy to melt any solid that forms in the vacuum adapter. It is very important to keep the vacuum adapter outlet free from solid because heating a closed system could build up pressure and cause the apparatus to shatter or fly apart. When distillation is complete, the solid is melted by heating the receiving flask and then transferred to a tared vial or other container. The last traces of solid can be transferred with a small amount of ethyl ether or another volatile solvent, which is then evaporated under a hood.

A burner can be used if no flammable liquids are in the vicinity.

A homemade" Dean–Stark trap can be constructed as described in J. Chem. Educ, 1963, 40, 349.

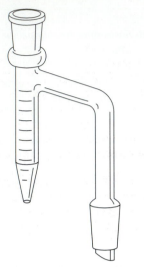

Figure E9 Dean–Stark trap

d. Water Separation

During some reactions that yield water as a by-product, it may be necessary to remove the water to prevent the decomposition of a water-sensitive product or to increase the yield. This can be accomplished with a simple distillation apparatus, but it is more convenient to use a *water separator,* such as the Dean–Stark trap shown in Figure E9. The water separator is inserted in the reaction flask, filled to the level of the sidearm with the reaction solvent (which must be less dense than water and immiscible with it), and fitted with a reflux condenser. As water forms during the reaction, its vapors and those of the reaction solvent condense inside the reflux condenser and drip down into the water separator. The organic solvent overflows through the sidearm and returns to the reaction flask, while the water collects at the bottom of the separator. The theoretical yield of water from the reaction can be calculated, so the volume of water in the separator is monitored to determine when the reaction is nearing completion.

The parts from an organic lab kit can be assembled as shown in Figure E10 so that the 25-mL flask, still head, and vacuum adapter together function like a Dean–Stark trap. When you assemble the apparatus in Figure E10, clamp both flasks securely and close the vacuum adapter outlet with a rubber bulb from a medicine dropper. Use a funnel to fill the water separator with the organic solvent to a level just below the bottom of the sidearm. Heat the reaction mixture under gentle reflux, being careful that the adapter drip tip does not become flooded with liquid. When disassembling the apparatus after the reaction, leave the water separator clamped while the other components are removed; then cautiously tilt the water separator to pour liquid from its sidearm into a beaker.

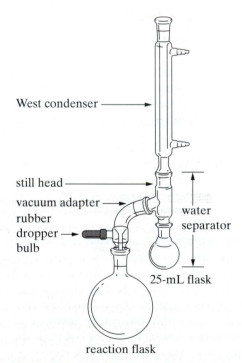

Figure E10 Apparatus for water separation

Vacuum Distillation

Principles and Applications

The boiling point of a liquid decreases when the external pressure is reduced, so under reduced pressure a liquid distills at a temperature that may be much lower than its normal boiling point. As shown in Table E3, a liquid that boils at 200°C at a pressure of 1 atmosphere (760 torr) will boil near 100°C at 25 torr. This reduces the likelihood that thermal decomposition and other high temperature reactions will occur during distillation. Distillation under reduced pressure is generally called *vacuum distillation.* As a rule, most liquids that boil around 200°C or above at atmospheric pressure should be purified by vacuum distillation, since above that temperature chemical transformations are more likely.

Vacuum distillation has certain inherent features that can cause potential hazards and experimental difficulties. According to Boyle's law $(V \propto 1/P)$ the volume of vapor generated by boiling a given amount of liquid is greater under reduced pressure. For example, a vapor bubble that occupies a volume of 0.10 mL at 760 torr will expand to 3.0 mL at 25 torr. This can cause excessive bumping in the boiling flask and mechanical carryover of liquid to the receiver. Special devices such as capillary bubblers are used to maintain smooth boiling during a vacuum distillation, and a Claisen connecting tube is added to reduce carryover. The vapor velocity is also greater because there are fewer molecules around to bump into; this can cause superheating of the vapor at the still head and a pressure differential throughout the system, so that the observed distillation temperature may be too high and the pressure reading too low. These problems can be circumvented by using the right kind of apparatus and by carrying out the distillation slowly. Even then, the separation attainable under vacuum distillation does not equal that possible at atmospheric pressure, and there is always the danger that the apparatus may implode because of the unbalanced external pressure on the system. Accordingly, a vacuum distillation must be performed with great care and attention to detail to obtain satisfactory results and prevent accidents.

Table E3 Approximate boiling points of liquids at 25 torr

normal b.p.	b.p. at 25 torr
150°C	60°C
200°C	100°C
250°C	140°C
300°C	180°C

a. Standard-Scale Vacuum Distillation

Experimental Considerations

Vacuum Sources. Most organic chemistry laboratories are provided with either water aspirators or vacuum lines that are connected to a central vacuum pump. Under optimum conditions, a water aspirator can attain a vacuum of 10–25 torr. In theory, an aspirator should be able to attain a pressure equal to the vapor pressure of the water flowing through it, which is a function of the water temperature, as shown in Table E4. In practice, the pressure is often 5–10 torr higher because of insufficient water pressure, leaks in the system, or deficiencies in the aspirator itself.

An aspirator must be provided with a solvent trap to prevent backup of water into the receiving flask due to changes of water pressure and to reduce pressure fluctuations throughout the system. A thick-walled filter

Table E4 Vapor pressure of water at 30°C and below

T, °C	P, torr	T, °C	P, torr
30	31.8	18	15.5
28	28.3	16	13.6
26	25.2	14.	12.0
24	22.4	12	10.5
22	19.8	10	9.2
20	17.5	8	8.0

Take Care! Never use a thin-walled container, such as an Erlenmeyer flask, as a trap; it may shatter under vacuum.

flask of the largest convenient size makes a suitable trap, if it is provided with a pressure-release valve and hooked up to the aspirator and distillation apparatus as illustrated in Figure E12 (p. 669).

If an in-house or other vacuum line is used, it may be necessary to place a cold trap between the vacuum line and vacuum adapter to keep organic vapors out of the vacuum pump. The cold trap can be a solvent trap immersed in an ice-salt bath, dry ice in acetone, or some other efficient coolant. If a pressure above the value attainable by a vacuum pump is desired, the pressure release valve can be replaced by a bleed valve consisting of the bottom part of a Bunsen burner (the pressure is varied by adjusting the needle valve of the burner).

Pressure Measurement. To know when the desired product is distilling, you need to know the pressure inside the system so that you can estimate its boiling point. If a manometer is attached to the system as shown in Figure E12, the pressure (in torr) is equal to the vertical distance between its two columns of mercury (in millimeters). The mercury in a manometer can present a hazard if the vacuum is broken suddenly—air rushing in can push the mercury column forcefully to the closed end of the tube, breaking it and releasing toxic mercury into the laboratory. For this and other reasons, the vacuum must always be released *slowly.* It is advisable to open the manometer to the system only when a pressure reading is being made.

If a manometer is not available, it may still be possible to carry out a vacuum distillation successfully if the boiling points of any impurities are quite different from that of the main fraction. By measuring the temperature of the aspirator water and referring to Table E4, you can estimate the *minimum* boiling temperature of the product as described in the next paragraph. The product should distill somewhat above that temperature; its actual boiling range will depend on the efficiency of the aspirator and the air tightness of your apparatus. You can also use a vacuum gauge to measure the pressure at the outlet of an aspirator or vacuum line, keeping in mind that the pressure in the system will be somewhat higher than the measured pressure.

Boiling Points under Reduced Pressure. If the boiling point of a substance at a given pressure is not known, it can be estimated using the vapor pressure-temperature nomograph in Figure E11. More precise estimates can be made using tables such as those in R. R. Dreisbach, *Pressure-Volume-Temperature Relationships of Organic Compounds* (New York: McGraw-Hill, 1952), or by using various empirical relationships. The main fraction should be collected over a range that brackets the expected boiling point, keeping in mind that pressure fluctuations and other factors may cause the boiling point value to vary by 10°C or more.

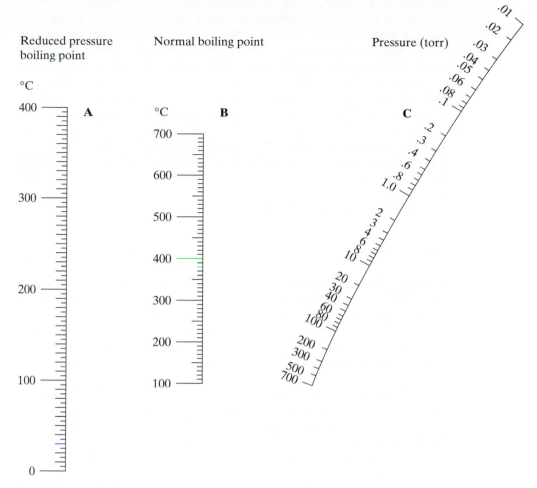

| Reduced pressure boiling point | Normal boiling point | Pressure (torr) |

To estimate the boiling point at pressure *P* given the boiling point at another pressure *P′*:
(a) connect pressure *P′* in **C** with the boiling point at that pressure in **A** using a ruler, and place a sharp pencil point where the ruler intersects line **B**; (b) pivot the ruler around the pencil point until it reaches the desired pressure (*P*) in **C**, then read the boiling point at that pressure from **A**. **Example:** To estimate the boiling point of dibutyl phthalate at 10 torr from its reported boiling point of 236° at 40 torr, place a ruler at 40 torr in **C** and 236° in **A**, causing it to intersect line **B** at about 345°. Then hold a pencil point at 345° on **B**, pivot the ruler about that point to 10 torr in **C**, and read the boiling point from **A**. This yields an estimated boiling point of 197° at 10 torr.

Figure E11 Reduced pressure boiling point nomograph

Heat Sources. The heat source should be capable of providing constant, uniform heating to prevent bumping and superheating and to maintain a constant distillation rate. An oil bath [OP-6a] is best, especially if it can be electrically heated and stirred. Heating mantles and heat lamps may also be satisfactory.

Smooth-Boiling Devices. You can reduce bumping and foaming during a vacuum distillation by using either a magnetic stirbar, microporous boiling chips, or a bubbler. Your lab kit may contain a narrow-tipped tube with a fine hole at the outlet, which can be used as a bubbler. Otherwise you can construct a flexible capillary bubbler, about as fine as a cat's whisker, by drawing it from a length of *thick-walled* capillary tubing—see your instructor for directions. (Capillary bubblers drawn from thin-walled tubing are very fragile and may break.) A short rubber tube with a screw clamp should be placed at the top of either kind of bubbler to control the rate of bubbling. The bubbler is inserted into the boiling flask through a thermometer adapter (or a rubber stopper if necessary) so that its tip extends to within a millimeter or two of the bottom. Under vacuum, this device should deliver a very fine stream of air bubbles, preventing development of the large bubbles that cause bumping. A bubbler has a few disadvantages: Air entering the system raises the pressure slightly and may oxidize the product at high temperatures, and a capillary bubbler may plug up when the vacuum is broken. Nevertheless, a well-constructed capillary bubbler often works better than the other devices.

Assembling the Apparatus.   The glassware used in assembling the apparatus for vacuum distillation must be free of cracks, star fractures, and other imperfections that might cause the apparatus to shatter under reduced pressure. Joints should be well lubricated with vacuum grease to prevent leaks [see OP-2], and all rubber tubing should be stretched and bent to make sure it is pliable and free of cracks. All connecting tubing should be heavy walled to prevent collapse and as short as possible to reduce the pressure differential between the vacuum source and the system. Connections between rubber and glass, as well as ground-joint connections, must be secure and airtight. If the distilland may contain a volatile impurity, have at least two receivers available to collect it and the main fraction. Commercially available rotating receivers (sometimes called "cows" because of their udder-like appearance) make it possible to switch receivers quickly; without one it is necessary to remove the heat source and break the vacuum before changing receivers.

General Directions for Vacuum Distillation

Safety Notes

> **Because of the possibility of an implosion, safety glasses must be worn during a vacuum distillation. It is advisable to work behind a hood sash or safety shield while the apparatus is under vacuum. A rapid pressure increase, accompanied by a thick fog in the distilling flask, indicates decomposition of the distilland. If this occurs, remove the heat source immediately and get away (warning others to do so also) until the flask cools. Report the incident to your instructor.**

Equipment and supplies:

- heat source
- boiling flask
- Claisen connecting tube
- stopper (if bubbler is not used)
- thermometer adapter
- vacuum adapter

- rings, clamps, supports
- smooth-boiling device
- three-way connecting tube
- thermometer
- condenser
- receiving flask(s)

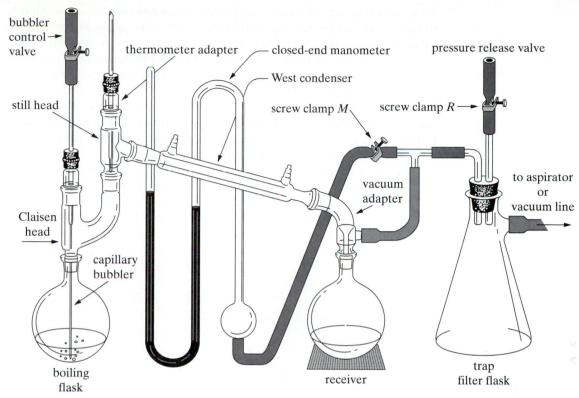

bubbler control valve

thermometer adapter

closed-end manometer

pressure release valve

still head

West condenser

screw clamp *M*

screw clamp *R*

Claisen head

vacuum adapter

to aspirator or vacuum line

capillary bubbler

boiling flask

receiver

trap filter flask

Figure E12 Apparatus for vacuum distillation

- joint clips (or wire, etc.)
- heavy-walled rubber tubing
- trap with pressure release valve
- glass tee (optional)
- condenser tubing
- screw clamps
- manometer (optional)

Inspect your glassware and rubber tubing for imperfections; if you have any doubt about their condition, see the instructor. Assemble the apparatus illustrated in Figure E12. If a manometer is not available, omit the glass tee and connect the vacuum adapter directly to the trap (also, disregard any instructions that refer to a manometer). The capillary bubbler (if you are using one) should be provided with a screw clamp on a short length of rubber tubing to control the bubbling rate. If you are using microporous boiling chips or a stirbar to prevent bumping, replace the capillary bubbler assembly with a ground-glass stopper. If you are using a vacuum-pump system rather than an aspirator, omit the solvent trap shown and use a cold trap, if necessary, as directed by your instructor.

Make sure that all joints and connections are tight; then add the material to be distilled through a funnel. If the liquid may contain a volatile solvent, remove it first by distilling at atmospheric pressure, then let the apparatus cool before proceeding. Add a few microporous boiling chips, drop in a stirbar, *or* open the screw clamp on the bubbler a turn or two. Raise the heat source into position, but do not begin heating yet. Turn on the condenser cooling water, open screw clamps *M* and *R*, and turn on the vacuum fully. Slowly close clamp *R* on the pressure-release valve and adjust the screw clamp on the bubbler (if you are using one) so that it emits a fine stream of bubbles. (If bumping and foaming occur, there may be some residual solvent in the distilland. Open clamp *R* and then close it down to a point where the solvent will evaporate without excessive bumping.) With clamp *R* completely closed, wait a minute or two until the pressure equilibrates, then read the manometer. If the observed pressure is more than ~10 torr above the estimated pressure, check the system for leaks caused by loose joints, cracked tubing, etc. If you find any, release the vacuum by opening clamp *R* and fix them. If the pressure is satisfactory, use the nomograph in Figure E11 to estimate the boiling range at that pressure.

Commence heating (and magnetic stirring, if used) until distillation begins, readjusting the bubbler clamp as necessary to maintain a very fine stream of bubbles. If excessive foaming occurs, try heating more slowly, using an antifoaming agent, or using a larger boiling flask. If bubbles form around any joint during the distillation, the joint is not tight enough; remove the heat source, break the vacuum, and regrease or otherwise repair it. Adjust the heat so that a distillation rate of about 1 drop per second is attained; then record the temperature and pressure readings. Close clamp *M* and leave it closed except when you need to make another pressure reading. If the temperature at the still head jumps up or down while the liquid is distilling, the pressure may be fluctuating due to changes in the aspirator flow rate; adjust the heating rate as necessary to maintain a suitable distillation rate. (To minimize such fluctuations, only a few students should use aspirators at the same time.)

If the initial distillation temperature is markedly lower than the estimated boiling range for the product, you are probably distilling a volatile forerun. Continue distilling until the temperature reaches the low end of the expected boiling range. Then change receivers without turning off the vacuum by the following procedure:

1 Lower the heat source and let the system cool down.
2 Open the bubbler clamp (if you are using one), then open clamp *R* until the system is at atmospheric pressure.
3 Replace the receiver with another one. If you are using boiling chips, add another chip or two.
4 Open clamp *M*, slowly close clamp *R*, adjust the bubbler, and record the pressure.
5 Commence heating until distillation begins.
6 Record the boiling temperature and pressure, close clamp *M*.

A capillary bubbler will sometimes plug up when the vacuum is broken; if that happens, you may have to replace it with a new one. Continue distilling until the upper end of the estimated boiling range is reached or until a significant drop in temperature indicates that the product is completely distilled. Stop the distillation before the boiling flask is completely dry.

When the distillation is completed, follow steps 1 and 2 for bringing the system back to atmospheric pressure, then turn off the vacuum. Disassemble the apparatus and clean the glassware promptly.

Summary

1 Inspect glassware and tubing, assemble apparatus, check connections and joints.
2 Add distilland and smooth-boiling device.

3 Position heat source, start cooling water, open clamps R and M, turn on vacuum.
4 Close R, let pressure equilibrate.
5 Read pressure, estimate boiling range.
   IF pressure is too high, check system for leaks and repair as necessary.
6 Adjust heat to attain desired distillation rate, adjust bubbler (if used).
7 Record temperature and pressure; close M.
   IF temperature is below estimated range, GO TO 8.
   IF temperature is within estimated range, GO TO 12.
8 Distill until temperature reaches lower end of expected boiling range.
9 Lower heat source, let cool, open bubbler clamp and R, change receivers.
10 Open M, close R, read pressure.
11 Adjust bubbler (if used), resume distillation, record temperature and pressure, close M.
12 Distill until upper end of temperature range is attained or only a little distilland remains.
13 Lower heat source, let cool, open R.
14 Turn off vacuum.
15 Disassemble and clean apparatus, dispose of residue and forerun.

b. Small-Scale Vacuum Distillation

The apparatus for small-scale simple distillation pictured in Figure E8 (p. 663) can be used under reduced pressure if the vacuum adapter sidearm is connected to a vacuum line through a trap, as illustrated in Figure E12. Except that a cooling bath is used instead of condenser water, the procedure is essentially the same as that used for standard-scale vacuum distillation. It may be necessary to insert a Claisen head between the boiling flask and the still head to prevent mechanical carryover of the distilland or to allow for insertion of a capillary bubbler.

Fractional Distillation

Principles and Applications

Although simple distillation is used to purify organic liquids containing small amounts of volatile impurities, it is not a very efficient way of separating the components of a mixture when each component makes up a substantial fraction of the mixture, unless the boiling points of the components are far apart. With mixtures of closer-boiling liquids, the distillate composition and boiling point will change continually throughout the distillation as illustrated in Figure E13, and most of the distillate will be a relatively impure mixture of the components.

The separation could be improved by redistilling portions of the initial distillate and subsequent distillates. Such a process is diagrammed in Figure E14, in which the initial distillate is delivered directly to a second distilling flask, which redistills the condensed vapors and delivers its distillate

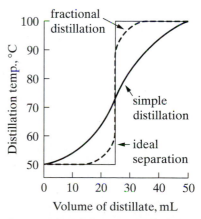

Figure E13 Separation efficiency of simple and fractional distillations

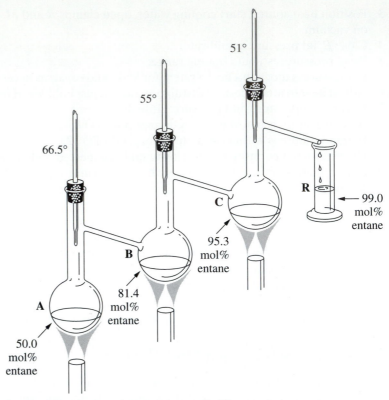

51°

55°

66.5°

R ← 99.0 mol% entane

C

95.3 mol% entane

B

81.4 mol% entane

A

50.0 mol% entane

Figure E14 Hypothetical multistage distilling apparatus

to a third distilling flask, which distills that liquid into a receiver. (Note that the apparatus is purely hypothetical; there are more practical ways of accomplishing the same result.) This process can be understood by referring to the temperature-composition diagram in Figure E15 for the imaginary entane-orctane mixture discussed in OP-25. If we boil a 50:50 mole percent mixture of entane (b.p. = 50°C) and orctane (b.p. = 100°C) in flask A (point *A* on the diagram in Figure E15), the vapor over that mixture will have a composition of 81.4 mol% entane and only 18.6 mol% orctane (point *A'*). This is because entane has a considerably higher vapor pressure at the boiling point of the mixture (66.5°C) than does orctane. The vapor is condensed into flask B (line *A'-B*), where it is boiled at 55°C to yield a vapor containing 95.3 mol% entane (line *B-B'*), which is condensed (line *B'-C*) into flask C. The condensed liquid in C boils at 51°C to yield a vapor that is 99.0 mol% entane (line *C-C'*), which is delivered to receiver R as a liquid of the same composition (line *C'-R*).

The boiling point decreases with each subsequent distillation because the distilland becomes richer in the more volatile component.

Only the first few drops of distillate will attain this degree of purity because as the more volatile entane is removed, the less volatile orctane will accumulate in the distilling flasks, reducing the proportion of entane in the vapor. The efficiency of the hypothetical apparatus would be improved considerably if some of the orctane-enriched liquid in each flask were drained back into the preceding flask through an overflow tube so that orctane would not accumulate as rapidly in the upper stages. Even then, the purity

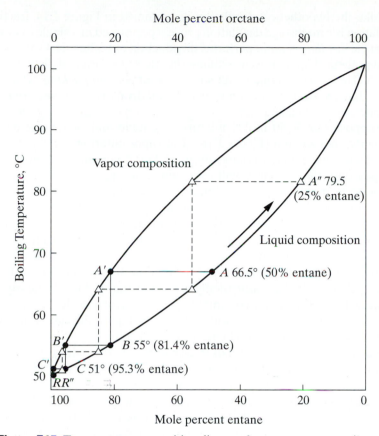

Figure E15 Temperature-composition diagram for entane-orctane mixture

of the distillate would decrease with time. For example, when the amount of entane in flask A had decreased to 25 mol% (point A''), the vapor condensing into the receiver (point R'') would be only 96.6 mol% entane, as shown by the broken lines in Figure E15. In other words, during a distillation the distilland climbs inexorably up the temperature-composition graph and the distillate composition changes accordingly. Nevertheless, the separation effected by several distillation stages is considerably better than by only one, as can be seen by comparing the curves for simple distillation and fractional distillation in Figure E13.

Distillation involving several concurrent vaporization-condensation cycles is called *fractional distillation*. During a fractional distillation the distillate is collected in several separate receivers, the contents of each receiver being a different *fraction*. The initial fraction is collected until the vapor temperature rises to a predetermined value, a second fraction is collected over a different temperature range, and so on. For a fractional distillation having the distillation curve illustrated by the broken line in Figure E13, the first 30 percent (15 mL) or so of distillate would be nearly pure entane, the last 30 percent would be nearly pure orctane, and the middle fraction would be a mixture of the two, which could be redistilled if desired.

Like the hypothetical distillation diagrammed in Figure E14, fractional distillation is a multistage distillation process performed in a single operation. A vertical *distilling column* performs the function of boiling flasks B and C in the hypothetical apparatus, redistilling (in effect) the original distillate from flask A. The column is filled with some type of *column packing,* which provides a large surface area from which repeated vaporization and condensation operations can take place.

Suppose our 50:50 mol% mixture of entane and orctane is distilled through such a column (Figure E16). The vapor entering the column from the pot will have the same composition (81.4 mol% entane) as before, but as it passes onto the column it will cool, condense onto the packing surface, and begin to trickle down the column on its way back to the pot. Since the temperature is higher near the pot, part of the condensate will vaporize on the way down, yielding a vapor richer in entane. This vapor will rise up the column until, at a higher level than before (since its boiling point is lower), it cools enough to recondense. This process of vaporization and condensation may be repeated a number of times on the way to the top of the column, so that when the vapor finally arrives at that point it is nearly pure entane. Although this is a continuous process involving the simultaneous upward flow of vapor and downward flow of liquid, the net result is the same as that produced by successive discrete distillations.

Each section of the distillation apparatus that provides a separation equivalent to one cycle of vaporization and condensation (one "step" on the temperature-composition graph) is called a *theoretical plate.* Since the first cycle occurs in the pot, it provides one theoretical plate; the column illustrated in Figure E16 contributes two more. Note that there is a continuous variation in both temperature and vapor composition as one proceeds up the column, and that the temperature is fixed by the liquid-vapor composition. For instance, the entane-rich liquid near the top of the column boils at a lower temperature than the original mixture, so the average temperature of plate 3 is lower than that of the plates below it.

In certain multistage columns, the "theoretical" plates are real. A Bruun column consists of a series of horizontal plates stacked at intervals inside a vertical tube, and one vaporization-condensation cycle occurs on each plate.

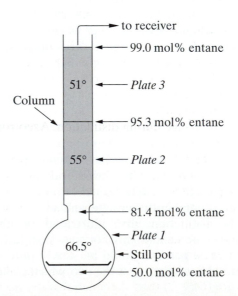

Figure E16 Operation of a fractionating column

The efficiency of a fractional distillation apparatus can be determined using the *Fenske equation,*

$$n = \frac{\log \dfrac{Z_A}{X_A} - \log \dfrac{Z_B}{X_B}}{\log \alpha}$$

where n is the total number of theoretical plates, X_A and X_B are the mole fractions of liquids A and B in the distilland, and Z_A and Z_B are the mole fractions of the same components in the vapor that emerges from the top of the column. The term α is the *relative volatility* of the two liquids, where the volatility of a liquid in a mixture is the ratio of its mole fraction in the vapor to its mole fraction in the liquid. The pot provides one theoretical plate, so the number of theoretical plates provided by the column is $n - 1$. The efficiency of a particular kind of column or column packing is given by its *HETP* (height equivalent to a theoretical plate), which is equal to the height of the column divided by the number of theoretical plates it provides. For example, a 24 cm column that provides 4 theoretical plates has an HETP of 6 cm. The lower its HETP, the more efficient is the column.

In practice, a number of factors limit the efficiency of a given column. Efficiency is highest under the equilibrium condition of *total reflux,* in which all of the vapors are returned to the pot. In practice, some of the vapors are continuously distilling into the receiver, which disturbs the equilibrium and reduces the column's efficiency. To maintain a reasonably high efficiency, the *reflux ratio (R),* the ratio of liquid returned to liquid distilled measured over the same time interval, must be kept reasonably high.

$$R = \frac{\text{liquid volume returning to pot}}{\text{liquid volume distilled}}$$

According to one rule of thumb, the reflux ratio should at least equal the number of theoretical plates for efficient operation. Reflux ratios of 5–10 are common for routine separations. To a lesser extent, the distillation rate and the *holdup* of a column (the amount of liquid that adheres to column's surface and packing) can also affect the efficiency of a distillation.

To this point we have considered only ideal liquids. No real liquid systems display ideal behavior, although some may come very close. The greatest deviations from ideal behavior occur with liquid mixtures that form azeotropes, where an *azeotrope* is a solution of two or more liquids whose composition does not change during distillation. Azeotrope formation can can make it difficult or impossible to purify certain liquids by distillation. For example, ethanol and water form a *minimum-boiling azeotrope* called 95% ethanol, which contains 95.6% ethanol and 4.4% water by mass and boils at 78.15°C—lower than the boiling points of both ethanol (78.5°C) and water. Distilling more dilute ethanol eventually yields 95% ethanol, but it is impossible to obtain pure ethyl alcohol by distilling 95% ethanol because the vapor has the same composition as the liquid in the pot. In this case, the problem caused by an azeotrope can be solved by a different azeotrope. Benzene and water form an azeotrope with ethanol that boils at a lower temperature (64.9°C) than the ethanol-water azeotrope, so by distilling 95% ethanol with some benzene, the benzene-water-ethanol azeotrope can be distilled off until all the water is removed and absolute (100%) ethanol is obtained.

Experimental Considerations

Heat Sources. For good results, the heat source must provide constant, uniform heating. An oil bath works best but a heating mantle will suffice if its heat output is carefully adjusted to bring about the desired distillation rate. Because of the potential fire hazard and the difficulty of maintaining a constant heating rate, a burner should not be used for fractional distillation.

Columns. The column is the most important part of any fractional distillation apparatus. One of the simplest is the Vigreux column, which has a series of indentations in a spiral arrangement down the length of the column. The surface area for condensation inside the column is comparatively small, so Vigreux columns are not very efficient, having HETP values in the 8–12 cm range. The most common type of column for fractional distillation is simply a straight tube (usually jacketed) with indentations at the bottom, filled with a suitable packing material. Packed columns are more efficient than Vigreux columns, but their holdup is higher and distillation rates are lower. Holdup losses of 3 mL or more can be expected from fractional distillation with the apparatus illustrated in Figure E18 (page 678), so it is not practical to distill less than about 20 mL from such a setup.

Packing materials include metal turnings and glass or porcelain beads, rings, helices, and saddles. Highly efficient packing materials such as Heli-Pak (glass helices) may provide HETP ratings of 1 cm or lower, but they are quite expensive. Glass beads and stainless steel "sponge" are more practical for use in undergraduate laboratories. Stainless steel sponge pads can be stretched and cut into 6–8 inch lengths for use in the distilling columns found in most organic lab kits. With this packing, the column is sometimes deliberately flooded by strong heating to wet the packing (see "Flooding"). Then the heat is reduced to drain the column before distillation is begun. Stainless steel packing should not be used to distill halogen compounds, which corrode it.

If any component of a distilland boils much above 100°C, the column and still head should be insulated to prevent heat losses that may reduce efficiency or prevent distillation entirely. They can be covered by one or more layers of crumpled aluminum foil, or glass wool can be wrapped around them and held in place by aluminum foil. The insulation should provide "windows" that can be opened to observe the vapors in the still head and the packing near the bottom of the column.

Flooding. One problem often encountered in a fractional distillation is *flooding,* in which the column becomes partly or entirely filled with liquid. Flooding is usually caused by an excessive heating rate, but it may also be caused by poor insulation, an unsuitable packing support, or improper packing. For example, sponge packing that is too tightly compressed or a glass-wool plug used as a packing support may hold up enough liquid to cause flooding. If flooding occurs, the heat source must be removed until all of the excess liquid has returned to the pot before the distillation is resumed. Flooding will greatly decrease the efficiency of a separation since it reduces the surface area of packing available for the separation.

Figure E17 Vigreux column

General Directions for Fractional Distillation

Equipment and supplies:

- heat source
- boiling flask
- distilling column
- insulating material (optional)
- thermometer
- condenser
- vacuum adapter
- joint clips or rubber bands

- clamps, rings, supports
- boiling chips or stirbar and stirrer
- column packing
- three-way connecting tube
- thermometer adapter
- condenser tubing
- receivers

Pack the column to within a centimeter or less of the upper ground joint. If you are using stainless steel sponge as a packing, *pull* it into the column using a copper wire bent into a hook at one end, making sure that it is as uniform as possible. To pack a column with glass beads, hold the column nearly horizontal with your hand over the bottom opening and place a few large beads in the top of the column. Then quickly pivot the column to an upright position so that (with a little luck) the beads will jam together at the constriction and support the remainder of the packing. (A small plug of stainless steel sponge can also serve as a support.) Slowly pour in the rest of the packing from a beaker, with continuous shaking, so that it is as uniform as possible. Other packing can be added similarly, except that glass helices should be dropped in one at a time, with shaking.

Assemble the apparatus illustrated in Figure E18, using large screw-cap vials or other appropriate containers, numbered and tared, as receivers. For very volatile distillates, small ground-joint flasks should be used as receivers. Use a boiling flask large enough that it will be no more than half full of distilland. The vacuum-adapter drip tube should extend into the receiver to reduce losses by evaporation. Make sure that all joints are tight, the column is perpendicular to the bench top, the boiling flask and receiver are supported properly, and none of the joints are under excessive strain. If the boiling temperature will rise to 100°C or higher during the distillation, insulate the apparatus from the bottom of the column to the top of the still head. Add the distilland to the distilling flask, and add a few boiling chips or a stirbar.

Raise the heat source into place, start the stirrer if you are using one, and begin heating to bring the mixture to the boiling point. When it boils, adjust the heating rate so that the reflux ring of condensing vapor passes up the column at a slow, even rate—it should take 5–10 minutes or more to reach the top of the column. Watch the packing at the bottom of the column closely for evidence of flooding. If flooding occurs, remove the heat source immediately and let the liquid drain into the boiling flask; then resume heating at a lower rate. If flooding is still a problem, you may need to reinsulate or repack the column. When the vapors rise above the column packing, adjust the heat to keep the reflux ring between the packing and the sidearm for a minute or so, giving the column time to equilibrate. Once distillation begins, read the thermometer when the temperature reading stabilizes. Distill at a rate of about 1 drop every 1–3 seconds, or at a rate that gives the desired reflux ratio. (Estimate the reflux ratio by counting the drops that

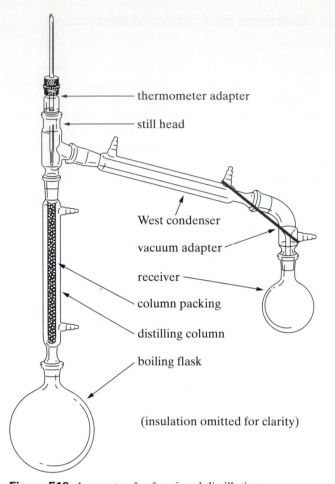

thermometer adapter

still head

West condenser

vacuum adapter

receiver

column packing

distilling column

boiling flask

(insulation omitted for clarity)

Figure E18 Apparatus for fractional distillation

drip into the pot and those that drip into the receiver during a short time period.) If the initial distillate is cloudy (due to dissolved water), change receivers when it becomes clear.

If the column is not very efficient or the components' boiling points are close together, the temperature may rise only gradually throughout the distillation. In that case, it is best to collect fractions continuously at regular temperature intervals and redistill them. Otherwise, continue distilling until the still-head temperature begins to rise sharply (or when a predetermined target temperature is reached), change receivers, and record the temperature. Change receivers again when the temperature begins to stabilize at a higher value (or when the next target temperature is reached) and record the temperature. Increase the heating rate as necessary to maintain a suitable distillation rate. Repeat this process as necessary, changing receivers each time the temperature begins to rise and each time it stabilizes at a higher value, until the pot is nearly dry *or* the temperature drops sharply *or* the final target temperature is reached. (Note that the temparature may also drop if the heating rate is not high enough to bring the vapors up to the still head.) Then lower and turn off the heat source and let the column drain. The fractions collected while the temperature was rising rapidly are impure;

unless you wish to redistill them, they should be placed in a solvent recovery container. Disassemble the apparatus and clean it promptly.

Summary

1 Pack column; assemble and insulate apparatus.
2 Add liquid and boiling chips or stirbar.
3 Turn on stirrer (if used) and heat source.
4 Adjust heat so that reflux ring passes slowly up the column.
5 Record temperature when distillation begins and thermometer reading stabilizes.
6 Distill until temperature rises sharply or target temperature is reached, change collectors, record temperature.
   IF more fractions are to be collected, REPEAT 6.
   IF you are distilling the last fraction, CONTINUE.
7 Distill until temperature drops sharply or final target temperature is reached.
8 Remove heat source, drain column.
9 Disassemble and clean apparatus, dispose of residue and impure fractions.

F. Measuring Physical Constants

Melting Point

Principles and Applications

The melting point of a pure substance is defined as the temperature at which the solid and liquid phases of the substance are in equilibrium at a pressure of 1 atmosphere. At a temperature slightly lower than the melting point, a mixture of the two phases solidifies; at a temperature slightly above the melting point, the mixture liquefies. Melting points can be used to characterize organic compounds and to assess their purity. The melting point of a pure compound is a unique property of that compound, which is essentially independent of its source and method of purification. This is not to say that no two compounds will have the same melting point; many compounds have melting points that differ by no more than a fraction of a degree. If two pure samples have *different* melting points, however, they are almost certainly different compounds.

The melting point of an organic solid is usually measured by grinding the solid to a powder and packing the powder inside a *melting-point tube,* a capillary tube that is closed at one end. The melting-point tube is then placed in an appropriate heating device and the *melting-point range* of the sample, the range of temperatures over which the solid is converted to a liquid, is observed and recorded. A pure substance usually melts within a range of no more than 1–2 °C; that is, the transition from a crystalline solid to a clear, mobile liquid occurs within a degree or two if the rate of heating is sufficiently slow and the sample is properly prepared.

The presence of impurities in a substance *lowers* its melting point and *broadens* its melting-point range. To better understand the effects of impurities on melting point, consider the phase diagram for phenol (P) and diphenylamine (D) in Figure F1. Pure phenol melts at 41°C and pure diphenylamine at 53°C. If a sample of phenol contains a small amount of diphenylamine as an impurity, its melting point will decrease in approximate proportion to the mole percent of diphenylamine present; likewise, the melting point of diphenylamine will decrease on addition of phenol. For example, the melting point of a mixture containing 10 mol% diphenylamine in phenol is given by the point M; that of 20 mol% phenol in diphenylamine is given by point N. Pure phenol and diphenylamine both have sharp melting points. Mixtures of the two (except the *eutectic mixture* at the minimum in the diagram) exhibit broader melting-point ranges that depend on their composition. The approximate melting-point range for a mixture is given by the distance between the broken line connecting points P and E or points E and D and the solid lines connecting the same points. For example, the melting-point range of a 10 mol% diphenylamine mixture is given in the figure by the distance between M' and M, and for an 80 mol% mixture by the distance between N' and N.

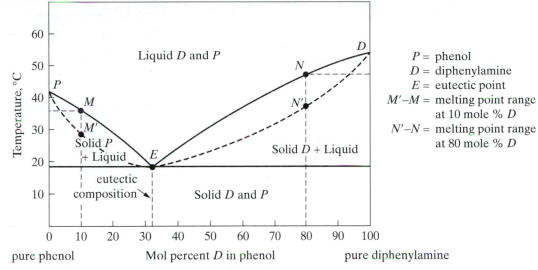

Figure F1 Phase diagram for the phenol-diphenylamine system

Since the melting point decreases in a nearly linear fashion as the amount of impurity increases (up to a point, at least), the difference between the observed and expected values may make it possible to estimate a compound's purity, as is done for camphor in Experiment 7. The melting point lowering effect can also be used to confirm the identity of a substance, such as the product of a reaction, when it is thought to be a certain known compound. The compound in question is mixed with a sample of the known compound and the melting point of the mixture is measured. If the two compounds are identical, the mixture melting point will be essentially the same as that for the known compound measured separately. If they are not identical, the known compound will act as an impurity in the unknown, so the melting point of the mixture will be lower and its range broader than for the known compound.

Experimental Considerations

Melting Behavior. The melting-point range of a sample is reported as the range between (1) the temperature at which the sample first begins to liquefy and (2), the temperature at which it is completely liquid, called the *liquefaction point.* When a single melting point is to be reported, the liquefaction point is generally used, although a compound may also be charactized by its *meniscus point,* the temperature at which the liquid meniscus is barely clear of the solid below it. Some automatic melting-point devices report melting points that are closer to the meniscus point than to the liquefaction point.

If traces of solvent remain in a sample due to insufficient drying or other causes, you may observe "sweating" of solvent from the sample or bubbles in the molten sample, which may resolidify when all of the solvent is driven off. If a sample does show this behavior, it should be dried and the melting point remeasured. Even a dry sample will tend to soften and shrink before it begins to liquefy; this process begins at the *eutectic temperature* (point *E* in Figure F1.) In any case, softening, shrinking, and sweating should

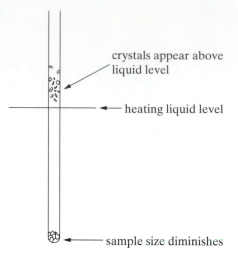

Figure F2 above: crystals appear above liquid level · heating liquid level · sample size diminishes

Figure F2 Sublimation of a sample in a melting-point tube

clamp here · heat

Stir with a gentle, up-and-down motion

Figure F3 Stirred heating bath

not be mistaken for melting behavior. The melting-point range does not begin until the first free liquid is clearly visible.

Some compounds sublime—change directly from the solid to the vapor state—when they are heated in an open container. Sublimation can be detected during a melting-point determination by a pronounced shrinking of the sample, accompanied by the appearance of crystals higher up inside the melting-point tube (see Figure F2). The melting point of a sample that sublimes at or below its melting temperature can be measured using a sealed capillary tube. An ordinary melting-point tube should be cut [see OP-3] short enough so that it does not project above the block of a melting-point apparatus such as the one in Figure F4, or so that it can be entirely immersed in a heating-bath liquid. The sample is introduced and the open end is sealed with a burner. Then the melting point is measured by one of the methods described in the General Directions.

If a sample becomes discolored and liquefies over an unusually broad range during a melting-point determination, it is probably undergoing thermal decomposition. Compounds that decompose on heating melt at temperatures that vary with the rate of heating. The approximate melting point (or *decomposition point*) of such a compound should be measured by heating the melting-point apparatus to within a few degrees of the expected melting temperature before inserting the sample, then raising the temperature at a rate of about 3–6°C per minute.

Apparatus for Measuring Melting Points. Melting points can be determined with good accuracy using a special melting-point tube (Thiele tube or Thiele–Dennis tube) filled with a heating-bath liquid such as mineral oil [see OP-6a], as illustrated in Figure F5 (page 686). The design of such tubes promotes good circulation of the heating liquid without stirring. A capillary tube containing the solid is secured to a thermometer, which is then immersed in the bath liquid. The solid is observed carefully for evidence of melting as the apparatus is heated. Melting points can also be measured in an ordinary beaker if the bath liquid is stirred constantly, either manually or with a magnetic stirrer (see Figure F3).

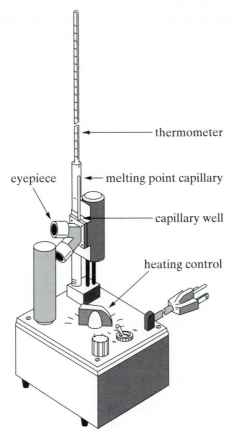

thermometer

eyepiece

melting point capillary

capillary well

heating control

Figure F4 Capillary melting-point apparatus (Mel-Temp®)

Commercial melting-point instruments that use capillary melting-point tubes are available, such as the Mel-Temp® illustrated in Figure F4. Such melting-point devices are quite accurate if operated properly, and they can be used to make several measurements at once. This feature is useful when a mixture melting point is being determined, since the melting points of the unknown compound, the known, and the mixture can be measured and compared at the same time. The heating rate is adjusted by a dial controlling the voltage input to a heating coil. The dial reading required to attain the desired heating rate at the melting point of the sample can be estimated from a heating-rate chart furnished with the instrument. The heating-rate dial is initially set higher than this to bring the temperature within 15–20°C of the expected melting point, then reduced to the estimated value.

Melting-Point Corrections. The observed melting point of a compound may be inaccurate because of defects in the thermometer used or because of the "emergent stem error" that results when a thermometer is not immersed to its intended depth in a heating bath. Many thermometers are designed to be immersed to the depth indicated by an engraved line on the stem, usually 76 mm from the bottom of the bulb; slight deviations from this depth will not result in serious error. Other thermometers are designed for total immersion; if they are used under other circumstances, the temperature readings will be in error. Such an error can be compensated for by adding an *emergent stem correction* calculated with the following equation. This requires that a second thermometer be held opposite the middle of the exposed part of the mercury column.

> emergent stem correction (to be added to t_1) = $0.00017 \cdot N(t_1 - t_2)$
> N = length in degrees of exposed mercury column
> t_1 = observed temperature
> t_2 = temperature at middle of exposed column

Melting points that have been corrected in this way should be reported as, for example, "m.p. 123–124.5° (corr.)."

Errors due to either thermometer defects or an emergent stem can be corrected by *calibrating* the thermometer under the conditions in which it is to be used. A thermometer used for melting-point determinations, for example, is calibrated by measuring the melting points of a series of known compounds, subtracting the observed melting point from the true melting point of each compound, and plotting this correction as a function of temperature. The correction at the melting point of any other compound is then read from the graph and added to its observed value. A list of pure compounds that can be used for melting-point calibrations is given in Table F1. Sets of pure calibration substances are available from chemical supply houses.

Mixture Melting Points. A mixture melting point is obtained by grinding together approximately equal quantities of two solids (a few milligrams of each) until they are thoroughly intermixed, then measuring the melting point of the mixture by the usual method. Usually one of the compounds (X) is an "unknown" that is believed to be identical to a known compound, Y. A sample of pure Y is mixed with X and the melting points of this mixture and of pure Y (and sometimes of X as well) are measured. If the melting-point ranges of pure Y and of the mixture are identical (within a degree or so), then X is probably identical to Y. If the mixture melts at a lower temperature and over a broader range than pure Y, then X and Y are different compounds.

Table F1 Substances used for melting-point calibrations

Substance	m.p., °C
ice	0
diphenylamine	54
m-dinitrobenzene	90
benzoic acid	122.5
salicylic acid	159
3,5-dinitrobenzoic acid	205

General Directions for Melting-Point Measurements

Safety Notes

> **Mineral oil begins to smoke and discolor below 200°C and can burst into flames at higher temperatures. Oil fires can be extinguished with solid-chemical fire extinguishers or powdered sodium bicarbonate.**

Equipment and supplies
(Starred items are for method *B* only):

- Mel-Temp® (method *A* only)
- 1-m length of glass tubing
- capillary melting-point tube
- *clamp, ring stand
- *burner
- *3-mm slice of rubber tubing
- flat-bottomed stirring rod or flat-bladed spatula
- watch glass
- thermometer
- *Thiele or Thiele–Dennis tube
- *heating oil
- *cut-away cork

Put a few milligrams of the dry solid on a small watch glass and grind it to a fine powder with a flat-bottomed stirring rod or a flat-bladed spatula. Press the open end of a capillary melting-point tube into a pile of the powder until enough has entered the tube to form a column 1–2 mm high. Using too much sample can result in a melting-point range that is too broad and a melting-point value that is too high. Tap the closed end of the tube gently on the bench top (or rub its sides with a small file); then drop it (open end up) through a 1-meter length of small-diameter glass tubing onto a hard surface, such as the bench top. Repeat the process several times to pack the sample firmly into the bottom of the tube. Use one of the following methods to measure the melting-point range of the sample. If you don't know the sample's expected melting point, determine its approximate value with rapid heating (6°C per minute or more), then carry out a more accurate melting-point measurement with a second sample as described here.

A melting-point tube can be constructed by sealing one end of an open-ended capillary tube that is approximately 10 cm long and 1 mm in diameter (I.D.).

A. Mel-Temp® method.

Place the melting-point tube (sealed side down) in one of the wells on the Mel-Temp's heating block. Use the heating-rate chart to estimate the dial setting that will cause the temperature to rise at a rate of 1–2°C per minute at the expected melting point of the sample. If the thermometer reading is well below the expected melting point, adjust the dial setting to raise the temperature quite rapidly until it is within 15°C of the expected value, then reduce the dial setting so that the temperature is rising no more than 2°C/minute (preferably 1°C/minute) by the time the temperature is within 5°C of the expected melting point. Continue heating at that rate as you observe the sample through the magnifying lens in the observation port, and record (as the limits of the melting point range) the temperatures (1) when the first free liquid appears in the melting-point tube and (2) when the sample is completely liquid. It is advisable to do at least two measurements on each compound. Let the block cool to 10–15° below the melting point before you do another measurement. Cooling can be accelerated by passing an air stream over the block.

B. Thiele-tube Method.

(You can use a similar method with other melting-point baths such as the one in Figure F3, but stirring and a hot plate or other flameless heat source will be required in that case.) Clamp the Thiele or Thiele–Dennis tube securely to a ring stand and add enough mineral oil (or other bath liquid) to just cover the top of the sidearm outlet, as shown in Figure F5A. Secure the melting-point tube at its top to a broad-range thermometer by means of a 3-mm-thick rubber band (cut from $\frac{1}{4}$-inch I.D. thin-walled rubber tubing) so that the sample is adjacent to the middle of the thermometer bulb (Figure F5C). Pinch the rubber band between your fingers when inserting the melting-point tube. Snap the thermometer into the center of a cut-away cork (Figure F5B) at a point above the rubber band, with the capillary tube and the degree markings on the same side as the opening in the cork. Insert this assembly into the bath liquid and move the thermometer, if necessary, so that its bulb is centered in the tube about 3 cm below the sidearm junction and the temperature can be read through the opening in the cork (Figure F5A). The rubber band should be 2–3 cm above the liquid level so that the bath liquid, as it expands on heating, will not contact it. If the hot oil covers the rubber band, it may soften and allow the capillary tube to drop out.

The cork is bored out to accommodate the thermometer, then cut with a single-edged razor blade or a sharp knife so that the thermometer can be snapped into place from the side rather than inserted from the top.

If the expected melting point of the compound is known, heat the bottom of the Thiele tube with a burner (or a microburner, for more precise control)

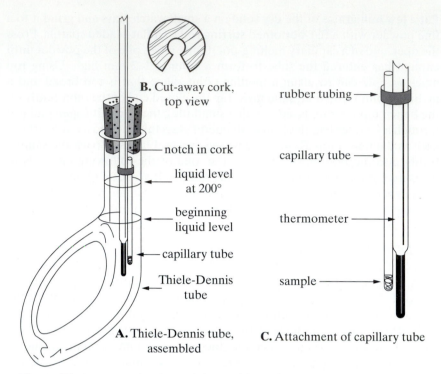

B. Cut-away cork, top view

notch in cork

liquid level at 200°

beginning liquid level

capillary tube

Thiele-Dennis tube

A. Thiele-Dennis tube, assembled

rubber tubing

capillary tube

thermometer

sample

C. Attachment of capillary tube

Figure F5 Apparatus for determining melting points

until the temperature is about 15°C below the expected value. Turn down the flame and apply it at the sidearm to reduce the heating rate so that the temperature is rising no more than 2°C/minute (preferably 1°C/minute) by the time the temperature is within 5°C of the expected melting point. Continue heating at that rate as you observe the sample closely, and record (as the limits of the melting-point range) the temperatures (1) when the first free liquid appears in the melting-point tube and (2) when the sample is completely liquid. It is advisable to do at least two measurements on each compound. Let the heating bath cool to 10–15° below the melting point before you do another measurement. Cooling can be accelerated by passing an air stream over the tube. Clean up the apparatus and place the oil in a container for recycling (if the oil is not too dark, it may be stored in the Thiele tube). Mineral oil can be removed from glassware by rinsing the glassware with petroleum ether, followed by acetone, then washing it with a detergent and water.

Summary

1 Assemble apparatus for melting-point determination, if necessary.
2 Grind solid to powder, fill capillary tube(s) to depth of 1–2 mm.
    IF you are using method *B*, GO TO 4.
3 Insert sample in Mel-Temp heating block. GO TO 5.
4 Secure capillary tube to thermometer; insert assembly in heating bath.
5 Heat rapidly to ~15°C below m.p.; then reduce heating rate to 1–2 °C/min.
6 Observe and record melting-point range.
7 Disassemble apparatus as needed, clean up.

Boiling Point

The boiling point of a liquid is defined as the temperature at which the vapor pressure of the liquid is equal to the external pressure at the surface of the liquid, and also as the temperature at which the liquid is in equilibrium with its vapor phase at that pressure. These definitions are the basis for various standard-scale and small-scale methods for measuring boiling points. For example, the boiling point of a liquid can be determined by distilling a small quantity of the liquid and observing the temperature at the still head, where the liquid and its vapors are assumed to be in equilibrium. It can also be determined by measuring the temperature at which the liquid's vapor, trapped inside a capillary tube immersed in the liquid, is balanced against the external pressure. Like the melting point of a solid, the boiling point of a liquid can be used to help identify it and assess its purity.

Boiling-Point Corrections. The *normal boiling point* of a liquid is its boiling point at an external pressure of 1 atmosphere (760 torr, 101.3 kPa). Since the atmospheric pressure at the time of a boiling-point determination is seldom exactly 760 torr, observed boiling points may differ somewhat from values reported in the literature and should be corrected. If a laboratory boiling-point determination is carried out at a location reasonably close to sea level, atmospheric pressure will rarely vary by more than 30 torr from 760. For deviations of this magnitude, a *boiling-point correction,* Δt, can be estimated using Equation **1**.

$$\Delta t \approx y(760 - P)(273.1 + t) \tag{1}$$

Δt = temperature correction, to be added to observed boiling point
P = barometric pressure, in torr
t = observed boiling point (°C)

The value 1.0×10^{-4} is used for the constant y if the liquid is water, an alcohol, a carboxylic acid, or another associated liquid; otherwise y is assigned the value 1.2×10^{-4}. For example, the boiling point of water at 730 torr is 98.9°C. Use of Equation **1** leads to a correction factor of $[(1.0 \times 10^{-4})(760 - 730)(273.1 + 98.9)]$ °C = 1.1°C, which yields the correct normal boiling point of 100.0°C.

At high altitudes, the atmospheric pressure may be considerably lower than 1 atmosphere, resulting in observed boiling points substantially lower than the normal values. For example, water boils at 93°C on the campus of the University of Wyoming (elevation 7520 feet) and at 81°C at the top of Mount Evans in Colorado (elevation 14,264 feet). For major deviations from atmospheric pressure, Equation **2** can be used in conjunction with approximate entropy of vaporization values obtained from the section "Correction of Boiling Points to Standard Pressure" in the *CRC Handbook of Chemistry and Physics.*

*Equation **2** is a simplified form of the Hass–Newton equation found in the CRC Handbook, 64th edition, p. D-189.*

$$\Delta t \approx \frac{(273 + t)}{\phi} \log \frac{760}{P} \tag{2}$$

ϕ = (entropy of vaporization at normal boiling point)/2.303 R

For example, hexane boils at 49.6°C at 400 torr; the value of ϕ for alkanes is found (from the *CRC Handbook*) to be about 4.65 at that temperature. Substituting into Equation **2**, we obtain a correction factor of:

$$\Delta t \approx \frac{(273°C + 49.6°C)}{4.65} \log \frac{760 \text{ torr}}{400 \text{ torr}} = 19.3°C$$

which gives a normal boiling point of 68.9°C. A second approximation, using the value of ϕ at the corrected boiling point, gives 68.8°C. This compares very favorably to the reported value of 68.7°C. As in melting-point determinations, it may be necessary to correct the boiling point for thermometer error, especially when working with high-boiling liquids. See OP-28 for details.

a. Distillation Boiling Point

During a carefully performed distillation of a pure liquid, the vapors surrounding the thermometer bulb are in equilibrium (or nearly so) with the liquid condensing on the bulb. Thus the still-head temperature recorded during the distillation of a pure liquid should equal its boiling point. If the liquid is contaminated by impurities, the distillation boiling point may be either too high or too low, depending on the nature of the impurity. Volatile impurities in a liquid lower its boiling point, whereas involatile impurities raise it; in either case, the distillation boiling-point range will be broadened. Therefore if there is any doubt about the purity of a liquid, it should be distilled or otherwise purified prior to a boiling-point determination.

General Directions for Distillation Boiling-Point Measurement

If special small-scale glassware is available, use it to assemble an apparatus for small-scale distillation. Otherwise assemble the apparatus pictured in Figure E8 [OP-25], using a 25-mL distilling flask and following the instructions given in that Operation. Correct thermometer placement is essential (see Figure E7). Be sure the still head is well insulated from the heat source to prevent superheating of the vapors. Add 5–10 mL of the pure liquid to the pot and drop in a boiling chip or two, or a stirbar. Distill the liquid slowly (1 drop per second or less), recording the temperature after the first 1–2 mL has distilled and again when only 1–2 mL remains in the pot. It is also advisable to record the *median boiling point,* the temperature at which approximately half of the liquid has distilled, since low- or high-boiling impurities may invalidate the readings at the beginning or end of a distillation. If the distillation temperature rises markedly during the distillation, collect the distillate in several fractions, changing collectors whenever the boiling point starts to rise and again when it levels off. Record the barometric pressure so that you can make a pressure correction, if necessary. If the boiling range is more than 2°C, it may be advisable to repeat the determination using the purest fraction collected.

b. Micro Boiling Point

When a liquid is heated to its boiling point, the pressure exerted by its vapor becomes just equal to the external pressure at the liquid's surface. If a tube that is closed at one end is filled with a liquid and immersed (open end

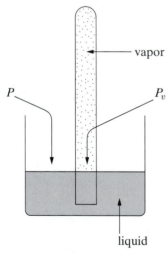

Figure F6 Micro boiling point principle. At the boiling point, $P_v = P$, where P_v = pressure exerted by vapor on liquid surface and P = pressure exerted by atmosphere on liquid surface.

down) in a reservoir containing the same liquid, the tube will begin to fill with vapor as the liquid is heated to its boiling point. At the boiling point, the vapor pressure inside the tube will be balanced by the pressure exerted on the liquid surface by the surrounding atmosphere, so that the liquid levels inside and outside the tube will be equal (see Figure F6). If the temperature is raised above the boiling point, vapor will escape in the form of bubbles; if the temperature is lowered below the boiling point, the tube will begin to fill with liquid. This behavior is the basis of a micro method for determining boiling points, which utilizes the apparatus in Figure F7.

General Directions for Micro Boiling Point Measurement

Equipment and supplies:

- boiling tube
- thermometer
- Thiele tube
- burner
- capillary tube
- rubber band
- mineral oil

Add 2–4 drops of the liquid to a boiling tube constructed of a 10-cm length of 4–5 mm O. D. glass tubing sealed at one end [see OP-3]. Insert a capillary melting-point tube (sealed at one end) into the boiling tube with its open end down, and secure the assembly to a thermometer by means of a rubber band cut from thin-walled rubber tubing, as illustrated in Figure F7. Insert this assembly into a Thiele tube as for a melting-point determination (see Figure F5, p. 686), with the rubber band 2 cm or more above the liquid level.

Heat the Thiele tube until a *rapid,* continuous stream of bubbles emerges from the capillary tube (a slow stream of bubbles may be caused by expansion of air rather than vaporization of the liquid). Avoid overheating or the liquid will boil away; it may be necessary to add more liquid if the sample is very volatile. Remove the heat source, let the bath cool slowly until the bubbling stops, and record the temperature when liquid just begins to enter the capillary tube. Let the temperature drop a few degrees so that the liquid partly fills the capillary tube. Then heat very slowly until the first bubble of vapor emerges from the mouth of the capillary tube. Record the temperature at that point also. The two temperatures represent the boiling-point range; they should be within a degree or two of each other. Cool the bath until liquid again fills the capillary tube; then repeat the determination. If repeated determinations on the same sample give appreciably different (usually higher) values for the boiling point, the sample is probably impure and should be distilled. Record the barometric pressure so that a pressure correction can be made. Clean and dry the boiling tube, saving it for future boiling-point measurements.

Summary

1. Add liquid to boiling tube, insert inverted m.p. tube.
2. Assemble Thiele tube apparatus.
3. Heat until continuous stream of bubbles emerges from m.p. tube, remove heat.
4. Record temperature when liquid enters capillary.
5. Heat slowly until first vapor bubble emerges from capillary; record temperature.

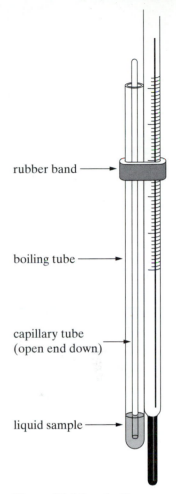

rubber band

boiling tube

capillary tube (open end down)

liquid sample

Figure F7 Micro boiling-point assembly

Sometimes the capillary tube will stick to the bottom of the boiling tube. This can be prevented by cutting a small nick in the open end of the capillary tube with a triangular file.

6 Let cool below boiling point.
IF another measurement is required, GO TO 3.
IF not, CONTINUE.
7 Record barometric pressure, correct observed boiling point.
8 Disassemble apparatus, clean up.

Refractive Index

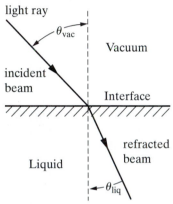

light ray

θ_{vac}

Vacuum

incident beam

Interface

refracted beam

Liquid

θ_{liq}

Figure F8 Refraction of light in a liquid

The *refractive index* (index of refraction) of a substance is defined as the ratio of the speed of light in a vacuum to its speed in the substance in question. When a beam of light passes into a liquid, its velocity is reduced, causing it to bend downward. The refractive index is related to the angles that the incident and refracted beams make with a line perpendicular to the liquid surface, according to Equation **1**.

$$n^t_\lambda = \frac{c_{vac}}{c_{liq}} = \frac{\sin\theta_{vac}}{\sin\theta_{liq}} \tag{1}$$

n^t_λ = refractive index at temperature t using light of wavelength λ
c = velocity of light

The refractive index is a unique physical property that can be measured with great accuracy (up to eight decimal places), and it is thus very useful for characterizing pure organic compounds. Refractive-index measurements are also used to assess the purity of known liquids and to determine the composition of solutions.

Experimental Considerations

It is much easier to make measurements in air than in a vacuum, so most refractive-index measurements are made in air, whose refractive index is 1.0003, and the small difference is corrected by the instrument. Refractive-index values depend on both the wavelength of the light used for their measurement and the density of the liquid, which varies with its temperature. Most values are reported with reference to light from the yellow D line of the sodium emission spectrum, which has a wavelength of 589.3 nm. Thus a refractive index value may be reported as $n^{20}_D = 1.3330$, for example, where the superscript is the temperature in °C and D refers to the sodium D line. Since most refractive index readings are made at or corrected to this wavelength, the D is sometimes omitted. If another light source is used, its wavelength is specified in nanometers.

The Abbe refractometer (Figure F9, p. 692) is a widely used instrument for measuring refractive indexes of liquids. It measures the *critical angle* (the smallest angle at which a light beam is completely reflected from a surface) at the boundary between the liquid and a glass prism and converts it to a refractive index value. The instrument employs a set of compensating prisms so that white light can be used to give refractive-index values corresponding to the sodium D line. A properly calibrated Abbe-3L refractometer should be reliable to ±0.0002. The calibration of the instrument can be checked by measuring the refractive index of distilled water, which should be 1.3330 at 20°C and 1.3325 at 25°C.

The sample block of an Abbe refractometer can be kept at a constant temperature of 20.0°C by water pumped through it from a thermostatted water bath. If a refractive index is measured at a temperature other than 20.0°C, the temperature should be read from the thermometer on the instrument and the refractive index corrected by using the following equation,

$$\Delta n = 0.00045 \times (t - 20.0)$$

where t is the temperature of the measurement in °C. The correction factor (including its sign) is added to the observed refractive index. For example, if the refractive index of an unknown liquid is found to be 1.3874 at 25.1°C, its refractive index at 20.0°C should be about $1.3874 + [0.00045 \times (25.1 - 20.0)]$ = 1.3897. Alternatively, the refractive index of a reference liquid similar in structure and properties to the unknown liquid can be measured at the same temperature (t) as the unknown, and a correction factor calculated from the following equation,

$$\Delta n = n^{20} - n^{t}$$

where n^{t} is the measured refractive index of the reference liquid at temperature t and n^{20} is the reported value of its refractive index at 20°C. The correction factor is then added to the measured refractive index of the unknown at temperature t. This method can correct for experimental errors (improper calibration of the instrument, etc.) as well as temperature differences.

Small amounts of impurities can cause substantial errors in the refractive index. For example, the presence of just 1% (by mass) of acetone in chloroform reduces the refractive index of the latter by 0.0015. Such errors can be critical when refractive-index values are being used for qualitative analysis, so it is essential that an unknown liquid be pure when its refractive index is measured.

Directions for Refractive Index Measurements

These directions apply to the B&L Abbe-3L refractometer; if you are using a different instrument your instructor will demonstrate its operation.

Equipment and supplies:

- Abbe-3L refractometer
- washing liquid
- dropper
- soft tissues

Raise the *hinged prism* of the Abbe refractometor and, using an eyedropper, place 2–3 drops of the sample in the middle of the *fixed prism* below it. Never allow the tip of a dropper or another hard object to touch the prisms—they are easily damaged. If the liquid is volatile and free flowing, it may be introduced into the channel alongside the closed prisms. With the prism assembly closed, switch on the *lamp* and move it toward the prisms to illuminate the visual field as viewed through the eyepiece. Rotate the *handwheel* until two distinct fields (light and dark) are visible in the eyepiece, and reposition the lamp for the best contrast and definition at the borderline between the fields.

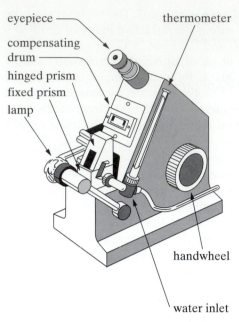

Figure F9 Abbe-3L Refractometer

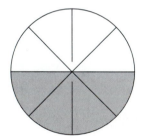

Figure F10 Visual field for properly adjusted refractometer.

Rotate the *compensating drum* on the front of the instrument until the borderline is sharp and achromatic (black and white) where it intersects an inscribed vertical line. If the borderline cannot be made sharp and achromatic, the sample may have evaporated. Rotate the handwheel (or the fine adjustment knob, if there is one) to center the borderline exactly on the crosshairs (Figure F10).

Depress and hold down the *display switch* on the left side of the instrument to display an optical scale in the eyepiece. Read the refractive index from this scale (estimate the fourth decimal place), and record the temperature if it is different from 20.0°C. Open the prism assembly and remove the sample by gently *blotting* it with a soft tissue (do not rub!). Wash the prisms by moistening a tissue or cotton ball with a suitable solvent (acetone, methanol, etc.) and blotting them gently. When the residual solvent has evaporated, close the prism assembly and turn off the instrument.

Summary

1   Insert sample between prisms.
2   Switch on and position lamp.
3   Rotate handwheel until two fields are visible; reposition lamp for best contrast.
4   Rotate compensating drum until borderline is sharp and achromatic.
5   Rotate handwheel or fine adjustment knob until line is centered on crosshairs.
6   Depress display switch to display optical scale, estimate refractive index to four decimal places, record temperature.
7   Clean and dry prisms, close prism assembly.
8   Make temperature correction, if necessary.

Optical Rotation

Principles and Applications

Light can be considered a wave phenomenon with vibrations occurring in an infinite number of planes perpendicular to the direction of propagation. When a beam of ordinary light passes through a *polarizer,* such as a Nicol prism, only the light-wave components that are parallel to the plane of the polarizer can pass through. The resulting beam of *plane-polarized light* has light waves whose vibrations are restricted to a single plane (see Figure F11). When a beam of plane-polarized light passes through an optically active substance, molecules of the substance interact with the light so as to rotate its plane of polarization. The angle by which a sample of an optically active substance rotates the plane of that beam is called its *observed rotation,* α. The observed rotation of a sample depends on the length of the light path through the sample and the concentration of the sample, as well as its identity. The first two are not intrinsic properties of the sample itself, so the observed rotation is divided by these factors to yield its *specific rotation,* $[\alpha]$.

$$[\alpha] = \frac{\alpha}{lc} \tag{1}$$

$[\alpha]$ = specific rotation
α = observed rotation
l = length of sample, in decimeters (dm)
c = concentration of solution, in grams of solute per milliliter solution (for a neat liquid, substitute the density in g/mL)

The specific rotation of a pure substance is an intrinsic property of the substance and can be used to characterize it.

The composition of a mixture of known optically active substances can be calculated from its specific rotation, using equation **2**.

$$[\alpha] = [\alpha]_A X_A + [\alpha]_B (1 - X_A) \tag{2}$$

In this equation $[\alpha]$ is the specific rotation of the mixture, $[\alpha]_A$ the specific rotation of component A, X_A the mole fraction of component A, and $[\alpha]_B$

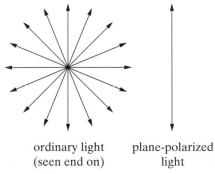

ordinary light   plane-polarized
(seen end on)   light

Figure F11 Schematic representations of ordinary and plane-polarized light

the specific rotation of component B. For example, an equilibrium mixture of α-D-glucose ($[\alpha] = 112°$) and β-D-glucose ($[\alpha] = 18.7°$) has a specific rotation of $52.7°$. The mole fraction of α-D-glucose in the mixture can be calculated by substituting these values into equation 2 and solving for X_A.

$$52.7° = (112°)\, X_A + (18.7°)(1 - X_A)$$
$$X_A = 0.364$$

Since both forms of glucose have the same molecular weight, the equilibrium mixture contains 36.4% α-D-glucose and 63.6% β-D-glucose by mass.

Experimental Considerations

The specific rotation of an optically active sample can be determined using an instrument called a *polarimeter*. A polarimeter consists of a *polarizer*, a *sample cell* to hold the sample, and an *analyzer*—a second polarizing prism that can be rotated at an angle to the polarizer. When the axis of the analyzer is perpendicular to that of the polarizer, it blocks out the plane-polarized light. In a precision polarimeter, the analyzer contains two or more prisms set at a small angle to each other, and it is rotated until the prisms bracket the point of minimum intensity, transmitting dim light of equal intensity. This feature is necessary because it is easier for the human eye to match two intensities than to estimate a point of minimum intensity. The angle by which the analyzer must be rotated to reach this point equals the angle at which the sample rotated the beam of plane-polarized light, its observed rotation, α. Using equation **1**, the observed rotation of the sample can be converted to its specific rotation given the sample's concentration and the length of the sample cell (usually 1 or 2 dm).

The optical rotation of a substance is usually measured in solution. Water and alcohol are common solvents for polar compounds, and dichloromethane can be used for less polar ones. Often a suitable solvent will be listed in the literature, along with the light source and temperature used for the measurement. The volume of solution required depends on the size of the polarimeter cell, but 10–25 mL is usually sufficient. A solution for polarimetry ordinarily contains about 1–10 g of solute per 100 mL of solution and should be prepared using an accurate balance and a volumetric flask. If the solution contains particles of dust or other solid impurities, it should be filtered [OP-11]. When possible, the concentration should be comparable to that reported in the literature. For example, the *CRC Handbook of*

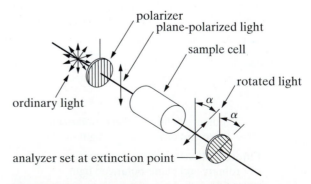

Figure F12 Schematic diagram of a polarimeter.

Chemistry and Physics reports the optical rotation of (+)-menthol as "+49.2 (al, $c = 5$)," where the concentration (c) is given in grams per 100 mL of an alcoholic (al) solution (c is defined differently here than in equation **1**). Thus a menthol solution for polarimetry can be prepared by accurately weighing about 1.25 g of (+)-menthol, dissolving it in ethyl alcohol (al) in a 25-mL volumetric flask, and adding more alcohol up to the calibration mark.

General Directions for Polarimetry

This procedure applies to a Zeiss-type polarimeter with a split-field image. Experimental details may vary somewhat for different instruments.

Remove the screw cap and glass end plate from one end of a clean 1- or 2-decimeter sample cell (do not get fingerprints on the end plate), and rinse the cell with a small amount of the solution to be analyzed. Stand the polarimeter cell vertically on the bench top and overfill it with the solution, rocking the tube if necessary to shake loose any air bubbles. Add the last milliliter or so with a dropper so that the liquid surface is convex. Carefully slide the glass end plate on so that there are no air bubbles trapped inside. If the tube has a bulge at one end, a small bubble can be tolerated; in that case the tube should be tilted so that the bubble migrates to the bulge and stays out of the light path. Screw on the cap just tightly enough to provide a leakproof seal—overtightening it may strain the glass and cause erroneous readings. If the light source is a sodium lamp, make sure it has had ample time to warm up (some require 30 minutes or more). Place the sample cell in the polarimeter trough, close the cover, see that the light source is oriented to provide maximum illumination in the eyepiece, and focus the eyepiece if necessary. Set the analyzer scale to zero and (if necessary) rotate it a few degrees in either direction until a dark and a light field are clearly visible. You may see a vertical bar down the middle and a background field on both sides, as shown in Figure F13, or two fields divided down the middle.

Focus the eyepiece so that the line(s) separating the fields are as sharp as possible. Starting from zero, rotate the analyzer scale about 10° (more if necessary) in either direction until you reach a point where both fields are of nearly equal intensity. Then back off a degree or so toward zero and use the fine adjustment knob (if there is one) to rotate the scale *away* from zero until the entire visual field is as uniform as possible and the dividing lines between fields have all but disappeared. (If the field is very bright and a slight turn of the knob has little effect on it, you are probably 90° off. Rotate the scale 90° back toward zero.) If you overshoot the final reading, move the scale back a few degrees so that you again approach it going away from zero. Read the rotation angle from the analyzer scale, using the Vernier scale (if there is one) to read fractions of a degree, and record the direction

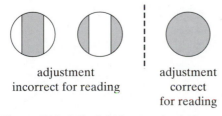

adjustment
incorrect for reading

adjustment
correct
for reading

Figure F13 Split-field image of polarimeter

of rotation, + for clockwise or − for counterclockwise. Obtain another reading of the rotation angle, this time approaching the final reading going *toward* zero. For accurate work you should take a half-dozen readings or more, reversing direction each time to compensate for mechanical play in the instrument, and average them.

Rinse the cell with the solvent used in preparing the solution, fill the cell with that solvent (or use a solvent blank provided), and determine its rotation angle by the same procedure as before. Remove the solvent and let the cell drain dry, or clean it as directed by your instructor. Subtract the average rotation angle of the solvent blank (note + and - signs) from that of the sample to obtain the observed rotation of the sample, and calculate its specific rotation using equation **1**. In reporting the specific rotation, specify the temperature, light source, solvent, and concentration.

Summary

1. Prepare solution of compound to be analyzed.
2. Rinse and fill sample cell with solution, place in polarimeter.
3. Adjust light source, focus eyepiece.
4. Rotate analyzer scale until optical field is uniform, read rotation angle.
5. Repeat step 4 several times, reversing direction each time, and average readings.
   IF solvent blank has been run, GO TO 7.
   IF not, GO TO 6.
6. Rinse and fill sample cell with solvent; place in polarimeter. GO TO 4.
7. Clean cell, drain dry, dispose of solution and solvent as directed.
8. Calculate observed rotation and specific rotation of sample.

G. Instrumental Analysis

Gas Chromatography

Gas chromatography (GC) is an invaluable tool for the separation and analysis of liquid mixtures. Like all chromatographic methods, its operation is based on the distribution of the sample components between a mobile phase and a stationary phase. *Gas-liquid chromatography* is the most common form of gas chromatography; *gas-solid chromatography* has limited application and will not be discussed here. The mobile phase for gas-liquid chromatography is an unreactive *carrier gas* such as helium, and the stationary phase consists of a high-boiling liquid on a solid support, contained within a heated column. The components of the sample must have a reasonably high vapor pressure so that their molecules will spend enough time in the vapor phase to travel through the column with the carrier gas. Thus gas chromatography can only be used to separate gases, liquids that are volatile enough to vaporize without decomposing when heated, and some volatile solids when in solution. Mixtures of solids and less volatile liquids can be separated by high performance liquid chromatography [OP-33].

See OP-16 for a general introduction to chromatographic methods.

    Preparative gas chromatography is used to separate the components of a mixture or purify a major component. Because even a large GC column will not accomodate sample sizes greater than about 0.5 mL, preparative GC cannot be used to separate large volumes of liquids. In the undergraduate organic chemistry lab, it is most useful for purifying the products of reactions conducted at the microscale level. *Analytical gas chromatography* is used for the qualititative and quantitative analysis of mixtures; that is, to identify the components of a mixture and find out how much of each component is present. An analytical gas chromatograph requires only a minute amount of sample, usually a microliter or so.

Principles and Applications

In gas-liquid chromatography the components of a mixture are distributed between a liquid stationary phase and a gaseous mobile phase, the carrier gas. The time it takes for a component to pass through the column, called its *retention time,* depends on its relative concentrations in the stationary and mobile phases. While the molecules of a component are in the gas phase, they pass through the column at the speed of the carrier gas. While they are in the liquid phase, they stay put. The more time a component spends in the vapor phase, the sooner it will get through the column, and the lower its retention time will be. The more time a component spends in the liquid phase, the longer its retention time will be. The time a component spends in the gas phase depends on its volatility (and thus on its boiling point) and the temperature of the column. The time it spends in the liquid phase depends on the strength of the attractive forces between its molecules and the molecules of the liquid phase.

    In order for any two components of a mixture to be separated sharply, they must have significantly different retention times, and their bands (the regions of the column they occupy) must be narrow enough that they do not overlap appreciably. The degree of separation depends on a number of factors, including the length and efficiency of the column, the carrier gas flow rate, and the temperature at which the separation is carried out. Modern gas chromatographs provide the means to vary these factors precisely, and they offer an almost endless variety of applications, from analyzing automobile emissions for noxious gases to detecting PCBs in lake trout.

There are several important differences between gas-liquid chromatography and liquid-solid chromatographic methods such as column chromatography [OP-16] and thin-layer chromatography [OP-17]. Unlike the mobile phase in a liquid-solid separation, the mobile phase in gas chromatography does not interact with the molecules of the sample; its only purpose is to carry them through the column. Thus the separation of the components of a mixture depend on (1) how strongly they are attracted to the stationary phase, and (2) how volatile they are. If the components have similar polarities, they will be attracted to the stationary phase to about the same extent, so they will tend to be separated in order of relative volatility, with the lower-boiling components having the shorter retention times. If the components have different polarities, their retention times will depend on their relative polarities and the polarity of the stationary phase. For example, a polar stationary phase will attract polar components more strongly than nonpolar ones, giving polar components longer retention times than nonpolar components of comparable volatility.

A column chromatography or TLC separation is usually conducted at room temperature and the temperature stays about the same throughout the separation, but a GC separation is usually carried out at an elevated temperature and the temperature can be changed throughout the separation according to a preset program. Increasing the temperature of the column increases the fraction of each component in the gas phase, thus decreasing its retention time. The column temperature also affects the separation efficiency, since at excessively high temperatures components will tend to spend most of their time in the vapor phase, causing their bands to overlap.

Instrumentation

In addition to the column and carrier gas, a gas chromatograph must have some means of vaporizing the sample, controlling the temperature of the separation, detecting each constituent of the sample as it leaves the column, and recording data from which the composition of the sample can be determined. A typical gas chromatograph includes the following components, as diagrammed in Figure G1.

Injection Port. The injection port is the starting point for a sample's passage through the column. The sample, in a microliter syringe, is injected into the column by inserting the needle through a rubber or silicone septum and depressing the plunger. The sample size for a packed column is typically about 1–2 μL, but may vary from a few tenths of a microliter to 20 μL. With an open tubular column (described later), a sample splitter delivers about 1 nL (10^{-3} μL) of the injected sample to the column; the rest is discarded. The injection port is heated to a temperature sufficient to vaporize the sample, usually about 50°C above the boiling point of its least volatile component.

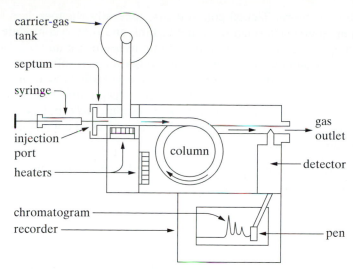

Figure G1 Schematic diagram of a gas chromatograph

Packed Columns. A packed column is a long tube packed with an inert solid support that is coated with the high-boiling liquid phase. The tube is usually made of stainless steel or glass. A typical column may be 1.5–3 meters long with an inner diameter of 2–4 mm ($\frac{1}{8}-\frac{1}{4}$-inch O.D.), and is usually bent into a coil to fit inside a column oven, where its temperature is controlled. Although such columns can be packed by the user, most columns are purchased ready-made. The column is packed with a finely ground support medium such as Chromosorb W, whose particles are previously coated with the liquid phase. Chromosorb W is made by crushing white diatomaceous earth, mixing it with sodium carbonate flux, and forming the mixture into bricks that are baked at over 900°C. The ground-up bricks are then separated by screens into particles of uniform diameter. Other supports include crushed firebrick, silica, alumina, and Teflon beads. The composition of the packing is described on a tag attached to each column. For example, a column packing described as "10% DEGS/Chromosorb W 80/100" contains 10% by mass of a diethylene glycol succinate liquid phase on a Chromosorb W support having a particle-size range from 80 to 100 mesh (0.17 to 0.15 mm). As for a distillation column [OP-27], the efficiency of a chromatography column can be measured in terms of the number of theoretical plates it provides. Most packed columns have efficiencies ranging from 500 to 1000 theoretical plates per meter.

The mesh number is the number of meshes per inch in the finest sieve that will let the particles pass through.

Open Tubular Columns. An *open tubular column* or *capillary column* consists of an open tube coated on the inside with the liquid phase. A very thin film of the liquid phase (ranging from 0.05 μm to 1.0 μm or more) is adsorbed on or chemically bonded to the inner wall of the column, which is a long capillary tube made of glass or fused silica. Fused silica open tubular (FSOT) columns are flexible enough to be bent into coils 15 cm or so in diameter. In size, the columns range from "microbore" columns that may have an inner diameter of 0.05–0.10 mm and a length of 20 meters to "megabore" columns having an I.D. of about 0.50 mm and a length of

100 meters or more. Packed columns are generally cheaper, less fragile, and easier to use than open tubular columns, and they accomodate much larger sample sizes. But open tubular columns are faster and less likely to react with the sample, and they provide unparalled resolution of sample components—a microbore column may have an efficiency of more than 10,000 theoretical plates per meter. Since a capillary column may be 20–50 times longer than a typical packed column, the capillary column is usually hundreds of times more efficient than a packed column containing the same liquid phase.

Column Oven. The column is mounted inside a heated chamber, the *column oven,* that controls its temperature. Under *isothermal operation,* the oven temperature is kept constant throughout a separation; under *program operation* it is varied according to a programmed sequence. As a general rule, the column temperature for isothermal operation should be approximately equal to or slightly above the average boiling point of the sample. If the sample has a broad boiling range, it may be separated by program operation. The more volatile components are ordinarily eluted during the low temperature end of the program, and less volatile components are eluted at its high temperature end.

Carrier Gas. A chemically unreactive gas is used to sweep the sample through the column. Helium is the most widely used carrier gas, although nitrogen and argon are also used with some detectors. Carrier gases usually come in pressurized gas cylinders, and various devices are used to measure and control their flow rates.

Detector. The detector is a device that detects the presence of each component as it leaves the column and sends an electrical signal to a recorder or other output device. A separate oven is used to heat the detector enclosure so that the sample molecules remain in the vapor phase as they pass the detector. The most widely used detectors for routine gas chromatography are *thermal conductivity* (*TC*) detectors and *flame ionization* (*FI*) detectors. A TC detector responds to changes in the thermal conductivity of the carrier gas stream as molecules of the sample pass through. With an FI detector, the component molecules are pyrolyzed in a hydrogen-oxygen flame, producing short-lived ions that are captured and used to generate an electrical current. Thermal conductivity detectors are comparatively simple and inexpensive, and they respond to a wide variety of organic and inorganic species. Flame ionization detectors are much more sensitive than thermal conductivity detectors, so they are used with open tubular columns (for which TC detectors are unsuited) as well as packed columns. An FI detector is somewhat inconvenient to use because it requires hydrogen and air to produce the flame, and it cannot be used for preparative gas chromatography because it destroys the sample.

Recorder. The recorder is an instrument that records a peak on moving chart paper as each component band passes the detector. The resulting *gas chromatogram* consists of a series of peaks of different sizes, each produced by a different component of the mixture. The recorder may be provided with an *integrator* that measures the area under each component peak. Many research-grade gas chromatographs are interfaced with computers that process their output signals and provide digital readouts of peak areas, retention times, and various instrumental parameters.

Liquid Phases

The factor that most often determines the success or failure of a gas chromatographic separation is the choice of the liquid phase. In general, polar liquid phases are best for separating polar compounds, and nonpolar liquid phases are best for separating nonpolar compounds. However, there are hundreds of liquid phases available, and choosing the right one for a particular separation may not be an easy task. Some typical liquid phases and their maximum operating temperatures are listed in Table G1 in order of polarity (lower to higher). General-purpose liquid phases such as dimethylpolysiloxane (dimethylsilicone) can separate a variety of compounds successfully, usually in approximate order of their boiling points. Other liquid phases have more specialized applications; for example, diethylene glycol succinate is a high-boiling ester used mainly to separate the esters of long-chain fatty acids, as in Experiment 54.

Heating a column above its maximum operating temperature will cause the liquid phase to vaporize.

The selection of a liquid phase can be facilitated by the use of *McReynolds numbers,* which indicate the affinity of a liquid phase for different types of compounds; the higher the number the greater the affinity. X' is a McReynolds number measuring the relative affinity of a liquid phase for aromatic compounds and alkenes; Y' measures its affinity for alcohols, phenols, and carboxylic acids; and Z' measures its affinity for aldehydes, ketones, ethers, esters, and related compounds. For example, the polar liquid phase DEGS has a much higher Z' value (590) than dimethylpolysiloxane (44), so it will retain an ester much longer. McReynolds numbers are particularly useful for selecting a liquid phase to separate compounds of different chemical classes, such as alcohols from esters. Thus a Carbowax 20M column should separate isopentyl acetate from isopentyl alcohol satisfactorily because its Y' and Z' values are very different, but an OV-17 column would not be a good choice for such a separation. Chromatography supply companies often provide detailed information about the kinds of separations their columns can accomplish. For example, the Alltech chromatography catalog reproduces a gas chromatogram of a mixture of 33 different drugs that were separated on one of its capillary columns, giving the conditions of the separation.

Table G1 Selected liquid phases for gas chromatography

Stationary phase	maximum T, °C	McReynolds numbers		
		X'	Y'	Z'
squalane	150	0	0	0
dimethylpolysiloxane (OV-1)	350	16	55	44
diphenyl/dimethylpolysiloxane (OV-17)	350	119	158	162
polyethylene glycol (Carbowax 20M)	250	322	536	368
diethylene glycol succinate (DEGS)	225	496	746	590
dicyanoallylpolysiloxane (OV-275)	275	629	872	763

Note: McReynold's numbers apply to the commercial stationary phase in parentheses.
Commercial designations for the same kind of stationary phase vary widely; thus AT-1, DB-1, Rtx-1, SE-30, and DC-200 are all similar to OV-1.

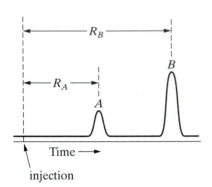

Figure G2 Retention times of two components

(1)

(2)

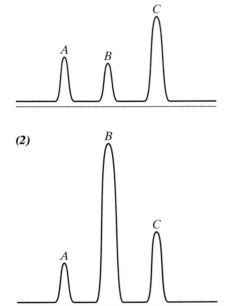

Figure G3 Use of gas chromatography for qualitative analysis: (1) chromatogram before adding ethanol; (2) chromatogram after adding ethanol

Qualitative Analysis

The retention time of a component is the time it spends on the column, from the time of injection to the time its concentration at the detector reaches a maximum. The retention time corresponds to the distance on the chromatogram, along a line parallel to the baseline, from the injection point to the top of the component's peak, as shown in Figure G2. This distance can be converted to units of time if the chart speed is known. If all instrumental parameters (temperature, column, flow rate, etc.) are kept constant, the retention time of a component will be characteristic of that compound and can be used to characterize it. You can seldom (if ever) identify a compound with certainty from its retention time alone, but you can sometimes use retention times to confirm the identity of a compound whose identity you suspect. For example, if a known compound and an unknown have identical retention times on two or more different stationery phases under the same operating conditions, the compounds are probably identical.

When you carry out a reaction and obtain a gas chromatogram of the product mixture, you may be able to guess which GC peak corresponds to which product, especially if the products have significantly different boiling points and the stationary phase is known to separate compounds in order of boiling point. It is always a good idea to back up your guess with proof, however. One way to identify the components of a product mixture is to "spike" the mixture with an authentic sample of a possible component and see which GC peak increases in relative area after spiking. For example, suppose you carry out the Fischer esterification of acetic acid by ethanol and record a gas chromatogram of the product mixture.

$$CH_3COOH + CH_3CH_2OH \rightleftharpoons CH_3COOCH_2CH_3 + H_2O$$

    acetic acid   ethanol   ethyl acetate

Since this is an equilibrium reaction, the product mixture would be expected to contain some unreacted starting materials as well as the product. To detect the presence of unreacted ethanol, you can add a little pure ethanol to the sample and record a second gas chromatogram. The peak that increases in relative area is the ethanol peak, as shown in Figure G3. The process can be repeated using different pure compounds to identify the other components, if necessary.

Another way to identify the components of a mixture using GC is to obtain spectra of the components after they pass through the column. As described in OP-37, a mass spectrometer coupled to a gas chromatograph can act as a "superdetector," identifying the components as they come off the column. When the GC column in use has a large enough capacity that individual components can be collected at the column outlet, they can often be identified from their IR or NMR spectra.

Quantitative Analysis

The use of gas chromatography for quantitative analysis is based on the fact that, over a wide range of concentrations, a detector's response to a given component is proportional to the amount of that component in the sample. Thus the area under a component's peak can be used to determine its mass percentage in the sample. If a peak on a gas chromatogram is symmetrical, its area can be calculated with fair accuracy by multiplying its height (h) in

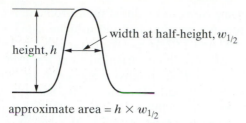

height, h
width at half-height, $w_{1/2}$

approximate area = $h \times w_{1/2}$

Figure G4 Measurement of approximate peak area

millimeters (measured from the baseline) by its width at a point exactly halfway between the top of the peak and the baseline ($w_{1/2}$).

$$\text{approximate peak area} = h \times w_{1/2}$$

Many gas chromatograms are equipped with an *integrator* that automatically calculates the peak areas and displays them digitally. Peak areas can also be measured by making a photocopy of the chromatogram, accurately cutting out each peak with a sharp knife or razor blade (use a ruler to cut a straight line along the baseline), and weighing the peaks to the nearest milligram (or tenth of a milligram) on an accurate balance.

From the peak areas for a sample, you can sometimes estimate the percentage of each component in the mixture by dividing its area by the sum of the areas and multiplying by 100%. For example, if a gas chromatogram has three peaks and the area of peak A is 1227 mm^2, the area of peak B is 214 mm^2, and the area of peak C is 635 mm^2, the sum of the areas is 2076 mm^2 so the mass percentages of the components are approximately:

$$\%A = \frac{1227}{2076} \times 100\% = 59.1\%$$

$$\%B = \frac{214}{2076} \times 100\% = 10.3\%$$

$$\%C = \frac{635}{2076} \times 100\% = \underline{30.6\%}$$

$$100.0\%$$

This kind of calculation is accurate only if the detector responds in the same way to each component. This may be true (or nearly so) if you are analyzing a mixture of very similar compounds, such as a series of long-chain fatty acid esters. In many cases, however, the same mass of two different components will not necessarily result in the same detector response. Flame ionization detectors respond mainly to ions produced by certain reduced carbon atoms, such as those in methyl and methylene groups. Because an ethanol molecule contains about three-fifths as much carbon by weight as a heptane molecule does, the response of an FI detector to a gram of ethanol is only about three-fifths as great as its response to a gram of heptane. Thermal conductivity detectors also respond differently to different substances, but the variations are usually not as large as those of FI detectors.

For accurate quantitative analysis of most mixtures you should multiply each peak area by a *detector response factor* to obtain a corrected area that is proportional to its mass. Detector response factors can be obtained from the literature in some cases, or by direct measurement. For example, detector response factors for cyclohexane and toluene using a TC detector are reported to be 0.942 and 1.02, respectively (relative to benzene). Suppose the chromatogram of a distillation fraction from Experiment 6 is recorded using such a detector, and has relative peak areas of 78.0 mm^2 for cyclohexane and 14.6 mm^2 for toluene. The corrected peak areas are then (0.942 × 78.0 mm^2) = 73.5 mm^2 for cyclohexane and (1.02 × 14.6 mm^2) = 14.9 mm^2 for toluene. Dividing each corrected area by the sum of the corrected areas and multiplying by 100 gives mass percentages of 83.1% for cyclohexane and 16.9% for toluene.

If detector response factors for the components of a mixture are not known, you can determine them by the following method, which is described for two components, A and R. We will call component R the reference compound and arbitrarily assign it a detector response factor of 1.00. (1) Obtain a gas chromatogram of a mixture containing carefully weighed amounts of A and R. (2) Measure the peak area for each component. (3) Calculate the detector response factor for A using the following relationship:

$$\text{detector response factor for A} = \frac{\text{mass}_A}{\text{area}_A} \times \frac{\text{area}_R}{\text{mass}_R}$$

This method can be used to calculate detector response factors for any number of components simultaneously, as long as their peaks are well resolved. Each component (including the reference compound) is carefully weighed, a gas chromatogram of the mixture is recorded, all of the peak areas are measured, and the detector response factors are calculated from the resulting masses and areas.

Preparative Gas Chromatography

Most preparative gas chromatographs are too expensive to be used routinely in undergraduate laboratories, but some inexpensive analytical gas chromatographs, such as the Gow-Mac 69-350, can be operated with preparative columns using a metal adapter that is connected to the exit port on the gas chromatograph. A special GC collection tube is inserted into the adapter just before the desired compound's peak begins to appear on the chromatogram, and removed just after the end of the component's peak has been recorded. The condensed liquid is then transferred to a small conical vial by centrifugation, and can be used to obtain an IR or NMR spectrum of the compound. Additional components can be collected in other collection tubes.

If an adapter and specialized collection glassware are not available, a 3-inch, 2 mm O.D. glass tube can be packed with glass wool and inserted into the GC exit port while the desired component's peak is being recorded. The component should condense on the surface of the glass wool, from which it can be washed off with a little solvent (if the component is quite volatile, the tube should be chilled first). An infrared or NMR spectrum of the compound can then be obtained in a solution of the solvent used, or the solvent can be evaporated.

General Directions for Recording a Gas Chromatogam

Do not attempt to operate the instrument without prior instruction and proper supervision. Consult the instructor if the instrument does not seem to be working properly or if you have questions about its operation. It will be assumed that all instrumental parameters have been preset, that an appropriate column has been installed, and that the column oven will be operated isothermally. If not, the instructor will show you what to do. Before you begin, be sure you have read the section in OP-5 about the use of syringes.

If the sample is a volatile solid, dissolve it in the minimum volume of a suitable low-boiling solvent; otherwise, use the neat liquid. Rinse a microsyringe with the sample a few times; then partially fill it with the sample. A microsyringe is a very delicate instrument, so handle it carefully and avoid using excessive force that might bend the needle or plunger. If there are air bubbles inside the syringe, tap the barrel with the needle pointing up, or eject the sample and refill the syringe more slowly. Hold the syringe with the needle pointing up and expel excess liquid until the desired volume of sample (usually 1–2 μL) is left inside. Wipe the needle dry with a tissue, and pull the plunger back a centimeter or so to prevent prevaporization of the sample.

Set the chart speed, if necessary, and switch on the recorder-chart drive (and the integrator, if there is one). Carefully insert the syringe needle into the injection port by holding the needle with its tip at the center of the septum and pushing the barrel slowly, but firmly, with the other hand until the needle is as far inside the port as it will go. Inject the sample by *gently* pushing the plunger all the way in just as the recorder pen crosses a chart line (the starting line); use as little force as possible to avoid bending the plunger. Withdraw the needle and mark the starting line, from which all retention times will be measured. Let the recorder run until all of the anticipated component peaks have appeared. Then turn off the chart drive and tear off the chart paper using a straight edge. Inspect the chromatogram carefully; if it is unsuitable because the significant peaks are too small, poorly resolved, or off-scale, repeat the analysis after taking measures to remedy the problem. If there is evidence of prevaporization (as indicated by an extraneous small peak preceding each major peak at a fixed interval), be sure the plunger is pulled back before you inject the sample for the next run. Injecting the sample immediately after you insert the needle in the injector port will also prevent prevaporization, but this technique requires good timing.

Before you analyze a different sample, clean the syringe and rinse it thoroughly with that sample. When you are finished, rinse the syringe with an appropriate low-boiling solvent such as methanol or dichloromethane, remove the plunger, and set it on a clean surface to dry.

If the gas chromatograph is interfaced with a computer, follow your instructor's directions for acquiring and printing the chromatogram.

Take Care! The injection port is hot! Don't touch it.

Summary

1. Rinse syringe, fill with recommended volume of sample.
2. Start chart drive and integrator (if applicable).
3. Insert syringe needle into injector port, inject sample when pen crosses a chart line.
4. Withdraw needle; mark starting line.
5. Stop chart drive when chromatogram is complete.
6. Remove chromatogram, clean and dry syringe.

High Performance Liquid Chromatography

Principles and Applications

High performance liquid chromatography (HPLC) can be regarded as a hybrid of column chromatography and gas chromatography, sharing some features of both methods. As in column chromatography, the mobile phase is a liquid that carries the sample through a column packed with fine particles that interact with the components of the sample to different extents, causing them to separate. As in gas chromatography, the sample may be injected onto the column and is detected as it leaves the column, and its passage through the column is recorded as a series of peaks on a chromatogram. The stationary phase may be a solid adsorbent, as in column chromatography, but it is more often an organic phase that is bonded to tiny beads of silica gel. The silica beads are, in effect, coated with a very thin layer of a liquid organic phase. Each component of the sample is partitioned between a liquid mobile phase and the liquid stationary phase according to a ratio—the *partition coefficient*—that depends on its solubility in each liquid. The components of a mixture generally have different partition coefficients in the liquid phases, so they pass down the column at different rates.

Instrumentation

HPLC was developed as a means of improving the efficiency of a column chromatographic separation by reducing the particle size. Most column chromatography packings contain particles with diameters in the 75–175 μm range, while most modern HPLC packings have particle diameters in the 3–10 μm range, increasing separation efficiency dramatically. But solvents will not flow easily through such small particles by gravity alone. Thus a powerful pump is needed to force eluents through the column at pressures up to 6000 psi ($\sim$400 atm).

The basic components of an HPLC system are diagrammed in Figure G5. The instrument ordinarily has several large solvent reservoirs, each of which can be filled with a different solvent. *Isocratic elution* is elution using a single solvent. *Gradient elution* utilizes two or more solvents of different polarity and varies the solvent ratio throughout a separation according to a programmed sequence. Gradient elution can greatly reduce the time needed for a separation and increase the separation efficiency.

Preparative HPLC systems, which require special wide-bore columns, are equipped with fraction collectors to collect the eluent as it comes off the column; the fractions are evaporated to yield the pure components. Analytical HPLC systems, which are used to determine the compositions of mixtures, require much smaller samples and the components are not recovered. A typical analytical HPLC column is constructed of stainless steel, with a length of 10–25 cm and an inner diameter of 2.1–4.6 mm. A microbore analytical column may have an inner diameter of 1 mm, while a large preparative column may have an inner diameter of up to 50 mm or so.

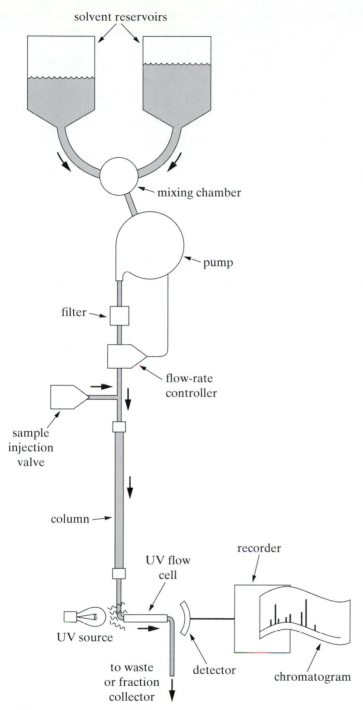

Figure G5 Schematic diagram of HPLC system

Samples are introduced onto the column by means of a syringe or sampling valve and are carried through it by the pressurized eluent mixture. Because the particles of the stationary phase are so small, a column can easily be plugged by particulate matter or adherent solutes introduced by either the eluent or the sample. To remove anything that might harm the column, the eluent is forced through one or more filters and the sample may be introduced through a short *guard column.*

As each component of a sample leaves the column, the detector responds to some property of the component, such as its ability to absorb ultraviolet radiation. The detector then sends an electrical signal to a recorder or computer, which traces the component's peak on a chart or monitor. The resulting chromatogram is a graph of some property of the components, such as UV absorbance, plotted against the volume of the mobile phase. Because different solutes may have greatly different UV absorptivities at the wavelength used by an *ultraviolet absorbance detector,* for example, a detector response factor (see OP-32) must be determined for each component before its percentage in the mixture can be calculated. Components that do not absorb ultraviolet radiation will not be detected by a UV detector; in that case a different type of detector can be used, or the components can be converted to derivatives that are UV-active. Any component whose refractive index is different from that of the eluent can be detected by a *refractive index (RI) detector,* but RI detectors have a lower sensitivity than UV detectors.

High performance liquid chromatography can be used to separate mixtures containing proteins, nucleic acids, steroids, antibiotics, pesticides, inorganic compounds, and many other substances whose volatilities are too low for gas chromatography [OP-32]. Because HPLC separations can be run at room temperature, there is little danger of decomposition reactions or other chemical changes that sometimes occur in the heated column of a gas chromatograph. These advantages have made HPLC the fastest growing separation technique in chemistry, but its use in undergraduate laboratories is limited because of the high cost of the instruments, columns, and high-purity solvents required (HPLC-grade water costs about $50 a gallon!).

Stationary Phases

Some HPLC columns contain a solid adsorbent such as silica gel or alumina; components of the stationary phase are adsorbed to different extents on the surface of the solid. Such a stationary phase is much more polar than the eluent, so nonpolar components are eluted faster than polar ones. Most modern HPLC columns contain a *bonded liquid phase*—an organic phase that is chemically bonded to particles of silica gel. Silica gel contains silanol ($-Si-OH$) groups to which long hydrocarbon chains, such as octadecyl groups, can be attached by reactions such as the following.

$$-Si-OH \xrightarrow{R_2SiCl_2} Si-O-\overset{\overset{\displaystyle R}{|}}{\underset{\underset{\displaystyle R}{|}}{Si}}-Cl \xrightarrow{H_2O} \xrightarrow{(CH_3)_3SiCl}$$

$$-Si-O-\overset{\overset{\displaystyle R}{|}}{\underset{\underset{\displaystyle R}{|}}{Si}}-O-Si(CH_3)_3 \qquad R = CH_3(CH_2)_{17}- \text{ (octadecyl)}$$

The bonded stationary phase in this example is *less* polar than the eluent, which may be a mixture of water with another solvent such as methanol, acetonitrile, or tetrahydrofuran. Thus polar components will spend more time in the eluent than in the stationary phase and will be eluted faster than nonpolar ones, reversing the usual order of elution. This mode of separation is called *reverse-phase chromatography*.

Other stationary phases operate by still different mechanisms. The stationary phase for *size-exclusion chromatography* is a porous solid that separates molecules based on their effective size and shape in solution. Small molecules can enter even the narrowest openings in the porous structure, larger molecules find fewer openings they can get into, and still larger molecules may be completely excluded from the solid phase. Thus large, bulky molecules pass down the column faster than smaller molecules. A stationary phase for *ion-exchange chromatography* has ionizable functional groups that carry a negative or positive charge over a suitable pH range, attracting ionic solutes from solution. In organic chemistry, such stationary phases are used mainly to separate ionizable organic compounds such as carboxylic acids, amines, and amino acids. Chiral stationary phases that can separate enantiomers and determine their optical purity are also available.

Hundreds of different stationary phases are used for HPLC separations. Some of the most popular reverse-phase packings consist of silica gel with bonded methyl ($-CH_3$), phenyl ($-C_6H_5$), octyl [$-(CH_2)_7CH_3$], octadecyl [$-(CH_2)_{17}CH_3$], cyanopropyl [$-(CH_2)_3CN$], and aminopropyl [$-(CH_2)_3NH_2$] groups. Size-exclusion stationary phases contain silica, glass, and polymeric gels of varying porosity. Ion-exchange stationary phases include (a) styrene-divinylbenzene copolymers to which ionizable functional groups (such as $-SO_3H$ and $-NH_4^+OH^-$) are attached; (b) beads with a thin surface layer of ion-exchange material; and (c) bonded phases on silica particles.

General Directions for HPLC

Do not attempt to operate the instrument without prior instruction and proper supervision. Because HPLC systems vary widely in construction and operation, only a very general outline of the procedure is provided here. It will be assumed that all instrumental parameters have been preset, that an appropriate reverse-phase column has been installed, and that the elution will be isocratic. If not, the instructor will provide additional directions.

Prepare an approximately 0.1% stock solution of the sample in the same solvent or solvent mixture as the one being used for the elution, and dilute an aliquot of this solution further, if necessary. Be certain you are using HPLC-grade solvents for preparing the solution and eluting it through the column. The solvents should be purged with helium prior to elution to remove dissolved gases. Filter the solution through a 1.0-μm membrane filter to remove any particulate matter. Degas it, if necessary, as directed by your instructor. Inject the sample (10 μm unless otherwise directed) through the injection port, or use a sampling valve to introduce it into the system. Using a guard column to protect the analytical column from particles is advisable. Once the sample is injected, a microprocessor will control all aspects of the separation. Wait until no more peaks appears on the chromatogram, then inspect the chromatogram to see if the components are well resolved. If they are not, repeat the determination using a higher percentage of the more polar solvent (usually water) in the solvent mixture.

OPERATION 34

Infrared Spectrometry

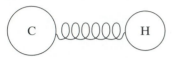

Figure G6 "Ball and spring" model of a chemical bond

Principles and Applications

The atoms of a molecule behave as if they were connected by flexible springs rather than by rigid bonds resembling the connectors of some ball and stick molecular models. A molecule's component parts can oscillate in different *vibrational modes,* which are vividly described by such terms as stretching, rocking, scissoring, twisting, and wagging. When infrared (IR) radiation is passed through a sample of a pure compound, its molecules can absorb radiation of the energy and frequency needed to bring about transitions between vibrational ground states and vibrational excited states. For example, if a molecule contains a C—H bond that vibrates 90 trillion times a second, the molecule must absorb infrared radiation of *just* that frequency (9.0×10^{13} Hz) to jump to a vibrational excited state in which the C—H bond vibrates twice as fast. The frequency of infrared radiation is usually given in wave numbers ($\bar{\nu}$). The wave number in cm^{-1} of a vibration is the number of peak-to-peak waves per centimeter. The relationship between wave number and frequency (ν) in hertz (s^{-1}) is given by:

$$\nu = c \cdot \bar{\nu}$$

where c is the speed of light ($\sim 3.0 \times 10^{10}$ cm/s). Thus the wave number of 9.0×10^{-13} Hz radiation is

$$\bar{\nu} = \frac{\nu}{c} = \frac{9.0 \times 10^{13}\ s^{-1}}{3.0 \times 10^{10}\ cm\ s^{-1}} = 3000\ cm^{-1}$$

The wave number of an infrared band is the inverse of its wavelength, which is generally measured in micrometers (μm); the two can be interconverted using the following relationship.

$$\text{wave number (in } cm^{-1}) = \frac{10^4\ \mu m/cm}{\text{wavelength (in } \mu m)}$$

So the wave number of a 5.85 μm C=O bond vibration is $(10{,}000/5.85)\ cm^{-1} = 1710\ cm^{-1}$.

The *infrared spectrum* of a compound, which graphs the amount of infrared radiation the compound absorbs over a broad frequency range (usually about $4000{-}600\ cm^{-1}$ or $2.5{-}17\ \mu$m), is obtained by analyzing a sample of the compound with an *infrared spectrometer.* On a typical infrared spectrum, wavelengths increase and wave numbers decrease going from left to right, so wave number ranges are usually reported with the higher value first. If the sample absorbs IR radiation of a given wave number, the recorder pen on the spectrometer moves downward a distance that depends on the amount of IR radiation absorbed. Thus an IR spectrum consists of a series of inverted peaks, with each peak corresponding to a different kind of bond vibration. Because vibrational transitions are usually accompanied by rotational transitions in the frequency region scanned, the inverted peaks appear as comparatively broad "valleys"

A spectrometer is an instrument that measures and records the components of a spectrum in order of wavelength, mass, or some other property.

called *IR bands,* rather than sharp peaks such as those seen in NMR spectra [OP-35].

Infrared spectrometry is most often used to detect the presence of specific functional groups and other structural features from band positions and intensities, and to show whether an unknown compound is identical to a known compound whose IR spectrum is reported in the literature. The *fingerprint region* of an infrared spectrum (1250–670 cm^{-1}) is best for establishing that two substances are identical, since the bands found in this region are often characteristic of the molecule as a whole and not of isolated bonds. Infrared spectrometry may also be used to assess the purity of a compound, monitor the rate of a reaction, measure the concentration of a solution, and study hydrogen bonding and other phenomena.

Instrumentation

In a conventional *dispersive infrared spectrometer,* infrared radiation is broken down into its component wavelengths by a diffraction grating or prism and beamed through the sample, one wavelength at a time. As the spectrum is scanned from lower to higher wavelength (higher to lower wave number), the radiation that is transmitted through the sample at each wavelength is detected and recorded on a chart. The resulting infrared spectrum is a graph of percent transmittance versus wavelength.

In a *Fourier-transform infrared (FTIR) spectrometer,* a beam containing all of the infrared wavelengths in the instrument's range passes through the sample at one time. The information contained in the resulting jumbled signal is then sorted out by a microprocessor and converted to an infrared spectrum that, in general appearance, resembles a scanned spectrum.

To see how an FTIR spectrometer works, refer to the diagram in Figure G7. The heart of the instrument is a *Michelson interferometer,* which causes two beams of infrared radiation to interfere with one another. The infrared radiation generated by a heated filament or other IR source passes through a *beam splitter* that sends half of the radiation to a fixed mirror and half to a moving mirror. The beam that reflects off the fixed mirror always

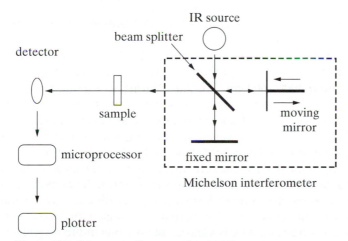

Figure G7 Schematic diagram of an FTIR spectrometer

travels the same distance from the source to the sample; the distance the other beam travels is varied continuously by the moving mirror. The beams from both mirrors are recombined before the infrared radiation passes through the sample. Suppose for a moment that the IR beam coming from the source is of a single wavelength, λ. When the moving and fixed mirrors are the same distance from the beam splitter, both beams will travel the same distance before they recombine. As a result, the peaks and troughs of their waveforms will be aligned and will interfere *constructively* with one another, sending a beam of high intensity to the sample. When the moving and fixed mirrors are not the same distance from the beam splitter, the beams will travel different distances and their waveforms will usually not be aligned peak to peak when they recombine. For example, if the peaks of one beam exactly align with the troughs of the other, the beams will interfere *destructively* with one another, sending no radiation to the sample. In other alignments the intensity of the combined beam will be somewhere between these two extremes.

As a result of constructive and destructive interference, the intensity of a light beam of wavelength λ will vary, with a sinusoidal wave pattern, at the frequency of the moving mirror. This pattern is called an *interferogram.* Because light beams of a different wavelength will align differently when they recombine, each different wavelength of IR radiation will generate a different interferogram. Interferograms are additive, so the interferogram that reaches the sample will be the sum of all the interferograms of all the wavelengths. If IR radiation of a particular wavelength is absorbed as it passes through the sample, the intensity of that wavelength's interferogram—and thus its contribution to the overall interferogram—will decrease, changing the waveform of the overall interferogram that exits the sample. Thus the interferogram that exits the sample will differ from the one that entered it in a way that depends on the amount by which the intensity of each of its component interferograms was reduced by the sample. In other words, this interferogram contains all of the intensity information for all the different wavelengths that passed through the sample—there is no need to break down the infrared radiation into its component wavelengths and measure the intensity of the transmitted radiation at each individual wavelength, as for a dispersive IR spectrometer. When the interferogram impinges on a detector, it generates an electrical signal that is sent to a microprocessor. The microprocessor uses a mathematical technique called Fourier-transform analysis to "decode" the interferogram and recover the intensity data for each wavelength. This data is then plotted as an FTIR spectrum.

Unlike a continuous-wave IR spectrometer, which takes five minutes or more to record a spectrum, an FTIR spectrometer can record a complete spectrum in a fraction of a second, so many spectra of the same sample can be run in a minute or less. The data from these spectra are averaged to yield a composite spectrum that is far cleaner and better resolved than a spectrum obtained by a dispersive instrument, because it lacks the instrumental "noise" that can distort IR bands. This averaging capability makes it possible to obtain good spectra using very small samples (1 mg or less). In addition, a helium-neon laser is used as a standard against which the frequencies of the radiation are measured, so the wave numbers obtained from an FTIR spectrum are much more accurate than those from a conventional spectrum.

Experimental Considerations

An infrared spectrometer is a precision instrument that may cost as much as a Jeep Grand Cherokee but does not respond as well to rough handling. Treat it with respect and follow instructions carefully to prevent damage and avoid the need for costly repairs. Never unplug or switch off the instrument unless directed otherwise—some infrared spectrometers must be turned on continually to prevent damage to their optical parts. Keep water and aqueous solutions away from an IR spectrometer since some of its components may be water sensitive. Never move the chart holder, drum, or pen carriage on a dispersive instrument while the instrument is in operation or before it has been properly reset at the end of a run—this can damage mechanical components. Do not leave objects lying on the bed of a flat-bed recorder, as they might jam the chart drive.

FTIR Spectrometers. A typical Fourier-transform infrared spectrometer is illustrated in Figure G8. To obtain an infrared spectrum using such an instrument, the user introduces a sample into the sample compartment and selects the desired number of scans (usually 4–16). Unless a large number of scans is selected, the spectrum should appear on the monitor in less than a minute. A copy of the spectrum can then be obtained by using a plotter or a computer printer. Some instruments print the wave number of each band directly on the spectrum; with other instruments you may need to determine the wave numbers by using a cursor, then write them on the spectrum.

Dispersive IR Spectrometers. A typical dispersive infrared spectrometer with a flat-bed recorder is shown in Figure G9. The chart paper on which spectra are recorded, which has wave number and wavelength values printed on it, is clipped to the recorder bed. With other instruments it may be wrapped around a moveable drum. The user may need to align the chart

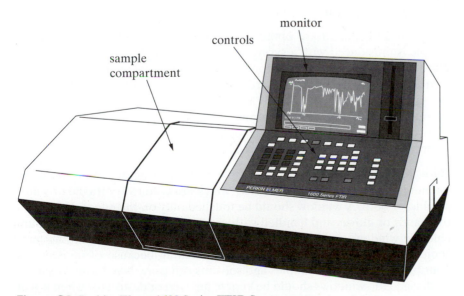

Figure G8 Perkin–Elmer 1600 Series FTIR Spectrometer

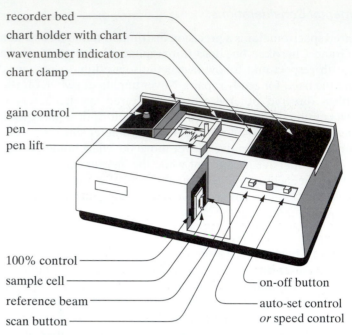

recorder bed
chart holder with chart
wavenumber indicator
chart clamp

gain control
pen
pen lift

100% control
sample cell
reference beam
scan button

on-off button
auto-set control
or speed control

Figure G9 Perkin–Elmer Model 710b infrared spectrometer

paper by matching a wave number on the chart (for instance, 4000 cm^{-1}) with an alignment mark on the recorder bed or drum. If it not aligned properly, the IR bands will appear at the wrong locations on the chart paper, and their wave numbers will be incorrect. The user then moves the recorder bed or drum to its initial position (with the pen at the far left of the chart paper), moves the pen to a point near the top of the chart paper with an attenuator control, and scans the spectrum.

Sample Preparation

Most infrared spectra are obtained by using sample cells into which the sample is introduced with a syringe or by some other means. Certain sampling accessories, such as horizontal ATR's (attenuated total reflectance devices), make it possible to record a spectrum without using a sample cell; the sample (if a liquid) is simply transferred to a sampling trough with a dropper or Pasteur pipet. Since sample cells and other sampling accessories vary widely in construction and application, most sampling techniques should be demonstrated by the instructor.

Care of Infrared Cells. Most sample cells use metal halide windows (sodium chloride, silver chloride, etc.), which are either very fragile or water soluble, or both. A window should be touched only on the edges with clean, dry hands or gloves and handled with great care to avoid damage. Sodium chloride windows must not be exposed to moisture; even breathing on a sodium chloride window can cause some etching because of moisture in your breath. Therefore samples and solvents run using NaCl windows must be dry, and the windows should be kept in an oven or desiccator when not in use. When assembling demountable cells, the cinch nuts must not be over-tightened, as this can fracture the windows.

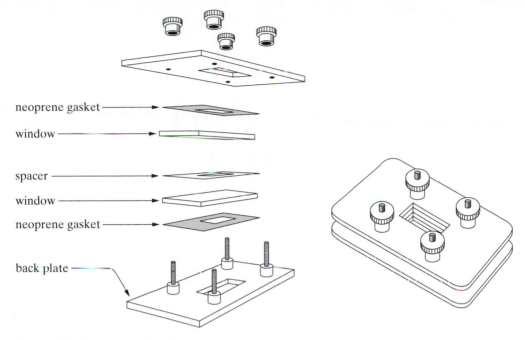

Figure G10 Demountable cell

Thin Films. Thin films cannot be used for very volatile liquids since they may evaporate before the spectrum is complete; with such liquids, use a spacer as described in the next paragraph. To prepare a thin film of a neat (undiluted) liquid using a *demountable cell,* disassemble the cell carefully according to the instructor's directions and place 1–2 drops of the liquid on the lower window. Position the upper window by touching an edge to the corresponding edge of the lower window and carefully lowering it into place. Press the plates together so that the liquid fills the space between them, taking care to exclude air bubbles, then assemble the cell as directed. After you have recorded a spectrum, disassemble the cell and rinse the windows with a *dry* volatile solvent (dichloromethane or NaCl-saturated absolute ethanol may be used). Let the solvent evaporate and return the cells to a desiccator or oven.

Volatile Neat Liquids. If a liquid is comparatively volatile, its spectrum can be run in a demountable, sealed, or sealed-demountable cell using a spacer approximately 0.015–0.030 mm thick. Fill a demountable cell by disassembling the cell, placing the spacer on the lower window, adding sufficient liquid to fill the cavity in the spacer, positioning the upper window, and reassembling the cell. After the spectrum has been run, disassemble and clean the cell as described previously.

A *sealed cell* or *sealed demountable cell* is filled by injecting the sample into one of its filling ports with a Luer-lock syringe body. Remove both plugs from the filling ports, draw about 0.5 mL of liquid into the syringe, and carefully insert the syringe tip into one of the ports with a half twist. Holding the cell upright with the syringe port at the bottom, depress the syringe's plunger until the space between the windows is filled and a little liquid appears at the upper port. If there is much resistance to filling, try

Take Care! The syringes are fragile and break easily.

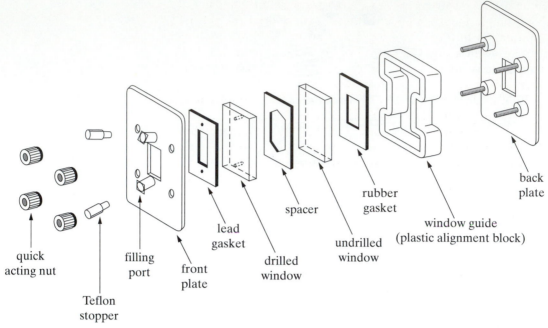

quick
acting nut

Teflon
stopper

filling
port

front
plate

lead
gasket

drilled
window

spacer

undrilled
window

rubber
gasket

window guide
(plastic alignment block)

back
plate

Figure G11 Sealed demountable cell

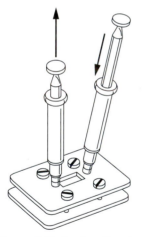

Figure G12 Push-pull technique for flushing a sealed or sealed demountable cell

A convenient solution cell for use with very small amounts of solute is described in J. Chem. Educ. **1991,** *68,* A124.

the push-pull technique described next. If there are air bubbles between the windows, they can sometimes be removed by tapping gently on the metal frame of the cell. Put the cell on a flat surface, remove the syringe, and insert a plug in the upper port with a slight twist. Remove any excess solvent in the lower port with a piece of tissue paper or cotton, and close that port with another plug.

After the spectrum has been run, clean the cell by removing most of the liquid with a syringe and flushing the cell several times with a volatile solvent. A convenient way of flushing a cell is to lay it on a flat surface, then insert a syringe filled with solvent in one port and an empty syringe in the other port. Twist the syringes slightly as you insert them so they will not pull out too easily. Slowly push on the one plunger while pulling on the other, as shown in Figure G12. This will draw washing liquid through the cell from one syringe to the other. Do this several times; then remove the excess liquid with an empty syringe and dry the cell by passing clean, dry air or nitrogen between the windows. An ear syringe or a special cell-drying syringe can be used to force air (gently!) through the cell, or it can be dried by attaching a trap and aspirator to one port and a drying tube filled with desiccant [OP-22b] to the other.

Solutions. Solutions of liquids or solids in a suitable solvent can be analyzed in a sealed or sealed demountable cell that has a spacer 0.1 mm or more in thickness. The solvent should be relatively nonpolar and must not react with the solute. It should ordinarily dissolve enough of the solute to yield a 5–10% solution; more dilute solutions can be used with FTIR spectrometers. It is best to use spectral-grade solvents so that impurities will not give rise to extraneous peaks. When the sample is run, the infrared spectrum of the solvent is subtracted from the solution spectrum to give the spectrum of the solute. Even then, strong solvent peaks will obscure certain portions of the spectrum. For example, the spectrum of a solute run in chloroform will yield no useful information in the 1250–1200 cm^{-1} and 800–650 cm^{-1} regions because the chloroform absorbs nearly all of the infrared radiation there. Sometimes a sample is run separately in two solvents, such as chloroform and carbon disulfide, to obtain a complete spectrum.

With an FTIR spectrometer, an IR cell is filled with the pure solvent and the solvent's spectrum is run as a *background*. The background spectrum is stored in the instrument until another background is recorded, so a number of samples can be run in the same cell using the same background. With a dispersive IR spectrometer, one cell is filled with the solution and another identical cell is filled with the solvent, using spacers of the same thickness. The spectrum is then run with the solvent cell in the reference beam.

With most FTIR spectrometers you can also record both spectra under the same conditions and subtract the solvent spectrum from the solution spectrum.

To prepare a sample cell for a solution spectrum, prepare a 5–10% solution of the substance to be analyzed and fill the cell using a Luer-lock syringe body, as described for volatile neat liquids. Usually 0.1–0.5 mL of solution will be required. A 0.1-mm spacer can be used unless the solution is quite dilute. When the spectrum has been recorded, clean the sample cell by removing the excess solution and flushing it (as described previously) with the pure solvent used in preparing the solution. If necessary, rinse the cell with a more volatile solvent before drying.

Mulls. A solid sample can be prepared as a *mull* in Nujol (a kind of mineral oil) or another mulling oil. The sample spectrum should be compared with a spectrum of the mulling agent so that peaks due to the oil can be identified and disregarded during interpretation. When Nujol is used, the aliphatic C—H stretching and bending regions (3000–2850, 1470, 1380 cm^{-1}) cannot be interpreted, but most functional groups and other structural features can be identified. If it is necessary to examine the entire spectrum, another sample can be prepared using a complementary mulling oil such as Fluorlube. While Nujol is essentially transparent at wave numbers lower than 1300 cm^{-1}, Fluorlube is transparent above 1300 cm^{-1}. Some scattering of infrared radiation by particles of the solid will reduce transmittance at the high wave number end of the spectrum, so the baseline of a dispersive IR spectrometer should not be set at that end, but wherever the transmittance is highest.

To prepare a mull, grind about 10–20 mg of the solid in an agate or mullite mortar until it coats the entire inner surface of the mortar and has a glassy appearance (about 5–10 minutes of grinding will be required). The particles must be ground to an average diameter of about 1 μm or less to avoid excessive radiation loss by scattering. Add a drop or two of mulling oil and grind the mixture until it has about the consistency of petroleum jelly. Transfer most of the mull to the lower window of a demountable cell using a rubber policeman. Spread the mull evenly with the top window, taking care to exclude air bubbles. Assemble the cell, run the spectrum, and clean the windows as for a neat liquid, using petroleum ether or another suitable solvent.

Potassium Bromide Discs. A potassium bromide disc is prepared by mixing a solid with *dry* spectral-grade potassium bromide and using a die to press the mixture into a more or less transparent wafer. Potassium bromide adsorbs moisture from the atmosphere, so it should be kept in a tightly capped container and stored in an oven or desiccator. It is advisable to dry it

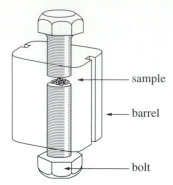

Figure G13 Potassium bromide Mini-Press

If another kind of press is to be used, follow the manufacturer's or your instructor's directions for preparing the disc.

See J. Chem. Educ. **1977**, *54*, 287 *for additional suggestions regarding the preparation of KBr discs.*

before use by heating it in a 110°C oven for several hours. Grind 0.5–2 mg of the solid very finely in a dry agate or mullite mortar as if you were preparing a mull; then add about 100 mg of the dry potassium bromide and mix it thoroughly with the sample. If you are using a Mini-Press (Figure G13), screw in the bottom bolt of the Mini-Press five full turns and introduce about half of the sample mixture into the barrel. Keeping the open end of the barrel pointed up, tap it gently against the bench top to level the mixture, brush down any material on the threads with a soft brush, and screw in the top bolt. Alternately tap the bottom bolt on the bench top and screw in the top bolt with your fingers to level the sample further. When the bolt is finger tight, clamp the bottom bolt in a vise and gradually tighten the top bolt as far as you can with a heavy wrench (or to 20 ft-lbs with a torque wrench). Leave the die under pressure for a minute or two. Remove both bolts, leaving the KBr disc in the center of the barrel, and check to see that it is reasonably transparent and homogeneous; if not, make up a new disc. A disc may be cloudy or inhomogeneous because the components are wet or not completely mixed, the pellet is too thick, the sample size is too large, or the die was not tightened enough.

Place the block containing the disc on a holder (provided with the Mini-Press) in the spectrometer sample beam. As for a mull, scattering will reduce transmittance at the left side of the spectrum, so for a dispersive instrument, set the baseline wherever the transmittance is highest. After you have run the spectrum, punch out the KBr disc using the eraser end of a pencil. Wash the barrel and bolts with water, then rinse them with acetone or methanol, and store the clean, dry press in a desiccator.

Melts. Infrared spectra of low-melting solids can sometimes be obtained by spreading a thin, uniform layer of the finely powdered solid on a silver chloride window, covering it with another silver chloride window, and heating the assembly *slowly* on a hot plate. As soon as the solid melts to form a uniform film between the plates, the windows should be removed and pressed together with forceps until solidification occurs. The windows are then installed in a cell and the spectrum is run as for a thin film. This method works well only when the solid crystallizes as a glassy film or very small crystals; larger crystals produce excessive light scattering.

General Directions for Recording an Infrared Spectrum

Do not attempt to operate the instrument without prior instruction and proper supervision. The construction and operation of commercial infrared spectrometers vary widely, so the following is meant only as a general guide to assist you in recording an infrared spectrum. Specific operating techniques must be learned from the instructor, the operating manual, or both. It will be assumed that the necessary operational parameters have been set beforehand; if not, your instructor will show you what to do. If you are using an FTIR spectrometer follow procedure *A*; for a dispersive IR spectrometer follow procedure *B*.

A. FTIR Spectrometer. Your instructor will tell you which keys to use for such operations as scanning a spectrum, using a cursor, and printing or plotting a spectrum. If a background spectrum has not been recorded recently, run the background with an empty sample compartment, unless your compound is in solution. In that case run the background with a cell containing the pure solvent in the sample compartment and use the same cell for the sample. Prepare a sample cell or KBr disk containing the sample by one of the methods described previously, and place the cell or KBr disc holder in the sample cell holder. Select the desired number of scans (four is usually sufficient for a routine spectrum) and start scanning the spectrum. Wait until a spectrum appears on the monitor. If the spectrum doesn't look right—for example, if the low or high wave number end is missing or you are seeing only a small part of the total spectrum, display the "normal" spectrum using the appropriate key(s) (some instruments use *Rerange* and *Rescale* keys for this purpose). The strongest bands should extend nearly to the bottom of your spectrum; if they do not, use a vertical-scale expansion key to improve the appearance of the spectrum. See that the printer or plotter is turned on and properly adjusted, then press the appropriate key to print or plot the spectrum. If the instrument you are using does not record wave numbers directly on the spectrum, use the cursor arrow keys to move the cursor to the significant bands, and write the displayed wave numbers on your spectrum. Reset the instrument to display a normal spectrum, if necessary. Then remove the sample cell, close the sample compartment, and clean the sample cell.

B. Dispersive IR Spectrometer. Prepare a sample cell or KBr disk containing the sample by one of the methods described previously, and place the cell or KBr disc holder in the sample cell holder. If you are running a solution spectrum, place an identical cell containing the solvent in the reference compartment; otherwise leave it empty. If necessary, place chart paper on the recorder bed (or wrap it around a drum), then align the paper properly and move the drum or carriage to the starting position. Set the 100% transmittance control so that the pen is at 85–90% T, lower the pen onto the chart paper, and scan the spectrum at "normal" or "fast" speed. Examine the spectrum to see that the absorption bands show satisfactory intensity and resolution. Ideally, the spectrum should be recorded so that the strongest absorption band has a maximum transmittance of 5–10%. If your first spectrum is not acceptable, try varying the following parameters, depending on the sample preparation method:

1 *Neat liquid:* Vary cell path length or film thickness.
2 *Solution:* Vary concentration or cell path length.
3 *KBr disc:* Vary amount of sample or thickness of disc.
4 *Mull:* Vary amount of sample or film thickness.

Then run another spectrum, using a slower speed if desired. Remove the sample and spectrum and reset the instrument, if necessary.

Interpretation of Infrared Spectra

Because most infrared bands are associated with specific chemical bonds, it is usually possible to deduce the functional class of an organic compound from its infrared spectrum. The stretching vibrations of chemical bonds resemble the vibrations of springs in that stronger bonds have higher vibrational energies and frequencies than weaker ones. Thus triple bonds generally absorb at higher wave numbers than double bonds, and double bonds absorb at higher wave numbers than single bonds. Because of the comparatively low mass of the hydrogen atom, however, single bonds to hydrogen (C—H, O—H, N—H, etc.) have higher vibrational frequencies than most double and triple

Take Care! Make sure the instrument has been reset correctly before attempting to move the drum or carriage.

bonds. Stretching bands involving single bonds to hydrogen occur at the high frequency (left) end of an infrared spectrum, in the region between 3700 and 2700 cm^{-1} (2.7–3.7 μm). Triple bonds usually absorb between 2700 and 1850 cm^{-1} (3.7–5.4 μm), and double bonds and aromatic bonds absorb between 1950 and 1450 cm^{-1} (5.1–6.9 μm). Most IR bands between 1500 and 600 cm^{-1} (6.7–16.7 μm) are produced by *bending* vibrations or single-bond stretching vibrations. It takes less energy to bend a bond than to stretch it, so bending vibrations tend to have comparatively low frequencies. An absorption band in the 1500–600 cm^{-1} region may be associated with more than one bond; for example, the so-called acyl-oxygen stretching band of an ester arises from the vibration of C—C—O units rather than of isolated C—O bonds.

What to Look for in an IR Spectrum. Consider the infrared spectrum of 2-methyl-1-propanol (isobutyl alcohol) in Figure G14. At first glance it may seem indecipherable—just a series of dips and rises in a graph. But each "dip" (IR band) arises from a stretching or bending vibration of one or more bonds in the 2-methyl-1-propanol molecule, and some of the bands can tell you a great deal about the molecules that gave rise to them. First look at the WAVENUMBERS scale at the bottom of the spectrum; from the numbers there you can read off the wave number, in cm^{-1}, of each band. For example, the first large band in the spectrum is between 3400 and 3200 cm^{-1}, and its minimum is at approximately 3330 cm^{-1}. A more accurate wave number, 3328.1 cm^{-1}, is shown above the right-hand corner of the spectrum. Bands designated by tick marks (short lines) at the bottom of the spectrum have their exact wave numbers listed on the IR spectra in this book, which are reproduced from the *Aldrich Library of FT-IR Spectra, Edition II*.

There is another numerical scale at the top of the spectrum that indicates the wavelength in microns (micrometers). The numbers on the left side of the spectrum are transmittance values; the minimum in the 3328 cm^{-1} band has a transmittance of ~10 percent, meaning that only 10 percent of the 3328 cm^{-1} IR radiation passed through the sample. The other 90 percent

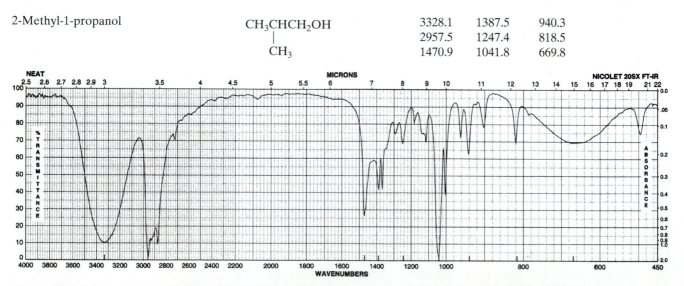

2-Methyl-1-propanol

$$CH_3CHCH_2OH$$
$$|$$
$$CH_3$$

3328.1	1387.5	940.3
2957.5	1247.4	818.5
1470.9	1041.8	669.8

Figure G14 Infrared spectrum of 2-methyl-1-propanol. (Spectrum from *The Aldrich Library of FT-IR Spectra, Edition II*, used with the permission of the Aldrich Chemical Company.)

was absorbed during a change in the bond vibrational frequency of some bond in the compound—but which bond? Since the band is at the left end of the spectrum, the bond must have a high vibrational energy, and we have already noted that bands in the 3700–2700 cm^{-1} region of the spectrum arise from vibrations of bonds to hydrogen atoms. There are only two bonds of this type in the molecule; C—H bonds and O—H bonds. And there are two bands in the 3700–2700 cm^{-1} region, the strong, broad, symmetrical one on the left and a rather ragged band with several minima (centered around 2900 cm^{-1}) on the right. The ragged band is actually composed of several overlapping bands, arising from vibrations of several different bonds of the same general type. Since there is only one O—H bond in the molecule while there are nine C—H bonds, it is reasonable to assume that the ragged band on the right arises from C—H stretching vibrations, and that the band on the left is the O—H band. Note that the area of a band has little to do with the number of bonds that give rise to it; the single O—H bond has a much larger band than the nine C—H bonds.

So the bond responsible for a given IR band can often be identified from its *location* on the spectrum (as indicated by its wave number), its *intensity* (strength), and its *shape*. Most O—H bands (like the one in the previous spectrum) are very broad and strong, where a *strong* band is one whose minimum is near the bottom of the spectrum. Most C—H bands are strong and ragged, like the one in the spectrum. Another very strong band appears at 1042 cm^{-1} in the Figure G14 spectrum. This band, which is in the wave number region for single-bond stretching vibrations (except those involving hydrogen), arises from stretching vibrations of the C—O single bond. The presence of both a C—O and an O—H band in the IR spectrum of a compound is good evidence that the compound is an alcohol (or possibly a phenol), since all alcohols contain a C—O—H grouping in their molecules.

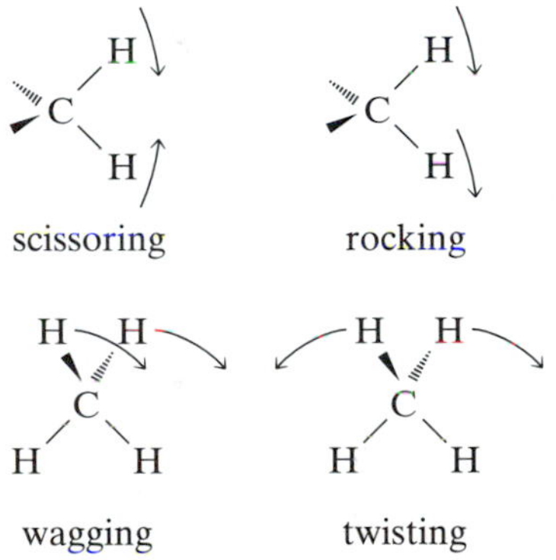

Although we have now located bands corresponding to every kind of bond in the 2-methyl-1-propanol molecule (except C—C bonds, which don't usually give prominent bands) there are still a number of bands left. This is because the same kind of bond can undergo different kinds of vibrations. For example, most of the bands to the left of the C—O band arise from scissoring, wagging, and twisting vibrations of CH$_2$ and CH$_3$ groups (Figure G15), and the very broad, weak band centered at 670 cm^{-1} arises from an O—H bending vibration.

Figure G15 Some carbon-hydrogen vibrations

Spectral Regions. You can see that some infrared bands, particularly the stretching bands we have discussed, are more easily recognized than others, and are more useful in revealing the presence of functional groups. Many of the other bands can be ignored for the time being, although they may provide useful information to a chemist skilled in spectral interpretation. The key to efficient IR spectral interpretation is *knowing where to look* for the more useful bands. Examining the following regions of the infrared spectrum will help you locate the most useful infrared bands quickly.

Region 1: 3600–3200 cm^{-1} (2.8–3.1 μm). Bands in this region can arise from O—H and N—H stretching vibrations of alcohols, phenols, amines, and amides. O—H bands are generally very strong and broad; N—H bands are somewhat weaker, and in the case of primary amines and amides, they have two peaks.

Region 2: 3100–2500 cm^{-1} (3.2–4.0 μm). This region contains most of the C—H stretching vibrations. A strong band in the 3000–2850 cm^{-1} region, arising from C—H bonds to sp³ carbon atoms, is present for most organic compounds. The sp² C—H bonds associated with aromatic hydrocarbons and alkenes absorb at higher frequencies (3100–3000 cm⁻¹), and the C—H bonds of aldehyde (CHO) groups absorb at lower frequencies. The O—H bond of a carboxylic acid gives rise to a very broad absorption band in this region.

Region 3: 1750–1630 cm⁻¹ (5.7–6.1 μm). This region contains most of the carbonyl (C=O) stretching bands of aldehydes, ketones, carboxylic acids, amides, and esters. The carbonyl band is usually strong and quite unmistakable. Unsaturated compounds may have a C=C stretching band in the 1670–1640 cm⁻¹ region, but this band is nearly always weaker and narrower than a carbonyl band.

Region 4: 1350–1000 cm⁻¹ (7.4–10.0 μm). This region is usually cluttered with many C—H bending bands and other bands, but it is often possible to identify the C—O stretching bands of alcohols, phenols, carboxylic acids, and esters, and some C—N stretching bands of amines and amides.

These four spectral regions are shaded on the infrared spectrum in Figure G16. In this spectrum, the absence of any band in Region 1 (or a broad band in Region 2) eliminates from consideration all compounds containing O—H and N—H bonds, including alcohols, phenols, and primary or secondary amines and amides. In Region 2, the appearance of a weak "shoulder" on the C—H band at 3050 cm⁻¹ indicates an sp² C—H bond associated with either an aromatic ring or a carbon-carbon double bond. Region 3 has a strong C=O band near 1720 cm⁻¹, and Region 4 shows a strong C—O band at 1276 cm⁻¹ as well as a weaker one at 1109 cm⁻¹. The absence of an O—H or N—H band and the presence of the C=O and two C—O bands suggest that the compound responsible for this spectrum is an ester. The compound is, in fact, the aromatic ester ethyl benzoate, whose structure is shown on the next page.

2981.9	1367.2	1108.5
1718.5	1275.8	1028.5
1451.4	1175.2	710.3

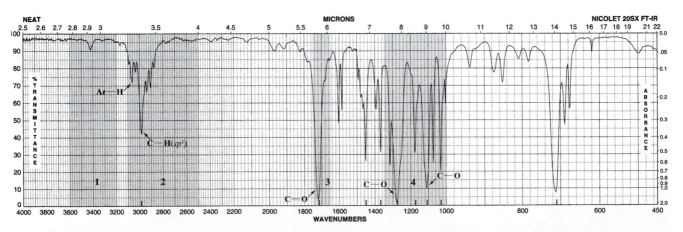

Figure G16 Classification of a compound from its IR spectrum (Spectrum from *The Aldrich Library of FT-IR Spectra, Edition II*, used with the permission of the Aldrich Chemical Company.)

Table G2 Important bands in regions 1–4 of infrared spectra

Region	Frequency range (cm^{-1})	Bond type	Family	Comments
1	3500–3200	N—H	amine, amide	Weak-medium. 1°: 2 bands; 2°: 1 band; 3°: no bands. See also region 3.
	3600–3200	O—H	alcohol, phenol	Broad, strong. See region 4.
2	3300–2500	O—H	carboxylic acid	Very broad, strong, centered around 3000. See regions 3 and 4.
	3100–3000	C—H	aromatic hydrocarbon, alkene	May be shoulder on stronger sp^3 C—H band.
	2850–2700	C—H	aldehyde	Weak to medium, usually two sharp bands. See region 3.
3	1740–1685	C=O	aldehyde	Strong. See region 2.
	1750–1660	C=O	ketone	Strong.
	1725–1665	C=O	carboxylic acid	Strong. See regions 2 and 4.
	1775–1715	C=O	ester	Strong. See region 4.
	1695–1615	C=O	amide	Strong. See region 1.
4	1350–1210	C—O	carboxylic acid	Medium-strong. See regions 1 and 3.
	1300–1180	C—O	phenol	Strong. See region 1.
	1200–1000	C—O	alcohol	Strong. See region 1. Frequency in order 3° > 2° > 1°.
	1310–1160	C—O	ester	Strong. See region 3. Accompanied by weaker C—O band as for alcohol.

Note: Tentative classifications must be confirmed by referring to the following descriptions of specific families.

Table G2 summarizes the locations and characteristics of important absorption bands from the four spectral regions, and tells you where to look for other bands that may help you confirm the presence of a particular functional group.

The best way to become proficient at identifying IR bands is to study spectra that contain those bands, such as the spectra in your lecture textbook, in the "Characteristic Infrared Bands" section that follows, and in collections of spectra described in section F of the Bibliography. When you have learned to recognize the most important IR bands, you can take some shortcuts that will help you identify functional groups rapidly—or at least eliminate the functional groups that aren't there. The following flow chart should help you do that. Just start at the top and work your way down, following the yes/no arrow that answers each question. This chart is intended as a rapid screening device and is *not* infallible; some bands (such as the C=C stretching band) are hard to identify with certainty, and the locations of other bands may vary widely. And some compounds may contain more than one functional group; thus hydroxyacetone (CH_3COCH_2OH) has both an O—H and a C=O band in its spectrum, but it is not a carboxylic acid, as you could tell from the location of its O—H band. When you arrive at a tentative conclusion about the nature of the compound responsible for an IR spectrum, you should refer to Table G2 to see whether other bands on the spectrum are consistent with your initial choice, then study the spectral characteristics of the appropriate class of compounds to confirm (or disprove) your tentative classification. The IR correlation chart on the back endpaper of this book may also help you identify some infrared spectral bands.

ethyl benzoate

Possible family   C=O present?   *Possible family*
                              yes /   \ no

carboxylic   ⟵ yes   O—H present?   O—H present?   ⟶ yes   alcohol
acid | no   no | or phenol

amide (1°, 2°) ⟵ yes N—H present?   N—H present?   ⟶ yes   amide (1°,2°)
                              | no   no |

ester   ⟵ yes   C—O present?   C=C present?   ⟶ yes   alkene
                              | no   no |

aldehyde   ⟵ yes   C—H at ~ 2700 cm⁻¹?   Ar—H present?   ⟶ yes   aromatic
                              | no   no | hydrocarbon

ketone   ⟵ yes   None of the above?   None of the above? ⟶ yes alkane
or 3° amide   or 3° amine

Flow Chart for Detecting Functional Classes from IR Bands.

For example, suppose the the flow chart suggests that your compound may be an alcohol or phenol. You can first check Table G2 to see if any other bands characteristic of alcohols and phenols appear in its spectrum, such as a C—O band. If so, you should read the "Characteristic Infrared Bands" sections about alcohols and phenols to find out whether your compound is an alcohol or phenol. If you find that your compound is an alcohol, you should then study its spectrum for clues to its structure. The frequency of its C—O band may tell you whether it is primary, secondary, or tertiary. By consulting the sections on aromatic hydrocarbons and alkenes (which also apply to other compounds that contain aromatic rings and C=C bonds) you can find out whether your alcohol is aromatic or contains a carbon-carbon double bond. Of course, if you find that your compound is *not* an alcohol or phenol, you should continue down the chart or start back at the beginning.

Characteristic Infrared Bands

This section contains information about the most useful infrared bands of compounds that can be identified as described in Part IV, Qualitative Organic Analysis—alcohols, aldehydes, ketones, amides, amines, carboxylic acids, esters, halides, aromatic hydrocarbons, and phenols. Since such compounds usually contain alkyl groups and may contain carbon-carbon double bonds, the IR bands of alkanes and alkenes are described as well. For each family of organic compounds, a summary of the main spectral features that characterize the family is followed by a description of individual bond vibrations and a representative infrared spectrum. The wave number ranges given are for solids (in Nujol mulls or KBr discs) or neat liquids; values for solutions may differ somewhat. Although the wave number ranges apply to most of the organic compounds in each class, compounds with certain structural features (such as highly strained rings) may have bands outside of the ranges indicated. On the spectra, absorption bands are designated either as stretching (ν) or bending (δ) bands; only those bands that are most useful for functional group classification are labeled. Note that the exact wave numbers of significant bands (designated by tick marks along the lower edge of the spectra) are listed with each spectrum.

Alkanes. Alkanes are identified primarily by the absence of any IR bands characteristic of functional groups. Their spectra are quite simple, containing only the C—H stretching and bending vibrations characteristic of sp³ hybridized carbon atoms. Since nearly all other organic compounds contain such C—H bonds, their spectra will also contain some or all of the bands described for alkanes. (See Experiment 16 for some alkane spectra and additional information.)

C—H *stretch:* 3000–2800 cm⁻¹ (multiple overlapping bands, strong to weak). CH₃ bands are near 2960 cm⁻¹ and 2870 cm⁻¹; CH₂ bands are near 2925 cm⁻¹ and 2850 cm⁻¹. Nearly always to the *right* of 3000 cm⁻¹.
C—H *bend:* (moderate to weak). Characteristic CH₃ band is near 1375 cm⁻¹; CH₂ band is near 1465 cm⁻¹.

Alkenes. Most alkenes contain the same kinds of bands as alkanes, plus additional bands associated with carbon-carbon double bonds and vinylic (=C—H) carbon-hydrogen bonds. The presence of one or two strong bands in the 1000–650 cm⁻¹ region and a sharp band near 1650 cm⁻¹ suggests an alkene functional group, especially if the compound is not aromatic. (See Figure G17.)

=C—H *stretch:* 3125–3030 cm⁻¹ (moderate to weak). May appear as a shoulder on a stronger sp³ C—H band, but nearly always to the *left* of 3000 cm⁻¹.
C=C *stretch:* 1675–1600 cm⁻¹ (moderate to weak; narrow). May be absent for symmetrical alkenes. Conjugation moves band to lower wavelengths.
=C—H *out-of-plane bend:* 1000–650 cm⁻¹ (usually strong). Position depends on type of substitution: RCH=CH₂ has bands at 995–985 and 915–905 cm⁻¹; *cis*-RCH=CHR a band at 730-665 cm⁻¹; *trans*-RCH=CHR a band at 980–960 cm⁻¹; and R₂C=CH₂ a band at 895–885 cm⁻¹.

Aromatic Hydrocarbons. Most compounds containing benzene rings are characterized by (1) aromatic C—H (Ar—H) stretching bands near

1-Hexene	CH₃CH₂CH₂CH₂CH=CH₂	2962.1	1466.1	909.2
		1821.3	1379.1	739.8
		1641.8	992.7	630.8

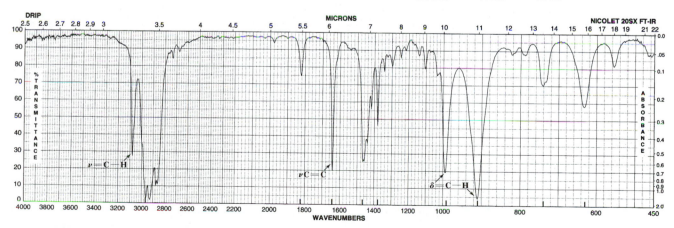

Figure G17 IR spectrum of an alkene, 1-hexene (Spectrum from *The Aldrich Library of FT-IR Spectra, Edition II,* used with the permission of the Aldrich Chemical Company.)

3070 cm^{-1}, (2) a distinctive pattern of weak bands in the 2000–1650 cm^{-1} region, (3) two sets of bands near 1600 cm^{-1} and 1515–1400 cm^{-1}, and (4) one or more strong absorption bands in the 900–675 cm^{-1} region. The presence of such bands and the absence of absorption bands characteristic of functional groups suggest an aromatic hydrocarbon. (See Figure G19.)

Ar—H *stretch:* 3100–3000 cm^{-1} (moderate to weak). May appear as a shoulder on a stronger sp^3 C—H band, but nearly always to the *left* of 3000 cm^{-1}.

Overtone-combination vibrations: 2000–1650 cm^{-1} (multiple bands, weak). The band pattern is related to the kind of ring substitution, as shown in Figure G18.

C⋯C *stretch:* 1615–1585 cm^{-1} and 1515–1400 cm^{-1} (variable).

Ar—H *out-of-plane bend:* 910–730 cm^{-1} (strong). The band frequency varies with the number of adjacent ring hydrogens:

> two adjacent hydrogens: 855–800 cm^{-1}
>
> three adjacent hydrogens: 800–765 cm^{-1}
>
> four or five adjacent hydrogens: 770–730 cm^{-1}

Monosubstituted, *meta*-disubstituted, and some trisubstituted benzenes show an additional ring-bending band around 715–680 cm^{-1}. For example, a *meta*-disubstituted benzene has three adjacent ring hydrogens, so it should have bands in the 800–765 cm^{-1} and 715–680 cm^{-1} regions.

Alcohols. The presence of a strong, broad band centered around 3300 cm^{-1} and a strong C—O band in the 1200–1000 cm^{-1} region is good evidence for an alcohol. A C—O band above 1200 cm^{-1} may suggest a phenol, particularly when it is accompanied by bands indicating an aromatic ring structure. (See Figure G20.)

monosubstituted

o-disubstituted

m-disubstituted

p-disubstituted

Figure G18 Typical absorption patterns of substituted aromatic compounds in the 2000–1670 cm^{-1} region

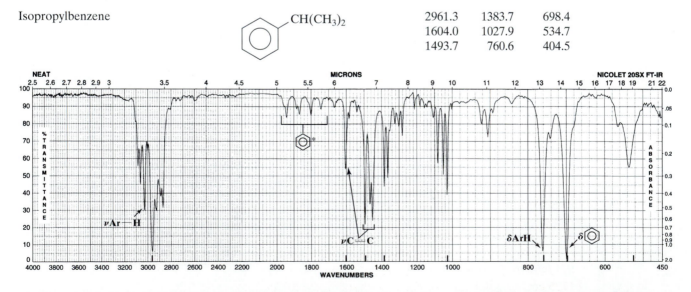

Isopropylbenzene   CH(CH$_3$)$_2$

2961.3	1383.7	698.4
1604.0	1027.9	534.7
1493.7	760.6	404.5

Figure G19 IR spectrum of an arene, isopropylbenzene. The bands marked ⬡* are aromatic overtone-combination bands. (Spectrum from *The Aldrich Library of FT-IR Spectra, Edition II,* used with the permission of the Aldrich Chemical Company.)

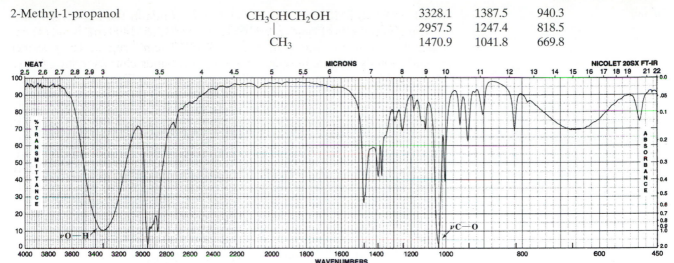

2-Methyl-1-propanol

$$CH_3CHCH_2OH$$
$$|$$
$$CH_3$$

3328.1	1387.5	940.3
2957.5	1247.4	818.5
1470.9	1041.8	669.8

Figure G20 IR spectrum of a primary alcohol, 2-methyl-1-propanol (Spectrum from *The Aldrich Library of FT-IR Spectra, Edition II*, used with the permission of the Aldrich Chemical Company.)

O—H *stretch:* 3600–3200 cm^{-1} (strong, broad). Usually centered near 3300 cm^{-1}.

C—O *stretch:* 1200–1000 cm^{-1} (strong to moderate). Most saturated aliphatic alcohols absorb near 1050 cm^{-1} if they are primary, near 1110 cm^{-1} if they are secondary, and near 1175 cm^{-1} if they are tertiary. Alicyclic alcohols, and alcohols with aromatic rings or vinyl groups on the carbon that is bonded to OH, absorb at wave numbers about 25–50 cm^{-1} lower than these.

Phenols. Phenols are characterized by a strong, broad band centered around 3300 cm^{-1} and a strong band near 1230 cm^{-1}, accompanied by bands indicating an aromatic structure (see "Aromatic Hydrocarbons"). (See Figure G21.)

O—H *stretch:* 3600–3200 cm^{-1} (strong, broad).
O—H *bend:* 1390–1315 cm^{-1} (moderate).
C—O *stretch:* 1300–1180 cm^{-1} (strong). Usually close to 1230 cm^{-1}. This band may be split, with several distinct peaks.

Aldehydes. The presence of a sharp, medium-intensity band near 2720 cm^{-1} and a strong carbonyl band near 1700 cm^{-1} is good evidence for an aldehyde. (See Figure G22.)

$$\overset{(O)}{\underset{||}{C}}$$—H *stretch:* 2850–2700 cm^{-1} (moderate to weak). From the carbonyl C—H bond. Most aldehydes have two bands near 2850 and 2720 cm^{-1}, with the low-frequency band well separated from other aliphatic C—H bands.

C=O *stretch:* 1740–1685 cm^{-1} (strong). Most unconjugated aldehydes absorb near 1725 cm^{-1}; conjugation of the carbonyl group with an aromatic ring or other unsaturated system shifts the band to the 1700–1685 cm^{-1} region. A weak overtone of this band may appear near 3400 cm^{-1}.

Phenol

OH

3372.6	1224.4	751.8
1595.3	1168.0	689.8
1498.9	809.8	506.0

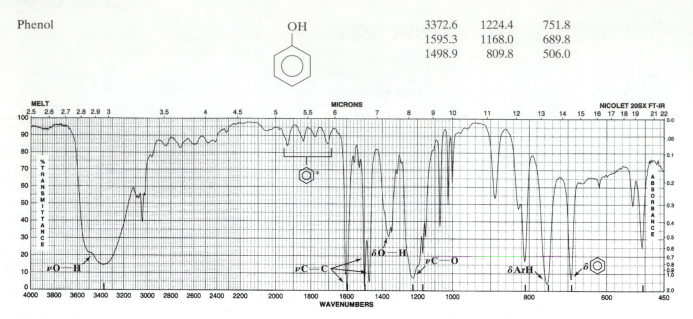

Figure G21 IR spectrum of phenol (Spectrum from *The Aldrich Library of FT-IR Spectra, Edition II*, used with the permission of the Aldrich Chemical Company.)

Note: The bands marked ⬡* are aromatic overtone-combination bands.

3-Methylbutanal

CH₃   O
CH₃CHCH₂CH

2960.1	1468.3	1016.6
2718.7	1368.8	898.9
1727.6	1170.8	524.1

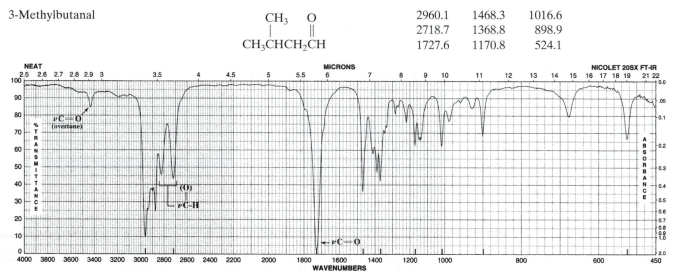

Figure G22 IR spectrum of an aldehyde, 3-methylbutanal (Spectrum from *The Aldrich Library of FT-IR Spectra, Edition II*, used with the permission of the Aldrich Chemical Company.)

Ketones. The presence of a strong carbonyl band around 1700 cm⁻¹ is good evidence for a ketone if other bands listed in Table G2 (O—H, N—H, C—O, and aldehyde C—H) are absent. Bands in the 1300-1100 cm⁻¹ region, which may be mistaken for C—O bands, arise from C—C—C vibrations involving the carbonyl carbon. (See Figure G23.)

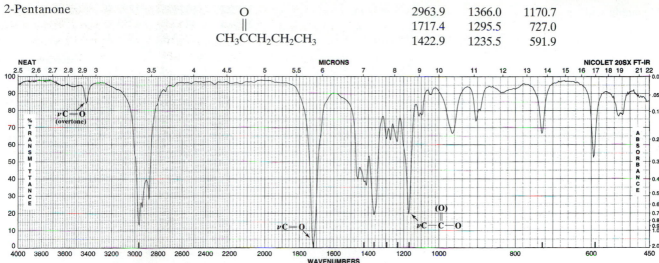

Figure G23 IR spectrum of a ketone, 2-pentanone (Spectrum from *The Aldrich Library of FT-IR Spectra, Edition II,* used with the permission of the Aldrich Chemical Company.)

C=O *stretch:* 1750–1660 cm^{-1} (strong). Most unconjugated aliphatic ketones absorb around 1715 cm^{-1}, and conjugated ketones absorb near 1670 cm^{-1}. A weak C=O overtone band is usually evident near 3400 cm^{-1}.

C—C—C *stretch/bend:* 1300–1100 cm^{-1} (moderate). Often multiple bands; unconjugated ketones absorb around 1230–1100 cm^{-1}, conjugated ketones around 1300–1230 cm^{-1}.

Carboxylic Acids. The presence of a very broad band centered near 3000 cm^{-1} and a carbonyl band around 1700 cm^{-1} is good evidence for a carboxylic acid. (See Figure G24.)

O—H *stretch:* 3300–2500 cm^{-1} (strong, very broad). C—H stretching bands are generally superimposed on this band.
C=O *stretch:* 1725–1665 cm^{-1} (strong). Unconjugated acids absorb around 1725–1700 cm^{-1}, conjugated acids around 1700—1665 cm^{-1}.
C—O *stretch:* 1350–1210 cm^{-1} (strong). May have a number of sharp peaks for long-chain acids.
O—H *bend:* 950–870 cm^{-1} (moderate, broad).

Esters. The presence of a strong carbonyl band around 1740 cm^{-1} and an unusually strong C—O band in the 1310–1160 cm^{-1} region is good evidence for an ester, especially if there is no O—H band. (See Figure G25.)

C=O *stretch:* 1775–1715 cm^{-1} (strong). Near 1770 cm^{-1} for phenyl esters (RCOOAr) and vinyl esters, 1740 cm^{-1} for most unconjugated esters, and 1730–1695 cm^{-1} for formates and conjugated esters.
C—O *stretch (acyl-oxygen):* 1310–1160 cm^{-1} (strong, broad). Occurs near 1310–1250 cm^{-1} for conjugated esters, 1240 cm^{-1} for acetates, and 1175 cm^{-1} for other unconjugated esters. Both the "acyl-oxygen" and "alkyl-oxygen" bands arise from coupled vibrations involving C—C—O groupings.

Hexanoic acid

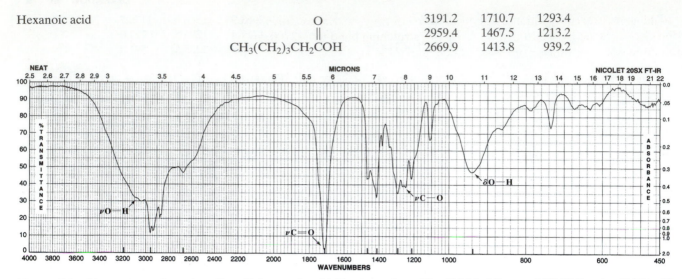

3191.2	1710.7	1293.4
2959.4	1467.5	1213.2
2669.9	1413.8	939.2

$$CH_3(CH_2)_3CH_2COH$$

Figure G24 IR spectrum of a carboxylic acid, hexanoic acid (Spectrum from *The Aldrich Library of FT-IR Spectra, Edition II,* used with the permission of the Aldrich Chemical Company.)

sec-Butyl acetate

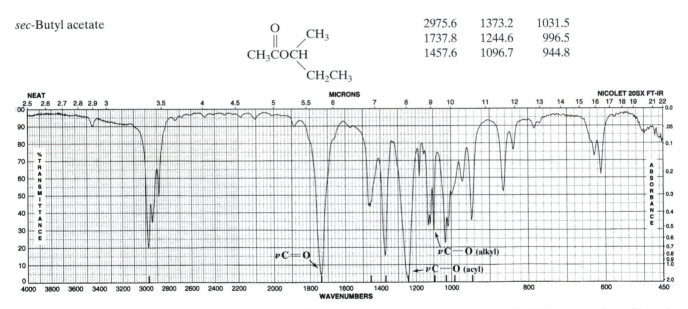

2975.6	1373.2	1031.5
1737.8	1244.6	996.5
1457.6	1096.7	944.8

$$CH_3COCH$$

Figure G25 IR spectrum of an ester, *sec*-butyl acetate (Spectrum from *The Aldrich Library of FT-IR Spectra, Edition II,* used with the permission of the Aldrich Chemical Company.)

C—O *stretch (alkyl-oxygen):* 1200–1000 cm^{-1} (moderate). Occurs in the same region as alcohol bands. Esters of phenols absorb at higher wave numbers.

Amines. Primary amines are characterized by a medium-intensity, two-pronged band near 3350 cm^{-1} and two medium-strong bands near 1615 and 800 cm^{-1}, the latter one very broad. Secondary amines have a single weak band near 3300 cm^{-1} and a broad band near 715 cm^{-1}. Tertiary amines may be distinguished by a shift of the methylene stretching band to ~2700 cm^{-1} and the presence of a C—N band. (See Figure G26.)

> N—H *stretch:* 3500–3200 cm^{-1} (moderate to weak, broad). Primary aliphatic amines give rise to a two-pronged band centered near 3350 cm^{-1}, secondary amines have one weak band near 3300 cm^{-1}, and tertiary amines have none. Aromatic primary and secondary amines absorb near 3400 and 3450 cm^{-1}, respectively.
>
> N—H *bend (scissoring):* 1650–1500 cm^{-1} (strong to moderate). Usually near 1615 cm^{-1} for primary amines. Seldom observed for secondary aliphatic amines; secondary aromatic amines absorb near 1515 cm^{-1}.

3-Methylbutylamine

CH_3
|
$CH_3CHCH_2CH_2NH_2$

3366.4	1384.1	847.7
2955.0	1066.7	815.5
1467.7	919.2	770.5

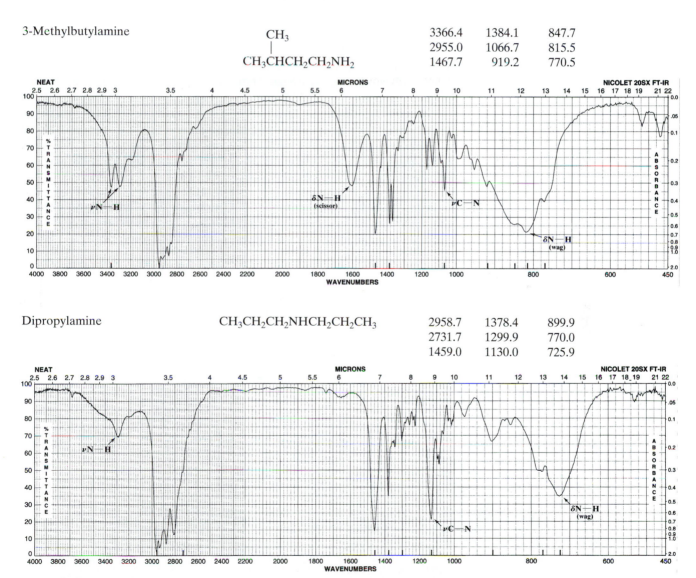

Dipropylamine

$CH_3CH_2CH_2NHCH_2CH_2CH_3$

2958.7	1378.4	899.9
2731.7	1299.9	770.0
1459.0	1130.0	725.9

Figure G26 IR spectra of a primary amine, 3-methylbutylamine, and a secondary amine, dipropylamine (Spectra from *The Aldrich Library of FT-IR Spectra, Edition II,* used with the permission of the Aldrich Chemical Company.)

N—H *bend (wagging):* 910–660 cm^{-1} (strong to moderate, broad). Often strong and very broad, around 910–770 cm^{-1} for primary amines. Closer to 715 cm^{-1} for secondary amines.

C—N *stretch:* 1340–1020 cm^{-1} (strong to moderate). Around 1340–1250 cm^{-1} for aromatic amines, 1250—1020 cm^{-1} for aliphatic amines. As for an alcohol C—O band, the frequency of an aliphatic C—N band varies with changes in the structure of the attached alkyl group.

Amides. The presence of a carbonyl band near 1640 cm^{-1} and two bands (or peaks) in the 3400–3000 cm^{-1} region is good evidence for an amide. (See Figure G27.)

N—H *stretch:* 3450–3300 cm^{-1} and 3225–3180 cm^{-1} (one or two bands, strong to moderate). Primary amides have two bands (or a two-pronged band) near 3400 and 3200 cm^{-1}. Secondary amides have a single N—H stretching band near 3340 cm^{-1}, with an N—H bending overtone near 3080 cm^{-1}.

C=O *stretch:* 1695–1615 cm^{-1} (strong). Usually centered near 1640 cm^{-1}.

N—H *bend:* 1655–1615 cm^{-1} (primary) or 1570–1515 cm^{-1} (secondary) (strong to moderate). This band usually overlaps the carbonyl band on spectra of primary amides obtained using KBr discs or mulls; it appears at lower frequencies on spectra obtained in solution. The band is near 1540 cm^{-1} for most secondary amides, and an overtone can sometimes be seen at about 3080 cm^{-1}.

Organic Halides. Alkyl chlorides and bromides show fairly strong absorption between 800 and 500 cm^{-1}. Additional chemical evidence is usually needed to characterize organic halides. (See Figure G28.)

2-Methylpropanamide

$$\underset{CH_3CH-CNH_2}{\overset{\displaystyle CH_3 \quad O}{}}$$

3352.3	1296.5	655.8
1640.2	1147.1	625.8
1401.0	1090.6	511.7

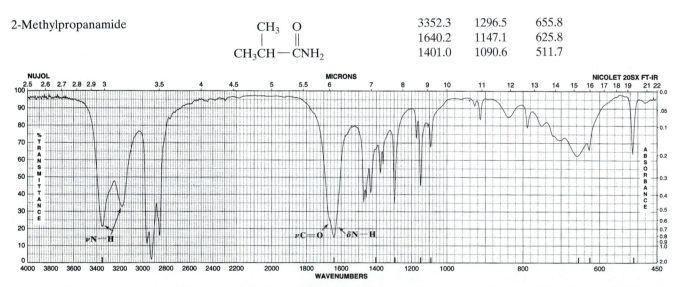

Figure G27 IR spectrum of an amide, 2-methylpropanamide (Spectrum from *The Aldrich Library of FT-IR Spectra, Edition II*, used with the permission of the Aldrich Chemical Company.)

1-Chloropentane $\qquad$ $CH_3(CH_2)_4Cl$

2959.3	1282.1	789.3
1467.0	1037.2	730.8
1380.3	925.7	653.7

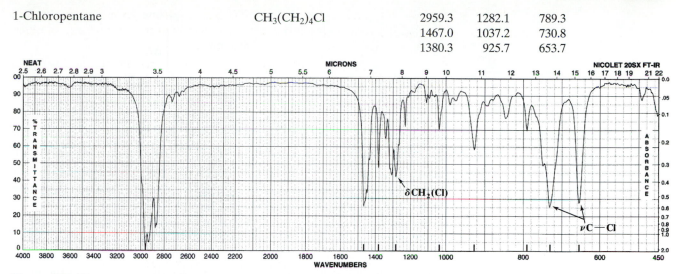

Figure G28 IR spectrum of an alkyl chloride, 1-chloropentane (Spectrum from *The Aldrich Library of FT-IR Spectra, Edition II,* used with the permission of the Aldrich Chemical Company.)

C(X)—H *bend:* 1300–1150 cm^{-1} (moderate). Observed only for halides with terminal halogen atoms ($-CH_2X$).

C—Cl *stretch:* 850–550 cm^{-1} (strong to moderate). Two bands near 725 and 645 cm^{-1} when the chlorine is terminal; below 625 cm^{-1} otherwise, unless several chlorine atoms are on the same or adjacent carbons. Ar—Cl bonds absorb around 1175–1000 cm^{-1}.

C—Br *stretch:* 760–500 cm^{-1} (strong to moderate). Near 645 cm^{-1} when the bromine is terminal. Ar—Br bonds absorb around 1175–1000 cm^{-1}.

Nuclear Magnetic Resonance Spectrometry

OPERATION 35

The theoretical principles underlying nuclear magnetic resonance (NMR) spectrometry can be found in most textbooks of organic chemistry and in appropriate sources listed in category F of the Bibliography in this book. Here we will review only those principles that are needed for an understanding of the operation of an NMR spectrometer.

Nuclear magnetic resonance (NMR) spectrometry is based on the magnetic properties of certain nuclei that possess a quality known as *spin*. The nucleus of a ^1H atom, which is a single proton, has spin. The nuclei of ^{13}C atoms also have spin, but the nuclei of ^{12}C atoms, which are nearly 100 times more abundant than ^{13}C atoms, do not. An atom with spin behaves like a tiny bar magnet. When placed in a strong magnetic field it tends to align

If you are not familiar with the principles and terminology of NMR spectrometry, read pp. 741–745 or consult your lecture text.

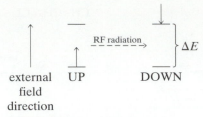

Figure G29 Spin transition of a magnetic nucleus

with the field. For convenience, we will refer to a nucleus in this orientation as being in an **up** spin state and a nucleus in the opposite orientation as being in a **down** spin state. If a sample containing magnetic nuclei is exposed to radio frequency (RF) radiation of just the right frequency, some of its **up** nuclei will flip over and become aligned against the external field. Such a transition is illustrated in Figure G29. Since a nucleus in the **down** state is less stable (contains more energy) than a nucleus in the **up** state, the spin transition results in an absorption of energy by the nucleus. Such a transition is possible only if the energy of a radio frequency photon, $h\nu$, is exactly equal to the energy of the transition, ΔE, so that $\nu = \Delta E/h$. When this is the case, the *resonance condition*—the condition under which nuclei of a given kind can undergo spin transitions—is fulfilled. The value of ΔE is directly proportional to the strength of the external magnetic field, H_o, so ν is also proportional to H_o. This means that the resonance condition for a nucleus can be fulfilled either by adjusting the frequency of the RF radiation or by adjusting the strength of the external field.

a. ^{1}H NMR Spectrometry

Instrumentation

There are two fundamentally different ways of obtaining an NMR spectrum. With a *continuous-wave (CW) NMR spectrometer,* the sample is irradiated continuously with radio frequency waves as the magnetic field (or sometimes the RF frequency) is varied, and the electromagnetic signals generated by nuclei as they change spins are converted to peaks on a moving chart. With a *Fourier-transform NMR (FT-NMR) spectrometer,* the sample is irradiated with intense pulses of full-spectrum RF radiation that displace the nuclei from their equilibrium distribution. Their response to the displacement is monitored, generating data that is converted by a computer to an NMR spectrum.

Continuous-Wave NMR. In a continuous-wave NMR spectrometer, a glass tube containing the sample is placed between the poles of a magnet and irradiated with RF radiation from a transmitter coil as the magnetic field is "swept" (varied continuously) over a preset range. In the instrument diagrammed in Figure G30, the magnetic field is swept from low to high field by varying the strength of an electric current passing through the sweep coils. When the resonance condition for a particular nucleus in the sample is met, nuclei of that type flip from the **up** state to the **down** state. As they do so, they generate a small fluctuating magnetic field that can be detected by a receiver coil encircling the sample tube. The receiver coil

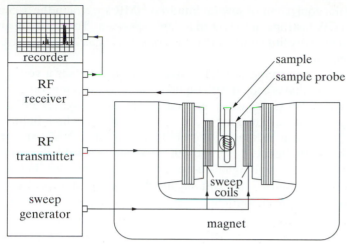

Figure G30 Schematic diagram of a continuous-wave NMR spectrometer

sends an electrical signal to an RF receiver, which amplifies and modifies the signal so that it can be displayed on a recorder as an NMR spectrum.

The NMR spectrum is a record of all the signals generated by all of the different kinds of nuclei in the sample that absorb RF radiation over the range swept by the instrument. If the sweep range is one in which the resonance conditions for ^{1}H nuclei (protons) are met, the spectrum should display a different signal for each kind of proton in a particular molecular environment. For example, protons on the benzene ring in *para*-xylene are in a different molecular environment than protons on the methyl groups, so the NMR spectrum of p-xylene will display two signals, one for each kind of proton.

A typical continuous-wave NMR spectrometer suitable for use by undergraduate students may operate at an RF frequency of 60 MHz and a magnetic field strength of approximately 1.4 tesla (14,000 gauss) for protons. When a ^{1}H NMR spectrum is recorded using a 60-MHz spectrometer, the magnetic field is swept over a range of about 1.4×10^{-5} tesla (0.14 gauss), which is only 10 millionths of the external field strength, or 10 parts per million (ppm). This sweep range must be extended to 15 ppm or so to detect certain protons, such as those in COOH groups.

Fourier-Transform NMR. A Fourier-Transform NMR (FT-NMR) spectrometer is capable of producing spectra with better resolution and a much higher signal-to-noise ratio than any CW instrument. In an FT-NMR instrument the nuclei are irradiated with a short ($\sim$10 μs) pulse of radio frequency radiation that covers the entire frequency range of interest. The pulse is so intense that it raises all of the absorbing nuclei into the **up** state. As the high-energy nuclei return to their equilibrium state, they generate a *free induction decay (FID)* signal that contains information about the nuclei whose resonance conditions were met by any of the RF frequencies in the pulse. The FID signal, which is equivalent in information content to a complete NMR spectrum, is picked up and sent to a computer that accumulates and averages the FID signals from a series of pulses. The computer then "decodes" the averaged FID signal by Fourier-transform analysis and converts it to a conventional NMR spectrum. The FID signal generated by one pulse takes less than a second to acquire, so an FT-NMR spectrometer can accumulate and

H_3C —⬡— CH_3

p-xylene

average the equivalent of several hundred NMR spectra in the 2–5 minutes it takes a CW instrument to record a single spectrum. The resulting averaged spectrum has very little noise and is thus much "cleaner" than a conventional CW spectrum.

Adding the data generated by successive pulses improves the quality of the NMR spectrum because a signal *increases in intensity with each addition, while instrumental* noise, *being random, tends to cancel out.*

A typical research-grade FT-NMR spectrometer uses a very powerful electromagnet that is cooled with liquid helium. At the temperature of liquid helium (4 K, −269°C) the wire coils that generate the magnetic field are electrical superconductors, making it possible to attain very high field strengths of 14 tesla or more. Increasing the field strength of an NMR spectrometer causes the NMR signals to spread out, reducing overlap between adjacent signals. This and the high signal-to-noise ratio make complex ^{1}H NMR spectra generated on an FT instrument much easier to interpret than those obtained with a CW instrument.

Chemical-Shift Reagents

The amount of structural information that can be obtained from a CW-NMR spectrum is often limited by the presence of overlapping signals. Increasing the magnetic field strength reduces overlapping by increasing the chemical shifts (in Hz) of all the signals by the same amount. Using a *chemical-shift reagent* also changes the chemical shifts of NMR signals, but it affects different signals differently; some are shifted more than others and some may not be shifted at all. Nevertheless, an appropriate chemical-shift reagent can often be used to separate the signals of interest and facilitate spectral interpretation.

Chemical-shift reagents are organometallic complexes of certain paramagnetic rare earth metals. These complexes can coordinate with the oxygen and nitrogen atoms of alcohols, amines, carbonyl compounds, and other Lewis bases. The local magnetic field produced by the paramagnetic metal atom shifts the signals of nearby protons to an extent that varies with distance; the closer a nucleus is to the metal atom, the more its chemical shift will change. Different chemical-shift reagents have different effects on a spectrum; thus tris(dipivaloylmethanato)europium(III) [Eu(dpm)$_3$] causes downfield chemical shifts, while the corresponding complex of praseodymium, Pr(dpm)$_3$, induces upfield chemical shifts.

Experimental Considerations

Sample Preparation. Most substances analyzed by NMR are first dissolved in a suitable solvent. Liquids that are no more viscous than water can sometimes be analyzed neat, but neat liquids may give broadened peaks and other spectral distortions due to intermolecular interactions. FT-NMR spectrometers yield good proton NMR spectra with solution concentrations as low as 0.1% (w/v), but CW-NMR instruments require concentrations on the order of of 10–20% (w/v) or higher. The liquid or solution is placed in a special thin-walled *NMR tube,* which is closed with a tight-fitting plastic cap to prevent evaporation. A typical NMR tube has an O.D. of 5 mm, a length of 17.5 cm, and is both fragile and expensive. The NMR tube should be straight and uniform; a tube that wobbles when it is rolled down a slightly inclined glass plate will give large spinning sidebands, as discussed under the heading "Sample Spinning." For a routine ^{1}H NMR analysis on a CW instrument, you can prepare the sample by (1) dissolving 50–100 mg of your compound in 0.5–0.8 mL of a suitable solvent; (2) filtering the solution directly into the NMR tube through a Pasteur pipet containing a small plug of tightly packed glass wool; (3) adding 5–15 μL of a reference standard, usually TMS (tetramethylsilane); (4) stoppering the NMR tube carefully; and (5) inverting the tube several times to mix the components thoroughly. Commercial NMR solvents may have 1–3% TMS added, in which case step (3) is omitted. The NMR tube should be filled to a depth of at least 2.5 cm but should be no more than three-fourths full. TMS boils near room temperature, so it should be kept in a refrigerator and added with a *cold* syringe or fine-tipped dropper. For very high-resolution spectra, the sample should be *degassed* by bubbling a fine stream of pure nitrogen through it for 1 minute; degassing is not necessary for routine spectra.

NMR Solvents. A solvent suitable for ^{1}H NMR analysis should ordinarily have no protons that produce intense signals of their own, since they might obscure signals from the sample. Thus hydrogen-containing solvents such as chloroform and acetone are used in their completely deuterated forms. Deuterium (^{2}H) undergoes resonance at about $6\frac{1}{2}$ times the field strength required for ^{1}H, so an isotopically pure deuterated solvent does not interfere with a proton NMR spectrum. Most deuterated solvents, however, contain a significant amount of the protic form, giving rise to one or more small signals. Thus the NMR spectrum of a deuterochloroform solution has a signal at 7.27 ppm, but this small signal usually doesn't interfere with the solute's signals. A solvent for FT-NMR analysis *must* contain deuterium, because the instrument locks onto the resonance signal of deuterium to help the user adjust the controls for maximum spectral resolution.

A good NMR solvent should also have a low viscosity, high solvent strength, and no appreciable interactions with the solute. Deuterochloroform (chloroform-d, $CDCl_3$) is the most widely used NMR solvent because its polarity is low enough to prevent significant solute-solvent interactions, and most organic compounds are sufficiently soluble in it for NMR analysis. When a more polar solvent is required, dimethyl sulfoxide-d_6 is often used, often in mixtures with deuterochloroform. It may be convenient to add 1–3% TMS to the bulk solvent so that it does not have to be added during sample preparation.

Table G3 compares the properties of some NMR solvents and gives the approximate chemical shifts (δ) of their ^{1}H NMR signals.

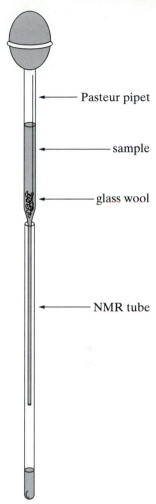

Figure G31 Filtering an NMR solution

Table G3 Properties of some NMR solvents

Solvent	δ, ppm	Solvent strength	Freedom from interactions	Viscosity
carbon disulfide	none	good	good	low
cyclohexane-d_{12}	1.4	poor	good	medium
acetonitrile-d_3	2.0	good	fair	low
acetone-d_6	2.1	good	poor	low
dimethyl sulfoxide-d_6	2.5	very good	poor	high
1,4-dioxane-d_8	3.5	good	fair	medium
deuterium oxide	~ 5.2 (v)	good	poor	medium
chloroform-d	7.3	very good	fair	low
pyridine-d_5	7.0–8.7	good	poor	medium
trifluoroacetic acid	~ 12.5 (v)	good	poor	medium

Note: δ is the chemical shift of the protic form of the solvent; v = variable; solvent strength refers to the ability to dissolve a broad spectrum of organic compounds.

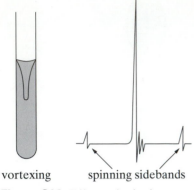

vortexing   spinning sidebands

Figure G32 Effects of spinning rate

Sample Spinning. To average out the effect of magnetic-field variations in the plane perpendicular to the axis of the NMR tube, an NMR sample is rotated at a rate of 30–60 revolutions per second while the NMR spectrum is being recorded. It is important to set the spinning rate correctly—excessively high rates can cause a vortex extending into the region of the receiver coil, and low rates can cause large *spinning sidebands* or signal distortion. Spinning sidebands are small peaks that are symmetrically spaced on either side of a main peak at a distance equal to the spinning rate; thus an NMR tube spun at 30 cycles per second may give rise to sidebands 30 Hz from each main peak. Spinning sidebands can be caused by field inhomogeneity and wobbling NMR tubes or sample spinners. To find out whether small signals are spinning sidebands or impurity peaks, change the spinning rate and scan again to see if their positions change.

Field Homogeneity. Recording a good NMR spectrum requires that the magnetic field be homogeneous (uniform) at the sample. The most important homogeneity control, usually called the Y control, is adjusted to produce a uniform field along the axis of the sample tube. For routine work on a previously tuned CW-NMR spectrometer, the Y control can be set by placing a blank sheet of paper over the chart paper and repeatedly scanning a strong peak in the spectrum of the sample (or of a standard acetaldehyde solution), each time making small adjustments in the Y control until the peak is as tall and narrow as possible and shows a good "ringing" (beat) pattern. Figure G33 shows an excellent ringing pattern for the quartet of acetaldehyde, characterized by the high amplitude, long duration, and exponential decay of the "wiggles" following the main peaks. An FT-NMR spectrum does not show a ringing pattern; magnetic field homogeneity is adjusted by maximizing the intensity of a deuterium lock signal. Ordinarily the instructor or a lab technician performs such adjustments.

Signal Amplitude. The amplitude (height) of the signals on an NMR spectrum is adjusted with two controls. The *spectrum amplitude* control changes the amplitude of both the signals and the baseline noise. The *RF power* con-

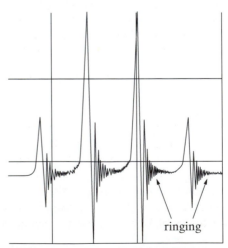

ringing

Figure G33 Ringing pattern for the quartet of acetaldehyde

trol increases signal height without increasing baseline noise up to the point where *saturation* begins, when the number of nuclei in both spin states is so nearly equal that increasing the intensity of the RF radiation no longer increases the number of transitions. Raising the RF power beyond that point causes distortion and reduces the signal size. Unless high sensitivity is required, the RF power level is usually set at about mid-range or at some other value where there is little likelihood of saturation, and the spectrum amplitude control is adjusted so that the strongest peak in the spectrum extends nearly to the top of the chart paper.

Sweep and Phasing Controls. There are four sweep controls on a typical continuous-wave NMR spectrometer; these control the reference point of the spectrum, the sweep rate (sweep width divided by sweep time), and the portion of the spectrum to be scanned. The *sweep zero* control is used to set the signal of the reference compound to the proper value (zero for TMS). The *sweep width* control sets the total chemical-shift range to be scanned, usually 600–1000 Hz when the entire spectrum is being scanned on a 60-MHz instrument. A 600-Hz (10-ppm) range can be used if the sample is known to contain no protons absorbing downfield of 10 δ. A CW-NMR spectrum is often scanned at a rate of 1 Hz per second, so the *sweep time* can can be set numerically equal to the sweep width (for example, 600 seconds for a 600-Hz sweep width). The *sweep offset* control is used when only a specific portion of the spectrum is to be scanned; it sets the upfield limit of the scan. For example, if a scan between 350 and 500 Hz is desired, the sweep offset should be 350 Hz and the sweep range 150 Hz.

The *phasing* control should be adjusted to obtain a straight baseline before and after a signal. Correct phasing is much more important when an NMR spectrum is being integrated than when it is being recorded.

Integration. When an NMR spectrum is *integrated,* the recorder pen traces a horizontal line until it reaches a signal; then it rises a distance that is proportional to the signal's area as it crosses the signal. Since the area of a signal on a proton NMR spectrum is proportional to the number of protons responsible for the signal, integrating the spectrum makes it possible to determine how many protons gave rise to each signal.

Following is an outline of the steps in the integration of a typical ^{1}H NMR spectrum on a CW instrument. If you will be expected to integrate your NMR spectrum, your instructor will provide more detailed directions.

1 The RF power is optimized to provide an acceptable signal-to-noise ratio.
2 The instrument is switched to the integral mode.
3 The integral amplitude control is set, while scanning the spectrum rapidly, so that the integrator trace spans the vertical axis of the chart.
4 With the sweep offset and sweep width controls set to scan a region free from NMR signals, the balance control is adjusted during a slow scan of that region to give a horizontal line.
5 The phasing control is adjusted while scanning over a signal to make the integrator traces before and after the signal as nearly horizontal as possible.
6 The integral over the entire spectrum is recorded (preferably once in each direction) using a sweep time about one-fifth to one-tenth that

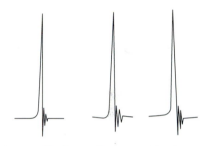

correct phasing   incorrect phasing

Figure G34 Effects of phasing on baseline

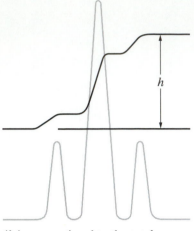

(*h* is proportional to the total area under the triplet)

Figure G35 Measuring signal areas

Take Care! Handle NMR tubes with great care; they are fragile and may break.

for the normal spectrum. The pen should be returned to the baseline after each scan.

7 The relative peak areas are determined by measuring the vertical distances between the integrator traces before and after each signal, and the results for successive scans are averaged (see Figure G35).

General Directions for Operating a CW-NMR Spectrometer

Equipment and supplies:

- sample
- tetramethylsilane (TMS)
- Pasteur pipet
- tissues
- NMR solvent
- NMR tube
- glass wool
- washing solvent

Do not attempt to operate the instrument without prior instruction and proper supervision. Do not make any adjustments other than the ones specified, except at the instructor's request and under his or her supervision. Some of the adjustments described here may be made in advance by the instructor or a lab technician. The following procedure applies to a typical 60-MHz continuous-wave NMR spectrometer; operating procedures for other instruments may vary considerably.

Fill the NMR tube to a depth of about 3 cm with a 10–20% (w/v) solution of the compound in a suitable solvent containing 1–3% TMS, and cap the tube. Wipe the outside of the NMR tube carefully with a tissue paper or lint-free cloth, insert it in the sample spinner using a depth gauge to adjust its position, wipe it again, and carefully place the assembly into the sample probe between the magnet pole faces. Adjust the air flow to spin the sample at 30–60 Hz. Align the chart paper on the recorder and cover it with a sheet of scrap paper. Set the sweep controls to scan the desired range at a suitable rate. (Typical settings for a 60-MHz instrument are: sweep offset, 0; sweep width, 600 Hz; sweep time, 600 s.) Set the RF power to about mid-range and the filter response time to 1 s or less. Set the spectrum amplitude control to about mid-range, and scan the spectrum to find the tallest peak. Readjust the spectrum amplitude to keep that peak on scale near the top of the chart during a scan. If necessary, optimize peak shape and ringing and adjust the phasing as directed by your instructor. Set the TMS peak to 0.0 δ ; you may have to sweep through the TMS signal several times, adjusting the control each time, until it is lined up with the zero on the chart paper. Set the recorder baseline, if necessary, at a convenient location near the bottom of the spectrum. Remove the blank paper and record the spectrum. If you are to integrate your spectrum, cover it with scrap paper while you are making the adjustments described previously, then remove the paper when you record the integral on the spectrum. Remove the chart paper and record the control settings and other relevant information on it.

Remove the sample tube as demonstrated by your instructor; follow directions carefully or the tube may break. Clean the sample tube immediately and thoroughly using a Pasteur pipet and an appropriate solvent. Generally the same solvent (in its protic form) is used for cleaning as was used for preparing the sample; thus if the solvent was $CDCl_3$, rinse the tube with $CHCl_3$—*not* the much more expensive deuterated solvent. Invert the tube in a suitable rack and let it drain dry. Before being reused, an NMR tube should be dried in an oven for several hours to remove all traces of solvent.

Interpretation of ¹H NMR Spectra

A ¹H NMR spectrum provides numerical data in the form of chemical shifts, signal areas, signal multiplicities, and coupling constants. Working out the structure of a molecule from these numbers is a fascinating mental exercise comparable to the work of a cryptographer who reconstructs meaningful messages from coded symbols.

The *chemical shift* (δ) is the distance, measured in hertz or parts per million, from the center of a signal to some reference signal, usually that of tetramethylsilane (TMS). The TMS signal occurs farther upfield (to the right) than nearly all other proton signals, so the chemical shift of a signal is usually measured as its distance downfield from (to the left of) that of TMS, as shown in Figure G36.

The *signal area,* which is the sum of the areas under all the peaks in a proton signal, is proportional to the number of protons giving rise to the signal. Signal areas are determined using an electronic integrator that traces a line across each proton signal after it is recorded. The area of the signal is proportional to the vertical rise of the integrator pen as it crosses the signal; that is, to the height of the "steps" drawn by the integrator pen, as shown in Figure G35. Integrated signal areas can be converted to proton numbers using the following relationship:

$$\text{number of protons responsible for signal} =$$
$$\text{total number of protons} \times \frac{\text{area under signal}}{\text{area under all signals}}$$

For example, suppose a compound with the molecular formula $C_{10}H_{14}$ has four signals with relative areas of 42, 7, 14, and 35. The sum of the areas is 98, so the number of protons responsible for the first signal is

$$14 \times \frac{42}{98} = 6$$

By similar calculations it can be shown that 1, 2, and 5 protons, respectively, are responsible for the other three signals. If the molecular formula of a compound is not known, relative proton numbers can be obtained by reducing the signal areas to the lowest ratio of integers.

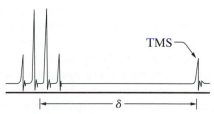

Figure G36 Chemical shift of a proton NMR signal

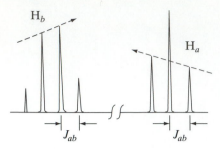

Figure G37 Signals of nearest-neighbor protons

The signal generated by a given set of protons may be split into several peaks as a result of *coupling* interactions with nearby proton sets. The *multiplicity* of a signal is simply the number of separate peaks it contains; its *coupling constant* is the distance between two adjacent peaks in the signal, measured in hertz (Hz). Figure G37 shows the signals of two sets of protons that are interacting with one another; the protons of set *a* have split the signal of the protons of set *b* into four peaks (a quartet) and the *b* protons have split the signal of the *a* protons into three peaks (a triplet). The coupling constant, which is equal for the two signals, is represented by J_{ab}. In the simplest case, the number of protons responsible for splitting the signal of a neighboring set of protons can be determined by subtracting 1 from the number of peaks in that signal. Thus the three peaks in the *a* signal can be produced by two neighboring *b* protons, and the four peaks in the *b* signal by three neighboring *a* protons. An interacting triplet-quartet grouping of this kind is good evidence for an ethyl (CH_3CH_2—) group.

Ideal triplets and quartets should be symmetrical, having relative peak area ratios of 1:2:1 and 1:3:3:1, respectively. As shown in Figure G37, however, the signals in an actual spectrum are often somewhat distorted, giving paired peaks of unequal height. Note that the two signals in the figure are not perfectly symmetrical but appear to "lean" toward one another, with the peaks on the side facing the other signal being higher than predicted. This and the fact that their coupling constants are equal are additional evidence that the protons responsible for the two signals are, in fact, coupling with one another and not with some other proton sets in the molecule.

The following general procedure should help you derive structural information from a proton NMR spectrum.

1 Measure the integrated area of each signal and use it to determine the number of protons responsible for the signal. Each set of equivalent protons (protons in the same molecular environment) gives rise to a signal, and the relative signal areas can tell you how many protons are in each set. For example, 3,3-dimethyl-2-butanone has nine hydrogen atoms on the three equivalent methyl groups to the left of the carbonyl group and three on the other methyl group, so its ^{1}H NMR spectrum has two signals with an area ratio of 3:1.

2 Determine the chemical shift of each signal on the delta scale by measuring the distance, in parts per million, from the center of the signal to the TMS reference peak. The chemical shift of a signal may indicate what kind of protons are responsible for the signal or may suggest their relative locations in the molecule. For example, alkyl hydrogen atoms that are remote from electron-withdrawing substituents should have a chemical shift of approximately 0.9 ppm if they are primary, 1.3 ppm if they are secondary, and 1.5 ppm if they are tertiary. Electron-withdrawing groups containing oxygen, nitrogen, or halogens tend to move ^{1}H NMR signals downfield, thereby increasing their chemical shifts. Benzene rings give rise to large downfield shifts, making it quite easy to recognize aromatic compounds from their ^{1}H NMR spectra. The correlation chart in Figure G38 summarizes chemical shift data for a number of proton types. Chemical shift values for compounds from the common families of organic compounds are given in Table G4 (p. 745).

$$CH_3 \quad O$$
$$CH_3 - C - C - CH_3$$
$$CH_3$$

3,3-dimethyl-2-butanone

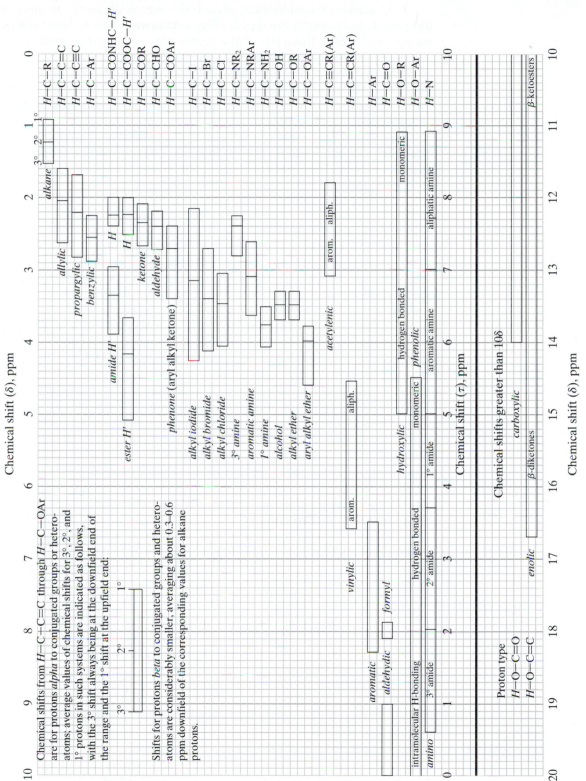

Figure G38 Correlation chart relating ¹H NMR chemical shifts to proton environments

3 Determine the multiplicity of each signal by counting the number of distinguishable peaks in the signal. (Very small peaks may be obscured by background noise.) If the signal of a proton set is reasonably symmetrical and contains evenly spaced peaks, it should be possible to estimate how many nearby protons are coupled with the protons in that set by subtracting 1 from the multiplicity of the signal. Irregular signals and signals that have been split by several dissimilar proton sets can be analyzed by more advanced methods.

4 Measure the coupling constant of each signal that contains more than one peak and try to determine from the resulting values and the way each signal "leans" what other signals might be coupled with it.

All of this information can be used to build up the structure of a molecule piece by piece. For example, consider the spectrum in Figure G39 of a ketone with the molecular formula $C_7H_{14}O$. Signals a and b have a relative area ratio of 6:1. Since the compound contains 14 protons, $\frac{6}{7}$ of them, or 12, must be responsible for signal a, and $\frac{1}{7}$ of them, or 2, for signal b. Signal a has only two peaks, indicating that the a protons have only one neighboring proton. The seven peaks in signal b (not all of which are clearly visible) indicate that the b protons have six neighbors, and the higher chemical shift of this signal suggests that these protons are close to the electron-withdrawing carbonyl group. The only alkyl group in which a single proton has six equivalent protons for neighbors is the isopropyl group, $(CH_3)_2CH-$. Two such groups provide the required total of 12 a and 2 b protons, and attaching them both to a carbonyl group gives the complete structure of the ketone, 2,4-dimethyl-3-pentanone. Note that this structure accounts for all the features of the ^1H NMR spectrum: the 6:1 area ratio for the a and b protons; the higher chemical shift of the signal for the b protons resulting from their proximity to the carbonyl group; the splitting of the a signal into two peaks by each neighboring b proton; the splitting of the b signal into seven peaks by each group of six neighboring a protons; and the equal coupling constants for the two signals.

2,4-dimethyl-3-pentanone

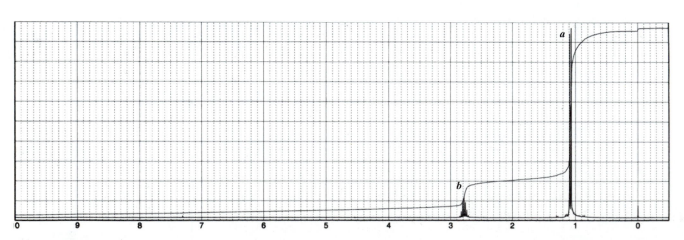

Figure G39 ^1H NMR spectrum of compound with molecular formula $C_7H_{14}O$. (Spectrum from *The Aldrich Library of ^{13}C and ^1H NMR Spectra* by C. J. Pouchert and J. Behnke, used with the permission of the Aldrich Chemical Company.)

Table G4 Approximate ^1H NMR chemical shift values for different types of protons

Family	Proton type	Chemical shift (δ, ppm)
Alcohol	H—O—R	1–5.5
	H—C—OH	3.4–4
Phenol	H—O—Ar	4–12
	H—Ar	6–8.5
Aldehyde	H—C=O	9–10
	H—C—CHO	2.2–2.7
Ketone	H—C—COR	2–2.5
Carboxylic acid	H—O—C=O	10.5–12
	H—C—COOH	2–2.6
Ester	H—C—COOR	2–2.5
	H—C—O—C=O	3.5–5
Amine	H—N—R (aliphatic)	1–3
	H—N—R (aromatic)*	3–5
	H—C—N	2.2–4
Amide	H—N—C=O*	5–9.5
	H—C—CO—N	2–2.4
	H—C—N—C=O	3–4
Halide	H—C—Br	2.5–4
	H—C—Cl	3–4
Aromatic hydrocarbon	H—Ar	6–8.5
	H—C—Ar	2.2–3

Note: Signals of proton types marked with an asterisk are often very broad

For further information about the interpretation of ^1H NMR spectra, refer to the textbook for your lecture course or to appropriate sources in category F of the Bibliography.

b. ^{13}C NMR Spectrometry

Because carbon-13 nuclei are much less abundant than hydrogen nuclei, signals from the ^{13}C nuclei in a typical sample are about 6000 times weaker than those from its ^1H nuclei. A continuous-wave NMR spectrometer can hardly distinguish such weak signals from background noise, so Fourier-transform NMR instruments must be used for recording carbon-13 NMR spectra. Since such instruments are highly complex and seldom available for use by undergraduate students, no effort will be made here to describe their operation.

A typical ^{13}C NMR spectrum is usually simpler and easier to interpret than the ^1H NMR spectrum of the same compound. Carbon-carbon splitting is unimportant because carbon-13 nuclei cannot couple with nonmagnetic carbon-12 nuclei, and there is only a slight chance that two carbon-13 atoms will be next to one another in the same molecule. Hydrogen nuclei couple strongly with carbon-13 nuclei, however—not only with the nuclei of the carbon atoms they are bonded to, but with those of more distant carbons as well. Since such coupling results in very complex spectra, it is usually prevented by various *decoupling* techniques. For *broad band proton decoupling* the sample is subjected to continuous broad-spectrum RF radiation covering the resonance frequencies of its protons, causing all of the protons to flip over (change spin states) so rapidly that their coupling effects on the adjacent carbon atoms average out to zero. In a broad band-decoupled ^{13}C NMR spectrum, each ^{13}C signal is a single peak rather than a multiplet. Another decoupling technique, *off-resonance decoupling,* yields *proton-coupled* spectra in which only those hydrogens that are attached directly to a carbon atom split the signal of that carbon atom. Thus the three protons of a methyl group will split the signal of the methyl carbon into a quartet, but will not split the signal of any other carbon in the molecule. Broad band-decoupled spectra are more common and easier to interpret than proton-coupled spectra, so we will restrict our discussion to them.

The area of a ^{13}C signal, unlike that of a ^{1}H signal, is not directly proportional to the number of carbon atoms responsible for the signal, so ^{13}C NMR spectra are not integrated. Consequently, three of the four parameters that can be obtained from ^{1}H NMR spectra are not present in a broad band-decoupled ^{13}C NMR spectra. Since there is no integration there are no signal areas, and since all of the signals are singlets there are no multiplicities or coupling constants to interpret. That leaves only the chemical shifts which, as it turns out, are extraordinarily useful.

The ^{13}C NMR spectrum of a compound gives us direct information about the compound's carbon "backbone;" information that is often not available from its ^{1}H NMR spectrum. A typical broad band-decoupled ^{13}C NMR spectrum of a compound contains one sharp peak for each kind of carbon atom in its molecules. Since many compounds have few, if any, magnetically equivalent carbon atoms, most of the peaks in a ^{13}C NMR spectrum may be one-carbon peaks, each arising from a different carbon atom. Thus the absence of a linear relationship between signal area and the number of carbon atoms is not a serious handicap.

Carbon-13 chemical shifts cover a much broader range than proton chemical shifts; about 250 ppm compared to 15 ppm or so for protons. As a result, the peaks on a ^{13}C NMR spectrum are usually well separated. The chemical shift of a carbon-13 atom is very sensitive to changes in its hybridization and molecular environment. Carbon atoms that are sp^2 hybridized have much higher chemical shifts ($100-160\ \delta$) than sp^3 carbons ($0-60\ \delta$), and the chemical shifts of sp carbons are somewhere in between ($65-105\ \delta$). As in ^{1}H NMR spectrometry, electron-withdrawing groups cause downfield chemical shifts on nearby carbon atoms, but they can cause upfield shifts at more distant carbon atoms. For example, a chlorine atom increases the chemical shift of an α-carbon atom by about 30 ppm and of a β-carbon atom by about 10 ppm, but it *decreases* the chemical shift of a γ-carbon atom by about 5 ppm.

$$\overset{\gamma}{C}-\overset{\beta}{C}-\overset{\alpha}{C}-Cl$$
$$\underset{-5}{}\quad\underset{+10}{}\quad\underset{+30}{}$$

Other electron-withdrawing groups have similar effects. Electron-donating groups have the opposite effect, increasing the chemical shifts of α and β carbon atoms and decreasing the chemical shifts of γ carbon atoms.

Table G5 ^{13}C NMR chemical shift ranges for different types of carbon atoms

Type of carbon atom	Chemical shift range (δ, ppm)
1° alkyl, RCH_3	0–40
2° alkyl, R_2CH_2	10–50
3° alkyl, R_3CH	15–50
alkene, $C{=}C$	100–160
alkyne, $C{\equiv}C$	60–90
aryl, $\langle\bigcirc\rangle C{-}$	100–170
alkyl halide, $C{-}X$ (X = Cl, Br)	5–75
alcohol or ether, $C{-}O$	40–90
amine, $C{-}N$	10–70
aldehyde or ketone, $C{=}O$	180–220
carboxylic acid or ester, $O{=}C{-}O$	160–185
amide, $O{=}C{-}N$	150–180

Table G5 shows chemical shift ranges for some kinds of ^{13}C atoms. Such tables can be used to assign the peaks in a ^{13}C NMR spectrum to specific carbon atoms. For example, the ^{13}C NMR spectrum of methyl methacrylate has five peaks with the chemical shifts shown here.

J. Chem. Educ. **1987**, *64,* 915 *describes a method for estimating* ^{13}C *chemical shifts.*

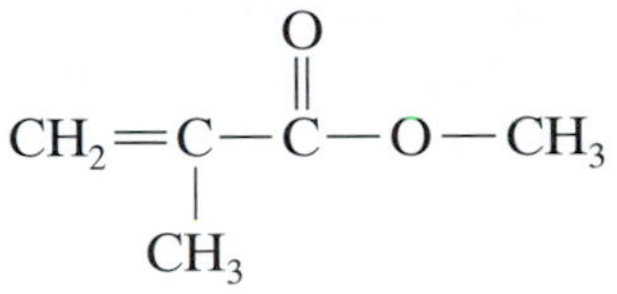

$$CH_2{=}C{-}\overset{\overset{\displaystyle O}{\|}}{C}{-}O{-}CH_3$$
$$|$$
$$CH_3$$

1	18 ppm	4	137 ppm
2	52 ppm	5	167 ppm
3	125 ppm		

methyl methacrylate

From Table G5 we find the following information for carbon atoms like those in methyl methacrylate.

ester carbonyl carbon: 160–185 ppm
alkene carbon: 100–160 ppm
carbon bonded to oxygen ($C{-}O$): 40–90 ppm
primary alkyl carbon: 0–40 ppm

Note that the $C{-}O$ chemical shift range from the table is for alcohols and ethers, but a carbon atom on the alcohol portion of an ester is in a similar environment. From this information it is easy to match the peak at 167 ppm with the carbonyl carbon, the peaks at 125 ppm and 137 ppm with the alkene carbons, the peak at 52 ppm with the OCH_3 methyl group, and the peak at 18 ppm with the remaining methyl group. Since the alkene carbon *beta* to the two oxygen atoms would be expected to have a higher chemical shift than the alkene carbon *gamma* to them, we can assign chemical shift values to the carbon atoms as shown. Such assignments can often be confirmed by a technique called *distortionless enhanced polarization transfer (DEPT),* which can pinpoint the carbon atom that is responsible for a particular peak.

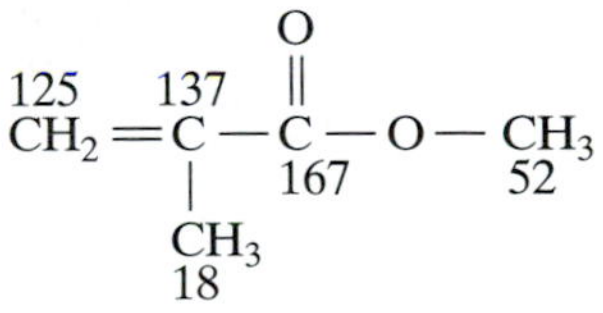

$$\underset{125}{CH_2}{=}\underset{137}{C}{-}\underset{167}{\overset{\overset{\displaystyle O}{\|}}{C}}{-}O{-}\underset{52}{CH_3}$$
$$\underset{18}{|}{CH_3}$$

OPERATION 36

Ultraviolet-Visible Spectrometry

Principles and Applications

1 nanometer (nm) = 10^{-9} m. The older unit "millimicrons" (mμ) is sometimes used for nm.

Ultraviolet-visible (UV-VIS) spectrometers, which detect the absorption of radiation in the visible (~400–800 nm) and near ultraviolet (~200–400 nm) regions of the electromagnetic spectrum, are useful for both qualitative and quantitative analysis of organic compounds. Radiation in these regions induces electronic transitions in which molecules are promoted from an electronic ground state to one or more excited states. An example of such a transition is illustrated in Figure G40, in which a pi electron in the ground state electron configuration of 1,3-butadiene jumps from its bonding molecular orbital (M.O.) to an unoccupied antibonding orbital. The energy required for electronic transitions is much greater than that needed to induce vibrational or nuclear magnetic transitions, so the wavelength of the radiation used is much shorter: 0.2–0.8 μm compared to about 2.5–50 μm for infrared spectrometry and several meters for NMR. The most important of these electronic transitions in organic compounds involve pi electrons in aromatic and conjugated aliphatic systems. Most single bonds and isolated double bonds absorb ultraviolet radiation with wavelengths shorter than 200 nm. Because oxygen in air also absorbs UV radiation below 200 nm, most UV-VIS spectrometers are not designed to scan this region.

An electronic spectrum in either the ultraviolet or visible region is usually quite featureless compared to an infrared or NMR spectrum; it may consist of only one or two broad *absorption bands*. The band structure is caused by rotational and vibrational transitions that accompany each electronic transition. Each different combination of vibrational and rotational transitions has a different energy, so collectively they span a broad range of wavelengths centered about the wavelength of the "pure" electronic transition. The height of an absorption band above the baseline of a UV-VIS spectrum is measured in units of *absorbance, A*. The position of an absorption band is given by its wavelength of maximum absorbance, λ_{max}, which is measured at the tip of the band's highest peak. For example, the absorption band illustrated in Figure G41 has an absorbance of 0.80 and a λ_{max} of 350 nm. Absorbance is related to *transmittance,* the fraction of incident radiation transmitted through a sample, by this equation:

$$A = \log(1/T) = -\log T \tag{1}$$

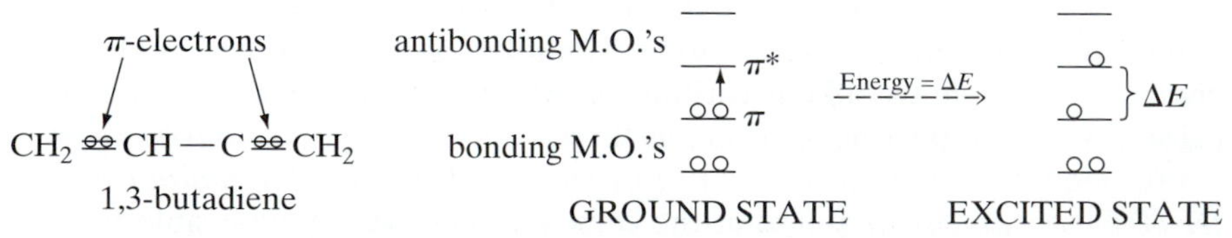

Figure G40 Electronic Energy Transition in 1,3-Butadiene

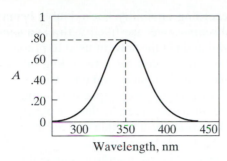

Figure G41 An ultraviolet absorption band

The transmittance of a UV-VIS band with $A = 0.80$ is $10^{-0.80} = 0.16$, meaning that about 16 percent of the light entering the sample passes through, and the remaining 84 percent is absorbed by the sample.

a. UV-VIS Spectra

Sample Preparation. Routine UV-VIS spectra are nearly always obtained in solution. The solvent must be transparent (or nearly so) in the regions to be scanned. Water, 95% ethanol, methanol, dioxane, acetonitrile, and cyclohexane are suitable down to about 210–220 nm; many other solvents can be used at higher wavelengths. The preferred solvent is 95% ethanol, in part because it does not require additional purification; most of the other solvents must be purified or purchased as spectral-grade solvents. The solvent must not, of course, react with the solute—for example, alcohols should not be used as solvents for aldehydes.

If possible, the solution to be analyzed should produce a maximum absorbance of about 1 when the solute's strongest absorption band is scanned. Using Beer's law (see equation **2**) we can show that the molar concentration of such a solution, if analyzed in a 1 cm sample cell, should be less than or equal to $1/\varepsilon_{max}$, where we define ε_{max} as the maximum molar absorptivity of the solute over the wavelength range to be scanned. For example, if the solute's strongest band has a molar absorptivity of 10,000 at its λ_{max} value, the solution concentration should be 1×10^{-4} M or less. To make up such dilute solutions accurately, it is usually necessary to prepare a stock solution that is too concentrated by several powers of 10, and then measure an aliquot of this solution and dilute it. For example, to prepare a $\sim 1.0 \times 10^{-4}$ M solution of cinnamic acid (M.W. = 148), you could measure 0.15 g (~ 1.0 mmol) of the solid into a 100-mL volumetric flask and make it up to volume with solvent; then pipet a 1-mL aliquot of this 0.010 M solution into another 100-mL volumetric flask and fill it to the mark with solvent. If the molar absorptivity of a sample is not known, it may be necessary to find the optimum concentration by trial and error, starting with a more concentrated solution and diluting it as needed to bring all of the absorption bands on scale.

Recording a Spectrum. The construction of ultraviolet-visible spectrometers varies widely. Both single- and double-beam instruments are available, with and without recording capability. For a double-beam recording instrument, two identical *sample cells* are filled with (1) a solution of the compound being analyzed and (2) the solvent used to prepare the solution. The most commonly used sample cell for a recording ultraviolet-visible spectrometer is a transparent rectangular container with a square cross section, having a path length of 1.00 cm and a capacity of about 3 mL. Fused quartz cells are used for the ultraviolet region; glass or plastic cells are suitable in the visible region. Two sides of a sample cell are nontransparent (usually frosted) and the other two are transparent.

Both sample cells are placed in appropriate cell holders in the instrument's sample compartment. A sample cell must be held by its nontransparent sides and inserted into its cell holder so that the light beam will pass through its transparent faces. Cells must be scrupulously cleaned; they should never be touched on the transparent sides, since even a fingerprint can yield a spectrum. Before a spectrum is recorded the user selects the wavelength region to be scanned and a radiation source appropriate for that region. A tungsten lamp can be used between 300 and 800 nm and a hydrogen lamp between 190 and 350 nm. Most modern instruments allow the operator to apply a baseline correction, which subtracts any absorbance differences between the sample and reference cells over the entire wavelength range to be scanned. A baseline is usually run with both cells filled with the solvent. The absorbance range can also be set; a typical range is from zero to one or two absorbance units. As the spectrum is scanned, the instrument automatically subtracts any absorption due to the solvent and records the spectrum of the solute on a chart paper. Chart papers, premarked with wavelength and absorbance values, usually come in the form of a roll that feeds onto a flat recorder bed during a scan.

Once a spectrum has been recorded, the data it contains can be presented as a tabulation of λ_{max} values giving either the absorbance, molar absorptivity (ε), or log ε at each wavelength specified. For example, the ultraviolet spectrum of cinnamic acid is reported in one reference book as "λ^{al} 210 (4.24), 215 (4.28), 221 (4.18), 268 (4.31)." The numbers in parentheses are log ε values for peaks having the λ_{max} values (in nanometers) given, and "al" indicates that the spectrum was run in ethyl alcohol.

General Directions for Operating a Recording UV-VIS Spectrometer

Do not attempt to operate the instrument without prior instruction and proper supervision. Ultraviolet-visible spectrometers vary widely in construction and operation, so the following is intended only as a general guide and may not be applicable to all instruments. Specific operating instructions should be learned from in-class demonstrations or the operator's manual.

Unless otherwise instructed, clean two sample cells by wiping their surfaces with a lens paper moistened with spectral grade methanol, then let the methanol evaporate. This should leave the cell surfaces free of contaminants that may have accumulated since they were last used. Be sure that the instrument, recorder, and source lamps have had sufficient time to warm up.

Select the appropriate radiation source for the desired wavelength range and set the absorbance range to 0–1 (or another appropriate value). Set the starting and ending wavelengths. If the instrument scans from high to low wavelength, the starting wavelength will be the highest wavelength of the range to be scanned. If there is no provision for setting the ending wavelength you may have to end the scan manually. Some instruments require manual adjustment of zero and 100% transmittance values (or infinite and zero absorbance values) before a spectrum or baseline is run; if so, make the adjustments as directed by your instructor. If the instrument provides for a baseline correction, fill two *clean* sample cells with the solvent, cap them, and place them in the sample and reference beams in the sample compartment. Then close the compartment door and run the baseline as directed by your instructor.

Leaving one cell in the reference beam, fill the other cell with the solution to be analyzed, cap it, and place it in the sample beam, then close the sample compartment door. Position the chart paper (if necessary) so that the scan starts on an ordinate (vertical) line. Label this line with the starting wavelength, lower the pen to the paper, and begin to scan the spectrum. If any absorption band goes off scale so that its top is "chopped off," change the absorbance range or dilute the sample. If both ultraviolet and visible regions are to be scanned, change the radiation source if necessary (many instruments do this automatically) and scan the spectrum in the other region. If the scan does not stop automatically when the end of the wavelength range has been reached, stop it manually. Raise the pen from the chart and tear off the chart paper. Write down the wavelength and absorbance ranges along with the wavelength interval between chart units; if you forget to record this information, it may be difficult or impossible to interpret the spectrum.

Rinse the sample cell several times with the solvent. If necessary, clean it further using a liquid detergent or a special cleaning solution. Never use a dry lens paper, an abrasive cleanser, or any scrubbing implement (such as a pipe cleaner with a wire core) that might scratch the cell. Drain both cells of excess solvent and dry them as directed by your instructor. Measure and write down the λ_{max} and absorbance values of any absorption bands of interest.

b. Colorimetery

Inexpensive nonrecording single-beam UV-VIS spectrometers, often called *colorimeters,* are frequently used for routine quantitative analysis of compounds in solution. The Spectronic 20 illustrated in Figure G42 is a widely used colorimeter, and most other colorimeters are operated similarly. The solution to be analyzed is prepared as described for a recording spectrometer, then placed in a *cuvette* that is inserted into the sample compartment. The wavelength control is set to a wavelength where the sample absorbs strongly and, after some preliminary adjustments, the absorbance (or percent transmittance) of the solution is read from the scale.

The concentration of a solution can be determined from its absorbance value using either Beer's law (equation **2**) or a calibration curve of absorbance versus molar concentration.

A cuvette looks like a small test tube— but it should never be used as one!

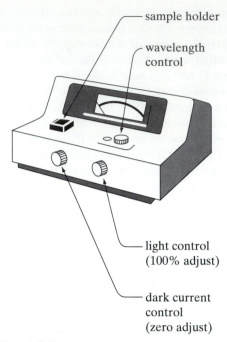

sample holder

wavelength control

light control (100% adjust)

dark current control (zero adjust)

Figure G42 Spectronic 20 spectrophotometer

Beer's law

$$c = \frac{A}{\varepsilon \cdot b}$$ (2)

c = molar concentration
A = absorbance
b = cell path length, in cm
ε = molar absorptivity, in L mol^{-1} cm^{-1}

Equation **2** is useful only when solutions of the solute obey Beer's law over the appropriate concentration range and when the absorptivity of the solute is known.

General Directions for Operating a Colorimeter

Do not attempt to operate the instrument without prior instruction and proper supervision. These directions are for operation of the B & L Spectronic 20 or a similar instrument and may not be applicable to all such instruments.

Make sure that the instrument has been switched on and that adequate time has been allowed for warmup. Set the wavelength to the desired value and rotate the zero adjust control until the pointer is at zero on the transmittance scale. To read the scale, position your eyes so that the pointer is directly over its reflection in the mirror; this prevents parallax errors. Insert a clean cuvette containing the pure solvent (the one used to prepare the solution being analyzed) into the sample holder, making sure that the alignment mark on the cuvette is opposite the mark on the cell holder. Adjust the 100% control until the transmittance reading is 100%. Rinse the cuvette with a little of the solution to be analyzed, then fill it with that solution. Replace it in the cell holder, being careful to position it in exactly the same alignment as before. Different cuvettes are sometimes used for the solvent and sample, but errors will result if the cuvettes are not well matched; for precise work, it is best to use the same cuvette for all measurements. Read the percent transmittance as accurately as possible. Note that it is more accurate to read $\%T$ and convert it to absorbance than to read the absorbance directly from the nonlinear absorbance scale.

If another solution containing the same solvent and solute is to be analyzed, drain the cuvette and rinse it with that solution before you make the next measurement. After the last measurement, rinse the cuvette with pure solvent, clean it, and let it air dry. Convert the percent transmittance values to absorbance values, if necessary, using the equation $A = \log(100\%/\%T)$.

Mass Spectrometry

The other kinds of spectrometry described in this Instrumental Analysis section use some kind of electromagnetic radiation—radio frequency, infrared, ultraviolet, or visible—to gently probe the molecules of a sample and induce them to reveal their secrets. No molecules are damaged; once the sample is removed from the spectrometer they return to their former states. In contrast, mass spectrometry takes a brute force approach to structure determination. Molecules entering a mass spectrometer are pummeled by high-energy electrons and shattered into fragments, which are pushed and pulled along a curved path until they smash into an ion collector at journey's end. The fragments cannot be put back together to form the original molecule, so mass spectrometry is a destructive method of analysis.

Mass spectrometry is not really a "spectral" method in the usual sense, in that no electromagnetic radiation is absorbed, but a mass spectrum does resemble a conventional spectrum in that it consists of a series of peaks of different amplitude plotted along a numerical scale. Although most mass spectrometers are costly, complex instruments and mass spectra are not easily interpreted, mass spectrometry has become an increasingly indispensible analytical tool for many kinds of research, medical, and industrial facilities.

Principles and Applications

When a compound is bombarded with a beam of high-energy electrons in a mass spectrometer, each of its molecules (M) can lose an electron and form a *molecular ion,* $M\bullet^+$.

$$M \rightarrow M\bullet^+ + e^-$$

Because a molecular ion contains both an unpaired electron and a positive charge, it is called a *radical cation.* If the energy of the electron beam is high enough, many of the molecular ions will have enough excess vibrational and electronic energy to break apart into fragments. Each pair of fragments consists of another positive ion (A^+), called a *daughter ion,* and a neutral molecule (X).

$$M\bullet^+ \rightarrow A^+ + X$$

The unpaired electron from the molecular ion may end up on either A^+ or X, depending on the kind of fragmentation. Each daughter ion may in turn break down, losing a neutral fragment to form yet another daughter ion, and so on.

$$A^+ \rightarrow B^+ + Y$$

For example, the fragmentation of an ammonia molecule takes place as shown, forming ions (cations and radical cations) with approximate masses of 17, 16, 15, and 14 atomic mass units.

| ammonia | molecular ion | daughter ions |

The positive ions are accelerated into an evacuated chamber where they are separated according to their mass to charge ratios (m/e), usually by means of strong electric and magnetic fields (see Figure G43). As a beam of ions with a given m/e value impinges on an *ion collector,* it gives rise to an electrical current that is amplified and displayed on a monitor as a peak whose amplitude is proportional to the number of ions striking the detector. A *mass spectrum* is a record, usually printed out as a table of data or a computer-generated bar graph, of the relative abundances of all the ions arranged in order their m/e values. Because most daughter ions have a charge of +1, the m/e value associated with a peak is nearly always equal to the mass of the ion that gave rise to that peak—or in rare cases to one-half of its mass. A large molecule may be fragmented into several hundred different ions with different m/e values and relative abundances, so mass spectrometers are provided with computers that record, store, and process the data.

Mass spectrometry is an extremely valuable tool for structural analysis; it can be used to identify or characterize a host of organic (and inorganic) compounds, including biologically active substances with very complex molecular structures. The mass spectrum of a compound ordinarily gives the mass of its molecular ion, which is essentially equal to its molecular weight. It also provides the masses of smaller pieces of its molecules, which can often be identified with the help of published tables of molecular fragments. A high-resolution mass spectrum provides data that can be used to determine a compound's molecular formula. Such information often makes it possible to piece together the compound's molecular structure, or at least to learn more about its structural features.

Different compounds yield distinctively different patterns of ion fragments, so an unknown compound can sometimes be identified by a comparison of its mass spectrum with mass spectral data from the literature. Unfortunately, peak intensities on mass spectra are very sensitive to instrumental parameters such as the energy of the electron beam. Thus there may be significant differences in the mass spectra recorded on different instruments, or even by different operators using the same instrument. Nevertheless, it is often possible to make a tentative identification from a literature comparison and then to confirm it by recording mass spectra of the unknown and the most likely known compounds under identical operating conditions. This process can be facilitated by using a computer to compare the spectrum of the unknown with the mass spectra in a memory bank, which may contain tens of thousands of such spectra.

Instrumentation

Mass spectrometers come in a wide variety of sizes and configurations, ranging from high-resolution mass spectrometers that may take up most of an instrument room to compact "tabletop" mass spectrometers that can be used as detectors for gas and liquid chromatographs. Although research-grade mass spectrometers are very expensive, tabletop mass spectrometers may be within the equipment budgets of some undergraduate chemistry programs.

In a conventional *single-focusing* mass spectrometer (diagrammed in Figure G43) the sample is introduced into a sample inlet system that is heated to keep some or all of its molecules in the vapor state. These molecules find their way into the *ionizing chamber* through an aperture called a *molecular leak,* which may be a tiny hole in a piece of gold foil. The ionizing chamber is kept at a pressure of about 10^{-6} torr to minimize collisions between particles and interference from ionized air. Molecules that wander into the path of the *electron beam* are ionized to molecular ions, which undergo fragmentation to yield daughter ions. All of these ions are pushed toward a slit by a positively charged *repeller plate,* then accelerated to a high velocity by electrically charged *accelerator plates* and directed into a *magnetic separator.* In the magnetic separator a powerful magnetic field deflects each kind of ion into a curved path whose radius depends on the ion's mass to charge ratio. Lighter (lower m/e) ions are deflected more than heavier (higher m/e) ions. At a given magnetic field strength only ions of a given mass (or m/e value) can pass through the slit that leads to the *ion collector.* As the field strength is increased, ions of progressively higher mass reach the ion collector and are detected. When a beam of ions strikes the ion collector, it generates an electrical signal whose intensity is proportional to the number of ions in the beam; that is, to their abundance. Each signal is amplified and displayed on a monitor screen or recorded on a mass spectrum.

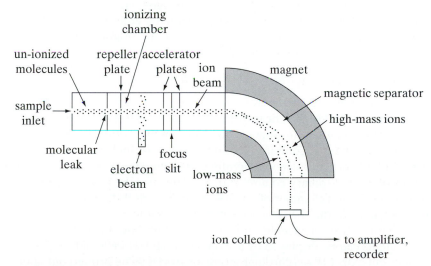

Figure G43 Schematic diagram of a single-focusing mass spectrometer

In a single-focusing mass spectrometer, variations in the kinetic energies of ions having the same mass cause their ion beam to broaden as it passes through the magnetic separator, reducing the instrument's resolving power. In a high-resolution *double-focusing* mass spectrometer the ions are initially directed along a curved path by an electrostatic field, which allows only particles of the same kinetic energy to pass through a slit leading to the magnetic separator. Double-focusing instruments can often resolve such ions as CH_2N^+ and N_2^+ whose mass numbers (28.0187 and 28.0061) differ by less than 0.05%.

In mass spectrometry, two adjacent peaks of equal amplitude are said to be resolved when the height of the valley between them is no more than 10 percent of their height.

Other kinds of mass spectrometers separate ions by very different methods. In a *quadrupole* mass spectrometer the ions are introduced between four parallel metal rods that create a rapidly oscillating magnetic field between them. Ions whose mass-to-charge ratio is compatible with the frequency of the field will oscillate along a straight path toward the ion collector; other ions will follow a different path and be removed. The frequency or intensity of the oscillating field is varied so that ions of all mass-to-charge ratios eventually reach the ion collector. In a *time-of-flight* mass spectrometer, ions beams are produced by brief pulses of electrons and accelerated by an electrical field pulse that gives all ions, regardless of mass, the same kinetic energy. Ions of a given mass (or *m/e* value) then pass through a *drift tube* at a speed that is inversely proportional to their mass; thus lighter ions reach the ion collector sooner than heavier ions. In a *Fourier-transform* mass spectometer (sometimes called an *ion trap* mass spectrometer), ions are forced into a circular path by a strong magnetic field and subjected to a radio frequency (RF) pulse. If the frequency of the radiation is equal to the frequency at which ions of a given *m/e* value move around their circular path (their *cyclotron frequency*), the ions will be accelerated and spiral outward. At the end of the RF pulse, they will be moving around a larger circle in a coherent packet. This packet of revolving ions generates an *image current* that decays with time after the pulse ends, producing a signal that is similar to the free induction decay signal observed with a Fourier-transform NMR spectrometer. The frequency of the RF pulse is varied to match the cyclotron frequencies of all the ions in a sample, causing each kind of ion to produce its own image current. The resulting image currents are then detected and converted to a conventional mass spectrum by Fourier-transform analysis. Some Fourier-transform mass spectrometers are capable of resolving ions whose masses differ by only 0.001% or so.

Gas Chromatography/Mass Spectrometry

Instruments in which mass spectrometers are interfaced with high performance liquid chromatographs (LC/MS) and even with other mass spectrometers (MS/MS) are also available.

Rapid-scan mass spectrometers, including some quadrupole, Fourier-transform, and time-of-flight instruments, can be interfaced with gas chromatographs and used to analyze the components of a mixture as they emerge from the GC column. This combination of instruments is, not surprisingly, called a *gas chromatograph/mass spectrometer (GC/MS)*. The output of a GC capillary column [OP-32] can often be introduced directly into the ionization chamber of a mass spectrometer. With a packed column, however, most of the carrier gas must be removed to maintain a sufficiently low pressure in the evacuated ionizing chamber. The mass spectra of the components can be displayed by a screen or recorder in "real time," as each component exits the gas chromatogram, or stored to be printed out later.

Gas chromatograph/mass spectrometers are particularly useful for identifying the components of natural products, biological systems, and ecosystems. For example, the flavor and odor components of essential oils, the physiologically active components of plants, and chemical pollutants in the environment can be characterized by GC/MS. Forensic chemists can use GC/MS to identify drug metabolites in the body fluids of a suspect, and physicans can even diagnose certain illnesses on the basis of a GC/MS analysis of a patient's breath.

Experimental Considerations

Samples prepared for mass spectrometry should be very pure since traces of impurities can make interpretation difficult. Sample sizes range from less than a microgram to several milligrams, and no special sample preparation is required. Liquids are inserted directly into the sample inlet with a hypodermic syringe, micropipet, or break-off device, and solids can be introduced by means of a melting-point capillary. The samples vaporize in the sample inlet system, after which their molecules flow through a molecular leak into the evacuated ionization chamber.

A typical single-focus mass spectrometer is prepared for operation by turning on the magnet current and adjusting controls that set the potential of the repeller plates, the accelerating voltage, the ionizing current (the number of electrons in the electron beam), and the energy of the electron beam. Increasing the repeller potential reduces the time that ions spend in the ionization chamber, while increasing the accelerating voltage increases the speed the ions attain by the time they enter the mass separator. Increasing the ionizing current increases the number of ions that are formed, and increasing the energy of the electron beam increases the amount of fragmentation. Since the settings of these controls determine the appearance of a mass spectrum, it is important to set them within ranges that are appropriate for a given analysis. The user also selects the range of masses to be scanned and the scan time. The mass spectrum is then scanned and recorded as a chart or computer printout.

Other types of instruments may operate quite differently, so no operating procedures will be given here. These must be learned by special instruction and by studying the manufacturer's operating manual. Computer programs that simulate the operation of specific mass spectrometers may also be available for your use.

Interpretation of Mass Spectra

Each ion recorded on a mass spectrum is characterized by its mass to charge ratio and relative abundance, the latter being proportional to the height of its peak. This information can be displayed by a variety of output devices including oscilloscopes, strip-chart recorders, and computer printers. The most intense peak in a mass spectrum, called the *base peak,* is assigned a relative intensity of 100 and the intensities of all other peaks are reported as percentages of the base-peak intensity. The molecular-ion peak may be the base peak but often it is not.

Molecular Weight and Molecular Formula. For most organic compounds, the molecular weight of the compound is nearly equal to the mass of its strongest molecular-ion peak. This molecular-ion peak is usually the last strong peak on the spectrum since no daughter ion should have a higher mass than the original molecular ion. Nevertheless, the molecular ion peak is ordinarily followed by at least two low-intensity peaks corresponding to isotopic variations of the molecular ion, because most of the elements in organic compounds have at least one isotope of higher mass number than the common form.

Just over one carbon atom in a hundred (1.08%) is a carbon-13 atom; the rest are carbon-12 atoms. Since benzene, for example, contains six carbon atoms, the chance that any one of the six will be carbon-13 is $6 \times 1.08\%$, or 6.48%. Thus the molecular-ion peak of benzene (C_6H_6, M.W. = 78) should be followed by a peak for a "heavy" form of benzene ($C_5{}^{13}CH_6$, M.W. = 79) with an intensity that is 6.48% of the parent peak intensity. Actually, the $m/e = 79$ peak, called the $M + 1$ *peak,* has a slightly higher intensity than this because benzene also contains minute quantities of benzene-d_1 (C_6H_5D), which also has an approximate molecular weight of 79. Likewise, oxygen-18 occurs naturally to the extent of about 0.20 atom for every 100 atoms of oxygen-16, so formaldehyde shows an $M + 2$ *peak* (corresponding to $CH_2{}^{18}O$) with an intensity that is 0.20% of the molecular-ion peak's intensity.

Intensities of the M + 1 and M + 2 peaks (relative to the molecular-ion peak's intensity) for a compound $C_wH_xN_yO_z$ can be calculated using Equations **1** and **2**:

$$\%(M + 1) = 1.08w + 0.015x + 0.37y + 0.037z \qquad \textbf{(1)}$$
$$\%(M + 2) = 0.006w(w - 1) + 0.0002wx + 0.004wy + 0.20z \quad \textbf{(2)}$$

For example, quinine (as the hydrate) has the molecular formula $C_{20}H_{30}N_2O_3$. Using equation **1** gives the intensity of its M + 1 peak as 22.9% of the M peak intensity, and equation **2** gives the intensity of of its M + 2 peak as 3.16% of the M peak intensity.

$$\%(M + 1) = 1.08(20) + 0.015(30) + 0.37(2) + 0.037(3) = 22.9$$
$$\%(M + 2) = 0.006(20)(19) + 0.0002(20)(30) + 0.004(20)(2)$$
$$+ 0.20(3) = 2.16$$

No two combinations of atoms with a given molecular mass are likely to yield M + 1 and M + 2 peaks of exactly the same relative intensity. Therefore if the intensities of these peaks in the mass spectrum of an unknown compound are measurable to at least two decimal places, its molecular formula (or several possible formulas) can be determined using published formula mass tables such as those in *Spectrometric Identification of Organic Compounds* [Bibliography, F20]. A general procedure for determining molecular formulas is to:

1 Locate the molecular-ion peak and determine its mass.
2 Measure the intensities of the M, M + 1, and M + 2 peaks and express the latter two as a percentage of the intensity of the M peak.
3 Find the formula (or formulas) listed under the M value that give M + 1 and M + 2 intensities close to the experimental values and that make sense from a chemical standpoint.

Some formulas can be eliminated immediately because they do not correspond to stable molecules or have the expected degree of hydrogen deficiency. The *index of hydrogen deficiency (I.H.D.)* of a compound, which is equal to the number of rings plus the number of pi bonds (or their aromatic equivalent) in a molecule of the compound, can be calculated from Equation **3**.

For compounds with the general formula $C_wH_xN_yO_z$,
$$\text{I.H.D.} = 1/2(2w - x + y + 2) \tag{3}$$

For example, the calculated I.H.D. of compound **1** is 7, which is consistent with its molecular structure—one ring and six pi bonds.

Example:

$$\text{I.H.D.} = {}^1\!/_2(18 - 7 + 1 + 2) = 7$$

A useful generalization that applies to most organic compounds is the *nitrogen rule,* which states that a stable compound whose molecular weight is an even number can have only zero or an even number of nitrogen atoms, whereas one whose molecular weight is an odd number can have only an odd number of nitrogen atoms. Finally, some molecular formulas can be eliminated because compounds with those formulas would be impossible to obtain from a given reaction or source.

Fragmentation Patterns. The use of fragmentation patterns to determine molecular structures is a complex subject covered in detail elsewhere; only a few generalizations will be given here. A molecular ion (or its daughter ions) often breaks down by eliminating small neutral molecules or free radicals such as CO, H_2O, HCN, C_2H_2, H•, or CH_3•, yielding ions with masses equal to $M - X$, where M represents the mass of the molecular ion or a daughter ion undergoing fragmentation, and X is the mass of the neutral species. Table G6 gives the masses and postulated structural formulas of some neutral species that are often lost by fragmentation.

To see how such data can be used to interpret mass spectra, consider the mass spectrum of methyl benzoate in Figure G44. The molecular ion peak in this spectrum is at $m/e = 136$, the base peak is at 105, and other strong peaks are at 77 and 51. The molecular ion is believed to be a radical cation with the structure shown in the margin. Since the mass numbers of the base peak and molecular-ion peak differ by 31 mass units, a neutral species with a mass number of 31 must have been lost from the molecular ion to produce the base-peak ion. Table G6 shows two neutral fragments having that mass number, CH_3O• and •CH_2OH. Since the molecular ion has a methoxyl group, it must have lost CH_3O• to form the base-peak ion, which must therefore be a benzoyl cation. The difference between the mass number of the benzoyl ion and that of the next major peak is $105 - 77 = 28$. The two neutral species having mass numbers of 28 are CO and C_2H_4; loss of CO from the benzoyl cation should yield a phenyl cation with the expected mass number of 77. Formation of the next major species ($m/e = 51$) requires a loss of 26 mass units from the phenyl ion. From Table G6, two species having that mass are •CN and C_2H_2 (acetylene), of which only the latter is a possibility. Loss of acetylene from the phenyl ion yields an ion with the formula

$$C_6H_5\overset{\overset{\displaystyle O}{\|}}{C}OCH_3 \xrightarrow{-CH_3O\bullet} C_6H_5C\equiv O+$$

molecular ion benzoyl ion
($m/e = 136$)   ($m/e = 105$)

$$C_6H_5C\equiv O+ \xrightarrow{-CO} C_6H_5{}^+$$

benzoyl ion phenyl ion
($m/e = 105$)   ($m/e = 77$)

$$C_6H_5^+ \xrightarrow{-C_2H_2} C_4H_3^+$$

phenyl ion　　$(m/e = 51)$
$(m/e = 77)$

$C_4H_3^+$; this species is often encountered in the mass spectra of aromatic compounds. There are a number of smaller peaks in the methyl benzoate spectrum that may provide additional structural information, but usually it is not possible (or necessary) to characterize all the peaks in a mass spectrum.

Table G6 Formulas of some neutral species lost from molecular ions

Mass of neutral species	Mass of resulting ion	Possible formulas
1	$M-1$	$H\cdot$
15	$M-15$	$\cdot CH_3$
16	$M-16$	$\cdot NH_2$
17	$M-17$	$\cdot OH, NH_3$
18	$M-18$	H_2O
26	$M-26$	$C_2H_2, \cdot CN$
27	$M-27$	$\cdot C_2H_3, HCN$
28	$M-28$	CO, C_2H_4
29	$M-29$	$\cdot CHO, \cdot C_2H_5$
30	$M-30$	$H_2CO\cdot, \cdot NO$
31	$M-31$	$CH_3O\cdot, \cdot CH_2OH$
35	$M-35$	$Cl\cdot$
36	$M-36$	HCl
42	$M-42$	CH_2CO
43	$M-43$	$CH_3CO\cdot, \cdot C_3H_7$
44	$M-44$	$CO_2, \cdot CONH_2$
45	$M-45$	$\cdot CO_2H, C_2H_5O\cdot$
46	$M-46$	NO_2
49	$M-49$	$\cdot CH_2Cl$
57	$M-57$	$CH_3COCH_2\cdot$
59	$M-59$	$\cdot CO_2CH_3$
77	$M-77$	$C_6H_5\cdot$
79	$M-79$	$Br\cdot$

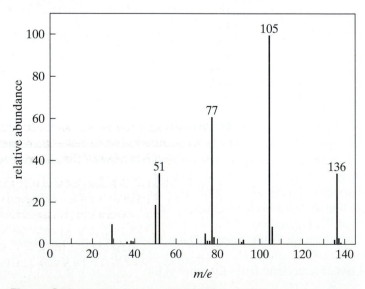

Figure G44 Mass spectrum of methyl benzoate

In the example, the interpretation was simplified because each major ion was produced from the preceding one along a single reaction path. Often there are fragmentation paths leading directly from the molecular ion to a number of different daughter ions. It is therefore a common practice to compare the mass number of the molecular ion with those of the significant daughter ions before attempting to compare the mass numbers of individual daughter ions.

The fragment ions described and other common fragment ions are listed in Table G7. Structures have been determined for some (but not all) of these cations. For example, the $C_7H_7^+$ ion with mass number 91, which results from the cleavage of alkylbenzenes, has been formulated as either a benzyl cation or a tropylium ion; in most cases, it appears to have the latter structure. Before you can interpret a mass spectrum proficiently using information from Tables G6, G7, and similar tables from other sources, you need to learn about the characteristic fragmentation patterns and mechanisms for different classes of organic compounds. This kind of information and some general rules for interpretation of mass spectra are given in some references on mass spectrometry listed in category F of the Bibliography.

Table G7 Some common fragment ions

m/e	Possible formulas
15	CH_3^+
17	OH^+
18	H_2O^+, NH_4^+
26	$C_2H_2^+$
27	$C_2H_3^+$
28	$CO^+, C_2H_4^+$
29	$CHO^+, C_2H_5^+$
30	$CH_2NH_2^+, NO^+$
31	CH_2OH^+, CH_3O^+
35	Cl^+ (also 37)
39	$C_3H_3^+$
41	$C_3H_5^+$
43	$CH_3CO^+, C_3H_7^+$
44	$CO_2^+, C_3H_8^+$
45	$CH_3OCH_2^+, CO_2H^+$
46	NO_2^+
49	CH_2Cl^+
51	$C_4H_3^+$
57	$C_4H_9^+, C_2H_5CO^+$
59	$COOCH_3^+$
65	$C_5H_5^+$
66	$C_5H_6^+$
71	$C_5H_{11}^+, C_3H_7CO^+$
76	$C_6H_4^+$
77	$C_6H_5^+$
78	$C_6H_6^+$
79	Br^+ (also 81)
91	$C_7H_7^+, C_6H_5N^+$
93	$C_6H_5O^+$
94	$C_6H_6O^+$
105	$C_6H_5CO^+$

Appendixes and Bibliography

Laboratory Equipment

Chemical Glassware

beaker

drying tube

evaporating dish

filter flask

sidearm
test tube

Erlenmeyer flask

Buchner funnel

Hirsch funnel

narrow-stemmed
funnel

powder (filling)
funnel

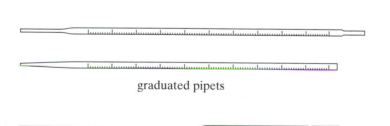

graduated pipets

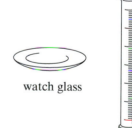

watch glass

graduated cylinder

Pasteur (capillary) pipet

Standard Taper Glassware

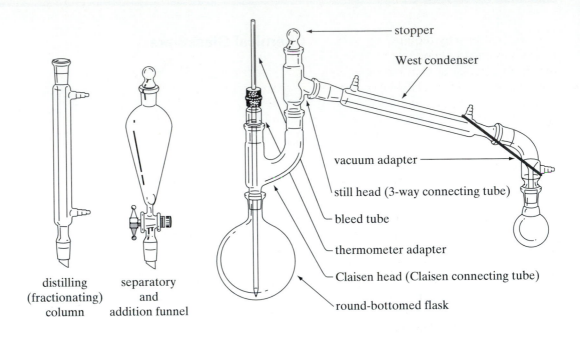

distilling
(fractionating)
column

separatory
and
addition funnel

stopper

West condenser

vacuum adapter

still head (3-way connecting tube)

bleed tube

thermometer adapter

Claisen head (Claisen connecting tube)

round-bottomed flask

Hardware

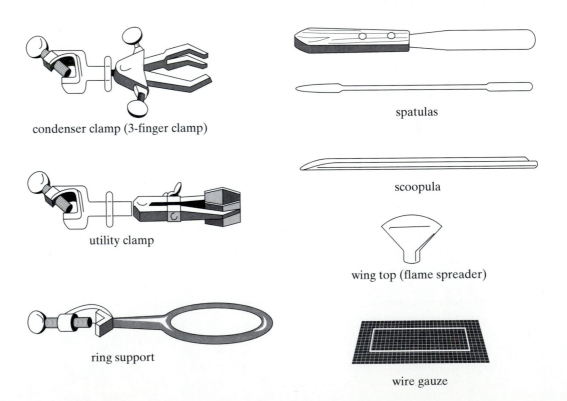

condenser clamp (3-finger clamp)

utility clamp

ring support

spatulas

scoopula

wing top (flame spreader)

wire gauze

Keeping a Laboratory Notebook

Your instructor may require you to maintain a laboratory notebook, write formal laboratory reports, or both. In either case you should write and submit an experimental plan before you begin an experiment. This appendix tells you how to keep a laboratory notebook. Information about writing lab reports and experimental plans will be found in Appendixes III and V.

A laboratory notebook is essentially a factual account of work performed in the laboratory. It may also include the writer's interpretation of the results. Your laboratory work may not require the degree of documentation expected of a research chemist, but you should at least be aware of the characteristics of a good laboratory notebook. Although any kind of notebook can be used to record data and information, a good laboratory notebook should be bound and have quadrilled (square-ruled), duplicate pages. A notebook with duplicate pages will permit you to turn in a copy of your notes to the instructor after each laboratory period. The first page or two should be reserved for a table of contents and the pages should be numbered sequentially so that you can find information quickly. Whenever possible, each entry should be written immediately after the work is performed and it should be dated and signed by the experimenter (notebooks of research chemists are usually signed by a witness as well). Each section of the notebook should have a clear, descriptive heading and the writing should be grammatically correct and sufficiently legible to be read and understood by any knowledgeable individual.

Before each experiment, you should write an experimental plan (see Appendix V), summarizing what you expect to do in the laboratory, how you intend to go about it, and other relevant information. Some or all of the following items should be included in your prelab writeup.

1 The experiment number and title
2 A clear, concise statement defining the scientific problem the experiment is designed to solve
3 A brief summary of the course of action to be followed in solving the problem
4 When appropriate, a working hypothesis regarding the outcome of the experiment
5 For a preparation, balanced equations for all significant reactions, including possible side reactions that might reduce the yield
6 A table listing relevant physical properties or other information (M.W., m.p., solubility, potential hazards, source, purity, etc.) about the reactants, products, solvents and any other chemicals involved in the experiment
7 Calculations of the quantities of reactants required and the theoretical yield of product
8 A list of the materials (chemicals, supplies, equipment) needed for the experiment
9 A checklist, flowchart (see Appendix V), or other kind of outline summarizing the experimental procedure

Some of this information, such as reaction equations and physical properties, will be given in the experiment itself, and certain items may not apply to some kinds of experiments such as kinetic studies or qualitative analyses.

During the experiment, you should keep a detailed account of your work, reporting everything of importance that you actually did and saw. Your data and observations contain the evidence you will need to arrive at a conclusion. Your notes should not simply restate the textbook procedure but should describe in your own words how you carried out the experiment. You should include all relevant data such as the quantities of materials that you actually used (not the theoretical quantities that you calculated, unless they are exactly the same) and the results of any analyses you performed. Raw data should be recorded with particular care; if you forget to record data at the time you measure it, or if you record it incorrectly or illegibly, the results of an entire experiment may be invalidated. As you gather evidence pertaining to the problem, you can write down one or more tentative hypotheses and describe how you tested them.

You may find it helpful to think of your lab notebook as telling a story about your accomplishments (or misadventures) in the laboratory. Although most professional journal articles are written in a dry, impersonal style, this need not be true of a lab notebook. The American Chemical Society's publication *Writing the Laboratory Notebook* [Bibliography, L35] suggests a more personal approach, stating that "The use of the active voice in the first person tells the story and clearly indicates who did the work." (*Example:* "I recorded the infrared spectrum of benzaldehyde ... ") Other scientists prefer to write in the passive voice, avoiding the use of personal pronouns. (*Example:* "The infrared spectrum of benzaldehyde was recorded ... ") If your instructor has a strong preference either way, you should follow his or her recommendation.

After you have finished an experiment summarize your results, then write down your conclusions and explain how you arrived at them. If your notebook pages are to be turned in as your laboratory report, include answers to assigned exercises and any other information requested by your instructor. See Appendix III for additional information about reporting your results.

APPENDIX III

Writing a Laboratory Report

You will be expected to submit a report for each experiment, either (1) in your laboratory notebook, (2) on a report form provided, or (3) on blank sheets of paper, bound or stapled together. Handwritten reports are usually acceptable if they are written legibly in ink. Your instructor may recommend a professional format such as the one described in Experiment 57 or suggest a different format. In any case, each report should include information under some or all of the following headings.

1 *Prelab Assignments:* Your experimental plan (see Appendix V), calculations, and any other material specified in the section "Before You Begin."

2 *Observations:* Any significant observations made during the course of the experiment. You should record all observations that might be of help to you (or another experimenter) if you were to repeat the experiment at a later time. These include the quantities of solvents or drying agents used, the reaction times, the distillation ranges, and a description of any experimental difficulties you encounter. You should also make note of phenomena that might provide clues about the nature of chemical or physical transformations taking place during the experiment such as color changes, phase separations, tar formation, and gas evolution. If you keep a detailed record of your observations in a laboratory notebook, it should not be necessary to include them in a parameters.

3 *Raw Data:* All numerical data obtained directly from an experiment, before it is graphed, used in calculations, or otherwise processed. This can include quantities of reactants and products, titration volumes, kinetic data, gas chromatographic retention times and integrated peak areas, and spectrometric parameters.

4 *Calculations:* Yield and stoichiometry calculations and any other calculations based on the raw data. If a number of repetitive calculations are required, one or two sample calculations of each type may suffice.

5 *Results:* A list, graph, tabulation, or verbal description of the significant results of the experiment. For a preparation, this should include a physical description of the product (color, physical state, evidence of purity, etc.), the percentage yield, and all significant physical constants, spectral data, and analytical data obtained for the product. For the qualitative analysis of an unknown compound, the results of all tests and derivative preparations should be included, along with spectral data and physical constants. The results of calculations based on the raw data should be reported in this section as well.

6 *Discussion:* A description of the scientific problem the experiment was designed to solve and your interpretation of the results as they pertain to the problem. In this section you should tell how you applied scientific methodology to solve the problem. You can include a description of any tentative hypotheses you formulated and how they were tested. Your discussion should also describe possible sources of error and, where relevant, discuss the significance of the results. This section may include your interpretation of spectra or chromatograms, which should be attached to the report.

7 *Conclusions:* Your final conclusions related to the problem and an explanation of how you arrived at your conclusions.

8 *Exercises:* Your answers to all exercises assigned by your instructor, showing your calculations and describing your reasoning when applicable.

The "Report" section following most experiments in this book lists specific items that should be addressed in an appropriate section of your report, but it is by no means complete. For example, you should calculate the percentage yield from a preparation whether or not you are specifically requested to. Each report should also include (on the first page or cover) the name and number of the experiment, your name, and the date on which the experiment was turned in to the instructor. The date(s) on which the experiment was performed may also be required. The following sample laboratory report illustrates the kind of writeup that might be prepared by a conscientious student. Note that the experimental plan is not included here; see Appendix V for directions on writing experimental plans.

Experiment 59: Preparation of Tetrahedranol

Name: Cynthia Sizer

Date: Jan. 25, 1999

(*Attached*: Experimental plan, IR spectrum, gas chromatograph, and worked exercises)

Prelab Assignment

Tetrahedryl acetate required:

$$\text{mass} = 0.0150 \text{ mol} \times \frac{110 \text{ g}}{1 \text{ mol}} = 1.65 \text{ g}$$

$$\text{volume} = 1.65 \text{ g} \times \frac{0.951 \text{ g}}{1 \text{ mL}} = 1.57 \text{ mL}$$

5 M sodium hydroxide required:

$$\text{volume} = 0.10 \text{ mol} \times \frac{1 \text{ L}}{5.0 \text{ mol}} \times \frac{1000 \text{ mL}}{1 \text{ L}} = 20 \text{ mL}$$

Observations

The reaction of tetrahedryl acetate with aqueous NaOH was carried out under reflux for 60 minutes, during which time the organic (top) layer slowly dissolved to give a homogeneous solution and the "fruity" odor of the ester disappeared. The organic layer was no longer visible after 45 minutes of heating. The acidified reaction mixture was extracted with two 10-mL portions of ethyl ether, the ether extracts were washed with two 10-mL portions of saturated aqueous sodium chloride, and the ether solution was dried over 0.70 g of anhydrous magnesium sulfate. The residue was distilled over a 4°C boiling range and the distillate solidified on cooling. Approximately 0.10 g of the dry product was dissolved in 0.50 mL of dichloromethane and gas chromatograms were recorded for (1) the original solution and (2) the solution spiked with an authentic sample of tetrahedryl acetate. The relative area of the second (72 s) product peak was considerably larger in the second chromatogram than in the first. The gas chromatograms were obtained using a 2-meter OV-101/Chromosorb W packed column. The infrared spectrum of the product was recorded using a thin film between heated silver bromide plates.

Data

Mass of tetrahedryl acetate: 1.644 g
Mass of dry product: 0.854 g
Distillation boiling range of product: 78–82°C
Micro boiling point of product: 81°C
Gas chromatography data:

  Column temperature: 110°C
  Injector temperature: 150°C
  Detector temperature: 150°C
  Helium flow rate: 25 cm^3/minute

Component	Retention time	Peak area
tetrahedranol	47 s	320 mm^2
tetrahedryl acetate	72 s	12 mm^2

Calculations

Percentage of tetrahedranol in product:

$$\frac{320\ mm}{332\ mm} \times 100\% = 96.4\%$$

Mass of tetrahedranol in product:

$$0.854\ g \times \frac{96.4\%}{100\%} = 0.823\ g$$

Theoretical yield of tetrahedranol (TetOAc = tetrahedryl acetate; TetOH = tetrahedranol):

$$1.644\ g\ TetOAc \times \frac{1\ mol\ TetOAc}{110.1\ g\ TetOAc} \times \frac{68.1\ g\ TetOH}{1\ mol\ TetOH} = 1.017\ g\ TetOH$$

Percent yield of tetrahedranol:

$$\frac{0.823\ g}{1.017\ g} \times 100\% = 80.9\%$$

Results

Tetrahedranol was obtained in 80.9% yield from the alkaline hydrolysis of tetrahedryl acetate. Tetrahedranol at room temperature was a colorless, almost transparent solid with a mild "spiritous" odor. It could easily be melted on a steam bath to form a clear, colorless liquid. It had a distillation boiling range of 78–82°C and a micro boiling point of 81°C. Its infrared spectrum contained significant absorption bands at 3370, 2975, 2885 and 1194 cm^{-1}.

Discussion

Statement of the problem: Can tetrahedranol with a purity of 95 percent or more be prepared by the alkaline hydrolysis of tetrahedryl acetate?

Working hypothesis 1: The alkaline hydrolysis of tetrahedryl acetate *will* yield tetrahedranol as one of the products.

I based this hypothesis on (1) our textbook's assertion that the alkaline hydrolysis of an ester yields the corresponding hydroxy compound (alcohol or phenol) and the salt of the corresponding carboxylic acid, and (2) the results of Experiment 4, in which I obtained salicylic acid (a phenol) from the alkaline hydrolysis of methyl salicylate. Observations in support of hypothesis 1 include the following:

1. The slow disappearance of the organic layer during the reaction suggests that the water-insoluble ester was being converted to a water-soluble product. According to Table 59.1 in the lab textbook, tetrahedranol is soluble in water.

2. The disappearance of the "fruity" odor of tetrahedryl acetate also suggests that the ester was reacting.

3. The solidification of the distillate and easy melting of the solid on a steam bath is consistent with the hypothesis, since Table 59.1 indicates that tetrahedranol has a melting point of 27°C.

Hypothesis 1 was tested by obtaining a micro boiling point and infrared spectrum of the product, with the following results:

1. The observed micro boiling point of 81°C is consistent with the hypothesis, since tetrahedranol has a reported boiling point of 82°C.

2. The infrared spectrum, suggesting a tertiary alcohol, is consistent with the hypothesis.

Wave number/cm^{-1}	Assignment	Interpretation
3370 (3368.4)	O—H stretch	Alcohol or phenol
2975, 2885 (2975.1, 2884.2)	C—H stretch	Absence of absorption above 3000 cm^{-1} eliminates a phenol as a possibility.
1194 (1193.6)	C—O stretch	Consistent with tertiary alcohol

The wave numbers of these bands are nearly identical to those on a published FTIR spectrum of tetrahedranol (shown in parentheses), and the band shapes and positions in the fingerprint region matched those of the published spectrum. My spectrum did have a weak band at 1739 cm^{-1} that was not present on the published spectrum, but this is probably the carbonyl (C=O) band of the tetrahedryl acetate impurity that was detected by gas chromatography.

Working hypothesis 2: The purity of the product *will not* be 95 percent or better.

I based this hypothesis on (1) the statement in the lecture textbook that some ester hydrolysis reactions, especially those involving esters of bulky alcohols, take more than an hour to reach completion, and (2) the fact that the ester and alcohol boiling points are only 15°C apart, suggesting that (as we learned in Experiment 5) simple distillation will not remove all of the impurity.

Hypothesis 2 was tested by obtaining a gas chromatogram of the product, identifying the peaks, and calculating the percentage of tetrahedranol in the product from the peak areas. Since the small 72 s peak became larger when the sample was spiked with tetrahedryl acetate, it must be the tetrahedryl acetate peak. Since the IR spectrum showed that the major product was tetrahedranol, the much larger peak at 47 s must be the tetrahedranol peak. The calculated mass percentage of tetrahedranol, 96.4%, is *not* consistent with my hypothesis. I am not disappointed with this result, since it was nice to know that the experiment came out better than expected! My assumptions that the reaction would not go to completion in an hour and that the impurity would not be completely removed from the product by simple distillation were both correct, but the reaction was more nearly complete than I had guessed.

The yield of tetrahedranol (0.823 g) was 0.194 g less than the theoretical value. Of this at least 0.019 g resulted from incomplete reaction of tetrahedryl acetate, based on the 0.031 g (0.854 g – 0.823 g) of ester in the distillate. The remaining losses could have arisen from (1) losses during transfers, (2) losses during the extraction and washing operations, and (3) losses during the distillation. During each transfer I used additional solvent (ether or water) to rinse out the vessel from which the product was transferred, so losses during transfers should not be a major factor. About 0.1 mL of residue remained in the distilling flask, which can account for no more than 0.1 g of the loss, since some of the residue must have been tetrahedryl acetate. Most of the remaining losses must have occurred during the extraction and washing operations, due to the partial solubility of tetrahedranol in water. Such losses might have been reduced by saturating the aqueous layer with potassium carbonate to salt out the alcohol or by carrying out several more ether extractions.

Conclusion

I conclude that tetrahedranol with a purity of 95 percent or more can be prepared by the hydrolysis of tetrahedryl acetate. I base this conclusion primarily on the infrared spectrum of the product, whose resemblance to the published spectrum leaves little doubt that the product is tetrahedranol, and on the gas chromatographic analysis, which indicates a purity of 96.4 percent. My conclusion that the product is tetrahedranol is supported by the other evidence cited in the discussion. My conclusion that the purity of the product is greater than 95 percent might possibly be in error, since the GC peak areas were not corrected by applying detector response factors.

Calculations for Organic Synthesis

In this book, the quantities of many reactants are given in units of "chemical amount," that is, in moles or millimoles. You will have to convert such quantities to units of mass or volume before you can begin a synthetic experiment. In any experiment, it is the relationship between *chemical* quantities that is significant, not the relationship between such *physical* quantities as mass and volume. Because there are no "mole-meters" that measure molar amounts directly, we are forced to use balances and volumetric glassware for that purpose. That should not obscure the fact that the chemical units are fundamental; only by knowing the chemical amounts of reactants involved in a preparation, for example, can you recognize the stoichiometric relationships between them or predict the yield of the expected product.

It is helpful to regard a chemical calculation as a process by which a given quantity is "converted" to the required quantity. This can be accomplished by multiplying the given quantity by a series of ratios used as unit or dimensional *conversion factors.* Unit conversions are carried out by using conversion factors, such as 454 g/lb, that are written as ratios between two quantities whose quotient is unity (for example, 454 g = 1 lb, so 454 g/1 lb = 1). Dimensional conversions are carried out by using conversion factors that are ratios of quantities in different *dimensions,* such as mass, volume, and chemical amount (amount of substance). For example, the density of a substance can be regarded as a conversion factor linking the two dimensions of mass and volume, and it is used to convert the mass of a given quantity of the substance to units of volume, or vice versa. A conversion factor can be inverted when necessary. For example, the molar mass of butyl acetate, written as 116 g/1 mol, will convert moles of butyl acetate to grams; the inverse ratio, 1 mol/116 g, will convert grams to moles. All calculations should be checked by making sure that the units involved cancel to yield the correct units in the answer. This does not ensure that your answer is correct, but if the units do *not* cancel, the answer is almost certainly wrong.

The following examples illustrate some fundamental types of calculations that you will encounter:

Chemical Amount and Mass: The chemical amount (in moles or millimoles) of a substance is converted to its mass by multiplying by its molar mass. Remember that the molar mass of a substance is obtained by simply appending the units g/mol to its molecular weight, which is a dimensionless quantity. For example, the mass of 15.0 mmol of butyl acetate (M.W. = 116) is

$$15.0 \text{ mmol} \times \frac{1 \text{mol}}{1000 \text{ mmol}} \times \frac{116 \text{ g}}{1 \text{ mol}} = 1.74 \text{ g}$$

Note that the chemical amount in millimoles must be converted to moles before the conversion factor is applied; otherwise, the units will not cancel. Mass can be converted to chemical amount by inverting the conversion factor before multiplying.

Chemical Amount and Volume: The chemical amount (moles or millimoles) of a pure liquid is converted to volume by multiplying by the liquid substance's molar mass and the inverse of its density. For example, the volume of 25.0 mmol of acetic acid (M.W. = 60.1; d = 1.049 g/mL) is

$$25.0 \text{ mmol} \times \frac{1 \text{ mol}}{1000 \text{ mmol}} \times \frac{60.1 \text{ g}}{1 \text{ mol}} \times \frac{1 \text{ mL}}{1.049 \text{ g}} = 14.4 \text{ mL}$$

The volume of a solution needed to provide a specified chemical amount of solute is calculated by multiplying the number of moles required by the inverse of the solution's molar concentration. For example, the volume of 6.0 M HCl (which contains 6.0 mol of HCl per liter of solution) needed to provide 18 mmol of HCl is

$$18 \text{ mmol} \times \frac{1 \text{ mol}}{1000 \text{ mmol}} \times \frac{1 \text{ L}}{6.0 \text{ mol}} \times \frac{1000 \text{ mL}}{1 \text{ L}} = 3.0 \text{ mL}$$

Note that concentrations expressed in mol/L and mmol/mL have the same numerical value. Thus a 6.0 M solution also has a concentration of 6.0 mmol/mL; using these units simplifies the previous calculation considerably:

$$18 \text{ mmol} \times \frac{1 \text{ mL}}{6.0 \text{ mmol}} = 3.0 \text{ mL}$$

Theoretical Yield: The maximum amount of product that could be attained from a reaction is called the *theoretical yield* of the reaction. Theoretical yields can be calculated using *stoichiometric factors*—ratios derived from the coefficients (expressed in moles) of the products and reactants in a balanced equation for the reaction. For example, the stoichiometric factors relating the chemical amount of the organic product to the chemical amounts of the two reactants in the following reaction are 1 mol dibenzalacetone/1 mol acetone and 1 mol dibenzalacetone/2 mol benzaldehyde.

$$2\text{PhCHO} + \text{CH}_3\overset{\overset{\text{O}}{\|}}{\text{C}}\text{CH}_3 \xrightarrow{\text{NaOH}} \text{PhCH}=\text{CHCCH}=\text{CHPh} + 2\text{H}_2\text{O}$$

benzaldehyde (B)   acetone (A)   dibenzalacetone (DBA)

Suppose you were trying to prepare dibenzalacetone (M.W. = 234.3) starting with 5.00 g of benzaldehyde (M.W. = 106.1) and 1.50 g of acetone (M.W. = 58.1). (For convenience, we will abbreviate the names as: dibenzalacetone = DBA, benzaldehyde = B, acetone = A.) You can calculate the maximum amount of product that could be formed from each reactant by converting the given quantity to moles, then applying the appropriate stoichiometric factor:

$$5.00 \text{ g B} \times \frac{1 \text{ mol B}}{106.1 \text{ g B}} \times \frac{1 \text{ mol DBA}}{2 \text{ mol B}} = 0.0236 \text{ mol DBA}$$

$$1.50 \text{ g A} \times \frac{1 \text{ mol A}}{58.1 \text{ g A}} \times \frac{1 \text{ mol DBA}}{1 \text{ mol A}} = 0.0258 \text{ mol DBA}$$

Since there is only enough benzaldehyde to produce 0.0236 mol of dibenzalacetone, it is impossible to obtain more than that from the specified quantities of reactants. Once that much product has been formed, the reaction mixture will have run out of benzaldehyde, and the *excess* (leftover) acetone will have nothing to react with. Therefore, benzaldehyde is the *limiting reactant* on which the yield calculations must be based. The theoretical yield of dibenzalacetone, in grams, is then

$$0.0236 \text{ mol DBA} \times \frac{234.3 \text{ g DBA}}{1 \text{ mol DBA}} = 5.53 \text{ g DBA}$$

Remember that the limiting reactant is always the one that would produce the least amount of product, which is not necessarily the one present in the lowest amount. In this example, benzaldehyde is the limiting reactant, although there is more than three times as much benzaldehyde (in grams) as there is acetone.

Percent Yield: It is seldom, if ever, possible to attain the theoretical yield from an organic preparation. The reaction may not go to completion, there may be side reactions that reduce the yield of product, and there are invariably material losses when the product is separated from the reaction mixture and purified. For example, in the reaction of benzaldehyde with acetone, some benzalacetone ($PhCH=CHCOCH_3$) would be formed as a by-product, reducing the yield of dibenzalacetone. The *percent yield* of a preparation compares the actual yield to the theoretical yield as defined here:

$$\text{Percent yield} = \frac{\text{actual yield}}{\text{theoretical yield}} \times 100\%$$

For example, if you prepared 4.09 g of dibenzalacetone from 5.00 g of benzaldehyde and 1.50 g of acetone (theoretical yield = 5.53 g), the percent yield of your synthesis would be:

$$\text{Percent yield} = \frac{4.09 \text{ g DBA}}{5.53 \text{ g DBA}} \times 100\% = 74.0\%$$

In many experiments, you will estimate the theoretical yield of a preparation based on the amounts of reactants given in the "Before You Begin" section. This will help you assess your performance by comparing your actual yield with an estimate of the "ideal" yield. But the percentage yield that you *report* should be based on the amounts of reactants that you actually used in the synthesis, not on the amounts given (unless they are exactly the same).

Planning an Experiment   **APPENDIX V**

Before starting any project, whether you are making a bookshelf, duck à l'orange, or isopentyl acetate, you must have a plan. An *experimental plan* should summarize what you expect to do in the laboratory and how you intend to go about it. You should state how the work is to be done in brief phrases, without excessive detail. You can always refer to the textbook procedure and operation descriptions for the details, but as you become more proficient in the laboratory, you should find yourself relying less on the textbook and more on your experimental plan. A good plan should give you quick access to the essential information you will need while performing the experiment: quantities of chemicals (including wash solvents, drying agents, etc.), reflux times, physical properties, hazard warnings, and other useful data should be included. You can list the supplies and equipment you will need for each operation (these are specified in most Operations) so that you can have them cleaned and ready when you need them. You may also wish to sketch the apparatus you will be using so that you can assemble it quickly in the laboratory.

An experimental plan should help you organize your time efficiently by listing tasks in the approximate order in which you expect to accomplish them. For example, whenever a reflux period is specified in a procedure, you will have some free time to set up the apparatus for the next step, reorganize your work area, review an operation, start a Minilab, take a melting point, record the spectrum of a previous product, or tie up other loose ends. Your plan should be flexible enough that you can alter it or deviate from it during the experiment, if there is good reason to do so. One simple and effective way of organizing your time is to use a laboratory checklist such as the one in Experiment 4. You should refer to relevant sections of the experiment (especially "Understanding the Experiment" and the directions) as you prepare your checklist, as well as the appropriate Operation descriptions. Leave enough space between the items on your checklist so that you can add new ones, as necessary, during the experiment. As you complete each task in the laboratory, simply check it off the list and go on to the next one.

A flow diagram such as the one in Figure 5.1 of Experiment 5 can help you organize your time by giving you a quick overview of the procedure, showing the purpose of each step. To create such a flow diagram, first list all the substances that you know to be present in the reaction mixture before the reaction starts (reactants, solvents, catalysts), as shown in the following general diagram. The reaction equation in the "Reactions and Properties" section tells you what substances will form as a result of the reaction. At the end of the reaction period, the reaction mixture will contain these products

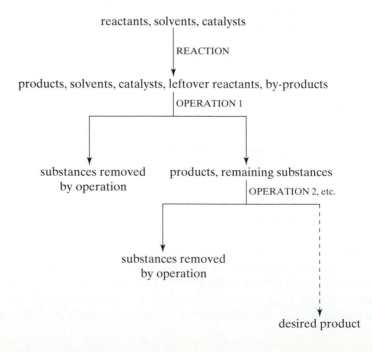

as well as the reaction solvent (if any), catalyst (if any), leftover reactants, and usually some by-products formed by various side reactions. The flow diagram should show how all of the unwanted substances in the reaction mixture are separated from the desired product. Each separation or purification operation is represented by a branch in the flow diagram, with the substance(s) being removed on one side and the desired product, along with the remaining substances, on the other side. After the last operation, the product stands alone, with all the impurities eliminated—on paper, at least!

Properties of Organic Compounds

The following tables are to be used in conjunction with the procedures described in Part IV, Qualitative Organic Analysis. Compounds that melt below ordinary ambient temperatures (about 25°C) are listed in order of increasing boiling point; those that (when pure) are generally solid at room temperature are listed in order of increasing melting point. Both melting and boiling points are specified for some borderline cases. Derivative preparations are described on pages 557–570 and are referred to by number at the heads of the appropriate columns. Melting points in parentheses are for derivatives that exist in more than one crystalline form or for which significantly different melting points have been reported in the literature. Sometimes the recrystallization solvent will determine the form in which a derivative crystallizes, so significant deviations from the listed melting points should not be considered conclusive proof that a sample is not the expected derivative. In cases where a given compound can form more than one product (for instance, nitro and bromo derivatives), the reaction conditions for the preparation may determine the derivative isolated (a mixture of derivatives may also result). A dash (—) in a derivative column indicates either that the derivative has not been reported in the literature or that it is not suitable for identification. See references from section G in the Bibliography for physical constants and derivative melting points not listed in these tables.

Most compounds are listed by their systematic (IUPAC) names except when those names would be too lengthy.

Abbreviations Used in Tables:
   d = decomposes on melting
   s = sublimes at or below melting point
   m = monosubstituted derivative (such as a mononitrated aromatic hydrocarbon)
   di = disubstituted derivative
   t = trisubstituted derivative
   tet = tetrasubstituted derivative

List of tables

Table 1 Alcohols

Compound	b.p.	m.p.	3,5-Dinitro-benzoate, D-1	4-Nitro-benzoate, D-1	1-Naphthyl-urethane, D-2	Phenyl-urethane, D-2
Methanol	65		108	96	124	47
Ethanol	78		93	57	79	52
2-Propanol	82		123	110	106	88
2-Methyl-2-propanol*	83	26	142	—	—	136
2-Propen-1-ol	97		49	28	108	70
1-Propanol	97		74	35	105	57
2-Butanol	99		76	26	97	65
2-Methyl-2-butanol	102		116	85	72	42
2-Methyl-1-propanol	108		87	69	104	86
3-Pentanol	116		101	17	95	48
1-Butanol	118		64	36	71	61
2-Pentanol	120		62	24	74	—
3-Methyl-3-pentanol	123		94 (62)	69	104	43
3-Methyl-1-butanol	132		61	21	68	57
4-Methyl-2-pentanol	132		65	26	88	143
1-Pentanol	138		46	11	68	46
Cyclopentanol	141		115	62	118	132
2-Ethyl-1-butanol	148		51	—	60	—
1-Hexanol	157		58	5	59	42
Cyclohexanol*	161	25	113	50	129	82
Furfuryl alcohol	172		80	76	130	45
1-Heptanol	177		47	10	62	60
2-Octanol	179		32	28	63	114
1-Octanol	195		61	12	67	74
1-Phenylethanol	202		95	43	106	92
Benzyl alcohol	205		113	85	134	77
2-Phenylethanol	219		108	62	119	78
1-Decanol	231		57	30	73	60
3-Phenylpropanol	236		45	47	—	92
1-Dodecanol*	259	24	60	45 (42)	80	74
1-Tetradecanol		39	67	51	82	74
(−)-Menthol		44	153	62	119	111
1-Hexadecanol		49	66	58	82	73
1-Octadecanol		59	77	64	89	79
Diphenylmethanol		68	141	132	136	139
Cholesterol		148	—	185	176	168
(+)-Borneol		208	154	137 (153)	132 (127)	138

*May be solid at or just below room temperature

Note: All temperatures are in °C

Table 2 Aldehydes

Compound	b.p.	m.p.	2,4-Dinitrophenyl-hydrazone, D-3	Semicarbazone, D-4	Oxime, D-5
Ethanal	21		168 (157)	162	47
Propanal	48		148 (155)	154	40
Propenal	52		165	171	—
2-Methylpropanal	64		187 (183)	125 (119)	—
Butanal	75		123	106	
3-Methylbutanal	92		123	107	48
Pentanal	103		106 (98)	—	52
2-Butenal	104		190	199	119
2-Ethylbutanal	117		95 (130)	99	—
Hexanal	130		104	106	51
Heptanal	153		106	109	57
2-Furaldehyde	162		212 (230)	202	91
2-Ethylhexanal	163		114 (120)	254d	—
Octanal	171		106	101	60
Benzaldehyde	179		239	222	35
4-Methylbenzaldehyde	204		234	234 (215)	80
3,7-Dimethyl-6-octenal	207		77	84 (91)	—
2-Chlorobenzaldehyde	213		213 (209)	229 (146)	76 (101)
4-Methoxybenzaldehyde	248		253d	210	133
Phenylethanal	*195*	33	121 (110)	153 (156)	99
2-Methoxybenzaldehyde		38	254	215	92
4-Chlorobenzaldehyde		48	265	230	110 (146)
3-Nitrobenzaldehyde		58	290	246	120
4-Nitrobenzaldehyde		106	320	221(211)	133 (182)

Note: All temperatures are in °C

Table 3 Ketones

Compound	b.p.	m.p.	2,4-Dinitrophenyl-hydrazone, D-3	Semicarbazone, D-4	Oxime, D-5
Acetone	56		126	187	59
2-Butanone	80		118	146	—
3-Methyl-2-butanone	94		124	113	—
2-Pentanone	102		143	112 (106)	58
3-Pentanone	102		156	138	69
3,3-Dimethyl-2-butanone	106		125	157	75 (79)
4-Methyl-2-pentanone	117		95 (81)	132	58
2,4-Dimethyl-3-pentanone	124		95 (88)	160	34
2-Hexanone	128		110	125	49
4-Methyl-3-penten-2-one	130		205	164 (133)	48
Cyclopentanone	131		146	210 (203)	56
4-Heptanone	144		75	132	—
2-Heptanone	151		89	123	—
Cyclohexanone	156		162	166	91
2,6-Dimethyl-4-heptanone	168		92	122	210
2-Octanone	173		58	124	—
Cycloheptanone	181		148	163	23
2,5-Hexanedione	194		257 (di)	185 (m); 224 (di)	137 (di)

continued

Table 3 *Continued*

Compound	b.p.	m.p.	2,4-Dinitrophenyl-hydrazone, D-3	Semicarbazone, D-4	Oxime, D-5
Acetophenone	202	*20*	238	198 (203)	60
2´-Methylacetophenone	214		159	205	61
Propiophenone	218	*21*	191	182 (174)	54
3´-Methylacetophenone	220		207	203	57
2-Undecanone	228		63	122	44
4-Phenyl-2-butanone	235		127	142	87
3´-Methoxyacetophenone	240		—	196	—
2´-Methoxyacetophenone	245		—	183	83 (96)
4´-Methylacetophenone	*226*	28	258	205	88
4´-Methoxyacetophenone		38	228	198	87
4-Phenyl-3-buten-2-one		42	227 (223)	187	117
Benzophenone		48	238	167	144
2-Acetonaphthone		54	262d	235	145
3´-Nitroacetophenone		80	228	257	132
9-Fluorenone		83	283	234	195
(±)-Camphor		179	177	237	118

Note: All temperatures are in °C

Table 4 Amides

Compound	b.p.	m.p.	Carboxylic acid, D-6	N-Xanthylamide, D-7
Formamide	195d		—	184
Propanamide		81	—	211
Ethanamide		82	17	240
Heptanamide		96	—	155
Nonanamide		99	12	148
Hexanamide		100	—	160
Hexadecanamide		106	63	142
Pentanamide		106	—	167
Octadecanamide		109	69	141
Butanamide		115	—	187
Chloroacetamide		120	61 (53)	209
4-Methylpentanamide		121	—	160
Succinimide		126	185	247
2-Methylpropanamide		129	—	211
Benzamide		130	122	223
3-Methylbutanamide		136	—	183
o-Toluamide		143	104 (108)	200
Furamide		143	133	210
Phenylacetamide		156	77s	195
p-Toluamide		159	180s	225
4-Nitrobenzamide		201	240	233
Phthalimide		238	210d	177

Table 5 Primary and secondary amines

Compound	b.p.	m.p.	Benzamide, D-8	p-Toluene-sulfonamide, D-9	Phenylthio-urea, D-10	Picrate, D-12
t-Butylamine	44		134	—	120	198
Propylamine	48		84	52	63	135
Diethylamine	56		42	60	34	155
sec-Butylamine	63		76	55	101	140
2-Methylpropylamine	69		57	78	82	150
Butylamine	77		42	—	65	151
Diisopropylamine	84		—	—	—	140
Pyrrolidine	89		—	123	—	112 (164)
3-Methylbutylamine	95		—	65	102	138
Pentylamine	104		—	—	69	139
Piperidine	106		48	96	101	152
Dipropylamine	109		—	—	69	75
Morpholine	128		75	147	136	146
Pyrrole	131		—	—	143	69d
Hexylamine	132		40	—	77	126
Cyclohexylamine	134		149	—	148	—
Diisobutylamine	139		—	—	113	121
N-Methylcyclohexylamine	147		86	—	—	170
Dibutylamine	159		—	—	86	59
N-Ethylbenzylamine	181		—	95	—	118
Aniline	184		163	103	154	198 (180)
Benzylamine	185		105	116 (185)	156	199 (194)
N-Methylaniline	196		63	94	87	145
o-Toluidine	200		144	108	136	213
m-Toluidine	203		125	114	104 (92)	200
N-Ethylaniline	205		60	87	89	138
2-Chloroaniline	209		99	105 (193)	156	134
2,6-Dimethylaniline	215		168	212	204	180
2-Methoxyaniline	225		60	127	136	200
2-Ethoxyaniline	229		104	164	137	—
3-Chloroaniline	230		120	138 (210)	124 (116)	177
4-Ethoxyaniline	250		173	106	136	69
Dicyclohexylamine	255d		153 (57)	—	—	173
N-Benzylaniline		37	107	149	103	48
p-Toluidine		44	158	118	141	182d
Diphenylamine		54	180 (109)	141	152	182
4-Methoxyaniline		58	154	114	157 (171)	170
4-Bromoaniline		66	204	101	148	180
2-Nitroaniline		71	110 (98)	142	—	73
4-Chloroaniline		72	192	95 (119)	152	178
1,2-Diaminobenzene		102	301 (di)	260 (di)	—	208
3-Nitroaniline		114	155 (di)	138	160	143
1,4-Diaminobenzene		142	300 (di)	266 (di)	—	—
4-Nitroaniline		147	199 (di)	191	—	100

Note: All temperatures are in °C. See other references in the Bibliography (Category G) for melting points of benzene-sulfonamides and 1-naphthylthioureas..

Table 6 Tertiary amines

Compound	b.p.	m.p.	Methiodide, D-11	Picrate, D-12
Triethylamine	89		280	173
Pyridine	115		117	167
2-Methylpyridine	129		230	169
3-Methylpyridine	143		92	150
4-Methylpyridine	143		152	167
Tripropylamine	157		207	116
2,4-Dimethylpyridine	159		113	183 (169)
N,N-Dimethylbenzylamine	183		179	93
N,N-Dimethylaniline	193		228d	163
Tributylamine	216 (211)		186	105
N,N-Diethylaniline	217		102	142
Quinoline	237		133 (72)	203
Isoquinoline	243	*26*	159	222
Tribenzylamine	—	91	184	190
Acridine	—	111	224	208

Note: All temperatures are in °C

Table 7 Carboxylic acids

Compound	b.p.	m.p.	Amide, D-13	*p*-Toluidide, D-14	Anilide, D-14	*p*-Nitrobenzyl Ester, D-15
Formic acid	101		43	53	50	31
Acetic acid	118		82	148	114	78
Propenoic acid	139		85	141	104	—
Propanoic acid	141		81	124	103	31
2-Methylpropanoic acid	154		128	107	105	—
Butanoic acid	164		115	75	95	35
3-Methylbutanoic acid	176		135	107	109	—
Pentanoic acid	186		106	74	63	—
2-Chloropropanoic acid	186		80	124		92 —
Dichloroacetic acid	194		985	153	118	—
2-Methylpentanoic acid	196		79	80	95	—
Hexanoic acid	205		101	75	95	—
2-Bromopropanoic acid	205d	*24*	123	125		99 —
Octanoic acid	239		107	70	57	—
Nonanoic acid	254		99	84	57	—
Decanoic acid		32	108	78	70	—
2,2-Dimethylpropanoic acid	*164*	35	178 (154)	—	129 (133)	—
Dodecanoic acid		44	110 (99)	87	78	—
3-Phenylpropanoic acid		48	105	135	98	36
Tetradecanoic acid		54	103	93	84	—
Hexadecanoic acid		62	106	98	90	42
Chloroacetic acid		63	120	162	137	—
Octadecanoic acid		70	109	102		95 —
trans-2-Butenoic acid		72	160	132	118	67
Phenylacetic acid		77	156	136	118	65
2-Methoxybenzoic acid		101	129	—	131	113

continued

Table 7 *Continued*

Compound	b.p.	m.p.	Amide, D-13	*p*-Toluidide, D-14	Anilide, D-14	*p*-Nitrobenzyl Ester, D-15
Oxalic acid (dihydrate)		101	419d (di)	268 (di)	254 (di)	204 (di)
2-Methylbenzoic acid		104	142	144	125	91
Nonanedioic acid		106	175 (di)	201 (di)	186 (di)	44
3-Methylbenzoic acid		112	94	118	126	87
Benzoic acid		122	130	158	163	89
Maleic acid		130	181 (m) 266 (di)	142 (di)	187 (di)	91
Decanedioic acid		133	170 (m) 210 (di)	201 (di)	122 (m) 200 (di)	73 (di)
Cinnamic acid		133	147	168	153	117
Propanedioic acid		135	50 (m) 170 (di)	86 (m) 253 (di)	132 (m) 230 (di)	86
2-Chlorobenzoic acid		140	140	131	118	106
3-Nitrobenzoic acid		140	143	162	155	141
Diphenylacetic acid		148	167	172	180	—
2-Bromobenzoic acid		150	155	—	141	110
Hexanedioic acid		152	125 (m) 224 (di)	238	151 (m) 241 (di)	106
4-Methylbenzoic acid		180s	160	160 (165)	145	104
4-Methoxybenzoic acid		184	167 (163)	186	169	132
Butanedioic acid		188	157 (m) 260 (di)	180 (m) 255 (di)	143 (m) 230 (di)	—
3,5-Dinitrobenzoic acid		205	183	—	234	157
Phthalic acid		210d	220 (di)	201 (di)	253 (di)	155
4-Nitrobenzoic acid		240	201	204	211	168
4-Chlorobenzoic acid		242	179	—	194	129
Terephthalic acid		>300s	—	—	337	263 (di)

Note: All temperatures are in °C

Table 8 Esters

Compound	b.p.	m.p.	Carboxylic acid, D-16	Alcohol or Phenol, D-16	*N*-Benzyl-amide, D-17	3,5-Dinitro-benzoate, D-18
Ethyl formate	54		8	—	60	93
Methyl acetate	57		17	—	61	108
Ethyl acetate	77		17	—	61	93
Methyl propanoate	80		—	—	43	108
Methyl acrylate	80		13	—	237	108
Isopropyl acetate	91		17	—	61	123
t-Butyl acetate	98		17	26	61	142
Ethyl propanoate	99		—	—	43	93
Methyl 2,2-dimethylpropanoate	101		35	—	—	108
Propyl acetate	102		17	—	61	74
Methyl butanoate	102		—	—	38	108
Ethyl 2-methylpropanoate	111		—	—	87	93
sec-Butyl acetate	112		17	—	61	76

continued

Table 8 *Continued*

Compound	b.p.	m.p.	Carboxylic acid, D-16	Alcohol or Phenol, D-16	N-Benzyl-amide, D-17	3,5-Dinitro-benzoate, D-18
Methyl 3-methylbutanoate	117		—	—	54	108
Isobutyl acetate	117		17	—	61	87
Ethyl butanoate	122		—	—	38	93
Butyl acetate	126		17	—	61	64
Methyl pentanoate	128		—	—	43	108
Ethyl 3-methylbutanoate	135		—	—	54	93
3-Methylbutyl acetate	142		17	—	61	61
Ethyl chloroacetate	145		63	—	—	93
Pentyl acetate	149		17	—	61	46
Ethyl hexanoate	168		—	—	53	93
Hexyl acetate	172		17	—	61	58
Cyclohexyl acetate	175		17	25	61	113
Dimethyl malonate	182		135	—	142	108
Diethyl oxalate	185		101*		223	93
Heptyl acetate	192		17	—	61	47
Phenyl acetate	197		17	42	61	146
Methyl benzoate	199		122	—	105	108
Diethyl malonate	199		135	—	142	93
o-Tolyl acetate	208		17	31	61	135
m-Tolyl acetate	212		17	12	61	165
Ethyl benzoate	213		122	—	105	93
p-Tolyl acetate	213		17	36	61	189
Methyl o-toluate	215		104	—	—	108
Benzyl acetate	217		17	—	61	113
Diethyl succinate	218		188	—	206	93
Isopropyl benzoate	218		122	—	105	123
Methyl phenylacetate	220		77s	—	122	108
Diethyl maleate	223		137	—	150	93
Ethyl phenylacetate	228		77s	—	122	93
Propyl benzoate	230		122	—	105	74
Diethyl adipate	245		152	—	189	93
Butyl benzoate	250		122	—	105	64
Ethyl cinnamate	271		133	—	226	93
Dimethyl phthalate	284		210d	—	179	108
(+)-Bornyl acetate	*226*	27	17	208	61	154
Methyl p-toluate		33	180s	—	133	108
Methyl cinnamate		36	133	—	226	108
Benzyl cinnamate		39	133	—	226	113
1-Naphthyl acetate		49	17	94	61	217
Ethyl p-nitrobenzoate		56	240	—	—	93
Phenyl benzoate		69	122	42	105	146
2-Naphthyl acetate		71	17	123	61	210
p-Tolyl benzoate		71	122	36	105	189
Methyl m-nitrobenzoate		78	140	—	101	108
Methyl p-nitrobenzoate		96	240	—	142	108

*dihydrate; the anhydrous acid melts at 190°C

Note: All temperatures are in °C. Additional derivatives of the acid and alcohol portions of most esters can be found in Tables 1 and 7.

Table 9 Alkyl halides

Compound	b.p.	Density d_4^{20}, C-10	S-Alkylthiuronium-picrate, D-19
Bromoethane	38	1.461	188
2-Bromopropane	60	1.314	196
1-Chloro-2-methylpropane	69	0.879	167 (174)
3-Bromopropene	71	1.398	155
1-Bromopropane	71	1.354	177
Iodoethane	72	1.936	188
1-Chlorobutane	78	0.884	177
2-Iodopropane	89	1.703	196
1-Bromo-2-methylpropane	93	1.264	167 (174)
1-Chloro-3-methylbutane	100	0.875	173
1-Bromobutane	101	1.274	177
I-Iodopropane	102	1.749	177
3-Iodopropene	102	1.848	155
1-Chloropentane	108	0.882	154
I-Bromo-3-methylbutane	119	1.207	173 (179)
2-Iodobutane	119	1.595	166
1-Iodo-2-methylpropane	120	1.606	167 (174)
1-Bromopentane	129	1.218	154
1-Iodobutane	131	1.617	177
1-Chlorohexane	134	0.876	157
1-Iodo-3-methylbutane	148	1.503	173
1-Iodopentane	155	1.516	154
1-Bromohexane	155	1.173	157
1-Iodohexane	181	1.439	157
1-Bromooctane	201	1.112	134
1-Iodooctane	225	1.330	134

Note: All temperatures are in °C; density is in g/mL

Table 10 Aryl halides

Compound	b.p.	m.p.	Density d_4^{20}, C-10	Nitro derivative, D-20	Carboxylic acid, D-21
Chlorobenzene	132		1.106	52	—
Bromobenzene	156		1.495	75 (70)	—
2-Chlorotoluene	159		1.083	63	140
3-Chlorotoluene	162		1.072	91	158
4-Chlorotoluene	162		1.071	38 (m)	242
1,3-Dichlorobenzene	173		1.288	103	—
1,2-Dichlorobenzene	181		1.306	110	—
2-Bromotoluene	182		1.423	82	150
3-Bromotoluene	184		1.410	103	155
Iodobenzene	188		1.831	171 (m)	—
2,6-Dichlorotoluene	199		1.269	50 (m)	139
2,4-Dichlorotoluene	200		1.249	104	164
3-Iodotoluene	204		1.698	108	187
2-Iodotoluene	211		1.698	103 (m)	162
1-Chloronaphthalene	259		1.191	180	—

continued

Table 10 *Continued*

Compound	b.p.	m.p.	Density d_4^{20}, C-10	Nitro derivative, D-20	Carboxylic acid, D-21
4-Bromotoluene	184	28	—	—	251
4-Iodotoluene		35	—	—	270
1,4-Dichlorobenzene		53	—	106,54 (m)	—
2-Chloronaphthalene		56	—	175	—
1,4-Dibromobenzene		89	—	84	—

Note: All temperatures are in °C, density is in g/mL. Nitro derivatives signified by (m) are mononitro compounds; all others are dinitro derivatives.

Table 11 Aromatic hydrocarbons

Compound	b.p.	m.p.	Nitro derivative, D-20	Carboxylic acid, D-21	Picrate, D-12
Benzene	80		89 (di)	—	84u
Toluene	111		70 (di)	122	88u
Ethylbenzene	136		37 (t)	122	96u
1,4-Xylene	138		139 (t)	300s	90u
1,3-Xylene	139		183 (t)	330s	91u
1,2-Xylene	144		118 (di)	210d	88u
Isopropylbenzene	152		109 (t)	122	—
Propylbenzene	159		—	122	103u
1,3,5-Trimethylbenzene	165		86 (di) 235 (t)	350 (t)	97u
t-Butylbenzene	169		62 (di)	122	—
4-Isopropyltoluene	177		54 (di)	300s	—
1,3-Diethylbenzene	181		62 (t)	330s	—
1,2,3,4-Tetrahydronaphthalene	206		96 (di)	210d	—
Diphenylmethane	262	26	172 (tet)	—	—
1,2-Diphenylethane		53	180 (di) 169 (tet)	—	—
Naphthalene		80	61 (m)	—	149
Triphenylmethane		92	206 (t)	—	—
Acenaphthene		96	101 (m)	—	161
Fluorene		114	199 (di) 156 (m)	—	87 (77)
Anthracene		216	—	—	138u

Note: All temperatures are in °C. Picrates designated u are unstable and cannot easily be purified by recrystallization.

Table 12 Phenols

Compound	b.p.	m.p.	Aryloxyacetic acid, D-22	Bromo deriv- ative, D-23	1-Naphthyl- urethane, D-24
2-Chlorophenol	176		145	49 (m) 76 (di)	120
3-Methylphenol	202		—	84 (t)	128
2-Methylphenol	*192*	31	152	56 (di)	142
4-Methylphenol	*232*	36	135	49 (di) 108 (tet)	146

continued

Table 12 *Continued*

Compound	b.p.	m.p.	Aryloxyacetic acid, D-22	Bromo deriv- ative, D-23	1-Naphthyl- urethane, D-24
Phenol	*182*	42	99	95 (t)	133
4-Chlorophenol		43	156	33 (m)	166
				90 (di)	
2,4-Dichlorophenol		43	135 (141)	68	—
2-Nitrophenol		45	158	117 (di)	113
4-Ethylphenol		47	97	—	128
5-Methyl-2-isopropylphenol		50	149	55 (m)	160
3,4-Dimethylphenol		63	163	171 (t)	142
4-Bromophenol		64	157	95 (t)	169
3,5-Dichlorophenol		68	—	189 (t)	—
2,5-Dimethylphenol		75	118	178 (t)	173
1-Naphthol		94	194	105 (di)	152
3-Nitrophenol		97	156	91 (di)	167
4-*t*-Butylphenol		100	86	50 (m)	110
1,2-Dihydroxybenzene		105	—	193 (tet)	175
1,3-Dihydroxybenzene		110	195	112 (di)	206
4-Nitrophenol		114	187	142 (di)	150
2-Naphthol		123	154	84 (m)	157
1,2,3-Trihydroxybenzene		133	198	158 (di)	—
1,4-Dihydroxybenzene		172	250	186 (di)	—

Note: All temperatures are in °C

The Chemical Literature

APPENDIX VII

The literature of chemistry consists of *primary, secondary,* and *tertiary* sources. Most primary sources in chemistry contain descriptions of original research carried out by professional chemists. They include scientific periodicals such as the *Journal of Organic Chemistry,* patents, dissertations, technical reports, and government bulletins. Secondary sources contain material from the primary literature that has been systematically organized, condensed, or restated to make it more accessible and understandable to users. Secondary sources include most monographs, textbooks, dictionaries, encyclopedias, reference works, review publications, and abstracting journals dealing with chemistry. Tertiary sources are intended to aid users of the primary and secondary sources or to provide facts about chemists and their work. Tertiary sources include guides to the chemical literature, directories of scientists and scientific organizations, bibliographies, trade catalogs, and publications devoted to the financial and professional aspects of chemistry. Some sources may combine several different functions; for example, *Chemical & Engineering News* prints articles about chemical research as well as financial and professional information and thus serves as both a secondary and a tertiary source.

Although a few secondary sources critically evaluate primary material and correct errors before printing it, primary sources are generally used when it is important to obtain the most accurate and detailed information available on a topic. Errors can always occur when the material reappears in a secondary source, and important information may be left out. Because primary sources are not organized in any systematic way, it is generally necessary to refer to other sources to determine where the desired information can be found. The Bibliography following this Appendix lists a number of secondary and tertiary sources that can be used to obtain information directly, to gain access to the primary literature, or both. For example, *Beilstein's Handbuch der organischen Chemie* (hereafter referred to as *Beilstein*) gives detailed, reliable information about organic compounds and also provides citations to the literature in which the information was first reported. In this Appendix references to the Bibliography will be given in the form (B6) where the letter indicates a section of the bibliography and the number indicates a specific work in that section.

Using *Chemical Abstracts* and *Beilstein*

Most of the reference works cited in the Bibliography are limited in scope; they make no attempt to cover the entire field of chemistry or to list all of the known organic compounds. The two major works that do attempt that kind of coverage are *Beilstein* (A3) and *Chemical Abstracts* (J3). *Beilstein* summarizes all the important information published about specific compounds, but it is many years behind the current literature in most areas. The earlier volumes of *Beilstein* are available only in German, so at least a rudimentary knowledge of that language is necessary to make good use of this resource. *Chemical Abstracts* (*CA*) prints *abstracts* (brief summaries) of scientific papers, patents, and other printed material related to chemistry shortly after publication. The coverage of *CA* is enormous; as of 1997 it provided world wide information on more than 15 million chemical substances and on more than 13 million articles and patents.

Before using *Chemical Abstracts* for the first time, read the introduction that appears in issue 1 of each volume (two volumes are published each year), which describes the layout of the abstracts. Each abstract of a scientific paper (or other article) contains an abstract number, the title and author of the paper, a citation that tells where the original paper can be located, and a concise summary of the important information in the paper. The contents of *CA* can be searched by computer (as described in "Online Searches") or by consulting the indexes. Although each weekly issue of *CA* has its own indexes, the semiannual and collective indexes are far more useful for literature searches. A Collective Index is published every five years; prior to 1957, these indexes came out every 10 years. In 1972, the Subject Index was divided into two parts: the Chemical Substances Index and the General Subject Index. Author, formula, and patent indexes are also published, along with ancillary materials such as the Ring Systems Handbook, Registry Handbook, Index Guide, and Service Source Index, which are updated periodically.

The first thing you must do before searching *Chemical Abstracts* for information about a particular compound or subject is to find the *CA index name* of the compound or the *CA index heading* for the subject. In some cases, it may be possible to derive (or guess) the index name, but that may be difficult since *CA* follows its own rules of nomenclature, which often differ from IUPAC rules. The Index Guide, which now appears with each Collective Index and at intervals in between, gives cross-references from alternative names of substances to the *CA* index name. Thus the entry under "aniline" lists the index name benzeneamine, followed by the *CA registry number* [62-53-3]. The Index Guide does not list every compound indexed or give every synonym for the compounds it does list, so you may have to try different approaches to find what you are looking for. If you can't locate a specific compound, try looking up a possible parent (unsubstituted) compound under its trivial name; the index name of this parent compound should begin with a root name under which you will find your compound listed in the Chemical Substances Index (or the Subject Index prior to 1972). For example, suppose you are searching for the following compound:

$$CH_3O \overset{}{\underset{}{\bigcirc}} CH = CHC \overset{O}{\underset{\|}{}} \bigcirc$$

The unsubstituted compound (PhCH=CHCOPh) is known by such names as chalcone and benzalacetophenone. Looking up "chalcone" in a recent Index Guide provides the *CA* index name "2-propen-1-one, 1,3-diphenyl," so you will find the substituted compound listed in the Chemical Substances Index under the same root name, as "2-propen-1-one, 3-(4-methoxy)phenyl-1-phenyl." Keep in mind that the index name of a compound may change from time to time. This compound was listed under "chalcone, 4-methoxy" during the 8th Collective Index period (1967–1971) and before. Between the 8th and 9th Collective index periods, some major changes were made in the *CA* nomenclature rules, which now require rigorously systematic names for most chemical substances.

If you have trouble finding the *CA* index name for a compound using the Index Guide, you might look for it in another source such as the *Merck Index* (A4) or the *Dictionary of Organic Compounds* (A11). The Registry Handbook gives index names for the compounds indexed by Chemical Abstracts, so if you can locate the registry number for a compound in another source you can easily find its index name in the Registry Handbook. if you have a fairly good idea what the index name for a compound might be, you may also be able to locate it in the formula indexes.

Once you locate the index name in use during a particular index period, you will find abstracts listed under that name in the Chemical Substances Index or Subject Index for that period. Abstract citations in indexes from 1967 on are given in the form **80**:12175e, where the first number is the *CA* volume number and the second is the abstract number. In an index from prior to 1967, an abstract citation such as **51**:4321[b] refers to the volume and column number (there are two columns on each page) in which the abstract appears; the superscript (either a letter from "a" through "i" or a number from 1 through 9) indicates the location of the abstract in that column. The information you are looking for may appear in the abstract itself, or you may have to read the original article cited in the abstract. Citations of such articles now appear in the form *Tetrahedron. Lett.* **1996,** 37(37), 6767–6770 (Eng.), where

The letter e in this citation is a check-letter. If you looked up abstract number 12715 by mistake, you would find that its check-letter is b.

the abbreviated name of the publication appears first, followed by the date, volume and issue number, page numbers, and language in which the paper is written. (Earlier citations were given with the volume number first, followed by the pages and year.) The full name of the publication will be found in the *Chemical Abstracts Service Source Index* (CASSI), which also provides a brief publication history of each source and a list of libraries that carry it.

To carry out a thorough index search of *Chemical Abstracts,* it is best to start with the most recent Collective Index and all semiannual indexes published since then and work your way back through the previous collective indexes, using the index guides or other sources to locate the appropriate index names. Trivial names are used more frequently in the earlier indexes, and these can often be recognized in the formula indexes or located in other sources. If you are looking for information about a specific compound, you may find it easier to search *Beilstein* through the most recent supplemental series that lists the compound, and *Chemical Abstracts* from that time to the present. For further information about the use of *Chemical Abstracts,* refer to one or more of the literature guides in category K of the Bibliography.

Beilstein's Handbook of Organic Chemistry (*Handbuch der Organischen Chemie*) is by far the most comprehensive source of organized information about organic compounds. *Beilstein* provides information on the structure, characterization, natural occurrence, preparation, purification, energy parameters, physical properties, and chemical properties of organic compounds. It also cites the primary sources from which the information was obtained. *Beilstein,* unlike *Chemical Abstracts,* evaluates its sources critically and doesn't hesitate to correct errors from previous series. The fifth supplemental series, covering the period from 1960 to 1979, is now being published in an English-language edition. The basic series (*Hauptwerke*) and four previous supplemental series (*Erganzungswerke* I-IV, abbreviated E I-IV) are published only in German and cover the literature through 1959.

To obtain all the information about a particular compound in *Beilstein,* you must search the basic series and all the available supplemental series. The enormous size of this so-called "handbook"—along with the language barrier—could make that a fearsome prospect, but there is really no reason to be intimidated by *Beilstein.* The amount of German you need to know is quite limited and can be learned quickly with the help of the *Beilstein Dictionary* (A2), a slim dictionary written specifically for *Beilstein* users. And *Beilstein* is so well organized that it is not difficult to find the information you seek. If a compound has been around for some time, you can locate its *Beilstein* entries by the following procedure:

1 Write the molecular formula of the compound with C and H first, followed by other elements in alphabetical order.
2 Locate the formula in volume (*Band*) 29 of the second supplemental series (*General-Formelregister, Zweites Erganzungswerke*) and look for the name of the compound among those listed under that formula. Although the names are in German, many are similar or identical to the English names. (If you need help, use a German-English dictionary.)
3 Write down the volume number and pages on which information about the compound appears in the basic series (H) and the first and second supplemental series (E I and E II), and look up the appropri-

ate entries in those volumes. (Each volume may include several individually bound subvolumes.)

4 Once you know the index name of the compound and its page number in the basic series, you can locate its entry in the corresponding volume of any later series. Alternatively, you can look it up in the cumulative subject index for that volume.

Each compound is also assigned a system number, which can be used to locate its entries in the same way. For example, the notation for indigo ($C_{16}H_{10}N_2O_2$) in volume 29 of E II reads "Indigo **24,** 417, I 370, II 233." So you will find entries for indigo on page 417 of volume 24 in the basic series and pages 370 and 233 of volume 24 in the first and second supplemental series, respectively. The page number in the basic series, written as "**H,** 417," is called its coordinating reference; to find indigo's entry in volume 24 of a later series, you can locate the pages with **H,** 417 printed at the top and leaf through them until you find the entry for indigo. Knowing the E II index name of a compound may also help you locate it in the current cumulative subject index (*Sachsregister*) for the appropriate volume. The cumulative indexes for some volumes are combined; thus the listing for indigo is found in the volume 23–25 subject index, and it reads "**24** 417 d, I 370 d, II 233 e, IV 469." The letters refer to the location of an entry on the page; thus "II 233 e" means that information about indigo will be found under the fifth entry on page 233 of E II. There is no listing for E III because supplementary series III and IV were issued jointly for volumes 17–27. Entries in the joint series are designated by E IV rather than by E III/IV.

Locating entries for a compound such as adamantane, which does not appear in the E II formula indexes, may take a little more time. All *Beilstein* entries are organized according to a detailed system, and learning that system is the best way to get complete access to the information contained in this source. However, you can usually locate such entries by either (1) finding the volume number in which a structurally similar compound appears and searching the cumulative indexes of that volume or (2) locating a *Beilstein* reference from another source. The E II indexes indicate that cyclohexane appears in volume 5, which contains all cyclic compounds lacking functional groups, so you will find adamantane listed in the cumulative indexes for that volume. You can also find *Beilstein* references for many compounds in certain reference books (A1, A4, A8, and A15, for example). The entry for adamantane in the *CRC Handbook* gives the notation "B5[4], 469," referring to a *Beilstein* entry on page 469 in volume 5 of E IV. If you look up that entry you will find a back reference (E III 393) and a coordinating reference (**H,** 165) that will help you find information about adamantane in the other series.

indigo

adamantane

The Beilstein *system is described in the "Notes for Users" (in English) at the beginning of each volume in the recent supplemental series.*

Online Searches

Both *Chemical Abstracts* and *Beilstein* can be searched electronically using a variety of online search services. Limited *Chemical Abstracts* searching capabilities are available through FirstSearch CA Student Edition, with coverage of the most commonly held journals at academic libraries. More sophisticated search options are available using the on-line service STN (Scientific & Technical Information Network) International, which is operated by the Chemical Abstracts Service (CAS) and provides a variety of scientific and technical databases. The fundamental CAS database, called

CAplus, includes all entries from the printed *Chemical Abstracts* since 1970 (and some back to 1967), plus additional bibliographic information. Other CAS databases available via STN International include *CAOLD,* which includes abstracts from *CA* prior to 1967; the *REGISTRY* File, a list of approximately 12 million substances with their *CA* registry numbers; *CJACS,* which gives the complete texts of articles published in selected American Chemical Society Journals since 1982; *CASREACT,* a database of recent organic reactions; *CHEMSOURCES,* a database of information on chemical products and their suppliers; and *CHEMLIST,* a listing of regulated and other hazardous substances. STN databases can be accessed with any computer that is connected to a telecommunications network, and are often available through university libraries. A simplified search option called STN*Easy* provides access to a variety of STN databases via the World Wide Web (WWW), and features a graphical interface that doesn't require special training to use. Basic and advanced searches are available on STN*Easy,* and online help is provided when needed. For a basic search, the user simply enters a category that determines the databases to be searched, types in the words to be searched, and selects a search strategy. Search strategies include "any of these terms," which retrieves references that contain any or all of the words listed, and "all of these words," which only retrieves references that contain all of them.

Access to *Beilstein* is available through the *BEILSTEIN* online database, which is provided by STN International, DIALOG, and other vendors. The database is intended to cover not only the contents of the printed work, but also information from Beilstein file cards and primary literature up to the current date. For detailed information about online searching of *Beilstein* and *CA,* see references K2 and K4 in the Bibliography.

Using the Bibliography

A number of books, articles, and other literature sources in organic chemistry are listed in the following Bibliography under 12 general categories:

A. Reference Works
B. Organic Reactions and Syntheses
C. Laboratory Safety
D. General Laboratory Techniques
E. Chromatography
F. Spectrometry
G. Qualitative Organic Analysis
H. Reaction Mechanisms and Advanced Topics
J. Reports of Chemical Research
K. Guides to the Chemical Literature
L. Sources on Selected Topics
M. Software for Organic Chemistry

Each source is referred to here by the category letter and its number within the category.

Category A: Reference Works

While *Beilstein* attempts to provide all the important information about the millions of organic compound mentioned in the chemical literature, the following reference books provide selected information about a much smaller number of compounds, usually numbering in the tens of thousands. The *CRC Handbook of Chemistry and Physics* (A15) and *Lange's Handbook of Chemistry* (A8) tabulate physical properties and other data for many common organic compounds and contain a large amount of useful information about chemistry. *The Merck Index* (A4) is an excellent source of information on approximately 10,000 organic and inorganic compounds. It describes their uses and hazardous properties, provides detailed physical and structural data, and gives literature references for the isolation and synthesis of many compounds. The *CRC Handbook of Data on Organic Compounds (HODOC)* (A9) contains data and references to published spectra for more than 27,000 organic compounds. The *Aldrich Catalog* (A1) describes the chemicals manufactured by the Aldrich Chemical Company, and gives their physical properties, hazard warnings, procedures for safe disposal, references to published Aldrich spectra, and references to listings in the *Merck Index, Beilstein,* and *Fieser* (B9). The *Dictionary of Organic Compounds (DOC)* (A11) is an important multi volume set, updated by annual supplements, that gives structures, physical constants, hazard descriptions, sources, uses, derivatives, and bibliographic references for more than 145,000 organic compounds. It is also available on CD-ROM. Figure 1 on the next page shows the level of information provided by three of these reference works.

Before you use a reference work to find information about organic compounds, always read the introduction or explanatory material at the beginning of the work or preceding the table you intend to use. The introductory section of a reference work may (1) describe the content and organization of the material, (2) list symbols and abbreviations, and (3) describe the system of nomenclature used. Different sources often use very different naming systems. For example, the *Merck Index* emphasizes therapeutic uses of compounds, so it lists aspirin under that name, but in *Lange's Handbook* you will find aspirin listed as acetylsalicylic acid, and in the *CRC Handbook of Chemistry and Physics* as salicylic acid acetate. To locate most compounds in the *CRC Handbook,* you must also know that they are entered under the name of the parent compound; thus 2,4-dinitrobenzene is listed as "benzene, 2,4-dinitro." Often an index is the quickest and most reliable means of access to a given entry in a reference work. When using the *Merck Index* or the *Dictionary of Organic Compounds,* you should ordinarily consult the name index first to find the name under which a compound is listed. If you can't find the compound in the name index, you may be able to locate it in a formula index. Carbon and hydrogen are listed first in most formula indexes, followed by the other elements in alphabetical order.

The *Beilstein Dictionary* (A2) is an invaluable aid to understanding the parts of *Beilstein* (A3) that are available only in German. *The Organic Chemist's Desk Reference* (A12) includes a user's guide to the *Dictionary of Organic Chemistry* as well as a discussion of nomenclature in *Chemical Abstracts,* a list of reference works in organic chemistry, and other useful information. *The Chemist's Companion* (A6) and *The Chemist's Ready Reference Handbook* (A13) provide practical information about a variety of theoretical and experimental topics. *Organic Chemistry: An Alphabetical Guide* (A10) discusses and defines many terms used in organic chemistry. *Kirk-Othmer* (A7) is an excellent source of comprehensive and up-to-date articles on a variety of chemical topics. It offers particularly good coverage of industrial chemistry and commercial products, but also includes entries on natural materials such as coffee, terpenoids, and vitamins.

Category B: Organic Reactions and Syntheses

Most of the works in this category are intended to help chemists and chemistry students design and carry out organic syntheses. The *Guidebook to Organic Synthesis* (B14), and *Organic Synthesis: The Disconnection Approach* (B28) describe strategies for planning an organic synthesis. At a more advanced level, *The Logic of Chemical Synthesis* (B8) deals with the analysis of complex synthetic problems, while *Principles of Organic Synthesis* (B17) discusses such topics as thermodynamics, kinetics, and stereochemistry as they apply to organic synthesis.

A number of works can help the experimenter select the type of reaction that will best accomplish a given synthetic transformation. *Synthetic*

Figure 1 Entries for benzoic acid from some reference works

No.	Name	Formula	Formula weight	Beilstein reference	Density	Refractive index	Melting point	Boiling point	Flash point	Solubility in 100 parts solvent
b55	Benzoic acid	C₆H₅COOH	122.12	9, 92	1.080		122.4	133^{10mm}	121	0.29 aq; 43 alc; 10 bz; 22 chl; 33 eth, acet

A. *Lange's Handbook of chemistry.* (*Reprinted with permission from Lange's Handbook of Chemistry,* 13th ed., by N. A. Lange, edited by J. A. Dean. Copyright McGraw-Hill, Inc., New York, 1985.)

1093. Benzoic Acid. Benzenecarboxylic acid; phenylformic acid; dracylic acid. C₇H₆O₂; mol wt 122.12. C 68.84%, H 4.95%, O 26.20%. Occurs in nature in free and combined forms. Gum benzoin may contain as much as 20%. Most berries contain appreciable amounts (around 0.05%). Excreted mainly as hippuric acid by almost all vertebrates, except fowl. Mfg processes include the air oxidation of toluene, the hydrolysis of benzotrichloride, and the decarboxylation of phthalic anhydride: Faith, Keyes & Clark's *Industrial Chemicals,* F. A. Lowenheim, M. K. Moran, Eds. (Wiley-Interscience, New York, 4th ed., 1975) pp 138-144. Lab prepn from benzyl chloride: A. I. Vogel, *Practical Organic Chemistry* (Longmans, Londona, 3rd ed., 1959) p 755; from benzaldehyde: Gattermann-Wieland, *Praxis des organischen Chemikers* (de Gruyter, Berlin, 40th ed, 1961) p 193. Prepn of ultra-pure benzoic acid for use as titrimetric and calorimetric standard: Schwab, Wicher, *J. Res. Nat. Bur. Standards* **25,** 747 (1940). *Review:* A. E. Williams in Kirk-Othmer *Encyclopedia of Chemical Technology* vol. 3 (Wiley-Interscience, New York, 3rd ed., 1978) pp 778-792.

Monoclinic tablets, plates, leaflets. d 1.321 (also reported as 1.266). mp 122.4°. Begins to sublime at around 100°. bp₇₆₀ 249.2°; bp₄₀₀ 227°; bp₂₀₀ 205.8°; bp₁₀₀ 186.2°; bp₆₀ 172.8°; bp₄₀ 162.6°; bp₂₀ 146.7°; bp₁₀ 132.1°. Volatile with steam. Flash pt 121-131°. K at 25°: 6.40 × 10⁻⁵; pH of satd soln at 25°: 2.8. Soly in water (g/l) at 0° = 1.7; at 10° = 2.1; at 20° = 2.9; at 25° = 3.4; at 40° = 6.0; at 50° = 9.5; at 60° = 12.0; at 70° = 17.7; at 80° = 27.5; at 90° = 45.5; at 95° = 68.0. Mixtures of excess benzoic acid and water form two liquid phases beginning at 89.7°. The two liquid phases unite at the critical soln temp of 117.2°. Composition of critical mixture: 32.34% benzoic acid, 67.66% water: see Ward, Cooper, *J. Phys. Chem.* **34,** 1484

(1930). One gram dissolves in 2.3 ml cold alc, 1.5 ml boiling alc, 4.5 ml chloroform, 3 ml ether, 3 ml acetone, 30 ml carbon tetrachloride, 10 ml benzene, 30 ml carbon disulfide, 23 ml oil of turpentine; also sol in volatile and fixed oils, slightly in petr ether. The soly in water is increased by alkaline substances, such as borax or trisodium phosphate, *see also* Sodium Benzoate.

Barium salt dihydrate, C₁₄H₁₀BaO₄.2H₂O, *barium benzoate.* Nacreous leaflets. *Poisonous!* Soluble in about 20 parts water; slightly sol in alc.

Calcium salt trihydrate, C₁₄H₁₀CaO₄.3H₂O, *calcium benzoate.* Orthorhombic crystals or powder. d 1.44. Soluble in 25 parts water; very sol in boiling water.

Cerium salt trihydrate, C₂₁H₁₅CeO₆.3H₂O, *cerous benzoate.* White to reddish-white powder. Sol in hot water or hot alc.

Copper salt dihydrate, C₁₄H₁₀CuO₄.2H₂O, *cupric benzoate.* Light blue, cryst powder. Slightly soluble in cold water, more in hot water; sol in alc or in dil acids with separation of benzoic acid.

Lead salt dihydrate, C₁₄H₁₀O₄Pb.2H₂O, *lead benzoate.* Cryst powder. *Poisonous!* Slightly sol in water.

Manganese salt tetrahydrate, C₁₄H₁₀MnO₄.4H₂O, *manganese benzoate.* Pale-red powder. Sol in water, alc. Also occurs with 3H₂O.

Nickel salt trihydrate, C₁₄H₁₀NiO₄.3H₂O, *nickel benzoate.* Light-green odorless powder. Slightly sol in water; sol in ammonia; dec by acids.

Potassium salt trihydrate, C₇H₅KO₂.3H₂O, *potassium benzoate.* Cryst powder. Sol in water, alc.

Silver salt, C₇H₅AgO₂, *silver benzoate.* Light-sensitive powder. Sol in 385 parts cold water, more sol in hot water; very slightly sol in alc.

Uranium salt, C₁₄H₁₀O₆U, *uranium benzoate, uranyl benzoate.* Yellow powder. Slightly sol in water, alc.

Toxicity: Mild irritant to skin, eyes, mucous membranes.

USE: Preserving foods, fats, fruit juices, alkaloidal solns, etc; manuf benzoates and benzoyl compds, dyes; as a mordant in calico printing; for curing tobacco. As standard in volumetric and calorimetric analysis.

THERAP CAT: Pharmaceutic aid (antifungal agent).

THERAP CAT (VET): Has been used with salicylic acid as a topical antifungal.

B. *The Merck Index,* 10th ed. (Reprinted with permission from The Merck Index, 10th ed., edited by M. Windholtz. Copyright Merck and Co., Inc., Rahway, N.J., 1983.)

Benzoic acid, ꭗCl   **B-00378**
[65-85-0]

PhCOOH

$C_7H_6O_2$ M 122

Preservative in the food industry. Used in manuf. of preservatives, plasticisers, alkyd resin coatings and caprolactam. Leaflets or needles (H_2O). V. spar. sol. H_2O. Mp 122° (ca. 100° subl.). Bp 249°, Bp_{10} 133°. pK_a 4.2 (25°). Steam-volatile.

▷Toxic by skin absorption. DG0875000.

Me ester: [93-58-3]. Used in perfumery and flavourings. Liq. d_{23}^{23} 1.09. Fp −12.3°. Bp 199.6°, Bp_{24} 96-8°.

▷DH3850000.

Et ester: [93-89-0]. Polymerisation catalyst. Used in perfumery and flavourings. Liq. d_4^{25} 1.04. Fp −34°. Bp 212.9°, Bp_{10} 87.2°.

▷Mod. toxic. DH0200000.

Propyl ester: [2315-68-6]. Flavour ingredient. d_{15}^{15} 1.03: Bp 230°.

Isopropyl ester: [939-48-0]. Polymerisation catalyst, flavour ingredient. d_{15}^{15} 1.02. Bp 218-9°.

▷DH3150000.

Butyl ester: [136-60-7]. Dye carrier; used in perfumery. d_{15}^{15} 1.01. Bp 248-9°.

▷DG4925000.

tert-Butyl ester: [774-65-2]. Bp_2 96°.

Benzyl ester: [120-51-4]. *Benzyl benzoate.* Contained in Peru balsam. Tobacco flavouring agent, insect repellant component. Leaflets. d^{18} 1.11. Mp 21°. Bp 323-4° (316-7°).

▷Mod. toxic. DG4200000.

Fluoride: [455-32-3]. *Benzoyl fluoride.* Fuming liq. Bp 159-61°. Hydrolysed by hot H_2O.

Chloride: [98-88-4]. *Benzoyl chloride.* Polymerisation catalyst, benzoylating agent. Can be used for synth. of aliphatic acid chlorides. Fuming liq. d_{15}^{15} 1.22. Fp −1°. Bp 197°.

▷Highly irritant, causes burns, violent reaction with dimethyl sulphoxide. DM6600000.

Bromide: [618-32-6]. *Benzoyl bromide.* Fuming liq. d^{15} 1.57. Fp −24°. Bp 218-9°, $Bp_{0.05}$ 48-50°.

Iodide: [618-38-2]. *Benzoyl iodide.* Needles. Mp 3°. Bp_{20} 128°.

Anhydride: [93-97-0]. *Benzoic anhydride.* Cross-linking agent for polymers. Acylation and decarboxylating agent, can be used in polymer-linked form. Rhombic prisms. d_4^{15} 1.99. Mp 42°. Bp 360°.

▷Mild irritant and allergen

Amide: see Benzamide, B-00140 [55-21-0].

▷CU8700000.

Ethylamide: [614-17-5]. Needles (H_2O). Mp 70-1°. Bp 298-300°.

▷CV4920000.

Anilide: Benzanilide. Benzoylaniline. Leaflets. Mp 163°. Bp_{10} 117-9°.

Toluidide: see 4-Methylaniline, M-00814

Xylidide: see 2,5-Dimethylaniline, D-05527

N-Chloroanilide: [5014-47-1]. *N-Chlorobenzanilide.* Needles (ligroin). Mp 81.5-82°.

Nitrile: [100-47-0]. *Phenyl cyanide. Cyanobenzene.* d_{15}^{15} 1.01. Fp −13°. Bp 190.7°, Bp_{10} 69°.

▷DI2450000.

Hydrazide: Plates. Mp 112.5°.

Hydroxamate: see N-Hydroxybenzamide, H-01254

Azide: Benzazide. Benzoylazimide. Plates. Mp 32°.

▷Explodes on heating

Org. Synth., Coll.Vol., 1, 75, 361 (*synth, deriv*)

Jesson, J.P. *et al, Proc. R. Soc. London, Ser. A,* 1962, **268**, 68 (*raman*)

Beynon, J.H. *et al, Z. Naturforsch., A,* 1965, **20**, 883 (*ms*)

Evans, H.B. *et al, J. Phys. Chem.,* 1968, **72**, 2552 (*pmr*)

Fauvet, G. *et al, Acta Crystallogr., Sect. B,* 1978, **34**, 1376 (*cryst struct, nitrile*)

Fieser, M. *et al, Reagents for Organic Synthesis,* Wiley, 1967-78, **1**, 49, 1004; **5**, 23, 24

Bretherick, L., *Handbook of Reactive Chemical Hazards,* 2nd Ed., Butterworths, London and Boston, 1979, 394, 622

Sax, N.I., *Dangerous Properties of Industrial Materials,* 5th Ed., Van Nostrand-Reinhold, 1979, 407, 408, 410, 651

Hazards in the Chemical Laboratory, (Bretherick, L., Ed.), 3rd Ed., Royal Society of Chemistry, London, 1981, 193

Figure 1 *Continued*

C. *Dictionary of Organic Compunds.* (Reprinted with permission from *Dictionary of Organic Compounds,* 5th ed., edited by J.R. A. Pollock and R. Stevens. Copyright Chapman & Hall, London, 1982. The 5th edition is updated by annual supplements that contain both revised versions of existing entries and completely new entries.)

Organic Chemistry (B32) and *Modern Synthetic Reactions* (B12) survey many important synthetic reactions, with references to the earlier literature. More comprehensive coverage of organic transformations, including the more recent synthetic reactions, can be found in *Some Modern Methods of Organic Synthesis* (B6), *Compendium of Organic Synthetic Methods* (B11), *Comprehensive Organic Transformations* (B13), *Comprehensive Organic Synthesis* (B31), and *Theilheimer's Synthetic Methods of Organic Chemistry* (B30). Reaction files from Theilheimer and other works can be searched online via the *REACCS* database from Molecular Design, Ltd.

General works on organic synthetic reactions vary from the *Reaction Guide for Organic Chemistry* (B15), a basic compilation of most reactions covered in the sophomore level organic chemistry course, to *Organic Reactions* (B19), a multivolume set containing very comprehensive monographs on a large variety of reactions. *Organic Reactions* describes each reaction's mechanism, scope, limitations, and experimental conditions. It also provides several detailed experimental procedures and a table that lists examples of each reaction with references to the original sources. *Name Reactions and Reagents in Organic Synthesis* (B16) describes those synthetic reactions that are identified by the names of one or more discoverers, such as the Dieckmann condensation. *Comprehensive Organic Chemistry* (B4) and *Rodd's Chemistry of Carbon Compounds* (B7) describe the reactions of different classes of organic compounds in considerable depth. Each volume of *Chemistry of Functional Groups* (B22) covers the chemistry of a different functional group. *Asymmetric Synthesis* (B3) deals with the synthesis of chiral compounds and *Stereoselective Synthesis* (B1) with the use of stereoselective reactions in organic synthesis. Other books cover only one type of synthetic reaction, such as cycloaddition (B5) or nitration (B18); a search of your library's catalog should reveal many similar works on specific reactions. *Protective Groups in Organic Synthesis* (B10) describes the use of protective groups in syntheses involving multifunctional reactants. In addition to the works dedicated to organic reactions, several advanced textbooks such as March (H5) and Carey/Sundberg (H1) describe the most important synthetic reactions and give literature references to specific synthetic procedures.

Two multivolume sources of information about chemical reagents are Fieser's *Reagents for Organic Synthesis* (B9) and the *Encyclopedia of Reagents for Organic Synthesis* (B21). Fieser provides information about the preparation, purification, handling, and hazards of many chemical reagents, as well as examples of their use, with literature citations. Only the individual volumes are indexed, but if you locate the entry for a reagent in a recent volume, it will provide back references to the previous volumes. The Encyclopedia of Reagents reviews nearly 3500 reagents, listed alphabetically, and gives a critical assessment of each reagent. *Borane Reagents* (B23) covers the applications of boranes in organic synthesis. *Organic Solvents* (B24) describes physical properties and purification methods for many solvents used in organic synthesis.

Although the primary chemical literature is the most important source of experimental procedures, a number of secondary sources (in addition to *Organic Reactions*) provide relatively detailed, reliable procedures. *Organic Syntheses* (B20) is a continuing series that contains an excellent selection of carefully tested synthetic procedures. Single-volume works that give experimental procedures include *Organicum* (B2), three works by Sandler and Karo (B25, B26, B27), and *Vogel's Textbook of Practical Organic Chemistry* (B29). Vogel is also an excellent source of information about laboratory techniques. *Houben-Weyl* (D4), which is listed under General Laboratory Techniques, includes many synthetic procedures (in German) as well.

Category C: Laboratory Safety

Working Safely with Chemicals (C2) is a short booklet about laboratory safety written for students. *Prudent Practices in the Laboratory* (C7) is an authoritative guide to safe laboratory practices including procedures for the safe handling and disposal of chemicals. *Hazards in the Chemical Laboratory* (C6) describes the toxic effects of hazardous substances and reviews recent developments in the safe design and operation of chemical laboratories. The *CRC Handbook of Laboratory Safety* (C1) deals with the recognition and control of hazards and compliance with safety regulations, and includes a chapter on responding to laboratory emergencies. The *First Aid Manual for Chemical Accidents* (C3) gives first aid procedures for accidents caused by specific chemicals and classes of chemicals. The *Sigma/Aldrich Library of Chemical Safety Data* (C4) and *Sax's Dangerous Properties of Industrial Materials* (C5) provide detailed health and safety data for many common chemicals. *Bretherick's Handbook of Reactive Chemical Hazards* (C8) describes the properties of chemicals that are hazardous by virtue of their instability or their tendency to react with other chemicals. The 9th edition of *The Merck Index* (A4) contains a section on first aid for poisoning and chemical burns and a listing of poison control centers in the United States; this section is not included in more recent editions, however.

Category D: General Laboratory Techniques

The Organic Chem Lab Survival Manual (D8) describes many of the lab techniques used by organic chemistry students and tells you what things *not* to do, such as plugging a heating mantle directly into a wall socket. *A Guide for the Perplexed Organic Experimentalist* (D3) deals with the practical aspects of laboratory work for anyone intending to do research in organic chemistry. *Advanced Practical Organic Chemistry* (D2) covers up-to-date laboratory techniques used by advanced students and professional organic chemists. Weissberger's *Technique of Organic Chemistry* (D6) and the more recent *Techniques of Chemistry* (D7) are multivolume sets that contain information about a wide variety of experimental methods. Information on classical laboratory techniques such as distillation and recrystallization can be found in Volume I of reference D6, which is subtitled *Physical Methods of Organic Chemistry*. The first four volumes of *Houben-Weyl* (D4) describe many laboratory methods for organic chemistry, in German. *Purification of Laboratory Chemicals* (D5) includes methods for the purification of more than 4000 common chemicals. *Natural Products* (D1) describes laboratory techniques and gives specific procedures for the isolation and structure determination of natural products.

Category E: Chromatography

References E5–E7 are good general sources of information on the theory and practice of all types of chromatography. *Gas Chromatography* (E1) and *High Performance Liquid Chromatography* (E3) are "open learning" texts designed for self-study. The remaining works provide up-to-date coverage of GC, HPLC, and TLC techniques and applications. *Principles of Instrumental Analysis* (F21) in the next section has chapters on instrumental chromatographic methods.

Category F: Spectrometry

Principles of Instrumental Analysis (F21) is an excellent source of information about the principles and applications of all important kinds of spectrometric methods. An article in the *Journal of Chemical Education* (F9) covers the basics of IR and NMR spectral interpretation. The works by Silverstein (F20), Feinstein (F5), Kemp (F11), and Pavia (F14) are good one-volume introductions to the interpretation of IR, NMR, MS, and UV-VIS spectra. More comprehensive coverage of specific spectrometric methods is provided for infrared spectrometry by references F4, F6 and F22; for nuclear magnetic resonance spectrometry by F1, F2, and F7; for mass spectrometry by F3, F8, F10, and F13; and for ultraviolet-visible spectrometry by F15. The *Sadtler Standard Spectra* (F19) consists of a large number of IR, NMR, and UV-VIS spectra in ring binders; although they are not arranged systematically, individual spectra can be located by using the index volumes. Spectra in the Aldrich collections (F16–F18) are arranged by functional classes and in order of increasing molecular complexity within a functional class, making it possible to observe the effect of various structural features on the spectra. The *Registry of Mass Spectral Data* (F12) contains data on 120,000 mass spectra, both in printed form and on a database.

Category G: Qualitative Organic Analysis

Organic Structure Determination (G3), *The Systematic Identification of Organic Compounds* (G5), and *Spectral and Chemical Characterization of Organic Compounds* (G1) cover the traditional "wet chemistry" methods for identifying organic compounds, but include chapters on spectrometric methods as well. *Qualitative Organic Analysis* (G2) emphasizes spectral methods of identification. The *CRC Handbook of Tables for Organic Compound Identification* (G4) lists the properties and derivative melting points for many organic compounds in the most important functional classes.

Category H: Reaction Mechanisms and Advanced Topics

Electron Flow in Organic Chemisty (H7) teaches an intuitive approach to organic chemistry by breaking down reaction mechanisms into elementary electron-flow pathways. *A Guidebook to Organic Reaction Mechanisms* (H8) by Sykes is an excellent survey of organic reaction mechanisms suitable for advanced students; his *Primer* (H9) is a more basic introduction to mechanisms based on a simplified classification scheme. *Mechanism and Theory in Organic Chemistry* (H4) and *Perspectives on Structure and Mechanism in Organic Chemistry* (H3) are advanced textbooks that present the theoretical aspects of organic chemistry and provide current information about important reaction mechanisms. The *Advanced Organic Chemistry* textbooks by Carey/Sundberg (H1) and March (H5) provide good coverage of the mechanisms and synthetic applications of a large number of organic reactions, giving numerous references to the primary literature. *Determination of Organic Reaction Mechanisms* (H2) describes experimental techniques for studying reaction mechanisms, and *Organic Reaction Mechanisms* (H6) is an annual survey of recent developments in the field.

Category J: Reports of Chemical Research

Chemical Abstracts (J3), previously described in detail, is the most comprehensive single source of information about research in chemistry. The *Science Citation Index* (J7) is an index of literature citations to papers, patents, and books published in the past. For example, if you find an interesting paper by Linus Pauling in a chemistry journal, you can look up the paper in the citation index to find later articles that were based, in part, on Pauling's original paper. In this way you can sometimes trace the development of an idea or a method from its origin to the present day. A similar index for chemistry, the *Chemistry Citation Index,* is available on CD-ROM. *Chemical Titles* (J4) and *Current Contents* (J5) reproduce the current tables of contents of the most important chemistry journals, to inform chemists quickly of recent research in their fields. *Index Chemicus* (J6), a weekly guide to new organic compounds and their chemistry, is available in print and on a searchable database. The other works in this category (J1, J2) provide annual summaries or reviews of research in organic chemistry.

Category K. Guides to the Chemical Literature

Information Sources in Chemistry (K1), *How to Find Chemical Information* (K3), and *Chemical Information* (K7) are general guides to the chemical literature. *The Beilstein Guide* (K6) and *The Beilstein Online Database* (K2) describe how to search and use *Beilstein's* print and online versions, respectively. *From CA to CAS Online* (K4) serves the same function for *Chemical Abstracts*. The three articles by Somerville (K5) describe the contents and uses of some major works on organic reactions and syntheses. A brief but useful guide to information sources for organic chemistry can also be found in March (H5), pp. 1239–1268.

Category L: Sources on Selected Topics

This category includes a number of books and articles that can be used as resources for library research papers or simply read for enjoyment and enlightenment. Most are about topics related to the experiments or minilabs, such as reference L10 on the birth control pill (see Experiment 24), L15 on sweetness (see Experiment 20), and L17 on indigo (see Minilab 42). Other references are listed here because they don't fit into any of the other categories. For example, L11 and L35 are guides for writing scientific papers and laboratory notebooks, respectively, and L20 describes organic nomenclature in depth.

Category M: Software for Organic Chemistry.

A number of software titles are designed to be used in preparation for or during an organic chemistry laboratory. *Organic Chemistry Laboratory* (M7) is a set of interactive tutorials intended to help students learn about selected lab techniques including distillation, extraction, melting point determination, and qualitative organic analysis. *SQUALOR* and *MacSQUALOR* (M5) allow the user to identify simulated unknowns for qualitative organic analysis. *Introduction to Spectroscopy* (M2) and *Organic Chemistry: Spectra of Compounds* (M6) help the user analyze and interpret spectral data. *Proton NMR Spectrum Simulator* (M1) generates a simulated NMR spectrum from a molecular structure entered by the user. *MassSpec* (M3) helps the user identify the structure fragments that correspond to peaks on a mass spectrum. *SynTree* (M4) helps the user work out retrosynthetic pathways leading from a selected target compound back to a readily available starting material.

Bibliography

A. Reference Works

1. *Aldrich Catalog Handbook of Fine Chemicals*. Milwaukee, WI: Aldrich Chemical Co., 1996–97 (and other years).
2. *Beilstein Dictionary: German-English: For the Users of the Beilstein Handbook of Organic Chemistry*. Ft. Worth, TX: Saunders, 1992.
3. *Beilstein's Handbook of Organic Chemistry*. New York: Springer-Verlag, 1918 to date.
4. Budavari, S., ed., *The Merck Index: An Encyclopedia of Chemicals, Drugs, and Biologicals*, 12th ed. Rahway, NJ.: Merck and Co., 1996.
5. *Chemical Abstracts Ring Systems Handbook*. Washington DC: American Chemical Society, 1993.
6. Gordon, A. J., and Ford, R. A., *The Chemist's Companion*. New York: Wiley-Interscience, 1972.
7. *Kirk-Othmer Encyclopedia of Chemical Technology*, 4th ed. New York: Wiley, 1993–.
8. Lange, N. A., *Lange's Handbook of Chemistry*, 14th ed., ed. by J. A. Dean. New York: McGraw-Hill, 1992.
9. Lide, D. R., and Milne, G. W. A., eds., *CRC Handbook of Data on Organic Compounds,* 3rd ed. Boca Raton, FL: CRC Press, 1994.
10. Mundy, B. P., and Ellerd, M. G., *Organic Chemistry: An Alphabetical Guide*. New York: Wiley, 1996.
11. Rhodes, P. H., ed., *Dictionary of Organic Compounds,* 6th ed. London: Chapman and Hall, 1995.
12. Rhodes, P. H., *The Organic Chemist's Desk Reference: A Companion Volume to the Dictionary of Organic Compounds,* 6th ed. London: Chapman and Hall, 1995.
13. Shugar, G. J., and Dean, J. A., *The Chemist's Ready Reference Handbook*. New York: McGraw-Hill, 1990.
14. Weast, R. C., and Astle, M. J., eds., *CRC Handbook of Data on Organic Compounds*. Boca Raton, FL: CRC Press, 1985.
15. Weast, R. C., ed., *CRC Handbook of Chemistry and Physics*, 78th ed. (and other editions). Boca Raton, FL: CRC Press, 1997.

B. Organic Reactions and Syntheses

1. Atkinson, R. S., *Stereoselective Synthesis*. New York: Wiley, 1995
2. Becker, H. et al., *Organicum: Practical Handbook of Organic Chemistry*, trans. by B. J. Hazzard. Reading, MA: Addison-Wesley Publishing Co., 1973.
3. Aitken, R.A. and Kilényi, eds., *Asymmetric Synthesis*. Glasgow, UK: Blackie, 1992.
4. Barton, D. H., and Ollis, W. D., eds., *Comprehensive Organic Chemistry: The Synthesis and Reactions of Organic Compounds*. New York: Pergamon Press, 1979.
5. Carruthers, W., *Cycloaddition Reactions in Organic Synthesis*. New York: Pergamon Press, 1990.
6. Carruthers, W., *Some Modern Methods of Organic Synthesis,* 3rd ed. New York: Cambridge, 1987.
7. Coffey, S. (1964–1989), Ansell, M. F. (1973–), Sainsbury (1991–), eds., *Rodd's Chemistry of Carbon Compounds,* 2nd ed. and supplements. New York: Elsevier, 1964–.
8. Corey, E. J. and Cheng, X.-M., *The Logic of Chemical Synthesis*. New York: Wiley, 1989.
9. Fieser, L. F., and Fieser, M. (1967–1986) Fieser, M., and Smith, J. G. (1988–), *Reagents for Organic Synthesis*. New York: Wiley, 1967–.
10. Greene, T. W. and Wuts, P.G.M., *Protective Groups in Organic Synthesis*, 2nd ed. New York: Wiley, 1991.
11. Harrison, I. T., and Harrison, S., *Compendium of Organic Synthetic Methods*. New York: Wiley, 1971–1977.
12. House, H. O., *Modern Synthetic Reactions,* 2nd ed. Menlo Park, CA: Benjamin, 1972.
13. Larock, R. C., *Comprehensive Organic Transformations: A Guide to Functional Group Preparations*. New York: VCH, 1989.
14. Mackie, R. K., et al., *Guidebook to Organic Synthesis*, 2nd ed. New York: Wiley, 1990.
15. Millam, M. J., *Reaction Guide for Organic Chemistry*. Lexington, MA: D. C. Heath, 1989.
16. Mundy, B. P., and Ellerd, M. G., *Name Reactions and Reagents in Organic Synthesis*. New York: Wiley, 1988.
17. Norman, R. O. C., and Coxon, J. M., *Principles of Organic Synthesis*, 3rd ed. Glasgow, UK: Blackie, 1993.
18. Olah, G. A., *Nitration: Methods and Mechanisms*. New York: VCH, 1989.
19. *Organic Reactions*. New York: Wiley, 1942–.
20. *Organic Syntheses, Collective Volumes*. New York: Wiley, 1941–.
21. Paquette, L. A., ed., *Encyclopedia of Reagents for Organic Synthesis*. New York: Wiley, 1995.
22. Patai, S., ed., *Chemistry of Functional Groups*. New York: Wiley, 1964–.
23. Pelter, A., Smith, K., and Brown, H.C., *Borane Reagents*. London: Academic Press, 1988.
24. Riddick, J. A., and Bunger, W. M*., Organic Solvents: Physical Properties and Methods of Purification,* 4th ed. New York: Wiley, 1986.
25. Sandler, S. R., and Karo, W., *Organic Functional Group Preparations*, 2nd ed. Orlando, FL: Academic Press, 1983, 1986, 1989.
26. Sandler, S. R., and Karo, W., *Polymer Syntheses*, 2nd ed. Orlando, FL: Academic Press, 1992, 1993.
27. Sandler, S. R., and Karo, W., *Sourcebook of Advanced Organic Laboratory Preparations*. San Diego: Academic Press, 1992.
28. Stuart, W., *Organic Synthesis: The Disconnection Approach*. New York: Wiley, 1982.
29. Tatchell, A. R., et al., *Vogel's Textbook of Practical Organic Chemistry*, 5th ed. New York: Wiley, 1989.
30. Theilheimer, W. (1948–81), Finch, A. F. (1982–), eds., *Theilheimer's Synthetic Methods of Organic Chemistry*. Basel, NY: Karger, 1946–.
31. Trost, B. M. et al., ed., *Comprehensive Organic Synthesis: Selectivity, Strategy & Efficiency in Modern Organic Chemistry*. Elmsford, NY: Pergamon Press, 1991.
32. Wagner, R. B., and Zook, H. D., *Synthetic Organic Chemistry*. New York: Wiley, 1953.

C. Laboratory Safety

1. Furr, A. K., ed., *CRC Handbook of Laboratory Safety*, 4th ed. Boca Raton, FL: CRC Press, 1995.
2. Gorman, C. E., ed., *Working Safely with Chemicals in the Laboratory,* 2nd ed. Schnectady, NY: Genium, 1995.
3. Lefèvre, M. J., *First Aid Manual for Chemical Accidents,* 2nd ed. New York: Van Nostrand Reinhold, 1989.
4. Lenga, R. E., ed., *The Sigma-Aldrich Library of Chemical Safety Data*, 2nd ed. Milwaukee, WI: Sigma-Aldrich, 1988.
5. Lewis, R. J., Sr., *Sax's Dangerous Properties of Industrial Materials*, 8th ed. New York: Van Nostrand Reinhold, 1992.
6. Luxon, S. G., ed., *Hazards in the Chemical Laboratory*, 5th ed. Cambridge: Royal Society of Chemistry, 1992.
7. National Research Council, *Prudent Practices in the Laboratory: Handling and Disposal of Chemicals*. Washington DC: National Academy Press, 1995.
8. Urben, P. G., ed., *Bretherick's Handbook of Reactive Chemical Hazards*. Stoneham, MA: Butterworths, 1995.

D. General Laboratory Techniques

1. Ikan, R., *Natural Products: A Laboratory Guide,* 2nd ed. San Diego: Academic Press, 1991.
2. Leonard, J., et al., *Advanced Practical Organic Chemistry*, 2nd ed. New York: Chapman and Hall, 1994.
3. Loewenthal, H. J. E., *Guide for the Perplexed Organic Experimentalist*, 2nd ed. New York: Wiley, 1992.
4. *Methoden der Organischen Chemie, Houben-Weyl,* 4th ed. Stuttgart: Georg Thieme, 1952–.
5. Perrin, D. D., and Armarego, W. L. F, *Purification of Laboratory Chemicals*, 3rd ed. New York: Pergamon Press, 1988.
6. Weissberger, A., ed., *Technique of Organic Chemistry*, 3rd ed. New York: Wiley, 1959–.
7. Weissberger, A., ed., *Techniques of Chemistry*. New York: Wiley, 1971–.
8. Zubrick, J. W., *The Organic Chem Lab Survival Manual: A Student's Guide to Techniques*, 3rd ed. New York: Wiley, 1992.

E. Chromatography

1. Fowlis, I. A., *Gas Chromatography,* 2nd ed. New York: Wiley, 1995.
2. Grob, R. L., ed., *Modern Practice of Gas Chromatography,* 3rd ed. New York: Wiley, 1995.
3. Lindsay, S., *High Performance Liquid Chromatography,* 2nd ed. New York: Wiley, 1992.
4. Meyer, V. R., *Practical High-Performance Liquid Chromatography*. New York: Wiley, 1994.
5. Miller, J. M., *Chromatography: Concepts and Contrasts*. New York: Wiley, 1988.
6. Poole, C. F., and Poole, S. K., *Chromatography Today*. New York: Elsevier, 1991.
7. Ravindranath, B., *Principles and Practice of Chromatography*. New York: Halsted, 1989.
8. Schomburg, G., *Gas Chromatography: A Practical Course*. New York: VCH, 1990.
9. Touchstone, J. C., *Practice of Thin Layer Chromatography*, 3rd ed. New York: Wiley, 1992.

F. Spectrometry

1. Bovey, F. A., *Nuclear Magnetic Resonance Spectroscopy,* 2nd ed. San Diego: Academic Press, 1988.
2. Breitmaier, E. *Structure Elucidation by NMR in Organic Chemistry: A Practical Guide*. New York: Wiley, 1993.
3. Chapman, J. R., *Practical Organic Mass Spectrometry: A Guide for Chemical and Biochemical Analysis*. New York: Wiley, 1993.
4. Colthup, N. B., Daly, L. H., and Wiberley, S. E., *Introduction to Infrared and Raman Spectroscopy*, 3rd ed. Orlando, FL: Academic Press, 1990.
5. Feinstein, K., *Guide to Spectroscopic Identification of Organic Compounds*. Boca Raton, FL: CRC Press, 1995.
6. George, W. O., and McIntyre, P. S., *Infrared Spectroscopy,* New York: Wiley, 1987.
7. Günther, H., *NMR Spectroscopy: Basic Principles, Concepts, and Applications in Chemistry*, 2nd ed. New York: Wiley, 1995.
8. Hoffmann, Edmond de, *Mass Spectrometry: Principles and Applications*. New York: Wiley, 1996.
9. Ingham, A. M., and Henson, R. C., "Interpreting Infrared and Nuclear Magnetic Resonance Spectra of Simple Organic Compounds for the Beginner." *J. Chem. Educ.*, **1984**, *61*, 704.
10. Johnstone, R. A. W., and Rose, M. E., *Mass Spectrometry for Chemists and Biochemists,* 2nd ed. New York: Cambridge, 1996.
11. Kemp, W., *Organic Spectroscopy,* 3rd ed. New York: W. H. Freeman, 1991.
12. McLafferty, F. W., and Stauffer, D. B., eds., *The Wiley/NBS Registry of Mass Spectral Data*. New York: Wiley, 1989.
13. McLafferty, F. W., and Turecek, F., *Interpretation of Mass Spectra*, 4th ed. Mill Valley, CA: University Science Books, 1993.

14. Pavia, D. L., et al., *Introduction to Spectroscopy: A Guide for Students of Organic Chemistry*, 2nd ed. Ft. Worth, TX: Saunders, 1995.
15. Perkampus, H.-H., *UV-VIS Spectroscopy and its Applications*. New York: Springer-Verlag, 1992.
16. Pouchert, C. J., and Behnke, J., *The Aldrich Library of ^{13}C and ^{1}H FT-NMR Spectra*, 2nd ed. Milwaukee, WI: Aldrich Chemical Co., 1992.
17. Pouchert, C. J., and Campbell, J. R., *The Aldrich Library of NMR Spectra*, 2nd ed. Milwaukee, WI: Aldrich Chemical Co., 1983.
18. Pouchert, C. J., *The Aldrich Library of FT-IR Spectra*. Milwaukee, WI: Aldrich Chemical Co., 1985, 1989.
19. *Sadtler Standard Spectra*. (Collections of infrared, ultraviolet, and NMR spectra.) Philadelphia: Sadtler Research Laboratories.
20. Silverstein, R. M., and Webster, F. X., *Spectrometric Identification of Organic Compounds*, 6th ed. New York: Wiley, 1998.
21. Skoog, D. A., *Principles of Instrumental Analysis*, 4th ed. Ft. Worth, TX: Saunders, 1992.
22. Smith, B. C., *Fundamentals of Fourier Transform Infrared Spectroscopy*. Boca Raton, FL: CRC Press, 1996.

G. Qualitative Organic Analysis

1. Criddle, W. J., *Spectral and Chemical Characterization of Organic Compounds: A Laboratory Handbook*, 3rd ed. New York: Wiley, 1990.
2. Kemp, W., *Qualitative Organic Analysis: Spectrochemical Techniques,* 2nd ed. New York: McGraw-Hill, 1986.
3. Pasto, D. J., and Johnson, C. R., *Organic Structure Determination*. Englewood Cliffs, NJ: Prentice-Hall, 1969.
4. Rappoport, Z., ed., *CRC Handbook of Tables for Organic Compound Identification*, 3rd ed. Cleveland: Chemical Rubber Co., 1967.
5. Shriner, R. L., et al., *The Systematic Identification of Organic Compounds*, 7th ed. New York: Wiley, 1998.

H. Reaction Mechanisms and Advanced Topics

1. Carey, F. A., and Sundberg, R. J., *Advanced Organic Chemistry*, 3rd ed. New York: Plenum Press, 1990.
2. Carpenter, B. K., *Determination of Organic Reaction Mechanisms*. New York: Wiley, 1984.
3. Carroll, F. A., *Perspectives on Structure and Mechanism in Organic Chemistry*. Belmont, CA: Brooks/Cole, 1998.
4. Lowry, T. H., and Richardson, K. S., *Mechanism and Theory in Organic Chemistry*, 3rd ed. New York: Harper & Row, 1987.
5. March, J., *Advanced Organic Chemistry: Reactions, Mechanisms and Structure*, 4th ed. New York: Wiley, 1992.
6. *Organic Reaction Mechanisms*. New York: Wiley, 1965–.
7. Scudder, P. H., *Electron Flow in Organic Chemistry*. New York: Wiley, 1992.
8. Sykes, P., *A Guidebook to Mechanism in Organic Chemistry*, 6th ed. Essex, UK: Longman, 1986.
9. Sykes, P., *A Primer to Mechanism in Organic Chemistry*. Essex, UK: Longman, 1995.

J. Reports of Chemical Research

1. *Annual Reports in Organic Synthesis*. New York: Academic Press, 1970–.
2. *Annual Reports on the Progress of Chemistry, Section B: Organic Chemistry*. London: Royal Society of Chemistry, 1904–.
3. *Chemical Abstracts*. Columbus, OH: CA Service, American Chemical Society, 1907–.
4. *Chemical Titles*. Columbus, OH: CA Service, American Chemical Society, 1961–.
5. *Current Contents: Physical, Chemical & Earth Sciences*. Philadelphia: ISI Press, 1967–.
6. *Index Chemicus*. Philadelphia: ISI Press, 1960–.
7. *Science Citation Index*. Philadelphia: ISI Press, 1961–.

K. Guides to the Chemical Literature

1. Bottle, R. T., and Rowland, J. F. B., eds., *Information Sources in Chemistry,* 4th ed. New Providence, NJ: Bowker-Saur, 1993.
2. Heller, S. R., ed., *The Beilstein Online Database: Implementation, Content, and Retrieval.* Washington, DC: American Chemical Society, 1990.
3. Maizell, R. E., *How to Find Chemical Information: A Guide for Practicing Chemists, Educators, and Students.* New York: Wiley, 1987.
4. Schulz, H, and Georgy, U., *From CA to CAS Online: Databases in Chemistry,* 2nd ed. New York: Springer-Verlag, 1994.
5. Somerville, A. N., "Information Sources for Organic Chemistry." *J. Chem. Educ.,* **1991,** *68,* 553, 843; **1992,** *69,* 379.
6. Weissbach, O., *The Beilstein Guide: A Manual for the Use of Beilstein's Handbuch der Organischen Chemie.* Berlin: Springer-Verlag, 1976.
7. Wolman, Y. *Chemical Information: A Practical Guide to Utilization,* 2nd ed. New York: Wiley, 1988.

L. Sources on Selected Topics

1. Agosta, W. C., "Medicines and Drugs from Plants." *J. Chem. Educ.,* **1997,** *74,* 857.
2. Agosta, W. C., *Chemical Communication: The Language of Pheromones.* New York: Scientific American Library, 1992.
3. Anastas, P. T., and Williamson, T. C., eds., *Green Chemistry: Designing Chemistry for the Environment.* Washington, DC: American Chemical Society, 1996.
4. Bauer, K, et al., *Common Fragrance and Flavor Materials: Preparation, Properties, and Uses,* 2nd ed. Deerfield Beach, FL: VCH, 1990.
5. Benfey, O. T., *From Vital Force to Structural Formulas.* Philadelphia: Beckman Center for the History of Chemistry, 1992.
6. Brewster, J. H., "Stereochemistry and the Origins of Life." *J. Chem. Educ.,* **1986,** *63,* 667.
7. Buxton, S. R., and Roberts, S. M., *Guide to Organic Stereochemistry: From Methane to Macromolecules.* Essex, UK: Addison Wesley Longman, 1996.
8. Cole, L. A., *The Eleventh Plague: The Politics of Biological and Chemical Warfare.* New York: W. H. Freeman, 1997.
9. Dehmlow, E. V., and Dehmlow, S. S., *Phase Transfer Catalysis,* 3rd ed. New York: VCH, 1993.
10. Djerassi, C., *From the Lab into the World: A Pill for People, Pets, and Bugs.* Washington, DC: American Chemical Society, 1994.
11. Dodd, J. S., ed., *The ACS Style Guide: A Manual for Authors and Editors.* Washington, DC: American Chemical Society, 1986.
12. Donnelly, T. H., "The Origins of the Use of Antioxidants in Foods." *J. Chem. Educ.,* **1996,** *73,* 159.
13. DuPré, D. B., "Blood or Taco Sauce? The Chemistry behind Criminalists' Testimony in the O. J. Simpson Murder Case." *J. Chem. Educ.,* **1996,** *73,* 60.
14. Eliel, E. L., and Wilen, S. H., *Stereochemistry of Organic Compounds.* New York: Wiley, 1993.
15. Ellis, J. W., et al., "Symposium: Sweeteners and Sweetness Theory." *J. Chem. Educ.,* **1995,** *72,* 671, 676, 680.
16. Farrell, K. T., *Spices, Condiments, and Seasonings,* 2nd ed. New York: Van Nostrand Reinhold, 1990.
17. Fernelius, W. C., and Renfrew, E. E., "Indigo." *Journal of Chemical Education,* **1983,** *60,* 633.
18. Fossey, J., et al., *Free Radicals in Organic Chemistry.* New York: Wiley, 1995.
19. French, L. G., "The Sassafrass Tree and Designer Drugs: from Herbal Tea to Ecstasy." *J. Chem. Educ.,* **1995,** *72,* 479.
20. Fresenius, Philipp, *Organic Chemical Nomenclature,* New York: Halsted Press, 1989.
21. Garratt, P. J., *Aromaticity.* New York: McGraw-Hill, 1986.

22. Gerber, S. M., ed., *Chemistry and Crime: from Sherlock Holmes to Today's Courtroom.* Washington, DC: American Chemical Society, 1983.
23. Gilchrist, T. L., *Heterocyclic Chemistry,* 2nd ed. New York: Wiley, 1992.
24. Glidewell, C., and Lloyd, D., "The Arithmetic of Aromaticity." *Journal of Chemical Education,* **1986,** *63,* 306.
25. Goldsmith, R. H., "A Tale of Two Sweeteners." *J. Chem. Educ.,* **1987,** *64,* 954.
26. Gribble, G. W., "Natural Organohalogens: Many More than You Think!" *J. Chem. Educ.,* **1994,** *71,* 907.
27. Gunstone, F. D., *Fatty Acid and Lipid Chemistry.* New York: Blackie, 1996.
28. Hammond, G. S., and Kuck, V. J., *Fullerenes: Synthesis, Properties, and Chemistry of Large Carbon Clusters.* Washington, DC: American Chemical Society, 1992.
29. Hill, J. W., and Jones, S. W., "Consumer Applications of Chemical Principles: Drugs." *J. Chem. Educ.,* **1985,** *62,* 328.
30. Hirsch, A., *The Chemistry of Fullerenes.* New York: Georg Thieme, 1994.
31. Hosler, D. M., and Mikita, M. A., "Ethnobotany: The Chemist's Source for the Identification of Useful Natural Products." *J. Chem. Educ.,* **1987,** *64,* 328.
32. Jacques, J., *The Molecule and Its Double.* New York: McGraw-Hill, 1993.
33. James, L. K., ed., *Nobel Laureates in Chemistry, 1901–1992.* Washington, DC: American Chemical Society, 1993.
34. Jandacek, R. J., "The Development of Olestra, a Noncaloric Substitute for Dietary Fat." *J. Chem. Educ.,* **1991,** *68,* 476.
35. Kanare, H. M., *Writing the Laboratory Notebook.* Washington: American Chemical Society, 1985.
36. Kauffman, G. B., "Wallace Hume Carothers and Nylon, the First Completely Synthetic Fiber." *J. Chem. Educ.,* **1988,** *65,* 803.
37. Kauffman, G. B., and Seymour, R. B., "Elastomers: I. Natural Rubber." *J. Chem. Educ.,* **1990,** *67,* 422.
38. Kent, J. A., ed., *Riegel's Handbook of Industrial Chemistry,* 9th ed. New York: Van Nostrand Reinhold, 1992.
39. Kikuchi, S., "A History of the Structural Theory of Benzene— the Aromatic Sextet Rule." *J. Chem. Educ.,* **1997,** *74,* 194.
40. Kimbrough, D. R., "Hot and Spicy vs. Cool and Minty as an Example of Organic Structure-Activity Relationships." *J. Chem. Educ.,* **1997,** *74,* 861.
41. Kimbrough, D. R., "The Photochemistry of Sunscreens." *J. Chem. Educ.,* **1997,** *74,* 51.
42. King, F. D., ed., *Medicinal Chemistry: Principles and Practice.* Cambridge: Royal Society of Chemistry, 1994.
43. Kopecky, Jan, *Organic Photochemistry: A Visual Approach.* New York: VCH, 1992.
44. Laing, M., "Beware—Fertilizer can EXPLODE!" *J. Chem. Educ.,* **1993,** *70,* 393.
45. Mann, J., et al., *Natural Products: Their Chemistry and Biological Significance.* New York: Wiley, 1994.
46. Morris, E. T., *Fragrance: The Story of Perfume from Cleopatra to Chanel.* New York: Scribners, 1990.
47. Ohloff, G., *Scent and Fragrances: The Fascination of Odors and Their Chemical Perspectives.* New York: Springer-Verlag, 1994.
48. Olah, G. A., and Molnar, A., *Hydrocarbon Chemistry.* New York: Wiley, 1995.
49. Panico, R., et al., *A Guide to IUPAC Nomenclature of Organic Compounds.* Cambridge, MA: Blackwell, 1993.
50. Pauli, G. H., "Chemistry of Food Additives." *J. Chem. Educ.,* **1984,** *61,* 332.
51. Sbrollini, M. C., "Olfactory Delights." *J. Chem. Educ.,* **1987,** *64,* 799.
52. Seymour, R. B., and Kauffman, G. B., "The Ubiquity and Longevity of Fibers." *J. Chem. Educ.,* **1993,** *70,* 449.
53. Seymour, R. B., and Carraher, C. E., Jr., *Polymer Chemistry: An Introduction,* 3rd ed. New York: Marcel Dekker, 1992.

54. Silverman, R. B., *The Organic Chemistry of Drug Design and Drug Action*. San Diego: Academic Press, 1992.

55. Sotheeswaran, S., "Herbal Medicine: The Scientific Evidence." *J. Chem. Educ.,* **1992**, *69,* 444.

56. Spessard, G. O., and Miessler, G. L., *Organometallic Chemistry*. Upper Saddle River, NJ: Prentice Hall, 1997.

57. Starks, C. M., et al., *Phase Transfer Catalysis: Fundamentals, Applications, and Industrial Perspectives*. New York: Chapman & Hall, 1994.

58. Sundberg, R. J., *Indoles*. San Diego: Academic Press, 1996.

59. Teranishi, R., et al., *Flavor Chemistry: Trends and Development*. Washington, DC: American Chemical Society, 1989.

60. Thayer, J. S., *Organometallic Chemistry: An Overview*. Deerfield Beach, FL: VCH, 1988.

61. Vartanian, P. F., "The Chemistry of Modern Petroleum Product Additives." *J. Chem. Educ.,* **1991**, *68,* 1015.

62. Vogler, A., and Kunkley, H., "Photochemistry and Beer.*" J. Chem. Educ.,* **1982**, *59,* 25.

63. Waddell, T. G., et al., "Legendary Chemical Aphrodisiacs." *J. Chem. Educ.,* **1980**, *57,* 341.

64. Walling, C., "The Development of Free Radical Chemistry." *J. Chem. Educ.,* **1986**, *63,* 99.

65. Walters, E. E., et al, eds., *Sweeteners: Discovery, Molecular Design, and Chemoreception*. Washington, DC: American Chemical Society, 1991.

66. Waring, D. R., and Hallas, G., eds., *The Chemistry and Application of Dyes*. New York: Plenum Press, 1990.

67. Watson, James D., *The Double Helix*. New York: Atheneum Publishers, 1968.

68. Weissermel, K., and Arpe, H.-J., *Industrial Organic Chemistry*, 2nd ed. New York: VCH, 1993.

69. Zanger, et al., "The Aromatic Substitution Game." *J. Chem. Educ.,* **1993**, *70,* 985.

M. Software for Organic Chemistry

1. Black, K., *Proton NMR Spectrum Simulator* (Macintosh). Madison, WI: JCE Software.

2. Clough, F. W., *Introduction to Spectroscopy v2.0: IR, NMR, CMR, and Mass Spec* (DOS or Macintosh). Campton, NH: Trinity Software.

3. Figueras, J., *MassSpec v3.0: A Graphics-Based Mass Spectrum Analyzer* (Windows or Macintosh). Campton, NH: Trinity Software.

4. Figueras, J., *SynTree 2.0: A Program for Exploring Organic Synthesis* (DOS/Windows or Macintosh). Campton, NH: Trinity Software.

5. Pavia, D. L., *SQUALOR: A Simulation of the Qualitative Organic Analysis Laboratory Experience,* and *MacSQUALOR,* (DOS or Macintosh) Campton, NH: Trinity Software.

6. Schatz, P. F., *Organic Chemistry: Spectra of Compounds* (Windows or Macintosh). Wellesley, MA: Falcon Software.

7. Smith, S. G., *Organic Chemistry Laboratory* (DOS). Wellesley, MA: Falcon Software.

Index

Infrared Spectrum-Structure Correlation Chart

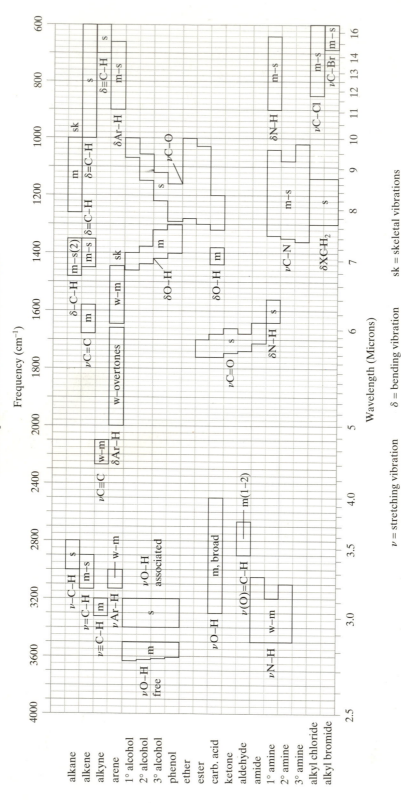

ν = stretching vibration δ = bending vibration   sk = skeletal vibrations

w = weak   m = medium intensity   s = strong